Probability, Random Processes, and Statistical Analysis

Together with the fundamentals of probability, random processes, and statistical analysis, this insightful book also presents a broad range of advanced topics and applications not covered in other textbooks.

Advanced topics include:

- Bayesian inference and conjugate priors
- Chernoff bound and large deviation approximation
- Principal component analysis and singular value decomposition
- Autoregressive moving average (ARMA) time series
- Maximum likelihood estimation and the Expectation-Maximization (EM) algorithm
- Brownian motion, geometric Brownian motion, and Ito process
- Black–Scholes differential equation for option pricing
- Hidden Markov model (HMM) and estimation algorithms
- Bayesian networks and sum-product algorithm
- Markov chain Monte Carlo methods
- Wiener and Kalman filters
- Queueing and loss networks

The book will be useful to students and researchers in such areas as communications, signal processing, networks, machine learning, bioinformatics, and econometrics and mathematical finance. With a solutions manual, lecture slides, supplementary materials, and MATLAB programs all available online, it is ideal for classroom teaching as well as a valuable reference for professionals.

Hisashi Kobayashi is the Sherman Fairchild University Professor Emeritus at Princeton University, where he was previously Dean of the School of Engineering and Applied Science. He also spent 15 years at the IBM Research Center, Yorktown Heights, NY, and was the Founding Director of IBM Tokyo Research Laboratory. He is an IEEE Life Fellow, an IEICE Fellow, was elected to the Engineering Academy of Japan (1992), and received the 2005 Eduard Rhein Technology Award.

Brian L. Mark is a Professor in the Department of Electrical and Computer Engineering at George Mason University. Prior to this, he was a research staff member at the NEC C&C Research Laboratories in Princeton, New Jersey, and in 2002 he received a National Science Foundation CAREER award.

William Turin is currently a Consultant at AT&T Labs Research. As a Member of Technical Staff at AT&T Bell Laboratories and later a Technology Consultant at AT&T Labs Research for 21 years, he developed methods for quantifying the performance of communication systems. He is the author of six books and numerous papers.

"This book provides a very comprehensive, well-written and modern approach to the fundamentals of probability and random processes, together with their applications in the statistical analysis of data and signals. It provides a one-stop, unified treatment that gives the reader an understanding of the models, methodologies, and underlying principles behind many of the most important statistical problems arising in engineering and the sciences today."
Dean H. Vincent Poor, Princeton University

"This is a well-written, up-to-date graduate text on probabilty and random processes. It is unique in combining statistical analysis with the probabilistic material. As noted by the authors, the material, as presented, can be used in a variety of current application areas, ranging from communications to bioinformatics. I particularly liked the historical introduction, which should make the field exciting to the student, as well as the introductory chapter on probability, which clearly describes for the student the distinction between the relative frequency and axiomatic approaches to probability. I recommend it unhesitatingly. It deserves to become a leading text in the field."
Professor Emeritus Mischa Schwartz, Columbia University

"Hisashi Kobayashi, Brian L. Mark, and William Turin are highly experienced university teachers and scientists. Based on this background, their book covers not only fundamentals, but also a large range of applications. Some of them are treated in a textbook for the first time. Without any doubt the book will be extremely valuable to graduate students and to scientists in universities and industry. Congratulations to the authors!"
Professor Dr.-Ing. Eberhard Hänsler, Technische Universität Darmstadt

"An up-to-date and comprehensive book with all the fundamentals in probability, random processes, stochastic analysis, and their interplays and applications, which lays a solid foundation for the students in related areas. It is also an ideal textbook with five relatively independent but logically interconnected parts and the corresponding solution manuals and lecture slides. Furthermore, to my best knowledge, similar editing in Part IV and Part V can't be found elsewhere."
Zhisheng Niu, Tsinghua University

Probability, Random Processes, and Statistical Analysis

HISASHI KOBAYASHI
Princeton University

BRIAN L. MARK
George Mason University

WILLIAM TURIN
AT&T Labs Research

CAMBRIDGE
UNIVERSITY PRESS

CAMBRIDGE
UNIVERSITY PRESS

University Printing House, Cambridge CB2 8BS, United Kingdom

One Liberty Plaza, 20th Floor, New York, NY 10006, USA

477 Williamstown Road, Port Melbourne, VIC 3207, Australia

314-321, 3rd Floor, Plot 3, Splendor Forum, Jasola District Centre, New Delhi - 110025, India

79 Anson Road, #06-04/06, Singapore 079906

Cambridge University Press is part of the University of Cambridge.

It furthers the University's mission by disseminating knowledge in the pursuit of education, learning and research at the highest international levels of excellence.

www.cambridge.org
Information on this title: www.cambridge.org/9780521895446

First published 2012

A catalogue record for this publication is available from the British Library

ISBN 978-0-521-89544-6 Hardback

Additional resources for this publication at www.cambridge.org/9781107024625

To
Masae, Karen, and Galina

Contents

Abbreviations and Acronyms

a.e.	(converge) almost everywhere
a.s.	(converge) almost surely
ANN	artificial neural network
ANOVA	analysis of variance
APP	a posteriori probability
AR	autoregressive
ARMA	autoregressive moving average
ARIMA	autoregressive integrated moving average
AWGN	additive white Gaussian noise
BCJR algorithm	Bahl-Cock-Jelinek-Raviv algorithm
BD process	birth-death process
BEC	binary erasure channel
BN	Bayesian network
BSC	binary symmetric channel
BUGS	Bayesian updating with Gibbs sampling
CDF	cumulative distribution function
CF	characteristic function
CLN	closed loss network
CLT	central limit theorem
CPD	conditional probability distribution
CPT	conditional probability table
CRLB	Cramér-Rao lower bound
CTCS	continuous-time, continuous space
CTMC	continuous-time Markov chain
D.	(converge) in distribution
DAG	directed acyclic graph
dB	decibel
DBN	dynamic Bayesian network
DTCS	discrete-time, continuous-space
DTMC	discrete-time Markov chain
ECM	exponential change of measure
EDA	exploratory data analysis
EM	expectation-maximization
EMC	embedded Markov chain

E-step	expectation step
FBA	forward-backward algorithm
FCFS	first come, first served
GBM	geometric Brownian motion
G/G/1	single server queue with general (independent) arrivals and general service times
G(K)/M/m	m server queue with K general sources and exponential service times (multiple repairmen model)
GLS	generalized loss station
GMM	Gaussian mixture model
GMP	Gauss-Markov process
HITS	hypertext induced topics search
HMM	hidden Markov model
HSMM	hidden semi-Markov model
i.i.d.	independent and identically distributed
IPP	interrupted Poisson process
IS	infinite server
k-NN	k nearest neighbor
LCFS	last come, first served
LCG	linear congruential generator
LFG	lagged Fibonacci generator
l.i.m.	limit in the mean
LRD	long range dependent
LSA	latent semantic analysis
LT	Laplace transform
M/D/1	single server queue with Poisson arrivals and deterministic service time
M/E_k/1	single server queue with Poisson arrivals and k-stage Erlang service times
M/G/1	single server queue with Poisson arrivals and general service times
M/G/∞	infinite-server queue with Poisson arrivals and general service times
M/G/$m(0)$	m server loss system with Poisson arrivals and general service times (generalized Erlang loss model)
M/H_2/1	single server queue with Poisson arrivals and 2-phase hyper-exponential service times
M/M/1	single server queue with Poisson arrivals and exponential service times
M/M/∞	infinite-server queue with Poisson arrivals and exponential service times
M/M/m	m server queue with Poisson arrivals and exponentially distributed service times
M/M/$m(0)$	m server loss system with Poisson arrivals and exponentially distributed service times (Erlang loss model)
M(K)/M/1	single server queue with K exponential sources and exponential service times (machine repairman model, finite-source model)
M(K)/M/m	m server queue with K exponential sources and exponential service times (machine repairmen model)

M(K)/M/m (0)	m server loss system with K exponential sources and exponential service times (Engset loss model)
M-step	maximization step
mod	modulo
m.s.	(converge) in mean square
MAP	maximum a posteriori probability
MA	moving average
MCMC	Markov chain Monte Carlo
MGF	moment generating function
MH algorithm	Metropolis-Hastings algorithm
MIMO	multiple-input, multiple-output
MLE	maximum-likelihood estimate
MLN	mixed loss network
MLSE	maximum-likelihood sequence estimate
MMBS	Markov modulated Bernoulli sequence
MMPP	Markov modulated Poisson process
MMPS	Markov modulated Poisson sequence
MMSE	minimum mean square error
MP test	most powerful test
MRG	multiple recursive generator
MRP	Markov renewal process
MSE	mean square error
MVUE	minimum variance unbiased estimator
OLN	open loss network
P.	(converge) in probability
PASTA	Poisson arrivals see time average
PCA	principal component analysis
PDF	probability density function
PGF	probability generating function
PMF	probability mass function
PRML	partial-response, maximum-likelihood
PS	processor sharing
ROC	receiver operating characteristic
RNG	random number generator
RR	round robin
RTS	Rauch-Tung-Striebel
RV	random variable
SLLN	strong law of large numbers
SMP	semi-Markov process
SNR	signal-to-noise ratio
SSS	strict-sense stationary
SVD	singular value decomposition
SVM	support vector machine
TCP	transmission control protocol

TPM	transition probability matrix
TPMF	transition probability matrix function
WLLN	weak law of large numbers
WSS	wide-sense stationary
w.p.1	(converge) with probability one

Preface

This book covers fundamental concepts in probability, random processes, and statistical analysis. A central theme in this book is the interplay between probability theory and statistical analysis. The book will be suitable to graduate students majoring in information sciences and systems in such departments as Electrical and Computer Engineering, Computer Science, Operations Research, Economics and Financial Engineering, Applied Mathematics and Statistics, Biology, Chemistry and Physics. The instructor and the reader may opt to skip some chapters or sections and focus on chapters that are relevant to their fields of study. At the end of this preface, we provide suggested course plans for various disciplines.

Organization of the book

Before we jump into a mathematical description of probability theory and random processes, we will provide in **Chapter 1, Introduction**, specific reasons why the subjects of this book pertain to study and research across diverse fields or disciplines: (i) communications, information and control systems, (ii) signal processing, (iii) machine learning, (iv) bioinformatics and related fields, (v) econometrics and mathematical finance, (vi) queueing and loss systems, and (vii) other applications. We will then provide a brief but fascinating historical review of the development of (a) classical probability theory, (b) modern probability theory, (c) random processes, and (d) statistical analysis and inference. This historical review also serves as an overview of various topics discussed in this volume.

The remainder of the volume consists of five parts (see Figure 0.1, Chapter Dependencies).

Part I: Probability, random variables and statistics, starting with *axiomatic probability theory* (**Chapter 2, Probability**) introduced by Kolmogorov, covers a broad range of basic materials on probability theory: *random variables* and *probability distributions* (**Chapter 3, Discrete random variables**, and **Chapter 4, Continuous random variables**). In Sections 4.4 and 4.5 we introduce the *canonical exponential family* and *conjugate priors*, which will be used in later chapters.

In Section 5.4 of **Chapter 5, Functions of random variables and their distributions**) we discuss how *random variates* with a specified distribution can be generated for Monte Carlo simulation.

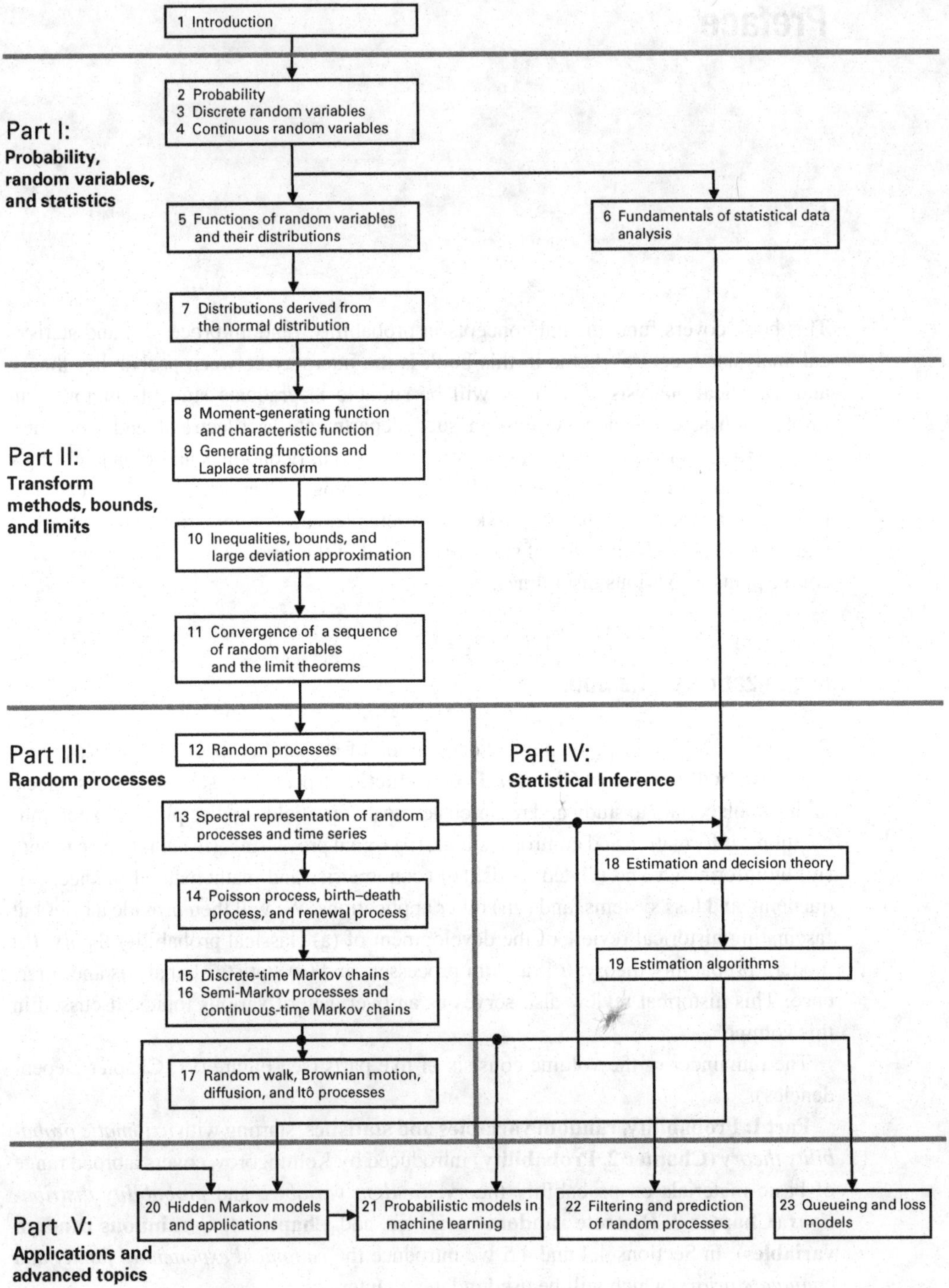

Figure 0.1 Chapter dependencies.

Chapter 6, Fundamentals of statistical data analysis, is rather unique to this volume: it relates the abstract notion of probability theory to real statistical data, with emphasis on *graphical presentations*, which are important in an exploratory stage of data analysis. **Chapter 7, Distributions derived from the normal distribution**, discusses important distributions related to the Gaussian distribution: *chi-squared, Student's t, Fisher's F, lognormal, Rayleigh*, and *Rice distributions*. *Complex-valued Gaussian* variables are especially useful to communication engineers.

Part II: Transform methods, bounds, and limits, **Chapter 8, Moment generating function and characteristic function**, and **Chapter 9, Generating functions and Laplace transform** discuss various transform methods. The reader will learn the relationships among these powerful techniques. We also describe numerical methods to invert the Laplace transform, which are not discussed in most textbooks. **Chapter 10, Inequalities, bounds, and large deviation approximation**, starts with derivations of several inequalities often encountered in probability and statistics: *Cauchy–Schwarz inequality*, *Jensen's inequality*, *log-sum inequality*, *Markov's inequality*, *Chebyshev's inequality*, and *Kolmogorov's inequalities* for martingales and submartingales. We then provide a detailed discussion of *Chernoff's bounds* and *large deviation theory*, which have been increasingly used in recent years. **Chapter 11, Convergence of a sequence of random variables**, may be mathematically challenging to some readers. But it is important to understand the *weak and strong laws of large numbers* and the *central limit theorem*, which are among the most important concepts in probability theory.

Part III: Random processes, **Chapter 12, Random processes**, introduces different classifications of random processes and serves as an introduction to a wide range of random processes, including Markov processes. The notions of *strict stationarity* and *wide-sense stationarity* and *ergodicity* are defined. The *complex-valued Gaussian process* is discussed in much more detail than can be found elsewhere. **Chapter 13, Spectral representation of random processes and time series**, focuses on spectral representation of random processes. The *power spectrum* of a wide-sense stationary process is defined, and its statistical counterpart, the *periodogram*, and its use in *time-series analysis* are presented. The *Karhunen–Loève expansion* is presented as a continuous-time counterpart of eigenvector expansion of the correlation matrix. The *principal component analysis (PCA)* and *singular value decomposition (SVD)* are presented in a unified setting. The chapter ends with a discussion of the autoregressive moving average (ARMA) time series and its spectrum.

Chapter 14, Poisson process, birth–death process, and renewal process, **Chapter 15, Discrete-time Markov chains**, and **Chapter 16, Semi-Markov processes and continuous-time Markov chains**, present standard materials found in most textbooks on random processes. But Section 15.2.2, Spectral expansion method, is probably unique to our book. In Section 16.3 we introduce the notion of reversible Markov chains, whose properties are used in the Markov chain Monte Carlo (MCMC) method in Chapter 21 on machine learning. Section 16.4, An application: phylogenetic tree and its Markov chain representation, illustrates how Markov chain theory is used in biostatistics and bioinformatics.

Chapter 17, Random walk, Brownian motion, diffusion and Itô processes, is also at the core of random process theory. Its application is diverse, ranging from physics, chemistry and material sciences, to operations research, and most recently to mathematical finance. We include two topics seldom covered in textbooks on random processes. They are *Einstein's diffusion equation* (Section 17.3.2) and the *Ornstein–Uhlenbeck process* (Section 17.3.4), which are important to research in areas of bioinformatics that deal with the motion of molecules in the three-dimensional space. Our discussion on *Itô's formula*, *geometric Brownian motion* and the *Black–Scholes differential equation* in Section 17.4 serves as an introduction to mathematical finance.

Part IV: Statistical Inference consists of three chapters. **Chapter 18, Estimation and decision theory**, contains three important subjects: 18.1 *Parameter estimation*, where the theory of *maximum-likelihood estimation* is developed and its theoretical lower bound, the *Cramér–Rao lower bound*, is derived; 18.2 *Hypothesis testing*, where we discuss the famous *Neyman–Pearson lemma* and derive the *likelihood ratio test*. Some of the concepts and techniques developed in radar detection and statistical communication theory, such as the *receiver operating characteristic (ROC) curve*, are now practiced in medical diagnosis and decision making, and also in data mining research; and 18.3 *Bayes decision and estimation theory*, built on *Bayes' theorem* (Section 2.4.2), serves as the foundation for the *Bayesian statistical approach* that is increasingly used in machine learning, econometrics, and other disciplines.

Chapter 19, Estimation algorithms, discusses classical numerical methods for estimation, including the *method of moments*, whose generalized version is used in econometrics, and the *Newton–Raphson* algorithm. We then discuss the *expectation-maximization (EM) algorithm*, a computationally efficient algorithm to obtain a maximum-likelihood estimate of a parameter. The EM algorithm is used widely in communications, speech recognition, machine learning, econometrics, and bioinformatics, especially for state estimation and model parameter estimation of a hidden Markov model, as discussed in Chapter 20.

Part V: Applications and Advanced Topics, covers three application topics. **Chapter 20, Hidden Markov models and applications**, will be valuable to students and researchers in a variety of fields. We provide a unified treatment of the forward–backward algorithm, the *Viterbi algorithm*, and the *BCJR (Bahl–Cocke–Jelinek–Raviv) algorithm*, using a *transition-based hidden Markov model (HMM) representation*, where the observable is a probabilistic function of the state transition. This approach is more concise than the conventional state-based HMM representation, where the observable is a function of the current state. The *EM algorithm* discussed in Section 19.2 is used to yield a maximum-likelihood estimate in an iterative fashion, and a forward–backward algorithm, known as the *Baum–Welch algorithm*, is derived in a simple and elegant fashion.

Chapter 21, Probabilistic models in machine learning. Probabilistic modeling and the Bayesian statistical approach are increasingly important to research in machine learning. This chapter serves as an introduction to this rapidly developing field. Applications such as speech recognition and biological sequence alignment are briefly discussed, and Bayesian networks are presented as a generalization of the HMM formulation. The advent of a computationally efficient simulation technique, called Markov chain Monte Carlo (MCMC), which can estimate a desired probability distribution as the stationary distribution of a Markov chain, now makes the Bayesian approach practical in machine learning, econometrics, and bioinformatics. The Metropolis–Hastings algorithm, the Gibbs sampler, simulated annealing, and other techniques are also discussed.

Chapter 22, Filtering and prediction of random processes, begins with a discussion on MMSE (minimum mean square error) estimation, conditional expectation, and regression analysis, followed by two major theories on filtering and prediction: the *Wiener filter* and the *Kalman filter*. The latter is applied in almost all disciplines, including econometrics and bioinformatics. The Wiener filter theory is important to gain insight in filtering and prediction problems, although it is not as widely practiced as the Kalman filter, because of its difficulty associated with the *Wiener–Hopf integral equation* or its discrete-time analog.

Chapter 23, Queueing and loss models, will be useful to performance modeling and analysis of computers and communication networks, transportation systems, manufacturing systems, and in other fields that involve scheduling and logistics. Several important topics are addressed here. The notion of *processor sharing*, originally developed as a mathematical limit of round-robin scheduling in a time-shared computer, has been recently found to be applicable to modeling of (i) statistically multiplexed traffic, (ii) Web servers, and (iii) links congested with TCP (transmission control protocol) traffic in the Internet. The *Erlang and Engset loss models*, traditionally presented as special cases of the *birth–death process* model, have been substantially generalized by the present authors in recent years, including general service time distributions, multiple classes of customers, and multiple types of servers. This generalized loss model, which we term a *generalized loss station (GLS)*, appears for the first time in a textbook. A *loss network*, which is also a recent development, is also presented in this chapter.

Suggested course plans

Familiarity with college-level calculus and matrix algebra is sufficient background. The book is suitable as a first-year graduate textbook on probability, statistics, and random processes.

In the Department of Electrical Engineering at Princeton, a first-year graduate course "ELE 525: Random Processes in Information Systems" covers the materials in Chapters 2 through 17 and Chapter 22, not including the Kalman filter. The Kalman filter, the topics in Chapter 18 and some subjects in state-space-based control theory form another semester course "ELE 530 Theory of Detection and Estimation." Chapter 23 was taught as part of another graduate course "ELE 531: Communication Networks," which includes network architectures and protocols, network performance and security.

In the Department of Electrical and Computer Engineering of George Mason University, a first-year graduate course "ECE 528: Introduction to Random Processes in

ECE" covers the materials in Chapters 2 through 9, together with selected materials from Chapters 10 through 14. A more advanced course, "ECE 728: Random Processes in ECE" covers the material in Chapters 11 through 17 and parts of Chapters 20 and 22. Chapter 23 was taught as part of another graduate course, "ECE 642: Design and Analysis of Computer Communication Networks." Figure 0.1 shows how a course may be designed by skipping some chapters, depending on the objective of the course.

We believe that the following combinations of chapters will pertain to courses or self-study plans, making this book useful as a reference book, if not a textbook, in various disciplines of science and engineering.

1. **Communications, information and control systems:**
   Chapters 1 through 13 and Chapter 22.
   Chapters 14 through 19 and 23 together with review of the preceding chapters may form a separate advanced graduate course.
2. **Signal processing:**
   Chapters 1 through 16, Chapters 18 through 20, plus Chapter 22 (excluding Section 22.2, Wiener filter theory) may constitute a workable course plan.
   Chapters 14, 17, and 23 will be of lesser importance to signal processing applications. Section 10.3, Large deviation theory, and Section 12.4, Complex-valued Gaussian process, may be skipped.
3. **Machine learning:**
   Chapters 1 through 16 and Chapters 18 through 21 should provide a good background required in machine learning, as far as probabilistic modeling and statistics are concerned. A book on machine learning principles and algorithms (e.g., [211]) should supplement the above mathematical topics.
   Chapter 11, Convergence of a sequence of random variables, Section 10.3, Large deviation theory, and Section 12.4, Complex-valued Gaussian process may be skipped.
4. **Biostatistics, bioinformatics and related fields:**
   Chapters 1–7 and Chapters 14–16 and 18–20 supplemented by "Classical Probability in Statistical Mechanics," (e.g., [32]) should provide a good background required in bioinformatics, as far as probability theory and statistics are concerned.
   Chapter 11 Convergence of a sequence of random variables may be skipped.
5. **Econometrics:**
   Chapters 1 through 10, 12 (excluding Section 12.4), 13, 22 (excluding Section 22.2 Wiener filter theory), and 18 through 20, supplemented by a Bayesian statistical approach (e.g., [128]), should provide a good background required in econometrics, as far as probability theory and statistics are concerned.
6. **Mathematical finance:**
   Chapters 1 through 11 and Chapters 14, 16, and 17. Game theoretic probability plays an important role in mathematical finance [299].
7. **Queueing and loss systems:**
   Chapters 1 through 17, and Chapter 23, supplemented by Kobayashi and Mark [203].

Supplementary materials

Owing to page limitations, this book does not include some important prerequisite materials or advanced subjects which might ordinarily be provided as appendices in more conventional books. Instead we are going to provide such supplementary materials on the book's website, which will be accessible online by interested readers. A preliminary list of topics to be included is as follows: (1) Selected topics of set theory, (2) Selected topics of linear algebra, (3) The Dirac delta function, (4) Stieltjes and Lebesgue integrals, and dF notation, (5) Selected topics in measure theory, (6) Interchanging limit and integral, (7) Differentiating integrals and sums, (8) Complex analysis: contour integral and the residue theorem, (9) Functional transformation and Jacobians, and (10) Stirling's (de Moivre's) approximation formula for a factorial.

Solution manuals

The solutions to problems with a ⋆ will be available online to all readers, whereas the solutions to the other problems will be accessible only to registered instructors authorized by Cambridge University Press.

Lecture slides

Lecture slides for all chapters will be available for registered instructors.

Matlab exercises and programs

MATLAB exercise problems and their programs will be available. The MATLAB programs used to generate numerical plots presented in the text will also be available online to registered users.

Acknowledgments

The first author would like to thank his Ph.D. thesis advisor **John B. Thomas**, Professor Emeritus, of Princeton University, who supervised his thesis (*Representation of complex-valued vector processes and their application to estimation and detection*, Ph.D. thesis, Princeton University, August, 1967) and whose excellent textbook (*An Introduction to Applied Probability and Random Processes*, John Wiley & Sons, Inc., 1971) has greatly influenced the structure and contents of this book. The second author would like to acknowledge the support of the National Science Foundation under Grant No. 0916568 and to thank Professor **Yariv Ephraim** of George Mason University for valuable discussions on HMMs and related topics. The third author would like to thank **Joseph Irgon**, a graduate student of Princeton University, for valuable discussions on applications of clustering, HMMs, and Bayesian networks in bioinformatics.

We thank Professor **Chee Wei Tan** and Dr. **Eugene Brevdo**, who served as teaching assistants for Kobayashi's graduate course "ELE 525: Random Processes in Information systems." They provided many figures and valuable suggestions when the manuscript of this book was used in class. Mr. **Qingwei Li** also did some MATLAB programming for some figures and numerical examples, and Dr. **Duanyi Wang** assisted in the manuscript preparation in the early phase of this effort. We also thank many graduate students at Princeton University and George Mason University who gave various useful comments on early drafts of this book.

A number of our colleagues have read portions of the manuscript and given us valuable comments and suggestions. They are (in alphabetical order of their last names) Professor **Erhan Çinlar** (ORFE, Princeton University), Professor **Paul Cuff** (EE, Princeton University), Dr. **Rodolphe Conan** (Research School of Astronomy & Astrophysics, Australian National University), Professor **Bradley Dickinson** (EE, Princeton University), Professor **Anthony Ephremides** (ECE, University of Maryland), Professor **Yariv Ephraim** (ECE, George Mason University), Dr. **Gerhard Fasol** (Eurotechnology, KK.), Mr. **Kaiser Fung** (Sirius XM Radio), Professor **David Goodman** (ECE, Polytechnic Institute of New York University), Professor **James Hamilton** (Economics, University of California at San Diego), Professor **Eberhard Hänsler** (EE, Technical University of Darmstadt), Professor **Takeo Hoshi** (Economics, University of California at San Diego), Professor **Sanjeev Kulkarni** (EE, Princeton University), Professor **S. Y. Kung** (EE, Princeton University), Dr. **Olivier Lardière** (LACISR, University of Victoria), Professor **Alexander Matasov** (Mechanics and Mathematics, Moscow Lomonosov State University), Professor **David McAllester** (CS, Toyota

Technological Institute at Chicago), Dr. **Taiji Oashi** (University of Maryland School of Pharmacy), Dean **Vincent Poor** (SEAS, Princeton University), Professor **Robert Schapire** (CS, Princeton University), Professor **Christoph Sims** (Economics, Princeton University), Dr. **Shigeki Sugii** (Singapore Bioimaging Consortium/Duke-NUS Graduate Medical School), Professor **Warren Powell** (ORFE, Princeton University), Professor **Shun-Zheng Yu** (CS, Sun Yat-Sen University), Dr. **Linda Zeger** (MIT Lincoln Laboratory), and Dr. **Xin-Ying Zhang** (Synopsis Inc.).

We thank Ms. **Sarah Matthews**, Mr. **Christopher Miller** and Dr. **Philip Meyler** of Cambridge University Press for their advice and encouragement, and especially for their patience despite our long delay in delivering the manuscript.

Finally, we thank our spouses, Masae Kobayashi, Karen Sauer, and Galina Turin, for their patience and understanding during the preparation of this book.

Hisashi Kobayashi
Brian L. Mark
William Turin

1 Introduction

1.1 Why study probability, random processes, and statistical analysis?

Many problems we face in daily life involve some degree of uncertainty and we need to use **probabilistic reasoning** in order to make a sound decision, be it an investment, medical, or social problem. In many cases "probability" may represent our personal judgment about how likely a particular event (e.g., price movement of a stock we own; positive or negative effect of some medicine we may choose to take when we are ill) is to occur. Probability attached to a given event is generally not based on any precise computation but is often a reasonable assessment based on our knowledge or experience. Such probability may be aptly called **subjective** or **qualitative** probability, and may not be scientifically estimated, unless the same event happens repeatedly.

The other type of probability we deal with is what we may call **objective** or **quantitative** probability, which can be estimated objectively based on empirical evidence from observable events. This philosophical question concerning subjective versus objective probability has been pondered by many probabilists and statisticians, as we will briefly discuss in Section 1.2. Philosophical discussion on subjective probability still continues today as a fascinating topic, as is found in arguments between two schools of statistics, i.e., **frequentist** statistics and **Bayesian statistics**, which will be discussed at the end of this chapter.

In this book we will discuss probability based on the modern probability theory established by Kolmogorov in the early twentieth century. The theory of statistical estimation and decision presented in Chapters 18 and 19 is largely based on results developed in the context of frequentist statistics, but we will embrace the Bayesian view of statistics and discuss many subjects recently developed and applied to such fields as machine learning, economics, and biology.

A given scientific or engineering problem that we wish to investigate or solve may often involve some unknown or unpredictable components. We then need to extract the essential part of the given complex system and cast it into some abstract model and solve it analytically or by simulation. Such a model is often **stochastic** or **probabilistic** in nature.

1.1.1 Communications, information, and control systems

Suppose that you are an electrical engineer assigned to a task of designing a mobile handset for a wireless communication system. The function of a receiver is to recover the sender's message, be it voice, data, or video, with acceptable signal quality or to keep the bit error probability below some specified value. There are at least five components that are probabilistic. First, the signal itself is not known *a priori*; all we can specify is its *statistical properties*. Second, the characteristic of a radio channel is not fixed and is often time varying. Third, there will be interfering signals coming from other users who may be using the same frequency band, operating in other cells. Fourth, there will be *thermal noise* introduced at the antenna and the front-end RF (radio frequency) amplifier. Last, the major challenge for you is to characterize these stochastic components as accurately as possible and then design receiver components and algorithms that allow you to recover the signal with a desired level of quality, which more frequently than not will be expressed in a probabilistic statement. An encouraging part is that recent advances made in computer and storage technologies allow you to apply sophisticated algorithms for channel estimation and adopt powerful error-correcting codes and complex decoding algorithms for reliable signal recovery. Signals as well as noise at the *physical layer* can be often represented by **Gaussian processes.** If these processes are band-pass processes, which is usually the case, they can be more conveniently represented in terms of their low-pass equivalents, which are **complex-valued** Gaussian processes.

Design and analysis of *communication and control systems* are often formulated as **statistical estimation and decision** problems (see Chapters 18 and 19) and/or **filtering and prediction** problems (Chapter 22). The **Kalman filter** was originally developed in the context of *stochastic control* and its use in the Apollo spacecraft (the first mission to land humans on the moon!) is among its earliest successful applications. In communication networks, arrivals of *data packets* (as in a packet-switched network), occurrences of *calls* (as in a circuit-switched network) or connections of *flows* (as in the Internet) are mathematically modelled as *point processes* such as **Poisson processes.**

Information theory pioneered by **Shannon**[1] is basically an *applied probability theory* and much of the material covered in Chapters 2 through 5, 10, and 11 is a prerequisite to the study of information theory.

1.1.2 Signal processing

Signal processing is concerned with manipulation of signals such as sound, images, video, biological signals, and data in storage media (such as magnetic or optical discs). Processing of such signals includes filtering, compression, feature extraction, and classification. It often involves characterization of such signals as *random processes* and use of *statistical inference* theory. A **Markov process** representation (Chapters 15, 16,

[1] Claude Elwood Shannon (1916–2001) was an American electrical engineer and mathematician, and has been called "the father of information theory."

and 20) of the underlying random process is sometimes found to be very powerful. It is interesting to note that **Markov**,[2] who introduced in 1907 [238] the concept of what we now call a **Markov chain** (Section 15.1), applied his Markov model representation to the first 20 000 letters of Pushkin's[3] *Eugene Onegin*. Shannon also discussed in his 1948 seminal paper [300] a Markov model representation of a written English text.

1.1.3 Machine learning

Probabilistic reasoning and the *Bayesian statistical approach* play an increasingly important role in **machine learning**. Machine learning (Chapter 21) refers to the design of computer algorithms for obtaining new knowledge or improving existing knowledge in the form of a *model* of some experimental data. A broad family of such models developed in this field include Bayesian networks (Section 21.5), which represent generalizations of hidden Markov models (HMMs) (Chapter 20), support vector machines (SVMs) [338], and artificial neural networks (ANNs) [149]. A major task of machine learning is to classify data using prior knowledge and/or statistical properties of data. Practical applications of machine learning include recognition of speech, image, and handwriting, robotics (such as pattern recognition), and bioinformatics.

The *speech recognition* technology practiced today is largely based on a statistical approach, in which a sequence of *phonemes* or sub-words is characterized as a random sequence drawn from an underlying (i.e., unobservable) *hidden Markov process*, with the number of states as large as several thousand. Such an HMM together with the **Viterbi algorithm** for sequence estimation, and the **expectation-maximization (EM) algorithm** (Chapter 19) for model parameter estimation, has been successfully used in speech recognition.

The EM algorithm is also used for data clustering in *machine learning* and *computer vision*. In natural language processing, two frequently used algorithms are the **forward–backward algorithm (FBA)**, also known as the **Baum–Welch algorithm (BWA)** (Section 20.6.2), and the *inside–outside algorithm*, a generalization of the FBA for unsupervised induction of probabilistic context-free grammars.

Search engines, such as Google search, make use of statistical inference applied to a *Markov chain*, where Markov states correspond to pages on the World Wide Web and state transitions are dictated by Web users randomly clicking on links to these pages.

In machine learning, complex probabilistic distributions are often represented in terms of *probabilistic graphical* models [206], in which the nodes represent variables and the edges correspond to probabilistic interactions between them. A Markov process on a phylogenetic tree (Section 16.4) and factor graphs (Section 21.6.1) for sum-product algorithms (Pearl's belief propagation) are such examples.

[2] Andrei A. Markov (1856–1922) was a Russian mathematician best known for the Markov chain.
[3] Alexander Sergeyevich Pushkin (1799–1837) was a Russian novelist/poet who is considered to be the founder of modern Russian literature.

1.1.4 Biostatistics, bioinformatics, and related fields

Biostatistics deals with applications of statistics in the design, analysis, and interpretation of biological experiments, including medicine and agriculture. *Bioinformatics* is an interdisciplinary study that applies statistics and information sciences to the field of biology. Biostatistics, bioinformatics, and related fields – such as *computational biology* and *epidemology* – make use of *probabilistic formulation* as well as *statistical analysis.* Such noted statisticians as **K. Pearson**[4] and **Fisher**[5] developed and applied statistics primarily to biology. Such notions as **survivor function**, **hazard function**, and **mean residual life**, originated with biostatistical applications of *survival analysis*, are used by the biotech industry, schools of public health, hospitals, etc.

Biological and chemical phenomena are often represented as **Gaussian processes**. The notion of **ergodicity** formulated for time-domain random processes can be extended, in the temporal–spatial domain, to *conformational sampling* of biological molecules.

The methods of **principal component analysis (PCA)**, **singular value decomposition (SVD)**, and **regression analysis** are routinely used in bioinformatics. **Maximum-likelihood estimation** is performed for constructing *phylogenetic trees* or *evolutionary trees* based on DNA and/or amino-acid sequences. **Bayesian inference** is used, for instance, to predict drug efficacy and adverse drug effects for patients. **Hypothesis testing** and the **statistical decision** approach are also found useful in biostatistics and epidemology. **Receiver operating characteristic (ROC)** curves, originally developed in radar detection theory, are now commonly used in medical decision making, and in recent years have been increasingly adopted in the *data mining* research community as well. A variant of the **Newton–Raphson algorithm**, called the "adopted basis Newton–Raphson (ABNR) algorithm," is often used for an energy minimization scheme for biological molecules. **Brownian motion** and **diffusion processes** are also important to bioinformatics. *Brownian dynamics simulation*, which is closely related to the **Langevin equation**, is used to study interactions between biological molecules and/or complex formation. The aforementioned **HMM** and related algorithms have been successfully used for sequence alignment for DNA and proteins, and also for prediction of receptor-binding ligands and protein structure. The HMM and related estimation algorithms are extensively discussed in Chapter 20.

1.1.5 Econometrics and mathematical finance

Econometrics is primarily concerned with developing and applying quantitative or statistical methods to better understand economic principles. In econometrics, a **time series** or a **discrete-time random process** is usually adopted to represent economic data (e.g., a country's gross domestic product (GDP) as a function of time). An

4 Karl Pearson (1857–1936) was a British statistician who applied statistics to biological problems of heredity and evolution.

5 Sir Ronald Aylmer Fisher (1890–1962) was a British statistician, evolutionary biologist, and geneticist.

autoregressive (AR) model or **autoregressive moving average (ARMA)** model is often adopted to take into account the temporal dependency. When multiple economic data (e.g., the inflation rate, unemployment rate, interest rate, and the GDP) are simultaneously considered, a multiple time-series representation with **vector autoregression (VAR)** [305] or **vector ARMA** is used.

The **principal component analysis (PCA)** and a **generalized method of moments**, as alternatives to *maximum-likelihood estimation* procedures such as the aforementioned *EM algorithm*, are among the statistical analysis tools available to deal with large cross-sectional data.

The **numerical Bayesian method** that adopts a **Markov chain Monte Carlo (MCMC)** simulation such as the **Metropolis–Hastings algorithm** is increasingly used in *Bayesian econometrics*. The aforementioned *HMM* was independently developed in econometrics and the term **regime-switching models** [142] is used synonymously with HMMs.

Mathematical finance is a branch of applied mathematics concerned with financial markets, and is closely related to *financial economics*. Unlike econometrics, it usually deals with continuous-time random processes, and its mathematical tools include **martingales**, **Brownian motion** (also known as the **Wiener process** or *Wiener–Lévy* process), and **stochastic differential equations**. These advanced theories of random processes have been much contributed to and used by researchers in statistical mechanics.

In the mid 1960s **Samuelson**[6] discovered the long forgotten Ph.D. thesis "Theory of speculation" [9, 10] of **Bachelier**[7] published in 1900. He recognized the applicability of the theory of **Brownian motion** to analysis of financial markets, by confirming that the history of stock markets' fluctuations fits quite well to what is known as the "square-root law" of Brownian motion. **Merton**,[8] **Scholes**,[9] and **Black**[10] made use of a new random process, called **geometric Brownian motion** (GBM) or *exponential Brownian motion*, introduced by Osborne [261] in 1959.

Geometric Brownian motion, denoted $S(t)$, has since been widely used to model the movement of a stock price – also bonds, commodity, and inflation. Black and Scholes [29] showed that if the derivative value function $F(t)$ is a function of $S(t)$ and time t, i.e., $F(t) = F(S, t)$, it should satisfy the following partial differential equation:

$$\frac{\partial}{\partial t}F(S,t) + bS\frac{\partial}{\partial S}F(S,t) + \frac{\sigma^2 S^2}{2}\frac{\partial^2}{\partial S^2}F(S,t) = rF(S,t), \quad t < T^*, \tag{1.1}$$

where r is the interest rate and d is the dividend yield of the stock; thus $b \triangleq r - d$ represents the cost of carrying the stock. The time T^* is the exercising time of the derivatives.

6 Paul Samuelson (1915–2009) was a US economist, and received the Nobel Prize in Economics in 1970.

7 Louis Bachelier (1870–1946) was a French mathematician, and is regarded as the founder of modern financial mathematics.

8 Robert C. Merton (1944–) is a US economist, Nobel Laureate, and Harvard University professor.

9 Myron S. Scholes (1941–) is a professor emeritus of Stanford Graduate School of Business, and a Nobel Laureate.

10 Fischer Black (1938–1995) was an American economist, taught at the University of Chicago and MIT Sloan School, and worked for Goldman Sachs. His premature death prevented him from receiving a Nobel Prize.

Equation (1.1), now widely known as the **Black–Scholes differential equation** for option pricing, was published in 1973. Scholes and Merton received a Nobel prize in Economics in 1997 for this equation and related work.

Today, the mathematical theory of finance uses advanced *probability theory* and *random processes*. So, scientists and engineers well versed in probability and random processes to the level of this textbook will find it not so difficult to comprehend additional topics such as the **stochastic differential equation** pioneered by **Itô**.[11] Chapter 17 serves as an introduction to mathematical finance.

1.1.6 Queueing and loss systems

When multiple users contend for a resource simultaneously, congestion develops at the resource. Then, either users must be put into a *queue* of some sort or must be rejected from accessing the resource. Such rejected users are said to be *lost*. Examples of the queued case are (i) a queue of automobiles at a toll booth of a highway and (ii) data packets stored temporally enqueued in a buffer at a network router. Examples of the lost case are (i) calls denied in the conventional telephone systems and (ii) calls denied in a cellular phone system. The mathematical theory for such queueing and loss systems is referred to as **queueing theory** or *traffic theory* and is a branch of applied probability theory.

Study of queueing and loss models dates back to the pioneering work by **Erlang**,[12] who published a significant paper [95] in 1917 in which he showed that arrivals of calls at a telephone exchange are distributed randomly and follow a Poisson distribution. The **Poisson process** (Section 14.1) plays an important role in queueing theory. Applications of queueing and loss models include not only the circuit- or packet-switched networks and cellular networks, but also an *Internet* model at the *TCP* (transmission control protocol) level. In Chapter 23, we discuss *queueing and loss system models*, including recent results on **loss network models** and a **generalized loss station (GLS)**, which generalizes the classical **Erlang** and **Engset** loss models in several aspects.

1.1.7 Other application domains

The **birth–death (BD) process** (Section 14.2) is used not only in queueing models, but also in social sciences, such as in the study of demography and population models in biology (in describing the population in biology such as the evolution of bacteria). **Renewal processes** and related concepts such as **hazard function** and **residual life** have immediate applications in *survival analysis*, *reliability theory*, and *duration modeling*, where survival analysis deals with death in biological organisms and failure in mechanical system, reliability theory is used in engineering, and duration modeling or analysis is used in economics or sociology.

[11] Kiyoshi Itô (1915–2008) was a Japanese mathematician and was a professor of Kyoto University.

[12] Agner Krarup Erlang (1878–1929) was a Danish mathematician and an engineer who founded traffic engineering.

Probability theory originally grew out of problems encountered by gamblers in the sixteenth century as described in Section 1.2. So, many examples in this book, such as **gambler's ruin** and **coin tossing**, are taken from betting problems. The term **martingale** originally referred to a class of betting strategies that were popular in France in the eighteenth century. **Martingale theory**, however, is a much more recent development, introduced in the mid-twentieth century by the aforementioned **Lévy**[13] and developed further by **Doob**.[14]

1.2 History and overview

1.2.1 Classical probability theory

It is known that the sixteenth-century Italian **Cardano**[15] wrote *Liber de Ludo Aleae*, a book about games of chance, in the 1560s but it was not published until 1663, long after his death. His book is said to contain the first systematic treatment of probability. But the origin of probability theory is often ascribed to the correspondence in 1654 between two famous French mathematicians: **Pascal**[16] and **Fermat**[17] [78]. They were inspired by questions on gambling posed by **Gombaud**.[18] In 1657, **Huygens**[19] published the first book on probability entitled *De Ratiociniis in Ludo Aleae* (*On Reasoning in Games of Chance*), a treatise on problems associated with gambling, motivated by Pascal and Fermat's correspondence. From this period, probability theory became a favorite subject of some great minds, including **Jacob Bernoulli**,[20] **Leibniz**,[21] **Daniel Bernoulli**,[22] **De Moivre**,[23] **Bayes**,[24] and **Lagrange**.[25]

In 1713, Daniel Bernoulli's book *Ars Conjectandi* (*The Art of Conjecture*) was published posthumously; it included not only the results by Fermat, Pascal, and Huygens, but also his own result: "If a very large number N of independent trials are

[13] Paul Pierre Lévy (1886–1971) was a French mathematician who investigated dependent variables and introduced *martingale theory*.

[14] Joseph Leo Doob (1910–2004) was an American mathematician who taught at the University of Illinois.

[15] Gerolamo Cardano (1501–1576) was an Italian mathematician, physician, astrologer, and gambler.

[16] Blaise Pascal (1623–1662) was a French mathematician, physicist, and philosopher.

[17] Pierre de Fermat (1601–1665) was a French lawyer and government official most remembered for his work in number theory; in particular for Fermat's last theorem. He also made important contributions to the foundations of the calculus.

[18] Antoine Gombaud (1607–1684), Chevalier de Méré, was a French writer. As a contemporary of Pascal, the two corresponded frequently on the calculation of probabilities.

[19] Christiaan Huygens (1629–1695) was a Dutch mathematician, astronomer, and physicist.

[20] Jacob Bernoulli (1654–1705) was a Swiss mathematician.

[21] Gottfried Wilhelm Leibniz (1646–1716) was a German mathematician and philosopher.

[22] Daniel Bernoulli (1700–1782) was a Dutch-born Swiss mathematician and scientist, and a nephew of Jacob Bernoulli. The Bernoulli family produced a number of famous mathematicians and scientists in the eighteenth century.

[23] Abraham de Moivre (1667–1754) was a French mathematician, who worked in London. He was an intimate friend of Isaac Newton.

[24] Thomas Bayes (1702–1761) was a British mathematician and Presbyterian minister.

[25] Joseph Lagrange (1736–1813) was a French mathematician and mathematical physicist.

made, the ratio of the number of successes n to N will, with probability close to 1, be very close to the theoretical probability of success." This result became generally known as **Bernoulli's theorem**. In 1733 De Moivre published *The Doctrine of Chances*, in which he sharpened Bernoulli's theorem and proved the first form of the **central limit theorem** (CLT), by introducing the **normal distribution** as an approximation to the **binomial distribution**. The normal distribution was subsequently used by **Laplace**[26] in 1783 to study measurement errors and by **Gauss**[27] in 1809 in the analysis of astronomical data. A special case of what is now known as **Bayes' rule** is stated in Bayes' "Essay towards solving a problem in the doctrine of chances," published posthumously in 1764 by his friend Richard Price in the *Philosophical Transactions of the Royal Society of London*.

By the 1770s, **Laplace** became the leading authority on mathematical probability, and his *Théorie Analytique des Probabilités* (*Analytic Theory of Probability*) published in 1812 remained authoritative for over half a century after his death. Laplace discussed an instance of the CLT for i.i.d. (independent and identically distributed) random variables (Theorem 8.2 of Section 8.2.5). **Poisson**[28] followed in Laplace's footsteps. In *Recherches sur la Probabilité des Jugements en Matière Criminelle et Matière Civile* (*Research on the Probability of Judgments in Criminal and Civil Matters*), published in 1837, the Poisson distribution first appears. In 1835, Poisson described Bernoulli's theorem under the name "La loi des grands nombres (The law of large numbers)."

The latter half of the nineteenth century witnessed the advancing empiricism in probability that equated the notion of probability merely as the limit of relative frequency, and this discouraged philosophical questions on probability, such as subjective vesus objective probability posed by Poisson and others. In England, **Venn**[29] wrote "The logic of chance" in 1866. His work emphasized the **frequency interpretation** of probability and influenced the development of the **theory of statistics**.

A renewal of mathematical and philosophical interest in probability came only towards the end of the nineteenth century, as probability began to play a fundamental role in **statistical mechanics** advanced by such theoretical physicists as **Maxwell**,[30] **Kelvin**,[31] **Boltzmann**,[32] and **Gibbs**.[33]

26 Pierre-Simon, Marquis de Laplace (1749–1827) was a French mathematician and astronomer whose work was pivotal to the development of mathematical astronomy.

27 Carl Friedrich Gauss (1777–1805) was a German mathematician and scientist who contributed to a wide variety of fields in both mathematics and physics, including number theory, analysis, differential geometry, magnetism, astronomy, and optics.

28 Siméon-Denis Poisson (1781–1840) was a French mathematician, geometer, and physicist.

29 John Venn (1834–1923) was a British logician and philosopher.

30 James Clerk Maxwell (1831–1879) was a Scottish physicist best known for his revolutionary work in electromagnetism and the kinetic theory of gases.

31 Lord William Thompson Kelvin (1824–1907) was a British mathematical physicist and engineer, best known for his work in the mathematical analysis of electricity and thermodynamics.

32 Ludwig Boltzmann (1844–1906) was an Austrian physicist famous for his founding contributions in the fields of statistical mechanics and statistical thermodynamics.

33 Josiah Willard Gibbs (1839–1903) was an American mathematical physicist, physical chemist, and contributed to thermodynamic and statistical mechanics.

By the late nineteenth century, Russian mathematicians became significant players in probability theory. The most notable among them were **Chebyshev**[34] and his student **Markov**, who introduced, as mentioned in Section 1.1.2, the theory of Markov chains [238] and generalized the law of large numbers to dependent variables.

1.2.2 Modern probability theory

By the early twentieth century, French mathematicians regained interest in mathematical probability: they included **Poincaré**,[35] **Borel**,[36] and the aforementioned Lévy.

At the beginning of the twentieth century, however, the status of probability was perceived as unsatisfactory by many mathematicians: They felt that it lacked clarity and rigor to be a respectable branch of mathematics. In 1900, at the second International Congress of Mathematicians held in Paris, **Hilbert**[37] lectured on what he believed to be the most important 23 open problems in mathematics. He listed **probability** as a subdiscipline of his sixth problem; i.e., **axiomatic foundations** for physics.

Hilbert's call for axiomatization of probability stimulated many mathematicians in the first two decades of the twentieth century. During this period, an increased emphasis on **measure theory** and generalized concepts of **integration** pioneered by Borel, **Lebesgue**,[38] **Hausdorff**,[39] **Fréchet**,[40] and others became notable. But it was not until 1933 that Hilbert's call was fully answered when **Kolmogorov**[41] published his 62-page monograph *Grundbegriffe der Wahrscheinlichkeitsrechnung* (*Basic Concepts of Probability Theory*) [208, 210].

Kolmogorov's measure theoretic approach to probability was a major breakthrough in the history of probability theory; his approach was soon accepted by a majority of mathematicians and statisticians. The "limiting frequency theory" approach, long advocated by **von Mises**[42] [342] and others, is no longer subscribed to by many. Kolmogorov himself, however, did not negate von Mises' approach, and in fact he wrote in 1963: " ... that the basis for the applicability of the results of the mathematical theory of probability to real 'random phenomena' must depend on some form of the frequency concept of

[34] Pafnuty Lvovich Chebyshev (1821–1894) is known for his work in the fields of probability, statistics, and number theory. Among his students were Markov and Aleksandr Mikhailovich Lyapunov (1857–1918).

[35] Henri Poincaré (1854–1912) was a French mathematician, theoretical physicist, and a philosopher of science.

[36] Émile Borel (1871–1956) was a French mathematician and pioneered measure theory.

[37] David Hilbert (1862–1943) was a German mathematician, known for his contributions to algebra, topology, geometry, number theory, and physics.

[38] Henri Léon Lebesgue (1875–1941) was a French mathematician who formulated measure theory and defined the Lebesgue integral that generalized the notion of the Riemann integral.

[39] Felix Hausdorff (1868–1942) was a German mathematician who founded the theory of topological and metric spaces. He also investigated set theory and introduced the concept of a partially ordered set.

[40] René Maurice Fréchet (1878–1973) was a French mathematician who contributed to real analysis and founded the theory of abstract spaces.

[41] Andrey Nikolaevich Kolmogorov (1903–1987) was a Soviet mathematician who made major advances in probability theory, topology, logic, turbulence, classical mechanics, and computational complexity.

[42] Richard von Mises (1887–1953) was an Austrian-born American mathematician who worked on fluid mechanics, aerodynamics, aeronautics, statistics, and probability theory.

probability, the unavoidable nature of which has been established by von Mises in a spirited manner." A majority of textbooks on probability theory today takes the **axiomatic approach**, and so does this textbook.

1.2.3 Random processes

In the axiomatic formulation of probability theory introduced in 1933, a **random variable** X is defined as a mapping $X(\omega)$ from a sample point ω in **sample space** Ω to a real number; i.e., $X(\omega) \in \mathbb{R}$. A **random process** (also called a **stochastic process**) is then defined as a collection of random variables. That is, for each t in the index set $\mathcal{T}$, $X(\omega; t)$ is a random variable. We will postpone a formal discussion of random processes until Chapter 12, but it should be noted that investigation of random processes began well before 1933. In the theory of random processes there are two fundamental processes: one is the Poisson process and the other is Brownian motion [186].

1.2.3.1 Poisson process to Markov process

The **Poisson process** is named after the aforementioned French mathematician Siméon-Denis Poisson. What the Poisson process is to a class of point processes (Chapter 14) is what the **Gaussian process** is to a class of continuous-time and continuous-valued processes (Chapter 12). The Poisson process can be considered as a special case of the *birth-death (BD) process*, which in turn is a special case of a *continuous-time Markov chain (CTMC)*, also called a *Markov process* (Chapter 16). Markov introduced (as mentioned already in Section 1.1.2 and towards the end of Section 1.2.2) what we call a *discrete-time Markov chain (DTMC)* in 1907 [238]. Chebyshev and Markov proved the **central limit theorem (CLT)** under conditions weaker than Laplace's proof, using *moments*. **Lyapunov**[43], another student of Chebyshev, later proved the same theorem using **characteristic functions**.

Markov process models have been used in numerous applications, including **statistical mechanics** in physics, *queueing theory* (Chapter 23), *information theory*, and many other scientific and engineering disciplines, including *signal processing* applications and *bioinformatics*, as discussed in the previous sections.

Another generalization of the Markov process led to **semi-Markov processes** (Section 16.1) and the **renewal process** (Section 14.3), which find applications in *simulation methodology* as well as the aforementioned reliability theory and queueing theory.

The concept of **martingale** in probability theory was originally introduced by Lévy, and much of the theory was developed by Doob, as remarked at the end of the previous section.

[43] Aleksandr Mikhailovich Lyapunov (1857–1918) contributed also to a study of stability conditions in dynamical systems.

1.2.3.2 Brownian motion to Itô process

The origin of **Brownian motion** goes back to in 1827, when **Brown**[44] observed the irregular motion of pollen particles suspended in water. Several decades later the aforementioned Bachelier gave a mathematical description of Brownian motion in his 1900 Ph.D. thesis "The theory of speculation" [9, 10]. His work presented a stochastic analysis of the stock and option markets. Five years later, in 1905, **Einstein,**[45] based on his independent work, published a mathematical analysis of Brownian motion [86], which showed that the concentration density of a large number of Brownian particles should satisfy a partial differential equation, now known as a **diffusion equation**. Almost at the same time, **Smoluchowski**[46] worked on the theory of Brownian motion [270, 307], independently of Einstein. The atomic nature of matter was still a controversial idea around the turn of the century. Einstein and Smoluchowski reasoned that, if the kinetic theory of fluids was right, the molecules of water would move randomly. Therefore, a small particle would receive a random number of impacts of random strength and from random directions in any short period of time. This random bombardment by the molecules of the fluid would cause a sufficiently small particle to move in exactly the way observed by Brown almost 80 years earlier.

Perrin[47] carried out experiments to verify the new mathematical models, and his results finally put an end to the century-long dispute about the reality of atoms and molecules. Perrin received the Nobel Prize in Physics in 1926.

The theory of Brownian motion was further investigated by Lévy, **Wiener,**[48] Kolmogorov, **Feller,**[49] and others. In Chapter 17 we will derive Brownian motion as a continuous limit of a **random walk**. But Wiener derived in 1923 the Brownian motion as a random real-valued continuous function $W(t)$ on $[0, \infty)$ [351]. We now call such a random function a **Wiener process**.

In 1944, the aforementioned **Itô** published his work on a theory of stochastic differential equations [157, 158]. The basic concepts of his theory, often called **Itô calculus**, consist of the **Itô integral** and **Itô's lemma**. **Stochastic differential equations** have been applied to many disciplines, including **Kalman–Bucy filtering** (Section 22.3) and the **Black–Scholes differential equation** for option pricing (Chapter 17.4.3). Itô became the first recipient of the Carl Friedrich Gauss Prize newly established at the 2006 International Congress of Mathematicians. We should

[44] Robert Brown (1773–1858) was a Scottish biologist.

[45] Albert Einstein (1879–1955). Scientists call year 1905 "Einstein's annus mirabilis" (a Latin phrase meaning "Einstein's year of miracles"). Within a few months, Einstein published three seminal papers in *Annalen der Physik* that profoundly transformed our understanding of physics. The first paper was on his "special theory of relativity" and the famous equation $E = mc^2$. The second paper was on the photoelectric effect by hypothesizing that light consisted of particles (called photons). For this work he received the Nobel Prize in Physics in 1921. The third paper was on Brownian motion.

[46] Marion Smoluchowski (1872–1917) was a Polish physicist.

[47] Jean-Baptiste Perrin (1870–1942) was a French physicist and a Nobel Prize winner.

[48] Norbert Wiener (1894–1964) was an American theoretical and applied mathematician. He was a pioneer in the study of stochastic and noise processes. Wiener is also the founder of cybernetics.

[49] William Feller (1906–1970) was a Croatian–American mathematician specializing in probability theory. He was a professor of mathematics at Princeton University from 1950 till 1970.

also note that **Stratonovich**[50] introduced a stochastic calculus, different from Itô calculus; the Stratonovich integration is most frequently used within physical sciences. He also solved the problem of optimal nonlinear filtering based on his theory of *conditional Markov process* [313], and the Kalman–Bucy linear filter is considered as a special case of Stratonovich's filter.

1.2.4 Statistical analysis and inference

The history of mathematical statistics is not separable from that of probability theory. Much of modern statistical theory and its application are founded on modern probability theory.

The **method of least squares** is often credited to **Gauss**, **Legendre**,[51] and **Adrain**.[52] Gauss also showed that among linear models where the errors are uncorrelated, having zero mean and equal variances, the *best linear unbiased estimate* of the linear coefficients is the *least-squares estimate*. This result is now known as the **Gauss–Markov theorem**. This result can be interpreted in the context of the least-squares approach to *regression analysis* introduced in the late nineteenth century.

Around the period "probabilists" in France were studying games of chance, some mathematicians in England were concerned with sampling, motivated by selling life insurance, financing pensions, and insuring ships at sea [20]. They contributed to the theory of **sampling** and reasoning from sets of data.[53] The concept of **regression** comes from genetics and was popularized by **Galton**,[54] who discovered "regression toward the mean" by experimenting with sweet peas. His sweet peas produced seeds with a normal variation of sizes that regressed from the distribution of their parents. The aforementioned K. Pearson is credited for the establishment of the discipline of statistics; he contributed to such classical statistical methods as **linear regression**, **correlation**, and the **chi-square test**. **Gosset**,[55] an employee of Guinness Brewery Co. in Dublin, studied in K. Pearson's biometric laboratory in 1906–07 and published in 1908 two papers on **Student's *t* distribution**, using the pseudonym 'Student,' which addressed the brewer's concern with small samples.

The notion of a **maximum likelihood estimator** (Section 18.1.2) for estimating a parameter of a distribution was introduced in the 1912–1922 period by the aforementioned Fisher, who is also known for his work on the **analysis of variance** (ANOVA), the ***F*-distribution**, **Fisher information**, and the **design of experiments**.

50 Ruslan Leont'evich Stratonovich (1930–1997) was a Russian mathematician, physicist, and engineer.
51 Adrien-Marie Legendre (1752–1833) was a French mathematician and made contributions to statistics, number theory, abstract algebra, and mathematical analysis.
52 Robert Adrain (1775–1843) was an American mathematician.
53 Personal communication with David Goodman.
54 Sir Francis Galton (1822–1911) was an English Victorian polymath, anthropologist, eugenicist, protogeneticist, and statistician.
55 William Sealy Gosset (1876–1937) was a British statistician, best known for Student's *t*-distribution.

The theory of **hypothesis testing** (Section 18.2) was also developed in the early part of the twentieth century. **Neyman**[56] and **E. Pearson**[57] are best known for the **Neyman–Pearson lemma** which states that, when performing a hypothesis test between two hypotheses, the **likelihood-ratio test** is the most powerful test (Section 18.2.2).

Estimation and decision theory provided foundations for radar detection theory, statistical communication theory, and control theory. Such notions as **unbiasedness**, **efficiency**, and **sufficiency** are important ingredients of estimation theory. The famous minimum variance unbiased estimator bound, known as the **Cramér–Rao lower bound** (Section 18.1.3), is due to **Cramér**[58] and **Rao**.[59]

Filter design theory was developed as part of circuit theory in the early twentieth century, but the birth of modern filter theory based on statistical estimation theory was in 1949, when **Wiener** published his work developed in the 1940s [350]. It is based on the minimum mean square error criterion (Chapter 22). The discrete-time version of the Wiener filter was obtained independently by Kolmogorov and published in 1941 [209]. Hence, the theory is sometimes called the Kolmogorov–Wiener theory.

Kalman[60] introduced in 1960 what is known as the **Kalman filter** [171], an efficient recursive filter that estimates the state of a dynamic system from a series of incomplete and noisy data (Section 22.3). There is a close relationship between the Kalman filter and the estimation problem for an **HMM**.

The **EM algorithm** (Section 19.2) developed by Dempster, Laird, and Rubin in 1977 [80] is an iterative algorithm to find **maximum likelihood estimates** of model parameters, when the model depends on **unobserved latent variables**, and has been widely applied to signal processing, communications, bioinformatics, machine learning, and econometrics, as mentioned in the previous section.

1.2.4.1 Frequentist statistics versus Bayesian statistics

As remarked earlier, the theory of mathematical statistics was founded from the late nineteenth century through the early twentieth century by such British statisticians as Galton, the two Pearsons, Gosset, and Fisher. Their work is based on the frequency interpretation of probability advocated by Venn in his 1866 book, as discussed in Section 1.2.1. This school of statisticians is often referred to as the **frequentists** in contrast with the "Bayesians" to be described below. Such statistical methods and tools as *regression*, *multiple regression*, *ANOVA*, *analysis of covariance (ANCOVA)*, *maximum likelihood*, *F test*, *experimental design*, *PCA*, and *time series analysis* all belong to **frequentist statistics**.

Use of probability to describe the degree of **belief** in states of a "system" of our interest forms the basis of the **Bayesian statistical approach** alluded to at the beginning

[56] Jerszy Neyman (1894–1981) was a Polish–American mathematician.
[57] Egon Sharpe Pearson (1895–1980) was a British statistician, a son of Karl Pearson.
[58] Carl Harald Cramér (1893–1985) was a Swedish probabilist and statistician.
[59] Calyampudi Radhakrishna Rao (1920–) is an Indian statistician.
[60] Rudolph Emil Kalman (1930–) is a Hungarian-born American system theorist.

of this chapter. It starts with some prior belief and updates it using observation data to give a posterior belief, which may be used to draw an **inference**. With the advent of computationally efficient methods such as **MCMC**, the Bayesian approach has recently become of real practical use, and is increasingly used in such fields as machine learning (e.g., Koller and Friedman [206] and Kononenko and Kukar [211]) and information theory (e.g., MacKay [234]), econometrics (e.g., Greenberg [128]), and bioinformatics (e.g., Wilkinson [352]).

1.3 Discussion and further reading

There exists a large body of literature concerning the history of probability and statistics; e.g., Todhunter [325] (1865), David [78], Maistrov [235], Hacking [135, 136], Stigler [312], Hald [140], and Franklin [111]. Shafer and Vovk [299] provide a rather comprehensive treatment on the history of probability and related bibliography, as well as their own view of probability. Fine [105] gives a brief but informative description of the history of probability in his introductory chapter.

Williams [355] (Appendix D) provides a well-selected list of books, together with his own witty remarks, on *probability* (including genetics, random walks, Markov chains, stochastic processes, martingales, and stochastic calculus), *frequentist statistics* (including regression, ANOVA and experimental design, time series), *Bayesian statistics* (including decision theory – frequentist and Bayesian, model choice), *quantum theory* and *quantum computing*.

Part I

Probability, random variables, and statistics

2 Probability

2.1 Randomness in the real world

2.1.1 Repeated experiments and statistical regularity

One way to approach the notion of probability is through the phenomenon of **statistical regularity**. There are many repeating situations in nature for which we can predict in advance, from previous experiences, *roughly* what will happen, but not *exactly* what will happen. We say in such cases that the occurrences are **random**. The reason that we cannot predict future events exactly may be that (i) we do not have enough data about the condition of the given problem, (ii) the laws governing a progression of events may be so complicated that we cannot undertake a detailed analysis, or possibly (iii) there is some basic indeterminacy in the physical world. Whatever the reason for the randomness, a definite average pattern of results may be observed in many situations leading to random occurrences when the situation is recreated a great number of times. For example, if a fair coin is flipped many times, it will turn up heads on about half of the flips.

Another example of randomness is the response time of a web (i.e., World Wide Web or WWW) access request you may send over the Internet in order to retrieve some information from a certain website. The amount of time you have to wait until you receive a response will not be precisely predictable, because the total round trip time depends on a number of factors. Thus, we say that the response time varies *randomly*. Although we cannot predict exactly what the response time of a given web access request will be, we may find experimentally that certain **average** properties do exhibit a reasonable regularity. The response time of small requests averaged over minutes will not vary greatly over an observation interval of several minutes; the response time averaged over a given day will not differ greatly from its value averaged over another day of similar system usage.

The tendency of repeated experiments to result in the convergence of the averages as more and more trials are made is what we refer to as *statistical regularity*. This statistical regularity of averages is an experimentally verifiable phenomenon in many situations that involve randomly varying quantities. We are therefore motivated to construct a mathematical model adequate for the study of such phenomena. This is the domain of probability and statistics.

2.1.2 Random experiments and relative frequencies

By an *experiment*, we mean a measurement procedure in which all conditions are predetermined to the limit of our ability or interest. We use the word *trial* to mean the making of the measurement. An experiment is called *random* when the conditions of the measurement are not predetermined with sufficient accuracy and completeness to permit a precise prediction of the result of a trial. Whether an experiment should be considered random depends on the precision with which we wish to distinguish **possible outcomes**. If we desire or are able to look closely enough, in some sense, any experiment is random.

We now discuss more precisely what we mean by statistical regularity. Let A denote one of the possible outcomes of some experiment, say, the "head" in coin tossing, and we repeat the experiment a large number of times N under uniform conditions. Denote by $N(A)$ the number of times that the outcome A occurs. The fraction

$$f_N(A) = \frac{N(A)}{N} \tag{2.1}$$

is called the **relative frequency** or simply the *frequency* of outcome A. If there is a practical certainty that the measured relative frequency will tend to a limit as N increases without limit, we would like to say that the outcome A has a definite probability of occurrence, and take $P[A]$ to be that limit; i.e.,

$$f_N(A) \to P[A] \text{ as } N \to \infty. \tag{2.2}$$

Unfortunately, this simple approach faces many difficulties. One obvious difficulty is that, strictly speaking, the limit may never be found, since an infinite number of repetitions of the experiment takes an infinite amount of time. Therefore, rather than defining a probability as the limit of a relative frequency, we will construct an abstract model of probability so that probabilities behave like the limits of relative frequencies. An important after-the-fact justification of this procedure is that it leads to the so-called **laws of large numbers**, according to which, in certain very general circumstances, the mathematical counterpart of an empirical relative frequency does converge to the appropriate probability. Hence, an empirical relative frequency may be used to estimate a probability.

2.2 Axioms of probability

A mathematical model will prove useful in predicting the results of experiments in the real world if the following two conditions are met. First, pertinent physical entities and their properties must be reflected in the model. Second, the properties of the model must be mathematically consistent and make analysis tractable. We begin by defining the following three abstract entities: **sample space**, **event**, and **probability measure**. We then develop a model by assigning them mathematically consistent properties that reflect constraints in the real world.

2.2.1 Sample space

The *sample space*[1] is a mathematical abstraction of the collection of all possible experimental outcomes. We denote this collection by the symbol Ω. An object in Ω is called a **sample point** and denoted ω. Each sample point, therefore, corresponds to a possible outcome of a real-world experiment.

Example 2.1: Tossing of two coins. Consider the experiment of tossing two coins, denoted coin i ($i = 1, 2$). We are interested in whether each coin falls heads (h) or tails (t), and denote the four possible outcomes of this experiment as (hh), (ht), (th), and (tt). Each outcome of the experiment corresponds to exactly one member of the sample space

$$\Omega = \{\omega_1, \omega_2, \omega_3, \omega_4\}, \tag{2.3}$$

where

$$\omega_1 = (\text{hh}), \qquad \omega_2 = (\text{ht}), \qquad \omega_3 = (\text{th}), \qquad \omega_4 = (\text{tt}). \tag{2.4}$$

□

Example 2.2: Response time to a web request. Consider the experiment in which we measure the response time for a web request that is sent over the Internet to a certain web server. In theory, the system response time can be anywhere between zero and plus infinity. Then Ω is the positive half-line:

$$\Omega = \{\omega \,:\, 0 \leq \omega < \infty\}. \tag{2.5}$$

□

2.2.2 Event

An *event* is a set of sample points. We usually denote events by capital letters, such as $A, B, \ldots$, or $A_1, A_2, \ldots$. An event is concisely defined by the expression

$$A = \{\omega : \text{ certain conditions on } \omega \text{ are satisfied}\}, \tag{2.6}$$

which reads "Event A is the set of all ω such that certain conditions on ω are satisfied." Clearly, an event is a subset of Ω.

Example 2.3: Tossing of two coins – continued. Consider again the experiment with two coins. If A is the subset of Ω defined by

[1] According to Feller [99], the notion of "sample space" comes from R. von Mises [341], who used the German word "Merkmalraum."

$$A = \{(\text{hh}), (\text{ht}), (\text{th})\}, \tag{2.7}$$

then A is the event that there is at least one head. Similarly,

$$B = \{(\text{hh})\} \tag{2.8}$$

is the event that there are two heads. Thus, an event may contain one or more sample points. An event like B, which contains only *one* sample point, is called a **simple event**.

In order to explain the significance of the notion of events, we may distinguish, in connection with a real-world experiment, between the terms **outcome** and **result**. By different outcomes we mean outcomes that are separately identifiable in an ultimate sense. On the other hand, by different results we mean sets of outcomes between which we choose to distinguish. Thus, *results* in the real world correspond to *events* in the mathematical model. For example, a result in our response-time experiment might be that the observed response time at the terminal is between 1.0 and 1.5 s. Such a result clearly embraces an infinite number of different possible response times or outcomes.

2.2.3 Probability measure

A **probability measure** is an assignment of real numbers to the events defined on Ω. The probability of an event A is denoted by $P[A]$.[2] The set of properties that the assignment must satisfy are sometimes called the **axioms of probability**. This axiomatic formulation of probability is the aforementioned great accomplishment by Kolmogorov [208] published in 1933. The probability assigned to an event corresponds to that value at which we expect the relative frequency of the associated result to stabilize in an infinitely long sequence of independent trials of the real-world experiment.

Example 2.4: Tossing of two coins – continued. If the sample space Ω is the one defined by (2.4) and (2.3), a *possible* probability assignment to the simple events is

$$P[\{\omega_1\}] = \frac{1}{2},\ P[\{\omega_2\}] = P[\{\omega_3\}] = \frac{1}{4}, \text{ and} P[\{\omega_4\}] = 0.$$

However, this assignment is not appropriate to reflect the coin-tossing experiment.

Example 2.5: Response time to the web request – continued. If Ω is the set of all response times given by (2.5) and we define the event E_t by

$$E_t = \{\omega :\ 0 \le \omega \le t\}, \tag{2.9}$$

[2] Note that some authors write $P(A)$ or $P\{A\}$ instead of $P[A]$, the notation we adopt in this textbook.

a possible probability assignment is

$$P[E_t] = 1 - e^{-t/T}, \tag{2.10}$$

where T is the average value of response time.

2.2.4 Properties of probability measure

Before we discuss the properties of the probability measure, we must know some elementary set theory, since the definition of sample space Ω and events implies the existence of certain other identifiable sets of points.

1. The **complement** of event A, denoted A^c, is the event containing all sample points that belong to Ω, but not to A:

$$A^c = \{\omega : \omega \text{ does not belong to } A\}. \tag{2.11}$$

2. The **union** of A and B, denoted $A \cup B$, is the event containing all sample points that belong to at least one of the two sets A, B:[3]

$$A \cup B = \{\omega : \omega \text{ belongs to } A \text{ or } B\}. \tag{2.12}$$

3. The **intersection** of A and B, denoted by $A \cap B$, is the event containing all sample points in both A and B:

$$A \cap B = \{\omega : \omega \text{ belongs to both } A \text{ and } B\}. \tag{2.13}$$

4. The event containing no sample points is called the **null event**, denoted $\emptyset$. For any event A, the intersection of A and A^c is the null event; i.e., $A \cap A^c = \emptyset$.
5. The event containing all points – that is, the sample space Ω – is called the **sure event** or *certain event* (an event that must occur). For any event A, the union of A and A^c is the sure event; that is, $A \cup A^c = \Omega$. It is clear that the null event and sure event are related by $\Omega^c = \emptyset$ and $\emptyset^c = \Omega$.
6. Two events A and B are called **disjoint** or *mutually exclusive* if they have no sample points in common; that is, if $A \cap B = \emptyset$.

The relations between the operations (i.e., complement, union, and intersection) are easily visualized in the schematic diagrams of Figure 2.1. Such drawings are called **Venn diagrams**.

[3] The word "or" is used in mathematics and logic in the inclusive sense. Thus, the statement "A or B" is the mathematical expression for "either A or B or both."

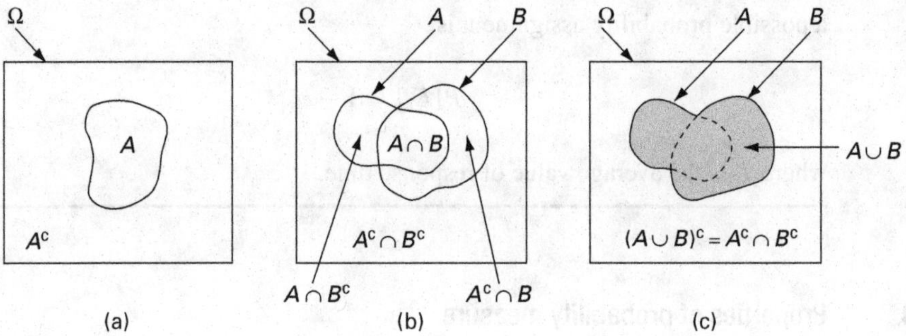

Figure 2.1 Venn diagrams.

The symbols $\cup$ and $\cap$ are operations between any two sets, just as $+$ and $\times$ are operations between any two numbers, and they obey similar laws:

$$\textit{commutative laws:} \quad \begin{array}{l} A \cup B = B \cup A \\ A \cap B = B \cap A \end{array}; \tag{2.14}$$

$$\textit{associative laws:} \quad \begin{array}{l} (A \cup B) \cup C = A \cup (B \cup C) \\ (A \cap B) \cap C = A \cap (B \cap C) \end{array}; \tag{2.15}$$

$$\textit{distributive laws:} \quad \begin{array}{l} A \cap (B \cup C) = (A \cap B) \cup (A \cap C) \\ A \cup (B \cap C) = (A \cup B) \cap (A \cup C) \end{array}. \tag{2.16}$$

Finally, we note two very useful identities:

$$(A \cup B)^c = A^c \cap B^c, \tag{2.17}$$

$$(A \cap B)^c = A^c \cup B^c, \tag{2.18}$$

which are sometimes called **de Morgan's laws**.[4] All of these laws can be readily verified by the Venn diagrams in Figure 2.1.

Having extended the notion of an event, we will now consider certain properties that a probability measure should satisfy. In a long sequence of N independent trials of a real-world experiment, the observed relative frequency $f_N(A)$ of the result A meets certain conditions:

1. The relative frequency $f_N(A)$ is always nonnegative:

$$f_N(A) \geq 0.$$

2. Every trial of an experiment is sure to have a result. Hence,

$$f_N(\Omega) = 1.$$

3. If two results A and B are *mutually exclusive*, then

$$f_N(A \text{ or } B) = f_N(A) + f_N(B).$$

[4] Augustus de Morgan (1806–1871) was an Indian-born British mathematician and logician.

Since we are going to use probability theory to predict the results of real-world random experiments, it is reasonable that we impose similar conditions on corresponding entities in our mathematical model. We therefore require our assignment of probability measure $P[\cdot]$ to satisfy the following axioms of probability:

Axiom 1. $P[A] \geq 0$ for all events A. (2.19)

Axiom 2. $P[\Omega] = 1$; that is, the probability of the *sure* event is 1. (2.20)

Axiom 3. If A and B are *mutually exclusive* events – i.e., if $A \cap B = \emptyset$ – then $P[A \cup B] = P[A] + P[B]$. (2.21)

These properties, motivated from real-world considerations, are self-consistent and are adequate for a formal development of probability theory, whenever the totality of events on Ω is *finite*.

A consequence of Axiom 3 is that if $A_1, A_2, \ldots, A_M$ are M mutually exclusive events, then their union $A_1 \cup A_2 \cup \cdots \cup A_M$, which we denote by $\bigcup_{m=1}^{M} A_m$, has the probability

$$P\left[\bigcup_{m=1}^{M} A_m\right] = \sum_{m=1}^{M} P[A_m]. \tag{2.22}$$

This is easily shown by successive applications of Axiom 3 or, more formally, by the method of mathematical induction. A consequence of Axioms 2 and 3 is

$$P[A] \leq 1 \tag{2.23}$$

for any event A, since

$$P[A] + P[A^c] = P(\Omega) = 1, \tag{2.24}$$

and $P[A^c] \geq 0$, where A^c is the complement of A. It also follows from Axioms 2 and 3 that the probability of the null event is zero; i.e.,

$$P[\emptyset] = 1 - P[\Omega] = 0. \tag{2.25}$$

Since the event $A \cup B$ is decomposable into a set of mutually exclusive events $A \cap B$, $A^c \cap B$, and $A \cap B^c$, we have

$$\begin{aligned} P[A \cup B] &= P[A \cap B] + P[A^c \cap B] + P[A \cap B^c] \\ &= P[A \cap B] + (P[B] - P[A \cap B]) + (P[A] - P[A \cap B]). \end{aligned} \tag{2.26}$$

Thus, we have

$$\boxed{P[A \cup B] = P[A] + P[B] - P[A \cap B].} \tag{2.27}$$

Hence,

$$P[A \cup B] \leq P[A] + P[B], \tag{2.28}$$

where the equality holds only when A and B are mutually exclusive. If the total number of possible events is infinite, then the above three properties alone are insufficient. It is necessary to extend Axiom 3 to include infinite unions of disjoint events.

Axiom 4. If $A_1, A_2, \ldots$ are mutually exclusive events, then their union $A_1 \cup A_2 \cup \cdots$, denoted $\bigcup_{m=1}^{\infty} A_m$, has the probability

$$P\left[\bigcup_{m=1}^{\infty} A_m\right] = \sum_{m=1}^{\infty} P[A_m]. \tag{2.29}$$

Since Axiom 4 subsumes Axiom 3, only three axioms are necessary to define probabilities for any events in Ω. Equation (2.25) implies that the probability of the null event is zero. However, the converse is not true: even if the probability of an event is zero, it does *not* imply that it is the null event. The null event is the mathematical counterpart of an impossible outcome; probability theory assigns probability zero to anything impossible, but does not imply that if an event has probability zero it is impossible. It is entirely conceivable that there is an event A such that $f_N(A) \to 0$ even though $N(A)$ does not remain zero. For example, the probability that you observe the response time of exactly 2 seconds is zero, yet the response time of 2 seconds is a possible event.

Because of Axiom 4, we require that the collection of events be **closed**[5] under the operation of taking countable unions. So we introduce the notion of σ-field.

DEFINITION 2.1 (σ-field).[6] *A collection $\mathcal{F}$ of subsets of Ω is called a σ-field, if it satisfies the following properties:*

$$\begin{array}{|l|}\hline \text{(a) } \emptyset \in \mathcal{F}; \\ \text{(b) if } A \in \mathcal{F}, \text{ then } A^{\mathrm{c}} \in \mathcal{F}; \\ \text{(c) if } A_1, A_2, \ldots \in \mathcal{F}, \text{ then } \bigcup_{m=1}^{\infty} A_m \in \mathcal{F}. \\ \hline \end{array} \tag{2.30}$$

□

Because of the property (c), a σ-field is indeed closed under the operation of taking countable unions. It is not difficult to see that it is also *closed under countable intersections* (Problem 2.13).

Example 2.6: Smallest σ-field. The smallest σ-field associated with Ω is $\mathcal{F} = \{\emptyset, \Omega\}$.

Example 2.7: If A is a subset of Ω, then $\mathcal{F} = \{\emptyset, A, A^{\mathrm{c}}, \Omega\}$ is a σ-field.

[5] In mathematics, a set is said to be *closed* under some *operation* if the operation on members of the set produces a member of the set. For example, the real numbers are closed under subtraction, but the natural numbers are not: 2 and 5 are both natural numbers, but the result of $2 - 5 = -3$ is not.

[6] Also called a σ-algebra.

Consider a collection $\mathcal{C}$ of subsets of Ω. We can find a smallest σ-field that contains all the elements of $\mathcal{C}$.

Example 2.8: Consider a sample space defined by $\Omega = \{a, b, c, d\}$. A set $\mathcal{C} = \{\{a\}, \{b\}\}$ is a subset of Ω, but it is not a field. The complement of a simple event $\{a\}$ is $\{a\}^c = \{b, c, d\}$. Similarly, $\{b\}^c = \{a, c, d\}$. We also find $\{a\} \cup \{b\} = \{a, b\}$ and $(\{a\} \cup \{b\})^c = \{c, d\}$. Then collecting all these subsets of Ω, we find that

$$\{\emptyset, \{a\}, \{b\}, \{a, b\}, \{c, d\}, \{b, c, d\}, \{a, c, d\}, \Omega\}$$

is the smallest σ-field containing all the elements of $\mathcal{C}$. □

We are ready to provide the following formal definition of probability measure and probability space.

DEFINITION 2.2 (Probability measure and probability space). *A* ***probability measure*** *P defined on $(\Omega, \mathcal{F})$ is a function that maps any element of $\mathcal{F}$ into* $[0, 1]$ *such that*

(a) $P[\emptyset] = 0$, $P[\Omega] = 1$;
(b) if $A_1, A_2, \ldots \in \mathcal{F}$ *and* $A_m \cap A_n = \emptyset$ $(m \neq n)$, *then*

$$P\left[\bigcup_{m=1}^{\infty} A_m\right] = \sum_{m=1}^{\infty} P[A_m]. \tag{2.31}$$

The triple $(\Omega, \mathcal{F}, P)$ is called a ***probability space.*** □

Example 2.9: Tossing of two coins and a product space. Let us consider again the experiment of tossing two coins. We now assume that the coins are possibly biased. The sample space of tossing the first coin (coin "1") is denoted $\Omega_1 = \{\mathrm{h}, \mathrm{t}\}$. Its σ-field is $\mathcal{F}_1 = \{\emptyset, \{\mathrm{h}\}, \{\mathrm{t}\}, \Omega_1\}$. A possible probability measure P_1 is given by

$$P_1[\emptyset] = 0,\ P_1[\{\mathrm{h}\}] = p_1,\ P_1[\{\mathrm{t}\}] = 1 - p_1,\ \text{and } P_1[\Omega_1] = 1,$$

where p_1 is a fixed real number in the interval $[0, 1]$. If $p_1 = \frac{1}{2}$, then we say that the coin 1 is unbiased or fair. Thus, the experiment of tossing "coin 1" has the probability space $(\Omega_1, \mathcal{F}_1, P_1)$. We can define the probability space $(\Omega_2, \mathcal{F}_2, P_2)$ for the experiment of tossing "coin 2" in a similar manner, except that $P_2[\{\mathrm{h}\}] = p_2$ and $P_2[\{\mathrm{t}\}] = 1 - p_2$, where p_2 may be different from p_1.

The sample space Ω of the experiment of tossing the two coins is the Cartesian product of the two sample spaces defined above:

$$\Omega = \Omega_1 \times \Omega_2 = \{(\omega_1, \omega_2) : \omega_1 \in \Omega_1, \omega_2 \in \Omega_2\}. \tag{2.32}$$

The σ-field is not so straightforward to construct. It certainly should contain all subsets of $\Omega_1 \times \Omega_2$ of the form $A_1 \times A_2 = \{(a_1, a_2) : a_1 \in A_1, a_2 \in A_2\}$, where A_i is an element of $\mathcal{F}_i$, $i = 1, 2$. Note, however, that the collection of all such sets $\mathcal{F}_1 \times \mathcal{F}_2 = \{A_1 \times A_2 : A_1 \in \mathcal{F}_1, A_2 \in \mathcal{F}_2\}$ is not generally a σ-field.

The smallest σ-field associated with Ω is found as

$$\begin{aligned}\mathcal{F} = \{&\emptyset, (\mathrm{hh}), (\mathrm{ht}), (\mathrm{th}), (\mathrm{tt}),\\ &(\mathrm{hh}) \cup (\mathrm{ht}), (\mathrm{th}) \cup (\mathrm{tt}), (\mathrm{hh}) \cup (\mathrm{th}),\\ &(\mathrm{ht}) \cup (\mathrm{tt}), (\mathrm{hh}) \cup (\mathrm{tt}), (\mathrm{ht}) \cup (\mathrm{th}),\\ &(\mathrm{ht}) \cup (\mathrm{th}) \cup (\mathrm{tt}), (\mathrm{hh}) \cup (\mathrm{th}) \cup (\mathrm{tt}),\\ &(\mathrm{hh}) \cup (\mathrm{ht}) \cup (\mathrm{tt}), (\mathrm{hh}) \cup (\mathrm{ht}) \cup (\mathrm{th}), \Omega\},\end{aligned}$$

where we write, for brevity of notation, simply (hh) to denote the simple event $\{(\mathrm{h}, \mathrm{h})\}$.

We now need to find a suitable probability measure P on $(\Omega, \mathcal{F})$. Since the two experiments (i.e., tossing coin 1 and then coin 2) are considered to be **statistically independent** (see the next section for a formal definition of statistical independence of two events) it is appropriate to define $P[\cdot]$ by

$$P[A_1 \times A_2] = P_1[A_1]P_2[A_2], \tag{2.33}$$

for any $A_i \in \mathcal{F}_i$, $i = 1, 2$. For example, $P[\{(\mathrm{t}, \mathrm{h})\}] = P_1[\{\mathrm{t}\}]P_2[\{\mathrm{h}\}] = (1 - p_1)p_2$. But the domain of the function $P[\cdot]$ defined above must be extended beyond the set $\{\mathcal{F}_1 \times \mathcal{F}_2\}$, since, for instance, the event $A = \{(\mathrm{hh}), (\mathrm{ht}), (\mathrm{th})\}$, is in the σ-field $\mathcal{F}$, but does not take the form $A_1 \times A_2$. It is easy to see, however, that its probability assignment can be found by proper interpretation:

$$\begin{aligned}P[A] &= 1 - P[A^c] = 1 - P[\{(\mathrm{tt})\}] = 1 - P[\{\mathrm{t}\} \times \{\mathrm{t}\}] = 1 - P_1[\{\mathrm{t}\}]P_2[\{\mathrm{t}\}]\\ &= 1 - (1 - p_1)(1 - p_2) = p_1 + p_2 - p_1 p_2.\end{aligned}$$

In a similar manner, we can extend the domain of $P[\cdot]$ to the whole σ-field $\mathcal{F}$. The resulting probability space $(\Omega, \mathcal{F}, P)$ is called the **product space** of the probability spaces $(\Omega_1, \mathcal{F}_1, P_1)$ and $(\Omega_2, \mathcal{F}_2, P_2)$, and the measure $P[\cdot]$ is sometimes called the product measure. There are, of course, other probability measures that can be assigned to $(\Omega, \mathcal{F}, P)$, but if the condition (2.33) is not met for all elements, then such measures may contradict the independence assumption. □

2.3 Bernoulli trials and Bernoulli's theorem

Repeated independent trials are called **Bernoulli trials** if there are only two possible outcomes for each trial and their probabilities remain the same throughout the trials. It is usual to refer to the two possible outcomes as "success" and "failure." The sample space of each individual trial is

$$\Omega = \{\mathrm{s}, \mathrm{f}\},$$

where s stands for success and f for failure. If we denote the probability of the simple event {s} by p, i.e.,

$$P[\{s\}] = p, \quad 0 \le p \le 1, \tag{2.34}$$

then the probability of the event {f} is given by[7]

$$P[\{f\}] = 1 - p \triangleq q, \tag{2.35}$$

since the event {f} is the complement of the event {s}. The probabilities (2.34) and (2.35) constitute the **Bernoulli distribution** for a single trial.

The sample space for an experiment consisting of two independent Bernoulli trials is the Cartesian product of the sample space Ω with itself:

$$\Omega^2 = \Omega \times \Omega = \{(\mathrm{ss}), (\mathrm{sf}), (\mathrm{fs}), (\mathrm{ff})\}.$$

Thus, the Bernoulli distribution for two trials is given by $\{p^2, pq, qp, q^2\}$. In general, the sample space for n Bernoulli trials is the nth-fold Cartesian product of Ω:

$$\Omega^n = \Omega \times \Omega \times \cdots \times \Omega = \{(\mathrm{ss}\ldots\mathrm{s}), (\mathrm{ss}\ldots\mathrm{f}), \ldots, (\mathrm{ff}\ldots\mathrm{s}), (\mathrm{ff}\ldots\mathrm{f})\}.$$

Each of the sample points is made up of a string of n symbols, s or f. Since the trials are independent, the probabilities multiply. For example, the probability of the outcome ssf ... fsf is given by

$$P[\{\mathrm{ssf}\ldots\mathrm{fsf}\}] = ppq \;\ldots\; qpq = p^k q^{n-k}, \tag{2.36}$$

where k is the number of successes and $n - k$ is that of failures in a given outcome of n Bernoulli trials. Equation (2.36) defines the *Bernoulli distribution* of n trials.

If the order in which the successes occur does not matter, then the number of sample points belonging to this event is equal to the number of *combinations* of n things taken k at a time. This number is referred to as the **binomial coefficient** and denoted as $\binom{n}{k}$ and is given by

$$\binom{n}{k} \triangleq \frac{n!}{k!(n-k)!} = \frac{n(n-1)\cdots(n-k+1)}{k \times (k-1)\cdots 2 \times 1}. \tag{2.37}$$

Since each of these sample points (simple events by themselves) has probability $p^k q^{n-k}$, the probability of k successes in n trials is given by

$$\boxed{B(k; n, p) \triangleq \binom{n}{k} p^k q^{n-k}, \quad k = 0, 1, 2, \ldots, n.} \tag{2.38}$$

The set of probabilities $B(k; n, p)$ is called the **binomial distribution**.

7 The notation $\triangleq$ is to be read "is defined as."

If we sum $B(k; n, p)$ over k from 0 to n, we find

$$\sum_{k=0}^{n} B(k; n, p) = \sum_{k=0}^{n} \binom{n}{k} p^k q^{n-k} = (p+q)^n = 1, \tag{2.39}$$

as it should be for the probability distribution.

If we take the ratio of $B(k; n, p)$ over $B(k-1; n, p)$ we have

$$\frac{B(k; n, p)}{B(k-1; n, p)} = \frac{n! p^k q^{n-k}}{(n-k)!k!} \frac{(n-k+1)!(k-1)!}{n! p^{k-1} q^{n-k+1}} = \frac{n-k+1}{k} \frac{p}{q}. \tag{2.40}$$

Therefore, $B(k; n, p) \geq B(k-1; n, p)$ if $k(1-p) \leq (n-k+1)p$ or $k \leq (n+1)p$. Thus, $B(k; n, p)$, as a function of k, increases until

$$k_{\max} = \lfloor (n+1)p \rfloor, \tag{2.41}$$

which is defined as the largest integer not exceeding $(n+1)p$. In other words, $k_{\max}$ represents the most likely value of k, or the most likely number of successes in n trials.

From (2.41) we find that $k_{\max}$ satisfies

$$(n+1)p - 1 < k_{\max} \leq (n+1)p, \tag{2.42}$$

from which we have

$$p - \frac{q}{n} < \frac{k_{\max}}{n} \leq p + \frac{p}{n}. \tag{2.43}$$

Therefore, by taking the limit $n \to \infty$, we find

$$\lim_{n\to\infty} \frac{k_{\max}}{n} = p. \tag{2.44}$$

In other words, the ratio of the most probable number of successes to the number of Bernoulli trials tends to p, the probability of success in a single trial. Equation (2.44) relates the relative frequency k/n to the axiomatic definition of p.

We can state a more general result along this line of argument, which is due to Swiss mathematician Jacob Bernoulli (1654–1705) in his book *Ars Conjectandi* (*The Art of Conjecture*) published in 1713, after his death, by his nephew Nicholas Bernoulli (1687–1759):

THEOREM 2.1 (Bernoulli's theorem (weak law of large numbers)).[8] *Let p denote the probability of success in a single trial, and let k be the number of successes that occur in n independent trials. Then, for any $\epsilon > 0$, the following inequality holds:*

$$P\left[\left|\frac{k}{n} - p\right| > \epsilon\right] \leq \frac{p(1-p)}{n\epsilon^2}. \tag{2.45}$$

[8] There is another Bernoulli's theorem (or Bernoulli's principle, Bernoulli's law) in fluid dynamics, named after Daniel Bernoulli (1700–1782), another nephew of Jacob Bernoulli.

Therefore, in the limit, we have the following ***weak law of large numbers:***[9]

$$\boxed{P\left[\left|\frac{k}{n} - p\right| > \epsilon\right] \to 0, \text{ as } n \to \infty.} \tag{2.46}$$

We say that k/n ***converges*** *to p* ***in probability.***[10]

Proof. First, we observe that $|\frac{k}{n} - p| > \epsilon$, if and only if $(k - np)^2 > n^2\epsilon^2$. Hence,

$$P\left[\left|\frac{k}{n} - p\right| > \epsilon\right] = P\left[(k - np)^2 > n^2\epsilon^2\right]. \tag{2.47}$$

By direct computation, we can find that (see Problem 2.14)

$$\sum_{k=0}^{n} (k - np)^2 B(k; n, p) = np(1 - p). \tag{2.48}$$

The above summation, which we denote by S, can be split into two components, one for those values of k which satisfy $(k - np)^2 > n^2\epsilon^2$ and the other for k such that $(k - np)^2 \leq n^2\epsilon^2$:

$$\begin{aligned}
S &= \sum_{(k-np)^2 > n^2\epsilon^2} (k - np)^2 B(k; n, p) + \sum_{(k-np)^2 \leq n^2\epsilon^2} (k - np)^2 B(k; n, p) \\
&> n^2\epsilon^2 \sum_{(k-np)^2 > n^2\epsilon^2} B(k; n, p) + \sum_{(k-np)^2 \leq n^2\epsilon^2} (k - np)^2 B(k; n, p) \\
&> n^2\epsilon^2 \sum_{(k-np)^2 > n^2\epsilon^2} B(k; n, p) \\
&= n^2\epsilon^2 P\left[\left|\frac{k}{n} - p\right| > \epsilon\right].
\end{aligned} \tag{2.49}$$

Therefore, from the last equation and (2.48), we have

$$np(1 - p) > n^2\epsilon^2 P\left[\left|\frac{k}{n} - p\right| > \epsilon\right], \tag{2.50}$$

which results in (2.45). By taking the limit $n \to \infty$, we prove the weak law of large numbers (2.46). □

The term "law of large numbers (la loi des grands nombres)" was christened by Siméon-Denis Poisson (1781–1840) in 1835 to describe Bernoulli's theorem. This theorem can be paraphrased as

$$\boxed{\lim_{n\to\infty} P\left[\left|\frac{k}{n} - p\right| < \epsilon\right] = 1,} \tag{2.51}$$

for any $\epsilon > 0$.

[9] The adjective "weak" is added to this theorem because stronger versions of the law of large numbers were later developed, as noted after this theorem.

[10] *Convergence in probability* and other modes of convergence, such as *convergence with probability 1*, will be duly discussed in Chapter 11.

Later, the French mathematician Émile Borel (1871–1956) in 1909 and the Italian mathematician Francesco Paolo Cantelli (1875–1966) in 1917 showed, using different approaches, stronger versions of the law of large numbers; i.e., k/n **converges** to p not only in probability, but also **with probability 1**. That is,

$$\boxed{P\left[\lim_{n\to\infty}\left|\frac{k}{n}-p\right|<\epsilon\right]=1,} \tag{2.52}$$

for any $\epsilon > 0$. Hence, Bernoulli's theorem is now called the **weak** law of large numbers. Note that in the stronger version $\lim_{n\to\infty}$ comes inside of $P\,[\,]$. We will further discuss weak and strong laws of large numbers in Chapter 11.

2.4 Conditional probability, Bayes' theorem, and statistical independence

2.4.1 Joint probability and conditional probability

So far we have been concerned primarily with the outcomes of a single experiment. In real-world problems, however, we often deal with the outcomes of combined experiments. A joint (or compound) experiment that consists of one experiment having possible outcomes A_m $(m = 1, 2, \cdots, M)$ and another having the possible outcomes B_n $(n = 1, 2, \cdots, N)$ can be considered as a single experiment having the set of possible outcomes (A_m, B_n). Probabilities relating to such a combined experiment are known as **joint** (or *compound*) probabilities. The joint probability of events A and B is often written as $P\,[A, B]$ instead of $P\,[A \cap B]$. From the axioms of probability and related results discussed in the previous section, it follows that

$$0 \le P\,[\text{A}, \text{B}] \le 1. \tag{2.53}$$

If the M possible events A_m are mutually exclusive and the same property holds for the N possible events B_n, we have

$$\sum_{m=1}^{M}\sum_{n=1}^{N} P\,[A_m, B_n] = 1. \tag{2.54}$$

Both (2.53) and (2.54) can be extended in an obvious fashion to cases in which we deal with more than two basic experiments.

Suppose that the combined experiment is repeated N times, out of which the result A_m occurs $N(A_m)$ times, the result B_n occurs $N(B_n)$ times, and the compound result (A_m, B_n) occurs $N(A_m, B_n)$ times. Then the relative frequency of the compound result is given by

$$f_N(A_m, B_n) = \frac{N(A_m, B_n)}{N}. \tag{2.55}$$

For the moment, let us focus our attention on those $N(A_m)$ trials in each of which the result A_m occurred. In each of these trials, one of the N_B possible results B_n,

$1 \leq n \leq N_B$, occurred; in particular, the result B_n occurred $N(A_m, B_n)$ times. Thus, the relative frequency of occurrence of the result B_n under the assumption that the result A_m also occurred is

$$f_N(B_n \mid A_m) = \frac{N(A_m, B_n)}{N(A_m)}. \tag{2.56}$$

This relative frequency is called the *conditional relative frequency* of B_n on the second experiment, given A_m on the first experiment. Alternatively, it may be expressed as

$$f_N(B_n \mid A_m) = \frac{f_N(A_m, B_n)}{f_N(A_m)}. \tag{2.57}$$

In accordance with (2.57), we define the following.

DEFINITION 2.3 (Conditional probability). *The **conditional probability** that event B occurs given that event A occurs is defined as*

$$\boxed{P[B \mid A] \triangleq \frac{P[A, B]}{P[A]},} \tag{2.58}$$

provided that $P[A] > 0$. *The conditional probability* $P[B \mid A]$ *is undefined if* $P[A] = 0$. □

We can rewrite (2.58) as

$$P[A, B] = P[B \mid A]P[A]. \tag{2.59}$$

Conditional probabilities possess essentially the *same* properties as the *unconditional* probabilities already discussed. The significance of (2.58) is that only the unconditional probabilities are primitive. Prior to Kolmogorov's work both conditional and unconditional probabilities were either defined in terms of equally probable cases or else taken as primitives. In that case, (2.58) was a theorem rather than a definition.

2.4.2 Bayes' theorem

A set of events $A_1, A_2, \ldots, A_n$ is called a **partition** of the sample space Ω if they are a set of mutually exclusive and *exhaustive* events in Ω; i.e., $A_1 \cup A_2 \cup \cdots \cup A_n = \Omega$ and $A_i \cap A_j = \emptyset$ for $i \neq j$. Then we obtain for any event B

$$\bigcup_{j=1}^{n} \{B \cap A_j\} = B, \tag{2.60}$$

and then

$$\sum_{j=1}^{n} P[B, A_j] = \sum_{j=1}^{n} P[A_j]P[B \mid A_j] = P[B]. \tag{2.61}$$

The last formula is sometimes referred to as the **total probability theorem**, meaning that the total probability $P[A]$ of event A is divided into the n components.

As a special case of the total probability theorem (2.61), for any events A and B we can write $P[B]$ as

$$P[B] = P[B \mid A]P[A] + P[B \mid A^c]P[A^c]. \tag{2.62}$$

This is simply because A and A^c form a partition of Ω.

We now state the following useful formula.

THEOREM 2.2 (**Bayes' theorem**). *Let B be an event in a sample space Ω and $A_1, A_2, \ldots, A_n$ be a partition of Ω. Then it can be shown that*

$$\boxed{P[A_j \mid B] = \frac{P[A_j]\,P[B \mid A_j]}{P[B]} = \frac{P[A_j]\,P[B \mid A_j]}{\sum_{i=1}^{n} P[B \mid A_i]P[A_i]}.} \tag{2.63}$$

□

The proof is straightforward, and is left to the reader as an exercise (Problem 2.17). The above theorem is called **Bayes' theorem** or **Bayes' rule**. The probability $P[A_j]$ is called the **prior** probability (or *a priori* probability) of event A_j (before event B occurs), whereas the conditional probability $P[A_j|B]$ is often called the **posterior** probability (or *a posteriori* probability) of event A_j after event B occurs.

Example 2.10: Medical test. Consider some disease and its medical diagnosis test. The following statistics are known about this disease and its medical test.

- For a person with this disease, the test yields a positive result 99% of the time and a negative result 1%.
- For a person without this disease, the test yields a negative result 99% of the time and a positive result 1%.
- Suppose that 1% of the population is infected by this disease and 99% of the population is not.

Suppose that you have taken this test and, unfortunately, the test result is positive. What is the chance that you are indeed infected by this disease?

Answer:

Let A represent a person's condition with respect to this disease and B represent their test result:

$$A = \text{"Not infected by the disease,"} \quad A^c = \text{"Infected by the disease,"}$$
$$B = \text{"Negative test result,"} \quad B^c = \text{"Positive test result."}$$

Let

$$P[A] = p, \ P[A^c] = 1 - p;$$
$$P[B|A] = \alpha, P[B^c|A] = 1 - \alpha;$$
$$P[B^c|A^c] = \beta, P[B|A^c] = 1 - \beta.$$

Then

$$P[A, B] = P[A]P[B|A] = p\alpha;$$
$$P[A, B^c] = p(1-\alpha);$$
$$P[A^c, B] = (1-p)(1-\beta);$$
$$P[A^c, B^c] = (1-p)\beta.$$

Then, the probability that a person is infected by the disease when the medical test is positive is obtained from Bayes' theorem as

$$P[A^c|B^c] = \frac{P[A^c, B^c]}{P[A, B^c] + P[A^c, B^c]} = \frac{(1-p)\beta}{p(1-\alpha) + (1-p)\beta}.$$

If we substitute $p = \alpha = \beta = 0.99$, then

$$P[A^c|B^c] = \frac{0.01 \times 0.99}{0.99 \times 0.01 + 0.01 \times 0.99} = \frac{0.0099}{0.0198} = 0.5.$$

That is, the probability that you have this disease is 50%.

If the test is less accurate, say $\alpha = 0.95$, then $P[A^c|B^c] = 0.167$; hence $P[A|B^c] = 0.833$. Hence, the probability that you are indeed sick is 16.7%.

If the test misses the sick patient more often, say $\beta = 0.95$, while α remains 0.99, then $P[A^c|B^c] = 0.4896$. Hence, the probability that you are sick decreases slightly from 50% to 48.96%.

Suppose the disease is much rarer and only 0.1% of the population is infected; i.e., $p = 0.999$. Then, while keeping $\alpha = \beta = 0.99$, we find $P[A^c|B^c] = 0.09016$, or $P[A|B^c] = 0.90984$. Hence, the probability that you are indeed sick is 9%. This means that the incidence of the disease is so low that the vast majority of people with a positive test result do not actually have the disease. Thus, retesting will improve the reliability of the result. □

2.4.2.1 Frequentist probabilities and Bayesian probabilities

As remarked in Chapter 1, a major controversy in probability theory that has been in existence since almost the birth of probability theory is regarding the types of statements to which probabilities can be assigned. The "frequentists" take the view that probabilities can be assigned only to the outcomes of an experiment that can be repeated a number of times, whereas the "Bayesians" take the subjective view of probability and believe that the notion of probability is applicable to any situation or event for which we attach some uncertainty or belief. In the medical test example discussed above, the frequentists will argue that the probability $P[A]$, the probability that a given individual is infected by this disease, should be given only if sufficient clinical data have been collected with a large number of individuals.

The Bayesians might argue that this kind of experiment – i.e., to find a certain individual is infected by the particular disease – is not repeatable. Even if we accept the existence of a *perfect* diagnostic method – which in reality is rare – and accept to estimate $P[A]$ by the proportion of infected people in individuals with backgrounds similar to that of this individual in question, the frequentists' method of probability assignments suffers from the problem that it requires a very large number of such individuals to be diagnosed. The Bayesians argue that a "subjective probability" can be and should be assigned to any situation for which we have uncertainty or belief. Calling the probabilities subjective does not imply, however, that they can be assigned arbitrarily. The assignment ought to be consistent with the axioms of the probability discussed in Section 2.2. This Bayesian view of probability is appealing to those who investigate **learning theory** from the probabilistic point of view. Referring to the above example, $P[A]$ is the prior probability or **belief** that the given individual is infected by the disease, and its posterior probability $P[A|B_i]$ is its updated belief based on **data** B_i, and this Bayes' theorem provides the fundamental principle of *learning* based on data or **evidence**.

Although this different view of probability assignment in the above example might appear as a mere philosophical argument, the significance of the difference between the two schools of thought will become clearer when we revisit Bayes' theorem in Section 4.5, where we deal with **parameters**, such as means and variances, associated with probability distributions.

2.4.3 Statistical independence of events

As interpreted above, $P[B \mid A]$ is the probability of occurrence of event B assuming the occurrence of event A. In general, this posterior probability is different from the prior probability $P[B]$, because knowing that event A has happened should help us sharpen our ability to infer about the occurrence of B. In other words, information about the result of one experiment should generally decrease our uncertainty about possible results of the other experiment. If not, we shall say that B is statistically independent of A.

DEFINITION 2.4 (Statistical independence). *Event B is said to be* statistically independent[11] *of event A if*

$$P[B \mid A] = P[B], \tag{2.64}$$

or, equivalently, if

$$\boxed{P[A, B] = P[A]P[B].} \tag{2.65}$$

Then the events A and B are said to be statistically independent. □

[11] Some authors (e.g., [99]) use the term "stochastically independent," while others (e.g., [131]) simply use the term "independent" without any adverb.

Equations (2.64) and (2.65) are also equivalent to

$$P[A \mid B] = P[A]; \tag{2.66}$$

that is, A is also statistically independent of B.

An interpretation of (2.64) and (2.65) is that if the events A and B are statistically independent, then knowledge of the occurrence of one event tells us no more about the probability of occurrence of the other event than what we know without that knowledge.

When more than two events are to be considered, the situation becomes more complicated. A set of M events A_m $(m = 1, 2, \cdots, M)$ is said to be *mutually independent* if and only if the probability of every intersection of M or fewer events equals the product of the probabilities of the constituents. For example, three events $A,\ B,\ C$ are mutually independent when

$$\begin{aligned} P[A, B] &= P[A]P[B], \\ P[B, C] &= P[B]P[C], \\ P[A, C] &= P[A]P[C], \end{aligned} \tag{2.67}$$

and

$$P[A, B, C] = P[A]P[B]P[C]. \tag{2.68}$$

No three of these relations necessarily implies the fourth. If only the equations in (2.67) are satisfied, we say that the events are *pairwise independent*. Pairwise independence does not imply mutual independence.

Example 2.11: Throwing two dice. Suppose that two true dice are thrown and that the dice are distinguishable. An outcome of this experiment is denoted by (m, n), where m and n are the faces of the dice. Let A and B be the following events of this experiment:

$$\begin{aligned} A &= \{m + n = 11\}, \\ B &= \{n \neq 5\}. \end{aligned}$$

Then we find

$$\begin{aligned} P[A] &= P[(5, 6)] + P[(6, 5)] \\ &= P[\{m = 5\}]P[\{n = 6\}] + P[\{m = 6\}]P[\{n = 5\}] = \frac{1}{18}, \\ P[B] &= 1 - P[\{n = 5\}] = \frac{5}{6}, \end{aligned}$$

and

$$P[A, B] = P[\{m = 5\} \cap \{n = 6\}] = \frac{1}{36}.$$

Therefore,

$$P[A, B] \neq P[A]P[B].$$

Thus, the events A and B are not statistically independent.

2.5 Summary of Chapter 2

Relative frequency of an event:	$f_N(A) = \frac{N(A)}{N}$	(2.1)
Properties of events:		
Complement of A	$A^c = \{\omega : \omega \text{ does not belong to } A\}$	(2.11)
Union of A and B	$A \cup B = \{\omega : \omega \text{ belongs to } A \text{ or } B\}$	(2.12)
Intersection of A and B	$A \cap B = \{\omega : \omega \text{ belongs to } A \text{ and } B\}$	(2.13)
Null event	$\emptyset = \text{empty event} = A \cap A^c$	
Sure event	$\Omega = \text{sample space} = A \cup A^c$	
A and B are disjoint	$A \cap B = \emptyset$	
Commutative laws	$A \cup B = B \cup A$	
	$A \cap B = B \cap A$	(2.14)
Associative laws	$(A \cup B) \cup C = A \cup (B \cup C)$	
	$(A \cap B) \cap C = A \cap (B \cap C)$	(2.15)
Distributive laws	$A \cap (B \cup C) = (A \cap B) \cup (A \cap C)$	
	$A \cup (B \cap C) = (A \cup B) \cap (A \cup C)$	(2.16)
DeMorgan's laws	$(A \cup B)^c = A^c \cap B^c$	(2.17)
	$(A \cap B)^c = A^c \cup B^c$	(2.18)
Axioms of probability:		
Axiom 1	$P[A] \geq 0$	(2.19)
Axiom 2	$P[\Omega] = 1$	(2.20)
Axiom 3	If $A \cap B = \emptyset$, $P[A \cup B] = P[A] + P[B]$	(2.21)
Axiom 4	If $A_1, A_2, \ldots, A_M$ are mutually exclusive,	
	$P\left[\bigcup_{m=1}^{M} A_m\right] = \sum_{m=1}^{M} P[A_m]$	(2.22)
Prob. of union of two events:	$P[A \cup B] = P[A] + P[B] - P[A \cap B]$	(2.27)
σ-field $\mathcal{F}$:	(a) $\emptyset \in \mathcal{F}$	
	(b) $A \in \mathcal{F} \Rightarrow A^c \in \mathcal{F}$	
	(c) $A_1, A_2, \ldots \in \mathcal{F} \Rightarrow \bigcup_{m=1}^{\infty} A_m \in \mathcal{F}$	(2.30)
Binomial distribution:	$B(k; n, p) \triangleq \binom{n}{k} p^k q^{n-k}, \quad k = 0, 1, \ldots, n$	(2.38)
Bernoulli's theorem:	$P\left[\left\lvert\frac{k}{n} - p\right\rvert > \epsilon\right] \to 0, \quad \text{as } n \to \infty$	(2.46)
Conditional probability:	$P[B \mid A] \triangleq \frac{P[A, B]}{P[A]}$	(2.58)
Bayes' theorem:	$P[A_j \mid B] = \frac{P[A_j]\, P[B \mid A_j]}{P[B]} = \frac{P[A_j]\, P[B \mid A_j]}{\sum_{i=1}^{n} P[B \mid A_i] P[A_i]}$	(2.63)
Statistical independence of events:	$P[A, B] = P[A] P[B]$	(2.65)

2.6 Discussion and further reading

The materials presented in this chapter can be found in most books on probability theory and its applications, e.g., Bertsekas and Tsitsiklis [24], Blake [31], Davenport and Root [77], Feller [99], Fine [105], Gray and Davisson [126], Grimmett and Stirzaker [131], Gubner [133], Leon-Garcia [222], Nelson [254], Papoulis and Pillai [262], Ross [289], Stark and Woods [310], Thomas [319], Trivedi [327], Wilks [353], Yates and Goodman [363]. Virtually all textbooks today take the axiomatic approach. When we deal with experiments that have equally probable outcomes (or simple events), such as tossing fair coins, throwing dice, or drawing cards, combinatorial analysis is a very powerful tool for probabilistic analysis. In the correspondence with Pascal in 1654, Fermat proposed a combinatorial method in answering the questions on gambling.

We assume that the reader is familiar with basic combinatorics such as those required in answering some questions of this chapter. For those who wish to review combinatorial mathematics or further study the subject, they are directed to Feller [99], Nelson [254], and Ross [289], who devote an entire chapter to combinatorics.

Discussion of subjective probability is found, for example, in Hacking [136].

2.7 Problems

Section 2.2: Axioms of probability

2.1* Tossing a coin three times. Consider the experiment of tossing a coin three times.

(a) What is the sample space Ω?
(b) Define event E_i as an outcome where exactly i tosses yield "heads," $i = 0, 1, 2, 3$. How many sample points does E_i contain, $i = 0, 1, 2, 3$?
(c) Define event F as an event in which at least two of the tosses yield "heads."

2.2* Tossing a coin until "head" or "tail" occurs twice in succession [99]. Consider an experiment in which a coin is tossed until "head (h)" or "tail (t)" appears twice in succession. Examples of simple events are {thh} and {hthtt}. Find the sample space Ω of this experiment.

2.3 Placing distinguishable particles in different cells [99]. Consider the experiment of placing three distinguishable particles (which we denote by a, b, and c) into three cells. Examples of sample points are $\omega_1 = (abc| \; - \; | -)$, $\omega_2 = (a\,c|\,b\,| -)$, etc.

(a) Write down the sample space Ω. How many sample points are there in Ω?
(b) Let events A, B, C be defined as

$$A = \text{multiple particles occupy a cell;}$$
$$B = \text{the first cell is not empty;}$$
$$C = \text{both } A \text{ and } B \text{ occur.}$$

How many sample points are there in events A, B, and C respectively?

2.4 Placing indistinguishable particles in different cells [99]. Consider the same experiment as Problem 2.3, but suppose that the three particles are not distinguishable. Examples of sample points are $\omega_1 = (***| - | -)$, $\omega_2 = (** \mid * \mid -)$, etc.

(a) Write down the sample space Ω. How many sample points are there?
(b) Let event $\tilde{A}$, $\tilde{B}$, $\tilde{C}$ be defined as

$$\tilde{A} = \text{multiple particles occupy a cell;}$$
$$\tilde{B} = \text{the first cell is not empty;}$$
$$\tilde{C} = \text{both } \tilde{A} \text{ and } \tilde{B} \text{ occur.}$$

How many sample points are there in events $\tilde{A}$, $\tilde{B}$, and $\tilde{C}$ respectively?

2.5* Probability assignment to the coin tossing experiment. Consider the coin tossing experiment of Problem 2.1. If the coin tossing is fair, what is the appropriate probability measure for this experiment? Find the probability of event E_i ($i = 0, 1, 2, 3$).

2.6* Probability assignment to the coin tossing experiment in Problem 2.2. Consider the coin tossing experiment described in Problem 2.2. Assume that the coin is a fair coin; i.e., "h" and "t" appear with probability 1/2 each.

(a) Find the appropriate probability assignment for each sample point in Ω.
(b) What is the probability that the experiment ends before the sixth toss?
(c) What is the probability that an even number of tosses is required?

2.7 Placing distinguishable particles in cells: Maxwell–Boltzmann statistics. Consider the experiment of Problem 2.3. Assume that a particle is equally probable to be in any of the cells.

(a) What should be the appropriate probability measure for this experiment?
(b) What is then the probability p that each of the three cells contains exactly one particle?
(c) Generalize the experiment and consider r distinguishable particles and n distinguishable cells, where $n \geq r$. Show that the probability that, in each of r preselected cells, one and only one particle is found is given by

$$\frac{r!}{n^r}. \tag{2.69}$$

In statistical mechanics, this is called the (classical) Maxwell–Boltzmann statistics.

2.8 Placing indistinguishable particles in cells: Bose–Einstein statistics. Consider the experiment of Problem 2.4. Assume that all distinguishable arrangements are equally likely.

(a) What should be the appropriate probability measure in this case?
(b) What is then the probability p that each of the three cells contains exactly one particle?

(c) Generalize the experiment and consider r indistinguishable particles and n distinguishable cells, where $n \geq r$. Show that the probability that, in each of r preselected cells, one and only one particle is found is given by

$$\binom{r+n-1}{r}^{-1}. \tag{2.70}$$

In statistical mechanics, this is called the **Bose**[12] – Einstein statistics.

2.9 Placing at most one particle in a cell: Fermi–Dirac statistics. Consider again the experiment of placing r indistinguishable particles in n distinguishable cells. Assume that (i) it is not possible for two or more particles to be in the same cell (this constraint is called the "**Pauli**[13] exclusion principle" in quantum mechanics) and (ii) all distinguishable arrangements satisfying the above constraint have equal probabilities.

Show that the probability that, in each of r preselected cells, one and only one particle is found is given by

$$\binom{n}{r}^{-1}, \tag{2.71}$$

which is known as **Fermi**[14]**–Dirac**[15] statistics.

Section 2.3: Bernoulli trials and Bernoulli's theorem

2.10* Distribution laws and Venn diagram. Prove the distribution laws (2.16) using Venn diagrams.

2.11* DeMorgan's law. Show that $(A \cap B)^c = A^c \cup B^c$; i.e., (2.18).

2.12 Axiom 3. Derive (2.22) using mathematical induction. Show

(a) the basis step: (2.22) is true for $M = 2$;
(b) the induction step: if (2.22) is true for $M = N(\geq 2)$, then it also holds for $M = N + 1$.

2.13 Closure property of σ-field. Show that a σ-field is closed under the operation of taking countable intersections:

[12] Satyendra Nath Bose (1894–1974) was an Indian mathematical physicist, best known for his work on quantum mechanics in the early 1920s, providing the foundation for Bose–Einstein statistics and the theory of the Bose–Einstein condensate. He is honored as the namesake of the boson.

[13] Wolfgang Ernst Pauli (1900–1958) was an Austrian theoretical physicist noted for his work on the theory of spin, and in particular the discovery of the exclusion principle.

[14] Enrico Fermi (1901–1954) was an Italian physicist, noted for his work on the development of the first nuclear reactor and for his contributions to the development of quantum theory, nuclear and particle physics, and statistical mechanics. Fermi was awarded the Nobel Prize in Physics in 1938 for his work on induced radioactivity.

[15] Paul Adrien Maurice Dirac (1902–1984) was a British theoretical physicist and a founder of the field of quantum mechanics. Dirac shared the 1933 Nobel Prize in Physics with Erwin Schrödinger, "for the discovery of new productive forms of atomic theory."

$$\text{if } A_1, A_2, \ldots \in \mathcal{F} \text{ then } \bigcap_{m=1}^{\infty} A_m \in \mathcal{F}.$$

2.14* Derivation of (2.48). Derive (2.48).

2.15 Multinomial coefficient. Let $k_1, k_2, \ldots, k_r$ be a set of nonnegative integers such that

$$n_1 + n_2 + \ldots + n_r = n.$$

Now consider the problem of partitioning n items into r groups. Show that the number of different ways of partitioning n items in such a way that k_i items are placed in group i $(i = 1, 2, \ldots, r)$ is given by

$$\frac{n!}{k_1!k_2!\cdots k_r!}. \tag{2.72}$$

This number is referred to as the **multinomial coefficient**. Needless to say, for $r = 2$, we have the binomial coefficient (2.37).

Section 2.4: Conditional probability, Bayes' theorem, and statistical independence

2.16* Joint probabilities. Interpret Eq. (2.54) in terms of the relative frequencies of compound results (A_m, B_n).

2.17* Proof of Bayes' theorem. Derive (2.63) of Bayes' theorem.

2.18* Independent events. Let A and B be two independent events such that with probability $\frac{1}{12}$ they will occur simultaneously, and with probability $\frac{1}{3}$ neither of them will occur. Obtain $P[A]$ and $P[B]$.

2.19* Medical test. Consider the medical test discussed in Example 2.10.

(a) Suppose

$$p = 0.999, \alpha = 0.95, \beta = 0.99.$$

Then what is the probability that a positive result is a false positive?

(b) Change the parameters to

$$p = 0.999, \alpha = 0.99, \beta = 0.95.$$

What is the probability that a positive result is a false positive?

2.20 Birthday problem.[16] In a group of r persons, what is the probability that each person has a distinct birthday? We assume that birth rates are constant throughout the year, and ignore complications due to leap years. Evaluate (approximately) this probability for $r = 23$ and $r = 56$.

Hint: Use the approximation $\ln(1 - x) \approx -x$ for $|x| \ll 1$.

[16] Feller [99] notes that this problem was first discussed by R. von Mises, "Über Aufteilungs- und Besetzungs-Wahrscheinlichkeiten," *Revue de la Faculté des Sciences de l'Université d'Istanbul, N.S.* vol. 4 (1938–1939), pp. 145–163.

2.21 Web access pattern. Consider a web server that receives web access requests from a large number of users. We assume that the request arrival mechanism is characterized by the following simple model. We divide the time axis into contiguous segments of Δ seconds, and Δ is chosen sufficiently small that the probability of receiving more than one request is negligibly small. We also assume that the arrivals in different segments are statistically independent events. Let p be the probability that a randomly chosen segment interval observes an arrival of a request.

(a) Assume an observation interval of $T = 5\Delta$, and let $p = 0.2$. What is the probability that at least two requests arrive during this observation interval?
(b) For any integer $n = 1, 2, 3, \ldots,$ find the probability that $T = n\Delta$ will be the observation interval required to see the first arrival.
(c) Suppose that you are informed that there are k successes in n Bernoulli trials. Obtain the conditional probability that any particular trial resulted in a success.

3 Discrete random variables

3.1 Random variables

In Section 2.2 we defined a sample space, sample points, events, and a probability measure assigned to the events. Recall that an event is a set of sample points. Now we introduce the notion of a *random variable*.[1] A real-valued function $X(\omega)$ defined on a sample space Ω of points ω is called a *random variable*[2], which we abbreviate as RV.[3] That is, the RV is an association of a real number with each sample point in the sample space. Thus, $X(\omega)$ may be regarded as a function that maps Ω into the real line: given any sample point ω, the function $X(\cdot)$ specifies a finite real number of $X(\omega)$. A simple example of such a mapping is illustrated in Figure 3.1.

For example, in the coin-tossing experiment, Ω contains only two sample points, the head and tail. Now we wish to associate 1 with the head and 0 with the tail. Then the mapping

$$X(\omega) = \begin{cases} 1, & \text{if } \omega = \text{head}, \\ 0, & \text{if } \omega = \text{tail}, \end{cases}$$

is clearly a random variable. In our example of measuring the response time of a request to a web server, Ω itself is the real line (the positive half-line). The function $X(\omega) = \omega$ is clearly a legitimate RV. So are $X(\omega) = 1/\omega$, $X(\omega) = \omega^2$, etc.

Two RVs X and Y defined on a probability space $(\Omega, \mathcal{F}, P)$ are said to be equal (everywhere), written $X = Y$, if $X(\omega) = Y(\omega)$ for all $\omega \in \Omega$; i.e., X and Y are equal if they are identical as functions mapping the sample space Ω to the real line. In probability theory and applications, a weaker type of equivalence between two RVs arises frequently. The RVs X and Y are said to be equal **almost surely (a.s.)** or **with probability one** if $P[\{\omega : X(\omega) = Y(\omega)\}] = 1$. Another way of expressing this is that the event that X and Y differ has zero probability; i.e., $P[\{\omega : X(\omega) \neq Y(\omega)\}] = 0$. We denote almost sure equivalence of X and Y by $X \overset{\text{a.s.}}{=} Y$ or write simply, $X = Y$, a.s. More generally, an event A is said to occur almost surely, or with probability one, if $P[A] = 1$.

[1] Although we discuss primarily discrete RVs in the present chapter, the definitions and properties we present in this section are applicable to both discrete and continuous RVs.

[2] Although we limit ourselves to real-valued functions here, we can generalize our treatment to complex-valued functions.

[3] The notation "RV" should be pronounced as "random variable." To avoid possible confusion, however, we write "a random variable" instead of "a RV."

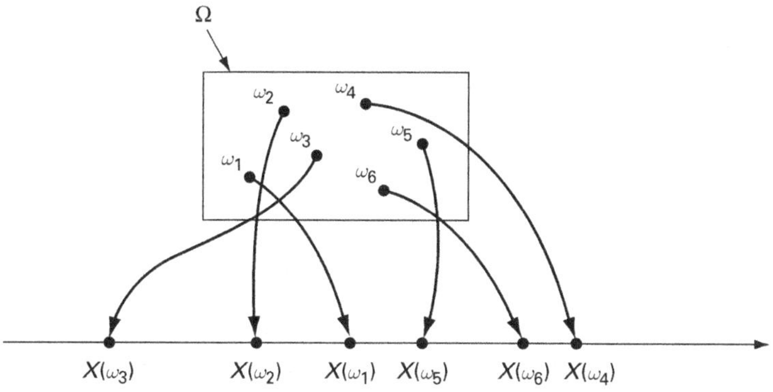

Figure 3.1 A random variable $X(\omega)$ as a mapping from Ω to the real line.

When a relation holds between two RVs almost surely, e.g., $X = Y$ or $X \leq Y$, it is common practice to omit the explicit designation "a.s.," and we will often follow this convention in this textbook.

Since many readers may find this kind of mathematical abstraction unexciting at best, we hasten to point out that it is sufficient here to think of a random variable as a symbol for a number that is going to be produced by a random experiment. Once produced, the number is, of course, no longer random and is called a *realization* or *instance* of the RV.

The word random applies to the process that produces the number, rather than to the number itself. Hereafter, in referring to these functions, we often delete empty parentheses and simply write X to denote the function $X(\cdot)$.

3.1.1 Distribution function

A random variable X is characterized by its *distribution function* $F_X(x)$:

$$\boxed{F_X(x) \triangleq P[\{\omega : X(\omega) \leq x\}],} \tag{3.1}$$

or simply

$$F_X(x) = P[X \leq x]. \tag{3.2}$$

The properties of distribution functions listed below follow directly from the definition in (3.1) or (3.2).

$$\boxed{\begin{array}{ll} \text{Property 1.} & F_X(x) \geq 0, \quad \text{for } -\infty < x < \infty. \\ \text{Property 2.} & F_X(-\infty) = 0. \\ \text{Property 3.} & F_X(\infty) = 1. \\ \text{Property 4.} & \text{If } b > a,\ F_X(b) - F_X(a) = P[a < X \leq b] \geq 0. \end{array}} \tag{3.3}$$

The first three properties follow, respectively, from the following facts: $F_X(x)$ is a probability, $P[\emptyset] = 0$, and $P[\Omega] = 1$. Property 4 follows from the fact that

$$\{\omega : X(\omega) \leq a\} \cap \{\omega : a < X(\omega) \leq b\} = \emptyset.$$

and

$$\{\omega : X(\omega) \leq a\} \cup \{\omega : a < X(\omega) \leq b\} = \{\omega : X(\omega) \leq b\}$$

(see Problem 3.1).

3.1.2 Two random variables and joint distribution function

Now we proceed to the case of two RVs. Given functions $X(\omega)$ and $Y(\omega)$ defined on the sample space Ω, we define the *joint distribution function* $F_{XY}(x, y)$ of the RVs X and Y by

$$F_{XY}(x, y) \triangleq P[\{\omega : X(\omega) \leq x, Y(\omega) \leq y\}] = P[X \leq x, Y \leq y]. \tag{3.4}$$

Thus, $F_{XY}(x, y)$ is the probability assigned to the set of all points ω that are associated with the region of the two-dimensional Euclidean space that is shaded in Figure 3.2.

The properties of joint distribution functions listed below follow directly from the definition in (3.4).

Property 1. $F_{XY}(x, y) \geq 0;$ for $-\infty < x < \infty, \ -\infty < y < \infty.$
Property 2. $F_{XY}(x, -\infty) = 0;$ for $-\infty < x < \infty,$
$F_{XY}(-\infty, y) = 0;$ for $-\infty < y < \infty.$
Property 3. $F_{XY}(\infty, \infty) = 1.$
Property 4. If $b > a$ and $d > c$,
$F_{XY}(b, d) \geq F_{XY}(b, c) \geq F_{XY}(a, c).$
Property 5. $F_{XY}(x, \infty) = F_X(x),$
$F_{XY}(\infty, y) = F_Y(y).$ (3.5)

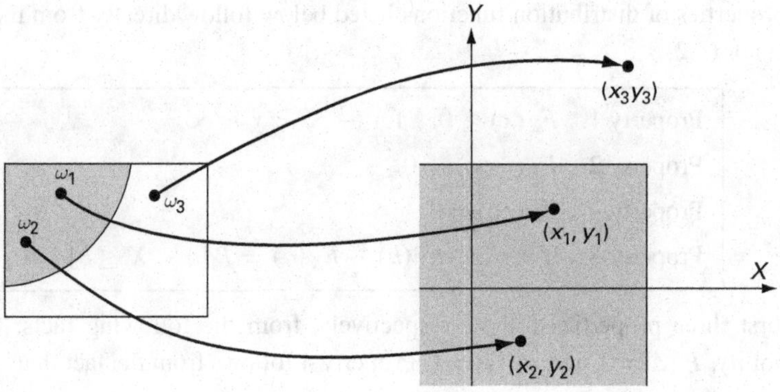

Figure 3.2 RVs $X(\omega)$ and $Y(\omega)$ as mappings from Ω to the two-dimensional Euclidean space.

Properties 1 through 4 are obvious extensions of the corresponding properties in (3.3). Property 5 is a consequence of the fact that

$$\begin{aligned}\{\omega : X(\omega) \leq x\} \cap \{\omega : Y(\omega) < \infty\} &= \{\omega : X(\omega) \leq x\} \cap \Omega \\ &= \{\omega : X(\omega) \leq x\}.\end{aligned}$$

Thus, $F_{XY}(x, \infty)$ and $F_{XY}(\infty, y)$ are both ordinary one-variable distribution functions. They are, respectively, the distributions of X and Y alone, and are usually designated as the *marginal distribution functions*. In summary, $F_{XY}(x, y)$ is a nondecreasing function of both arguments and is always bounded by zero and one.

These definitions and results are extendable in a more or less obvious manner to the case of multidimensional RVs: let $X_1, X_2, \ldots, X_m$ be RVs defined on Ω, and let $\boldsymbol{X}$ denote the m-tuple $(X_1, X_2, \ldots, X_m)$. We then define the m-dimensional joint distribution function $F_{\boldsymbol{X}}(\boldsymbol{x})$ as

$$\begin{aligned}F_{\boldsymbol{X}}(\boldsymbol{x}) &= P\left[\{\omega : X_1(\omega) \leq x_1, X_2(\omega) \leq x_2, \ldots, X_m(\omega) \leq x_m\}\right] \\ &= P\left[X_1 \leq x_1, X_2 \leq x_2, \ldots, X_m \leq x_m\right],\end{aligned} \tag{3.6}$$

where $\boldsymbol{x} = (x_1,\ x_2,\ \ldots,\ x_m)$. We refer to $\boldsymbol{X}$ as an m-dimensional *vector* of RVs or, simply, as a *random vector*.

3.2 Discrete random variables and probability distributions

Random variable X is called a *discrete random variable* (discrete RV) if the range of the function $X(\omega)$ consists of isolated points on the real line; that is, if X can take on only a finite or countably infinite number of values $\{x_1, x_2, x_3, \ldots\}$. For example, the number of heads appearing in N tosses of a coin is a discrete RV.

For a discrete RV X, we denote by $p_X(x_i)$, the *probability* that X takes the value x_i:

$$\boxed{p_X(x_i) \triangleq P[X = x_i],\quad i = 1, 2, \ldots} \tag{3.7}$$

The complete set of probabilities $\{p_X(x_i)\}$ associated with the possible values x_i of X is called the **probability distribution** of the discrete RV X. The probability distribution and the distribution function defined by (3.1) are related by

$$F_X(x) = \sum_{x_i \leq x} p_X(x_i). \tag{3.8}$$

Therefore, the distribution function is often referred to as the **cumulative distribution function** (CDF). The function, $p_X : \mathbb{R} \to [0, 1]$, defined by

$$\boxed{p_X(x) = P[X = x],\quad x \in \mathbb{R},} \tag{3.9}$$

is called the **probability mass function** (PMF). Since $p_X(x) = 0$ whenever $x \notin \{x_1, x_2, \ldots\}$, the PMF provides an equivalent characterization of the discrete RV X as its probability distribution $\{p_X(x_i)\}$.

Alternatively, we can write

$$F_X(x) = \sum_i p_X(x_i)u(x - x_i), \quad -\infty < x < \infty, \tag{3.10}$$

where $u(t)$ is the *unit step function* defined by

$$u(x) = \begin{cases} 1, & \text{for } x \geq 0, \\ 0, & \text{for } x < 0. \end{cases} \tag{3.11}$$

The formal derivative of this last equation is

$$f_X(x) = \frac{dF_X(x)}{dx} = \sum_i p_X(x_i)\delta(x - x_i), \quad -\infty < x < \infty, \tag{3.12}$$

where $\delta(t)$ is the *Dirac delta function* [81] or the *impulse function* defined by

$$\delta(x) \triangleq \frac{du(x)}{dx} = 0, \quad \text{for } x \neq 0, \tag{3.13}$$

and

$$\int_{-\infty}^{\infty} \delta(x)\, dx = \int_{-\epsilon}^{\epsilon} \delta(x)\, dx = 1 \tag{3.14}$$

for any $\epsilon > 0$. A formal definition of the delta function is given in terms of its *sampling property* with regard to an arbitrary continuous function $g(t)$:

$$\int_{-\infty}^{\infty} g(x)\delta(x - a)\, dx \triangleq g(a). \tag{3.15}$$

The function $f_X(x)$ of (3.12) is called the *probability density function* (PDF) of the discrete RV X. As we shall see in Section 4.1, continuous RVs are characterized naturally by their PDFs. Hence, the representation (3.12) allows both discrete and continuous RVs to be considered in terms of their PDFs. By letting x go to infinity in (3.10), we find that

$$F_X(\infty) = \sum_{\text{all } i} p_X(x_i) = 1. \tag{3.16}$$

An example of a probability distribution and the corresponding distribution function is shown in Figure 3.3.

3.2.1 Joint and conditional probability distributions

In a similar manner we define the **joint probability distribution** of two discrete RVs X and Y as the set of probabilities $\{p_{XY}(x_i, y_j)\}$ for all possible values of the pairs (x_i, y_j):

$$p_{XY}(x_i, y_j) \triangleq P[X = x_i, Y = y_j]. \tag{3.17}$$

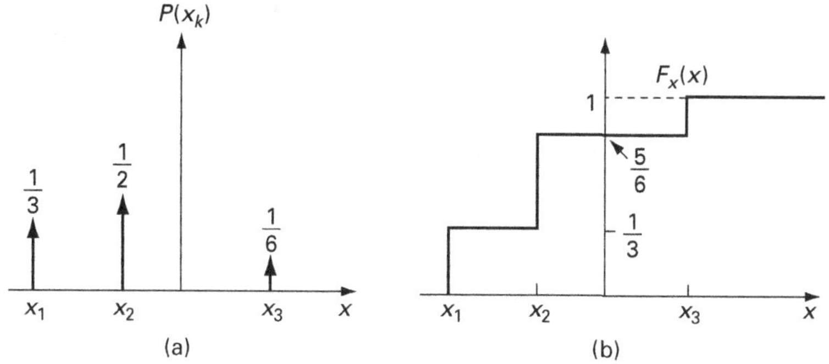

Figure 3.3 (a) The probability distribution and (b) the distribution function of a discrete RV.

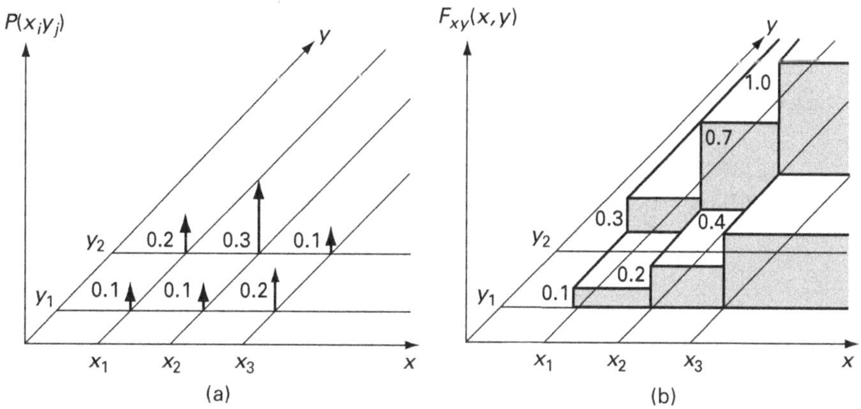

Figure 3.4 (a) The joint probability distribution and (b) the joint distribution function.

The corresponding joint distribution function defined by (3.4) is given by

$$F_{XY}(x, y) = \sum_{x_i \le x} \sum_{y_j \le y} p_{XY}(x_i, y_j). \tag{3.18}$$

Therefore,

$$F_{XY}(\infty, \infty) = \sum_{\text{all } i} \sum_{\text{all } j} p_{XY}(x_i, y_j) = 1. \tag{3.19}$$

An example of the joint probability distribution and the associated joint distribution function of two discrete RVs is illustrated in Figure 3.4.

The set of values that a discrete RV (or a random vector) takes on can be regarded as the set of distinct sample points or distinct simple events. The various results obtained in Section 2.4 for a sample space and events are therefore directly translatable to the case of discrete RVs. Thus, (3.19) follows directly from (2.54).

Similarly, we define the conditional probability distributions:

$$p_{Y|X}(y_j|x_i) \triangleq P[Y = y_j|X = x_i] \text{ and } p_{X|Y}(x_i|y_j) \triangleq P[X = x_i|Y = y_j]. \tag{3.20}$$

Then, the relations among the joint probability distribution, the conditional probability distribution, and the marginal distributions for discrete RVs are now self-evident:

$$p_{XY}(x_i, y_j) = p_X(x_i)p_{Y|X}(y_j|x_i) = p_Y(y_j)p_{X|Y}(x_i|y_j), \tag{3.21}$$

$$\sum_{\text{all } j} p_{Y|X}(y_j|x_i) = \sum_{\text{all } i} p_{X|Y}(x_i|y_j) = 1, \tag{3.22}$$

$$p_X(x_i) = \sum_{\text{all } j} p_{XY}(x_i, y_j) \text{ and } p_Y(y_j) = \sum_{\text{all } i} p_{XY}(x_i, y_j), \text{ etc.} \tag{3.23}$$

The **conditional distribution function** of X given Y is defined by

$$F_{X|Y}(x|y) \triangleq P[X \le x|Y = y] = \sum_{x_i \le x} p_{X|Y}(x_i|y), \tag{3.24}$$

when $P[Y = y] > 0$; i.e., $y = y_j$ for some j.

The notion of *statistical independence* of two events is also directly applicable to two discrete RVs.

DEFINITION 3.1 (**Independent random variables**). *We say that random variables X and Y are* independent *or* statistically independent *if and only if*

$$\boxed{p_{XY}(x_i, y_j) = p_X(x_i)p_Y(y_j), \quad \text{for all values } (x_i, y_j),} \tag{3.25}$$

or, equivalently, if and only if

$$F_{XY}(x_i, y_j) = F_X(x_i)F_Y(y_j), \quad \text{for all values of } x_i \text{ and } y_j. \tag{3.26}$$

□

Equivalence between (3.25) and (3.26) can be shown easily by using the relations (3.8) and (3.18) (Problem 3.3).

Similarly, the discrete RVs $X, Y, \ldots, Z$ are said to be independent RVs if and only if

$$p_{XY\cdots Z}(x_k, y_l, \ldots, z_m) = p_X(x_k)p_Y(y_l)\cdots p_Z(z_m) \tag{3.27}$$

is satisfied for all values $(x_k, y_l, \ldots, z_m)$ or, equivalently,

$$F_{XY\cdots Z}(x_k, y_l, \ldots, z_m) = F_X(x_k)F_Y(y_l)\cdots F_Z(z_m) \tag{3.28}$$

for all values of $x_k, y_l, \ldots, z_m$.

Consider, for instance, an experiment in which we toss an ordinary die n times. Let X_n denote the result of the nth toss, $n = 1, 2, \ldots, N$. Clearly, X_n is a discrete RV that takes on only integers between 1 and 6. The *empirical average* (also called the *sample mean*) of the N results, denoted by $\overline{X}_N$, is

$$\overline{X}_N = \frac{1}{N}\sum_{n=1}^{N} X_n. \tag{3.29}$$

Let $N(i)$ denote the number of tosses that result in the integer i, $1 \leq i \leq 6$, then the summation of (3.29) can be rewritten as

$$\overline{X}_N = \frac{1}{N}\sum_{i=1}^{6} iN(i) = \sum_{i=1}^{6} if_N(i), \tag{3.30}$$

where $f_N(i) = N(i)/N$ is the *relative frequency* of the outcome i (cf. (2.1) of Chapter 2). Since the X_n are RVs, so is their sample mean $\overline{X}_N$.[4] But when N becomes sufficiently large, $f_N(i)$ will tend to the probability $p_X(i)$. Thus, for large N, we expect $\overline{X}_N$ to stabilize at the value $E[X]$ defined by

$$E[X] \triangleq \sum_{1 \leq i \leq 6} ip_X(i). \tag{3.31}$$

We call $E[X]$ the *expectation* of X. We often write the expectation of X as μ_X for conciseness.

DEFINITION 3.2 (**Expectation**). *The* expectation, *the* expected value, *or the* mean *of a discrete RV X with probability distribution $\{p_X(x_i)\}$ is defined as*

$$\boxed{\mu_X = E[X] \triangleq \sum_{\text{all } i} x_i p_X(x_i),} \tag{3.32}$$

provided the sum converges absolutely.[5] □

The notation $E[X]$, $E(X)$, or μ_X is commonly used in mathematics and statistics. In physics, $\langle X \rangle$, $\langle X \rangle_{\text{Av}}$, or $\overline{X}$ are common substitutes for $E[X]$. The definition of expectation extends straightforwardly to a function $h(X)$ of the RV X (Problem 3.6):

$$\boxed{E[h(X)] \triangleq \sum_{\text{all } i} h(x_i)p_X(x_i).} \tag{3.33}$$

Discrete RVs X and Y are independent if and only if

$$\boxed{E[h(X)g(Y)] = E[h(X)]E[g(Y)]} \tag{3.34}$$

for arbitrary real-valued functions $h(\cdot)$ and $g(\cdot)$ (see Problem 3.7).

Most of the RVs we deal with have a finite expectation or mean. There are, however, some exceptions, as we shall see later in this chapter (cf. Zipf's distribution, Cauchy distribution). If $E[X] = \infty$, we usually say that X does not possess an expectation; we may say, instead, that X has an infinite expectation. One important property of the expectation is that $E[\,\cdot\,]$ is a *linear operator*. Therefore, the expectation of a weighted sum of many RVs is the weighted sum of their expectations:

[4] In general, a function $Y = f(\boldsymbol{X})$ of an n-tuple $\boldsymbol{X}(\omega) = (X_1(\omega), X_2(\omega), \ldots, X_N(\omega))$ is a random variable, since Y can be written as a function of $\omega \in \Omega$; i.e., $Y = Y(\omega)$. See Chapter 5 for further details.

[5] A series $\sum_n a_n$ is said to converge absolutely if and only if $\sum_n |a_n| < \infty$. Absolute convergence implies that the value of the sum is independent of the order in which the sum is performed.

$$E\left[\sum_i a_i X_i\right] = \sum_i a_i E[X_i]. \tag{3.35}$$

This is true whether the RVs X_i are independent or not.

Given a conditional probability distribution $\{p_{X|Y}(x_i|y_j)\}$, the **conditional expectation** of X conditioned on the event $\{Y = y_j\}$ is defined by

$$\boxed{E[X|Y = y_j] \triangleq \sum_{\text{all } i} x_i p_{X|Y}(x_i|y_j).} \tag{3.36}$$

It is often of interest to consider the conditional expectation of a random variable X with respect to alternative realizations of the RV Y. The **conditional expectation of** X **given** Y, denoted $E[X|Y]$, can be thought of as a random variable that takes the value $E[X|Y = y_j]$ with probability $P[Y = y_j] = p_Y(y_j)$.

DEFINITION 3.3 (**Conditional expectation**). *Let* $\{p_{X|Y}(x_i|y_j)\}$ *be the conditional probability distribution of a discrete RV* X *conditioned on another discrete RV* Y *to be equal to* y_j *and define a function* $\psi : \{y_j\} \to \mathbb{R}$, *by* $\psi(y_j) = E[X|Y = y_j]$. *Then the conditional expectation of* X *given* Y *is given by*

$$\boxed{E[X|Y] \triangleq \psi(Y).} \tag{3.37}$$

□

It should be emphasized that $E[X|Y]$ is a function of the RV Y. Since $E[X|Y]$ is itself a (discrete) RV, we can take its expectation. We can easily show (Problem 3.8) that

$$\boxed{E[E[X|Y]] = E[X].} \tag{3.38}$$

This basic property of conditional expectation is called the **law of iterated expectations** and is also known as the **law of total expectation**, or the **tower property**.

3.2.2 Moments, central moments, and variance

If X is a random variable, so are its kth power X^k and $(X - \mu_X)^k$. We now define the expectation of these random variables.

DEFINITION 3.4 (**Moments and central moments**). *For a positive integer k,*

$$E[X^k] = \sum_{\text{all } i} x_i^k p_X(x_i), \tag{3.39}$$

is called the kth moment *of X, provided the series converges absolutely. Similarly,*

$$E\left[(X - \mu_X)^k\right] = \sum_{\text{all } i} (x_i - \mu_X)^k p_X(x_i) \tag{3.40}$$

is called the kth central moment *of X.* □

Clearly, the first moment is equal to the expectation μ_X. The second central moment is given the special name *variance*, and is usually denoted as σ_X^2.

DEFINITION 3.5 (**Variance and standard deviation**). *Let X be a RV with finite second moment $E[X^2]$ and mean μ_X. We define the* variance *of X as*

$$\boxed{\sigma_X^2 = \text{Var}[X] \triangleq E[(X-\mu_X)^2] = E[X^2] - \mu_X^2.} \tag{3.41}$$

The square root of the variance, σ_X, is called the standard deviation. □

The concept of **conditional variance** can be defined in terms of the conditional expectation.

DEFINITION 3.6 (**Conditional variance**). *Let X and Y be discrete RVs. The conditional variance of X given Y is defined as*

$$\boxed{\text{Var}[X|Y] \triangleq E[(X - E[X|Y])^2|Y].} \tag{3.42}$$

□

3.2.3 Covariance and correlation coefficient

Let X and Y be two RVs, and define another RV Z by

$$Z = X + Y. \tag{3.43}$$

Then its mean or expectation is

$$\mu_Z = E[Z] = E[X] + E[Y] = \mu_X + \mu_Y. \tag{3.44}$$

The variance of Z is

$$\begin{aligned}\text{Var}[Z] &= E[(Z-\mu_Z)^2] = E[(X-\mu_X+Y-\mu_Y)^2] \\ &= E[(X-\mu_X)^2] + E[(Y-\mu_Y)^2] + 2E[(X-\mu_X)(Y-\mu_Y)] \qquad (3.45)\\ &= \sigma_X^2 + \sigma_Y^2 + 2\sigma_{X,Y}, \qquad (3.46)\end{aligned}$$

where

$$\boxed{\sigma_{X,Y} \triangleq \text{Cov}[X,Y] \triangleq E[(X-\mu_X)(Y-\mu_Y)]} \tag{3.47}$$

is called the *covariance* between X and Y. Expanding the above expression gives

$$\sigma_{X,Y} = \text{Cov}[X,Y] = E[XY] - \mu_X\mu_Y. \tag{3.48}$$

Let us *normalize* X and Y by their standard deviations:

$$X^* = \frac{X}{\sigma_X} \text{ and } Y^* = \frac{Y}{\sigma_Y}. \tag{3.49}$$

The covariance between X^* and Y^* is called the *correlation coefficient* of X and Y, and is denoted by $\rho(X, Y)$:

$$\boxed{\rho(X, Y) \triangleq \mathrm{Cov}[X^*, Y^*] = \frac{\sigma_{X,Y}}{\sigma_X \sigma_Y}.} \tag{3.50}$$

By the well-known Cauchy–Schwarz inequality, the correlation coefficient can be seen to satisfy the following inequality (see Problem 3.10):

$$|\rho(X, Y)| \leq 1,$$

where equality holds if and only if $X^* \overset{\text{a.s.}}{=} cY^*$ for some scalar $c \neq 0$; i.e., if and only if X^* and Y^* are linearly dependent in the almost sure sense.[6] In particular, if $\rho(X, Y) = 1$, the constant $c > 0$, and X and Y are said to have **perfect positive correlation**, whereas if $\rho(X, Y) = -1$, we then have $c < 0$ and X and Y are said to have **perfect negative correlation** (see Problem 6.17). In this sense, $\rho(X, Y)$ is a measure of the degree of correlation between X and Y.

DEFINITION 3.7 (**Uncorrelated random variables**). *We say X and Y are uncorrelated if*

$$\boxed{\mathrm{Cov}[X, Y] = \rho(X, Y) = 0.} \tag{3.51}$$

□

Suppose that variables X and Y are *independent*; i.e.,

$$p_{XY}(x_i, y_j) = p_X(x_i) p_Y(y_j), \quad \text{for all } x_i, y_j. \tag{3.52}$$

Then

$$E[XY] = \sum_i \sum_j x_i y_j p_{XY}(x_i, y_j) = \left(\sum_i x_i p_X(x_i) \right) \left(\sum_j y_j p_Y(y_j) \right) = \mu_X \mu_Y, \tag{3.53}$$

where the rearrangement from the second expression to the third can be justified since the series converges absolutely. We state this as a theorem.

THEOREM 3.1 (**Expectation of the product of independent random variables**). *If random variables X and Y are* statistically independent *with finite expectations μ_X and μ_Y, their product XY is a random variable with expectation $\mu_X \mu_Y$:*

$$E[XY] = E[X]E[Y] = \mu_X \mu_Y. \tag{3.54}$$

□

Then in view of Definition 3.7 and (3.47), we readily have the following theorem.

[6] The concept of almost sure equivalence of two RVs is discussed in Section 3.1.

THEOREM 3.2 (**Independence implies uncorrelatedness**). *If X and Y are independent, then they are uncorrelated. However, the converse is not true.* □

Furthermore, we readily have the following theorem from (3.46).

THEOREM 3.3 (**Variance of sum of independent variables**). *If X and Y are independent, the variance of $Z = X + Y$ is given by*

$$\sigma_Z^2 = \sigma_X^2 + \sigma_Y^2. \tag{3.55}$$

□

We generalize the above results to the sum of $n > 2$ variables in Theorem 3.4.

THEOREM 3.4 (**Sum of n random variables**). *Let $X_1, X_2, \ldots, X_n$ be random variables with finite means $\mu_1, \mu_2, \ldots, \mu_n$ and variances $\sigma_1^2, \sigma_2^2, \ldots, \sigma_n^2$. Consider the sum variable*

$$S_n = X_1 + X_2 + \ldots + X_n.$$

Then

$$E[S_n] = \mu_1 + \mu_2 + \ldots + \mu_n, \tag{3.56}$$

$$\text{Var}[S_n] = \sum_{i=1}^n \sigma_i^2 + 2\sum_{i<j} \text{Cov}[X_i, X_j], \tag{3.57}$$

where the last sum extends over the $n(n-1)/2$ pairs (X_i, X_j) with $i < j$. In particular, if all X_i are pairwise independent, then

$$\text{Var}[S_n] = \sigma_1^2 + \sigma_2^2 + \ldots + \sigma_n^2. \tag{3.58}$$

Proof. Equation (3.56) is immediate from the definition of $E[\cdot]$, which is a linear operation (cf. (3.35)). Write $E[S_n] = \mu_S$. Then $S_n - \mu_S = \sum_{i=1}^n (X_i - \mu_i)$ and

$$(S_n - \mu_S)^2 = \left[\sum_{i=1}^n (X_i - \mu_i)\right]\left[\sum_{j=1}^n (X_j - \mu_j)\right]$$
$$= \sum_{i=1}^n (X_i - \mu_i)^2 + 2\sum_{i<j}(X_i - \mu_i)(X_j - \mu_j). \tag{3.59}$$

Taking the expectation, we get (3.57). □

3.3 Important probability distributions

In this section we will discuss several important probability distributions: the binomial, geometric, hypergeometric, Poisson, negative-binomial (or Pascal) and Zipf (or zeta) distributions.

3.3.1 Bernoulli distribution and binomial distribution

We already defined and discussed the Bernoulli and binomial distributions in Section 2.3, so we only briefly summarize them here.

Let p and $q = 1 - p$ be the probabilities of success and failure in a Bernoulli trial. A Bernoulli random variable B is defined by

$$B = \begin{cases} 1, & \text{if success occurs,} \\ 0, & \text{if failure occurs.} \end{cases} \tag{3.60}$$

Then the probability distribution of Y is given by the Bernoulli distribution (cf. (2.34) and (2.35)) defined by

$$\boxed{P[B = 1] = p, \quad P[B = 0] = q.} \tag{3.61}$$

It is easy to show that $E[B] = p$ and $\text{Var}[B] = pq$.

Next, consider a random variable X defined by

$$X = \text{Number of successes in } n \text{ independent Bernoulli trials.}$$

Then the probability distribution of X is given by the **binomial distribution** defined by (2.38):

$$\boxed{P[X = k] = B(k; n, p) \triangleq \binom{n}{k} p^k q^{n-k}, \quad k = 0, 1, 2, \ldots, n.} \tag{3.62}$$

The expectation of the RV X can be computed as

$$E[X] = \sum_{k=0}^{n} k \binom{n}{k} p^k q^{n-k} = \sum_{k=1}^{n} np \binom{n-1}{k-1} p^{k-1} q^{n-k}, \tag{3.63}$$

where we used the identity

$$k \binom{n}{k} = \frac{n!}{(k-1)!(n-k)!} = n \binom{n-1}{k-1}.$$

Then, by setting $k - 1 = i$ and $n - 1 = m$, and using the identity

$$\sum_{i=0}^{m} \binom{m}{i} p^i q^{m-i} = (p+q)^m = 1,$$

we find

$$\boxed{\mu_X = E[X] = np.} \tag{3.64}$$

In order to find $E[X^2]$ we first note the following relation:

$$k^2 \binom{n}{k} = [k(k-1) + k] \binom{n}{k} = n(n-1) \binom{n-2}{k-2} + n \binom{n-1}{k-1}.$$

Thus,

$$
\begin{aligned}
E[X^2] &= \sum_{k=1}^{n} k^2 \binom{n}{k} p^k q^{n-k} \\
&= n(n-1)p^2 \sum_{k=2}^{n} \binom{n-2}{k-2} p^{k-2} q^{n-k} + np \sum_{k=1}^{n} \binom{n-1}{k-1} p^{k-1} q^{n-k} \\
&= n(n-1)p^2 + np = n^2p^2 + np(1-p) = npq + (np)^2. \qquad (3.65)
\end{aligned}
$$

Since the variance of X is $\text{Var}[X] = E[X^2] - (E[X])^2$, we find

$$\boxed{\sigma_X^2 = \text{Var}[X] = npq.} \qquad (3.66)$$

We could have obtained the above mean μ_X and variance σ_X^2 more readily by noting that X is the sum of n independent Bernoulli RVs B_i, $i = 1, \ldots, n$ (Problem 3.11):

$$X = B_1 + B_2 + \cdots + B_n. \qquad (3.67)$$

The binomial distribution can easily be generalized to what is called the **multinomial distribution** associated with an experiment where each trial can have one of $r (\geq 2)$ outcomes (see Problem 3.12).

3.3.2 Geometric distribution

Consider a sequence of Bernoulli trials with probability of success p. Let

$$X = \text{Number of trials needed to achieve the } \textit{first} \text{ success.}$$

Then, the probability distribution of X is given by the following *geometric distribution*:

$$\boxed{P[X = k] = q^{k-1}p, \quad k = 1, 2, \ldots,} \qquad (3.68)$$

where $q = 1 - p$ is the probability of failure per trial. The probability that more than k trials are needed to get the first success is given by

$$P[X > k] = q^k, \quad k = 0, 1, 2, \ldots. \qquad (3.69)$$

Suppose that m trials have resulted in all failures. What is the probability that more than k additional trials will be needed to achieve the first success? This conditional probability is readily found to be

$$P[X > m + k | X > m] = \frac{P[X > m + k]}{P[X > m]} = \frac{q^{k+m}}{q^m} = q^k. \qquad (3.70)$$

The last probability is independent of m, the number of failures in the past. This property is called the *memoryless* property of the geometric distribution. This is an immediate result of the property of Bernoulli trials in which the outcomes of the trials are independent of each other; i.e., success or failure in the past does not influence possible outcomes of the present or future trials.

The expectation of X can be evaluated as

$$E[X] = \sum_{k=1}^{\infty} kq^{k-1}p = p\sum_{k=1}^{\infty} kq^{k-1}. \tag{3.71}$$

Recall the following formula for a geometric series:

$$\sum_{k=1}^{\infty} x^k = \frac{x}{1-x}, \quad \text{for } |x| < 1. \tag{3.72}$$

Differentiating both sides, we have

$$\sum_{k=1}^{\infty} kx^{k-1} = \frac{1}{(1-x)^2}. \tag{3.73}$$

By setting $x = q = 1 - p$ in the above and substituting it into (3.71), we find

$$\boxed{E[X] = \frac{1}{p}.} \tag{3.74}$$

Similarly, we can compute the variance of X (Problem 3.13):

$$\boxed{\mathrm{Var}[X] = \frac{q}{p^2}.} \tag{3.75}$$

The expectation (3.74) can be obtained somewhat more quickly, if we use a special formula for nonnegative random variables (Problem 3.5) and the result (3.69):

$$E[X] = \sum_{k=0}^{\infty} P[X > k] = \sum_{k=0}^{\infty} q^k = \frac{1}{1-q}. \tag{3.76}$$

3.3.3 Poisson distribution

As we remarked in Chapter 1, Siméon-Denis Poisson (1781–1840) derived this distribution to approximate the binomial distribution when the probability of occurrence p is small. In 1898 **Bortkiewicz**[7] published "Das Gesetz der kleinen Zahlen (The law of small numbers)," in which he analyzed data of the number of soldiers in the Prussian cavalry corps who were killed by being kicked by a horse, and showed that those numbers followed a Poisson distribution.

[7] Ladislaus Josephovich Bortkiewicz (1868–1931) was a Russian economist and statistician of Polish descent.

Let us consider the limiting case of the binomial distribution $B(k; n, p)$ of (2.38) when $n \to \infty$ and $p \to 0$, while keeping

$$np = \lambda,$$

where λ is a fixed parameter. Then, as will be shown below, the limit distribution becomes

$$\boxed{P(k; \lambda) \triangleq \frac{\lambda^k}{k!} e^{-\lambda}, \quad k = 0, 1, 2, \ldots,} \tag{3.77}$$

which is the *Poisson distribution* with *mean* λ.

In order to derive the above distribution, let S_n denote the total number of successes in n Bernoulli trials:

$$S_n = X_1 + X_2 + \cdots + X_n, \tag{3.78}$$

where X_i is a binary RV such that

$$X_i = \begin{cases} 1, & \text{if the } i\text{th trial is success,} \\ 0, & \text{otherwise.} \end{cases} \tag{3.79}$$

Then for given p and finite n, the RV S_n has the binomial distribution, as discussed in Section 2.3:

$$P[S_n = k] = B(k; n, p) = \binom{n}{k} p^k (1-p)^{n-k}, \tag{3.80}$$

which can be rearranged, using $p = \lambda/n$, as

$$\begin{aligned} B(k; n, p) &= \frac{n!}{k!(n-k)!} p^k (1-p)^{n-k} \\ &= \frac{1}{k!}\left(1 - \frac{1}{n}\right)\left(1 - \frac{2}{n}\right)\cdots\left(1 - \frac{k-1}{n}\right)\lambda^k (1-p)^{n-k}. \end{aligned} \tag{3.81}$$

Substituting the relation $p = \frac{\lambda}{n}$ in the last term, we have

$$(1-p)^{n-k} = \left[\left(1 - \frac{\lambda}{n}\right)^n\right]^{(n-k)/n}. \tag{3.82}$$

Then using the formula $\lim_{\delta \to 0}(1 - a\delta)^{1/\delta} = e^{-a}$, we have

$$\lim_{n \to \infty} (1-p)^{n-k} = e^{-\lambda}. \tag{3.83}$$

Thus, in the limit

$$\lim_{n \to \infty} B(k; n, p) = \frac{\lambda^k}{k!} e^{-\lambda}, \quad k = 0, 1, 2, \ldots. \tag{3.84}$$

Therefore, the two-parameter binomial distribution becomes the one-parameter distribution of (3.77), which we often write as $P(k; \lambda)$. The expectation, the second moment, and variance of the Poisson RV can be obtained (Problem 3.18) as

$$\boxed{\begin{aligned} E[X] &= \lambda, \\ E[X^2] &= \lambda^2 + \lambda, \\ \mathrm{Var}[X] &= \lambda. \end{aligned}} \tag{3.85}$$

In Figure 3.5 we plot the Poisson distribution for $\lambda = 0.5, 1, 2, 4, 8,$ and 16. As λ increases, the distribution appears to become normal (or Gaussian) distribution-like in shape, but of course this distribution is defined for nonnegative discrete integers, whereas the normal distribution is a continuous distribution defined over $(-\infty, \infty)$.

The cumulative Poisson distribution is often denoted as $Q(k; \lambda)$:

$$\boxed{Q(k; \lambda) = \sum_{i=0}^{k} P(i; \lambda), \quad k = 0, 1, 2, \ldots.} \tag{3.86}$$

The following set of formulas relating $P(k; \lambda)$ and $Q(k; \lambda)$ are useful (Problem 3.20):

$$\sum_{k'=0}^{k} P(k - k'; \lambda_1) Q(k'; \lambda_2) = Q(k; \lambda_1 + \lambda_2); \tag{3.87}$$

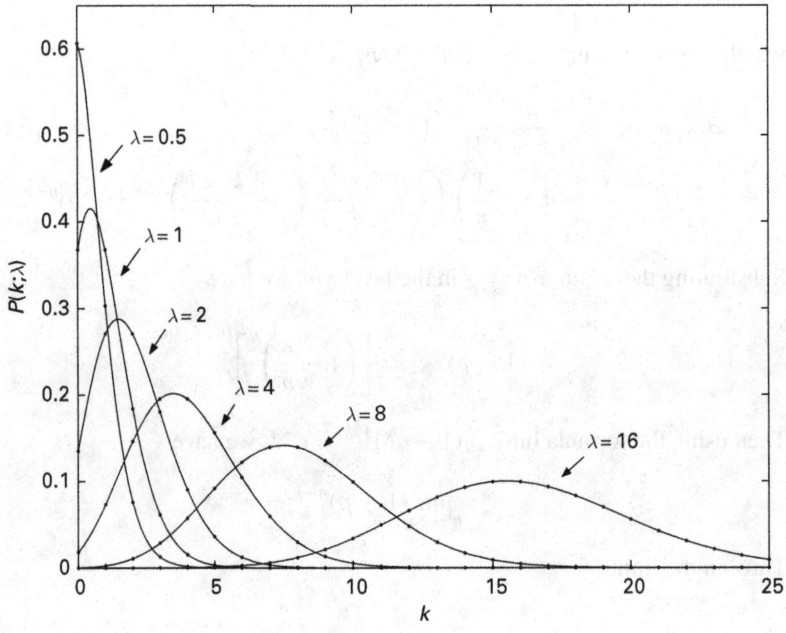

Figure 3.5 The Poisson distribution with different values of λ.

$$Q(k;\lambda) = \int_{\lambda}^{\infty} P(k;y)\,dy; \tag{3.88}$$

$$Q(k;\lambda) = \frac{\lambda Q(k-1;\lambda) + (k+1)Q(k+1;\lambda)}{k+\lambda+1}; \tag{3.89}$$

$$\sum_{j=0}^{k-1} Q(j;\lambda) = kQ(k;\lambda) - \lambda Q(k-1;\lambda); \tag{3.90}$$

$$\int_{\lambda}^{\infty} Q(k-1;y)\,dy = kQ(k;\lambda) - \lambda Q(k-1;\lambda). \tag{3.91}$$

Since the binomial distribution is an approximation to the hypergeometric distribution (see Problems 3.15 and 3.16) for large N, and the Poisson distribution is an approximation to the binomial distribution for large n and small p, one might expect the Poisson distribution to be an approximation to the hypergeometric distribution under certain conditions. As a matter of fact, roughly speaking, this is the case if (i) n is large, (ii) p is small (with $np = \lambda$), and (iii) N is much larger than n.

Like the binomial distribution, the Poisson distribution is a very important one and has been applied to many problems. For instance, the number of defective items in industrial products, the number of telephone calls initiated per unit time, the distribution of bacterial colonies per unit volume of a culture, the number of bomb fragments striking a target per unit area, etc. In all these situations we have a large number n of independent or nearly independent "trials," a small probability p of a "success" in any given trial, and we ask for the probability of getting k successes in n trials [353]. Feller [99] discusses several application examples in which observed data fit the Poisson distribution.

3.3.4 Negative binomial (or Pascal) distribution

Consider a succession of Bernoulli trials. How many trials are needed to achieve r successes ($r \geq 1$)?

Let us define a random variable Y_r:

$$Y_r \triangleq \text{Number of trials needed to achieve } r \text{ successes.} \tag{3.92}$$

Then we readily find

$$\begin{aligned} P[Y_r = k] &= P[r-1 \text{ successes in } k-1 \text{ trials and a success at the } k\text{th trial}] \\ &= \binom{k-1}{r-1} p^{r-1} q^{k-r} p = \binom{k-1}{r-1} p^r q^{k-r}, \end{aligned} \tag{3.93}$$

where $q = 1 - p$. Thus, we have

$$\boxed{P[Y_r = k] = \binom{k-1}{r-1} p^r q^{k-r}, \quad k \geq r.} \tag{3.94}$$

The last expression could have been obtained directly, by observing that $\binom{k-1}{r-1} = \binom{k-1}{k-r}$ and that the event $Y_r = k$ is obtained if and only if $k - r$ failures occur in the first $k - 1$ trials (hence $r - 1$ successes), immediately followed by a success. Since (3.94) defines the probabilities, they should add up to one; i.e.,

$$\sum_{k=r}^{\infty} \binom{k-1}{r-1} p^r q^{k-r} = 1, \tag{3.95}$$

from which we obtain the following identity:

$$\sum_{k=r}^{\infty} \binom{k-1}{r-1} q^{k-r} = p^{-r} = (1-q)^{-r}. \tag{3.96}$$

It is suggested that the reader prove the above identity directly (Problem 3.21).

Using the identity

$$\begin{aligned} \binom{-n}{i} &= \frac{(-n)!}{i!(-n-i)!} = (-1)^i \frac{n(n+1)\cdots(n+i-1)}{i!} \\ &= (-1)^i \binom{n+i-1}{i}, \end{aligned} \tag{3.97}$$

we may write (3.94) as

$$\boxed{P[Y_r = k] = \binom{-r}{k-r} p^r (-1)^{k-r} (1-p)^{k-r} = \binom{-r}{k-r} p^r (-q)^{k-r}, \quad k \geq r,} \tag{3.98}$$

where $q = 1 - p$. Therefore, the above distribution (3.94) or (3.98) is called the *negative binomial distribution* with parameters (r, p). It is also known as the *Pascal distribution* or sometimes referred to as the *binomial waiting time distribution* [353]. When $r = 1$, this distribution becomes the geometric distribution.

Recall that the binomial distribution is concerned with the distribution of the number of successes r in a specified number k of trials, whereas the negative binomial (or Pascal) distribution is concerned with the number of trials k needed to achieve a specified number of successes r.

Alternatively, we could derive the above distribution, observing that Y_r is the sum of r geometrically distributed RVs:

$$Y_r = X_1 + X_2 + \cdots + X_r, \tag{3.99}$$

where each X_i $(1 \leq i \leq r)$ is distributed according to (3.68). In order to show that Y_r defined above has the distribution (3.94), we need to study additional properties of the sum of independent RVs, so we defer this derivation until a later section.

Yet another derivation of the negative binomial distribution is as follows. Note that k or fewer trials are needed to achieve r successes, if and only if the number of successes achieved in k Bernoulli trials is at least r. Therefore, the probabilities of the above two equivalent events are equal; i.e.,

$$P[Y_r \leq k] = \sum_{i=r}^{k} B(i; k, p), \tag{3.100}$$

where $B(i; k, p) = \binom{k}{i} p^i (1-p)^{k-i}$. Then

$$P[Y_r = k] = P[Y_r \leq k] - P[Y_r \leq k-1] \tag{3.101}$$

$$= \sum_{i=r}^{k} \binom{k}{i} p^i q^{k-i} - \sum_{i=r}^{k-1} \binom{k-1}{i} p^i q^{k-1-i}. \tag{3.102}$$

It is left for an exercise to show that the above expression is equivalent to (3.94) (Problem 3.22).

The jth moment of Y_r can be computed as

$$E[Y_r^j] = \sum_{k=r}^{\infty} k^j \binom{k-1}{r-1} p^r q^{k-r} = \frac{r}{p} \sum_{k=r}^{\infty} k^{j-1} \binom{k}{r} p^{r+1} q^{k-r}$$

$$= \frac{r}{p} \sum_{i=r+1}^{\infty} (i-1)^{j-1} \binom{i-1}{r} p^{r+1} q^{i-(r+1)}$$

$$= \frac{r}{p} E\left[(Y_{r+1} - 1)^{j-1}\right], \tag{3.103}$$

where Y_{r+1} is a negative binomial RV, i.e., the number of trials needed to achieve $r+1$ successes. Setting $j = 1$ in the last expression yields

$$\boxed{E[X] = \frac{r}{p}.} \tag{3.104}$$

Similarly, by setting $j = 2$,

$$E[Y_r^2] = \frac{r}{p} E[Y_{r+1} - 1] = \frac{r}{p}\left(\frac{r+1}{p} - 1\right). \tag{3.105}$$

Thus,

$$\boxed{\mathrm{Var}[Y_r] = \frac{rq}{p^2}.} \tag{3.106}$$

3.3.4.1 Shifted negative binomial distribution

Analogous to the *shifted* geometric distribution defined in Problem 3.14, let us define a random variable Z_r by

$$Z_r = \text{Number of } \textit{failures} \text{ needed to achieve } r \text{ successes},$$

which is related to Y_r simply by

$$Z_r = Y_r - r. \tag{3.107}$$

The probability distribution of Z_r is given by (Problem 3.23)

$$f(k; r, p) \triangleq P[Z_r = k] = \binom{r+k-1}{r-1} p^r q^k, \quad k = 0, 1, 2, \ldots. \tag{3.108}$$

This distribution is obtained by *shifting* the negative binomial distribution of Y_r to the left by r. Thus, the *support* of this distribution is $[0, \infty)$ in contrast to $[r, \infty)$ of the distribution (3.94). This shifted form of negative binomial distribution is useful in modeling count data like the Poisson distribution discussed in the previous example. It is more flexible than the Poisson distribution in fitting empirical count data, because the negative binomial is a two-parameter distribution (i.e., p and r), whereas the Poisson distribution involves only one parameter (i.e., λ).

The distribution (3.108) can be generalized for a noninteger parameter r. Then the distribution (3.108) takes the following form:

$$f(k; r, p) = \frac{\Gamma(r+k)}{\Gamma(r)\Gamma(k+1)} p^r (1-p)^k, \quad k = 0, 1, 2, \ldots, \tag{3.109}$$

where $\Gamma(r)$ is the gamma function:

$$\Gamma(r) = \int_0^\infty x^{r-1} e^{-x} \, dx, \tag{3.110}$$

which will be discussed in detail in Section 4.2.3.

Figure 3.6 is a plot of the distribution (3.109) for $r = 0.5, 1, 2, 4, 8, 16$, and 32, while p is set to $p = 0.5$. We can also show (Problem 3.24) that this distribution approaches the Poisson distribution in the limit $r \to \infty$ and $(1-p) \to 0$, while keeping $(1-p)r = \lambda$.

3.3.5 Zipf's law and zeta distribution

In the 1930s, the American linguist **Zipf**[8] examined the frequency of occurrences of words in various natural languages (English, Chinese, etc.) and found that while a small number of words are used very often, there are many words that are rarely used. By ranking the words based on their frequency of occurrences,[9] he found that f_n, the frequency of the nth ranked word, can be expressed as

$$f_n \propto \frac{1}{n^\alpha}, \quad n = 1, 2, 3, \ldots, \tag{3.111}$$

[8] George Kingsley Zipf (1902–1950) was a linguist and philologist at Harvard University.

[9] In English, the top ranked words are "the," "of," "and," and "to."

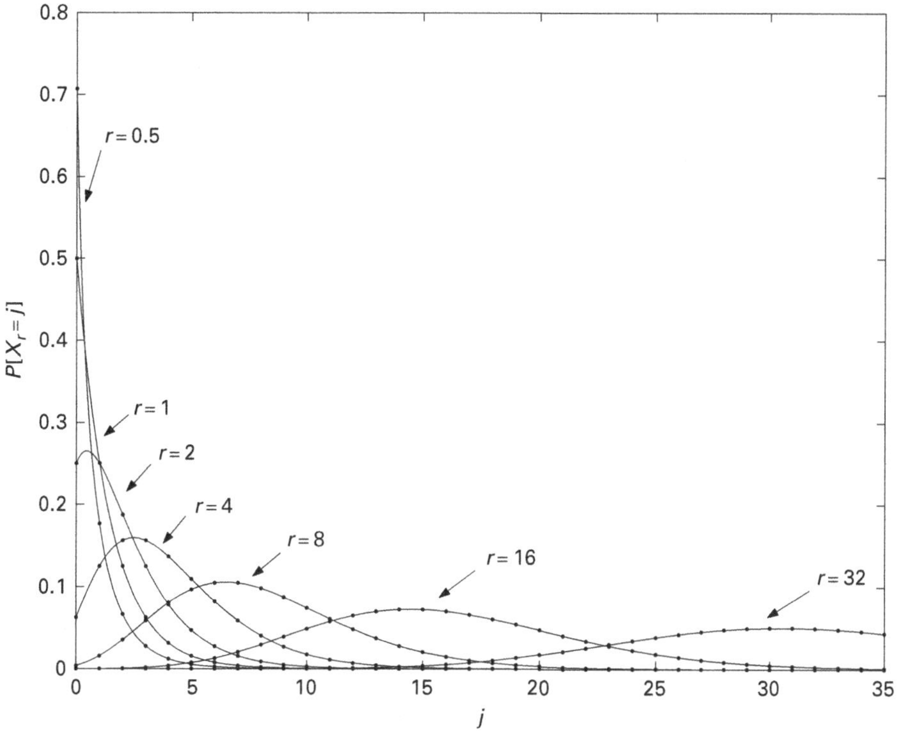

Figure 3.6 The *shifted* negative binomial distribution (3.108) or (3.109); i.e., the probability distribution of the number of *failures* needed to achieve r successes in Bernoulli trials with $p = 0.5$ and for $r = 0.5, 1, 2, 4, 8, 16, 32$.

where the *power exponent* α (> 1) is close to unity.[10] Later, many empirical distributions encountered in economics and other realms have been found to exhibit a similar "power-law" behavior, which is called *Zipf's law*. For example, consider request traffic to a certain website or a web proxy server, and let $X = n$ denote the nth most frequently accessed web page. Some recent studies ([37] and references therein) show that the generalized Zipf law fits the empirical data obtained at several proxy server sites. Zipf's law can be verified by plotting $\log f_n$ versus $\log n$. If the scatter diagram is close to a single straight line of slope $-\alpha$, the distribution of (3.111) may apply.

We define the Zipf distribution with parameter α, often referred to as the *zeta distribution*, for a discrete random variable X which takes on integers (that often represent the rank orders in occurrences) $1, 2, 3, \ldots$, as follows:

$$\boxed{p_n = P[X = n] = \frac{n^{-\alpha}}{\zeta(\alpha)}, \quad n = 1, 2, 3, \ldots,} \tag{3.112}$$

[10] When $\alpha = 1$, the series becomes the so-called harmonic series and does not converge (Problem 3.25).

where $\zeta(s)$ is defined by

$$\boxed{\zeta(s) = \sum_{n=1}^{\infty} n^{-s}, \quad s > 1,} \tag{3.113}$$

which is known as the **Euler**[11] zeta function, which was extended in 1859 to the complex numbers s ($s \neq 1$) by **Riemann**.[12] Thus, the function (3.113) is often called the Riemann zeta function.

3.3.5.1 Euler and Riemann zeta functions

We list some interesting results about the Euler and Riemann zeta function.

- Euler discovered the connection between the zeta function and prime numbers p:

$$\begin{aligned}\zeta(s) &= \prod_{p\geq 2:\text{prime}} \frac{1}{1-p^{-s}} \\ &= \left(1+\frac{1}{2^s}+\frac{1}{4^s}+\cdots\right)\left(1+\frac{1}{3^s}+\frac{1}{9^s}+\cdots\right)\cdots\left(1+\frac{1}{p^s}+\frac{1}{p^{2s}}+\cdots\right)\cdots,\end{aligned} \tag{3.114}$$

which can be alternatively expressed by the identity

$$1+\frac{1}{2^s}+\frac{1}{3^s}+\frac{1}{4^s}+\frac{1}{5^s}+\cdots = \frac{2^s}{2^s-1}\frac{3^s}{3^s-1}\frac{5^s}{5^s-1}\frac{7^s}{7^s-1}\cdots. \tag{3.115}$$

- The harmonic series, obtained when $s = 1$ in the left-hand side of the Euler formula (3.114) or (3.115), diverges (Problem 3.25 (a)). This implies that there are infinitely many prime numbers.
- The zeta function $\zeta(\alpha)$ is a decreasing function for real values $\alpha > 1$ and is convex. Except for $\alpha =$ even integer, the value of $\zeta(\alpha)$ must be approximated, since it is a transcendental function.[13]

$$\begin{aligned} &\zeta(1) = \infty, && \zeta(1.5) \approx 2.612, \\ &\zeta(2) = \tfrac{\pi^2}{6} \approx 1.645, && \zeta(2.5) \approx 1.341, \\ &\zeta(3) \approx 1.202, && \zeta(3.5) \approx 1.127, \\ &\zeta(4) = \tfrac{\pi^4}{90} \approx 1.0823, && \\ &\lim_{\alpha\to\infty} \zeta(\alpha) = 1. && \end{aligned}$$

[11] Leonhard Euler (1707–1783) was a Swiss mathematician and physicist who spent most of his life in Saint Petersburg, Russia, and Berlin, Germany. Euler's contributions are in the fields of number theory, calculus, geometry, graph theory, mechanics, optics, astronomy, etc. Euler's work on Latin squares laid the groundwork for today's Sudoku puzzles.

[12] Georg Friedrich Bernhard Riemann (1826–1866) was a German mathematician who studied under Gauss. He made important contributions to complex analysis, differential geometry, and mathematical physics.

[13] A transcendental function is a function which is not an algebraic function. Examples include the exponential function, the trigonometric functions, and their inverse functions.

3.3.5.2 Mean, variance, and moments of the zeta distribution

The mth moment of the random variable X with the probability distribution (3.112) can be obtained as follows:

$$E[X^m] = \sum_{n=1}^{\infty} n^m p_n = \sum_{n=1}^{\infty} n^m \frac{n^{-\alpha}}{\zeta(\alpha)}. \tag{3.116}$$

Thus,

$$\boxed{E[X^m] = \frac{\zeta(\alpha - m)}{\zeta(\alpha)}, \quad \alpha > m + 1.} \tag{3.117}$$

The mean and variance are given as

$$E[X] = \frac{\zeta(\alpha - 1)}{\zeta(\alpha)}, \quad \text{if } \alpha > 2; \tag{3.118}$$

$$\text{Var}[X] = \frac{\zeta(\alpha - 2)}{\zeta(\alpha)} - \left[\frac{\zeta(\alpha - 1)}{\zeta(\alpha)}\right]^2, \quad \text{if } \alpha > 3. \tag{3.119}$$

3.4 Summary of Chapter 3

(Cumulative) distribution function (CDF):	$F_X(x) \triangleq P[\{\omega : X(\omega) \le x\}]$	(3.1)
Probability distribution, $\{p_X(x_i)\}$	$p_X(x_i) \triangleq P[X = x_i], i = 1, 2, \ldots$	(3.7)
Probability mass function (PMF):	$p_X(x) \triangleq P[X = x], x \in \mathbb{R}$	(3.9)
Dirac delta function $\delta(x)$:	$\int_{-\infty}^{\infty} g(x)\delta(x - a)\, dx \triangleq g(a)$	(3.15)
Probability density function (PDF):	$f_X(x) = \sum_i p_X(x_i)\delta(x - x_i)$	(3.12)
Joint probability distribution:	$p_{XY}(x_i, y_j) = P[X = x_i, Y = y_j]$	(3.17)
Conditional probability distribution:	$p_{Y\|X}(y_j\|x_i) = P[Y = y_j\|X = x_i]$	(3.20)
Independent RVs:	$p_{XY}(x_i, y_j) = p_X(x_i)p_Y(y_j)$	(3.25)
Independent RVs:	$E[h(X)g(Y)] = E[h(X)]E[g(Y)]$	(3.34)
Expectation:	$\mu_X = E[X] \triangleq \sum_{\text{all}i} x_i p_X(x_i)$	(3.32)
	$E[h(X)] = \sum_{\text{all}i} h(x_i)p_X(x_i)$	(3.33)
Conditional expectation:	$E[X\|Y] = \psi(Y),\ \psi(y_j) = E[X\|Y = y_j]$	(3.37)
	$E[X\|Y = y_j] = \sum_{\text{all}i} x_i p_{X\|Y}(x_i\|y_j)$	(3.36)
Variance:	$\sigma_X^2 = \text{Var}[X] \triangleq E[(X - \mu_X)^2]$	(3.41)
Conditional variance:	$\text{Var}[X\|Y] = E[(X - E[X\|Y])^2\|Y]$	(3.42)
Covariance:	$\sigma_{X,Y} \triangleq \text{Cov}[X, Y] = E[(X - \mu_X)(Y - \mu_Y)]$	(3.47)
Correlation coefficient:	$\rho(X, Y) \triangleq \frac{\sigma_{X,Y}}{\sigma_X \sigma_Y}$	(3.50)

Uncorrelated RVs:	$\text{Cov}[X, Y] = \rho(X, Y) = 0$	(3.51)
Bernoulli distribution:	$P[B = 1] = p,\ P[B = 0] = q$	(3.61)
Binomial distribution:	$B(k; n, p) \triangleq \binom{n}{k} p^k q^{n-k},\quad 0 \leq k \leq n$	(3.62)
Geometric distribution:	$P[X = k] = q^{k-1} p,\quad k = 1, 2, \ldots$	(3.68)
Poisson distribution:	$P(k; \lambda) \triangleq \frac{\lambda^k}{k!} e^{-\lambda},\quad k = 0, 1, 2, \ldots$	(3.77)
Cumulative Poisson dist.:	$Q(k; \lambda) = \sum_{i=0}^{k} P(i; \lambda),\quad k = 0, 1, 2, \ldots$	(3.86)
Negative binomial dist.:	$P\,[Y_r = k] = \binom{k-1}{r-1} p^r q^{k-r},\quad k \geq r$	(3.94)
Negative binomial dist.:	$P\,[Y_r = k] = \binom{-r}{k-r} p^r (-q)^{k-r},\quad k \geq r$	(3.98)
Zipf's law:	$1/n^{\alpha},\quad n \geq 1$	(3.111)
Zeta distribution:	$p_n = n^{-\alpha}/\zeta(\alpha),\ n \geq 1$	(3.112)
Zeta function:	$\zeta(s) = \sum_{n=1}^{\infty} n^{-s},\quad s > 1$	(3.113)

3.5 Discussion and further reading

In this chapter we introduced the notion of a random variable as a mapping from the sample space Ω to the real line $\mathbb{R}$, or in the case of two joint variables as a mapping from Ω to $\mathbb{R} \times \mathbb{R}$. We then discussed several important discrete RVs, most of which are also found in other textbooks on probability, such as Feller [99], Grimmett and Stirzaker [131], Gubner [133], Nelson [254], Papoulis and Pillai [262], and Ross [289]. We discussed Zipf's law (zeta distribution) more than typical textbooks owing to its increasing importance in recent years, such as its use in the stochastic modeling of web server systems.

3.6 Problems

Section 3.1: Random Variables

3.1* Property 4 of (3.3). Derive Eq. (3.3).
Hint: Use Axiom 3 in Section 2.2.4.

Section 3.2: Discrete random variables and probability distributions

3.2* A nonnegative discrete RV. Consider a discrete RV X whose range is the set of nonnegative integers. Let the probability distribution of X be of the form

$$p_i = P[X = i] = k\rho^i,\ \ i = 0, 1, 2, \ldots,$$

where ρ is a given parameter, $0 < \rho < 1$.

(a) Determine the constant k.
(b) Obtain the distribution function of X.

3.3* Statistical independence. Show that (3.25) and (3.26) are equivalent. Prove also the equivalence of (3.27) and (3.28).

3.4 Discrete RVs, distribution function, and conditional probability. A sample space Ω consists of the four points

$$\Omega = \{\omega_1, \omega_2, \omega_3, \omega_4\}$$

and the probabilities of the simple events are

$$P[\{\omega_1\}] = \tfrac{1}{2}, \quad P[\{\omega_2\}] = \tfrac{1}{4}, \quad P[\{\omega_3\}] = \tfrac{1}{8}, \quad P[\{\omega_4\}] = \tfrac{1}{8}.$$

Define RVs X and Y by

$$\begin{aligned} X(\omega_1) = 1, \quad X(\omega_2) = 1, \quad X(\omega_3) = 2, \quad X(\omega_4) = 3; \\ Y(\omega_1) = 3, \quad Y(\omega_2) = 3, \quad Y(\omega_3) = 1, \quad Y(\omega_4) = 1. \end{aligned}$$

(a) Find the probability distribution and the distribution function of X. Do the same for Y.
(b) Find the conditional probability $P[Y = y_j | X = x_i]$ for all possible pairs $X = x_i$, $Y = y_j$.
(c) Are the RVs X and Y statistically independent?

3.5 Expectation of nonnegative random variable. Let X be a random variable that takes on integers 0, 1, 2, ... Show that the following formula holds:

$$E[X] = \sum_{k=0}^{\infty} P[X > k]. \tag{3.120}$$

3.6 Expectation of a function of a random variable. Show that (3.33) holds by defining a random variable $Z = h(X)$.

3.7 Statistical independence and expectation. Show that two discrete RVs X and Y are independent if and only if (3.34) for arbitrary functions $h(\cdot)$ and $g(\cdot)$.

3.8* Properties of conditional expectation. Let X and Y be discrete RVs. Show the following:

(a) $E[E[X|Y]] = E[X]$ (cf. (3.38)).
(b) $E[h(Y)g(X)|Y] = h(Y)E[g(X)|Y]$, where h and g are scalar functions.
(c) $E[\cdot\,|Y]$ is a linear operator.

3.9 Conditional variance. Show that

$$E[\mathrm{Var}[X|Y]] = \mathrm{Var}[X] - \mathrm{Var}[E[X|Y]].$$

3.10* Correlation coefficient and Cauchy–Schwarz inequality. For random variables X and Y, a version of the Cauchy–Schwarz inequality states the following (see also Problem 10.21):

$$(E[XY])^2 \le E[X^2]E[Y^2],$$

with equality holding if and only if $X^* = X - E[X]$ and $Y^* = Y - E[Y]$ are linearly dependent; i.e., $X^* \overset{\text{a.s.}}{=} cY^*$ for some constant c. Show that the Cauchy–Schwarz inequality implies that the correlation coefficient satisfies $|\rho(X, Y)| \le 1$ and furthermore that $\rho(X, Y) = 1$ implies perfect positive correlation of X and Y, while $\rho(X, Y) = -1$ implies perfect negative correlation.

Section 3.3: Important probability distributions

3.11* Alternative derivation of the expectation and variance of binomial distribution. Derive μ_X and σ_X^2 by using the representation of X as in (3.67).

3.12* Trinomial and multinomial distributions.

(a) Suppose that the outcome of a trial is one of three different events E_1, E_2, and E_3. Let the probability of these outcomes be p, q, and $1 - p - q$, respectively. Suppose that we make n independent trials. Show that the probability that E_1 will occur k_1 times, E_2 will occur k_2 times, and E_3 will occur $n - k_1 - k_2$ times is

$$\frac{n!}{k_1!k_2!(n - k_1 - k_2)!} p^{k_1} q^{k_2} (1 - p - q)^{n-k_1-k_2}. \tag{3.121}$$

This distribution is called a *trinomial distribution*.

(b) Generalize the above result to the case where the outcome of a trial is one of m different events $\{E_i : 1 \le i \le m\}$, which are mutually exclusive. Let the probability of event E_i be p_i, with $\sum_{i=1}^m p_i = 1$. Suppose that we make n independent trials. Show that the probability that E_i will occur k_i times $(1 \le i \le m)$, where $\sum_{i=1}^m k_i = n$ is given by

$$p_{\boldsymbol{k}} = \frac{n!}{k_1!k_2!\cdots k_m!} p_1^{k_1} p_2^{k_2} \cdots p_m^{k_m}, \tag{3.122}$$

where $\boldsymbol{k}$ stands for the vector $[k_1, k_2, \ldots, k_m]$. This distribution is called a *multinomial distribution*.

3.13 Variance of geometric distribution. Show that the variance of the geometric RV is given by (3.75).

3.14 Shifted geometric distribution. Referring to the experiment in Section 3.3.2 on the geometric distribution, define the random variable $Y = X - 1$, which represents the number of *failures* before the first success is achieved. Show that

$$P[Y = k] = pq^k, \; k = 0, 1, 2, \ldots, \tag{3.123}$$

which is often called the *shifted* geometric distribution, because Y starts from 0, while X starts from 1. Find the mean and variance of Y.

3.15 Hypergeometric distribution. Assume that there are N items contained in a box, out of which N_1 items are from class-1 (e.g., "good") and the remainder $N_2 = N - N_1$ are from class-2 (e.g., "bad"). Suppose that we take at random a sample of n items *without replacement* (i.e., we do not return the selected item to the box) and without regard to order. Let

$$X = \text{Number of class-1 items selected.}$$

(a) Show that the probability $p_k = P[X = k]$ is given by

$$\boxed{p_k \triangleq P[X = k] = \frac{\binom{N_1}{k}\binom{N_2}{n-k}}{\binom{N}{n}},} \tag{3.124}$$

where k must be in the range

$$\max\{0, n - N_2\} \leq k \leq \min\{n, N_1\}.$$

The probability distribution is known as the *hypergeometric distribution*, and is often used in the study of the quality control problem in production systems.

(b) Derive the following alternative expression of the hypergeometric distribution:

$$\boxed{p_k = \frac{\binom{n}{k}\binom{N-n}{N_1-k}}{\binom{N}{N_1}}.} \tag{3.125}$$

(c) Show that the mean is given by

$$E[X] = \frac{nN_1}{N}\sum_{k=1}^{n-1}\frac{\binom{N_1-1}{k-1}\binom{N_2}{n-k}}{\binom{N-1}{n-1}}. \tag{3.126}$$

(d) Suppose that the population sizes N and N_1 are both decreased by one to $N - 1$ and $N_1 - 1$ respectively and the sample size is also decreased by one to $n - 1$. Let X^* be the number of class-1 items that are selected. Find the distribution $\{p_k^*\}$ of X^*.

(e) Show that $E[X]$ of (3.126) can be simplified to

$$\boxed{E[X] = \frac{nN_1}{N} = np.} \tag{3.127}$$

Here, the ratio $p = N_1/N$ represents the fraction of class-1 items in the population before sampling is done. The formula (3.127) says that the expected number of class-1 items when we take n samples *without replacement* is given by np.

(f) Show that the jth moment $E[X^j]$ can be given as

$$E[X^j] = \frac{nN_1}{N}E\left[(X^* + 1)^{j-1}\right], \tag{3.128}$$

where X^* is the hypergeometric RV defined in part (d).

(g) Show that the variance of X is given by

$$\mathrm{Var}[X] = \frac{N-n}{N-1}npq, \tag{3.129}$$

where $p = N_1/N$ is defined in part (e), and similarly, $q = 1 - p = N_2/N$.

3.16 Relation between hypergeometric and binomial distributions.

(a) **Limit of the hypergeometric distribution.** Show that the hypergeometric distribution converges to the binomial distribution in the limit $N_1, N_2 \to \infty$ while keeping $N_1/N \to p$ and $N_2/N \to 1 - p$.

(b) **Generalization of the hypergeometric distribution.** Assume that the population of size N contains $R = 3$ classes of sizes N_1, N_2, and $N_3 (= N - N_1 - N_2)$ respectively. Suppose that a size n sample is taken. What is the probability that it contains k_1 from class-1, k_2 from class-2, and $n - k_1 - k_2$ from class-3?

3.17 Bridge game. The 52 cards consist of four classes: club, diamond, heart, and spade. Find the probability that a hand of 13 cards consists of five clubs, four diamonds, two hearts, and two spades.

3.18* Mean, second moment, and variance of the Poisson distribution. Show that the mean, second moment, and variance of the Poisson RV X are given by (3.85).

3.19 Tail of the Poisson distribution.

(a) For the Poisson-distributed variable S, show that

$$\frac{\lambda^k}{k!}e^{-\lambda} < P[S \ge k] < \frac{\lambda^k}{k!}e^{-\lambda}\frac{1}{1-\frac{\lambda}{k+1}}. \tag{3.130}$$

(b) Show further for sufficiently large k that

$$\frac{\Delta^k}{\sqrt{2\pi k}} < P[S \ge k] < \frac{\Delta^k}{\sqrt{2\pi k}}\frac{1}{1-\frac{\lambda}{k+1}}, \tag{3.131}$$

where $\Delta = \frac{\lambda}{k}e^{1-(\lambda-k)}$. Thus, as long as $\Delta < 1$, the probabilities such as $P[S = k]$ and $P[S \ge k]$ can be evaluated accurately.
Hint: Use Stirling's formula: $k! \approx \sqrt{2\pi k}k^k e^{-k}$, as $k \to \infty$.

3.20* Identities relating $P(k;\lambda)$ and $Q(k;\lambda)$. Derive the identities (3.87) through (3.91).
Hint: To show (3.88), write $P(k; y) = \frac{d}{dy}\left(-e^{-y}\right)\frac{y^k}{k!}$, apply the integration by parts, and express the left-hand side in terms of $\int_\lambda^\infty P(k-1; y)\,dy$ and $P(k;\lambda)$.

3.21* Derivation of the identity (3.96). Show the identity (3.96) without making use of the binomial distribution.
Hint: Consider $f(x) = (1-x)^{-r}$ and expand it into a series in powers of x.

3.22* **Equivalence of two expressions for the negative binomial distribution.** Show that the probability distributions (3.94) and (3.101) are equivalent.

3.23 **Another negative binomial distribution.** By referring to the negative-binomial variable Y_r defined in the text, let us define $Z_r = Y_r - r$, which represents the number of failures that precede the rth success. Show that the probability distribution of the RV is given by

$$P\left[Z_r = k\right] = \binom{r+k-1}{r-1} p^r (1-p)^k, \quad k = 0, 1, 2, \ldots. \tag{3.132}$$

This distribution is sometimes also referred to as the negative binomial distribution.

3.24 **Negative binomial and Poisson distribution.** Show that the negative binomial distribution of the form (3.109) approaches the Poisson distribution with mean λ if we let $r \to \infty$ and $(1-p) \to 0$ in such a manner that $(1-p)r = \lambda$. *Hint*: $r \ln(1/p)$ also converges to λ.

3.25 **Zipf's law with $\alpha = 1$.** The simplest case of Zipf's law may be the case $\alpha = 1$.

(a) Show, however, that the harmonic series does not converge; i.e.,

$$\zeta(1) = \sum_{n=1}^{\infty} \frac{1}{n} = \infty.$$

Hint: Consider a continuous function $f(x) = 1/x, x > 0$, and use the fact $\int_1^y f(x)\,dx = \ln y$.

(b) So, let us limit the maximum value that X takes on to be finite, say N, and consider the probability distribution

$$p_n = \frac{n-1}{z_N}, \quad n = 1, 2, 3, \ldots, N, \tag{3.133}$$

where z_N is the normalization constant

$$z_N = \sum_{n=1}^{N} \frac{1}{n}. \tag{3.134}$$

Show that

$$\ln(N+1) < z_N < \ln N + 1. \tag{3.135}$$

3.26 **Generalized Zipf's law.**

(a) Show that

$$\zeta(\alpha) = \sum_{n=1}^{\infty} \frac{1}{n^\alpha} = \infty, \quad \text{for } 0 < \alpha < 1.$$

(b) Limit the range of the distribution to N. Find the upper bound and lower bound of the normalization constant z_N.

4 Continuous random variables

4.1 Continuous random variables

Many RVs that we encounter in mathematical models of real-world problems are *continuous*. For example, the inter-arrival times of packets from terminals to a server may assume *any* value between 0 and ∞. Similarly, the response time of a web request sent over the Internet may take on any positive value. A random variable is called a **continuous RV** if its range is a continuum or, equivalently, if its distribution function is everywhere continuous.

4.1.1 Distribution function and probability density function

For a given continuous RV X, the probability that X lies in a small interval $(x, x+\Delta]$ is

$$P[x < X \leq x + \Delta] = F_X(x+\Delta) - F_X(x) = \Delta \frac{[F_X(x+\Delta) - F_X(x)]}{\Delta}. \tag{4.1}$$

If Δ is small and the distribution function $F_X(x)$ is differentiable everywhere except possibly a finite set of points, then the following approximation holds:

$$P[x < X \leq x + \Delta] \cong \Delta F'_X(x), \tag{4.2}$$

in which the prime denotes the derivative of F_X. Whenever it exists, the derivative of F_X is called the **probability density function (PDF)** of X and is denoted by f_X. Thus,

$$\boxed{f_X(x) = \frac{dF_X(x)}{dx},} \tag{4.3}$$

or, in differential notation, we write formally

$$f_X(x)\,dx = dF_X(x) = P[x < X \leq x + dx]. \tag{4.4}$$

Therefore, given the PDF of X, its distribution function is computable as

$$F_X(x) = \int_{-\infty}^{x} f_X(u)\,du. \tag{4.5}$$

Since the distribution function is a nondecreasing function, the PDF is always nonnegative:

$$f_X(x) \geq 0. \tag{4.6}$$

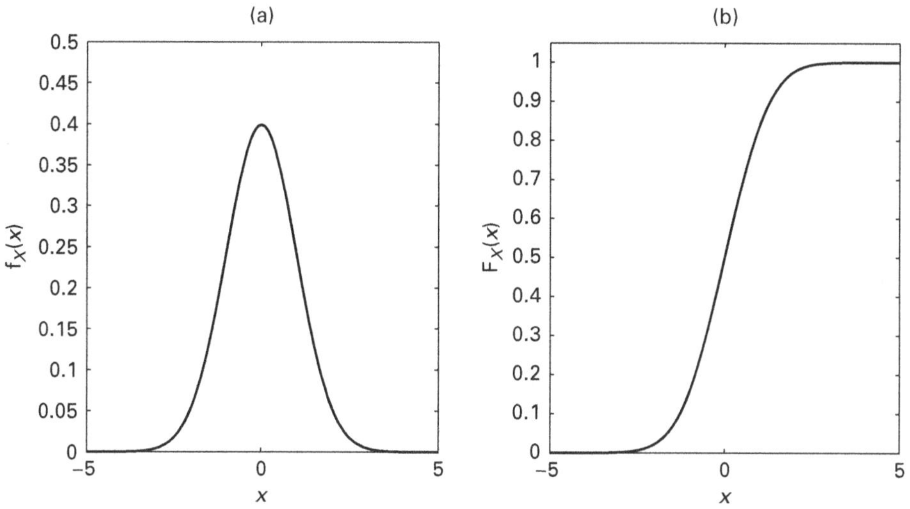

Figure 4.1 (a) The PDF and (b) the distribution function of a unit normal RV.

Figure 4.1 shows an example of the distribution function and the corresponding PDF of a continuous RV, in particular a unit normal random variable; i.e., a normally distributed random variable with zero mean and unit variance (see Section 4.2.4).

From Property 4 of (3.3) and (4.5) it follows that the probability $P[a < X \leq b]$ is given by the integral of the PDF over that interval:

$$P[a < X \leq b] = \int_a^b f_X(x)\,dx. \tag{4.7}$$

In particular, when $a = -\infty$ and $b = \infty$, we obtain

$$\int_{-\infty}^{\infty} f_X(x)\,dx = 1. \tag{4.8}$$

This result simply reflects the fact that the probability of the sure event is unity (see Property 2 in Section 2.2.4).

4.1.2 Expectation, moments, central moments, and variance

The notions of expectation, moments, central moments, and variance that we defined for discrete RVs in the previous chapter carry over directly to continuous RVs in a straightforward manner. The PMF should be replaced by the PDF and summation by integration.

Let μ_X denote the expectation of a continuous RV X, with PDF $f_X(x)$:

$$\boxed{\mu_X = E[X] = \int_{-\infty}^{\infty} x f_X(x)\,dx,} \tag{4.9}$$

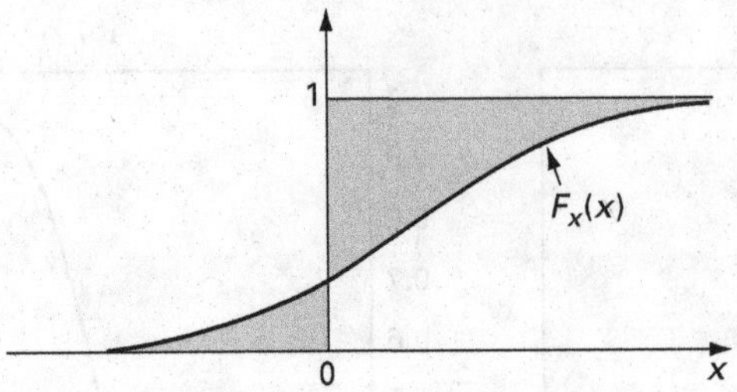

Figure 4.2 The expectation of the RV X is the difference of the shaded regions.

provided the integral exists. We may also write

$$\mu_X = \int_0^\infty x f_X(x)\,dx + \int_{-\infty}^0 x f_X(x)\,dx$$
$$= \int_0^\infty [1 - F_X(x)]\,dx - \int_{-\infty}^0 F_X(x)\,dx. \tag{4.10}$$

The last expression was obtained (see Problem 4.1) by applying integration by parts and by assuming that $E[|X|] < \infty$.[1] The result (4.10) says that the mean value of the RV X is equal to the difference of the right-hand and left-hand shaded areas in Figure 4.2.

If X is a nonnegative RV, the second term of (4.10) disappears and we obtain the following formula:

$$\mu_X = \int_0^\infty F_X^c(x)\,dx, \tag{4.11}$$

where

$$F_X^c(x) = 1 - F_X(x), \quad x \geq 0, \tag{4.12}$$

is called the **complementary distribution function** or the **survivor function** of the RV X. The formulas (4.10) and (4.11) hold for a *discrete random variable* as well (Problem 4.2).

The mth **moment** and mth **central moment** of a continuous RV X are defined as

$$E[X^m] = \int_{-\infty}^\infty x^m f_x(x)\,dx, \tag{4.13}$$

$$E[(X - \mu_X)^m] = \int_{-\infty}^\infty (x - \mu_X)^m f_X(x)\,dx. \tag{4.14}$$

The first moment is the expectation $E[X] = \mu_X$ and the second central moment is the variance $\text{Var}[X] = \sigma_X^2$, just like in the discrete RV case:

[1] An example of a distribution that does not satisfy this condition is the **Cauchy distribution** defined by (8.103):

$$f(x) = \frac{1}{\pi\alpha\left[1 + \frac{(x-\mu)^2}{\alpha^2}\right]}, \quad -\infty < x < \infty.$$

$$\sigma_X^2 = \mathrm{Var}[X] = \int_{-\infty}^{\infty} (x - \mu_X)^2 f_X(x)\, dx$$
$$= E[X^2] - \mu_X^2. \tag{4.15}$$

If we think of a one-dimensional PDF $f_X(\cdot)$ as a mass distribution along a rod, the moments $E[X^n]$ have direct physical analogs. The expectation μ_X corresponds to the center of gravity, $E[X^2]$ is the moment of inertia around the origin, and σ_X^2 is the central moment of inertia. Another type of analogy is found in electrical circuits. If the RV X represents a voltage or current, the mean μ_X gives the dc (direct current) component, the second moment $E[X^2]$ gives the average power carried by X, and σ_X is the root-mean-square value of the ac (alternating current) component.

We saw that the expression for the expectation μ_X is somewhat simplified when X is a nonnegative RV. The expression for the second moment may also be simplified for a nonnegative RV:

$$E[X^2] = 2 \int_0^{\infty} x F_X^c(x)\, dx. \tag{4.16}$$

Therefore, the variance of a positive RV is given by

$$\sigma_X^2 = E[X^2] - \mu_X^2 = 2 \int_0^{\infty} x F_X^c(x)\, dx - \mu_X^2. \tag{4.17}$$

The notions and properties of **covariance** and the **correlation coefficient** between two continuous RVs are exactly the same as those for discrete RVs. We will review these subjects in Section 4.3, where we give a formal discussion of **joint PDF** and **conditional PDF**.

4.2 Important continuous random variables and their distributions

4.2.1 Uniform distribution

A continuous RV X is said to be **uniformly distributed** on $[a, b]$ if

$$F_X(x) = \begin{cases} 0, & \text{if } x < a, \\ \frac{x-a}{b-a}, & \text{if } a \le x \le b, \\ 1, & \text{if } x > b. \end{cases} \tag{4.18}$$

The corresponding PDF is given by

$$f_X(x) = \begin{cases} \frac{1}{b-a}, & \text{if } a \le x \le b, \\ 0, & \text{elsewhere.} \end{cases} \tag{4.19}$$

The expectation, second moment, and variance can be readily found (Problem 4.9) as

$$\mu_X = \frac{b+a}{2}, \tag{4.20}$$

$$E[X^2] = \frac{b^2+ba+a^2}{3}, \tag{4.21}$$

$$\sigma_X^2 = \frac{(b-a)^2}{12}. \tag{4.22}$$

The **unit uniform RV** U defined over the unit interval [0, 1] has distribution and PDF given, respectively, by

$$\boxed{\begin{aligned} F_U(u) &= u, \quad 0 \le u \le 1, \\ f_U(u) &= 1, \quad 0 \le u \le 1. \end{aligned}} \tag{4.23}$$

4.2.2 Exponential distribution

Consider a nonnegative RV X that has the distribution function

$$\boxed{F_X(x) = 1 - e^{-\lambda x}, \quad x \ge 0,} \tag{4.24}$$

and the density function

$$\boxed{f_X(x) = \lambda e^{-\lambda x}, \quad x \ge 0.} \tag{4.25}$$

This distribution is called the **exponential distribution function** with **rate** λ and is shown in Figure 4.3. It frequently appears in many applications, including queueing models and reliability theory.

The exponential distribution is the continuous counterpart of the geometric distribution discussed in Section 3.3.2, which is associated with a Bernoulli trial sequence, a Bernoulli sequence for short. A random process that is associated with the exponential distribution is a **Poisson process**, which we will study in more detail in later chapters.

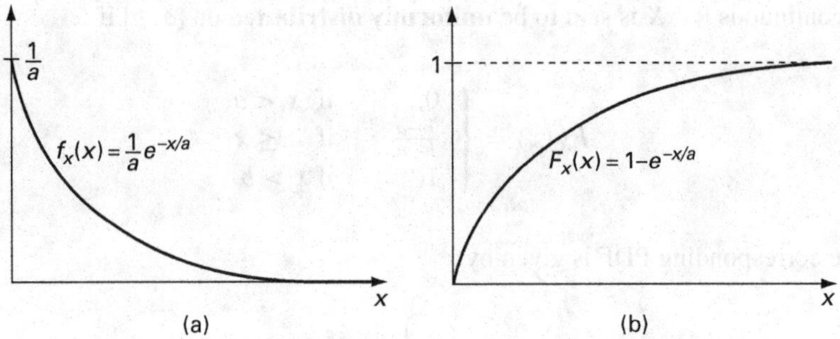

Figure 4.3 (a) The PDF and (b) the distribution function of an exponential RV.

Similar to the property (3.70) of the geometric distribution, the exponential RV X also possesses the **memoryless property**:

$$P[X > t + s | X > s] = \frac{P[X > t + s]}{P[X > s]} = \frac{e^{-\lambda(t+s)}}{e^{-\lambda s}} \\ = e^{-\lambda t} = P[X > t]. \tag{4.26}$$

Suppose that X represents the *lifetime* that is exponentially distributed. Then λ [s^{-1}] is the death rate in the sense that the probability that the life will end in the next δt [s] is $\lambda\delta t$, and λ^{-1} is the life expectancy. The above memoryless property implies that whatever the present age s is, the *residual lifetime* t is unaffected by the current age and has the same distribution of the lifetime itself; i.e., the exponential distribution with mean λ^{-1}.

As another example, suppose that X represents the interval between successive arrivals of buses in some city and assume that this RV is exponentially distributed with rate λ [min^{-1}], which is to say that buses arrive according to a **Poisson process** with rate λ. You have already waited for s [min], and a bus has not arrived yet. Then the conditional probability that you have to wait for at least t more minutes is the same as the original distribution; i.e., the exponential distribution with mean λ^{-1} [min]. In other words, the probability distribution of the additional waiting time does not depend on how long you have already waited! This counterintuitive property is called the "waiting time paradox of the Poisson process" (e.g., see Feller [99], pp. 10–15).

The expectation can be computed by applying (4.11) to the complementary distribution $F_X^c(x) = e^{-\lambda x}$:

$$\mu_X = \int_0^\infty F_X^c(x)\,dx = \int_0^\infty e^{-\lambda x}\,dx = \frac{1}{\lambda}. \tag{4.27}$$

Similarly, the second moment can be found by using (4.17):

$$\sigma_X^2 = 2\int_0^\infty x\,e^{-\lambda x} - \mu_X^2 = \frac{2}{\lambda^2} - \frac{1}{\lambda^2} = \frac{1}{\lambda^2}. \tag{4.28}$$

Thus, the standard deviation σ_X is the same as the mean μ_X.

The ratio of the standard deviation σ_X to the mean μ_X,

$$c_X = \frac{\sigma_X}{\mu_X}, \tag{4.29}$$

is called the **coefficient of variation** of the RV X, and indicates in a rough way the degree of departure from the exponential distribution. For an exponential distribution we have $c_X = 1$ from (4.27) and (4.28).

4.2.3 Gamma distribution

By generalizing the exponential distribution of rate λ to a distribution with two positive parameters (λ, β), we define the **gamma distribution**[2] by its PDF

$$f_Y(y; \lambda, \beta) \triangleq \frac{\lambda(\lambda y)^{\beta-1} e^{-\lambda y}}{\Gamma(\beta)}, \quad y \geq 0; \beta, \lambda > 0, \tag{4.30}$$

where $\Gamma(\beta)$ is the **gamma function**

$$\Gamma(\beta) = \int_0^\infty t^{\beta-1} e^{-t} dt = (\beta - 1)\Gamma(\beta - 1), \tag{4.31}$$

and the last recursive expression was obtained by applying integration by parts (Problem 4.13). This recursion is analogous to that of factorials; i.e., $n! = n(n-1)!$

The case where $\lambda = 1$ is called the **standard gamma distribution**:

$$f_X(x; \beta) \triangleq \frac{x^{\beta-1} e^{-x}}{\Gamma(\beta)}, \quad x \geq 0; \beta > 0. \tag{4.32}$$

The gamma function can be viewed as the normalization constant for the gamma PDF (4.32) (and for (4.30), as well). The cumulative distribution function (CDF) of the standard gamma distribution is given by

$$F_X(x; \beta) = \int_0^x f_X(t; \gamma)\, dt = 1 - \frac{\Gamma(\beta, x)}{\Gamma(\beta)} = \frac{\gamma(\beta, x)}{\Gamma(\beta)}, \tag{4.33}$$

where $\Gamma(\beta, x)$ and $\gamma(\beta, x)$ are called the **upper incomplete gamma function** and the **lower incomplete gamma function**, respectively, defined by

$$\Gamma(\beta, x) = \int_x^\infty t^{\beta-1} e^{-t}\, dt, \tag{4.34}$$

$$\gamma(\beta, x) = \int_0^x t^{\beta-1} e^{-t}\, dt. \tag{4.35}$$

It should be apparent that

$$\gamma(\beta, x) + \Gamma(\beta, x) = \Gamma(\beta), \quad \text{for all } x \geq 0,$$

and $\Gamma(\beta, 0) = \Gamma(\beta)$, $\Gamma(\beta, \infty) = 0$, $\gamma(\beta, 0) = 0$, and $\gamma(\beta, \infty) = \Gamma(\beta)$.

The mean of the standard gamma variable X of (4.32) is

$$E[X] = \frac{1}{\Gamma(\beta)} \int_0^\infty \lambda x\, e^{-x} x^{\beta-1}\, dx = \frac{\Gamma(\beta+1)}{\Gamma(\beta)} = \beta, \tag{4.36}$$

where the last expression was obtained using the recursive relation (4.31). Similarly, we find

$$E[X^2] = \beta(\beta + 1). \tag{4.37}$$

[2] Some authors use the parameters (α, β) or (β, α) instead of (λ, β).

Thus, the variance is

$$\mathrm{Var}[X] = E[X^2] - (E[X])^2 = \beta. \tag{4.38}$$

Then the mean, second moment, and variance of the RV Y defined by (4.30) are readily obtainable from the simple relation $X = \lambda Y$:

$$E[Y] = \frac{\beta}{\lambda}, \quad E[Y^2] = \frac{\beta(\beta+1)}{\lambda^2}, \quad \text{and } \mathrm{Var}[Y] = \frac{\beta}{\lambda^2}. \tag{4.39}$$

Figure 4.4 shows a family of the gamma distributions of (4.30) for three different cases, while the mean β/λ is kept to be unity. The parameter λ is just a scaling factor, but the shape of the PDF critically depends on the value of β, as seen in the figure:

- $\beta < 1$. The PDF $f_Y(y; \gamma, \lambda)$ with $\beta < 1$ is suitable when we want to fit a distribution with a **long tail**, as illustrated by the dashed curve that corresponds to the case $\beta = 0.2$. However, it goes to infinity as $y \to 0$.
- $\beta = 1$. The special case $\beta = 1$ reduces to the **exponential distribution** with mean $1/\lambda$.
- $\beta > 1$. The PDF (4.32) is zero at the origin and has a single maximum (i.e., its **mode** at $X = \beta - 1$) and then decreases towards zero as $x \to \infty$. In general, the *mode* of a probability distribution is defined as the value at which its PMF or PDF takes its maximum value. The mode of (4.30) is, therefore, at $Y = (\beta - 1)/\lambda$.
  If $\beta = k$, a positive integer, the random variable Y with the PDF (4.30) is equivalent to the sum of k i.i.d. exponential variables with mean $1/\lambda$; hence, $E[Y] = k/\lambda$.

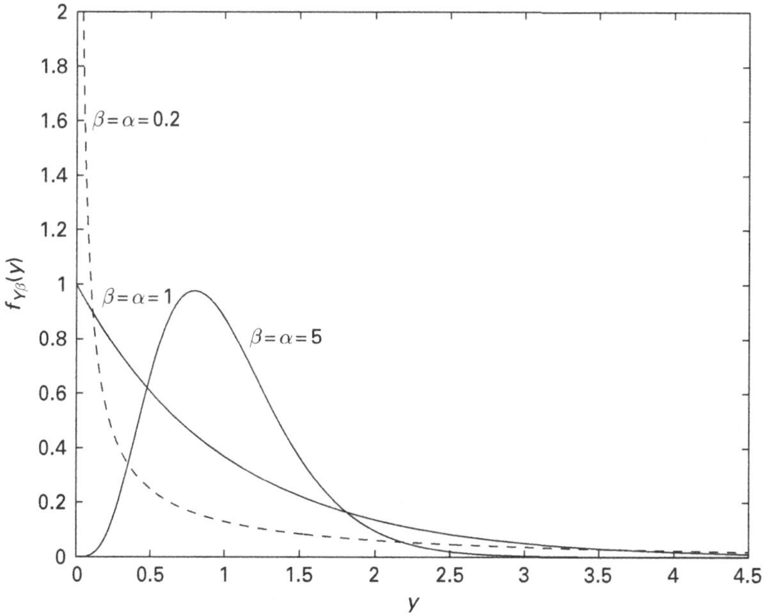

Figure 4.4 Gamma distributions for different parameters (λ, β), where the mean is kept constant (=1); i.e., $\beta/\lambda = 1$.

This distribution is known as the k-stage **Erlang distribution** (see Problem 4.11) and is widely used in queueing and teletraffic theory. Thus, the gamma variable can be viewed as the sum of β i.i.d. exponential random variables, where β is not necessarily an integer. Then, by virtue of the **central limit theorem (CLT)** (see Section 8.2.5), as β becomes large, the gamma distribution should approach the normal (or Gaussian) distribution (see Section 4.2.4) with mean β/λ and variance β/λ^2. Thus, if a positive RV exhibits a bell-shape distribution, a gamma distribution with suitable β and λ may provide a good fit.

If $\beta = n/2$ for an integer n in (4.32) (i.e., $\lambda = 1$), then it becomes the **chi-squared distribution** with n degrees of freedom, which will be discussed in Section 7.1.

The gamma distribution has been applied to many fields, including waiting time in queueing systems, the lifetime of devices as in reliability theory, the load on web servers, etc. It has also been applied to fit the distribution of rainfall in climatology and the distribution of insurance claims in finance services.

4.2.4 Normal (or Gaussian) distribution

A real RV X having the PDF

$$f_X(x) = \frac{1}{\sqrt{2\pi\sigma^2}} \exp\left[-\frac{(x-\mu)^2}{2\sigma^2}\right] \tag{4.40}$$

is called a **normal** or **Gaussian random variable**[3] with *mean* μ and *variance* σ^2. Often, the distribution is denoted by $N(\mu, \sigma^2)$.

Many RVs in physical situations are distributed in such a way as to have (at least approximately) the normal distribution. The main rationale for the prevalent use of the normal distribution in stochastic modeling is the CLT, which states that, under very weak restrictions, the sum of independent samples from any distribution with finite mean and variance tends, with proper scaling, to the normal distribution as the sample size becomes large.

The most convenient form of the normal distribution for tabulation is the one that corresponds to a random variable U defined by $U = (X - \mu)/\sigma$. The PDF of U, usually denoted $\phi(u)$,

$$\boxed{\phi(u) \triangleq \frac{1}{\sqrt{2\pi}} \exp\left(-\frac{u^2}{2}\right),} \tag{4.41}$$

is called the **standard** (or **unit**) **normal distribution**, and this distribution is often denoted $N(0, 1)$, where the arguments 0 and 1 represent the mean and variance respectively.

[3] Although the normal distribution is often called the Gaussian distribution, it was used in probability theory earlier by De Moivre and Laplace [99].

In order to verify that the functions (4.40) and (4.41) are indeed the PDFs with specified mean and variance (see Problem 4.15), we need first to show that (4.41) is a bona fide PDF; i.e., we must show that

$$I \triangleq \int_{-\infty}^{\infty} e^{-u^2/2}\, du \tag{4.42}$$

is equal to $\sqrt{2\pi}$. Toward this end, we consider I^2:

$$\begin{aligned} I^2 &= \int_{-\infty}^{\infty} e^{-u^2/2}\, du \int_{-\infty}^{\infty} e^{-v^2/2}\, dv \\ &= \int_{-\infty}^{\infty} \int_{-\infty}^{\infty} e^{-(u^2+v^2)/2}\, du\, dv. \end{aligned} \tag{4.43}$$

Changing the variables (u, v) to polar coordinates (r, θ):

$$u = r\cos\theta, \quad v = r\sin\theta; \quad \text{hence,} \quad du\, dv = r\, d\theta\, dr. \tag{4.44}$$

Then

$$\begin{aligned} I^2 &= \int_0^{\infty} \int_0^{2\pi} e^{-r^2/2} r\, d\theta\, dr = 2\pi \int_0^{\infty} r\, e^{-r^2/2}\, dr \\ &= -2\pi\, e^{-r^2/2}\Big|_0^{\infty} = 2\pi. \end{aligned} \tag{4.45}$$

Thus, $I = \sqrt{2\pi}$, and we have proved that $\phi(u)$ is a PDF. To show that its mean and variance are indeed zero and one is rather straightforward; thus, it is left to the reader as an exercise (see Problem 4.15). The distribution function of U,

$$\boxed{\Phi(u) = \int_{-\infty}^{u} \phi(t)\, dt = \frac{1}{\sqrt{2\pi}} \int_{-\infty}^{u} \exp\left(-\frac{t^2}{2}\right) dt,} \tag{4.46}$$

is widely tabulated, and is available in MATLAB, Mathematica, and other programming libraries. The PDF of (4.40) can be written as $\phi\left(\frac{x-\mu}{\sigma}\right)$ and the corresponding distribution function is $F_X(x) = \Phi\left(\frac{x-\mu}{\sigma}\right)$.

As we saw earlier in this chapter, Figure 4.1 shows the density and distribution functions of the standard normal distribution. The following properties of the distribution function $\Phi(x)$ are important:

1. For any $\infty < x < \infty$,

$$\Phi(-x) = 1 - \Phi(x). \tag{4.47}$$

2. For any $x > 0$,

$$\frac{e^{-x^2/2}}{\sqrt{2\pi}}\left(\frac{1}{x} - \frac{1}{x^3}\right) < 1 - \Phi(x) < \frac{e^{-x^2/2}}{\sqrt{2\pi}}\frac{1}{x}. \tag{4.48}$$

3. For large x,

$$1 - \Phi(x) \approx \frac{e^{-x^2/2}}{\sqrt{2\pi}\,x}, \quad x \gg 1. \tag{4.49}$$

Property 1 is due to the fact that $\phi(x)$ is symmetric around $x = 0$. In order to prove the second inequality in Property 2, we verify by differentiation the following identity [99]:

$$\frac{e^{-x^2/2}}{\sqrt{2\pi}\,x} = \frac{1}{\sqrt{2\pi}} \int_x^\infty e^{-y^2/2}\left(1 + \frac{1}{y^2}\right) dy. \tag{4.50}$$

The integrand on the right side is greater than the integrand of

$$1 - \Phi(x) = \frac{1}{\sqrt{2\pi}} \int_x^\infty e^{-y^2/2}\,dy, \tag{4.51}$$

which proves the second inequality. The first inequality can be derived by using the following identity:

$$\frac{e^{-x^2/2}}{\sqrt{2\pi}}\left(\frac{1}{x} - \frac{1}{x^3}\right) = \frac{1}{\sqrt{2\pi}} \int_x^\infty e^{-y^2/2}\left(1 - \frac{3}{y^4}\right) dy. \tag{4.52}$$

The integrand is smaller than $e^{-y^2/2}$. Property 3 readily follows from Property 2. This approximation is quite accurate for $x \gtrsim 4$.

4.2.4.1 Moments of the unit normal distribution

Since the density function of the standard normal distribution is an even function of u, the nth moment is zero for odd values of n:

$$E[U^n] = 0 \quad (n \text{ odd}). \tag{4.53}$$

When $n \geq 2$ is even, we obtain

$$E[U^n] = 1 \times 3 \times 5 \cdots (n-1) \quad (n \text{ even}), \tag{4.54}$$

either by direct evaluation of the integral or through the **characteristic function** (see Section 8.2). In particular,

$$E[U] = 0 \ \text{ and } \ \sigma_U^2 = E[U^2] = 1. \tag{4.55}$$

The normal RV X with the density function (4.40) can be written as

$$X = \sigma U + \mu. \tag{4.56}$$

Therefore, its expectation and variance are given by

$$E[X] = \sigma E[U] + \mu = \mu \ \text{ and } \ \sigma_X^2 = E[(X-\mu)^2] = \sigma^2 E[U^2] = \sigma^2. \tag{4.57}$$

One important property of the normal variable is its **reproductive property**. Suppose $X_i,\ i = 1,\ 2,\ \ldots,\ n$, are independent RVs having distributions $N(\mu_i, \sigma_i^2)$, and let Y be a random variable defined by

$$Y = \sum_{i=1}^{n} a_i X_i, \tag{4.58}$$

where the a_i are real constants. Then the distribution of Y is also normal:

$$N\left(\sum_{i=1}^{n} a_i \mu_i, \sum_{i=1}^{n} a_i^2 \sigma_i^2\right).$$

4.2.4.2 The normal approximation to the binomial distribution and the De Moivre–Laplace limit theorem[4]

In *The Doctrine of Chances*, published by Abraham de Moivre in 1733, Bernoulli's theorem was sharpened by introducing the normal distribution approximation to the binomial distribution for the special case $p = \frac{1}{2}$. Subsequently, in 1812, de Moivre's result was generalized to any p by Laplace. Their result, known as the **De Moivre–Laplace limit theorem**, is a special case of the CLT. It states that the binomial distribution of the number of "successes" S_n in n independent Bernoulli trials with probability p of success on each trial converges to a normal distribution as n goes to infinity.

We are interested in the probability of the event that the number of successes lies between α and β ($\alpha < \beta$); i.e.,

$$P[\alpha \le S_n \le \beta] = B(\alpha; n, p) + B(\alpha + 1; n, p) + \cdots + B(\beta; n, p), \tag{4.59}$$

where $B(k; n, p)$ is defined in (2.38).

It is convenient to introduce the new variable $\delta_k = k - np$. Then

$$k = np + \delta_k, \quad n - k = nq - \delta_k. \tag{4.60}$$

By applying Stirling's approximation formula[5] to the factorials, we obtain

$$\begin{aligned} B(k; n, p) &\sim \left[\frac{n}{2\pi k(n-k)}\right]^{1/2} \left(\frac{np}{k}\right)^k \left(\frac{nq}{n-k}\right)^{n-k} \\ &= \left[\frac{n}{2\pi(np + \delta_k)(nq - \delta_k)}\right]^{1/2} \frac{1}{(1 + \delta_k/np)^{np+\delta_k}(1 - \delta_k/nq)^{nq-\delta_k}}, \end{aligned} \tag{4.61}$$

where the symbol $\sim$ means here that the ratio of both sides tends to unity. If we assume $\delta_k^3/n^2 = (k - np)^3/n^2 \to 0$, then it can be shown (e.g., see Feller [99], p. 169) that (4.61) takes on the simpler form

$$B(k; n, p) \sim \left[\frac{n}{2\pi(np + \delta_k)(nq - \delta_k)}\right]^{1/2} e^{-\delta_k^2/2npq}. \tag{4.62}$$

[4] Readers may skip the remainder of this subsection at their first reading.

[5] Stirling's approximation for a factorial $n!$ is given by

$$n! \sim \sqrt{2\pi}\, n^{n+(1/2)} e^{-n}.$$

See Section 4.2.3 for a similar approximation to the gamma function $\Gamma(x)$.

However, $np + \delta_k \sim np$ and $nq - \delta_k \sim nq$, which further simplify (4.62), yielding

$$B(k; n, p) \sim \frac{1}{\sqrt{2\pi npq}} e^{-\delta_k^2/2npq} = \frac{1}{\sqrt{npq}} \Phi\left(\frac{k - np}{\sqrt{npq}}\right). \tag{4.63}$$

This is the desired asymptotic formula.

Now let

$$\sigma \triangleq \sqrt{npq} \tag{4.64}$$

and define x_k, which is a function of the variable k:

$$x_k = \frac{k - np}{\sigma} = \frac{\delta_k}{\sqrt{npq}}. \tag{4.65}$$

Then we can rewrite (4.63) in the form

$$B(k; n, p) \sim \sigma^{-1} \phi(x_k), \tag{4.66}$$

provided that (i) $\delta_k n^{-1} \to 0$ and (ii) $\delta_k^3 n^{-2} \to 0$, as $n \to \infty$ and $k \to \infty$. Condition (ii) implies condition (i), and is equivalent, in view of (4.65), to assuming that $x_k^3 n^{-1/2} \to 0$.

Thus, if

$$\frac{x_\alpha^3}{\sigma} \to 0, \text{ and } \frac{x_\beta^3}{\sigma} \to 0, \tag{4.67}$$

then (4.66) holds for all terms in (4.59), and thus

$$P[\alpha \le S_n \le \beta] \sim \sigma^{-1}[\phi(x_\alpha) + \phi(x_{\alpha+1}) + \cdots + \phi(x_\beta)]. \tag{4.68}$$

By interpreting the right side as a Riemann sum approximating an integral, we can derive the following theorem.[6]

THEOREM 4.1 **(De Moivre–Laplace limit theorem)**. *If α and β are such that $\sigma^{-1}x_\alpha^3 \to 0$ and $\sigma^{-1}x_\beta^3 \to 0$, then*

$$P[\alpha \le S_n \le \beta] \sim \Phi(x_{\beta+\frac{1}{2}}) - \Phi(x_{\alpha-\frac{1}{2}}), \tag{4.69}$$

where $\sigma = \sqrt{npq}$ and $x_t = \frac{t-np}{\sigma}$. □

If we normalize S_n and define

$$S_n^* = \frac{S_n - np}{\sqrt{npq}}, \tag{4.70}$$

then the inequality $\alpha \le S_n \le \beta$ is the same as $x_\alpha \le S_n^* \le x_\beta$, and (4.69) implies

$$P[x_\alpha \le S_n^* \le x_\beta] \sim \Phi\left(x_\beta + \frac{1}{2\sigma}\right) - \Phi\left(x_\alpha - \frac{1}{2\sigma}\right). \tag{4.71}$$

[6] See Feller [99], pp. 171–172 for details. An alternative proof is to use the characteristic function (CF) used in proving Theorem 8.2 of page 201.

Since $\frac{1}{\sigma} = \frac{1}{\sqrt{npq}} \to 0$ as $n \to \infty$, the right side tends to $\Phi(x_\beta) - \Phi(x_\alpha)$. Thus, we have the following corollary.

COROLLARY 4.1 (Corollary to the limit theorem). *For any $a < b$,*

$$\boxed{P\left[a \le \frac{S_n - np}{\sqrt{npq}} \le b\right] \to \Phi(b) - \Phi(a)} \tag{4.72}$$

as $n \to \infty$. □

Example 4.1: This example is taken from Feller [99], p. 174.

(a) Let

$$p = \frac{1}{2}, n = 200, \alpha = 95, \beta = 105.$$

The probability $P\,[95 \le S_{200} \le 105]$ may be interpreted as the probability that, in 200 tosses of a coin, the number of heads deviates from the expectation $E[S_{200}] = 100$ by at most 5. We have

$$\sigma^{-1} = \frac{1}{\sqrt{npq}} = 0.141\,421\ldots \text{ and } -x_{\alpha-\frac{1}{2}} = x_{\beta+\frac{1}{2}} = 0.777\,82\ldots$$

From a table[7] of the normal distribution, we find $\Phi(x_{\beta+\frac{1}{2}}) - \Phi(x_{\alpha-\frac{1}{2}}) = 0.563\,31\ldots$. The true value (obtainable from a table of the binomial distribution) is $0.563\,25\ldots$. The error is surprisingly small.

(b) Let

$$p = \frac{1}{10}, n = 500, \alpha = 50, \beta = 55.$$

The correct value is $P[50 \le S_{500} \le 55] = 0.317\,573\ldots$ We have $\sigma^{-1} = (45)^{-1/2} = 0.149\,071\,2\ldots$, and we get the approximation $\Phi(5.5\sigma^{-1}) - \Phi(-0.5\sigma^{-1}) = 0.3235\ldots$ Hence, the error is about 2%.

□

The two-dimensional and multidimensional Gaussian distributions are often referred to as the **bivariate** and **multivariate normal distributions** respectively. We shall further discuss the normal distribution in Section 4.3.1. A number of other important distributions are derived from, or reduced to, the normal distribution, as we discuss throughout this book, especially in Chapter 7.

[7] Or the normal CDF function in MATLAB.

4.2.5 Weibull distributions

The **Weibull**[8] distributions are often used in reliability engineering. There are three different, but related, distributions under the name Weibull distribution.

The **three-parameter Weibull distribution** is defined by

$$\boxed{F_X(x) = \begin{cases} 0, & x \leq \gamma, \\ 1 - e^{-[(x-\gamma)/\beta]^\alpha}, & x > \gamma, \end{cases}} \tag{4.73}$$

where $\alpha > 0$ determines the shape or the slope of the distribution; $\beta > 0$ is the scale parameter; and γ is the shifting or translation parameter.

The corresponding PDF $f_X(x)$ is given by

$$f_X(x) = \begin{cases} 0, & x \leq \gamma, \\ \frac{\alpha}{\beta}\left(\frac{x-\gamma}{\beta}\right)^{\alpha-1} e^{-[(x-\gamma)/\beta]^\alpha}, & x > \gamma. \end{cases} \tag{4.74}$$

The **two-parameter Weibull distribution** is defined by setting the shifting parameter $\gamma = 0$ in the above distribution:

$$\boxed{F_X(x) = 1 - e^{-(x/\beta)^\alpha}, \ x > 0,} \tag{4.75}$$

and its PDF is given by

$$f_X(x) = \frac{\alpha}{\beta^\alpha} x^{\alpha-1} e^{-(x/\beta)^\alpha}, \ x \geq 0. \tag{4.76}$$

If $\alpha = 1$, then the above PDF reduces to

$$f_X(x) = \frac{1}{\beta} e^{-x/\beta}, \ x \geq 0, \tag{4.77}$$

which is the exponential distribution discussed in Section 4.2.2.

The **standard Weibull distribution** is a one-parameter distribution, obtained by setting the scaling parameter $\beta = 1$:

$$\boxed{F_X(x) = 1 - e^{-x^\alpha}, \ x > 0,} \tag{4.78}$$

and its PDF is given by

$$f_X(x) = \alpha x^{\alpha-1} e^{-x^\alpha}, \ x \geq 0. \tag{4.79}$$

When $\alpha = 1$, the standard Weibull distribution reduces to the exponential distribution with mean one. For $\alpha = 2$, the distribution function and the PDF become

$$F_X(x) = 1 - e^{-x^2} \text{ and } f_X(x) = 2x\, e^{-x^2}. \tag{4.80}$$

[8] Ernst Hjalmar Waloddi Weibull (1887–1979) was a Swedish engineer, scientist, and mathematician.

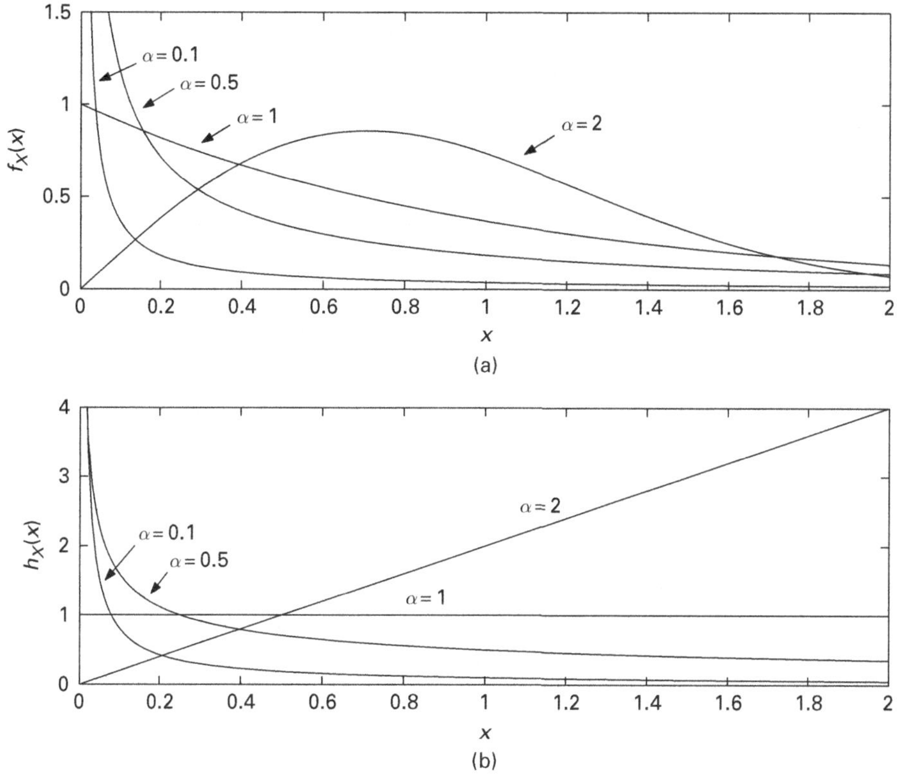

Figure 4.5 (a) The PDF of the standard Weibull distributions for $\alpha = 0.5, 1$, and 2. (b) The **hazard function** (see Problem 6.10 part (d)) of the standard Weibull distributions.

This distribution is equivalent to the **Rayleigh distribution** with $\sigma = 1/\sqrt{2}$, which will be discussed in Section 7.5.

Figure 4.5 (a) plots the PDF for $\alpha = 0.1, 0.5, 1, 2$, and 5 for the standard Weibull distribution. When $\alpha \ll 1$, the tail of the distribution decays very slowly. Therefore, a Weibull distribution with $\alpha \ll 1$ is a **heavy-tailed** distribution. The expectation, second moment, and variance of the standard Weibull distribution are given respectively by (Problem 4.17)

$$E[X] = \Gamma\left(\frac{1}{\alpha} + 1\right), \tag{4.81}$$

$$E[X^2] = \Gamma\left(\frac{2}{\alpha} + 1\right), \tag{4.82}$$

$$\sigma_X^2 = \left[\Gamma\left(\frac{2}{\alpha} + 1\right) - \Gamma^2\left(\frac{1}{\alpha} + 1\right)\right], \tag{4.83}$$

where $\Gamma(\cdot)$ is the gamma function defined in (4.31). Similarly, it is not difficult to calculate other statistics of interest, such as the **median** and the **mode** (Problem 4.18).

Once we have found the expectation and variance for the standard Weibull distribution, the corresponding statistics for the two-parameter and three-parameter distributions are easy to obtain, since it is a matter of scaling and shifting (Problem 4.19).

4.2.6 Pareto distribution

In Section 3.3.5 we discussed Zipf's law or the zeta distribution, which decays according to a power law $n^{-\alpha}$, where $\alpha > 1$. The **Pareto**[9] distribution is also a power law distribution found in the distribution of incomes (before modern industrial capitalism created the vast middle class) and in a large number of real-world situations. In fact, Zipf's distribution can be viewed as the discrete version of the Pareto distribution.

The **Pareto distribution** is a two-parameter distribution with $\alpha > 0$ and $\beta > 0$, and defined by

$$F_X(x) = \begin{cases} 0, & 0 \le x < \beta, \\ 1 - (x/\beta)^{-\alpha}, & \beta \le x < \infty. \end{cases} \tag{4.84}$$

The PDF is readily computed as

$$f_X(x) = \begin{cases} 0, & 0 \le x < \beta, \\ \frac{\alpha}{\beta}\left(\frac{x}{\beta}\right)^{-(\alpha+1)}, & \beta \le x < \infty. \end{cases} \tag{4.85}$$

Noting that the survivor function is given by

$$F_X^c(x) = \begin{cases} 1, & 0 \le x < \beta, \\ \left(\frac{x}{\beta}\right)^{-\alpha}, & \beta \le x < \infty, \end{cases} \tag{4.86}$$

we can readily calculate the expectation as

$$E[X] = \int_0^\infty F_X^c(x)\,dx = \beta + \beta^\alpha \int_\beta^\infty x^{-\alpha}\,dx = \frac{\alpha\beta}{\alpha - 1}, \quad \text{for } \alpha > 1, \tag{4.87}$$

whereas $E[X]$ does not exist for $\alpha \le 1$. Similarly, we find the second moment as

$$E[X^2] = 2\int_0^\infty x F_X^c(x)\,dx = 2\int_0^\beta x\,dx + 2\beta^\alpha \int_\beta^\infty x^{-\alpha+1}\,dx = \frac{\alpha\beta^2}{\alpha - 2}, \quad \text{for } \alpha > 2. \tag{4.88}$$

Thus, $E[X^2]$ does not exist for $\alpha \le 2$. The variance is then given by

$$\text{Var}[X] = \sigma_X^2 = \frac{\alpha\beta^2}{\alpha - 2} - \frac{\alpha^2\beta^2}{(\alpha - 1)^2} = \frac{\alpha\beta^2}{(\alpha - 1)^2(\alpha - 2)}, \quad \text{for } \alpha > 2. \tag{4.89}$$

[9] Vilfredo Federico Damaso Pareto (1848–1923) was a French–Italian sociologist and economist.

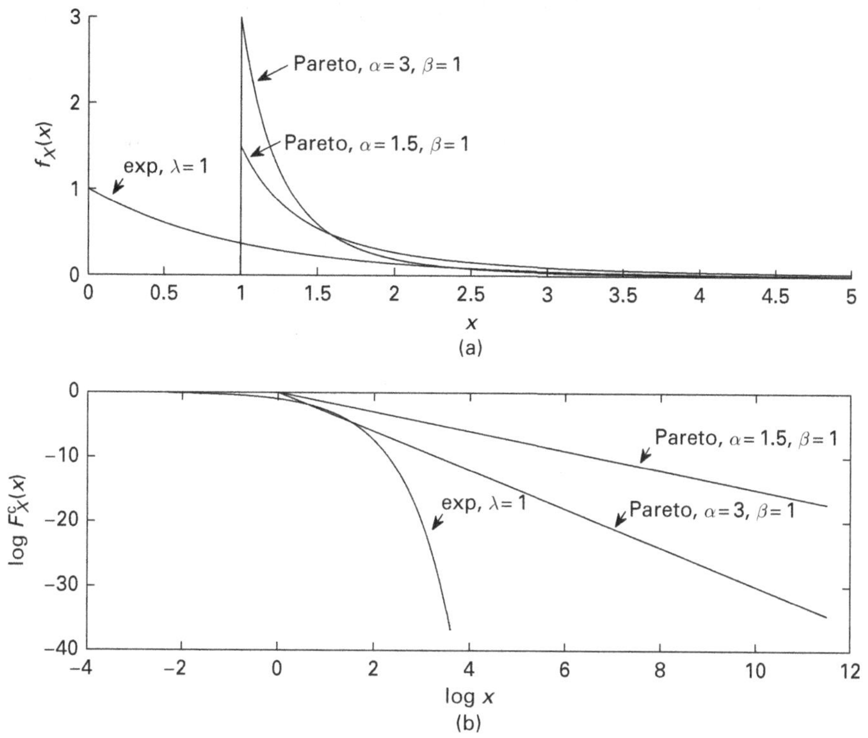

Figure 4.6 (a) The PDFs of the Pareto distributions with $\beta = 1, \alpha = 1.5$ and 3, together with the PDF of the exponential distribution $\lambda = 1$. (b) The log-survivor function $\log F_X^c(x)$ versus $\log x$ of the Pareto distributions and the exponential distribution.

For $1 < \alpha \leq 2$, the variance is infinite, although the mean is finite. Proceeding in a similar manner, we find the nth moment:

$$E[X^n] = \begin{cases} \frac{\alpha\beta^n}{\alpha - n}, & \text{if } \alpha > n, \\ \infty, & \text{if } \alpha \leq n. \end{cases} \tag{4.90}$$

Figure 4.6 (a) shows some plots of the PDF of the Pareto distributions. The tail of the distribution decays much more slowly than that of an exponential distribution. Their difference is more pronounced if we plot the **log-survivor function**, i.e., $\log F_X^c(x)$ versus $\log x$, as given in Figure 4.6 (b). For the Pareto distribution it is a straight line with slope $-\alpha$, whereas the exponential distribution grows exponentially in the negative direction. This is because the log-survivor of the exponential distribution with rate parameter λ is $\ln e^{-\lambda x} = -\lambda x = -\lambda\, e^{\ln x}$. Thus, the Pareto distribution, which follows a power law is said to have a long or **heavy-tailed** distribution.

If we translate the variable X to Y by $Y = X - \beta$, then we have

$$F_Y(y) = 1 - \left(\frac{\beta}{y + \beta}\right)^{\alpha}, \quad y \geq 0, \tag{4.91}$$

which is called the **translated Pareto distribution**.

4.3 Joint and conditional probability density functions

The various definitions and results given above can be extended to the case of two RVs X and Y. If the joint probability distribution $F_{XY}(x, y)$ is everywhere continuous and possesses a second partial derivative everywhere (except possibly on a finite set of curves), we define the **joint PDF** by

$$\boxed{f_{XY}(x, y) = \frac{\partial^2 F_{XY}(x, y)}{\partial x \partial y}.} \tag{4.92}$$

Then

$$F_{XY}(x, y) = \int_{-\infty}^{x} \int_{-\infty}^{y} f_{XY}(u, v)\, du\, dv. \tag{4.93}$$

Thus, $f_{XY}(x, y)\, dx\, dy$ may be interpreted as the probability that the point (X, Y) falls in an incremental area $dx\, dy$ about the point (x, y) in a two-dimensional Euclidean space. Since the joint probability distribution is a nondecreasing function of its arguments, it follows that

$$f_{XY}(x, y) \geq 0. \tag{4.94}$$

By letting x and y both approach infinity in (4.93), we obtain

$$\int_{-\infty}^{\infty} \int_{-\infty}^{\infty} f_{XY}(x, y)\, dx\, dy = 1. \tag{4.95}$$

If we instead let only one of the upper limits approach infinity, we obtain, on application of Property 5 in (3.5), the **marginal distribution** of X:

$$\int_{-\infty}^{x} \int_{-\infty}^{\infty} f_{XY}(u, v)\, du\, dv = F_X(x) \tag{4.96}$$

Similarly, the marginal distribution of Y is obtained as

$$\int_{-\infty}^{\infty} \int_{-\infty}^{y} f_{XY}(u, v)\, du\, dv = F_Y(y). \tag{4.97}$$

By differentiating both sides of (4.96) and (4.97), we obtain the following relation between the joint PDF and the marginal PDFs of X and Y:

$$\int_{-\infty}^{\infty} f_{XY}(x, v)\, dv = f_X(x) \tag{4.98}$$

and

$$\int_{-\infty}^{\infty} f_{XY}(u, y)\, du = f_Y(y). \tag{4.99}$$

Let us consider now the probability that the RV Y is less than or equal to y, subject to the hypothesis that the RV X has a value falling in $(x, x + \Delta]$. It follows from the definition of conditional probability that

$$P[Y \le y | x < X \le x + \Delta] = \frac{P[x < X \le x + \Delta, Y \le y]}{P[x < X \le x + \Delta]} = \frac{\int_x^{x+\Delta} \int_{-\infty}^{y} f_{XY}(u, v)\, du\, dv}{\int_x^{x+\Delta} f_X(u)\, du}. \quad (4.100)$$

The denominator can be replaced by $f_X(x)\Delta$ and the numerator by $\Delta \int_{-\infty}^{y} f_{XY}(x, v)\, dv$, as $\Delta \to 0$. Thus, we get

$$\boxed{F_{Y|X}(y|x) = \frac{\int_{-\infty}^{y} f_{XY}(x, v)\, dv}{f_X(x)},} \quad (4.101)$$

which is the **conditional distribution function** of the RV Y subject to the hypothesis $X = x$. Assuming that the usual continuity requirements are met for $F_{Y|X}(y|x)$, we define the *conditional PDF* $f_{Y|X}(y|x)$ by

$$f_{Y|X}(y|x) \triangleq \frac{\partial F_{Y|X}(y|x)}{\partial y} = \frac{f_{XY}(x, y)}{f_X(x)}. \quad (4.102)$$

Then

$$F_{Y|X}(y|x) = \int_{-\infty}^{y} f_{Y|X}(v|x)\, dv. \quad (4.103)$$

The concept of **conditional expectation** for continuous RVs is analogous to that for discrete RVs (see Section 3.2). Let X and Y be continuous RVs. The conditional expectation of X given Y is defined by

$$\boxed{E[X|Y] \triangleq \psi(Y),} \quad (4.104)$$

where

$$\psi(y) = E[X|Y = y] = \int_{-\infty}^{\infty} x f_{X|Y}(x|y) dx. \quad (4.105)$$

The basic properties of conditional expectation in the case of discrete RVs also hold for continuous RVs (see Problem 4.8). In particular, the **law of iterated expectations** holds:

$$E[E[X|Y]] = E[X]. \quad (4.106)$$

The definition of **conditional variance** for continuous RVs X and Y is formally the same as (3.42) for discrete RVs, namely

$$\mathrm{Var}[X|Y] \triangleq E[(X - E[X|Y])^2|Y]. \quad (4.107)$$

As we shall discuss in Section 22.1.3, the conditional expectation $E[X|Y]$ has an important interpretation as the best estimate of X as a function of Y in the minimum mean square error (MMSE) sense.

4.3.1 Bivariate normal (or Gaussian) distribution

By extending the standard normal distribution (4.41) of Section 4.2.4, we define the **standard bivariate normal (or Gaussian) distribution** as follows. Two normal RVs U_1 and U_2 that have the joint PDF given by

$$\phi_\rho(u_1, u_2) \triangleq \frac{1}{2\pi\sqrt{(1-\rho^2)}} \exp\left[-\frac{1}{2(1-\rho^2)}(u_1^2 - 2\rho u_1 u_2 + u_2^2)\right] \tag{4.108}$$

are called standard bivariate normal variables, where ρ is the **correlation coefficient** between U_1 and U_2, i.e.,

$$\rho \triangleq \text{Cov}[U_1, U_2] = \int_{-\infty}^{\infty}\int_{-\infty}^{\infty} u_1 u_2 \phi_\rho(u_1, u_2)\, du_1\, du_2, \tag{4.109}$$

and $-1 \le \rho \le 1$. However, when $\rho = 1$, or $\rho = -1$, the joint PDF cannot be used: when $\rho = 1$, the two RVs are completely correlated; i.e., $U_2 = U_1$. Similarly, $\rho = -1$ implies that $U_2 = -U_1$. Thus, the two RVs degenerate to a single variable; hence, the joint PDF (4.108) does not exist for $\rho = \pm 1$. Figure 4.7 shows $\phi_\rho(u_1, u_2)$ when $\rho = -0.75$. The contour lines (or level curves) of this surface are all ellipses (Problem 4.21).

When $\rho = 0$, the RVs U_1 and U_2 are said to be **uncorrelated** and (4.108) reduces to

$$\phi_0(u_1, u_2) = \frac{1}{\sqrt{2\pi}} \exp\left(-\frac{u_1^2}{2}\right) \frac{1}{\sqrt{2\pi}} \exp\left(-\frac{u_2^2}{2}\right) = \phi(u_1)\phi(u_2), \tag{4.110}$$

where $\phi(u_1)$ and $\phi(u_2)$ are the marginal PDFs of the variables U_1 and U_2 respectively and are both the standard normal distribution of (4.41). Equation (4.110) also implies that U_1 and U_2 are independent, since the joint PDF is the product of the marginal

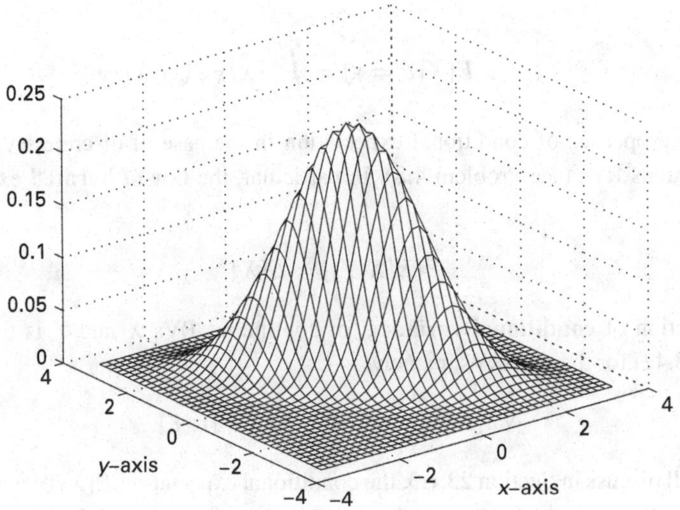

Figure 4.7 The standard bivariate normal distribution $\phi_\rho(u_1, u_2)$ of (4.108) with $\rho = -0.75$.

PDFs. Therefore, bivariate normal variables are independent when they are uncorrelated. Note that two RVs that are uncorrelated are not necessarily independent, unless they are normal RVs.

The conditional PDF of U_2 given $U_1 = u_1$ can be easily computed as

$$f_{U_2|U_1}(u_2|u_1) = \frac{\phi_\rho(u_1, u_2)}{\phi(u_1)} = \frac{1}{\sqrt{2\pi(1-\rho^2)}} \exp\left[-\frac{(u_2 - \rho u_1)^2}{2(1-\rho^2)}\right], \tag{4.111}$$

which is also a normal distribution, but its mean is ρu_1, not zero, and the variance is reduced from unity to $1 - \rho^2$.

Let us define RVs X_1 and X_2 that are related to the above U_1 and U_2 according to the following linear transformations:

$$U_1 = \frac{X_1 - \mu_1}{\sigma_1} \text{ and } U_2 = \frac{X_2 - \mu_2}{\sigma_2}. \tag{4.112}$$

Then, we find that the joint PDF of (X_1, X_2), sometimes denoted as $N(\mu_1, \mu_2, \sigma_1^2, \sigma_2^2, \rho)$, is given by

$$f_{X_1,X_2}(x_1, x_2) = \frac{1}{2\pi\sigma_1\sigma_2\sqrt{1-\rho^2}} \exp\left[-\frac{Q(x_1, x_2)}{2}\right], \tag{4.113}$$

where the function $Q(x_1, x_2)$ is

$$Q(x_1, x_2) = \frac{1}{1-\rho^2}\left[\left(\frac{x_1-\mu_1}{\sigma_1}\right)^2 - 2\rho\left(\frac{x_1-\mu_1}{\sigma_1}\right)\left(\frac{x_2-\mu_2}{\sigma_2}\right) + \left(\frac{x_2-\mu_2}{\sigma_2}\right)^2\right]. \tag{4.114}$$

If we adopt a matrix and vector representation, the PDF of the vector variable

$$\boldsymbol{X} \triangleq \begin{bmatrix} X_1 \\ X_2 \end{bmatrix} \tag{4.115}$$

is given as

$$f_{\boldsymbol{X}}(\boldsymbol{x}) = \frac{1}{2\pi|\det \boldsymbol{C}|^{1/2}} \exp\left[-\frac{1}{2}(\boldsymbol{x} - \boldsymbol{\mu})^\top \boldsymbol{C}^{-1}(\boldsymbol{x} - \boldsymbol{\mu})\right], \tag{4.116}$$

where $\top$ denotes matrix transpose, $\boldsymbol{x} = (x_1, x_2)^\top$ and $\boldsymbol{\mu} = (\mu_1, \mu_2)^\top$ are 2×1 column vectors, $\boldsymbol{C}$ is a 2×2 matrix called the **covariance matrix**, given by

$$\boldsymbol{C} = \begin{bmatrix} \sigma_1^2 & \rho\sigma_1\sigma_2 \\ \rho\sigma_1\sigma_2 & \sigma_2^2 \end{bmatrix}, \tag{4.117}$$

and $|\det \boldsymbol{C}|$ is the **determinant** of $\boldsymbol{C}$:

$$|\det \boldsymbol{C}| = \sigma_1^2\sigma_2^2(1 - \rho^2). \tag{4.118}$$

By noting that the inverse of C is given as

$$C^{-1} = \frac{1}{|\det C|}\begin{bmatrix} \sigma_2^2 & -\rho\sigma_1\sigma_2 \\ -\rho\sigma_1\sigma_2 & \sigma_1^2 \end{bmatrix} = \frac{1}{1-\rho^2}\begin{bmatrix} \frac{1}{\sigma_1^2} & -\frac{\rho}{\sigma_1\sigma_2} \\ -\frac{\rho}{\sigma_1\sigma_2} & \frac{1}{\sigma_2^2} \end{bmatrix}, \tag{4.119}$$

we can readily see that the PDF (4.116) is equivalent to the joint PDF (4.113).

4.3.2 Multivariate normal (or Gaussian) distribution

The PDF (4.116) applies more generally to the case of a **multivariate normal (or Gaussian)** variable $X = (X_1, \cdots, X_M)^\top$, for which the corresponding mean vector $\mu = E[X]$ is also M-dimensional, and the covariance matrix C is an $M \times M$ matrix, defined by

$$C = E[(X-\mu)(X-\mu)^\top]. \tag{4.120}$$

In this case, we write $X \sim N(\mu, C)$.

A multivariate normal variable X has the property that the conditional distribution of a subset of the component random variables given the remaining random variables is also a multivariate normal distribution. For example, suppose that X is partitioned as follows:

$$X = \begin{bmatrix} X_a \\ X_b \end{bmatrix}, \tag{4.121}$$

where $X_a = (X_1, \ldots, X_m)^\top$, $X_b = (X_{m+1}, \ldots, X_M)^\top$, and $1 \le m < M$. Assume that the mean vector μ and covariance matrix C of X are partitioned correspondingly as follows:

$$\mu = \begin{bmatrix} \mu_a \\ \mu_b \end{bmatrix}, \quad C = \begin{bmatrix} C_{aa} & C_{ab} \\ C_{ba} & C_{bb} \end{bmatrix}, \tag{4.122}$$

where $\mu_a = (\mu_1, \ldots, \mu_m)^\top$, $\mu_b = (\mu_{m+1}, \ldots, \mu_M)^\top$, and $C_{xy} = E[(X_x - \mu_x)(X_y - \mu_y)^\top]$ for all $x, y \in \{a, b\}$. Let $x_a = (x_1, \ldots, x_m)^\top$ and $x_b = (x_{m+1}, \ldots, x_M)^\top$ denote realizations of X_a and X_b respectively. Then the conditional density of X_a given $X_b = x_b$ is also given by a multivariate normal distribution:

$$f_{X_a|X_b}(x_a|x_b) = \frac{1}{2\pi|\det \tilde{C}|^{1/2}} \exp\left[-\frac{1}{2}(x_a - \tilde{\mu})^\top \tilde{C}^{-1}(x_a - \tilde{\mu})\right], \tag{4.123}$$

where

$$\tilde{\mu} = \mu_a + C_{ab}C_{bb}^{-1}(x_b - \mu_b), \tag{4.124}$$

$$\tilde{C} = C_{aa} - C_{ab}C_{bb}^{-1}C_{ba}. \tag{4.125}$$

In other words, $X_a|X_b = x_b \sim N(\tilde{\mu}, \tilde{C})$ (see Problem 4.22).

4.4 Exponential family of distributions

A family of PDFs (or PMFs in the case of discrete RVs) parameterized by a vector $\boldsymbol{\eta}$, representable in the form

$$\boxed{f_X(\boldsymbol{x};\boldsymbol{\theta}) = h(\boldsymbol{x}) \exp\left[\boldsymbol{\eta}^\top(\boldsymbol{\theta})\boldsymbol{T}(\boldsymbol{x}) - A(\boldsymbol{\theta})\right],} \tag{4.126}$$

with $\boldsymbol{x} \in \mathbb{R}^n$ and $\boldsymbol{\theta} \in \mathbb{R}^M$, is called an **exponential family**. Here, the functions $A(\boldsymbol{\theta})$ and $h(\boldsymbol{x})$ are scalar valued, whereas $\boldsymbol{\eta}(\boldsymbol{\theta})$ and $\boldsymbol{T}(\boldsymbol{x})$ may both be vector-valued functions of dimension M. The function $\boldsymbol{T}(\boldsymbol{x})$ is called the **sufficient statistic**, which will be further discussed in Section 18.1. If the dimension of $\boldsymbol{\theta}$ is smaller than that of $\boldsymbol{\eta}$, the family is called a **curved exponential family**.

If, by means of a mapping $\boldsymbol{\eta}(\boldsymbol{\theta}) = \boldsymbol{\eta}$, an exponential family of the form (4.126) can be transformed into the form

$$\boxed{f_X(\boldsymbol{x};\boldsymbol{\eta}) = h(\boldsymbol{x}) \exp[\boldsymbol{\eta}^\top\boldsymbol{T}(\boldsymbol{x}) - A(\boldsymbol{\eta})],} \tag{4.127}$$

then $\boldsymbol{\eta}$ is called the **canonical** or **natural parameter**, and an exponential family of the form (4.127) is called a **canonical exponential family** or **natural exponential family**. The canonical form of an exponential family is often more convenient than the form (4.126).

The exponential family of distributions contains a large class of distributions, including many of the standard ones we have already seen, including the exponential, gamma, normal, Poisson, binomial, etc. Example 4.2 shows that Poisson distributions belong to the exponential family, while Example 4.3 shows that normal distributions also belong to the exponential family. On the other hand, the family of uniform distributions does *not* belong to the exponential family (Problem 4.27). The three-parameter Weibull distributions with PDF given by (4.74) belongs to the exponential family if and only if the shape parameter α is assumed known (Problem 4.28). Similarly, the Pareto distributions with PDF given by (4.85) belong to the exponential family if and only if the parameter β is fixed (Problem 4.29).

Example 4.2: **M independent Poisson variables.** Let $\boldsymbol{X} = (X_1, X_2, \ldots, X_M)^\top$ represent M independent Poisson variables $X_1, \ldots, X_M$, with corresponding means $\theta_1, \theta_2, \ldots, \theta_M$. The PMF takes the product form:

$$p_X(\boldsymbol{x};\boldsymbol{\theta}) = \prod_{i=1}^{M} P(x_i;\theta_i) = \prod_{i=1}^{M} \frac{{\theta_i}^{x_i}}{x_i!} e^{-\theta_i}, \quad x_i = 0, 1, 2, \ldots, \quad i = 1, 2, \ldots, M,$$

which can be written as

$$p_X(\boldsymbol{x};\boldsymbol{\theta}) = \left(\prod_{i=1}^{M} \frac{1}{x_i!}\right) \exp\left(\sum_{i=1}^{M} x_i \log\theta_i - \sum_{i=1}^{M} \theta_i\right). \tag{4.128}$$

Defining $\boldsymbol{\eta} = (\eta_1, \eta_2, \ldots, \eta_M)^\top$ by

$$\eta_i = \log \theta_i \text{ or } \theta_i = e^{\eta_i}, \tag{4.129}$$

we can write the PDF of $\boldsymbol{X}$ as

$$p_{\boldsymbol{X}}(\boldsymbol{x}; \boldsymbol{\eta}) = \left(\prod_{i=1}^{M} \frac{1}{x_i!}\right) \exp\left(\sum_{i=1}^{M} x_i \eta_i - \sum_{i=1}^{M} e^{\eta_i}\right). \tag{4.130}$$

Thus, we see that this distribution belongs to the canonical exponential family with

$$h(\boldsymbol{x}) = \prod_{i=1}^{M} \frac{1}{x_i!}, \quad \boldsymbol{T}(\boldsymbol{x}) = \boldsymbol{x}, \text{ and } A(\boldsymbol{\eta}) = \sum_{i=1}^{M} e^{\eta_i}. \tag{4.131}$$

□

Example 4.3: Normal distribution. Consider a normal RV $X \sim N(\mu, \sigma^2)$. With $\boldsymbol{\theta} = (\mu, \sigma)$ we write the PDF of each sample x_i $(i = 1, 2, \ldots)$ as

$$\begin{aligned} f(x_i; \boldsymbol{\theta}) &= \frac{1}{\sqrt{2\pi}\sigma} \exp\left[-\frac{(x_i - \mu)^2}{2\sigma^2}\right] \\ &= \frac{1}{\sqrt{2\pi}} \exp\left(-\frac{x_i^2}{2\sigma^2} + \frac{x_i \mu}{\sigma^2} - \frac{\mu^2}{2\sigma^2} - \log \sigma\right), \quad i = 1, 2, \ldots, n. \end{aligned}$$

As in the previous example, we can present the normal distribution in the canonical exponential family form by identifying

$$\boldsymbol{\eta} = \begin{bmatrix} \eta_1 \\ \eta_2 \end{bmatrix} = \begin{bmatrix} \frac{1}{\sigma^2} \\ \frac{\mu}{\sigma^2} \end{bmatrix}, \quad \boldsymbol{T}(x) = \begin{bmatrix} -\frac{x^2}{2} \\ x \end{bmatrix},$$

$$h(\boldsymbol{X}) = \frac{1}{\sqrt{2\pi}}, \quad A(\boldsymbol{\eta}) = \frac{\mu^2}{2\sigma^2} + \log \sigma.$$

We can write the original parameter as $\boldsymbol{\theta} = (\mu, \sigma^2)$, where $\mu = \eta_2/\eta_1$ and $\sigma^2 = 1/\eta_1$. Hence,

$$A(\boldsymbol{\eta}) = \frac{\eta_2^2}{2\eta_1} - \frac{\log \eta_1}{2}.$$

We will return to this example in Chapter 18 (see Example 18.4), where we discuss maximum-likelihood estimation. □

4.5 Bayesian inference and conjugate priors

In Bayesian statistics, probability distributions are used to describe unknown quantities of interest. Suppose, for example, that an observed sample X is assumed to be drawn from a certain family of distributions specified in terms of some parameter. In the frequentist approach to inference, this parameter is assumed to take on a fixed value θ. By contrast, the Bayesian paradigm treats this parameter as a random variable Θ, which is assigned a **prior** PDF $\pi(\theta) = f_\Theta(\theta)$, if Θ is a continuous RV, or a prior PMF $\pi(\theta) = p_\Theta(\theta)$, if Θ is a discrete RV, before observing the data x. We adopt the notation $\pi(\theta)$ for the prior PDF or prior PMF for notational brevity and its popular use in the Bayesian statistics literature.

Recall Bayes' theorem (2.63) discussed in Section 2.4.2:

$$P[A \mid B] = \frac{P[B \mid A]\, P[A]}{P[B]}. \tag{4.132}$$

By setting $A = \{\Theta = \theta\}$ and $B = \{X = x\}$, we have for discrete RVs X and Θ

$$\boxed{\pi(\theta|x) = \frac{p(x|\theta)\pi(\theta)}{p(x)},} \tag{4.133}$$

where $p(x) = \sum_\theta p(x|\theta)\pi(\theta)$.[10] If Θ is a continuous RV and X is a discrete RV, (4.133) still holds formally, but in this case, $p(x) = \int_\theta p(x|\theta)\pi(\theta)d\theta$. If Θ and X are both continuous RVs, the analog of (4.133) is

$$\boxed{\pi(\theta|x) = \frac{f(x|\theta)\pi(\theta)}{f(x)},} \tag{4.134}$$

where $f(x) = \int_\theta f(x|\theta)\pi(\theta)\, d\theta$. If Θ is discrete and X is continuous, (4.133) holds formally, but in this case $f(x) = \sum_\theta f(x|\theta)\pi(\theta)$. Equations (4.133) and (4.134) are the basis of Bayesian statistics.

When the conditional PDF $f(x|\theta)$ is viewed as a function of θ with x given, it is called the **likelihood function** and is denoted as

$$L_x(\theta) = f(x|\theta) \text{ or } L_x(\theta) = p(x|\theta), \tag{4.135}$$

according to whether X is continuous or discrete, respectively. The likelihood function $L_x(\theta)$ is defined similarly in the frequentist paradigm, namely

$$L_x(\theta) = f(x;\theta) \text{ or } L_x(\theta) = p(x;\theta), \tag{4.136}$$

but here θ is viewed as an unknown but fixed parameter.

Of central interest in Bayesian inference is the calculation of the **posterior** distribution $\pi(\theta|x)$, given by (4.133) or (4.134). Note that the frequentist approach does not postulate the existence of a posterior distribution, since θ is assumed to be a constant.

[10] We write $p(x)$ and $p(x|\theta)$ instead of $p_X(x)$ and $p_{X|\Theta}(x|\theta)$ to simplify the notation.

When the likelihood $L_x(\theta)$ is given, the form of the posterior distribution depends on the prior distribution $\pi(\theta)$. Since the denominator of (4.133) or (4.134) is independent of θ, we can write the posterior distribution as

$$\pi(\theta|x) \propto L_x(\theta)\pi(\theta). \tag{4.137}$$

Thus, the posterior PDF or PMF $\pi(\theta|x)$ has the same shape as the likelihood function times the prior $\pi(\theta)$, and this information may be sufficient, if, for instance, we are interested in finding θ that maximizes $\pi(\theta|x)$. In general, computation of the posterior distribution requires numerical integration or summation of the product $L_x(\theta)\pi(\theta)$ over the range of values of the parameter θ.

For certain choices of the prior distribution, the posterior distribution has the same mathematical form as the prior distribution. In this case, the prior distribution is known as a **conjugate prior distribution** or simply a *conjugate prior* of the given likelihood function. The use of conjugate priors can lead to convenient closed-form expressions for the posterior distribution.

Example 4.4: The Bernoulli distribution and its conjugate prior, the beta distribution. Consider the Bernoulli trials discussed in Section 2.3. Here we write the probability of success as θ instead of p defined there. Defining the binary variable X_i as taking on $X_i = 1$ when the ith trial succeeds and $X_i = 0$ when it fails, we have

$$P[X_i = 1|\theta] = p(1|\theta) = \theta \text{ and } P[X_i = 0|\theta] = p(0|\theta) = 1 - \theta.$$

Since the result of the ith trial, x_i, is either 1 or 0, we can write

$$p(x_i|\theta) = \theta^{x_i}(1-\theta)^{1-x_i}. \tag{4.138}$$

For n independent trials, we observe the data $\boldsymbol{x} \triangleq (x_1, x_2, \ldots, x_n)^\top$ and the likelihood function of θ given $\boldsymbol{x}$ is

$$\begin{aligned} L_{\boldsymbol{x}}(\theta) = p(\boldsymbol{x}|\theta) &= \prod_{i=1}^{n} p(x_i|\theta) = \prod_{i=1}^{n} \theta^{x_i}(1-\theta)^{1-x_i} \\ &= \theta^{\sum_{i=1}^n x_i}(1-\theta)^{n-\sum_{i=1}^n x_i}. \end{aligned} \tag{4.139}$$

To complete the specification of this experiment from the Bayesian point of view, we need a prior PDF $\pi(\theta)$. The prior distribution $\pi(\theta)$ should allow θ to take any value in the interval [0, 1], and not outside. A possible choice is the **beta distribution**

$$\pi(\theta) = \text{Beta}(\theta; \alpha, \beta) \triangleq \frac{\theta^{\alpha-1}(1-\theta)^{\beta-1}}{B(\alpha, \beta)}, \quad 0 \le \theta \le 1,\ \alpha > 0,\ \beta > 0, \tag{4.140}$$

where

$$B(\alpha, \beta) = \int_0^\infty \theta^{\alpha-1}(1-\theta)^{\beta-1}\, d\theta \tag{4.141}$$

is the **beta function**, which is related to the gamma function by

$$B(\alpha, \beta) = \frac{\Gamma(\alpha)\Gamma(\beta)}{\Gamma(\alpha + \beta)}, \tag{4.142}$$

where α and β are called **prior hyperparameters** to distinguish them from the model parameter θ. Its mean and variance are given by

$$E[\Theta] = \frac{\alpha}{\alpha + \beta} \text{ and } \mathrm{Var}[\Theta] = \frac{\alpha\beta}{(\alpha + \beta)^2(\alpha + \beta + 1)}. \tag{4.143}$$

A nice thing about the beta distribution is that it can produce an enormous variety of distribution shapes depending on the choice of α and β. In Figure 4.8 (a) we plot the beta PDF with four different values 0.5, 1, 5, 10, for $\alpha = \beta$, and in Figure 4.8 (b) for four different pairs for $\alpha \neq \beta$; i.e., $(\alpha, \beta) = (1, 2), (1, 4), (1, 10), (10, 5)$, and $(5, 2)$. From these curves and (4.143) we see that, when $\alpha = \beta$, the beta PDF is symmetric around 0.5, giving more weight to the region around 0.5, as the common value $\alpha = \beta$ increases; for $\alpha > \beta$, the PDF is skewed towards one; and the variance decreases as α and/or β increase.

With the conjugate prior $\pi(\theta)$ of (4.140) associated with the likelihood function (4.138) of the Bernoulli trials, we evaluate the posterior probability after the trial data $\boldsymbol{x} = (x_1, x_2, \ldots, x_n)$ is obtained, as

$$\begin{aligned}\pi(\theta|\boldsymbol{x}) &\propto p(\boldsymbol{x}|\theta)\pi(\theta) \propto \theta^{\sum_{i=1}^n x_i}(1-\theta)^{n-\sum_{i=1}^n x_i}\theta^{\alpha-1}(1-\theta)^{\beta-1} \\ &\propto \theta^{(\alpha+\sum_{i=1}^n x_i)-1}(1-\theta)^{(\beta+n-\sum_{i=1}^n x_i)-1},\end{aligned} \tag{4.144}$$

where the normalization constant $B(\alpha, \beta)$ of the beta distribution given by (4.141) is subsumed into the proportionality constant, because it does not depend on θ. Thus, we have found that the posterior distribution is also a beta distribution $\mathrm{Beta}(\theta; \alpha_1, \beta_1)$, where

$$\alpha_1 = \alpha + \sum_{i=1}^n x_i \text{ and } \beta_1 = \beta + n - \sum_{i=1}^n x_i, \tag{4.145}$$

which may be termed the **posterior hyperparameters**. The mean of the posterior distribution can be readily found from (4.143) and (4.145) as (Problem 4.30)

$$\begin{aligned}E[\Theta|\boldsymbol{x}] &= \left(\frac{\alpha+\beta}{\alpha+\beta+n}\right)\frac{\alpha}{\alpha+\beta} + \left(\frac{n}{\alpha+\beta+n}\right)\overline{x}_n, \\ &= \left(\frac{\alpha+\beta}{\alpha+\beta+n}\right)E[\Theta] + \left(\frac{n}{\alpha+\beta+n}\right)\hat{\theta}_{\mathrm{MLE}}(\boldsymbol{x}),\end{aligned} \tag{4.146}$$

where $E[\Theta]$ is the prior estimate of the probability of success as defined in (4.143) and

$$\hat{\theta}_{\mathrm{MLE}}(\boldsymbol{x}) = \overline{x}_n \triangleq \frac{x_1 + x_2 + \cdots + x_n}{n}$$

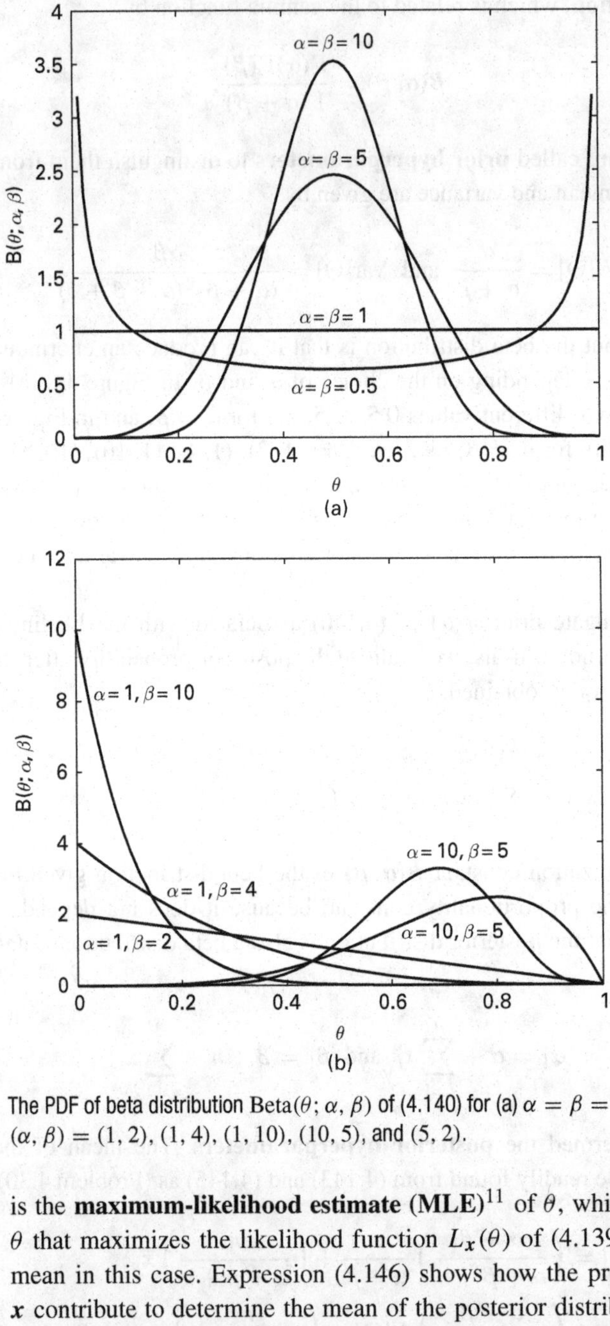

Figure 4.8 The PDF of beta distribution Beta$(\theta; \alpha, \beta)$ of (4.140) for (a) $\alpha = \beta = 0.5, 1.0, 5,$ and 10; (b) $(\alpha, \beta) = (1, 2), (1, 4), (1, 10), (10, 5)$, and $(5, 2)$.

is the **maximum-likelihood estimate (MLE)**[11] of θ, which is defined as the value of θ that maximizes the likelihood function $L_{\boldsymbol{x}}(\theta)$ of (4.139), and is simply the sample mean in this case. Expression (4.146) shows how the prior distribution and the data $\boldsymbol{x}$ contribute to determine the mean of the posterior distribution. As the sample size n increases, the weight on the prior mean diminishes, whereas the weight on the MLE approaches one; i.e., $E[\Theta|\boldsymbol{x}] \to \hat{\theta}_{\mathrm{MLE}}(\boldsymbol{x})$ as $n \to \infty$. This behavior illustrates how Bayesian inference generally works. □

[11] See Section 18.1.2 for a detailed discussion on the MLE.

Conjugate priors can be determined for most of the well-known families of distributions. For a likelihood function belonging to the regular exponential family, i.e.,

$$L_{\boldsymbol{x}}(\boldsymbol{\theta}) = h(\boldsymbol{x}) \exp[\boldsymbol{\eta}^\top(\boldsymbol{\theta})\boldsymbol{T}(\boldsymbol{x}) - A(\boldsymbol{\theta})], \tag{4.147}$$

a family of conjugate prior distributions can be constructed as follows:

$$f(\boldsymbol{\theta}; \boldsymbol{\alpha}, \beta) \propto \exp[\boldsymbol{\eta}^\top(\boldsymbol{\theta})\boldsymbol{\alpha} - \beta A(\boldsymbol{\theta})], \tag{4.148}$$

where $\boldsymbol{\alpha}$ and β are *prior hyperparameters*. In this case, the posterior distribution has the form

$$f(\boldsymbol{\theta}|\boldsymbol{x}; \boldsymbol{\alpha}, \beta) \propto \exp\{\boldsymbol{\eta}^\top(\boldsymbol{\theta})[\boldsymbol{\alpha} + \boldsymbol{T}(\boldsymbol{x})] - (1+\beta)A(\boldsymbol{\theta})\}, \tag{4.149}$$

which clearly also belongs to the regular exponential family. Here, $\boldsymbol{\alpha}_1 = \boldsymbol{\alpha} + \boldsymbol{T}(\boldsymbol{x})$ and $\beta_1 = 1 + \beta$ are the *posterior hyperparameters* corresponding to the prior hyperparameters $\boldsymbol{\alpha}$ and β respectively.

Example 4.5: Conjugate prior for the exponential distribution. The likelihood function for the exponential distribution has the form (cf. (4.25))

$$L_x(\lambda) = \lambda \exp(-\lambda x), \quad x \geq 0, \tag{4.150}$$

where λ is the model parameter. We choose a conjugate prior having the form of a gamma distribution (cf. (4.30)):

$$f(\lambda; \alpha, \beta) = \frac{\alpha(\alpha\lambda)^{\beta-1} e^{-\alpha\lambda}}{\Gamma(\beta)}, \quad \lambda \geq 0, \tag{4.151}$$

where α and β are the prior hyperparameters. Using (4.137), the posterior distribution is computed as

$$f(\lambda|x; \alpha, \beta) = \frac{\alpha(\alpha\lambda)^{\beta} e^{-(\alpha+x)\lambda}}{\Gamma(\beta+1)}, \quad \lambda \geq 0, \tag{4.152}$$

which is a gamma distribution such that the posterior hyperparameters are $\alpha_1 = \alpha + x$ and $\beta_1 = \beta + 1$. If M independent samples $x_1, \ldots, x_M$, are drawn from an exponential distribution, the likelihood function for the vector $\boldsymbol{x} = (x_1, \ldots, x_M)^\top$ has the form

$$L_{\boldsymbol{x}}(\lambda) = \lambda^M \exp\left(-\lambda \sum_{i=1}^{M} x_i\right), \quad \boldsymbol{x} \geq \boldsymbol{0}. \tag{4.153}$$

Using the conjugate prior given by (4.151), we find that the posterior distribution is a gamma distribution with posterior hyperparameters $\alpha_1 = \alpha + M$ and $\beta_1 = \beta + \sum_{i=1}^{M} x_i$. □

Example 4.6: Conjugate prior for a normal distribution with fixed variance σ^2. The likelihood function for a normal family of distributions with fixed variance σ^2 has the form (cf. (4.25))

$$L_x(\mu) = \frac{1}{\sqrt{2\pi\sigma^2}} \exp\left[-\frac{(x-\mu)^2}{2\sigma^2}\right], \tag{4.154}$$

where μ is the model parameter. Choosing a normal distribution as the conjugate prior, we have

$$f(\mu; \mu_0, \sigma_0^2) = \frac{1}{\sqrt{2\pi\sigma_0^2}} \exp\left[-\frac{(\mu-\mu_0)^2}{2\sigma_0^2}\right], \tag{4.155}$$

with prior hyperparameters μ_0 and σ_0^2. Applying (4.137), we find that the posterior distribution has the form

$$f(\lambda|x; \mu_0, \sigma_0^2) \propto \exp\left\{-\frac{1}{2}\left[\frac{(\mu-x)^2}{\sigma^2} + \frac{(\mu-\mu_0)^2}{\sigma_0^2}\right]\right\}. \tag{4.156}$$

After some algebraic manipulations, we obtain

$$f(\lambda|x; \mu_0, \sigma_0^2) \propto \exp\left[-\frac{1}{2}\left(\frac{1}{\sigma^2} + \frac{1}{\sigma_0^2}\right)\left(\mu - \frac{\frac{x}{\sigma^2} + \frac{\mu_0}{\sigma_0^2}}{\frac{1}{\sigma^2} + \frac{1}{\sigma^2}}\right)^2\right]. \tag{4.157}$$

Hence, the posterior hyperparameters are

$$\mu_1 = \frac{\frac{x}{\sigma^2} + \frac{\mu_0}{\sigma_0^2}}{\frac{1}{\sigma^2} + \frac{1}{\sigma_0^2}} \quad \text{and} \quad \sigma_1^2 = \left(\frac{1}{\sigma^2} + \frac{1}{\sigma_0^2}\right)^{-1}.$$

Generalizing to the case of n independent samples, i.e., $\boldsymbol{x} = (x_1, x_2, \ldots, x_n)^\top$, we can show that the posterior hyperparameters are given by

$$\mu_1 = \frac{\frac{\sum_{i=1}^n x_i}{\sigma^2} + \frac{\mu_0}{\sigma_0^2}}{\frac{n}{\sigma^2} + \frac{1}{\sigma_0^2}} \quad \text{and} \quad \sigma_1^2 = \left(\frac{n}{\sigma^2} + \frac{1}{\sigma_0^2}\right)^{-1}. \tag{4.158}$$

The second posterior hyperparameter σ_1^2 in the last expression is the harmonic mean of the prior σ_0^2 and the variance of data. For notational conciseness, the inverse of the variance, $h \triangleq \sigma^{-2}$, called the **precision**, is often used in the Bayesian statistics literature. From the last expression, for instance, the posterior precision is simply given by $h_1 = nh + h_0$, where $h_0 = \sigma_0^{-2}$ is the precision of the prior distribution. Use of precision instead of variance eliminates most of the inversions in the equations presented above. □

4.6 Summary of Chapter 4

Probability density function:	$f_X(x) = dF_X(x)/dx$	(4.3)
Expectation of continuous RV:	$E[X] = \int_{-\infty}^{\infty} x f_X(x)\, dx$	(4.9)
Unit uniform dist.:	$F_U(u) = u, \quad 0 \le u \le 1$	(4.23)
Exponential dist.:	$F_X(x) = 1 - e^{-\lambda x}, \quad x \ge 0$	(4.24)
Memoryless property:	$P[X > t + s \mid X > s] = P[X > t]$	(4.26)
Mean of exponent. dist.:	$\mu_X = 1/\lambda$	(4.27)
Variance of exponent. dist.:	$\sigma_X^2 = 1/\lambda^2$	(4.28)
Coeff. of variation:	$c_X = \sigma_X / \mu_X$	(4.29)
Gamma distribution:	$f_{\lambda,\beta}(y) \triangleq \frac{\lambda(\lambda y)^{\beta-1}}{\Gamma(\beta)} e^{-\lambda y}$	(4.30)
Gamma function:	$\Gamma(\beta) = \int_0^{\infty} x^{\beta-1} e^{-x}\, dx = (\beta - 1)\Gamma(\beta - 1)$	(4.31)
PDF of the unit normal variable:	$\phi(u) \triangleq \frac{1}{\sqrt{2\pi}} \exp\left(-\frac{u^2}{2}\right)$	(4.41)
Moments of U:	$E[U^n] = 1 \times 3 \times 5 \cdots (n-1) \quad (n \text{ even})$	(4.54)
Normal approx. of binomial dist.:	$B(k; n, p) \sim \frac{1}{\sqrt{npq}} \Phi\left(\frac{k-np}{\sqrt{npq}}\right)$	(4.63)
Standard Weibull distribution:	$F_X(x) = 1 - e^{-x^{\alpha}}, \quad x > 0$	(4.78)
Pareto distribution:	$F_X(x) = \begin{cases} 0, & 0 \le x < \beta \\ 1 - (x/\beta)^{-\alpha}, & \beta \le x < \infty \end{cases}$	(4.84)
Joint PDF:	$f_{XY}(x, y) = \frac{\partial^2 F_{XY}(x,y)}{\partial x \partial y}$	(4.92)
Conditional dist. function:	$F_{Y\mid X}(y\mid x) = \frac{\int_{-\infty}^{y} f_{XY}(x,v)\, dv}{f_X(x)}$	(4.101)
	$= \int_{-\infty}^{y} f_{Y\mid X}(v\mid x)\, dv$	(4.103)
Conditional PDF:	$f_{Y\mid X}(y\mid x) = \frac{f_{XY}(x,y)}{f_X(x)}$	(4.102)
Conditional expectation:	$E[X\mid Y] \triangleq \psi(Y)$	(4.104)
	$\psi(y) = E[X \mid Y = y] = \int_{-\infty}^{\infty} x f_{X\mid Y}(x\mid y)$	(4.105)
Standard bivariate normal dist.:	$\frac{1}{2\pi\sqrt{(1-\rho^2)}} \exp\left[-\frac{1}{2(1-\rho^2)}(u_1^2 - 2\rho u_1 u_2 + u_2^2)\right]$	(4.108)
Conditional PDF of U_2:	$f_{U_2\mid U_1}(u_2\mid u_1) = \frac{1}{\sqrt{2\pi(1-\rho^2)}} \exp\left[-\frac{(u_2-\rho u_1)^2}{2(1-\rho^2)}\right]$	(4.111)
Bivariate normal dist.:	$\frac{1}{2\pi \lvert\det \boldsymbol{C}\rvert^{1/2}} \exp\left[-\frac{1}{2}(\boldsymbol{x} - \boldsymbol{\mu})^{\top} \boldsymbol{C}^{-1} (\boldsymbol{x} - \boldsymbol{\mu})\right]$	(4.116)
where	$\boldsymbol{C} = \begin{bmatrix} \sigma_1^2 & \rho\sigma_1\sigma_2 \\ \rho\sigma_1\sigma_2 & \sigma_2^2 \end{bmatrix}$	(4.117)
Exponential family:	$f_X(\boldsymbol{x}; \boldsymbol{\theta}) = h(\boldsymbol{x}) \exp[\boldsymbol{\eta}^{\top}(\boldsymbol{\theta}) \boldsymbol{T}(\boldsymbol{x}) - A(\boldsymbol{\theta})]$	(4.126)
Canonical exponential family:	$f_X(\boldsymbol{x}; \boldsymbol{\eta}) = h(\boldsymbol{x}) \exp[\boldsymbol{\eta}^{\top} \boldsymbol{T}(\boldsymbol{x}) - A(\boldsymbol{\eta})]$	(4.127)
Posterior dist. (discrete x):	$\pi(\theta\mid x) = \frac{p(x\mid\theta)\pi(\theta)}{p(x)}$	(4.133)
Posterior dist. (continuous x):	$\pi(\theta\mid x) = \frac{f(x\mid\theta)\pi(\theta)}{f(x)}$	(4.134)
Likelihood function:	$L_x(\theta) = f(x\mid\theta)$ or $L_x(\theta) = p(x\mid\theta)$	(4.135)
Posterior dist. calculation:	$\pi(\theta\mid x) \propto L_x(\theta)\pi(\theta)$	(4.137)

4.7 Discussion and further reading

In this chapter we focused on continuous RVs and the associated concept of PDFs. We then discussed several important examples of continuous RVs, most of which are also found in other textbooks on probability such as Feller [99], Grimmett and Stirzaker [131], Gubner [133], Nelson [254], Papoulis and Pillai [262], and Ross [289]. Our treatment of the Weibull distribution and Pareto distribution is more substantial, because these so-called heavy-tailed distributions are increasingly used in stochastic modeling today. Exponential families of distributions are important in statistical inference, which we will discuss in more detail in Chapter 18. The material of Section 4.5 forms the basis for Bayesian statistics, which is covered in more depth in books such as Berger [18], Bernardo and Smith [19], and Gelman *et al.* [117].

4.8 Problems

Section 4.1: Continuous random variables

4.1* Expectation of a nonnegative continuous RV. Show the derivation of (4.10). *Hint:* Use the following property: if the kth absolute moment of X is finite, then

$$\lim_{x\to\infty} x^k[1 - F_X(x)] = 0.$$

4.2* Properties of discrete RVs. Show that (4.10) and (4.11) hold for a discrete RV as well.

4.3 Mixture models and mixed RVs. In a **mixture model**, the distribution function of a random variable Z is defined as a *mixture* of a set of RVs $\{X_1, X_2, \ldots, X_n\}$ in the sense that Z takes on the value of X_i with probability p_i, where $p_i \geq 0$, $i = 1, \ldots, n$ and $\sum_{i=1}^n p_i = 1$.

(a) Show that the distribution function of Z is given by

$$F_Z(z) = \sum_{i=1}^n p_i F_{X_i}(z).$$

(b) Show that the expectation of Z is given by

$$E[Z] = \sum_{i=1}^n p_i E[X_i].$$

(c) Verify the following statements:

(i) If all of the X_i are continuous RVs, then Z is a continuous RV.

(ii) If all of the X_i are discrete RVs, then Z is a discrete RV.

(iii) If some of the X_i are continuous RVs and some are discrete RVs, Z is neither a continuous RV nor a discrete RV. In this case, Z is said to be a **mixed random variable**.

4.4* Expectation and the Riemann–Stieltjes integral. The concept of expectation for a discrete RV X is defined by (3.32) in terms of the PMF $p_X(x)$ and for a continuous RV X by (4.9) in terms of the PDF $f_X(x)$.

(a) For a discrete RV X, show that the definition (4.9) is equivalent to (3.32) when the PDF of X is expressed in terms of the Dirac delta function as in (3.12). Thus, the definition (4.9) is applicable to mixed RVs if the PDF $f_X(x)$ is allowed to include Dirac delta functions.

(b) A definition of expectation for mixed RVs that avoids the use of Dirac delta functions can be formulated using the **Riemann–Stieltjes integral** (also called Stieltjes integral)[12] a generalization of the standard Riemann integral:

$$E[X] = \int_{-\infty}^{\infty} x \, dF_X(x), \tag{4.159}$$

where $F_X(x)$ is the distribution function of the RV X.

(i) Show that (4.159) is equivalent to (3.32) if X is a continuous RV.

(ii) Show that (4.159) is equivalent to (4.9) if X is a discrete RV.

(iii) Using integration by parts for the Stieltjes integral, show that (4.10) and (4.11) follow from (4.159).

4.5 Expectation of functions of RVs. Let X and Y be continuous RVs.

(a) Show that

$$E[h(X)] = \int_{-\infty}^{\infty} h(x) f_X(x) \, dx, \tag{4.160}$$

where $h(\cdot)$ is an arbitrary function.

(b) Show that X and Y are independent if and only if

$$E[h(X)g(Y)] = E[h(X)]E[g(Y)], \tag{4.161}$$

for arbitrary functions $h(\cdot)$ and $g(\cdot)$.

4.6 Second moment of a continuous RV. Show that the second moment of a continuous RV X is given by

$$E[X^2] = 2\int_0^{\infty} x F_X^c(x) \, dx - 2\int_{-\infty}^{0} x F_X(x) \, dx.$$

4.7 Joint PDF of two continuous RVs. Consider a pair of continuous RVs (X, Y) that have a joint PDF of the form

$$f_{XY}(x, y) = \begin{cases} k \exp(-\lambda x - \mu y), & x \geq 0, y \geq 0, \\ 0, & \text{elsewhere,} \end{cases}$$

where $\lambda > 0$, $\mu > 0$.

[12] Thomas Joannes Stieltjes (1856–1894) was a Dutch mathematician whose main contributions were in analysis, number theory, and continued fractions.

(a) Obtain the joint distribution function $F_{XY}(x, y)$ and determine the normalization constant k.
(b) Find the distribution functions $F_X(x)$ and $F_Y(y)$ and the conditional distribution function $F_{Y|X}(y|x)$.

4.8 Conditional expectation of two RVs. Show the following properties of conditional expectation for two continuous RVs X and Y:

(a) $E[E[X|Y]] = E[X]$ (cf. (4.106)).
(b) $E[h(Y)g(X)|Y] = h(Y)E[g(X)|Y]$, where h and g are scalar functions.
(c) $E[\cdot|Y]$ is a linear operator.

Section 4.2: Important continuous random variables and their distributions

4.9* Expectation, second moment, and variance of the uniform RV. Derive (4.20) through (4.22).

4.10* Moments of uniform RV. Consider a random variable X that is uniformly distributed in the interval $[a, b]$.

(a) Find its nth moment $E[X^n]$.
(b) Find its nth central moment $E[(X - \mu_X)^n]$.

4.11 Erlang distribution.[13] Consider the continuous-time analog of the negative binomial distribution. Let RV Y_r be the sum of r independent RVs X_i:

$$Y_r = X_1 + X_2 + \cdots + X_r, \tag{4.162}$$

where each X_i $(1 \le i \le r)$ is an exponential variable with mean $1/\lambda$.

(a) Find the mean and variance of the RV Y_r.
(b) The distribution function of Y_r is given by

$$F_{Y_r}(y) = 1 - e^{-\lambda y} \sum_{j=0}^{r-1} \frac{(\lambda y)^j}{j!}, \quad y \ge 0, \tag{4.163}$$

Find the corresponding PDF.
(c) Define a random variable S_r by the **sample mean** of the r RVs X_i; i.e.,

$$S_r = \frac{1}{r} \sum_{i=1}^{r} X_i = Y_r / r. \tag{4.164}$$

Show that its distribution function is given by

$$F_{S_r}(t) = 1 - \mathrm{e}^{-r\lambda t} \sum_{j=0}^{r-1} \frac{(r\lambda t)^j}{j!}, \quad t \ge 0. \tag{4.165}$$

Find its PDF.

[13] Also called Erlangian distribution.

The distributions of Y_r given by (4.163) and of S_r given by (4.165) are both referred to as the r**-stage Erlang distributions**, and often denoted E_r.

(d) Program the expression for the PDF of S_r for $r = 1, 2, 4, 8, 16$ and plot the curves.

4.12 Hyperexponential (or mixed exponential) distribution. Suppose that a random variable S is drawn from k types of exponential distributions of mean $1/\mu_i$, with probability π_i, $i = 1, 2, \ldots, k$. Then the distribution function of S takes the following **mixed exponential distribution**:

$$F_S(t) = \sum_{i=1}^{k} \pi_i(1 - e^{-\mu_i t}) = 1 - \sum_{i=1}^{k} \pi_i e^{-\mu_i t}. \tag{4.166}$$

This mixed exponential distribution is often referred to as the k-stage **hyperexponential distribution**, and often denoted H_k.

(a) Find its PDF and its mean.

(b) Plot a two-stage hyperexponential distribution, $F_S(t) = 1 - \pi_1 e^{-\mu_1 t} - \pi_2 e^{-\mu_2 t}$ and its PDF, where $\pi_1 = 0.0526$, $\mu_1 = 0.1$, and $\mu_2 = 2.0$. In the same figure plot the exponential distribution with the same mean.

4.13* Recursive formula for the gamma function. Derive the recursive expression (4.31).

4.14 Poisson distribution and the gamma distribution. Show that the Poisson distribution $P(k; \lambda)$ can be written in terms of the gamma distribution $f_{\lambda,\beta}(y)$ of (4.30) as follows:

$$P(k; \lambda) = f_{1,k+1}(\lambda). \tag{4.167}$$

4.15* Mean and variance of the normal distribution. Show that the mean and variance of the normal distribution (4.40) are indeed μ and σ^2 given as the model parameters.

4.16* $\Gamma(1/2)$. Show

$$\Gamma\left(\frac{1}{2}\right) = \sqrt{\pi}.$$

Hint: Use that $\phi(u)$ of (4.41) is a PDF.

4.17 Mean, second moment, and variance of the Weibull distribution. Show that the mean, second moment, and variance of the standard Weibull distribution are given by (4.81) through (4.83).

4.18 Median and mode of the Weibull distribution.

(a) Show that the median of the Weibull distribution is given by

$$\beta\,(\ln 2)^{1/\alpha}.$$

(b) Show that the mode of the Weibull distribution is given by

$$\beta\left(1-\frac{1}{\alpha}\right)^{1/\alpha}.$$

4.19 Expectation and variance of two-parameter and three-parameter Weibull distributions.

(a) Show that the mean and variance of the two-parameter Weibull distribution are given by

$$E[X]=\beta\Gamma\left(\frac{1}{\alpha}+1\right) \text{ and } \mathrm{Var}[X]=\beta^2\left[\Gamma\left(\frac{2}{\alpha}+1\right)-\Gamma^2\left(\frac{1}{\alpha}+1\right)\right]. \tag{4.168}$$

(b) Show that the mean and variance of the three-parameter Weibull distribution are given by

$$E[X]=\beta\Gamma\left(\frac{1}{\alpha}+1\right)+\gamma \text{ and } \mathrm{Var}[X]=\beta^2\left[\Gamma\left(\frac{2}{\alpha}+1\right)-\Gamma^2\left(\frac{1}{\alpha}+1\right)\right]. \tag{4.169}$$

4.20 Residual lifetime of Weibull distribution. Suppose that a standard Weibull RV X with parameter α represents the lifetime of some item (e.g., an electric bulb). If it has lasted already t [h], what is the probability that it will fail (or finish) in the next infinitesimal interval $[t, t+dt]$?

Section 4.3: Joint and conditional probability density functions

4.21* Joint bivariate normal distribution and ellipses. Show that all level curves (or contour lines) in Figure 4.7b are ellipses. *Hint*: An ellipse centered at the origin (0, 0) and having its major axis parallel to the x-axis may be specified by the equation

$$\frac{x^2}{a^2}+\frac{y^2}{b^2}=1.$$

4.22* Conditional multivariate normal distribution. Derive (4.123).

4.23 Circularly symmetric and independent RVs [262]. We say that the joint PDF $f_{XY}(x, y)$ of two RVs X and Y is **circularly symmetric**, if

$$f_{XY}(x,y)=g(r), \text{ where } r=\sqrt{x^2+y^2}.$$

Show that if the RVs are not only circularly symmetric, but also independent, then they are normal with zero mean and equal variance.

4.24 Buffon's needle problem.[14] Suppose that we have a tabletop with a number of parallel lines drawn on it, which are equally spaced, one inch apart. Suppose we have a needle that is one inch long and drop the needle randomly on the table. Show that the probability that the needle crosses or touches one of the lines is given by

$$\frac{2}{\pi} \approx 0.636\,619\,7.$$

Hint: Let us introduce the (X, Y) coordinates and let the parallel lines be represented by $Y = n$ $(n = 0, \pm 1, \pm 2, \ldots)$. Let the two ends of the needle be positioned at (X, Y) and (X', Y'). Let Θ be the angle of the needle from the X-axis; i.e.,

$$X' = X + \cos\Theta, \quad Y' = Y + \sin\Theta;$$

hence,

$$\tan\Theta = \frac{Y' - Y}{X' - X}.$$

Without loss of generality we assume that $Y' \geq Y$; if not, switch the labels of the two ends of the needle. Thus, we need to consider the range $0 < \Theta \leq \pi$. We also assume without loss of generality that $0 < Y \leq 1$; we label the line just below the needle as $Y = 0$. Figure 4.9 (a) shows a case where the needle does not touch any of the lines, whereas Figure 4.9 (b) shows a case when it intersects the line $Y = 1$.

4.25 Modifications of Buffon's needle experiment. Consider the following variations of the Buffon's needle experiment of Problem 4.24.

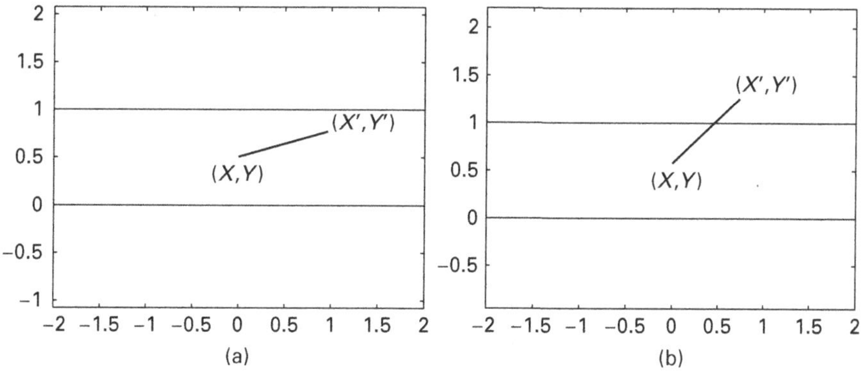

Figure 4.9 "Buffon's needle" experiment: (a) the needle does not touch any line; (b) the needle crosses the line $Y = 1$.

[14] Georges Louis Leclerc, Comte de Buffon (1707–1788) was a French mathematician and natural historian, who discovered in 1777 this simple way to estimate π.

(a) Let the line spacing be a [inches] and the needle length be ℓ [inches], with $\ell \leq a$. Find the probability that the needle hits a line.
(b) Now consider a floor on which rectangular tiles of a [inches] by b [inches] are placed. Thus, the vertical lines are at $X = ma$ ($m = 0, \pm 1, \pm 2, \ldots$) and the horizontal lines are at $Y = nb$ ($n = 0, \pm 1, \pm 2, \ldots$). Let the needle length be $\ell \leq \min\{a, b\}$. Find the probability that the needle will land within one tile.

Section 4.4: Exponential family of distributions

4.26* Exponential families of distributions. Show that the following families of distributions belong to the exponential family in canonical form:

(a) exponential distributions with PDF given by (4.25), parameterized by λ;
(b) gamma distributions with PDF given by (4.30), parameterized by (λ, β);
(c) binomial distributions given by (3.62), parameterized by (n, p);
(d) negative binomial (Pascal) distributions given by (3.98), parameterized by (r, p).

4.27 Uniform distribution. Show that the family of uniform distributions with PDF given by (4.19) does not belong to the exponential family.

4.28 Weibull distribution. Show that the family of three-parameter Weibull distributions with PDF given by (4.74) belongs to the exponential family if and only if the shape parameter α is fixed.

4.29 Pareto distribution. Show that the family of Pareto distributions with PDF given by (4.85) belongs to the exponential family if and only if the parameter β is fixed.

Section 4.5: Bayesian inference and conjugate priors

4.30* Posterior hyperparameters of the beta distribution associated with the Bernoulli distribution in Example 4.4.

(a) Derive the posterior mean (4.146).
(b) Find the posterior variance $\mathrm{Var}[\Theta|\boldsymbol{x}]$.

4.31 Conjugate prior for a geometric distribution. Consider a random variable X that is geometrically distributed (cf. (3.68)):

$$p_X(x|p) = p^{x-1}(1-p), \quad x \in \mathbb{Z}_+ = \{1, 2, \ldots\}, \tag{4.170}$$

where $p \in [0, 1]$ is the model parameter. Choose a beta distribution $\mathrm{Beta}(\alpha_0, \beta_0)$ as the conjugate prior $\pi(p)$. Suppose that you have a vector of n independent data samples, $x_i \in \mathbb{Z}_+$, given by $\boldsymbol{x} = (x_1, x_2, \ldots, x_n)$. Show that the posterior distribution $\pi(p|\boldsymbol{x})$ is also a gamma distribution. Find the posterior hyperparameters.

4.32 Conjugate prior for a multinomial distribution. Consider a random vector $\boldsymbol{X} = (X_1, \ldots, X_n)$ that is distributed according to a multinomial distribution (cf. (3.68)):

$$p(\boldsymbol{x}|\boldsymbol{p}) = \frac{n!}{n_1!n_2!\cdots n_m!} p_1^{n_1} p_2^{n_2} \cdots p_m^{n_m}, \tag{4.171}$$

where $\boldsymbol{x} = (x_1, \ldots, x_n)$, $\boldsymbol{p} = (p_1, \ldots, p_m)$ is a probability vector such $\sum_{i=1}^{m} p_i = 1$, and the n_i are nonnegative integers satisfying $\sum_{i=1}^{m} n_i = n$. An appropriate conjugate prior for the likelihood function $L_{\boldsymbol{x}}(\boldsymbol{p}) = p(\boldsymbol{x}|\boldsymbol{p})$ is the so-called **Dirichlet distribution**:[15]

$$\pi(\boldsymbol{p}) = \frac{1}{B(\boldsymbol{\alpha})} \prod_{i=1}^{m} p_i^{\alpha_i - 1}, \tag{4.172}$$

where $\boldsymbol{p} = (p_1, \ldots, p_m)$ is a probability vector, $\boldsymbol{\alpha} = (\alpha_1, \ldots, \alpha_m)$ with $\alpha_i > 0$, $i = 1, \ldots, m$, and $B(\boldsymbol{\alpha})$ is the multinomial beta function defined in terms of the gamma function by

$$B(\alpha) = \frac{\prod_{i=1}^{m} \Gamma(\alpha_i)}{\Gamma\left(\sum_{i=1}^{m} \alpha_i\right)}. \tag{4.173}$$

Show that the posterior distribution $\pi(\boldsymbol{p}|\boldsymbol{x})$ is also a Dirichlet distribution and find the corresponding posterior hyperparameters.

[15] Johann Peter Gustav Lejeune Dirichlet (1805–1859) was a German mathematician credited with the modern formal definition of a function.

5 Functions of random variables and their distributions

In many engineering applications, the input X to a given system (e.g., a receiver) is a random variable, and thus the corresponding output Y is also a random variable. The input–output relation is characterized by a known deterministic function $Y = g(X)$. Then, given the PDF $f_X(x)$ (or PMF if X is a discrete RV), we wish to find the PDF $f_Y(y)$ (or PMF) of the output RV. In a more general setting of *multiple-input, multiple-output* (MIMO) system, we may have a set of RVs denoted as $(X_1, X_2, \ldots, X_M)$ that are related to another set of RVs $(Y_1, Y_2, \ldots, Y_N)$ through the N known functions:

$$Y_n = g_n(X_1, X_2, \ldots, X_M), n = 1, 2, \ldots, N. \tag{5.1}$$

We start with the case in which a random variable Y is a single-valued function of another RV X.

5.1 Function of one random variable

Let X be a random variable with PDF $f_X(x)$ and let $g(\cdot)$ be a function that maps from $\mathbb{R}$ to $\mathbb{R}$. Then $Y = g(X)$ is also a random variable. One way to find the PDF is first to calculate the distribution function $F_Y(y)$:

$$\begin{aligned} F_Y(y) &= P[Y \le y] = P[g(X) \le y] \\ &= P[X \in \mathcal{D}_y], \end{aligned} \tag{5.2}$$

where $\mathcal{D}_y$ is the domain in the real line $\mathbb{R} = \{-\infty < x < \infty\}$ that is mapped to the range $\{-\infty < g(x) \le y\}$; i.e.,

$$\mathcal{D}_y = \{x : g(x) \le y\}. \tag{5.3}$$

Then we can write the last expression of (5.2) as

$$\boxed{F_Y(y) = \int_{-\infty}^{\infty} I(x \in \mathcal{D}_y) f_X(x)\, dx,} \tag{5.4}$$

where $I(A)$ is the indicator function; i.e.,

$$I(A) = \begin{cases} 1, & \text{if } A \text{ is true}, \\ 0, & \text{otherwise}. \end{cases} \tag{5.5}$$

Example 5.1: Linear transformation of a random variable. Let us consider the simplest mapping; i.e., a linear transformation $Y = g(X) = aX + b$. Then

$$\mathcal{D}_y = \begin{cases} \left(-\infty, \frac{y-b}{a}\right], & \text{if } a > 0, \\ \left[\frac{y-b}{a}, \infty\right), & \text{if } a < 0. \end{cases} \tag{5.6}$$

Therefore, we have

$$F_Y(y) = \begin{cases} F_X\left(\frac{y-b}{a}\right), & \text{if } a > 0, \\ 1 - F_X\left(\frac{y-b}{a}\right), & \text{if } a < 0. \end{cases} \tag{5.7}$$

By differentiating the above expression, we find

$$f_Y(y) = \begin{cases} \frac{1}{a} f_X\left(\frac{y-b}{a}\right), & \text{if } a > 0, \\ -\frac{1}{a} f_X\left(\frac{y-b}{a}\right), & \text{if } a < 0, \end{cases} \tag{5.8}$$

or equivalently

$$f_Y(y) = \frac{1}{|a|} f_X\left(\frac{y-b}{a}\right) \text{ for } a \neq 0. \tag{5.9}$$

If $a = 0$, then $Y = b$ with probability one; therefore,

$$F_Y(y) = u(y-b) \text{ and } f_Y(y) = \delta(y-b), \tag{5.10}$$

where $u(\cdot)$ and $\delta(\cdot)$ are the *unit step function* and the *Dirac delta function*, respectively (see Section 3.2).

□

Example 5.2: Square-law detector. Next, let us consider a simple nonlinear mapping $Y = g(X) = X^2$, which is sometimes called a square-law detector, in which X represents the input signal and Y represents the detector output. Noting that $g(x) = x^2$ and $\mathcal{D}_y = [-\sqrt{y}, \sqrt{y}\,]$, we find

$$F_Y(y) = P[X^2 \leq y] = P[-\sqrt{y} \leq X \leq \sqrt{y}] = F_X(\sqrt{y}) - F_X(-\sqrt{y}) \text{ for } y \geq 0. \tag{5.11}$$

By differentiating the above equation with respect to y, we obtain

$$f_Y(y) = \frac{1}{2\sqrt{y}}[f_X(\sqrt{y}) + f_X(-\sqrt{y})]. \tag{5.12}$$

An alternative way to derive the above $f_Y(y)$ is as follows. Note that $y = g(x)$ has two solutions $x_1 = \sqrt{y}$ and $x_2 = -\sqrt{y}$ for each $y > 0$. Then the RV Y falls in the interval $(y, y + \delta y)$ if and only if X falls in either of the following two mutually

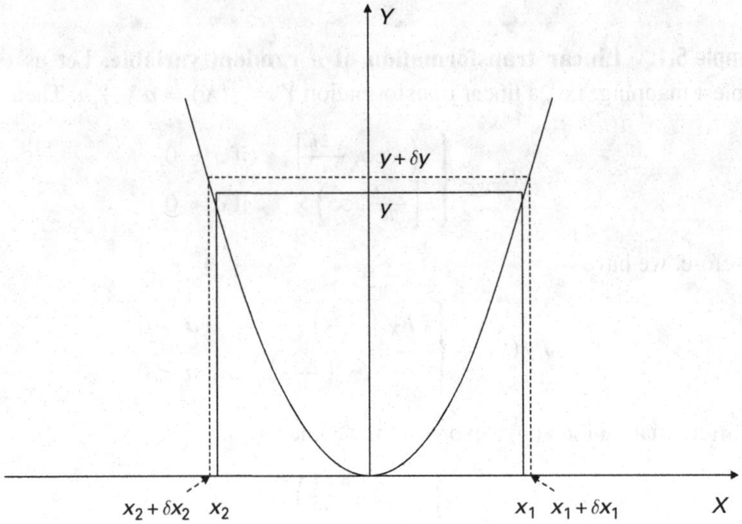

Figure 5.1 The nonlinear function $Y = X^2$ and mapping of the interval $(y, y + \delta y)$ into the two disjoint intervals $(x_1, x_1 + \delta x_1)$ and $(x_2 + \delta x_2, x_2)$ $(\delta x_2 < 0)$.

exclusive intervals: $(x_1, x_1 + \delta x_1)$ and $(x_2 + \delta x_2, x_2)$. Hence, for a given $\delta y (> 0)$ and the corresponding $\delta x_1 (> 0)$ and $\delta x_2 (< 0)$ (see Figure 5.1) we have

$$P[y < Y \le y + \delta y] = P[x_1 < X \le x_1 + \delta x_1] + P[x_2 + \delta x_2 \le X < x_2]. \quad (5.13)$$

For sufficiently small δy and δx_i $(i = 1, 2)$, we can write the above as

$$f_Y(y)\delta y \approx f_X(x_1)\delta x_1 + f_X(x_2)(-\delta x_2). \quad (5.14)$$

By dividing both sides by δy and taking the limit $\delta y \to 0$, we find

$$f_Y(y) = \frac{f_X(x_1)}{g'(x_1)} + \frac{f_X(x_2)}{-g'(x_2)}. \quad (5.15)$$

Because $g'(x) = 2x$, $g'(x_1) = 2\sqrt{y}$ and $g'(x_2) = -2\sqrt{y}$, the last equation reduces to (5.12).

□

We can easily extend the result of the last example to a general class of functions $g(x)$. Suppose that, for given y, the mapping $y = g(x)$ has multiple solutions $x_1, x_2, \ldots, x_m$, where the integer m, in general, depends on y, and thus it may be more appropriate to write it as $m(y)$.

$$\boxed{x_i = g^{-1}(y), i = 1, 2, \ldots, m(y),} \quad (5.16)$$

Assuming that $g(x)$ is continuous at all these $m(y)$ points, we find that the PDF of $Y = g(X)$ is given by

$$\boxed{f_Y(y) = \sum_{i=1}^{m(y)} \frac{f_X(x_i)}{\mid g'(x_i) \mid}.} \tag{5.17}$$

5.2 Function of two random variables

Let us now consider the case where Z is a function of two RVs X and Y, which have joint PDF $f_{XY}(x, y)$:

$$Z = g(X, Y). \tag{5.18}$$

How should we go about obtaining the PDF of Z? As discussed in the previous section, one way to solve this problem is to find the distribution function first and then differentiate it to obtain the PDF. So we begin with

$$\begin{aligned} F_Z(z) &= P[g(X, Y) \leq z] = P[(X, Y) \in \mathcal{D}_z] \\ &= \int\int I((x, y) \in \mathcal{D}_z) f_{XY}(x, y)\, dx\, dy, \end{aligned} \tag{5.19}$$

where $\mathcal{D}_z$ represents the domain in the (X, Y) plane that is mapped to the range $g(X, Y) \leq z$,

$$\mathcal{D}_z = \{(x, y) : \ g(x, y) \leq z\}, \tag{5.20}$$

and $I(\cdot)$ is the indicator function defined in (5.5).

Example 5.3: Sum of two random variables. Consider the simplest example of a two-variable function; i.e., $Z = X + Y$. Then the region $\mathcal{D}_z = \{(X, Y) : X + Y \leq z\}$ can be represented as $\mathcal{D}_z = \bigcup_{-\infty<y<\infty} \mathcal{H}_y$, where $\mathcal{H}_y = \{(X, Y) : y < Y < y + dy, \infty < X < z - Y\}$ is a horizontal strip of width dy (see Figure 5.2). Thus, we can rewrite the integration (5.19) as

$$F_Z(z) = \int\int I(x + y \leq z) f_{XY}(x, y)\, dxdy = \int_{-\infty}^{\infty} \left[\int_{-\infty}^{z-y} f_{XY}(x, y)\, dx\right] dy. \tag{5.21}$$

By differentiating the above with respect to z, we obtain

$$f_Z(z) = \int_{-\infty}^{\infty} \left[\frac{\partial}{\partial z} \int_{-\infty}^{z-y} f_{XY}(x, y)\, dx\right] dy. \tag{5.22}$$

By applying Leibniz's rule (5.94) (see Problem 5.6) to the expression inside the square brackets, we find

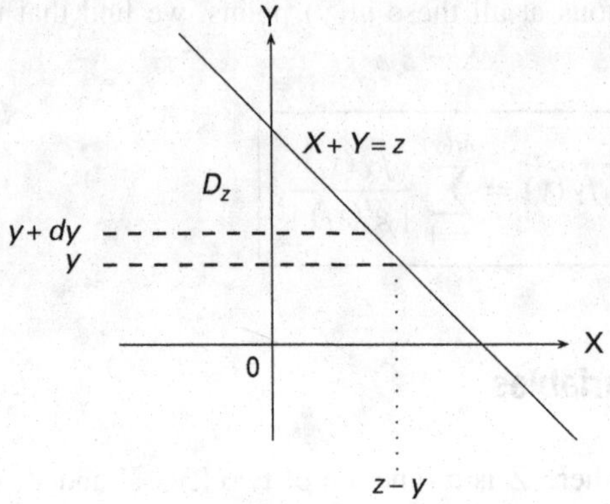

Figure 5.2 The region $\mathcal{D}_z = \{(X, Y) : X + Y \leq z\}$ and a horizontal strip of width dy extending horizontally over the interval $\{-\infty < X < z - y\}$.

$$
\begin{aligned}
\frac{\partial}{\partial z}\int_{-\infty}^{z-y} f_{XY}(x, y)\, dx &= f_{XY}(z-y, y) \times 1 - f_{XY}(-\infty, y) \times 0 \\
&\quad + \int_{-\infty}^{z-y} \frac{\partial f_{XY}(x, y)}{\partial z}\, dx \\
&= f_{XY}(z-y, y).
\end{aligned} \tag{5.23}
$$

Thus,

$$
\boxed{f_Z(z) = \int_{-\infty}^{\infty} f_{XY}(z-y, y)\, dy.} \tag{5.24}
$$

Since X and Y are symmetrical, we can interchange the roles of X and Y. Thus, an alternative procedure is to integrate the joint PDF over the vertical strip of width dx extending over the interval $\{-\infty < Y \leq z - x\}$ first, followed by the integration along the x-axis. Then we arrive at the formula

$$
f_Z(z) = \int_{-\infty}^{\infty} f_{XY}(x, z-x)\, dx. \tag{5.25}
$$

If, in particular, X and Y are statistically independent, i.e.,

$$
f_{XY}(x, y) = f_X(x) f_Y(y), \tag{5.26}
$$

then the formulas (5.24) and (5.25) reduce to

$$
f_Z(z) = \int_{-\infty}^{\infty} f_X(z-y) f_Y(y)\, dy = \int_{-\infty}^{\infty} f_X(x) f_Y(z-x)\, dx. \tag{5.27}
$$

The above integration formula is called the **convolution** (or **convolution integral**) of functions $f_X(\cdot)$ and $f_Y(\cdot)$. Thus, we have shown that if two RVs are statistically independent, the PDF of their sum is equal to the convolution of their PDFs, and we write

$$\boxed{f_Z(z) = f_X(z) \circledast f_Y(z).} \tag{5.28}$$

The above result will be derived again in Section 8.2.2, but using the **characteristic function** approach.

In the case where X and Y are independent, we can express the CDF of Z in terms of the CDFs of X and Y by rewriting (5.21) using the Stieltjes integral:

$$F_Z(z) = \int\int I(x+y \leq z) dF_X(x) dF_Y(y) = \int_{-\infty}^{\infty}\int_{-\infty}^{z-y} dF_X(x)\, dF_Y(y),$$

where we obtained the first equality by formally replacing $f_X(x)dx$ by $dF_X(x)$ and $f_Y(Y)dy$ by $dF_Y(y)$. Hence, we have

$$\boxed{F_Z(z) = \int_{-\infty}^{\infty} F_X(z-y)\, dF_Y(y) = \int_{-\infty}^{\infty} F_Y(z-x)\, dF_X(x),} \tag{5.29}$$

where the second equality is due to the symmetry between X and Y. In (5.29), if we formally replace $dF_Y(y)$ by $f_Y(y)dy$ and $dF_X(x)$ by $f_X(x)dx$, we obtain convolution formulas for the CDF $F_Z(z)$; i.e.,

$$\boxed{F_Z(z) = F_X(z) \circledast f_Y(z) = f_X(z) \circledast F_Y(z).} \tag{5.30}$$

□

Example 5.4: $Z = X^2 + Y^2$. Given the joint PDF $f_{XY}(x, y)$ of RVs X and Y, let us find the PDF of $Z = X^2 + Y^2$. This type of problem arises, for instance, when a signal of the form $S(t) = X\cos(\omega t - \phi) + Y\cos(\omega t - \phi)$ is received (usually corrupted with noise) and the phase ϕ of the signal is not known to the receiver. It is known in signal detection theory that the best strategy to detect such a signal is to compute the **power** of the signal, which is defined as the sum of the squares of X and Y.

We first seek to find the distribution function

$$F_Z(z) = P[X^2 + Y^2 \leq z] = \int\int I(x^2 + y^2 \leq z) f_{XY}(x, y)\, dx\, dy, \tag{5.31}$$

where $I(A)$ is the indicator function defined by (5.5). The region $\{(x, y) : x^2 + y^2 \leq z\}$ represents the area surrounded by a circle with radius $\sqrt{z}$ (see Figure 5.3); hence, we can write

$$F_Z(z) = \int_{-\sqrt{z}}^{\sqrt{z}} \left[\int_{-\sqrt{z-y^2}}^{\sqrt{z-y^2}} f_{XY}(x, y)\, dx\right] dy. \tag{5.32}$$

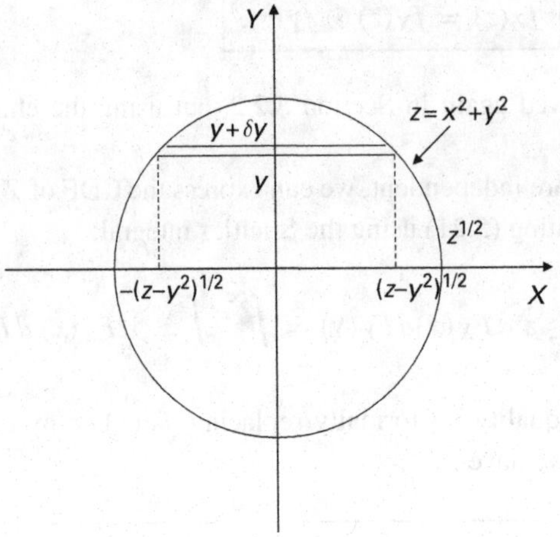

Figure 5.3 The circle of radius $\sqrt{z}$ and its decomposition into horizontal strips of width dy.

Differentiate $F_Z(z)$ with respect to z and exchange the order of differentiation and integration. By applying Leibniz's rule (5.94) again, the differentiation of the expression in the square brackets yields

$$f_{XY}(\sqrt{z-y^2}, y)\frac{1}{2\sqrt{z-y^2}} - f_{XY}(-\sqrt{z-y^2}, y)\left(-\frac{1}{2\sqrt{z-y^2}}\right) + 0. \tag{5.33}$$

We then finally obtain

$$f_Z(z) = \frac{d}{dz}F_Z(z) = \int_{-\sqrt{z}}^{\sqrt{z}} \frac{1}{2\sqrt{z-y^2}}\left[f_{XY}(\sqrt{z-y^2}, y) + f_{XY}(-\sqrt{z-y^2}, y)\right] dy. \tag{5.34}$$

An interesting case that often arises in signal detection problems, and for which we have a closed-form solution, is when X and Y are independent normal variables with zero mean and common variance (Problem 5.13). □

Example 5.5: $R = \sqrt{X^2 + Y^2}$**.** Let us set $Z = R^2$ in the previous example. In the context of detecting a signal of the form $S(t) = X\cos(\omega t - \phi) + Y\sin(\omega t - \phi)$, the RV $R = \sqrt{X^2 + Y^2}$ represents the **envelope** of the signal, i.e., $S(t) = R\cos(\omega t - \theta)$, where $\theta - \phi = \tan^{-1}\frac{Y}{X}$.

The distribution function of R is given by

$$F_R(r) = \int_{-r}^{r}\left[\int_{-\sqrt{r^2-y^2}}^{\sqrt{r^2-y^2}} f_{XY}(x, y)\, dx\right] dy. \tag{5.35}$$

Differentiation of the expression inside the square brackets leads, using Leibniz's rule again, to the following expression:

$$f_{XY}(\sqrt{r^2-y^2}, y)\frac{1}{2}\frac{2r}{\sqrt{r^2-y^2}} - f_{XY}(-\sqrt{r^2-y^2}, y)\left(-\frac{1}{2}\frac{2r}{\sqrt{r^2-y^2}}\right) + 0. \quad (5.36)$$

Thus, we obtain

$$f_R(r) = \frac{dF_R(z)}{dr} = \int_{-r}^{r} \frac{r}{\sqrt{r^2-y^2}}\left[f_{XY}(\sqrt{r^2-y^2}, y) + f_{XY}(-\sqrt{r^2-y^2}, y)\right] dy. \quad (5.37)$$

Again, an important and useful case is found when X and Y are independent normal variables with common variance (see Section 7.5.1). □

5.3 Two functions of two random variables and the Jacobian matrix

Let us continue the discussion of the previous section and consider two functions of two RVs. Let us denote, as before, the two RVs by X and Y and their joint distribution by $F_{XY}(x, y)$. Now consider two functions $g(x, y)$ and $h(x, y)$ that transform (X, Y) into a pair of new RVs (U, V) according to

$$U = g(X, Y) \text{ and } V = h(X, Y). \quad (5.38)$$

From the discussion of the previous section, we already know how to compute the **marginal** distribution functions of U and V and their PDFs. But how should we find the **joint** distribution function $F_{UV}(u, v)$ and the joint PDF $f_{UV}(u, v)$?

Let us define domain $\mathcal{D}_{u,v}$ as the region in the X–Y plane such that

$$\mathcal{D}_{u,v} = \{(x, y) : \; g(x, y) \le u, h(x, y) \le v\}. \quad (5.39)$$

Then we have

$$\begin{aligned} F_{UV}(u, v) &= P[g(X, Y) \le u, h(X, Y) \le v] = P[(X, Y) \in \mathcal{D}_{u,v}] \\ &= \int\int I((x, y) \in \mathcal{D}_{u,v}) f_{XY}(x, y)\, dx\, dy, \end{aligned} \quad (5.40)$$

where $I(A)$ is the indicator function defined earlier. Thus, the joint PDF is obtained by differentiating the CDF $F_{UV}(u, v)$ (see Problem 5.16 for a special example of such a procedure).

If $g(x, y)$ and $h(x, y)$ are continuous and differentiable functions, then, as in the case of one function of one random variable, we can derive a closed-form formula for the joint PDF directly. Let us set $(U, V) = (u, v)$ in (5.38). Then there are in general multiple solutions $(X, Y) = (x_i, y_i)$, $i = 1, 2, \dots, m$, such that

$$u = g(x_i, y_i) \text{ and } v = h(x_i, y_i), \; i = 1, 2, \dots, m, \quad (5.41)$$

where the number of solutions m generally depends on (u, v).

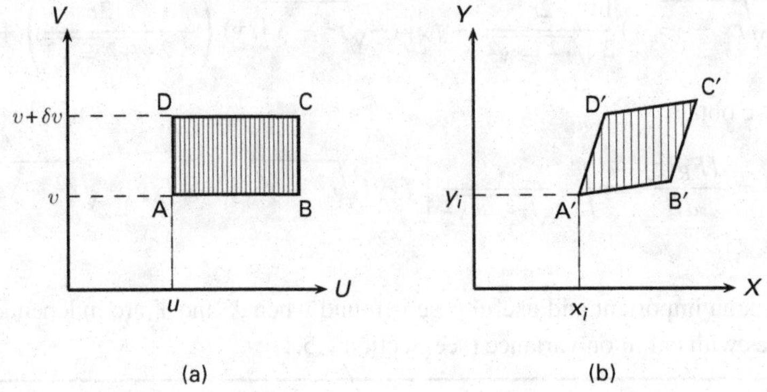

Figure 5.4 (a) The rectangle $ABCD$ at (u, v) in the U–V plane; (b) the ith parallelogram $A'B'C'D'$ at (x_i, y_i) in the X–Y plane, where $x_i = p_i(u, v)$ and $y_i = q_i(u, v)$.

Let the inverse mapping of (5.41) be p_i and q_i i.e.,

$$x_i = p_i(u, v) \text{ and } y_i = q_i(u, v). \tag{5.42}$$

By referring to Figure 5.4 (a), the functions p_i and q_i accomplish the following mapping:

$$\begin{aligned}
A &= (u, v) & &\rightarrow & A' &= (x_i, y_i);\\
B &= (u + \delta u, v) & &\rightarrow & B' &= \left(x_i + \tfrac{\partial}{\partial u} p_i(u, v)\delta u,\ y_i + \tfrac{\partial}{\partial u} q_i(u, v)\delta u\right);\\
C &= (u, v + \delta v) & &\rightarrow & C' &= \left(x_i + \tfrac{\partial}{\partial v} p_i(u, v)\delta v,\ y_i + \tfrac{\partial}{\partial v} q_i(u, v)\delta v\right);\\
D &= (u + \delta u, v + \delta v) & &\rightarrow & D' &= C' + (B' - A').
\end{aligned} \tag{5.43}$$

The probability that (U, V) falls in the rectangle $ABCD$ is given by

$$P[u < U \leq u + \delta u, v < V \leq v + \delta v] = f_{UV}(u, v)\delta u\, \delta v, \tag{5.44}$$

and this probability should be equal to the sum of probabilities that (X, Y) falls in one of the m corresponding parallelograms of sizes Δ_i, $i = 1, 2, \ldots, m$:

$$\sum_{i=1}^{m} f_{XY}(x_i, y_i)\Delta_i, \tag{5.45}$$

where Δ_i corresponds to the area $A'B'C'D'$ of Figure 5.4 (b), which can be calculated, by using the well-known result in **analytic geometry**, as follows.

The area S of a triangle with vertices at $P_1 = (x_1, y_1)$, $P_2 = (x_2, y_2)$, and $P_3 = (x_3, y_3)$ is given (Problem 5.17) by

$$S = \left| \frac{1}{2} \det \begin{bmatrix} 1 & 1 & 1 \\ x_1 & x_2 & x_3 \\ y_1 & y_2 & y_3 \end{bmatrix} \right|. \tag{5.46}$$

If we set the point to be the origin, i.e., $P_1 = O = (0, 0)$, then the formula (5.47) is simplified to

$$S = \left| \frac{1}{2} \det \begin{bmatrix} 1 & 1 & 1 \\ 0 & x_2 & x_3 \\ 0 & y_2 & y_3 \end{bmatrix} \right| = \left| \frac{1}{2} \det \begin{bmatrix} x_2 & x_3 \\ y_2 & y_3 \end{bmatrix} \right|. \tag{5.47}$$

Now in referring to Figure 5.4, let vectors $\overrightarrow{A'B'}$ and $\overrightarrow{A'C'}$ be interpreted as vectors $\overrightarrow{P_1 P_2} = \overrightarrow{O P_2}$ and $\overrightarrow{O P_3}$ respectively. Then, the area of triangle $A'B'C'$, which is one half of Δ_i that we are interested in, is given by

$$\begin{aligned} \frac{\Delta_i}{2} &= \left| \frac{1}{2} \det \begin{bmatrix} \frac{\partial}{\partial u} p_i(u, v)\delta u & \frac{\partial}{\partial v} p_i(u, v)\delta v \\ \frac{\partial}{\partial u} q_i(u, v)\delta u & \frac{\partial}{\partial v} q_i(u, v)\delta v \end{bmatrix} \right| \\ &= \left| \frac{1}{2} \det \begin{bmatrix} \frac{\partial}{\partial u} p_i(u, v) & \frac{\partial}{\partial v} p_i(u, v) \\ \frac{\partial}{\partial u} q_i(u, v) & \frac{\partial}{\partial v} q_i(u, v) \end{bmatrix} \delta u \, \delta v \right|. \end{aligned} \tag{5.48}$$

We define a 2×2 matrix

$$\boxed{\boldsymbol{J}\left(\frac{p_i, q_i}{u, v}\right) \triangleq \begin{bmatrix} \frac{\partial p_i}{\partial u} & \frac{\partial p_i}{\partial v} \\ \frac{\partial q_i}{\partial u} & \frac{\partial q_i}{\partial v} \end{bmatrix},} \tag{5.49}$$

which is called the **Jacobian matrix**[1] of the transformation functions $p_i(x, y)$ and $q_i(x, y)$. Then we can write

$$\Delta_i = \left| \boldsymbol{J}\left(\frac{p_i, q_i}{u, v}\right) \right|, \tag{5.50}$$

where we adopt a simplifying notation

$$|\mathbf{A}| \triangleq |\det \mathbf{A}|. \tag{5.51}$$

By equating (5.44) and (5.45) we find that

$$\boxed{f_{UV}(u, v) = \sum_{i=1}^{m} \left| \boldsymbol{J}\left(\frac{p_i, q_i}{u, v}\right) \right| f_{XY}(x_i, y_i).} \tag{5.52}$$

The determinant $\det \boldsymbol{J}$ is called the **Jacobian** or **Jacobian determinant**[2] of the transformation. If we define the Jacobian matrix of the original mapping by

$$\boldsymbol{J}\left(\frac{g, h}{x, y}\right) = \begin{bmatrix} \frac{\partial g}{\partial x} & \frac{\partial g}{\partial y} \\ \frac{\partial h}{\partial x} & \frac{\partial h}{\partial y} \end{bmatrix}, \tag{5.53}$$

[1] Named after the German mathematician Carl Gustav Jacob Jacobi (1804–1851).

[2] The term "Jacobian" is often a shorthand for the Jacobian matrix $\boldsymbol{J}$ as well as for its determinant.

we can show (Problem 5.18) the following identity:

$$\left| J\left(\frac{p_i, q_i}{u, v}\right) \right| = \left| J\left(\frac{g, h}{x, y}\right) \right|^{-1}. \tag{5.54}$$

Example 5.6: Two linear transformations. Let the two transformations be linear functions:

$$g(X, Y) = aX + bY \text{ and } h(X, Y) = cX + dY, \tag{5.55}$$

where $ad - bc \neq 0$. Then we can write the above in matrix form:

$$\begin{bmatrix} U \\ V \end{bmatrix} = \begin{bmatrix} a & b \\ c & d \end{bmatrix} \begin{bmatrix} X \\ Y \end{bmatrix}. \tag{5.56}$$

For a given $(U, V) = (u, v)$, there is only one solution, hence $m = 1$ and the solution (x_1, y_1) is readily found from the above matrix equation as

$$\begin{bmatrix} x_1 \\ y_1 \end{bmatrix} = \frac{1}{\Delta} \begin{bmatrix} d & -b \\ -c & a \end{bmatrix} \begin{bmatrix} u \\ v \end{bmatrix}. \tag{5.57}$$

Thus, we find the inverse mapping

$$p_1(u, v) = \frac{1}{\Delta}(du - bv) \text{ and } q_1(u, v) = \frac{1}{\Delta}(-cu + av), \tag{5.58}$$

where

$$\Delta = ad - bc. \tag{5.59}$$

The Jacobian is then found to be

$$J\left(\frac{p_1, q_1}{u, v}\right) = \begin{bmatrix} \frac{d}{\Delta} & \frac{-b}{\Delta} \\ \frac{-c}{\Delta} & \frac{a}{\Delta} \end{bmatrix}, \tag{5.60}$$

which gives

$$\left| J\left(\frac{p_1, q_1}{u, v}\right) \right| = |\Delta|^{-1}. \tag{5.61}$$

Therefore, we finally obtain the joint PDF of the transformed variables:

$$\begin{aligned} f_{UV}(u, v) &= \left| J\left(\frac{p_1, q_1}{u, v}\right) \right| f_{XY}(x_1, y_1) \\ &= |\Delta|^{-1} f_{XY}\left(\Delta^{-1}(du - bv), \Delta^{-1}(-cu + av)\right). \end{aligned} \tag{5.62}$$

We can verify the formula (5.54) by computing

$$J\left(\frac{g, h}{x, y}\right) = \begin{bmatrix} a & b \\ c & d \end{bmatrix}. \tag{5.63}$$

As a special case, let $a = b = c = 1$ and $d = 0$. Then $U = X + Y$ and $V = X$. Noting that $\Delta = -1$, we readily find the joint PDF of (U, V) as

$$f_{UV}(u, v) = |-1| f_{XY}(v, u - v). \tag{5.64}$$

By integrating over v, we obtain

$$f_U(u) = \int_{-\infty}^{\infty} f_{XY}(v, u - v)\, dv, \tag{5.65}$$

which agrees with (5.25) of Example 5.3. If we set instead $c = 0$ and $a = b = d$, we find that $U = X + Y$, $V = Y$, and $\Delta = 1$. In this case we have

$$f_{UV}(u, v) = f_{XY}(u - v, v), \tag{5.66}$$

from which we obtain

$$f_U(u) = \int_{-\infty}^{\infty} f_{XY}(u - v, v)\, dv, \tag{5.67}$$

which is equivalent to (5.22). □

5.4 Generation of random variates for Monte Carlo simulation[3]

Monte Carlo simulation usually refers to a numerical technique for solving a nonprobabilistic mathematical problem (e.g., a certain integration expression) by introducing a random variable whose mean or distribution is related to the solution of the original problem. A Monte Carlo simulation can be viewed as a way to estimate the expected value of some *response variable* Y:

$$E[Y] = \int_0^1 \cdots \int_0^1 Y(\boldsymbol{R}) f_{\boldsymbol{R}}(\boldsymbol{r})\, d\boldsymbol{r},$$

where $\boldsymbol{R}$ is a **random vector** representing a stream of random numbers of length m, which is the total number of **uniform variates**[4] to be generated during a simulation run: $\boldsymbol{R} = (R_1, R_2, \ldots, R_m)^\top$ (e.g., see [203]: Chapters 16 and 17). A Monte Carlo simulation is usually adopted when the functional form $Y(\boldsymbol{R})$ is not known explicitly; otherwise a simulation would be unnecessary.

In Chapter 21 of this book, we will discuss a computationally efficient method, MCMC, to numerically compute the probability distribution functions, PDFs or PMFs, when analytic expressions are difficult to come by or too difficult to evaluate. As we remarked in Chapter 1, MCMC is becoming a very important tool for Bayesian econometricians, machine learning researchers, and researchers in the field of bioinformatics.

[3] The reader may opt to postpone this section until the study of Chapter 21.

[4] A particular outcome or sample value of a random variable is often called a **variate**.

The term Monte Carlo simulation is also often used to mean *self-driven simulation* of some stochastic system (e.g., a queueing system) (e.g., see [203]). In either case, Monte Carlo simulation requires a mechanism for generating **random variates** whose PDF or PMF is specified. Generation of variates from any specified distribution is possible once we know how to generate a sequence of independent variates drawn from the **uniform distribution**. This assertion will be demonstrated in Section 5.4.2 by using the results we studied in this chapter. We will first concentrate on the generation of the uniform variates; i.e., real numbers drawn from the uniform distribution $U(0, 1)$.

5.4.1 Random number generator (RNG)

In a digital computer a real number is expressed with only finite accuracy, and thus we normally generate *integers* Z_i between zero and some positive integer m. Then, the fraction

$$Y_i = Z_i/m \tag{5.68}$$

lies between 0 and 1; i.e., $Y_i \in (0, 1)$. The most common method of generating such a sequence of **random numbers** Z_i is by means of a simple *recurrence* relation such as

$$\boxed{Z_i \equiv aZ_{i-1} + c \pmod{m},} \tag{5.69}$$

where a, the *multiplier*, is a positive integer and c, the *increment*, is a nonnegative integer, and m, the **modulus**, is also a positive integer. Two integers a and b are said to be **congruent modulo** m if their difference is an integral multiple of m, and this congruence relation is expressed as

$$a \equiv b \pmod{m}. \tag{5.70}$$

For instance, $5 \equiv 2 \pmod 3$, $13 \equiv 1 \pmod 3$, etc.

The generator of a sequence of numbers according to a congruence relationship (5.69) is thus called the **linear congruential generator (LCG)**. For the special case when $c = 0$ is chosen, this RNG is called a **multiplicative congruential generator**, and the case $c \neq 0$ is said to give a **mixed congruential generator**. Advantages of using such a simple recurrence formula of LCG are: (i) statistical properties of the resulting sequence can be reasonably well understood, so that we can choose optimal values of a, m, c, and the initial value Z_0, called a *seed*; (ii) it requires little computational time and memory space; and (iii) the sequence can be easily reproduced by just saving the seed Z_0. Similarly, the sequence generation process can be interrupted and restarted by saving the last number.

The sequence $\{Z_i\}$ generated by (5.69) is not a random sequence in the true sense of the word "random." But for all practical purposes, we may be content with accepting such sequences as random sequences if they *appear* to be sufficiently random; that is, if no important statistical tests reveal a significant discrepancy from the behavior that a truly random sequence is supposed to demonstrate. Thus, a sequence generated in a deterministic way such as (5.69) is often called a **pseudorandom sequence** or *quasi-random sequence*.

Other methods for random number generation include **lagged Fibonacci generator (LFG)**, **multiple recursive generator (MRG)**, **combined generators**, and **Marsaglia–Zamam's AWC (add-with-carry)** and **SWB (subtract-with-borrow)** algorithms. For detailed discussions of these RNG algorithms, the reader is directed to [203] (Chapter 16), for example, and references therein.

5.4.2 Generation of variates from general distributions

Anyone of the random number generation methods discussed above typically generates integers $\{Z_i\}$ between 0 and $m-1$ (or between 1 and $m-1$ for the case of the multiplicative congruential method) with a uniform frequency. Thus, the fractions $\{Y_i\}$ of (5.68) are uniformly distributed between zero and one. How can we then generate a sequence of random observations from a given probability distribution?

5.4.2.1 Transform method

Let X be the RV of our concern and $F(x)$ be the distribution function; that is,

$$F(x) = P[X \leq x]. \tag{5.71}$$

Set $F(X) = Y$; then Y is defined over the range zero and one. Now we show that if Y is a random variable uniformly distributed between zero and one, the variable X defined by

$$\boxed{X = F^{-1}(Y), 0 \leq Y \leq 1,} \tag{5.72}$$

has the cumulative distribution $F(x)$ (see Figure 5.5):

$$P[X \leq x] = P[Y \leq F(x)] = F(x). \tag{5.73}$$

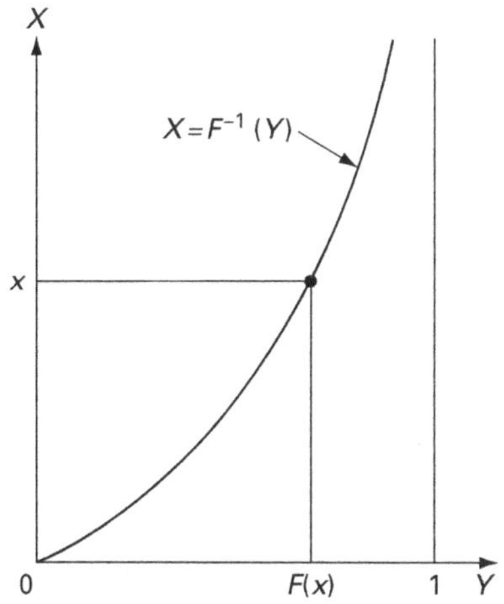

Figure 5.5 Transformation of the uniform RV into RV X with the distribution $F(x)$.

The inverse mapping $F^{-1}(\cdot)$ can be performed by writing the equation for this function, or by developing a table giving the values of X for a finite (but sufficiently dense) set of points of Y from zero to one.

Example 5.7: Transform method for exponential variates. Consider a random variable X with the *exponential* distribution $F(x) = 1 - e^{-\mu x}, x \geq 0$. Applying the procedure outlined above, set this function equal to a random decimal number $Y = 1 - e^{-\mu x}$ so that

$$X = F^{-1}(Y) = -[\ln(1 - Y)]/\mu. \tag{5.74}$$

Since $1 - Y$ is itself a random decimal number between zero and one, we can use a simpler transformation

$$X = F^{-1}(1 - Y) = -(\ln Y)/\mu. \tag{5.75}$$

Thus, one can generate a sequence of random observations from an exponential distribution by applying the transformation (5.74) or (5.75) to a random decimal sequence.

□

Although the algorithm based on the logarithm transformation is easy to program, it is not the fastest method. Other algorithms that do not use a natural logarithm subroutine are often adopted; e.g., the **rectangle-wedge-tail** method (e.g., see [191, 203]).

5.4.2.2 Acceptance–rejection method

It is often possible to calculate the PDF $f_X(x)$ but difficult to evaluate its integral $F_X(x)$ or the inverse $F_X^{-1}(x)$. A technique called the **acceptance–rejection method**[5] has been developed to deal with this situation. Let $f_X(x)$ be bounded by M and have a finite range (or support), say $a \leq x \leq b$, as shown in Figure 5.6. This method can be described in three steps as given in Algorithm 5.1.

The number of trials before an acceptable x is found is a random variable N with geometric distribution: $P[N = n] = \rho(1 - \rho)^{n-1}, n \geq 1$, where ρ is the probability that the inequality (5.76) is satisfied for a given pair (u_1, u_2). It should simply be given as the ratio of the *white area* under the function $f_X(x)$, which is unity since $f_X(x)$ is a PDF, to the *rectangle of size* $(b - a) \times M$. Hence, ρ is given by $\rho = 1/M(b - a)$. The mean value of N is $E[N] = 1/\rho = M(b - a)$. This implies that the method may not be efficient for PDFs with large $M(b - a)$.

[5] This method is sometimes called the **rejection** method.

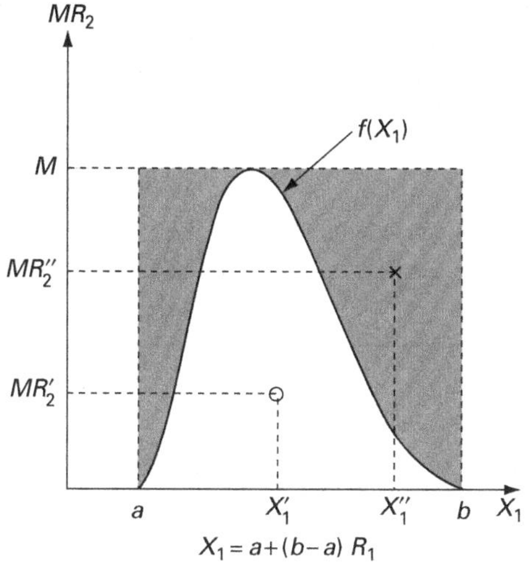

Figure 5.6 The acceptance–rejection method to generate a random variable with the PDF $f_X(x)$. (The point "$\circ$" is to be accepted, whereas the point "$\times$" is to be rejected.)

Algorithm 5.1 Acceptance–rejection algorithm

1: Generate a uniform variate $u_1 \in [0, 1]$, and set $x = a + (b - a)u_1$, which is in $[a, b]$.
2: Generate another uniform variate $u_2 \in [0, 1]$.
3: If

$$Mu_2 \leq f_X(x), \tag{5.76}$$

accept x, and reject otherwise.
4: Stop when the number of accepted variates x has reached a prescribed number. The PDF corresponding to the accepted x will then be $f_X(x)$. Otherwise, return to step 1.

The acceptance–rejection method described above, although simple, has two limitations: first, the PDF $f_X(x)$ has to have finite support $[a, b]$; second, $M(b - a)$ cannot be too large. In order to overcome these restrictions, we generalize the method as follows. For a given PDF $f_X(x)$, called the *target density*, we find another PDF $f_Y(y)$, called the *proposal density*, with readily available $F_Y^{-1}(y)$, and a constant $c > 0$ such that

$$\boxed{f_X(x) \leq c f_Y(x), \text{ for all } x.} \tag{5.77}$$

A generalized and improved algorithm takes the steps given in Algorithm 5.2. The expected number of iterations to generate one acceptable x is $E[N] = c$, whereas in the original algorithm this quantity was $E[N] = M(b - a)$, which can be quite large. The original acceptance–rejection method corresponds to choosing the proposal density $f_Y(y) = 1/(b - a)$ for $y \in [a, b]$. Thus, $x = F_Y^{-1}(u_1) = a + (b - a)u_1$.

Algorithm 5.2 Generalized and improved acceptance–rejection algorithm

1: Generate a uniform variate $u_1 \in [0, 1]$, and set $x = F_Y^{-1}(u_1)$.
2: Generate another uniform variate $u_2 \in [0, 1]$, and set $y = cf_Y(x)u_2$, which is uniformly distributed over $[0, cf_Y(x)]$.
3: If $y \leq f_X(x)$, accept x. Otherwise, reject and go to step 1.
4: Stop when the number of accepted variates x has reached a prescribed number. Otherwise, return to step 1.

Example 5.8: Acceptance–rejection method to generate a gamma variate. Consider the gamma distribution discussed in Section 4.2.3. Choose the parameters $\lambda = 1$ and $\beta = 3/2$:[6]

$$f_{X_{1,3/2}}(x) \triangleq \frac{x^{1/2}e^{-x}}{\sqrt{\pi}/2}, \; x \geq 0. \tag{5.78}$$

We choose an exponential distribution:

$$f_{Y_\lambda}(y) = \lambda\, e^{-\lambda y}, \; y \geq 0, \tag{5.79}$$

where λ will be determined below. The ratio

$$\frac{f_{X_{1,3/2}}(x)}{f_{Y_\lambda}(y)} = \frac{2}{\sqrt{\pi}\lambda} x^{1/2} e^{-(1-\lambda)x}$$

is maximum when $x = 1/2(1-\lambda)$; thus, we set

$$c = \max\left\{\frac{f_X(x)}{f_Y(x)}\right\} = \frac{1}{\lambda}\sqrt{\frac{2}{\pi\, e(1-\lambda)}}.$$

$E[N] = c$ can be minimized by choosing $\lambda = 2/3$.

Thus, a gamma variate with $(\lambda, \beta) = (1, 3/2)$ can be generated by Algorithm 5.3. The expected number of iterations needed is

$$E[N] = c = \frac{3\sqrt{3}}{\sqrt{2\pi\, e}}.$$

□

[6] With the translation $x = z/2$, this distribution becomes the χ^2 distribution with three degrees of freedom:

$$f_{\chi_3^2}(z) = \frac{z^{1/2}e^{-z/2}}{\sqrt{2\pi}}, \; z \geq 0.$$

Algorithm 5.3 Acceptance–rejection algorithm to generate gamma variates

1: Generate a uniform variate $u_1 \in [0, 1]$ and set $x = -\frac{3}{2} \ln u_1$.
2: Generate a uniform variate $u_2 \in [0, 1]$.
3: If $u_2 < \sqrt{\frac{2e}{3}} x^{1/2} e^{-x/3}$, accept x. Otherwise, reject.
4: Stop when the number of accepted variates x has reached a prescribed number. Otherwise return to step 1.

5.4.2.3 Composition methods

There are two types of "composition methods:" the first type applies to a case where the desired distribution function is composed of simpler distribution functions; the second is a case where the RV in question can be decomposed into simpler RVs.

Type 1: Suppose that the distribution $F(x)$ is represented as the mixture of two distributions $F_1(x)$ and $F_2(x)$:

$$F(x) = pF_1(x) + (1 - p)F_2(x). \tag{5.80}$$

We can obtain a variate X drawn from the distribution $F(x)$, by taking the following two steps:

1. Generate a uniform variate u.
2. If $u < p$, set x equal to a variate drawn from the distribution $F_1(x)$; otherwise, draw x from the other distribution $F_2(x)$.

This composite procedure is directly applicable to, for example, a mixed Gaussian distribution and a hyperexponential distribution.

Type 2: Suppose that the distribution $F(x)$ of RV X is such that the variable is representable in terms of one or more independent RVs with simple distributions. A notable example is a k-stage Erlang variate with mean $1/\mu$, which can be represented as the sum of k independent exponential random variates with mean $1/k\mu$:

$$X = X_1 + X_2 + \cdots + X_k. \tag{5.81}$$

A more efficient procedure is obtained (Problem 5.20) by noting the following representation of x:

$$x = -\frac{\ln\left(\prod_{i=1}^{k} u_i\right)}{k\mu}, \tag{5.82}$$

where $u_1, u_2, \ldots, u_k$ are k independent uniform variates between zero and one. Negative binomial variates and chi-squared variates can be generated by this type of composition method.

5.4.3 Generation of normal (or Gaussian) variates[7]

Since the normal or Gaussian distribution is of sufficient importance in many simulation studies, we will discuss two basic methods to generate normal variates.

5.4.3.1 Sum method

The distribution of the RV X of (5.81) approaches the normal or Gaussian distribution by virtue of the CLT, as $k \to \infty$. This property of *asymptotic normality* holds for any distributional forms of the component RVs $\{X_i\}$ so long as their means and variances are finite. Let us define a random variable S by

$$S = U_1 + U_2 + \cdots + U_n, \tag{5.83}$$

where the U_i are independent RVs uniformly distributed in [0, 1]. Then, for large n, the variable S is approximately normally distributed with mean $n/2$ and variance $n/12$. Then the variable X defined by

$$X = \frac{(S - \frac{n}{2})\sigma}{\sqrt{n/12}} + \mu \tag{5.84}$$

approximates the normal variable with mean μ and variance σ^2. A convenient choice is $n = 12k^2$ with some integer k, since it eliminates the square-root term from the last expression. However, this value of n truncates the distribution at $\pm 6k\sigma = \pm\sqrt{3n}\sigma$ limits. Some of the old subroutine programs under the name "Gaussian random generator" are based on this method with a fairly small size of n; thus, they are not appropriate when one is concerned with the tail of the distribution.

5.4.3.2 Box–Muller method

As we shall show in Problem 5.22 (see also Section 7.5.1), if u_1 and u_2 are independent uniform variates in [0, 1],

$$x = (-2 \ln u_1)^{1/2} \cos 2\pi u_2 \tag{5.85}$$

and

$$y = (-2 \ln u_1)^{1/2} \sin 2\pi u_2 \tag{5.86}$$

are independent normal variates sampled from the unit normal distribution $N(0, 1)$.

The Box–Muller method is based on the above transformations and is much superior to the simple sum method for generation of normal variates, but it takes considerably more computation time, because of the logarithm, sine, and cosine functions involved. A few clever techniques have been devised to reduce the computational steps required of the original Box–Muller method. See [203] (pp. 653–657) for a further discussion on improved methods for generation of normal variates and other general variates.

[7] Some authors (e.g., Knuth [191]) use the term *normal deviate* instead of *normal variate*.

5.5 Summary of Chapter 5

CDF of $Y = g(X)$:	$F_Y(y) = \int_{-\infty}^{\infty} I(x \in \mathcal{D}_y) f_X(x)\, dx$	(5.4)
where	$\mathcal{D}_y = \{x : g(x) \le y\}$	(5.3)
PDF of $Y = g(X)$:	$f_Y(y) = \sum_{i=1}^{m(y)} \frac{f_X(x_i)}{\lvert g'(x_i)\rvert}$	(5.17)
where	$x_i = g^{-1}(y), i = 1, 2, \ldots, m(y)$	(5.16)
PDF of $Z = X + Y$:	$f_Z(z) = \int_{-\infty}^{\infty} f_{XY}(z - y, y)\, dy$	(5.24)
	$= \int_{-\infty}^{\infty} f_{XY}(x, z - x)\, dx$	(5.25)
Sum of independent RVs:	$f_Z(z) = f_X(z) \circledast f_Y(z)$	(5.28)
	$F_Z(z) = \int_{-\infty}^{\infty} F_X(z - y)\, dF_Y(y)$	
	$= \int_{-\infty}^{\infty} F_Y(z - x)\, dF_X(x)$	(5.29)
	$F_Z(z) = F_X(z) \circledast f_Y(z) = f_X(z) \circledast F_Y(z)$	(5.30)
Two functions of two RVs:	$U = g(X, Y)$ and $V = h(X, Y)$	(5.38)
Inverse mapping:	$X_i = p_i(U, V)$ and $Y_i = q_i(U, V)$	(5.42)
Joint PDF of U and V:	$f_{UV}(u, v) = \sum_{i=1}^{m} \left\lvert \boldsymbol{J}\left(\frac{p_i, q_i}{u, v}\right)\right\rvert f_{XY}(x_i, y_i)$	(5.52)
Jacobian matrix:	$\boldsymbol{J}\left(\frac{p_i, q_i}{u, v}\right) = \begin{bmatrix} \frac{\partial p_i}{\partial u} & \frac{\partial p_i}{\partial v} \\ \frac{\partial q_i}{\partial u} & \frac{\partial q_i}{\partial v} \end{bmatrix}$	(5.49)
Jacobian of the inverse:	$\left\lvert \boldsymbol{J}\left(\frac{p_i, q_i}{u, v}\right)\right\rvert = \left\lvert \boldsymbol{J}\left(\frac{g, h}{x, y}\right)\right\rvert^{-1}$	(5.54)
Linear congruential generator:	$Z_i = aZ_{i-1} + c (\mathrm{mod}\ m)$	(5.69)
Transform method for random variate:	$X = F^{-1}(Y), 0 \le Y \le 1$	(5.72)
Acceptance–rejection method:	Algorithm 5.1	
Generalized A–R method:	Algorithm 5.2	
Box–Muller method for $N(0, 1)$:	$x = (-2 \ln u_1)^{1/2} \cos 2\pi u_2$	(5.85)
	$y = (-2 \ln u_1)^{1/2} \sin 2\pi u_2$	(5.86)

5.6 Discussion and further reading

The topic of this chapter, namely the transformation of RVs, is treated to varying degrees by different textbooks. Thomas [319] and Fine [105] devote one chapter as we do, whereas Papoulis and Pillai [262] spend two chapters, and discuss more examples. A majority of other books on probability treat this subject more tersely.

Discussion on random number generation can be found in [191] and many books on simulation. Tezuka [318] provides a comprehensive introduction to this subject and statistical tests that assess the quality of random number generation. The 1992 article by Ferrenberg *et al.* [104] suggests that *trinomial-type* generators, which include a large class of RNGs, are not adequate in large-scale simulation tasks. Some of these algorithms are used as "most advanced" generators. We should not overreact, however, to these findings, because the "defects" of the criticized generators may not be a real problem in a small simulation task: a periodic pattern associated with such bad generators may not appear in a small simulation because the simulation run may be substantially shorter than such a period. Random number generation functions or programs available in MATLAB and other packages should be adequate for the majority of simulation experiments.

5.7 Problems

Section 5.1: Function of one random variable

5.1* Half-wave rectifier. Let $g(x) = x \cdot u(x)$ represent the function of a half-wave rectifier; i.e.,

$$g(x) = \begin{cases} x, & x \geq 0, \\ 0, & x < 0. \end{cases} \tag{5.87}$$

Find the distribution function $F_Y(y)$ and the PDF $f_Y(y)$ of the rectifier output $Y = g(X)$ in terms of the inputs $F_X(x)$ and $f_X(x)$.

5.2 Square law detector – continued. Continue to consider Example 5.2, in which $Y = g(X) = X^2$.

(a) Let X have a uniform distribution over $[-1, +1]$. Find the distribution function and PDF of the square-law detector output Y.

(b) Let X be a Gaussian variable with zero mean and variance σ^2; i.e.,

$$F_X(x) = \Phi\left(\frac{x}{\sigma}\right) \text{ and } f_X(x) = \frac{1}{\sigma}\phi\left(\frac{x}{\sigma}\right), \tag{5.88}$$

where $\Phi(u)$ and $\phi(u)$ are the distribution function and PDF of the unit normal variable U defined by (4.46) and (4.41) in Section 4.2.4. Show that the distribution function of the square-law detector output Y is given by

$$F_Y(y) = \left[2\Phi\left(\frac{\sqrt{y}}{\sigma}\right) - 1\right], y \geq 0, \tag{5.89}$$

and the corresponding PDF is given by

$$f_Y(y) = \frac{1}{\sigma\sqrt{y}}\phi(\sqrt{y}/\sigma) = \frac{1}{\sqrt{2\pi y}\sigma}\exp\left(-\frac{y}{2\sigma^2}\right), y \geq 0. \tag{5.90}$$

Further show that the PDF of the normalized output defined by $\chi_1^2 = Y/\sigma^2$ is given by

$$f_{\chi_1^2}(x) = \frac{x^{-1/2}e^{-x/2}}{\sqrt{2\pi}}, \quad x \geq 0, \tag{5.91}$$

which is known as the **chi-square distribution**[8] with degree of freedom $n = 1$, as will be discussed in Section 7.1.

5.3 Exponential-law detector. Consider a nonlinear detector that follows an exponential law; i.e., $g(x) = e^x$.

(a) Find the distribution function $F_Y(y)$ and the PDF $f_Y(y)$ of the detector output in terms of the distribution function and PDF of the input variable X.
(b) Let X have an exponential distribution

$$F_X(x) = \begin{cases} 1 - e^{-\lambda x}, & x \geq 0, \\ 0, & x < 0. \end{cases}$$

Find $F_Y(y)$ and $f_Y(y)$.

5.4 Cauchy distribution. Let X be uniformly distributed between $[-\frac{\pi}{2}, \frac{\pi}{2}]$. Let $Y = \tan X$. Show that the PDF of Y is given by

$$f_Y(y) = \frac{1/\pi}{y^2 + 1}, \quad -\infty < y < \infty, \tag{5.92}$$

which is known as the **Cauchy distribution** with parameter unity.

5.5 Inverse of a random variable and the Cauchy distribution. Consider the case where Y is the inverse of X; i.e., $Y = g(X) = 1/X$.

(a) Find the PDF $f_Y(y)$ in terms of $f_X(x)$.
(b) Suppose that X is a random variable with the following Cauchy distribution with parameter α:

$$f_X(x) = \frac{1}{\pi\alpha\left(1 + \frac{x^2}{\alpha^2}\right)} = \frac{\alpha/\pi}{x^2 + \alpha^2}, \quad -\infty < x < \infty. \tag{5.93}$$

Derive the PDF of Y.

Section 5.2: Function of two random variables

5.6* Leibniz's rule.[9] In deriving (5.23), we used a special case of Leibniz's rule for differentiation under the integral sign.

THEOREM 5.1 (Leibniz's rule). *The following rule holds for differentiation of a definite integral, when the integration limits are functions of the differential variable:*

$$\boxed{\frac{d}{dz}\int_{a(z)}^{b(z)} h(z, y)\, dy = h(z, b(z))b'(z) - h(z, a(z))a'(z) + \int_{a(z)}^{b(z)} \frac{\partial}{\partial z} h(z, y)\, dy.} \tag{5.94}$$

8 Treat χ_n^2 as one symbol that represents a nonnegative RV.
9 Gottfried Wilhelm Leibniz (1646–1716) was a German mathematician. His last name is occasionally spelled as Leibnitz.

In particular, if h is a function of y only, the rule reduces to

$$\boxed{\frac{d}{dz}\int_{a(z)}^{b(z)} h(y)\,dy = h(b(z))b'(z) - h(a(z))a'(z).} \tag{5.95}$$

□

(a) Define

$$\int_{-\infty}^{y} h(x)\,dx \triangleq H(y).$$

Then prove (5.95).

(b) Define

$$\int_{-\infty}^{y} h(z,x)\,dx \triangleq H(z,y) \text{ and } \frac{\partial H(z,y)}{\partial y} \triangleq g(z,y).$$

Then prove (5.94).

(c) Alternative proof of (5.94). Consider a function $G(a, b, c)$, where a, b, and c stand for $a(z)$, $b(z)$, and $c(z)$ respectively. By applying the *chain rule* to the function G, we have

$$\frac{dG(a,b,c)}{dz} = \frac{\partial G}{\partial a}a'(z) + \frac{\partial G}{\partial b}b'(z) + \frac{\partial G}{\partial c}c'(z). \tag{5.96}$$

Consider a special case

$$c(z) = z \text{ and } G(a,b,c) \triangleq \int_a^b h(z,y)\,dy.$$

Then prove (5.94).

5.7 Sum of uniform variables. Let X and Y be independent uniform RVs.

(a) Let both RVs be uniformly distributed in the unit interval $(0, 1]$. Find the PDF of $Z = X + Y$.

(b) Suppose that X is uniformly distributed over $(0, a]$ and Y is uniformly distributed over $(0, b]$, $0 < a \le b$. Find the PDF of Z.

5.8 Sum of exponential variables. Let X and Y be independent exponential variables with rate parameters λ and μ, respectively:

$$f_X(x) = \lambda\, e^{-\lambda x} u(x) \text{ and } f_Y(y) = \mu\, e^{-\mu y}\cdot u(y),$$

where $u(\cdot)$ is the unit step function defined earlier.

(a) Find the PDF of $Z = X + Y$.

(b) What form does $f_Z(z)$ take when $\lambda = \mu$?

5.9 Difference of two RVs. Let $Z = X - Y$. Express the PDF of Z in terms of the joint PDF $f_{XY}(x, y)$. How can this be simplified when X and Y are statistically independent?

5.10 Ratio of two RVs. Given two RVs X and Y with the joint PDF $f_{XY}(x, y)$, define $Z = X/Y$.

(a) Show that the PDF of Z is given by

$$f_Z(z) = \int_{-\infty}^{\infty} |y| f_{XY}(yz, y)\, dy, \quad -\infty < z < \infty. \tag{5.97}$$

(b) Assume that X and Y are both nonnegative variables. Show that the PDF of $Z = X/Y$ is given by

$$f_Z(z) = \begin{cases} \int_0^{\infty} f_{X,Y}(yz, y)\, dy, & z > 0, \\ 0, & z \leq 0. \end{cases} \tag{5.98}$$

5.11 Product of two RVs. Show that $Z = XY$ has the PDF given by

$$f_Z(z) = \int_{-\infty}^{\infty} |x|^{-1} f_{XY}(x, z/x)\, dx. \tag{5.99}$$

5.12 Bivariate normal distribution and Cauchy distribution. Consider the bivariate normal variables $\mathbf{X} = (X_1, X_2)$ defined by (4.113) of Section 4.3.1:

$$f_{X_1,X_2}(x_1, x_2) = \frac{1}{2\pi\sigma_1\sigma_2\sqrt{1-\rho^2}} \exp\left[-\frac{1}{2}Q(x_1, x_2)\right], \tag{5.100}$$

where

$$Q(x_1, x_2) = \frac{1}{1-\rho^2}\left[\left(\frac{x_1-\mu_1}{\sigma_1}\right)^2 - 2\rho\left(\frac{x_1-\mu_1}{\sigma_1}\right)\left(\frac{x_2-\mu_2}{\sigma_2}\right) + \left(\frac{x_2-\mu_2}{\sigma_2}\right)^2\right]. \tag{5.101}$$

(a) Show that the PDF of $Z = (X_1 - \mu_1)/(X_2 - \mu_2)$ is given by the Cauchy distribution

$$f_Z(z) = \frac{1}{\pi\alpha\left[1 + \frac{(z-\mu)^2}{\alpha^2}\right]}, \quad -\infty < z < \infty, \tag{5.102}$$

where $\mu = \rho\sigma_1/\sigma_2$ and $\alpha = \sigma_1\sqrt{1-\rho^2}/\sigma_2$. See also Problem 5.4 and Section 8.2.5 for a discussion on the above Cauchy distribution.

Hint: Start with the simplest case first and then generalize the result to more general cases; i.e., (i) $\mu_1 = \mu_2 = \rho = 0$ and $\sigma_1 = \sigma_2 = \sigma$; (ii) $\mu_1 = \mu_2 = \rho = 0$ and $\sigma_1 \neq \sigma_2$; (iii) $\mu_1 = \mu_2 = 0$, $\rho \neq 0$, and $\sigma_1 \neq \sigma_2$; (iv) no restriction.

(b) Show that the distribution function is given by

$$F_Z(z) = \frac{1}{2} + \frac{1}{\pi}\tan^{-1}\frac{z-\mu}{\alpha}. \tag{5.103}$$

5.13 Independent normal distribution and exponential distribution. Let X_1 and X_2 be independent normal variables with zero mean and common variance σ^2. Show that $Z = X_1^2 + X_2^2$ is exponentially distributed with mean $2\sigma^2$:

$$f_Z(z) = \frac{1}{2\sigma^2} e^{-z/2\sigma^2} u(z). \tag{5.104}$$

5.14 Maximum of two RVs. Let

$$Z = \max\{X, Y\} = \begin{cases} X, & \text{if } X \geq Y, \\ Y, & \text{otherwise.} \end{cases} \tag{5.105}$$

(a) Find the domain $\mathcal{D}_z$ and sketch the region in the X–Y plane.
(b) Show that the distribution function of Z is given by

$$F_Z(z) = F_{XY}(z, z). \tag{5.106}$$

(c) Show that if X and Y are independent, the PDF of Z is

$$f_Z(z) = F_X(z) f_Y(z) + f_X(z) F_Y(z). \tag{5.107}$$

5.15 Minimum of two RVs. Let

$$Z = \min\{X, Y\} = \begin{cases} X, & \text{if } X \leq Y, \\ Y, & \text{otherwise.} \end{cases} \tag{5.108}$$

(a) Find the domain $\mathcal{D}_z$ and sketch the region in the X–Y plane.
(b) Show that the distribution function of Z is given by

$$F_Z(z) = F_X(z) + F_Y(z) - F_{XY}(z, z). \tag{5.109}$$

(c) Find the PDF $f_Z(z)$ when X and Y are independent.
(d) Let X and Y be both exponentially distributed with rate parameters λ and μ respectively. Show that Z is also exponentially distributed.

Section 5.3: Two functions of two random variables and the Jacobian matrix

5.16* Maximum and minimum of two RVs. Let $U = \min\{X, Y\}$ and $V = \max\{X, Y\}$.

(a) Find the domain $\mathcal{D}_{u,v}$ and sketch the region in the X–Y plane.
(b) Show that

$$F_{UV}(u, v) = \begin{cases} F_{XY}(u, v) + F_{XY}(v, u) - F_{XY}(u, u), & v \geq u, \\ F_{XY}(v, v), & v < u. \end{cases} \tag{5.110}$$

(c) Find the marginal distribution functions $F_U(u)$ and $F_V(v)$ from the above and verify the solutions obtained in Problems 5.14 and 5.15.
(d) Assume X and Y are statistically independent and are uniformly distributed in the intervals $(0, a)$ and $(0, b)$ respectively. Find the PDF $f_{UV}(u, v)$.

5.17 Area of a triangle. Derive the formula (5.47).
Hint: Consider the simpler case first by setting $P_1 = (0, 0)$, and represent the area of the triangle OP_2P_3 as that of a rectangle minus three triangles.

5.18 Inverse of Jacobian matrix. Prove the relationship (5.54) between the Jacobian matrices of the forward mapping and the inverse mapping.

Section 5.4: Generation of random variates for Monte Carlo simulation

5.19* Use of a rejection method. Consider the following PDF:

$$f_X(x) = \begin{cases} 2x, & 0 \le x \le 1, \\ 0, & \text{elsewhere.} \end{cases}$$

Use the acceptance–rejection method to generate variates according to this distribution.

5.20* Erlang variates. Show that the relation (5.82) holds. Write a program to generate Erlang variates.

5.21 Poisson variates. Suppose that we generate a sequence of uniform variates $U_1, U_2, \ldots$. Let X be defined as an integer variable such that

$$\prod_{i=1}^{X+1} U_i < e^{-\lambda} \le \prod_{i=1}^{X} U_i.$$

Show that X is Poisson distributed with mean λ.

5.22* The polar method for generating the Gaussian variate. Derive formulas (5.85) and (5.86) by following the following steps. Transform the pair (X_1, X_2) into the polar coordinates (R, Θ) according to

$$X_1 = R\cos\Theta,\ X_2 = R\sin\Theta,$$

where X_1 and X_2 are independent RVs, both of which are from $N(0, 1)$.

(a) Show that the distribution function of R is given by

$$F_R(r) = 1 - \exp\left(-\frac{r^2}{2}\right)$$

and Θ is a uniform RV from $U[0, 2\pi)$.

(b) Set

$$Y_1 = 1 - \exp\left(-\frac{R^2}{2}\right),\ Y_2 = \frac{\Theta}{2\pi}.$$

What are the distributions of the RVs Y_1 and Y_2?

6 Fundamentals of statistical data analysis

The **theory of statistics** involves interpreting a set of finite observations as a *sample point* drawn at random from a sample space. The study of statistics has the following three objectives: (i) to make the *best estimate* of important parameters of the population; (ii) to assess the *uncertainty* of the estimate; and (iii) to reduce a bulk of data to understandable forms. In much the same way as an examination of the properties of probability distribution functions forms the basic theory of probability, the foundation of statistical analysis is to examine the empirical distributions and certain descriptive measures associated with them. This chapter provides basic concepts of statistical data analysis.

6.1 Sample mean and sample variance

Let us consider a situation where we select randomly and independently ***n* samples** from a **population** whose distribution has mean μ and variance σ^2. The set of such samples, denoted as $(x_1, x_2, \ldots, x_n)$, is referred to as a **random sample** of size n. Random samples are an important foundation of statistical theory, because a majority of the results known in mathematical statistics rely on assumptions that are consistent with a random sample. Let us start with a simple question: How can we estimate the population mean μ_X and population variance σ_X^2 from the random sample $(x_1, x_2, \ldots, x_n)$?

The **sample mean** (also called the **empirical average**) $\overline{x}$ is defined as

$$\boxed{\overline{x} = \frac{1}{n}\sum_{i=1}^{n} x_i.} \tag{6.1}$$

Each sample x_i can be viewed as an **instance** or *realization* of the **associated RV** X_i. Thus, the sample mean $\overline{x}$ is an instance of the **sample mean variable** $\overline{X}$ defined by

$$\overline{X} = \frac{1}{n}\sum_{i=1}^{n} X_i. \tag{6.2}$$

In fact the term "sample mean" is often used in statistical theory to describe the variable $\overline{X}$, but the quantity we can actually observe is its instance $\overline{x}$.

Taking the expectation of both sides in (6.2), we have

$$E[\overline{X}] = \frac{1}{n}\sum_{i=1}^{n} E[X_i]. \tag{6.3}$$

Since $X_1, X_2, \ldots, X_n$ are independent and identically distributed (i.i.d.) RVs, we have $E[X_1] = E[X_2] = \cdots = E[X_n] = \mu_X$. Substituting these values into (6.3), we find that the **expectation** of the *sample mean variable* satisfies

$$E[\overline{X}] = \mu_X, \tag{6.4}$$

which asserts that $\overline{x}$ of (6.1) is an **unbiased estimate** of μ_X. An unbiased estimate is one that is, on the average, right on target.

Consider the variance of $\overline{X}$:

$$\text{Var}\,[\overline{X}] = E[(\overline{X} - E[\overline{X}])^2], \tag{6.5}$$

where $\overline{X} - E[\overline{X}] = \overline{X} - \mu_X$ can be rewritten as

$$\overline{X} - E[\overline{X}] = \frac{1}{n}\sum_{i=1}^{n}(X_i - \mu_X) = \frac{1}{n}\sum_{i=1}^{n} Y_i, \tag{6.6}$$

where

$$Y_i \triangleq X_i - \mu_X, \quad i = 1, 2, \ldots, n.$$

Therefore,

$$\text{Var}\,[\overline{X}] = E\left[\left(\frac{1}{n}\sum_{i=1}^{n} Y_i\right)^2\right] = \frac{1}{n^2}\sum_{i=1}^{n} E[Y_i^2] + \frac{1}{n^2}\sum_{i=1}^{n}\sum_{j=1(j\neq i)}^{n} E[Y_i Y_j]. \tag{6.7}$$

Since the random variables $\{Y_i;\ 1 \le i \le n\}$ are statistically independent with zero mean and variance σ_X^2, we have

$$\boxed{\text{Var}\,[\overline{X}] = \frac{\sigma_X^2}{n}.} \tag{6.8}$$

Thus, the **variance** of the *sample mean variable* is the population variance divided by the sample size.

The deviations of the individual observations from the sample mean provide information about the dispersion of the x_i about $\overline{x}$. We define the **sample variance** s_x^2 by

$$\boxed{s_x^2 \triangleq \frac{1}{n-1}\sum_{i=1}^{n}(x_i - \overline{x})^2.} \tag{6.9}$$

This quantity can be viewed as an *instance* of the **sample variance variable**

$$S_X^2 \triangleq \frac{1}{n-1}\sum_{i=1}^{n}(X_i - \overline{X})^2, \tag{6.10}$$

which is also commonly called the sample variance. We find, after some rearrangement (Problem 6.1),

$$S_X^2 = \frac{1}{n}\sum_{i=1}^{n} Y_i^2 - \frac{1}{n(n-1)}\sum_{i=1}^{n}\sum_{j=1(j\neq i)}^{n} Y_i Y_j. \tag{6.11}$$

Taking expectations, we have

$$\boxed{E[S_X^2] = \frac{1}{n}\sum_{i=1}^{n} E[Y_i^2] = \sigma_X^2.} \tag{6.12}$$

The reason for using $n-1$ rather than n as the divisor in (6.9) is to make $E[S_X^2]$ equal to σ_X^2; that is, to make s^2 an unbiased estimate of σ_X^2. The positive square root of the sample variance, s_x, is called the **sample standard deviation**.

6.2 Relative frequency and histograms

When the observed data takes on discrete values, we can just count the number of occurrences for the individual values. Suppose that the sample size n is given and $k(\leq n)$ **distinct values** exist. Let n_j be the number of times that the jth value is observed, $1 \leq j \leq k$. Then the fraction

$$\boxed{f_j = \frac{n_j}{n}, \quad j = 1, 2, \ldots, k,} \tag{6.13}$$

is, as defined in (2.1), the **relative frequency** of the jth value.

When the underlying random variable X is a continuous variable, we often adopt the method of "grouping" or "classifying" the data: the range of observations is divided into k intervals, called **class intervals**, at points $c_0, c_1, c_2, \ldots, c_k$. Let us designate the interval $(c_{j-1}, c_j]$ as the jth class, $1 \leq j \leq k$. Note that the lengths of the class intervals

$$\Delta_j \triangleq c_j - c_{j-1}, \quad j = 1, 2, \ldots, k,$$

need not be equal. Let n_j denote the number of observations that fall in the jth class interval. Then the relative frequency of the jth class takes the same form as (6.13). The grouped distribution may be represented graphically as the following "staircase function" in an (x, h)-coordinate system:

$$\boxed{h(x) = \frac{f_j}{\Delta_j} = \frac{n_j}{n\Delta_j}, \quad \text{for } x \in (c_{j-1}, c_j], \ \ j = 1, 2, \ldots, k.} \tag{6.14}$$

Such a diagram is called a **histogram** and can be regarded as an estimate of the PDF of the population. If the class lengths Δ_j are all the same, the shape of the histogram remains unchanged whether we use the relative frequency of the classes $\{f_j\}$ or the frequency counts of the classes $\{n_j\}$ as the ordinate. Such diagrams are also called histograms.

The choice of the class intervals in the histogram representation is by no means trivial. Certainly, we should choose them in such a way that the characteristic features of the distribution are emphasized and chance variations are obscured. If the class lengths are too small, chance variations dominate because each interval includes only a small number of observations. On the other hand, if the class lengths are too large, a great deal of information concerning the characteristics of the distribution will be lost.

Let $\{x_k : 1 \le k \le n\}$ denote n observations in the order observed and let $\{x_{(i)} : 1 \le i \le n\}$ denote the same observations ranked in order of magnitude. The frequency $H(x)$ of observations that are smaller than or equal to x is called the **cumulative relative frequency**, and is given by

$$H(x) = \begin{cases} 0, & \text{for } x < x_{(1)}, \\ \frac{i}{n}, & \text{for } x_{(i)} \le x < x_{(i+1)}, \quad i = 1, 2, \ldots, n-1, \\ 1, & \text{for } x \ge x_{(n)}, \end{cases} \tag{6.15}$$

which is the empirical analog of the CDF $F_X(x)$. If we use the unit step function

$$u(x) = \begin{cases} 1, & \text{for } x \ge 0, \\ 0, & \text{for } x < 0, \end{cases} \tag{6.16}$$

the above $H(x)$ can be more concisely written as

$$H(x) = \frac{1}{n}\sum_{i=1}^{n} u(x - x_{(i)}) = \frac{1}{n}\sum_{k=1}^{n} u(x - x_k), \quad -\infty < x < \infty. \tag{6.17}$$

Interestingly enough, use of the unit step function makes it unnecessary to obtain the rank-ordered data in order to find $H(x)$. The graphical plot of $H(x)$ is a nondecreasing step curve, which increases from zero to one in "jumps" of $1/n$ at points $x = x_{(1)}$, $x_{(2)}, \ldots, x_{(n)}$. If several observations take on the same value, the jump is a multiple of $1/n$.

When **grouped data** are presented as a *cumulative relative frequency distribution*, it is usually called the **cumulative histogram**. The cumulative histogram is far less sensitive to variations in class lengths than the histogram. This is because the accumulation is essentially equivalent to integration along the x-axis, which *filters out* the chance variations contained in the histogram. The cumulative relative frequency distribution or the cumulative histogram is, therefore, quite helpful in portraying the gross features of data.

6.3 Graphical presentations

Reducing primary data to the sample mean, sample variance, and histogram can reveal a great amount of information concerning the nature of the population distribution. But sometimes important features of the underlying distribution are obscured or hidden by the data reduction procedures. In this section we will discuss some graphical methods that are often valuable in an exploratory analysis of measurement data. They are: (a) histograms on the probability or log-normal probability papers; (b) the survivor functions on log-linear and log-log papers; and (c) the dot diagram and correlation coefficient.

6.3.1 Histogram on probability paper

6.3.1.1 Testing the normal distribution hypothesis

As we stated in Section 4.2.4, RVs occurring in physical situations often have the normal (or Gaussian) distribution, or can at least be treated approximately as normal RVs. As we shall see in subsequent sections, most statistical analysis techniques are based on the assumption of normality of measured variables. Thus, when we collect measurement data and obtain some empirical distribution, the first thing we might do is to examine whether the underlying distribution is normal. A **fractile diagram**[1] (Hald [139]) is useful for this purpose. For a given distribution function $F(x)$

$$P = F(x) \tag{6.18}$$

provides the dependence of the cumulative distribution on the variable x. The inverse function

$$x_P = F^{-1}(P) \tag{6.19}$$

gives the value of the variable x that corresponds to the given cumulative probability P. The value x_P is called the ***P*-fractile**. Some authors use the terms **percentile** or **quantile** instead of the term fractile.

The distribution function of the **standard normal distribution** $N(0, 1)$ is often denoted by $\Phi(\cdot)$ as defined in (4.46):

$$\Phi(u) = \frac{1}{\sqrt{2\pi}} \int_{-\infty}^{u} \exp\left(-\frac{t^2}{2}\right) dt. \tag{6.20}$$

Then the fractile, u_P, of the distribution $N(0, 1)$ is derived as

$$u_P = \Phi^{-1}(P). \tag{6.21}$$

Suppose that for a given cumulative relative frequency $H(x)$ we wish to test whether this empirical distribution resembles a normal distribution; that is, to test whether

$$H(x) \cong \Phi\left(\frac{x-\mu}{\sigma}\right) \tag{6.22}$$

holds for some parameters μ and σ, where the symbol $\cong$ means "to have the distribution of." Testing this relation is equivalent to testing the relation

$$u_{H(x)} \approx \frac{x-\mu}{\sigma}. \tag{6.23}$$

[1] This term should not be confused with a similar term "fractal diagram" known in fractal geometry.

According to the definition of $H(x)$, the plot of $u_{H(x)}$ versus x forms a step (or staircase) curve:

$$u_{H(x)} = \begin{cases} -\infty, & \text{for } x < x_{(1)}, \\ u_{i/n}, & \text{for } x_{(i)} \leq x < x_{(i+1)}, \quad i = 1, 2, \ldots, n-1, \\ \infty, & \text{for } x \geq x_{(n)}. \end{cases} \tag{6.24}$$

Therefore, the staircase function plot

$$u = u_{H(x)} \tag{6.25}$$

provides an estimate of the straight line

$$u = \frac{x - \mu}{\sigma} \tag{6.26}$$

in the same way that the cumulative frequency distribution $y = H(x)$ forms an estimate of the CDF $y = F(x)$. The graphical plot of the function (6.25) in an (x, u)-coordinate system is called the **fractile diagram**.

Instead of plotting (x, u_P) on ordinary graph paper, we may plot (x, P) directly on special graph paper called **probability paper**. On the ordinate axis of a probability paper, the corresponding values of $P = \Phi(u)$ are marked, rather than the u values. Probability paper is used in the same manner as other special graph papers, such as logarithmic paper. Figure 6.1 (a) shows a probability paper with step curve $u = u_{H(x)}$, based on $n = 50$ sample points drawn from a normal distribution with zero mean and unit variance. Instead of the step curve, we often plot n points $(x_{(i)}, (i - \frac{1}{2})/n)$ which are situated at the midpoints of the vertical parts of the step curve. The advantages are that it is easier to plot n points than to draw a step curve, and that possible systematic deviations from a straight line are more easily detected from this dot diagram. The result of this procedure is shown in Figure 6.1 (b).

If the distribution in question is normal, the points of the fractile diagrams should vary randomly about a straight line. In a small sample, say $n < 20$, the permissible random variation of points in the fractile diagram is so large that it is generally difficult to examine whether systematic deviations from a straight line exist.

6.3.1.2 Testing the log-normal distribution hypothesis

Some random variables we deal with are often modeled by a **log-normal distribution** (see Section 7.4). In order to test whether a log-normal distribution fits given empirical data, we should plot the step curve or dot diagram on **log-normal paper**, which is a simple modification of the above probability paper. The ordinate axis is the same as in the probability paper, i.e., $u_P = \Phi^{-1}(P)$, whereas the horizontal axis is changed from the linear scale (in the probability paper) to the logarithmic scale, i.e., $\log_{10} x = \log x / \log 10$.[2] If the empirical data exhibits a straight line on this log-normal probability paper, then a log-normal distribution should be a good candidate to represent this variable. In Figure 6.2 (a) and (b) we plot the step curve and dot diagram respectively

[2] We use log to mean the natural logarithm; i.e., $\log_e$ or ln.

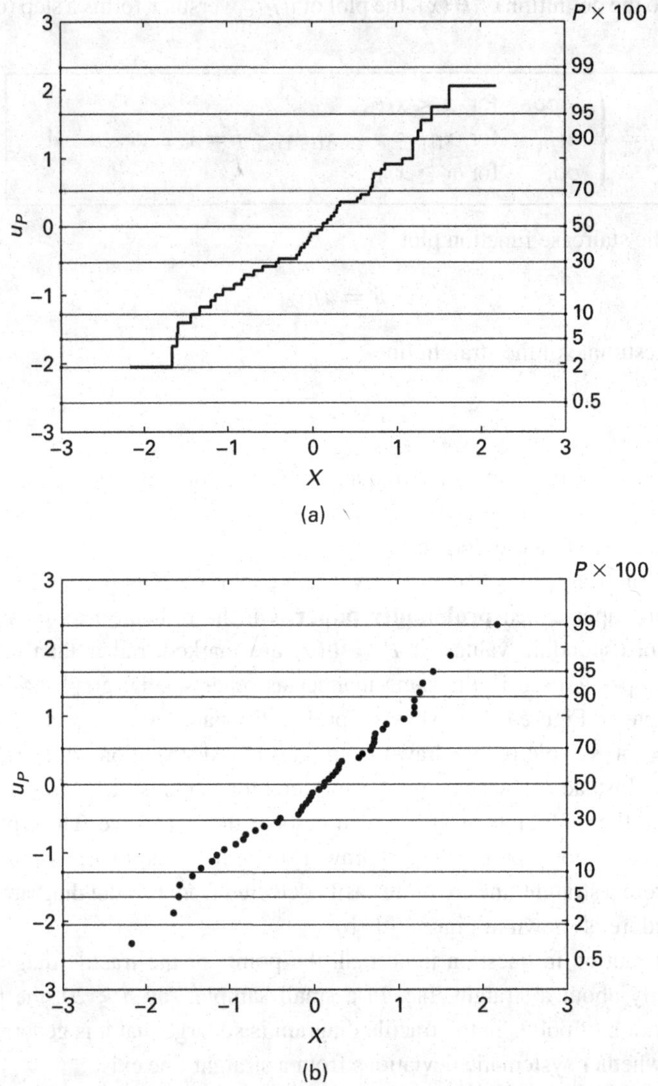

Figure 6.1 The fractile diagram of normal variates: (a) step curve; (b) dot diagram.

of a simulated set of $n = 50$ sample points x_i, where $x_i = e^{y_i}$ and y_i is drawn from $N(2, 4)$; i.e., $\mu_Y = 2$ and $\sigma_Y = 2$. From the results to be discussed in Section 7.4, we find that $\mu_X = e^{\mu_Y + (\sigma_Y^2/2)} = e^4$ and $\sigma_X^2 = e^{2\mu_Y+\sigma_Y^2}\left(e^{\sigma_Y^2} - 1\right) = e^8(e^4 - 1)$.

6.3.2 Log-survivor function curve

Suppose that a random variable X represents the life of some item (e.g., light-bulb) or the interval between failures of some machine. Given the distribution function $F_X(x)$ of the RV X, the probability that X survives time duration t,

$$S_X(t) \triangleq P[X > t] = 1 - F_X(t), \tag{6.27}$$

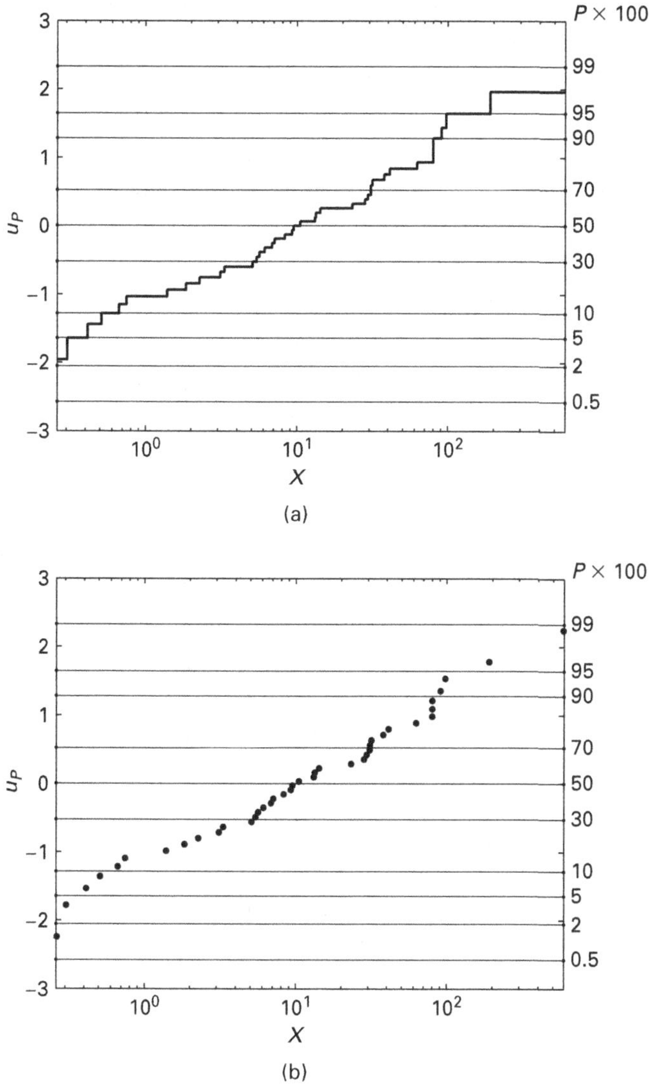

Figure 6.2 The fractile diagram of log-normal variates: (a) step curve; (b) dot diagram.

is often called the **survivor function**, or the **survival function** in reliability theory. It is equivalent to the complementary distribution function $F^c_X(t)$ defined earlier.

The natural logarithm of (6.27) is known as the **log-survivor function** or the **log-survival function** (Cox and Lewis [71]):

$$\boxed{\log S_X(t) = \log(1 - F_X(t)).} \tag{6.28}$$

The log-survivor function will show the details of the tail end of the distribution more effectively than the distribution itself.

If, for instance, $F_X(x)$ is an exponential distribution with mean $1/\alpha$, then its log-survivor function is a straight line: $S_X(t) = \log e^{-\alpha x} = -\alpha x$.

If $F_X(x)$ is a mixed exponential distribution (or hyperexponential distribution)

$$F_X(x) = \pi_1(1 - e^{-\alpha_1 x}) + \pi_2(1 - e^{-\alpha_2 x}), \quad \alpha_1 > \alpha_2, \quad \pi_1 + \pi_2 = 1, \tag{6.29}$$

then its log-survivor function has two asymptotic straight lines, since

$$\begin{aligned} \log S_X(t) &= \log(\pi_1 e^{-\alpha_1 t} + \pi_2 e^{-\alpha_2 t}) \\ &\approx \begin{cases} -\alpha_1 t + \log \pi_1, & \text{for small } t, \\ -\alpha_2 t + \log \pi_2, & \text{for large } t. \end{cases} \end{aligned} \tag{6.30}$$

The **sample log-survivor function** or **empirical log-survivor function** is similarly defined as

$$\boxed{\log[1 - H(t)],} \tag{6.31}$$

where $H(t)$ represents the cumulative relative frequency (ungrouped data) or the cumulative histogram (grouped data). In the ungrouped case we find from (6.15) that

$$\log\left(1 - \frac{i}{n}\right), \quad 1 \le i \le n, \tag{6.32}$$

should be plotted against $x_{(i)}$, where the subscript (i) represents the rank as in (6.15). In order to avoid difficulties at $i = n$, we may sometimes modify (6.32) into

$$\log\left(1 - \frac{i}{n+1}\right), \quad 1 \le i \le n. \tag{6.33}$$

As an example, Figure 6.3 plots the log-survivor function using a sample of size 1000 drawn from the above **hyperexponential distribution** with parameters

$$\pi_1 = 0.0526, \quad \pi_2 = 1 - \pi_1, \quad \alpha_1 = 0.1, \quad \text{and} \quad \alpha_2 = 2.0. \tag{6.34}$$

Out of the 1000 samples taken, 18 sample points that exceed $x = 10$ fall outside the scale of the figure; hence they are not shown. The asymptotes of (6.30) can be easily recognized from this log-survivor function.

Characteristically, the log-survivor function of the mixed exponential distribution (6.29) is convex with a linear tail. Observations of (or departures from) such characteristic shapes are used to postulate a functional form for a distribution. See Gaver *et al.* [115] and Lewis and Shedler [225].

6.3.2.1 Testing the Pareto distribution hypothesis

As discussed in Section 4.2.6, a simple way to examine whether the tail of an empirical distribution fits the **power law** of the **Pareto distribution** is to plot the log-survivor function on paper with the log-log scale, whereas the log-survivor function curve discussed above is plotted in the log-linear scale.

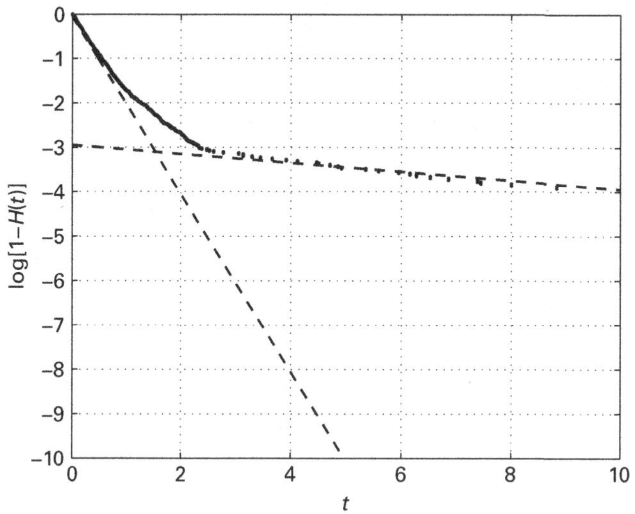

Figure 6.3 The log-survivor function of a mixed-exponential (or hyperexponential) distribution with $\pi_1 = 0.0526$, $\pi_2 = 1 - \pi_1$, $\alpha_1 = 0.1$, and $\alpha_2 = 2.0$.

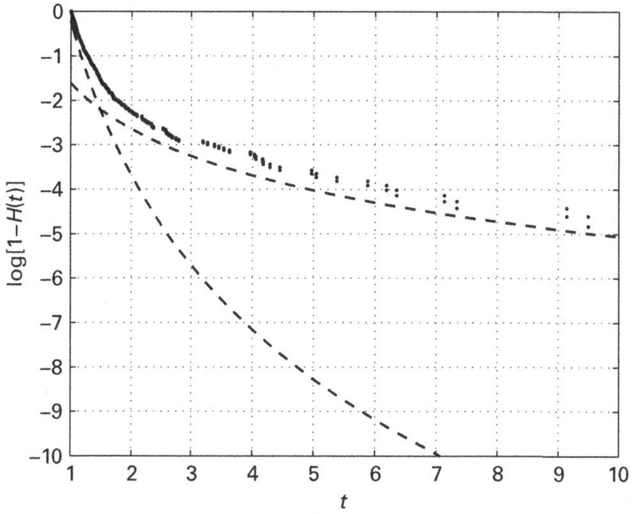

Figure 6.4 The log-survivor function of a mixed Pareto distribution, $\beta_1 = \beta_2 = 1$, $\pi_2 = 1 - \pi_1$, $\alpha_1 = 1.5$, $\alpha_2 = 5$, and $\pi_1 = 0.2$.

Analogous to the mixed exponential distribution, a **mixed Pareto distribution** is considered:

$$S_X(t) = \pi_1 \frac{\beta_1^{\alpha_1}}{t^{\alpha_1}} + (-\pi_1)\frac{\beta_2^{\alpha_2}}{t^{\alpha_2}}, \quad 0 < \max\{\beta_1, \beta_2\} \le t. \tag{6.35}$$

As an example, Figure 6.4 plots the log-survivor function of 500 samples drawn from the mixed Pareto distribution with $\beta_1 = \beta_2 = 1$, $\alpha_1 = 1.5$, $\alpha_2 = 5$, and $\pi_1 = 0.2$.

6.3.3 Hazard function and mean residual life curves

Other graphical plots that can be derived from the histogram or distribution function are the **hazard function curve** and the **mean residual life curve.** These notions are also related to reliability theory and **renewal process** theory, which will be briefly discussed in Section 14.3.

Suppose that X represents the life of some item, with the distribution function $F_X(x)$. The function defined by

$$\boxed{h_X(t) = \frac{f_X(t)}{S_X(t)} = \frac{f_X(t)}{1 - F_X(t)}} \tag{6.36}$$

is called the **hazard function** or the **failure rate**, because $h_X(t)\,dt$ represents the probability that the life will end in the interval $(t, t + dt]$, given that X has survived up to age t; i.e., $X \geq t$. If X represents the service time of a customer, as in queueing theory, $h_X(t)$ is called the **completion rate function.**

The hazard functions of the exponential, **Weibull**, **Pareto**, and **log-normal** distributions are given as follows:

$$h_X(t) = \begin{cases} \lambda, & t \geq 0, \quad \text{for exponential,} \\ \frac{\alpha}{\beta}\left(\frac{t}{\beta}\right)^{\alpha-1}, & t \geq 0, \quad \text{for Weibull,} \\ \frac{\alpha}{t}, & t \geq \beta, \quad \text{for Pareto,} \\ \dfrac{t^{-1}\exp\left[-\frac{(\log t - \mu_Y)^2}{2\sigma_Y^2}\right]}{\int_{\log t}^{\infty}\exp\left[-\frac{(u-\mu_Y)^2}{2\sigma_Y^2}\right]du}, & t > 0, \quad \text{for log-normal,} \end{cases} \tag{6.37}$$

where in the log-normal distribution the parameters μ_Y and σ_Y are given as

$$\mu_Y = \log \mu_X - \frac{1}{2}\log\left(1 + \frac{\sigma_X^2}{\mu_X^2}\right)$$

and

$$\sigma_Y^2 = \log\left(1 + \frac{\sigma_X^2}{\mu_X^2}\right).$$

From (6.36) we can express the survivor function in terms of the hazard function:

$$\boxed{S_X(x) = e^{-\int_0^x h_X(t)\,dt}, \quad x \geq 0,} \tag{6.38}$$

from which we have

$$h_X(t) = -\frac{d \log S_X(t)}{dt}, \quad t \geq 0. \tag{6.39}$$

The last equation, of course, could have been readily derived from (6.36).

Given that the service time variable X is greater than t, we call the difference

$$R = X - t \tag{6.40}$$

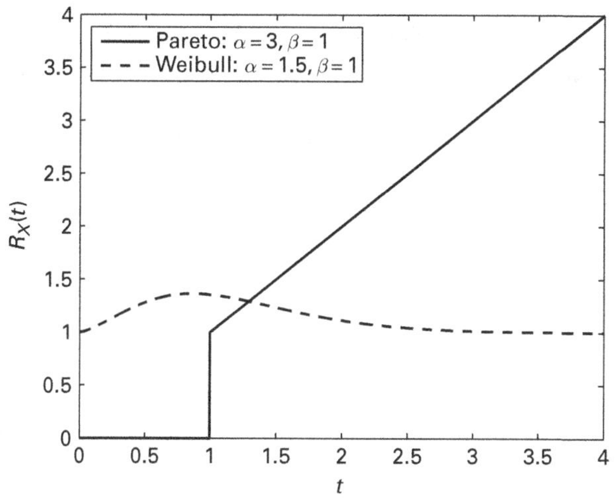

Figure 6.5 The mean residual life curves of a Pareto distribution with $\alpha = 3.0$ and $\beta = 1.0$, and a Weibull distribution with $\alpha = 1.5$ and $\beta = 1.0$.

the **residual life** conditioned on $X > t$. Then the **mean residual life function** is given by

$$\boxed{R_X(t) = E[R|X > t] = \frac{\int_t^\infty S_X(u)\,du}{S_X(t)}.} \tag{6.41}$$

At $t = 0$, the mean residual life becomes

$$R_X(0) = \int_0^\infty S_X(u)\,du = E[X], \tag{6.42}$$

as expected. Figure 6.5 shows mean residual life curves of a Pareto distribution and a Weibull distribution.

6.3.4 Dot diagram and correlation coefficient

In analyzing a simulation model or an operational system, we usually measure a number of variables, and we wish to find possible statistical associations among them. Thus, the search for **correlations** between two or more quantities is one of the most important functions in the *output analysis* of the measurement and evaluation process. A typical method of graphically examining correlations between two variables X and Y based on n observations of the pair

$$(x_i, y_i),\ 1 \le i \le n, \tag{6.43}$$

is to plot the points (x_i, y_i) one by one as coordinates. Such a diagram is called a **dot** or **scatter diagram**. The density of dots in a given region is proportional to the relative frequency of the pairs (X, Y) in the region.

Example 6.1: Scatter diagram of Internet distances [134, 364]. The approximate geographic distance between a pair of Internet hosts can be inferred by sending probe packets between the two hosts and measuring the round-trip delays experienced by the probes. The relationship between geographic distance g and round-trip delay d from a given Internet host to Internet hosts can be characterized by a scatter diagram consisting of points (g, d). Owing to the inherent randomness in round-trip delays over the Internet, delay measurements taken between a given pair of hosts separated by a fixed geographic distance g at different times yield different delays d.

The scatter diagram in Figure 6.6 was obtained by sending probe packets from a host at Stanford University to 79 other hosts on the Internet across the USA [364]. The line labeled *baseline* provides a lower bound on the d as a function of g based on the observation that the packet propagation speed over the Internet is at most the speed of light through an optical fiber. If the refractive index of the fiber is denoted by η, the propagation speed of the optical signal is $v = c/\eta$, where c is the speed of light in vacuo. Typically, the value of η is slightly less than 1.5, so we make the approximation $v \approx 2c/3$. If the round-trip delay between a pair of hosts is measured to be d, the corresponding (one-way) geographical distance is upper bounded by $\hat{g} = vd/2 \approx cd/3$. When the unit of time is milliseconds and the unit of geographical distance is kilometers, $c \approx 300$ km/ms, so d and $\hat{g}$ can be related approximately by

$$d \approx \frac{1}{100}\hat{g}, \tag{6.44}$$

which is the equation of *baseline* in Figure 6.6.

Since packets generally traverse multiple hops between two hosts and experience queueing and processing delays at each hop, the measured round-trip delay will

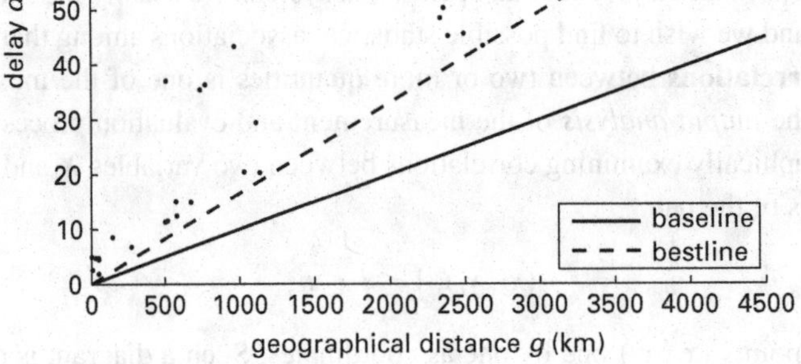

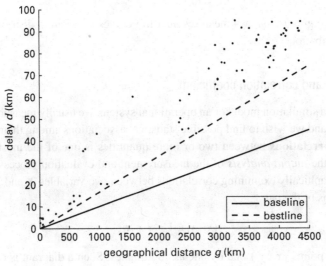

Figure 6.6 Scatter diagram of delay measurements from Internet host at Stanford University to 79 other hosts across the USA [364].

typically be much larger than the delay predicted by the equation of the baseline in (6.44). Gueye *et al.* [134] propose a tighter linear bound determined by solving a linear programming problem that minimizes the slope and y-intercept of the line subject to the constraints imposed by the set of scatter points. This deterministic bound corresponds to the line labeled *bestline* in Figure 6.6. An alternative approach that retains more of the statistical information captured by the scatter points is discussed in [364].

The most frequently used measure of statistical association between a pair of variables is the **correlation coefficient**. For a given pair of random variables X and Y, the covariance of X and Y, written $\mathrm{Cov}[X, Y]$ or σ_{XY}, is defined as

$$\sigma_{XY} \triangleq \mathrm{Cov}[X, Y] = E[(X - \mu_X)(Y - \mu_Y)] = E[XY] - \mu_X \mu_Y. \tag{6.45}$$

We say X and Y are **uncorrelated** if $\sigma_{XY} = 0$.

If X and Y are statistically independent, then they are uncorrelated, but **the converse is not true**: the condition $\sigma_{XY} = 0$ does not imply that X and Y are independent (see Problem 6.15). The correlation coefficient ρ_{XY} between X and Y is defined as

$$\boxed{\rho_{XY} = \frac{\sigma_{XY}}{\sigma_X \sigma_Y}.} \tag{6.46}$$

The correlation coefficient always satisfies the condition

$$-1 \leq \rho_{XY} \leq 1. \tag{6.47}$$

We say that X and Y are **properly linearly dependent** if there exist nonzero constants a and b such that $aX - bY$ is a constant c; that is,

$$P[aX - bY = c] = 1. \tag{6.48}$$

Therefore,

$$\mathrm{Var}[aX - bY - c] = 0, \tag{6.49}$$

from which we have

$$\rho_{XY} = +1 \quad \text{or} \quad -1 \tag{6.50}$$

depending on whether ab is positive or negative. Conversely, if $\rho = \pm 1$, then it implies (Problem 6.17) that

$$P\left[\mp\frac{(X - \mu_X)}{\sigma_X} + \frac{Y - \mu_Y}{\sigma_Y} = 0\right] = 1. \tag{6.51}$$

The **sample covariance** of the two variables based on observations $\{(x_i, y_i); 1 \le i \le n\}$ is defined as

$$\boxed{\begin{aligned} s_{xy} &= \frac{1}{n-1}\sum_{i=1}^{n}(x_i - \overline{x})(y_i - \overline{y}) \\ &= \frac{1}{n-1}\sum_{i=1}^{n} x_i y_i - \frac{n\overline{x}\,\overline{y}}{n-1}, \end{aligned}} \tag{6.52}$$

where $\overline{x}$ and $\overline{y}$ are the sample means of $\{x_i\}$ and $\{y_i\}$ respectively. The **sample correlation coefficient** is defined accordingly:

$$\boxed{r_{xy} = \frac{s_{xy}}{s_x s_y},} \tag{6.53}$$

where s_x^2 and s_y^2 are the *sample variances* of $\{x_i\}$ and $\{y_i\}$ respectively.

6.4 Summary of Chapter 6

Sample mean:	$\overline{x} = \frac{1}{n}\sum_{i=1}^{n} x_i$	(6.1)
Variance of the sample mean:	$\mathrm{Var}[\overline{X}] = \frac{\sigma_X^2}{n}$	(6.8)
Sample variance:	$s_x^2 \triangleq \frac{1}{n-1}\sum_{i=1}^{n}(x_i - \overline{x})^2$	(6.9)
Unbiasedness of the sample variance:	$E[S_X^2] = \sigma_X^2$	(6.12)
Relative frequency:	$f_j = n_j/n, \quad j = 1, 2, \ldots, k$	(6.13)
Histogram:	$h(x) = \frac{f_j}{\Delta_j}, \quad \text{for } x \in (c_{j-1}, c_j]$	(6.14)
Fractile diagram:	$(x, u_{H_X(x)})$	(6.24)
Log-survivor function:	$\log S_X(t) = \log(1 - F_X(t))$	(6.28)
Sample log-survivor:	$\log(1 - H_X(t))$	(6.31)
Hazard function:	$h_X(t) = \frac{f_X(t)}{S_X(t)} = \frac{f_X(t)}{1-F_X(t)}$	(6.36)
Mean residual life function:	$R_X(t) = E[R \mid X > t] = \frac{\int_t^\infty S_X(u)\,du}{S_X(t)}$	(6.41)
Dot diagram:	$(x_i, y_i); 1 \le i \le n$	(6.43)
Covariance:	$\sigma_{XY} = E[XY] - \mu_X \mu_Y$	(6.45)
Uncorrelated if:	$\sigma_{XY} = 0$	
Correlation coefficient:	$\rho_{XY} = \frac{\sigma_{XY}}{\sigma_X \sigma_Y}$	(6.46)
Sample correlation coefficient:	$r_{xy} = \frac{s_{xy}}{s_x s_y}$	(6.53)

6.5 Discussion and further reading

Most textbooks on probability theory and mathematical statistics do not seem to deal with graphical presentations of real data. We consider that this is an unfortunate state of affairs. Various types of graphical presentations of collected data should be explored before we can narrow down proper directions of mathematical modeling or analysis of the system in question.

Hald [139] seems to be one of the few textbooks that discusses the fractile diagram. A monograph by Cox and Lewis [71] presents several empirical log-survivor functions as well as scatter diagrams. Much of the material given in this chapter is taken from the first author's earlier book [197] on system modeling and analysis, in which additional examples of graphical plots based on computer performance data are found.

The **exploratory data analysis** (EDA) approach developed by Tukey [328] and others indeed exploits various graphical techniques as well as quantitative techniques in analyzing data to formulate plausible hypotheses. Two graphical techniques introduced by Tukey are the box plot and the stem-and-leaf diagram. A box plot, also known as a box-and-whiskers plot, graphically depicts the sample minimum, lower quartile, medium, upper quartile, and sample maximum, and may also indicate outliers of a data set. A stem-and-leaf plot, also called a stemplot, tabulates the data in ascending order in two columns. The first consists of the *stems* of the data set in ascending order, while the second consists of the *leaves* corresponding to each stem. Typically, a leaf contains the last digit of the associated sample value while the stem contains the remaining digits. Exploratory data analysis complements the conventional statistical theory, which places more emphasis on formal testing of a hypothesis and estimation of model parameters, two subjects to be studied in Chapter 18.

6.6 Problems

Section 6.1: Sample mean and sample variance

6.1* Derivation of (6.11). Derive (6.11)

6.2 Recursive formula for sample mean and variance. Let $\overline{x}_i$ and s_i^2 be the sample mean and sample variance based on data $(x_1, x_2, \ldots, x_i)$, where $i \leq n$. Then the last value of the sequence – that is, $\overline{x}_n$ and s_n^2 – are the desired quantities:

$$\overline{x} = \overline{x}_n \text{ and } s^2 = s_n^2.$$

(a) Derive the following recursive formula for the sample mean:

$$\overline{x}_i = \overline{x}_{i-1} + \frac{x_i - \overline{x}_{i-1}}{i}, \quad i \geq 1$$

with the initial value

$$\overline{x}_0 = 0.$$

(b) Similarly, show the recursive formula for the sample variance:

$$s_i^2 = \left(\frac{i-2}{i-1}\right) s_{i-1}^2 + \frac{(x_i - \overline{x}_{i-1})^2}{i}, \quad i > 1,$$

with the initial values

$$s_0^2 = s_1^2 = 0.$$

Section 6.2: Relative frequency and histograms

6.3 Expectation and variance of the histogram. Consider the histogram value $h(x)$ in the jth class interval $x \in (c_{j-1}, c_j]$ given by (6.14), which we denote as $h_j(\boldsymbol{x})$, where $\boldsymbol{x} = (x_1, x_2, \ldots, x_n)$ is the n random samples. Then $h_j(\boldsymbol{X})$ is a random variable, where the argument $\boldsymbol{x}$ is replaced by the corresponding RV $\boldsymbol{X} = (X_1, X_2, \ldots, X_n)$.

(a) Show that the expectation of the RV $h_j(\boldsymbol{X})$ is given by

$$E[h_j(\boldsymbol{X})] = \frac{F_X(c_j) - F_X(c_{j-1})}{\Delta_j} \approx f_X(c_j).$$

(b) Show that the variance of $h_j(\boldsymbol{X})$ is

$$\text{Var}[h_j(\boldsymbol{X})] = \frac{[F_X(c_j) - F_X(c_{j-1})][1 - F_X(c_j) + F_X(c_{j-1})]}{n\Delta_j^2} \approx \frac{f_X(c_i)}{n\Delta_j}.$$

6.4 Expectation and variance of the cumulative histogram. Find expressions for the expectation and variance of H_j (the cumulative histogram in the jth interval) in terms of the underlying distribution function $F_X(x)$. Explain why the shape of the cumulative histogram is rather insensitive to the choice of class lengths $\{\Delta_j\}$.

Section 6.3: Graphical presentations

6.5 Log-survivor function curve of Erlang distributions. Plot the sample log-survivor function by generating 1000 values of a random variable X that has the two-stage Erlang distribution of mean one. Do the same for the four-stage Erlang distribution.
Hint: To generate samples drawn from the k-stage Erlang distribution, apply the transform method of Example 5.7 in Section 5.4.2 to generate k samples drawn from an exponential distribution.

6.6* Log-survivor functions and hazard functions of a constant and uniform RVs. Find the expression for the log-survivor function and the completion rate function, when the service time is

(a) constant a;
(b) uniformly distributed in $[a, b]$.

6.7 Hazard function and distribution functions. Show that the distribution function $F_X(x)$ is given in terms of the corresponding hazard function $h_X(x)$ as follows:

$$\boxed{F_X(x) = 1 - e^{-\int_0^x h_X(t)\,dt}, \quad x \geq 0,} \tag{6.54}$$

and hence

$$f_X(x) = h_X(x)e^{-\int_0^x h_X(t)\,dt}. \tag{6.55}$$

6.8 Hazard function of a k-stage hyperexponential distribution. Consider the k-stage hyperexponential (or mixed exponential) distribution defined in (4.166) of Chapter 4. Show that its hazard function $h_X(t)$ is monotone decreasing. Find $\lim_{t\to\infty} h_X(t)$.

6.9 Hazard function of the Pareto distribution. Find the hazard function of the Pareto distribution.

6.10 Hazard function of the Weibull distribution. The Weibull distribution is often used in modeling reliability problems.

(a) Find the hazard function $h_X(t)$ of the standard Weibull distribution. What functional form does $h_X(t)$ take for $\alpha = 1$ and $\alpha = 2$?
(b) Plot the hazard function of the standard Weibull distribution for $\alpha = 0.1, 0.5, 1, 2,$ and 5, and confirm that they agree with the curves of Figure 4.5.

6.11* Mean residual life function and the hazard function. Show that the mean residual life function $R_X(t)$ is a monotone-decreasing function if and only if the hazard function $h_X(t)$ is monotone increasing.
Hint: Consider the **conditional survivor function** of $R = X - t$, given that X is greater than t, defined by

$$S_X(r|t) \triangleq P[R > r|X > t], \tag{6.56}$$

and find its relations with the hazard function $h_X(t)$ and the mean residual life function.

6.12* Conditional survivor and mean residual life functions for standard Weibull distribution.

(a) Find the conditional survivor function $S_X(r|t)$ (see Problem 6.11) of the standard Weibull distribution.
(b) Find the mean residual life function $R_X(t)$ for the standard Weibull distribution.

6.13 Mean residual life functions.

(a) For the hyperexponential distribution (6.29), show that

$$\lim_{t\to\infty} R_X(t) = \frac{1}{\alpha_2}.$$

(b) Consider the **standard gamma distribution** defined in (4.32):

$$f_X(x;\beta) \triangleq \frac{x^{\beta-1}e^{-x}}{\Gamma(\beta)}, \quad x \geq 0; \beta > 0.$$

Show that $R_X(t)$ is a monotone-increasing (decreasing) function if $\beta < 1$ ($\beta > 1$). Find $R_X(0)$ and $\lim_{t\to\infty} R_X(t)$.

6.14 Mean residual life functions – continued. Find an expression for the mean residual life function $R_X(t)$ for each of the following distributions:

(a) Pareto distribution with parameters $\alpha > 1$ and $\beta > 0$.
(b) Two-parameter Weibull distribution with parameters α and β.

6.15* Covariance between two RVs. Suppose that RVs X and Y are functionally related according to

$$Y = \cos X.$$

Let the probability density function of X be given by

$$f_X(x) = \begin{cases} \frac{1}{2\pi}, & -\pi < x < \pi, \\ 0, & \text{elsewhere.} \end{cases}$$

Find $\mathrm{Cov}[X, Y]$.

6.16 Correlation coefficient. Given two RVs X and Y, define a new RV

$$Z = \left[t \left(\frac{X - \mu_X}{\sigma_X} \right) + \frac{Y - \mu_Y}{\sigma_Y} \right]^2,$$

where t is a real constant.

(a) Compute $E[Z]$.
(b) Show that $-1 \leq \rho_{XY} \leq 1$, where ρ_{XY} is the correlation coefficient between X and Y.

6.17 Correlation coefficient – continued. Show that if $\rho = \pm 1$, then (6.51) holds.

6.18* Sample covariance. Show that the sample covariance s_{XY} defined by (6.52) is an unbiased estimate of the covariance σ_{XY}.

6.19 Recursive formula for sample covariance. Generalize the recursive computation formula of Problem 6.2 to the sample covariance.

7 Distributions derived from the normal distribution

In Sections 4.2.4 and 4.3.1 we defined the normal (or Gaussian) distributions for both single and multiple variables and discussed their properties. The normal distribution plays a central role in the mathematical theory of statistics for at least two reasons. First, the normal distribution often describes a variety of physical quantities observed in the real world. In a communication system, for example, a received waveform is often a superposition of a desired signal waveform and (unwanted) noise process, and the amplitude of the noise is often normally distributed, because the source of such noise is usually what is known as *thermal noise* at the receiver front. The normality of thermal noise is a good example of manifestation in the real world of the CLT, which says that the sum of a large number of independent RVs, properly scaled, tends to be normally distributed. In Chapter 3 we saw that the binomial distribution and the Poisson distribution also tend to a normal distribution in the limit. We also discussed the CLT and asymptotic normality.

The second reason for the frequent use of the normal distribution is its mathematical tractability. For instance, sums of independent normal RVs are themselves normally distributed. Such reproductivity of the distribution is enjoyed only by a limited class of distributions (that is, binomial, gamma, Poisson). Many important results in the theory of statistics are founded on the assumption of a normal distribution.

7.1 Chi-squared distribution

An important distribution that is derived from the normal distribution is what is known as the chi-squared distribution, often denoted χ^2-distribution. Let U_i, $1 \le i \le n$, be n i.i.d. RVs with the *standard normal distribution* defined by (4.40) in Section 4.2.4. Denote the sum of n independent standard normal variables squared by χ_n^2:

$$\chi_n^2 = \sum_{i=1}^{n} U_i^2. \tag{7.1}$$

The distribution of the RV χ_n^2 is solely determined by n, which is called the *degree of freedom (d.f.)* of this distribution. The notation χ^2, introduced by K. Pearson (1900) [266], may be somewhat confusing when the reader encounters this notation for the first time, since χ^2 instead of χ represents the RV defined above.

The derivation of the PDF of χ_n^2 is left to the reader as an exercise (Problem 7.2). It is given by

$$f_{\chi_n^2}(x) = \frac{x^{(n/2)-1}e^{-x/2}}{2^{n/2}\Gamma\left(\frac{n}{2}\right)}dx, \quad 0 \le x < \infty, \tag{7.2}$$

where $\Gamma(z)$ denotes the *gamma function* defined by

$$\Gamma(z) = \int_0^\infty t^{z-1}e^{-t}dt. \tag{7.3}$$

For the argument $z = n/2$, where n is a positive integer, we find

$$\Gamma(1) = 1, \quad \Gamma\left(\frac{1}{2}\right) = \sqrt{\pi}; \tag{7.4}$$

and for $n > 2$,

$$\Gamma\left(\frac{n}{2}\right) = \begin{cases} \left(\frac{n}{2}-1\right)!, & \text{for } n \text{ even}, \\ \left(\frac{n}{2}-1\right)\left(\frac{n}{2}-2\right)\cdots\frac{3}{2}\times\frac{1}{2}\sqrt{\pi}, & \text{for } n \text{ odd}. \end{cases} \tag{7.5}$$

Figure 7.1 gives the PDF curves for several values of n. For $n = 1$, we have

$$f_{\chi_1^2}(x) = \frac{x^{-1/2}e^{-x/2}}{\sqrt{2\pi}}, \quad x > 0. \tag{7.6}$$

The PDF curve is monotonically decreasing, the abscissa axis forming the asymptote, as shown in Figure 7.1. For $n = 2$, we have

$$f_{\chi_2^2}(x) = \frac{e^{-x/2}}{2}, \quad x \ge 0, \tag{7.7}$$

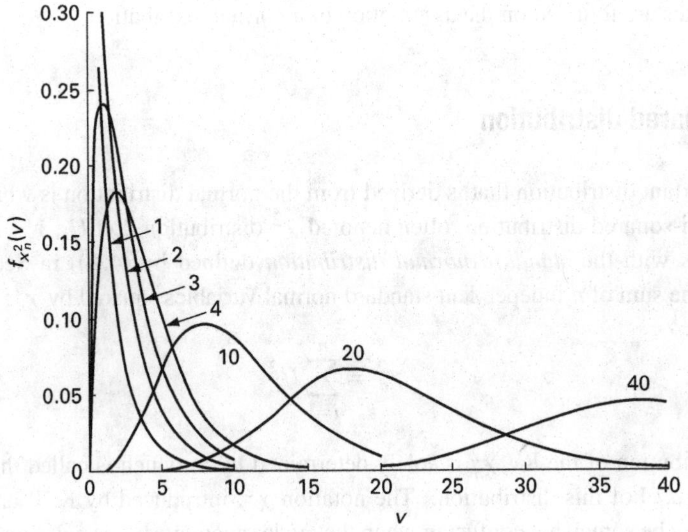

Figure 7.1 The χ_n^2 distribution with degree of freedom n.

which is the exponential distribution with mean $E[\chi_2^2] = 2$. For $n = 3$, we have

$$f_{\chi_3^2}(x) = \frac{x^{1/2}e^{-x/2}}{\sqrt{2\pi}}, \quad x \geq 0. \tag{7.8}$$

The PDF curve originates at the point (0, 0) and increases until $x = 1$, beyond which it decreases monotonically. For values $n > 3$, the distribution curve takes a course similar to that for $n = 3$. The expectation of χ_n^2 is equal to the number of d.f.:

$$E[\chi_n^2] = n, \tag{7.9}$$

which is immediate from the definition of the χ^2 variable. The *mode*, the abscissa of the maximum of the curve, is equal to $n - 2$ and the variance

$$\text{Var}[\chi_n^2] = 2n. \tag{7.10}$$

The χ^2 distribution is related to a number of well-studied distribution functions: the gamma, Erlang, Poisson, and Rayleigh distributions. In order to demonstrate this, we make the change of variable

$$Y_n = \frac{\chi_n^2}{2} \tag{7.11}$$

in the PDF of the χ_n^2 distribution. We then obtain

$$f_{Y_n}(y) = \frac{y^{(n/2)-1}e^{-y}}{\Gamma\left(\frac{n}{2}\right)}, \tag{7.12}$$

which is a special case of the *gamma distribution*

$$f(y) = \begin{cases} \frac{e^{-\lambda y}}{\Gamma(\beta)}\lambda(\lambda y)^{\beta-1}, & y \geq 0, \\ 0, & y < 0, \end{cases} \tag{7.13}$$

in which the parameters λ and β are

$$\lambda = 1 \quad \text{and} \quad \beta = \frac{n}{2}. \tag{7.14}$$

If we consider the case in which n is an even integer, i.e.,

$$n = 2k, \tag{7.15}$$

then (7.12) is reduced to

$$f_{Y_{2k}}(y) = \frac{y^{k-1}e^{-y}}{(k-1)!}, \tag{7.16}$$

which is the *k-stage Erlang distribution with mean k*. The above result is easily understood if we recognize that the sum of the squares of two independent standard normal variables is distributed exponentially with mean two (see Problem 5.13). Hence,

Y of (7.11) is equivalent to the sum of k independent exponential variables, each of which has mean unity.

From the set of results shown above, we find the following relation between the chi-squared distribution and the Poisson distribution:

$$P[\chi^2_{2k} > 2\lambda] = \int_\lambda^\infty \frac{y^{k-1}e^{-y}}{(k-1)!}dy$$
$$= \int_\lambda^\infty P(k-1; y)dy = Q(k-1; \lambda), \quad (7.17)$$

where the function $\{P(k; \lambda);\ k = 0, 1, 2, \ldots\}$ is the Poisson distribution with mean λ and $\{Q(k; \lambda);\ k = 0, 1, 2, \ldots\}$ is its cumulative distribution, as defined in (3.77) and (3.86) respectively.

Example 7.1: Independent observations from $N(\mu, \sigma^2)$. Let $X_1,\ X_2,\ \ldots,\ X_n$ be independent observations from a population distributed according to $N(\mu, \sigma^2)$.
Case 1: Suppose that the population mean μ is known, which is very seldom the case. Then an estimate of σ^2 should be given, not by the sample variance, but by

$$\tilde{s}^2 = \frac{1}{n}\sum_{i=1}^n (X_i - \mu)^2. \quad (7.18)$$

The n variables $X_1 - \mu,\ X_2 - \mu,\ \ldots,\ X_n - \mu$ are independent and identically distributed (i.i.d.) with the common distribution $N(0, \sigma^2)$. By normalizing these variables by σ, we find

$$\tilde{s}^2 = \frac{\sigma^2}{n}\sum_{i=1}^n U_i^2, \quad (7.19)$$

where

$$U_i = \frac{X_i - \mu}{\sigma},\quad 1 \le i \le n, \quad (7.20)$$

are i.i.d. RVs with the *standard normal distribution* $N(0, 1)$. Thus, $\tilde{s}^2$ may be written as

$$\tilde{s}^2 = \frac{\sigma^2}{n}\chi_n^2. \quad (7.21)$$

Case 2: Suppose the population mean μ is unknown, as in most cases. Then an estimate of σ^2 should be given by the sample variance s^2 we defined earlier, i.e.,

$$s^2 = \frac{1}{n-1}\sum_{i=1}^n (X_i - \bar{X})^2. \quad (7.22)$$

If we write s^2 in a manner similar to $\tilde{s}^2$, we have

$$s^2 = \frac{\sigma^2}{n-1}\chi^2, \quad (7.23)$$

where

$$\chi^2 = \sum_{i=1}^{n} \left(\frac{X_i - \bar{X}}{\sigma} \right)^2. \tag{7.24}$$

However, the n variables $\{(X_i - \bar{X})/\sigma;\ 1 \le i \le n\}$ are *linearly dependent*, since they must satisfy the relation

$$\sum_{i=1}^{n} \frac{X_i - \bar{X}}{\sigma} = 0. \tag{7.25}$$

Thus, the above χ^2 is not the χ_n^2 RV with n d.f. Instead, it can be transformed to a sum of $n - 1$ independent standard normal variables (Problem 7.1). In other words, χ^2 turns out to be χ_{n-1}^2. Hence,

$$s^2 = \frac{\sigma^2}{n-1} \chi_{n-1}^2. \tag{7.26}$$

□

Let $X_1, X_2, \ldots, X_n$ be independent RVs, with X_k being distributed according to $N(\mu_k, 1)$; i.e., $X_k = U_k + \mu_k$, where the U_k are independent standard normal variables. Let us define a new RV $\chi_n^2(\mu^2)$ by

$$\chi_n^2(\mu^2) \triangleq \sum_{k=1}^{n} X_k^2 = \sum_{k=1}^{n} (U_k + \mu_k)^2, \tag{7.27}$$

which is referred to as the *noncentral chi-square variable* with n d.f. and noncentrality parameter μ^2, where

$$\mu^2 = \sum_{k=1}^{n} \mu_k^2. \tag{7.28}$$

The characteristic function (to be discussed in Section 8.2) of $\chi_n^2(\mu^2)$ is given (Problem 8.20) by

$$\phi_{\chi_n^2(\mu^2)}(u) = \frac{1}{(1 - 2iu)^{n/2}} \exp\left(\frac{iu\mu^2}{1 - 2iu} \right), \tag{7.29}$$

where $i = \sqrt{-1}$.

7.2 Student's *t*-distribution

From the reproductive property of the normal distribution, we can show that the sample mean $\bar{X}$ of n independent observations $\{X_1, X_2, \ldots, X_n\}$ from the population $N(\mu, \sigma^2)$ is normally distributed according to $N(\mu, \sigma^2/n)$. Thus, the variable U defined by

$$U = \frac{(\bar{X} - \mu)\sqrt{n}}{\sigma} \tag{7.30}$$

is a standard normal variable.

If σ is known and μ is to be estimated from the sample mean, then we can use tables of the standard normal distribution to test whether U is significantly different from zero. In practice, however, we may not know the population variance σ^2 either. Hence, we must replace σ^2 by its estimate s^2; i.e., the sample variance defined by (7.22). Thus, the statistic we should use, instead of U, for a significance test is the following variable defined by the sample mean and sample variance:

$$t_{n-1} = \frac{(\bar{X} - \mu)\sqrt{n}}{s}. \tag{7.31}$$

By using the relation between s^2 and χ^2_{n-1} given by (7.26), we can write

$$t_{n-1} = \frac{(\bar{X} - \mu)\sqrt{n}/\sigma}{s/\sigma} = \frac{U}{\sqrt{\chi^2_{n-1}/(n-1)}}. \tag{7.32}$$

Thus, the distribution of t_{n-1} depends only on $n - 1$ (the d.f. for s^2), not on the population mean μ nor on the variance σ^2. The distribution of the variable t_{n-1} is called the *Student's t-distribution* (or simply the t-distribution) with $(n - 1)$ d.f. "Student" is the pseudonym of Gosset[1] in his 1908 paper [314].

We can obtain (Problem 7.6) the PDF of the t-distribution for k degrees of freedom as

$$f_{t_k}(t) = \frac{\Gamma\left(\frac{k+1}{2}\right)}{\Gamma\left(\frac{k}{2}\right)\sqrt{\pi k}}\left(1 + \frac{t^2}{k}\right)^{-(k+1)/2}, \quad -\infty < t < \infty. \tag{7.33}$$

For $k = 1$, the distribution reduces to

$$f_{t_1}(t) = \frac{1}{\pi(1+t^2)}, \tag{7.34}$$

which is called Cauchy's distribution. For $k = 2$, we have

$$f_{t_2}(t) = (2 + t^2)^{-3/2}, \tag{7.35}$$

which has zero mean but infinite variance. As one may expect, the t-distribution is more dispersed than the normal distribution is, since the use of s rather than σ introduces additional uncertainty. Moreover, while there is one standard normal distribution, there is a whole family of t-distributions. With a small sample size, this distribution is considerably more spread out than the normal distribution; but as the sample size increases, the t-distribution approaches the normal distribution. Figure 7.2 shows the distribution curves for various values of k.

[1] William S. Gosset (1876–1937) was a statistician of the Guinness brewing company.

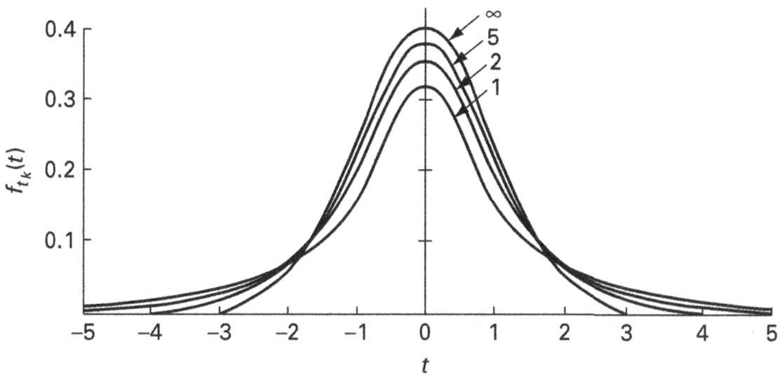

Figure 7.2 Student's t-distribution with k degrees of freedom ($k = 1, 2, 5, \infty$).

Since the t-distribution is symmetric around $t = 0$, all odd moments of the distribution (7.33) that exist are zero. As for the even moments that exist, we have from (7.32)

$$E[t_k^{2r}] = E[(\chi_1^2)^r]E\left[\left(\frac{\chi_k^2}{k}\right)^{-r}\right], \tag{7.36}$$

since U^2 is equivalent to the χ_1^2 variable. Then using the result of Problem 7.3 (b), we find

$$E[t_k^{2r}] = \frac{k^r\Gamma\left(\frac{1}{2}+r\right)\Gamma\left(\frac{k}{2}-r\right)}{\Gamma\left(\frac{1}{2}\right)\Gamma\left(\frac{k}{2}\right)}, \tag{7.37}$$

which shows that the $2r$th moment exists if and only if $-1 < 2r < k$. The *mean* and *variance* of the t_k-distribution can, therefore, be defined for $k > 1$ and $k > 2$ respectively and have values

$$E[t_k] = 0, \qquad \text{Var}\,[t_k] = \frac{k}{k-2}. \tag{7.38}$$

The t-distribution will be further discussed in Chapter 9 with respect to the confidence interval of an estimate based on simulation or real experimental data.

7.3 Fisher's *F*-distribution

Suppose that RVs V_1 and V_2 are statistically independent, having the chi-squared distributions with n_1 and n_2 d.f., respectively. We define the variable F_{n_1,n_2}, or simply F, by

$$F = \frac{V_1/n_1}{V_2/n_2}. \tag{7.39}$$

Then we can show that the probability that F falls in the interval $(x, x+dx)$ is given by

$$f_F(x)dx = \frac{\Gamma\left(\frac{n_1+n_2}{2}\right)\left(\frac{n_1}{n_2}\right)^{n_1/2}}{\Gamma\left(\frac{n_1}{2}\right)\Gamma\left(\frac{n_2}{2}\right)} x^{(n_1/2)-1}\left(1+\frac{n_1 x}{n_2}\right)^{-(n_1+n_2)/2} dx, \quad x > 0, \tag{7.40}$$

which is called the *F-distribution* (F stands for Fisher[2]) with (n_1, n_2) d.f. The F-distribution is often referred to as the *variance-ratio distribution* or the *Snedecor distribution.*[3] The above distribution can be derived in a manner similar to the derivation of the t-distribution: start with the joint PDF of V_1 and V_2; then obtain the joint PDF of F and V_2, and take the marginal distribution of F.

The rth moment of the F-distribution is given by (Problem 7.7)

$$E[F^r] = \frac{\left(\frac{n_2}{n_1}\right)^r \Gamma\left(\frac{n_1}{2}+r\right)\Gamma\left(\frac{n_2}{2}-r\right)}{\Gamma\left(\frac{n_1}{2}\right)\Gamma\left(\frac{n_2}{2}\right)}, \tag{7.41}$$

which exists only for $-n_1 < 2r < n_2$. Thus, the mean and variance of F are given by

$$E[F] = \frac{n_2}{n_2-2} \quad \text{for } n_2 > 2 \tag{7.42}$$

and

$$\text{Var}\,[F] = \frac{2n_2^2(n_1+n_2-2)}{n_1(n_2-2)^2(n_2-4)} \quad \text{for } n_2 > 4. \tag{7.43}$$

The *mode*, the value for which F is maximum, is given by

$$\text{mode } F = \frac{n_2(n_1-2)}{n_1(n_2+1)}. \tag{7.44}$$

Figure 7.3 shows the F-curves for several pairs of (n_1, n_2). Suppose that we have two normal populations $N(\mu_i, \sigma_i^2)$, $i = 1, 2$. Assume that we have independent observations of sample size n_i, with s_i^2 as their sample variances. In a previous section we have seen that s_i^2/σ_i^2 is χ^2-distributed with $(n_i - 1)$ d.f., $i = 1, 2$. Then the ratio of the sample variances is

$$\frac{s_1^2}{s_2^2} = \frac{\sigma_1^2}{\sigma_2^2} F_{n_1-1, n_2-1}. \tag{7.45}$$

Therefore, the F-distribution is used to test the equality of two sample variances. It is extensively used in the analysis of variance (ANOVA).

[2] However, the RV originally proposed by Fisher [106] was z defined by $z = \frac{1}{2} \ln F$.

[3] George Waddel Snedecor (1881–1974) was an American statistician.

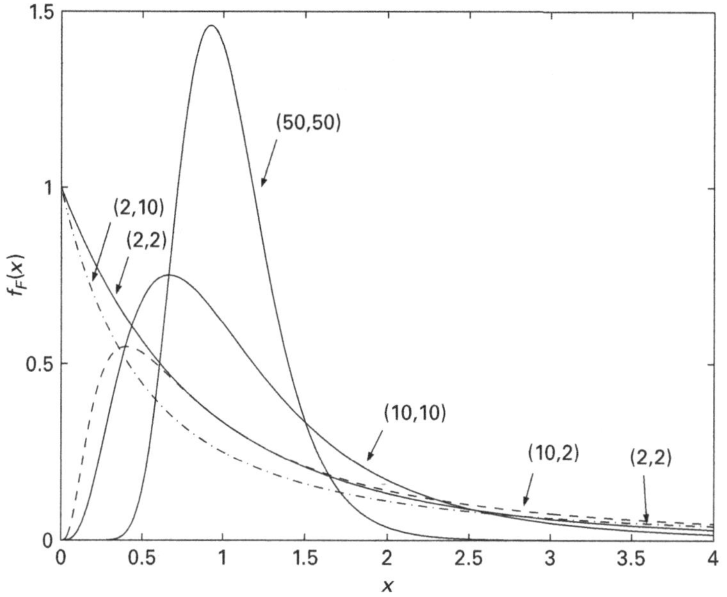

Figure 7.3 The F-distributions for various degrees of freedom (n_1, n_2).

7.4 Log-normal distribution

The log-normal distribution is often used by economists in modeling; e.g., the distribution of incomes. In mathematical finance, the price movement of a stock is often characterized by *geometric Brownian motion*, which exhibits a log-normal distribution (see Section 17.4.2). The log-normal distribution has a long or heavy-tailed distribution, like some other distributions such as the gamma (with $\beta \ll 1$), Pareto, and Weibull distributions. The log-normal distribution is also used in reliability engineering and in radio communication modeling. The latter example is associated with the so-called *shadowing effect* on signal propagation over a wireless channel. At the end of this section we will present a physical argument to justify use of the log-normal distribution in modeling radio signal propagation. But we first give its mathematical definition and discuss key properties of the distribution. As the name indicates, a *positive* RV X is said to have the log-normal distribution if its logarithm

$$Y = \ln X$$

has the normal distribution; i.e.,

$$f_Y(y) = \frac{1}{\sqrt{2\pi}\sigma_Y} \exp\left[-\frac{(y-\mu_Y)^2}{2\sigma_Y^2}\right], \quad -\infty < y < \infty. \tag{7.46}$$

Then, by using $dy = dx/x$, and $f_Y(y)\,dy = f_X(x)\,dx$, we readily find the PDF of the log-normal RV X:

$$f_X(x) = \frac{1}{\sqrt{2\pi}\sigma_Y x} \exp\left[-\frac{(\ln x - \mu_Y)^2}{2\sigma_Y^2}\right], \quad x > 0. \tag{7.47}$$

In order to obtain the expected value of X, we make use of the **moment-generating function (MGF)** (see Section 8.1 of Chapter 8) of the normal RV Y given by

$$M_Y(t) = E[e^{tY}] = \exp\left(\mu_Y t + \frac{\sigma_Y^2 t^2}{2}\right). \tag{7.48}$$

Then

$$\mu_X = E[X] = E[e^Y] = M_Y(1) = \exp\left(\mu_Y + \frac{\sigma_Y^2}{2}\right). \tag{7.49}$$

Similarly, the second moment of X is

$$E[X^2] = E[e^{2Y}] = M_Y(2) = \exp\left(2\mu_Y + 2\sigma_Y^2\right) = \mu_X^2 e^{\sigma_Y^2}. \tag{7.50}$$

Therefore, the variance of the log-normal variable is given by

$$\begin{aligned}\sigma_X^2 &= \exp\left(2\mu_Y + 2\sigma_Y^2\right) - \left[\exp\left(\mu_Y + \frac{\sigma_Y^2}{2}\right)\right]^2 \\ &= \exp\left(2\mu_Y + \sigma_Y^2\right)\left[\exp\left(\sigma_Y^2\right) - 1\right] = \mu_X^2\left[\exp\left(\sigma_Y^2\right) - 1\right].\end{aligned} \tag{7.51}$$

From (7.49) and (7.51) we find expressions for the mean and variance of Y in terms of those of X:

$$\mu_Y = \ln \mu_X - \frac{1}{2}\ln\left(1 + \frac{\sigma_X^2}{\mu_X^2}\right), \tag{7.52}$$

and

$$\sigma_Y^2 = \ln\left(1 + \frac{\sigma_X^2}{\mu_X^2}\right). \tag{7.53}$$

Now let us discuss how the log-normal RV appears in the signal propagation in a radio channel. Consider the signal power (or signal strength) at the receiver. It should be the signal power sent from the transmitter divided by the attenuation or loss factor L (>1) due to propagation loss. If the propagation is in free space, then $L = 4\pi d^2$, where d is the distance between the transmitter and the receiver. In practice, there are additional components such as absorption of signals in trees, buildings, and other objects, and these lossy components will vary. Thus, it is proper to treat L as a random variable. Furthermore, if we divide the path between the transmitter and receiver into contiguous and disjoint segments, then the overall loss L is the product of the loss within each segment:

$$L = \prod_{i=1}^{n} L_i \tag{7.54}$$

It is reasonable to assume that in most cases these RVs' L_i are statistically independent. Of course, the mean values of the L_i may be commonly affected by such factors as the temperature, precipitation, and so forth, but the variation of L_i from its mean should be unrelated to that of L_j; hence, L_i and L_j are statistically independent for $j \neq i$. Taking the logarithm of (7.54), we have

$$Y = \sum_{i=1}^{n} Y_i, \tag{7.55}$$

where we set

$$Y = \ln L \text{ and } Y_i = \ln L_i, \text{ for } i = 1, 2, \ldots, n.$$

The transformed RVs $Y_1, Y_2, \ldots, Y_n$ are statistically independent because L_i are independent. We do not require the assumption that they are statistically identical to each other, because a generalized version of the CLT, as stated in Theorem 11.23 of Section 11.3.4, does not require the identical distribution assumption. Assume that the Y_i have finite mean μ_i and variance σ_i^2. Then, from the CLT, we can show that Y is asymptotically (i.e., as $n \to \infty$) normally distributed according to $N(\mu_Y, \sigma_Y^2)$, where $\mu_Y = \sum_{i=1}^{n} \mu_i$ and $\sigma_Y^2 = \sum_{i=1}^{n} \sigma_i^2$, as long as none of the σ_i^2 represent a significant portion of their sum σ_Y^2. Therefore, the overall attenuation factor is log-normally distributed.

It is common practice in communication engineering to use the so-called decibel (dB) representation; i.e., use

$$Z = 10 \log_{10} L = 10 \frac{\ln L}{\ln 10} = (10 \log_{10} e) Y \text{ [dB]}. \tag{7.56}$$

Then its expected value is, using (7.52), given as

$$\begin{aligned} E[Z] &= \frac{10}{\ln 10} \mu_Y = \frac{10}{\ln 10} \left[\ln \mu_L - \frac{1}{2} \ln \left(1 + \frac{\sigma_L^2}{\mu_L^2} \right) \right] \\ &= 10 \left[\log_{10} \mu_L - \frac{1}{2} \log_{10} \left(1 + \frac{\sigma_L^2}{\mu_L^2} \right) \right] \text{ [dB]}. \end{aligned} \tag{7.57}$$

The standard deviation of the log-normal variable Z is

$$\sigma_Z = \frac{10}{\ln 10} \sigma_Y = 10 \sqrt{\log_{10} e \log_{10} \left(1 + \frac{\sigma_L^2}{\mu_L^2} \right)} \text{ [dB]}. \tag{7.58}$$

7.5 Rayleigh and Rice distributions

The Rayleigh and Rice distributions are primarily used by communication engineers, and they can be viewed as special cases of the chi-squared and non-central chi-squared distributions, respectively. The Rayleigh distribution is also a special case of the Weibull distribution defined earlier. Because of their prevalent use in communication engineering, we provide some discussion of these distributions.

7.5.1 Rayleigh distribution

Let us assume that X and Y are independent normal variables with zero mean and common variance σ^2. We define a new RV

$$R = \sqrt{X^2 + Y^2}, \quad R \geq 0. \tag{7.59}$$

Then the PDF of the RV R is

$$f_R(r) = \frac{r}{\sigma^2} e^{-r^2/2\sigma^2}, \quad r \geq 0. \tag{7.60}$$

This is known as the **Rayleigh distribution.**[4]

The derivation of the above expression is as follows. By writing $X = \sigma U_1$ and $Y = \sigma U_2$, where U_i are from $N(0, 1)$, we readily recognize $R^2 = \sigma^2 \chi_2^2$. Then from the general expression of the PDF of χ_n^2, we readily have

$$f_{\chi_2^2}(x) = \frac{e^{-x/2}}{2\Gamma(1)} = \frac{1}{2} e^{-x/2}, \quad x \geq 0. \tag{7.61}$$

Then from the relation $r^2 = \sigma^2 v$ (r and v are the values that the RVs R and χ^2 take respectively), we have $2r\,dr = \sigma^2\,dv$. Then equating $f_{\chi_2^2}(v)\,dv = f_R(r)\,dr$, we readily obtain (7.60).

The variable $S = R^2 = \sigma^2 \chi_2^2$ is exponentially distributed. This can be shown by applying the transformation $s = \sigma^2 v$ to the formula $f_{\chi_2^2}(v) = \frac{1}{2} e^{-v/2}$:

$$f_S(s)\,ds = f_{\chi_2^2}(v)\,dv = \frac{1}{2\sigma^2} e^{-s/2\sigma^2}\,ds, \quad s > 0. \tag{7.62}$$

A direct way to derive the Rayleigh distribution, without recourse to the chi-squared distribution, is to transform the bivariate normal RVs (X, Y) into the polar coordinate variables (R, Θ) by

$$X = R\cos\Theta \text{ and } Y = R\sin\Theta, \quad R \geq 0, \quad \Theta \in [0, 2\pi]. \tag{7.63}$$

The joint PDF of (X, Y) is given by

$$f_{XY}(x, y) = \frac{1}{2\pi\sigma^2} \exp\left(-\frac{x^2 + y^2}{2\sigma^2}\right) = \frac{1}{2\pi\sigma^2} \exp\left(-\frac{r^2}{2\sigma^2}\right). \tag{7.64}$$

[4] John W. Strutt, 3rd Baron Rayleigh (1842–1919) was an English physicist who won the Nobel Prize in Physics in 1904 for co-discovering the element argon.

Since the Jacobian of the above transformation is given by

$$\left|J\left(\frac{x,y}{r,\theta}\right)\right| = \left|\det\begin{bmatrix} \frac{\partial x}{\partial r} & \frac{\partial x}{\partial \theta} \\ \frac{\partial y}{\partial r} & \frac{\partial y}{\partial \theta}\end{bmatrix}\right| = \left|\det\begin{bmatrix} \cos\theta & -r\sin\theta \\ \sin\theta & r\cos\theta\end{bmatrix}\right| = r, \tag{7.65}$$

we find

$$f_{R\Theta}(r,\theta) = J\left(\frac{x,y}{r,\theta}\right) f_{X,Y}(x,y) = \frac{1}{2\pi}\frac{r}{\sigma^2}\exp\left(-\frac{r^2}{2\sigma^2}\right) = f_\Theta(\theta) f_R(r), \tag{7.66}$$

from which it is apparent that R and Θ are independent, and the marginal PDF of R is given by (7.60) and that of Θ is

$$f_\Theta(\theta) = \frac{1}{2\pi}, \quad 0 \le \theta < 2\pi, \tag{7.67}$$

which is the uniform distribution.

The MGF of the RV R^2 is readily obtainable from that of χ_2^2 derived in an exercise of this section, but the MGF of R does not have a simple closed-form expression. Thus, expressions for the mean and higher moments of the Rayleigh distribution are also complicated (Problem 7.10).

Let us define a new RV T by

$$T = \frac{Y}{X}, \quad -\infty < T < \infty. \tag{7.68}$$

By using (7.63), we find

$$T = \tan\Theta. \tag{7.69}$$

But the mapping from Θ to T is a two-to-one mapping, because for any real number t there are two θ: if $\tan\theta_1 = t$, then so is $\tan(\theta_1 + \pi)$. Thus, we change (7.67) to the uniform distribution within the interval $(-\frac{\pi}{2}, \frac{\pi}{2})$:

$$f_\Theta(\theta) = \frac{1}{\pi}, \quad -\frac{\pi}{2} < \theta < \frac{\pi}{2}. \tag{7.70}$$

Then, from the relation $f_T(t)\,dt = f_\Theta(\theta)\,d\theta$, and $dt = \sec^2\theta\,d\theta = (1+\tan^2\theta)\,d\theta$, we have

$$f_T(t) = f_\Theta(\theta)\frac{1}{1+t^2} = \frac{1}{\pi(t^2+1)}, \quad -\infty < t < \infty, \tag{7.71}$$

which is the Cauchy distribution defined earlier.

7.5.1.1 Rayleigh distribution and the Weibull distribution

As we remarked in Section 4.2.5, the Rayleigh distribution can be derived as a special case of the two-parameter Weibull distribution. By setting $\alpha = 2$ and $\beta = \sqrt{2}\sigma$ in the Weibull PDF of (4.76), and denoting the random variable X as R, we obtain the PDF

$$f_R(r) = \frac{1}{\sigma^2} r\, e^{-r^2/2\sigma^2}, \quad r \ge 0, \tag{7.72}$$

which is indeed the Rayleigh distribution.

It is instructive to note that if we modify the variable R of (4.76) to

$$R = (X^2 + Y^2)^{1/\alpha}, \tag{7.73}$$

then the bivariate normal distribution (7.64) leads to

$$f_R(r) = \frac{\alpha r^{\alpha-1}}{2\sigma^2} e^{-r^\alpha/2\sigma^2}, \quad r \geq 0, \tag{7.74}$$

which is the two-parameter Weibull distribution with parameters α and $\beta = (2\sigma^2)^{1/\alpha}$.

7.5.2 Rice distribution

Now let us assume that the independent normal RVs X and Y have nonzero means μ_X and μ_Y. We still retain the assumption that they have a common variance σ^2. Then the PDF of $R = \sqrt{X^2 + Y^2}$ is given by

$$f_R(r) = \frac{r\, e^{-(r^2+\mu^2)/2\sigma^2}}{2\pi\sigma^2} I_0\left(\frac{r\mu}{\sigma^2}\right), \quad r \geq 0, \tag{7.75}$$

which is known as **Rice distribution** or **Rician distribution** because it was originally derived by **S.O. Rice**[5] [280]. The parameter μ is the distance between the center of the bivariate normal distribution $(X, Y) = (\mu_X, \mu_Y)$ and the origin $(0, 0)$, i.e.,

$$\mu = \sqrt{\mu_X^2 + \mu_Y^2}, \tag{7.76}$$

and the function $I_0(x)$ is

$$I_0(x) = \frac{1}{\pi} \int_0^\pi e^{x\cos\phi}\, d\phi, \quad -\infty < x < \infty, \tag{7.77}$$

which is the **modified Bessel function of the first kind and zeroth order**.

The Rice distribution (7.75) can be derived by transforming the (X, Y) into the polar coordinates defined by (7.63). Using the Jacobian given by (7.65), we find the joint PDF of (R, Θ) given by

$$\begin{aligned} f_{R\Theta}(r, \theta) = r f_{XY}(x, y) &= \frac{r}{2\pi\sigma^2} \exp\left[-\frac{(x-\mu_X)^2 + (y-\mu_Y)^2}{2\sigma^2}\right] \\ &= \frac{r}{2\pi\sigma^2} \exp\left[-\frac{r^2+\mu^2}{2\sigma^2} - \frac{r(\mu_X\cos\theta + \mu_Y\sin\theta)}{\sigma^2}\right] \\ &= \frac{r}{2\pi\sigma^2} \exp\left[-\frac{r^2+\mu^2}{2\sigma^2} - \frac{r\mu\cos(\theta-\psi)}{\sigma^2}\right], \end{aligned} \tag{7.78}$$

where $\psi = \tan^{-1}(\mu_Y/\mu_X)$. Then by writing $\theta - \psi = \phi$, and integrating the above joint PDF with respect to the RV Θ, which is uniformly distributed over $[0, 2\pi]$, we find the marginal PDF of the RV R as

[5] Stephen O. Rice (1907–1986) was an American communication theorist.

$$
\begin{aligned}
f_R(r) &= \int_0^{2\pi} f_{R\Theta}(r,\theta)\,d\theta \\
&= \frac{r}{\sigma^2}\exp\left(-\frac{r^2+\mu^2}{2\sigma^2}\right)\frac{1}{2\pi}\int_0^{2\pi}\exp\left[-\frac{r\mu}{\sigma^2}\cos(\theta-\psi)\right]d\theta. \qquad (7.79)
\end{aligned}
$$

Since $\cos(\theta-\psi)$ is a periodic function of θ, we can write

$$
\begin{aligned}
\frac{1}{2\pi}\int_0^{2\pi}\exp\left[-\frac{r\mu}{\sigma^2}\cos(\theta-\psi)\right]d\theta &= \frac{1}{2\pi}\int_0^{2\pi}\exp\left(-\frac{r\mu}{\sigma^2}\cos\phi\right)d\phi \\
&= \frac{1}{\pi}\int_0^{\pi}\exp\left(-\frac{r\mu}{\sigma^2}\cos\phi\right)d\phi, \qquad (7.80)
\end{aligned}
$$

where we used the property that $\cos\phi$ is a symmetric function around $\phi=\pi$. Then using the definition of the modified Bessel function of order zero given in (7.77), the Rice distribution (7.75) follows.

An alternative way to derive this is to use the formula we obtained for the noncentral chi-squared distribution (7.112) (Problem 7.11). If we set $\sigma=1$, i.e., we normalize the amplitude by σ, the distribution of this normalized amplitude RV $V=R/\sigma$ is

$$
f_V(v) = \frac{v\,e^{-(v^2+m^2)/2}}{2\pi} I_0(vm), \quad v \geq 0, \qquad (7.81)
$$

where $m=\mu/\sigma$.

Figure 7.4 shows the plot of this normalized Rice distribution for $m = 0, 1, 2, 4$, and 6. The case $m=0$ corresponds to the normalized Rayleigh distribution. When $m>2$, the distribution resembles a normal distribution with mean slightly larger than m. But

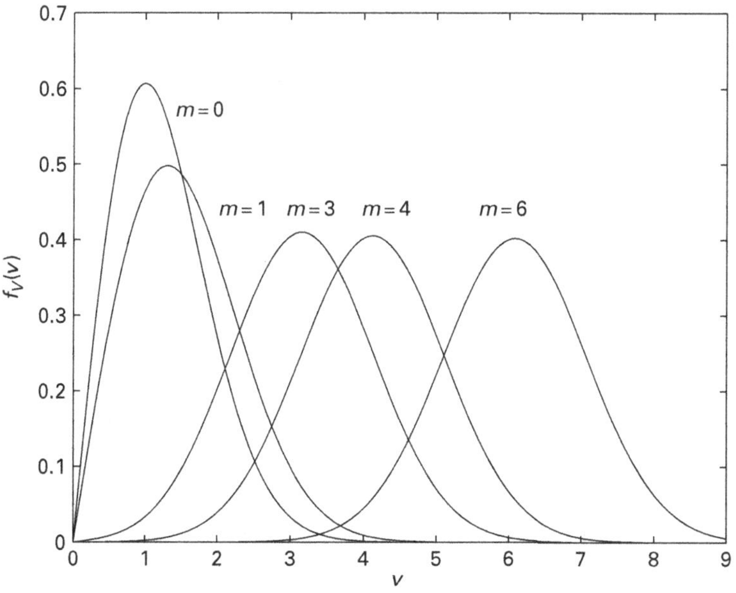

Figure 7.4 The normalized Rice distribution for $m = 0, 1, 3, 4$, and 6.

the RV V (and R as well) is a nonnegative RV, whereas a normal variable can take on negative values. The above result is not difficult to see, once we recognize that the joint variable (X, Y) has a circular normal distribution around the mean (μ_X, μ_Y). If we rotate the coordinates (X, Y) by $-\psi$, where $\psi = \tan^{-1}(\mu_Y/\mu_X)$, the joint PDF under these new coordinates is given by (Problem 7.12)

$$f_{X'Y'}(x', y') = \frac{1}{\sqrt{2\pi}\sigma} e^{-(x'-\mu)^2/2\sigma^2} \frac{1}{\sqrt{2\pi}\sigma} e^{-y'^2/2\sigma^2}. \tag{7.82}$$

By applying the Taylor series expansion to $e^{x\cos\phi}$ in the integration formula (7.77), and noting the fact that $I_0(x)$ is an even function, we obtain the following series representation:

$$I_0(x) = \sum_{m=0}^{\infty} \left[\frac{(x/2)^m}{m!}\right]^2. \tag{7.83}$$

If $x \ll 1$, then

$$I_0(x) \approx 1 + \frac{x^2}{4}, \quad \text{for } x \approx 0, \tag{7.84}$$

For large values of $x \gg 1$, we have the following asymptote:

$$I_0(x) \approx \frac{e^x}{\sqrt{2\pi x}}, \quad \text{for } x \gg 1. \tag{7.85}$$

7.6 Complex-valued normal variables[6]

7.6.1 Complex-valued Gaussian variables and their properties

Let Z be a complex-valued RV

$$Z = X + iY, \tag{7.86}$$

where X and Y are real-valued RVs. Then

$$E[Z^2] = E[X^2] - E[Y^2] + 2i\,E[XY]. \tag{7.87}$$

It is apparent that

$$E[Z^2] = 0 \text{ if and only if } E[X^2] = E[Y^2] \triangleq \sigma^2 \text{ and } E[XY] = 0, \tag{7.88}$$

and if this condition is met, then

$$E[ZZ] = E[Z^*Z^*] = 0 \text{ and } E[ZZ^*] = E[X^2] + E[Y^2] = 2\sigma^2. \tag{7.89}$$

[6] The reader may skip this section on first reading.

If in addition the RVs X and Y are normal and $E[X] = E[Y] = 0$, then we can write the joint PDF as

$$f_{XY}(x, y) = \frac{1}{2\pi\sigma^2} e^{-(x^2+y^2)/2\sigma^2}. \tag{7.90}$$

The normal RVs X and Y with this joint PDF are called *circularly symmetric*. If we transform the coordinates (X, Y) into the polar coordinates (R, Θ) by

$$X = R\cos\Theta, \quad Y = R\sin\Theta,$$

then we obtain, as we did in earlier sections, the joint PDF

$$f_{R\Theta}(r, \theta) = f_R(r) f_\Theta(\theta), \tag{7.91}$$

where

$$f_R(r) = \frac{r\, e^{-r^2/2\sigma^2}}{\sigma^2}, \quad r \geq 0, \tag{7.92}$$

$$f_\Theta(\theta) = \frac{1}{2\pi}, \quad 0 \leq \theta \leq 2\pi; \tag{7.93}$$

i.e., $f_R(r)$ is the Rayleigh distribution discussed in Section 7.5.1.

Transformation of the RVs (X, Y) to (Z, Z^*) yields its Jacobian

$$\left| \boldsymbol{J}\left(\frac{z, z^*}{x, y}\right) \right| = \left| \det \begin{bmatrix} 1 & 1 \\ i & -i \end{bmatrix} \right| = |-2i| = 2. \tag{7.94}$$

Therefore,

$$dx\, dy = \frac{1}{2} dz\, dz^*.$$

Thus, the joint PDF of Z and Z^* is given by

$$\boxed{f_{ZZ^*}(z, z^*) = \frac{1}{4\pi\sigma^2} e^{-(zz^*)/2\sigma^2}.} \tag{7.95}$$

7.6.2 Multivariate Gaussian variables

Let us now consider a multidimensional case of the **circularly symmetric** Gaussian variables. Possible applications include (i) a sequence of complex-valued RVs and (ii) multiple complex-valued signals received by an array of multiple antennas.

Let $\boldsymbol{X} = (X_1, X_2, \ldots, X_M)^\top$ and $\boldsymbol{Y} = (Y_1, Y_2, \ldots, Y_M)^\top$ be real-valued multivariate Gaussian variables (see Section 4.3.1) that satisfy

$$\boxed{\begin{aligned} &E[\boldsymbol{X}] = E[\boldsymbol{Y}] = \boldsymbol{0}, \\ &E[\boldsymbol{X}\boldsymbol{X}^\top] = E[\boldsymbol{Y}\boldsymbol{Y}^\top] = \boldsymbol{A}, \\ &E[\boldsymbol{Y}\boldsymbol{X}^\top] = -E[\boldsymbol{X}\boldsymbol{Y}^\top] = \boldsymbol{B}, \\ &\boldsymbol{A} = \boldsymbol{A}^\top, \quad \boldsymbol{B} = -\boldsymbol{B}^\top. \end{aligned}} \tag{7.96}$$

We define a $2M$-dimensional normal variable $\boldsymbol{W}$ by

$$\boldsymbol{W} \triangleq \begin{bmatrix} \boldsymbol{X} \\ \boldsymbol{Y} \end{bmatrix} = (X_1, X_2, \ldots, X_M, Y_1, Y_2, \ldots, Y_M)^\top. \tag{7.97}$$

Then, it satisfies the following properties:

$$\boxed{\begin{aligned} E[\boldsymbol{W}] &= \boldsymbol{0}, \\ E[\boldsymbol{W}\boldsymbol{W}^\top] &= \begin{bmatrix} \boldsymbol{A} & -\boldsymbol{B} \\ \boldsymbol{B} & \boldsymbol{A} \end{bmatrix} \triangleq \boldsymbol{\Sigma}. \end{aligned}} \tag{7.98}$$

We now define an M-dimensional complex-valued normal variable $\boldsymbol{Z}$:

$$\boldsymbol{Z} \triangleq \boldsymbol{X} + i\boldsymbol{Y}, \tag{7.99}$$

which satisfies

$$\boxed{\begin{aligned} & E[\boldsymbol{Z}] = \boldsymbol{0}, \\ & E[\boldsymbol{Z}\boldsymbol{Z}^\top] = E[\boldsymbol{X}\boldsymbol{X}^\top] - E[\boldsymbol{T}\boldsymbol{Y}^\top] + iE[\boldsymbol{Y}\boldsymbol{X}^\top] + iE[\boldsymbol{X}\boldsymbol{Y}^\top] \\ & \qquad\qquad = \boldsymbol{A} - \boldsymbol{A} + i\boldsymbol{B} + i(-\boldsymbol{B}) = \boldsymbol{0}, \\ & E[\boldsymbol{Z}^*\boldsymbol{Z}^{\mathrm{H}}] = \boldsymbol{0}, \\ & E[\boldsymbol{Z}\boldsymbol{Z}^{\mathrm{H}}] = 2[\boldsymbol{A} + \mathrm{i}\boldsymbol{B}] \triangleq 2\boldsymbol{C}, \end{aligned}} \tag{7.100}$$

where the superscript operator H, called Hermitian conjugate or conjugate transpose, represents the transpose of the complex conjugate of a complex-valued matrix; i.e., $\boldsymbol{Z}^{\mathrm{H}} = \boldsymbol{Z}^{*\top}$

Then we can establish the following relations between the $2M \times 2M$ matrix $\boldsymbol{\Sigma}$ of (7.98) and the $M \times M$ matrix $\boldsymbol{C}$ of (7.100).

1. The multiplication of matrices $\boldsymbol{\Sigma}$ is *isomorphic* to the multiplication of complex-valued matrices $\boldsymbol{C}$; i.e.,

$$\boxed{\begin{aligned} & \begin{bmatrix} \boldsymbol{A}_1 & -\boldsymbol{B}_1 \\ \boldsymbol{B}_1 & \boldsymbol{A}_1 \end{bmatrix} \cdot \begin{bmatrix} \boldsymbol{A}_2 & -\boldsymbol{B}_2 \\ \boldsymbol{B}_2 & \boldsymbol{A}_2 \end{bmatrix} = \begin{bmatrix} \boldsymbol{A}_1\boldsymbol{A}_2 - \boldsymbol{B}_1\boldsymbol{B}_2 & -(\boldsymbol{A}_1\boldsymbol{B}_2 + \boldsymbol{B}_1\boldsymbol{A}_2) \\ \boldsymbol{A}_1\boldsymbol{B}_2 + \boldsymbol{B}_1\boldsymbol{A}_2 & \boldsymbol{A}_1\boldsymbol{A}_2 - \boldsymbol{B}_1\boldsymbol{B}_2 \end{bmatrix} \\ & [\boldsymbol{A}_1 + i\boldsymbol{B}_1] \cdot [\boldsymbol{A}_2 + i\boldsymbol{B}_2] = [\boldsymbol{A}_1\boldsymbol{A}_2 - \boldsymbol{B}_1\boldsymbol{B}_2 + i(\boldsymbol{A}_1\boldsymbol{B}_2 + \boldsymbol{B}_1\boldsymbol{A}_2)] \end{aligned}} \tag{7.101}$$

2. The matrix $\boldsymbol{\Sigma}$ is symmetric, if and only if the sub-matrix $\boldsymbol{A}$ is symmetric and $\boldsymbol{B}$ is skew-symmetric. The matrix $\boldsymbol{C}$ is Hermitian, if and only if $\boldsymbol{\Sigma}$ is symmetric.
3. The determinants of $\boldsymbol{\Sigma}$ and $\boldsymbol{C}$ are related by

$$\boxed{\det \boldsymbol{\Sigma} = |\det \boldsymbol{C}|^2.} \tag{7.102}$$

4. If the matrix $\boldsymbol{C}$ is nonsingular, with the inverse

$$\boldsymbol{C}^{-1} = [\boldsymbol{A} + i\boldsymbol{B}]^{-1} \triangleq \boldsymbol{P} + i\boldsymbol{Q}, \tag{7.103}$$

then $\boldsymbol{\Sigma}$ is also nonsingular, with the inverse

$$\boldsymbol{\Sigma}^{-1} = \begin{bmatrix} \boldsymbol{P} & -\boldsymbol{Q} \\ \boldsymbol{Q} & \boldsymbol{P} \end{bmatrix}, \tag{7.104}$$

and conversely.
5. If $\boldsymbol{\Sigma}$ is symmetric (hence $\boldsymbol{C}$ is Hermitian[7]), then the corresponding quadratic forms involving $\boldsymbol{x}$, $\boldsymbol{y}$ and $\boldsymbol{z} = \boldsymbol{x} + i\boldsymbol{y}$ are related by

$$\boxed{[\boldsymbol{x}^\top \boldsymbol{y}^\top]\boldsymbol{\Sigma}^{-1}\begin{bmatrix} \boldsymbol{x} \\ \boldsymbol{y} \end{bmatrix} = \boldsymbol{z}^\top \boldsymbol{C}^{-1}\boldsymbol{z}^*.} \tag{7.105}$$

The PDF of $\boldsymbol{W}$ is given by

$$f_{\boldsymbol{W}}(\boldsymbol{w}) = \frac{1}{(2\pi)^M |\det \boldsymbol{\Sigma}|^{1/2}} \exp\left(-\frac{1}{2}\boldsymbol{w}^\top \boldsymbol{\Sigma}^{-1}\boldsymbol{w}\right). \tag{7.106}$$

Thus, the joint PDF of $\boldsymbol{X}$ and $\boldsymbol{Y}$ is given by

$$f_{\boldsymbol{XY}}(\boldsymbol{x}, \boldsymbol{y}) = \frac{1}{(2\pi)^M |\det \boldsymbol{\Sigma}|^{1/2}} \exp\left(-\frac{1}{2}[\boldsymbol{x}^\top \boldsymbol{y}^\top]\begin{bmatrix} \boldsymbol{P} & -\boldsymbol{Q} \\ \boldsymbol{Q} & \boldsymbol{P} \end{bmatrix}\begin{bmatrix} \boldsymbol{x} \\ \boldsymbol{y} \end{bmatrix}\right). \tag{7.107}$$

We now transform $(\boldsymbol{X}, \boldsymbol{Y})$ to $(\boldsymbol{Z}, \boldsymbol{Z}^*)$. Then its Jacobian is obtained, similar to (7.94), as

$$\left|J\left(\frac{\boldsymbol{z}, \boldsymbol{z}^*}{\boldsymbol{x}, \boldsymbol{y}}\right)\right| = \left|\det\begin{bmatrix} \boldsymbol{I} & \boldsymbol{I} \\ i\boldsymbol{I} & -i\boldsymbol{I} \end{bmatrix}\right| = |-2i|^M = 2^M, \tag{7.108}$$

where $\boldsymbol{I}$ is the $M \times M$ identity matrix. Then, by using (7.102), (7.105), (7.107), and (7.108), the joint PDF of $(\boldsymbol{Z}, \boldsymbol{Z}^*)$ is obtained as

$$f_{\boldsymbol{ZZ}^*}(\boldsymbol{z}, \boldsymbol{z}^*) = \frac{1}{(4\pi)^M |\det \boldsymbol{C}|} \exp\left(-\frac{1}{2}\boldsymbol{z}^\top \boldsymbol{C}^{-1}\boldsymbol{z}^*\right). \tag{7.109}$$

In Section 8.2.6 we will discuss the characteristic function of the multivariate complex-valued normal distribution.

[7] A complex-valued square matrix $\boldsymbol{C}$ is called *Hermitian* or *self-adjoint* if its conjugate transpose $\boldsymbol{C}^{*\top}$, sometimes written as $\boldsymbol{C}^{\mathrm{H}}$, is equal to $\boldsymbol{C}$.

7.7 Summary of Chapter 7

Chi-squared	$\chi_n^2 = \sum_{i=1}^n U_i^2$	(7.1)
PDF	$f_{\chi_n^2}(x) = \frac{x^{(n/2)-1}e^{-x/2}}{2^{n/2}\Gamma(\frac{n}{2})}dx$	(7.24)
Gamma function	$\Gamma(z) = \int_0^\infty t^{z-1}e^{-t}dt$	(7.3)
Expectation, variance	$E[\chi_n^2] = n,\ \mathrm{Var}[\chi_n^2] = 2n$	(7.9), (7.10)
Noncentral chi-squared	$\chi_n^2(\mu^2) = \sum_{k=1}^n (U_k + \mu_k)^2,$ $\mu^2 = \sum_{k=1}^n \mu_k^2$	(7.27)
Student's t	$t_{n-1} = \frac{(\bar{X}-\mu)\sqrt{n}}{s} = \frac{U}{\sqrt{\chi_{n-1}^2/(n-1)}}$	(7.31), (7.32)
PDF	$f_{t_k}(t) = \frac{\Gamma\left(\frac{k+1}{2}\right)}{\Gamma\left(\frac{k}{2}\right)\sqrt{\pi k}}\left(1+\frac{t^2}{k}\right)^{-(k+1)/2}$	(7.33)
Moments	$E[t_k^{2r+1}] = 0,\ E[t_k^{2r}] = \frac{k^r\Gamma\left(\frac{1}{2}+r\right)\Gamma\left(\frac{k}{2}-r\right)}{\Gamma\left(\frac{1}{2}\right)\Gamma\left(\frac{k}{2}\right)}$	(7.37)
	$E[t_k] = 0,\ k>1;\ \mathrm{Var}\ [t_k] = \frac{k}{k-2},\ k>2$	(7.38)
Fisher's F	$F_{n_1,n_2} = F = \frac{V_1/n_1}{V_2/n_2},\ V_1 \sim \chi_{n_1}^2,$ $V_2 \sim \chi_{n_2}^2$	(7.39)
Moments	$E[F^r] = \frac{\left(\frac{n_2}{n_1}\right)^r \Gamma\left(\frac{n_1}{2}+r\right)\Gamma\left(\frac{n_2}{2}-r\right)}{\Gamma\left(\frac{n_1}{2}\right)\Gamma\left(\frac{n_2}{2}\right)}$	(7.41)
	$\mathrm{Var}[F] = \frac{2n_2^2(n_1+n_2-2)}{n_1(n_2-2)^2(n_2-4)},\ n_2>4$	(7.43)
Mode	$\mathrm{mode}\ F = \frac{n_2(n_1-2)}{n_1(n_2+1)}$	(7.44)
Log-normal	$Y = \ln X,\ Y \sim N(\mu_Y, \sigma_Y^2)$	(7.46)
Moments	$\mu_X = E[X] = \exp\left(\mu_Y + \frac{\sigma_Y^2}{2}\right)$	(7.49)
	$\sigma_X^2 = \exp\left(2\mu_Y + 2\sigma_Y^2\right) - \left[\exp\left(\mu_Y + \frac{\sigma_Y^2}{2}\right)\right]^2$	(7.51)
	$\mu_Y = \ln\mu_X - \frac{1}{2}\ln\left(1 + \frac{\sigma_X^2}{\mu_X^2}\right)$	(7.52)
	$\sigma_Y^2 = \ln\left(1 + \frac{\sigma_X^2}{\mu_X^2}\right)$	(7.53)
Rayleigh	$R = \sqrt{X^2+Y^2},\ X, Y \sim N(0, \sigma^2)$	(7.59)
PDF	$f_R(r) = \frac{r}{\sigma^2}e^{-r^2/2\sigma^2}$	(7.60)
Rice	$R = \sqrt{X^2+Y^2},\ X \sim N(\mu_X, \sigma^2),$ $Y \sim N(\mu_Y, \sigma^2)$	
PDF	$f_R(r) = \frac{r\,e^{-(r^2+\mu^2)/2\sigma^2}}{2\pi\sigma^2} I_0\left(\frac{r\mu}{\sigma^2}\right)$	(7.75)
Bessel function	$I_0(x) = \frac{1}{\pi}\int_0^\pi e^{x\cos\phi}\,d\phi$	(7.77)
Complex-valued Gaussian	$Z = X + iY;\ \ X, Y \sim N(0, \sigma^2)$	
Joint PDF	$f_{ZZ^*}(z, z^*) = \frac{1}{4\pi\sigma^2}e^{-(zz^*)/2\sigma^2}$	(7.95)

7.8 Discussion and further reading

The χ^2, Student's t, and F distributions are discussed in most textbooks on probability theory and statistics, although the depth of their coverage differs significantly from book to book.

The Rayleigh and Rice (sometimes written as Ricean or Rician) distributions are seldom discussed by authors who are not electrical engineers. These distributions and the Nakagami distribution (see Problem 7.13) are especially important to communication engineers in characterizing wireless (i.e., radio and optical) channels. These three distributions, and the Weibull distribution as well, are used in characterizing *fast fading* (or *rapid* fading), which is caused by reflections of radio signals at local surfaces and motion of objects. If the receiver's position is on a *line of sight* (LOS) vis-à-vis the transmitter, there will be an LOS signal component in the received signal. Then the Rice distribution is appropriate, whereas the Rayleigh distribution applies when there is no LOS component in the received signal.

The log-normal distribution also increasingly appears in the literature on wireless communications, because so-called *slow fading*, which occurs due to *shadowing* by buildings, mountains, trees, and other objects, is often a limiting factor in recovering information at the receiver end. The log-normal distribution is applied to non-communication areas as well, where its heavy-tailed distribution aptly represents an empirically obtained distribution, since its two-parameter characterization is easy to deal with.

The complex-valued normal variables are also predominantly used by communication engineers. In digital communications, complex-valued representations of signals and/or noise are often adopted because they are more concise than the real-valued representations. There are two situations where complex-valued RVs or processes will be useful. One is when a signal (typically with additive Gaussian noise) goes through a narrowband filter at the receiver. The filtered signal can be conveniently characterized as the real part of a complex-valued process [150]. The other situation occurs, as remarked above with regard to the Rayleigh and Rice distributions, when a signal is sent over a radio (i.e., wireless) channel. In a typical situation, however, there are many reflecting and scattering objects between the transmitter and the receiver. Then, the received signal consists of an LOS component, if any, plus a myriad of tiny replicas of the transmitted signal, where their amplitudes and phases vary randomly. By virtue of the CLT, the sum of these NLOS (non-line-of-sight) components can be represented as the real part of a complex-valued process with Gaussian amplitude.

As for multivariate complex-valued normal RVs and their application, see, for example, Wooding [360], Turin [329], Grettenberg [130], Wainstein and Zubakov [343], and Kobayashi [192] and references therein. We will discuss the characteristic function of the complex-valued normal distribution in Section 8.2.6.

7.9 Problems

Section 7.1: Chi-squared distribution

7.1* Sample variance and chi-squared variable. Show that χ^2 of (7.24) is a sum of $(n-1)$ independent standard normal variables.

7.2 Derivation of the χ_n^2 distribution. The region consisting of $\mathbf{U} = (U_1, U_2, \ldots, U_n)^\top$ such that

$$\chi < \sqrt{U_1^2 + U_2^2 + \cdots + U_n^2} < \chi + d\chi, \quad \chi > 0,$$

is a hypershell with inner radius χ and outer radius $\chi + d\chi$. The volume dV of such a shell is proportional to its thickness $d\chi$ and to the $(n-1)$th power of its radius χ:

$$dV = A\chi^{n-1}\, d\chi,$$

where the constant A will be determined below.

(a) Show that the PDF of χ_n is given by

$$f_{\chi_n}(\chi)\, d\chi = \frac{A}{(2\pi)^{n/2}} \chi^{n-1} e^{-\chi^2/2}\, d\chi, \quad \chi > 0.$$

(b) Show that the PDF of χ_n^2 is given by

$$f_{\chi_n^2}(v)\, dv = \frac{A}{2(2\pi)^{n/2}\sqrt{v}} v^{(n-1)/2} e^{-v/2}\, dv, \quad v > 0.$$

(c) Show that the constant A is given by

$$A = \frac{2(2\pi)^{n/2}}{\Gamma\left(\frac{n}{2}\right) 2^{n/2}}.$$

7.3* Moments of gamma and χ^2-distributions.

(a) Consider the gamma distribution

$$f_X(x) = \frac{x^{\beta-1} e^{-x}}{\Gamma(\beta)}.$$

Show that the mth moment is

$$E[X^m] = \frac{\Gamma(\beta + m)}{\Gamma(\beta)}, \quad m = 1, 2, 3, \ldots$$

(b) Show that the χ_n^2 has the mth moment

$$E[(\chi_n^2)^m] = \frac{2^m \Gamma\left(\frac{n}{2} + m\right)}{\Gamma\left(\frac{n}{2}\right)}, \quad m = 1, 2, 3, \ldots$$

7.4 χ^2 distribution and exponential distribution. Show (without having recourse to (7.7)) that if X_1 and X_2 are independent standard normal variables, their squared sum

$$Y = X_1^2 + X_2^2$$

is exponentially distributed with mean 2.

7.5 Noncentral chi-squared distribution.

(a) Show that the PDF $f_{\chi_n^2(\mu^2)}(x)$ can be expressed in terms of the following weighted sum of the PDFs of the regular chi-squared distributions $f_{\chi^2_{n+2j}}(x),\ j = 0, 1, 2, \ldots$, with the weights being equal to the Poisson distribution with parameter $\mu^2/2$:

$$f_{\chi_n^2(\mu^2)}(x) = e^{-\mu^2/2} \sum_{j=0}^{\infty} \frac{(\mu^2/2)^j}{j!} f_{\chi^2_{n+2j}}(x). \tag{7.110}$$

(b) Show that for $n = 2$, the (7.110) reduces to

$$f_{\chi_2^2(\mu^2)}(x) = e^{-\mu^2/2} \sum_{j=0}^{\infty} \frac{(\mu^2/2)^j}{j!} f_{\chi^2_{2+2j}}(x). \tag{7.111}$$

(c) Show that (7.111) can further be expressed as

$$f_{\chi_2^2(\mu^2)}(y) = e^{-\mu^2/2 - y} I_0(\mu^2 y), \tag{7.112}$$

where $I_0(x)$ is the modified Bessel function of the first kind with zero d.f., and is defined by (7.83).

Section 7.2: Student's t-distribution

7.6 Derivation of the t-distribution.

(a) Show that the joint PDF of U and χ^2_{n-1} is given by

$$f_{U,\chi^2_{n-1}}(u, v) = \frac{1}{2\sqrt{2\pi}\,\Gamma\left(\frac{n-1}{2}\right)} \left(\frac{v}{2}\right)^{(n-3)/2} e^{-(u^2+v)/2}.$$

(b) By applying the transformation (7.32) and noting that the Jacobian of this transformation is $\sqrt{v/n-1}$, show that the joint PDF of t_{n-1} and χ^2_{n-1} is

$$f_{t_{n-1}\chi^2_{n-1}}(t, v) = \frac{1}{2\sqrt{\pi(n-1)}\,\Gamma\left(\frac{n-1}{2}\right)} \left(\frac{v}{2}\right)^{(n-2)/2} e^{-(v/2)[1+t^2/(n-1)]}.$$

(c) Taking the marginal distribution of t, show that the distribution of t_{n-1} is

$$f_{t_{n-1}}(t) = \frac{\Gamma\left(\frac{n}{2}\right)}{\sqrt{\pi(n-1)}\,\Gamma\left(\frac{n-1}{2}\right)} \left(1 + \frac{t^2}{n-1}\right)^{-n/2}.$$

Section 7.3: Fisher's F-distribution

7.7* Moments of the F-distribution. Show that the rth moment of the F-distribution is given by (7.41). *Hint:* Use the result of Problem 7.3 on the χ^2-distribution.

7.8 The F-distribution and the t-distribution. Show that when the d.f. in the numerator of F is one, the F- and t-distributions have a simple relation.

Section 7.4: Log-normal distribution

7.9* Median and mode of the log-normal distribution.

(a) Show that the median of the log-normal distribution $F_X(x)$, where $Y = \ln X$ is $N(\mu_Y, \sigma_Y^2)$, is given by

$$e^{\mu_Y} = \frac{\mu_X}{\sqrt{1 + \frac{\sigma_X^2}{\mu_X^2}}}.$$

(b) Show that the mode of the log-normal distribution $F_X(x)$ is

$$e^{\mu_Y - \sigma_Y^2} = \frac{\mu_X}{\left(1 + \frac{\sigma_X^2}{\mu_X^2}\right)^{\frac{3}{2}}}.$$

Section 7.5: Rayleigh and Rice distributions

7.10* MGF of R^2 and R variables in the Rayleigh distribution.[8]

(a) Derive the MGF of a random variable $Z \triangleq X^2 + Y^2$, where X and Y are independent normal RVs from $N(0, \sigma^2)$. Obtain the mean and variance of Z.

(b) Show that the MGF of the Rayleigh variable $R = \sqrt{Z}$ is given by

$$M_R(t) = 1 + \sqrt{2\pi}\sigma t \Phi(\sigma t) e^{(\sigma^2 t^2)/2}. \tag{7.113}$$

(c) Show that the mean and variance are

$$E[R] = \sqrt{\frac{\pi}{2}}\sigma \text{ and } \mathrm{Var}[R] = \left(2 - \frac{\pi}{2}\right)\sigma^2.$$

7.11 Alternative derivation of the Rice distribution. Derive the Rice distribution (7.75) using the result on the noncentral chi-squared distribution.

7.12 Rotation of the Rice distribution. Derive Eq. (7.82).

8 This exercise should be tried after studying Chapter 8.

7.13* **Nakagami *m*-distribution.** Consider a random variable Z as the sum of m independent Rayleigh variables R_i:

$$Z = \sum_{i=1}^{m} R_i^2, \tag{7.114}$$

or equivalently

$$Z = \sum_{i=1}^{2m} X_i^2, \tag{7.115}$$

where the X_i are independent normal variables from $N(0, \sigma^2)$.

(a) Show that the PDF of Z is given by

$$f_Z(z) = \frac{m^m}{\Omega^m \Gamma(m)} z^{m-1} e^{-mz/\Omega}, \quad z \geq 0, \tag{7.116}$$

where $\Omega = 2m\sigma^2$ and $\Gamma(m)$ is the gamma function defined in (4.31), and $\Gamma(m) = (m-1)!$ when m is an integer.

(b) Define the *envelope* of the signal as its square root; i.e.,

$$R = \sqrt{Z} = \left(\sum_{i=1}^{m} R_i^2 \right)^{1/2}, \quad R \geq 0. \tag{7.117}$$

Show that the PDF of R is given by

$$f_R(r) = \frac{2m^m}{\Omega^m \Gamma(m)} r^{2m-1} e^{-\frac{mr^2}{\Omega}}, \quad r \geq 0, \tag{7.118}$$

which is known as the **Nakagami-*m*** distribution. The parameter m is referred to as the *fading figure*. The Nakagami distribution, like the Rice and Weibull distributions, can be seen as another generalization of the Rayleigh distribution.

(c) Show that the expectation and the variance of the envelope R are given by

$$E[R] = \frac{\Gamma(m+\frac{1}{2})}{\Gamma(m)} \left(\frac{\Omega}{m} \right)^{1/2}, \tag{7.119}$$

$$\mathrm{Var}[V] = \Omega \left\{ 1 - \frac{1}{m} \left[\frac{\Gamma\left(m+\frac{1}{2}\right)}{\Gamma(m)} \right]^2 \right\}. \tag{7.120}$$

7.14 **The CDF of the Nakagami-*m* distribution.**

(a) Show that when m is a positive integer, the Nakagami-m CDF is given by

$$\boxed{F_R(r) = 1 - Q\left(m-1; \frac{r^2}{2\sigma^2}\right), \quad r \geq 0,} \tag{7.121}$$

where $Q(k;\lambda)$ is the cumulative Poisson distribution:

$$Q(k;\lambda) = Q(k-1;\lambda) + P(k;\lambda), \quad k = 0, 1, 2, \ldots \tag{7.122}$$

(b) By differentiating (7.121), verify that the Nakagami PDF obtained is equivalent to (7.118).

(c) Show that when m is a positive real number, the Nakagami-m CDF is given by

$$F_R(r) = \frac{1}{\Gamma(m)} \int_0^{mr^2/\Omega} y^{m-1} e^{-y}\, dy = \frac{\gamma\left(m, \frac{mr^2}{\Omega}\right)}{\Gamma(m)}, \tag{7.123}$$

where $\gamma(\beta, \lambda)$ is the *lower incomplete gamma function* defined by

$$\gamma(\beta, \lambda) \triangleq \int_0^{\lambda} y^{\beta-1} e^{-y}\, dy. \tag{7.124}$$

Hence, show that

$$\boxed{F_R(r) = 1 - \frac{\Gamma\left(m, \frac{mr^2}{\Omega}\right)}{\Gamma(m)}, \quad r \geq 0.} \tag{7.125}$$

7.15 Upper incomplete gamma functions.

(a) Define the *upper incomplete gamma function* by

$$\Gamma(\beta, \lambda) \triangleq \int_{\lambda}^{\infty} y^{\beta-1} e^{-y}\, dy. \tag{7.126}$$

Show that we can alternatively write the CDF of R as

$$F_R(r) = 1 - \frac{\Gamma\left(m, \frac{mr^2}{\Omega}\right)}{\Gamma(m)}, \quad r \geq 0. \tag{7.127}$$

(b) Show that, for an integer $m = k$, the upper incomplete gamma function and the cumulative Poisson distribution are related by

$$Q(k;\lambda) = \frac{\Gamma(k+1;\lambda)}{k!}, \tag{7.128}$$

which implies

$$\Gamma(k;\lambda) = (k-1)!\,Q(k-1;\lambda). \tag{7.129}$$

Section 7.6: Complex-valued normal variables

7.16 Relations between Σ and *C*. Show Properties 1–5.

7.17* Joint PDF of $(\mathbf{Z}, \mathbf{Z}^*)$. Derive (7.109).

Part II

Transform methods, bounds, and limits

8 Moment-generating function and characteristic function

In this chapter and the next we will discuss four different types of transforms (or functions) that we can apply to probability distributions or PDFs. They are the MGF, the *characteristic function* (CF; equivalent to the Fourier transform), the *probability-generating function* (PGF; equivalent to the Z-transform), and the *Laplace transform* (LT). They are closely related to each other, as shown in Table 8.1. It is often more convenient to deal with one of these transforms than to work directly with the original probability distributions or PDFs, when we wish to calculate the moments of a given RV or obtain the probability distribution of the sum of two or more RVs.

8.1 Moment-generating function (MGF)

8.1.1 Moment-generating function of one random variable

For a given RV X, its **MGF** $M_X(t)$ is defined by

$$\boxed{M_X(t) = E[e^{tX}] = \int_{-\infty}^{\infty} e^{tx}\, dF_X(x), t \in I,} \tag{8.1}$$

where $I \subseteq \mathbb{R} = (-\infty, \infty)$ is an interval in which $M_X(t)$ is finite. As for the argument of the MGF, we may use another symbol, say ξ, lest t be confused with the time index. The exponential function e^{tx} has the following Taylor series expansion:

$$e^{tx} = 1 + tx + \frac{t^2x^2}{2!} + \cdots + \frac{t^n x^n}{n!} + \cdots \tag{8.2}$$

Thus, we may rewrite (8.1) as

$$M_X(t) = 1 + tE[X] + \frac{t^2}{2!}E[X^2] + \cdots + \frac{t^n}{n!}E[X^n] + \cdots \tag{8.3}$$

If we differentiate the above equation n times with respect to t, all terms involving $E[X^k]$ for $k < n$ disappear. If we then set $t = 0$ in the resultant expression, all terms involving $E[X^k]$ for $k > n$ will disappear, leaving only one term, i.e., $E[X^n]$. Thus,

$$\left.\frac{d^n}{dt^n} M_X(t)\right|_{t=0} = \frac{n!}{n!}E[X^n]; \tag{8.4}$$

Table 8.1. Four different types of transform methods

Transform	Definition	Inverse transform
MGF	$M_X(t) = E\left[e^{tX}\right]$	$E\left[X^n\right] = M_X^{(n)}(0)$
CF	$\phi_X(u) = E\left[e^{iuX}\right]$	$f_X(x) = \mathcal{F}\{\phi_X(u)\}$
PGF	$P_X(z) = E\left[z^X\right]$	$p_k = \frac{P_X^{(k)}(0)}{k!}$
LT	$\Phi_X(s) = E\left[e^{-sX}\right]$	$f_X(x) = \mathcal{L}^{-1}\{\Phi_X(s)\}$

i.e.,

$$\boxed{E[X^n] = M_X^{(n)}(0), n = 0, 1, 2, \ldots,} \tag{8.5}$$

where the case $n = 0$ should be interpreted as $M_X(0) = 1$. Thus, the function $M_X(t)$ generates all the moments of X simply by its differentiation, hence the name *moment-generating function* (MGF).

The natural logarithm of $M_X(t)$ is denoted by $m_X(t)$:

$$\boxed{m_X(t) = \ln M_X(t), t \in I,} \tag{8.6}$$

which is referred to as the **logarithmic MGF** (log-MGF) or the **cumulant MGF**. It is easy to show that (Problem 8.1)

$$m'_X(0) = \left.\frac{M'_X(t)}{M_X(0)}\right|_{t=0} = E[X] \tag{8.7}$$

and

$$m''_X(0) = \left.\frac{M''_X(t)M_X(t) - (M'_X(t))^2}{M_X^2(t)}\right|_{t=0} = E[X^2] - (E[X])^2 = \sigma_X^2. \tag{8.8}$$

The last two formulas are sometimes simpler to deal with than the MGF $M_X(t)$.

Example 8.1: MGF of the binomial distribution. Consider the binomial distribution defined in (2.38):

$$B(k; n, p) = \binom{n}{k} p^k q^{n-k}, \; k = 0, 1, 2, \ldots, n, \tag{8.9}$$

where $q = 1 - p$. The corresponding MGF is

$$\boxed{M_X(t) = \sum_{k=0}^{n} e^{tk} \binom{n}{k} p^k q^{n-k} = \left(p e^t + q\right)^n, -\infty < t < \infty;} \tag{8.10}$$

where the interval of convergence is $I = \mathbb{R}$. The log-MGF is

$$m_X(t) = n \ln \left(p e^t + q\right), -\infty < t < \infty. \tag{8.11}$$

We can readily find the expectation

$$E[X] = M'_X(0) = n(p e^t + q)^{n-1} p e^t \Big|_{t=0} = np \tag{8.12}$$

and the second moment

$$E[X^2] = M''_X(0) = n^2 p^2 + npq. \tag{8.13}$$

Thus, the variance of X is found as

$$\text{Var}[X] = E[X^2] - E^2[X] = npq. \tag{8.14}$$

Alternatively, we can use the log-MGF, obtaining

$$E[X] = m'_X(0) = n \frac{p e^t}{p e^t + q}\Big|_{t=0} = np \tag{8.15}$$

and

$$\sigma_X^2 = m''_X(0) = n \frac{p e^t (p e^t + q) - (p e^t)^2}{(p e^t + q)^2}\Big|_{t=0} = npq. \tag{8.16}$$

□

Example 8.2: MGF of Poisson distribution. Suppose that X is a Poisson RV whose distribution is defined by (3.77):

$$P(k; \lambda) = \frac{\lambda^k}{k!} e^{-\lambda}, k = 0, 1, 2, \ldots. \tag{8.17}$$

Its MGF is

$$M_X(t) = E[e^{tX}] = e^{-\lambda} \sum_{k=0}^{\infty} \frac{(\lambda e^t)^k}{k!} = e^{\lambda(e^t - 1)}. \tag{8.18}$$

Then, the log-MGF is readily given as

$$\boxed{m_X(t) = \lambda(e^t - 1), -\infty < t < \infty,} \tag{8.19}$$

from which we find

$$E[X] = m'_X(0) = \lambda e^t \Big|_{t=0} = \lambda \tag{8.20}$$

and

$$\mathrm{Var}[X] = m''_X(0) = \lambda e^t \Big|_{t=0} = \lambda. \tag{8.21}$$

□

Example 8.3: MGF of the normal distribution. Let RV U be the unit normal RV:

$$f_U(u) = \frac{1}{\sqrt{2\pi}} e^{-u^2/2}, \; -\infty < u < \infty. \tag{8.22}$$

Its MGF is

$$M_U(t) = E[e^{tU}] = \frac{1}{\sqrt{2\pi}} \int_{-\infty}^{\infty} e^{tu} e^{-u^2/2} \, du = e^{t^2/2} \left[\frac{1}{\sqrt{2\pi}} \int_{-\infty}^{\infty} e^{-(u-t)^2/2} \, du \right]. \tag{8.23}$$

The last term in square brackets has a value of unity, as is readily seen if the change of variables $x = u - t$ is made. Thus, we have

$$M_U(t) = e^{t^2/2} \tag{8.24}$$

and the log-MGF is given by

$$\boxed{m_U(t) = \frac{t^2}{2}, \; -\infty < t < \infty,} \tag{8.25}$$

from which we confirm

$$E[U] = m'_U(0) = 0 \tag{8.26}$$

and

$$\mathrm{Var}[U] = m''_U(0) = 1. \tag{8.27}$$

By applying the transformation $U = (X - \mu)/\sigma$ or

$$X = \mu + \sigma U, \tag{8.28}$$

we find the MGF of the **normal (or Gaussian) distribution** $N(\mu, \sigma^2)$ as

$$M_X(t) = E\left[e^{t(\mu+\sigma U)}\right] = e^{t\mu} E[e^{t\sigma U}] = e^{t\mu} M_U(t\sigma) = \exp\left[t\mu + \frac{(t\sigma)^2}{2}\right]. \tag{8.29}$$

The log-MGF is

$$\boxed{m_X(t) = t\mu + \frac{(t\sigma)^2}{2}, \; -\infty < t < \infty,} \tag{8.30}$$

from which it is apparent that $E[X] = m'_X(0) = \mu$ and $\mathrm{Var}[X] = m''_X(0) = \sigma^2$.

The nth **central moment** of the normal RV X is given by

$$\boxed{E[(X-\mu)^n] = \sigma^n E[U^n], n = 1, 2, \ldots.} \tag{8.31}$$

By applying the Taylor-series expansion to $M_U(t) = e^{t^2/2}$, we have

$$M_U(t) = 1 + \frac{t^2}{2} + \frac{t^4}{8} + \cdots + \frac{1}{2^k k!} t^{2k} + \cdots. \tag{8.32}$$

Similarly, the Taylor series of the MGF is given, from (8.3), as

$$M_U(t) = 1 + tE[U] + \frac{E[U^2]}{2} t^2 + \cdots + \frac{E[U^n]}{n!} t^n + \cdots. \tag{8.33}$$

By equating the coefficients of like powers of t in these equations, we find

$$E[U^n] = \begin{cases} 0, & n \text{ odd}, \\ \frac{n!}{2^{n/2}(n/2)!}, & n \text{ even}, \end{cases} \tag{8.34}$$

or

$$\boxed{E[U^n] = \begin{cases} 0, & n \text{ odd}, \\ 1 \times 3 \times 5 \cdots (n-3)(n-1), & n \text{ even}. \end{cases}} \tag{8.35}$$

In particular, we find from (8.31) and the last equation,

$$E[(X-\mu)^4] = 3\sigma^4 \text{ and } E[(X-\mu)^6] = 15\sigma^6. \tag{8.36}$$

□

8.1.2 Moment-generating function of sum of independent random variables

Consider the sum of two independent RVs, say, $Y = X_1 + X_2$. The MGF of Y is

$$M_Y(t) = E[e^{tY}] = E[e^{tX_1}]E[e^{tX_2}] = M_{X_1}(t)M_{X_2}(t), \tag{8.37}$$

where we use the property that the expectation of the product of functions of independent RVs is equal to the product of the expectations of these functions (see (3.34) and (4.161)). By mathematical induction, we can generalize the result (8.37) to more than two RVs.

Let $X_i, i = 1, 2, \ldots, m$, be m independent RVs with corresponding MGFs $M_{X_i}(t)$. Define the RV Y to be the sum of the X_i's:

$$Y = \sum_{i=1}^{m} X_i. \tag{8.38}$$

Then the MGF of Y is given by the product of $M_{X_i}(t)$'s

$$\boxed{M_Y(t) = \prod_{i=1}^{m} M_{X_i}(t), t \in I,} \tag{8.39}$$

where I is an interval where all the MGFs $M_{X_i}(t)$ exist.

8.1.3 Joint moment-generating function of multivariate random variables

Suppose that $\boldsymbol{X} = (X_1, X_2, \ldots, X_m)^\top$ is an m-dimensional **vector RV** or **random vector**. We define the **joint MGF $M_{\boldsymbol{X}}(\boldsymbol{t})$**:

$$\boxed{M_{\boldsymbol{X}}(\boldsymbol{t}) \triangleq E\left[e^{t_1X_1+t_2X_2+\cdots+t_mX_m}\right] = E\left[e^{\langle \boldsymbol{t},\boldsymbol{x}\rangle}\right], \text{ for } \boldsymbol{t} \in \boldsymbol{I},} \tag{8.40}$$

where $\boldsymbol{I} \in \mathbb{R}^m$ is a region for which $M_{\boldsymbol{X}}(\boldsymbol{t})$ is finite and $\langle \boldsymbol{t}, \boldsymbol{x}\rangle$ represents the **inner product** (scalar product) of the two vectors.

The joint moment $E[X_1^{n_1} \cdots X_m^{n_m}]$, if it exists, can be obtained by differentiating the joint MGF n_1 times with respect to t_1, n_2 times with respect to t_2, …, n_m times with respect to t_m, and setting $t_1 = t_2 = \cdots = t_m = 0$:

$$\boxed{E\left[X_1^{n_1} \cdots X_m^{n_m}\right] = \left.\frac{\partial^{n_1+\cdots+n_m} M_{\boldsymbol{X}}(\boldsymbol{t})}{\partial t_1^{n_1} \cdots \partial t_m^{n_m}}\right|_{\boldsymbol{t}=\boldsymbol{0}}.} \tag{8.41}$$

It should be observed that the joint MGF of any subset of the components of the random vector $\boldsymbol{X}$ is obtained by setting equal to zero those t_i that correspond to the RVs not included in the subset. For example, the joint MGF of the random vector $(X_1, X_2)^\top$ is given by $M_{\boldsymbol{X}}(t_1, t_2, 0, \ldots, 0)$.

Example 8.4: Joint MGFs of bivariate and multivariate normal distributions. Let us consider the bivariate normal variables $\boldsymbol{X} = (X_1, X_2)^\top$ whose PDF is given by (4.108) of Section 4.3.1. Define a scalar variable Y by

$$Y = t_1X_1 + t_2X_2 = \boldsymbol{t}^\top \boldsymbol{X} \triangleq \langle \boldsymbol{t}, \boldsymbol{X}\rangle. \tag{8.42}$$

Since X_1 and X_2 are both normal RVs, so is Y with mean

$$\mu_Y = t_1\mu_1 + t_2\mu_1 = \langle \boldsymbol{t}, \boldsymbol{\mu}\rangle \tag{8.43}$$

and variance

$$\begin{aligned}\sigma_Y^2 &= E[(Y - \mu_Y)^2] \\ &= t_1^2E[(X_1 - \mu_1)^2] + 2t_1t_2E[(X_1 - \mu_1)(X_2 - \mu_2)] + t_2^2E[(X_2 - \mu_2)^2] \\ &= t_1^2\sigma_1^2 + 2\rho t_1t_2\sigma_1\sigma_2 + t_2^2\sigma_2^2 = \langle \boldsymbol{t}, \boldsymbol{t}\boldsymbol{C}\rangle = \boldsymbol{t}^\top \boldsymbol{C}\boldsymbol{t},\end{aligned} \tag{8.44}$$

where $\boldsymbol{t}^\top = (t_1, t_2)$, ρ is the **correlation coefficient** between X_1 and X_2, defined by

$$\rho = \frac{E[(X_1 - \mu_1)(X_2 - \mu_2)]}{\sigma_1\sigma_2}, \tag{8.45}$$

and $\boldsymbol{C}$ is the **covariance matrix** of the bivariate normal distribution defined by (4.117):

$$\boldsymbol{C} = \begin{bmatrix} \sigma_1^2 & \rho\sigma_1\sigma_2 \\ \rho\sigma_1\sigma_2 & \sigma_2^2 \end{bmatrix}. \tag{8.46}$$

By writing $Y = \sigma_Y U + \mu_Y$, with U being the unit normal variable, the MGF of Y can be expressed[1] as

$$\begin{aligned} M_Y(\xi) &= E[e^{\xi Y}] = e^{\xi\mu_Y} E[e^{\sigma_Y \xi U}] \\ &= e^{\xi\mu_Y} M_U(\sigma_Y\xi) = e^{\xi\mu_Y} e^{(\sigma_Y\xi)^2/2}, \end{aligned} \tag{8.47}$$

where we used the formula (8.24). If we set $\xi = 1$ in the last equation, we have

$$M_Y(1) = E[e^Y] = e^{\mu_Y + (\sigma_Y^2/2)}. \tag{8.48}$$

From the definition of joint MGF of $\boldsymbol{X}$, we have

$$E[e^Y] = E[e^{\langle \boldsymbol{t}, \boldsymbol{X}\rangle}] = M_{\boldsymbol{X}}(\boldsymbol{t}). \tag{8.49}$$

Thus, we find the MGF of $\boldsymbol{X}$:

$$M_{\boldsymbol{X}}(\boldsymbol{t}) = e^{\mu_Y + (\sigma_Y^2/2)} = \exp\left(\boldsymbol{t}^\top \boldsymbol{\mu} + \frac{\boldsymbol{t}^\top \boldsymbol{C}\boldsymbol{t}}{2}\right). \tag{8.50}$$

We can generalize the results of the bivariate normal distribution to a general multivariate normal distribution, where the vector RV $\boldsymbol{X}$ is now

$$\boldsymbol{X} = (X_1, X_2, \ldots, X_m)^\top, \tag{8.51}$$

which has the joint PDF (cf. (4.116))

$$\boxed{f_{\boldsymbol{X}}(\boldsymbol{x}) = \frac{1}{(2\pi)^{m/2}|\det \boldsymbol{C}|^{1/2}} \exp\left[-\frac{(\boldsymbol{x}-\boldsymbol{\mu})^\top \boldsymbol{C}^{-1}(\boldsymbol{x}-\boldsymbol{\mu})}{2}\right],} \tag{8.52}$$

where the covariance matrix $\boldsymbol{C}$ is now an $m \times m$ matrix:

$$\boldsymbol{C} = E[(\boldsymbol{X}-\boldsymbol{\mu})(\boldsymbol{X}-\boldsymbol{\mu})^\top] = \begin{bmatrix} \sigma_1^2 & \rho_{12}\sigma_1\sigma_2 & \cdots & \rho_{1m}\sigma_1\sigma_m \\ \rho_{21}\sigma_2\sigma_1 & \sigma_2^2 & \cdots & \rho_{2m}\sigma_2\sigma_m \\ \vdots & \vdots & \ddots & \vdots \\ \rho_{m1}\sigma_m\sigma_1 & \rho_{m2}\sigma_m\sigma_2 & \cdots & \sigma_m^2 \end{bmatrix}. \tag{8.53}$$

[1] We use ξ instead of t for fear the latter might be confused as $\boldsymbol{t}$.

The corresponding joint MGF is obtained (Problem 8.7) as

$$M_X(t) = \exp\left(t^\top \mu + \frac{t^\top C t}{2}\right), t \in \mathbb{R}^m. \tag{8.54}$$

□

8.2 Characteristic function (CF)

Another function that we find useful for characterizing random variables is the **characteristic function (CF)**. We begin with the CF of a single RV.

8.2.1 Characteristic function of one random variable

If X is a random variable and u is a real parameter, $e^{iuX} = \cos(uX) + i\sin(uX)$ is a complex-valued RV, where $i = \sqrt{-1}$. This interpretation requires that we extend our definition of RV to include mappings from Ω into the complex plane. The expectation of e^{iuX},

$$\phi_X(u) = E[e^{iuX}] = \int_{-\infty}^{\infty} e^{iux}\, dF_X(x), -\infty < u < \infty, \tag{8.55}$$

is called the **CF** of the RV X. In the terminology of Fourier analysis, $\phi_X(u)$ is the **Fourier–Stieltjes transform** of $F_X(x)$. If $F_X(x)$ is a continuous function and its derivative $F'_X(x) = f_X(x)$ exists, the above definition can be replaced by

$$\phi_X(u) = E[e^{iuX}] = \int_{-\infty}^{\infty} e^{iuX} f_X(x)\, dx, \tag{8.56}$$

which means that $\phi_X(u)$ is the **Fourier transform**[2] of the PDF $f_X(x)$.

For a discrete RV,

$$\phi_X(u) = E[e^{iuX}] = \sum_k e^{iux_k} p_X(x_k). \tag{8.57}$$

Formally, the CF $\phi_X(u)$ is obtained from the MGF $M_X(t)$ by the substitution $t = iu$, where $i = \sqrt{-1}$. There are two main advantages in working with the CF instead of the MGF.

1. The MGF $M_X(t)$ may not exist, since the integral of (8.1) may not converge absolutely, whereas the absolute convergence of the CF $\phi_X(u)$ is guaranteed, as shown in (a) of Theorem 8.1.

[2] In most engineering books, the Fourier transform of a function $s(t)$ is defined as $\tilde{S}(\omega) = \int_{-\infty}^{\infty} s(t)e^{-i\omega t}\, dt$, with the minus sign in the exponent of the integrand. But this is just a matter of definition. We may call the complex conjugate $\tilde{S}^*$, instead, the Fourier transform of $s(t)$, if we so wish.

2. The CF may be inverted to obtain the corresponding distribution function $F_X(x)$ (or PDF $f_X(x)$) by using the Fourier transform method. Knowledge of either the CF or the distribution function is equivalent to the knowledge of the other. Such a relationship does not exist between the MGF and the distribution function.

Theorem 8.1 lists the main properties of the CF.

THEOREM 8.1 (Properties of the CF). *The CF satisfies the following:*

(a) ***Absolute convergence.*** *The CF $\phi(u)$ always exists and its absolute magnitude is less than or equal to one:*

$$|\phi(u)| \leq \phi(0) = 1. \tag{8.58}$$

(b) ***Uniform continuity.*** *The CF is* ***uniformly continuous*** *on the real line $-\infty < u < \infty$.*

(c) ***Fourier inversion.*** *Let $\phi(u)$ be the CF of the distribution function $F(x)$ and suppose that $\phi(u)$ is absolutely integrable over $(-\infty, \infty)$. Then $F(x)$ has a bounded continuous PDF given by*

$$\boxed{f(x) = \frac{1}{2\pi}\int_{-\infty}^{\infty} \phi(u)e^{-iux}\,du, -\infty < x < \infty.} \tag{8.59}$$

(d) ***Self-adjoint property.***

$$\phi(-u) = \phi^*(u), \tag{8.60}$$

where $*$ indicates the complex conjugate.

(e) ***Nonnegative definiteness.*** *For any set of real numbers $u_1, u_2, \ldots, u_n$ and complex numbers $z_1, z_2, \ldots, z_n$:*

$$\sum_{j=1}^{n}\sum_{k=1}^{n} \phi(u_j - u_k)z_j z_k^* \geq 0. \tag{8.61}$$

Proof.

(a)

$$|\phi(u)| \leq \int_{-\infty}^{\infty} |e^{iux}|\,dF(x) = \int_{-\infty}^{\infty} dF(x) = 1$$

(b) We need to show that, for given $\epsilon > 0$, there exists a $\delta > 0$ such that whenever $|u_1 - u_2| < \delta$, then $|\phi(u_1) - \phi(u_2)| < \epsilon$.

$$\begin{aligned}|\phi(u+\delta) - \phi(u)| &= |E[e^{iX(u+\delta)} - e^{iXu}]| \\ &\leq E[|e^{iuX}(e^{i\delta X} - 1)|] \leq E[|h(\delta)|],\end{aligned}$$

where $h(\delta) = e^{i\delta X} - 1$. From the identity $|e^{i\delta x} - 1| = 2\left|\sin\frac{\delta x}{2}\right| \leq 2$, we have $|h(\delta)| \leq 2$, *almost surely*.[3] Therefore, $h(\delta) \to 0$, as $\delta \to 0$, and so $E[|h(\delta)|] \to 0$ by applying Lebesgue's dominated convergence theorem (see Theorem 11.10).

(c) This is the inverse Fourier transform theorem, and the reader is referred to an advanced textbook on the Fourier transform (e.g., [45, 323, 349]). A sufficient, but not necessary condition for a function $\phi(u)$ to be the CF of a continuous RV is

$$\int_{-\infty}^{\infty} |\phi(u)|\, dx < \infty.$$

(d) This property is due to the fact that the distribution function is real-valued:

$$\phi(-u) = \int_{-\infty}^{\infty} e^{-iux}\, dF(x) = \phi^*(u).$$

(e) This property comes from the fact that the distribution function is nonnegative.

$$\sum_j \sum_k \phi(u_j - u_k) z_j z_k^* = \sum_j \sum_k \int_{-\infty}^{\infty} \left(z_j e^{iu_j x}\right)\left(z_k^* e^{-iu_k x}\right) dF(x)$$
$$= E\left[\left|\sum_j z_j e^{iu_j X}\right|^2\right] \geq 0.$$

We can generalize the result to a complex function $z(u)$:

$$\int_{-\infty}^{\infty}\int_{-\infty}^{\infty} \phi(u_1 - u_2) z(u_1) z^*(u_2)\, du_1\, du_2 \geq 0,$$

if the integral exists.

□

Another important property is that the CF can also be used for moment generation. We will discuss this and other properties after the next example.

Example 8.5: Characteristic function of the normal distribution. Consider first the unit normal RV X defined by (8.22):[4]

$$f_X(x) = \frac{1}{\sqrt{2\pi}} e^{-x^2/2}, \qquad -\infty < x < \infty. \tag{8.62}$$

Then one may be tempted to find its CF by the simple substitution $t = iu$ in the MGF (8.24). Although it would yield the same result $\phi_X(u) = e^{-u^2/2}$, a mathematically correct derivation of this result should involve the use of complex analysis; note that the CF is defined in terms of the real parameter u.

[3] The concept of almost sure equivalence is discussed in Section 3.1.

[4] We use X here instead of U, since the lower case u is used as the argument of the the characteristic function.

$$\phi_X(u) = E[e^{iuX}] = \int_{-\infty}^{\infty} e^{iux} \frac{e^{-x^2/2}}{\sqrt{2\pi}}\,dx = \frac{e^{-u^2/2}}{\sqrt{2\pi}} \int_{-\infty}^{\infty} \exp\left[-\frac{(x-iu)^2}{2}\right] dx. \tag{8.63}$$

We make the change of variable $s = x - iu$ and integrate in the complex plane, obtaining

$$\phi_X(u) = \frac{e^{-u^2/2}}{\sqrt{2\pi}} \int_{-\infty-iu}^{\infty-iu} e^{-s^2/2}\,ds, \tag{8.64}$$

where the integration should be made along a line parallel to the real axis.

In order to obtain the **line integral** of (8.64), we consider integration of $e^{-s^2/2}$ along the rectangular contour sketched in Figure 8.1 (a) for the case $u > 0$ and in Figure 8.1 (b) for the case $u < 0$. Since the function $e^{-s^2/2}$ is **analytic** (i.e., possessing no poles), the integral around the entire contour of Figure 8.1 (a) is zero. This is known as the **Cauchy–Goursat integral theorem** in **complex analysis**; i.e., the *theory of functions of complex variables* (e.g., see [243, 302]):

$$\int_{-\alpha-iu}^{\alpha-iu} e^{-s^2/2}\,ds + \int_{\alpha-iu}^{\alpha} e^{-s^2/2}\,ds + \int_{\alpha}^{-\alpha} e^{-x^2/2}\,dx + \int_{-\alpha}^{-\alpha-iu} e^{-s^2/2}\,ds = 0 \tag{8.65}$$

The second and fourth terms approach zero as $\alpha \to \infty$, because the integral in both cases contains a factor $e^{-\alpha^2/2}$, which will exponentially approach zero as $\alpha \to \infty$.[5]

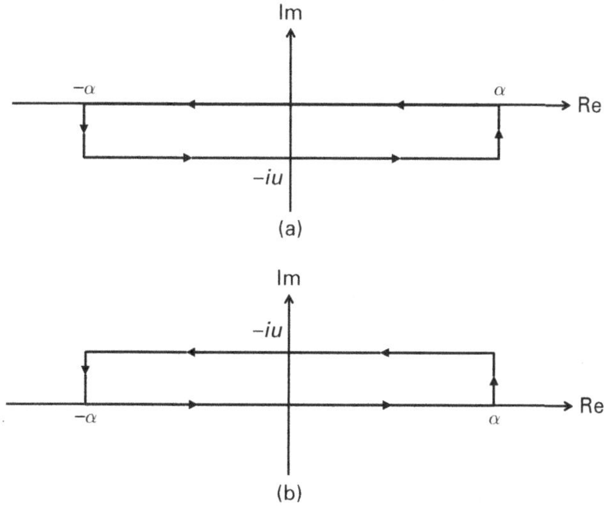

Figure 8.1 Contours for complex integral to obtain the CF of the normal distribution: (a) for $u > 0$; (b) $u < 0$.

[5] For $s = \alpha + iy$; $-u < y < 0$, we have $e^{-s^2/2} = e^{-\alpha^2/2} e^{-i\alpha y + (y^2/2)}$. Similarly, for $s = -\alpha + iy$; $-u < y < 0$, we have $e^{-s^2/2} = e^{-\alpha^2/2} e^{i\alpha y + (y^2/2)}$.

Then by letting α go to infinity and reversing the direction of integration in the third term, we find

$$\int_{-\infty-iu}^{\infty-iu} e^{-s^2/2}\,ds - \int_{-\infty}^{\infty} e^{-x^2/2}\,dx = 0. \tag{8.66}$$

From the definition of the PDF of the unit normal distribution, we know that the second term is $\sqrt{2\pi}$; thus, the first term in (8.66) must also be $\sqrt{2\pi}$. Substituting this result into (8.64), we find

$$\boxed{\phi_X(u) = e^{-u^2/2},\, u > 0.} \tag{8.67}$$

For the case $u < 0$, the contour integral in Figure 8.1 (b) will lead to the same result (Problem 8.14). Thus, (8.67) holds for for both $u > 0$ and $u < 0$. Clearly the above expression holds for $u = 0$ as well, since $\phi_Y(0) = 1$. Hence, (8.67) holds for $-\infty < u < \infty$.

By applying the transformation $Y = (X - \mu)/\sigma$ we find the CF of the normal distribution $N(\mu, \sigma^2)$ to be

$$\begin{aligned}\phi_X(u) &= E[e^{iu(\mu+\sigma Y)}] = e^{iu\mu} E[e^{i(u\sigma)Y}] \\ &= e^{iu\mu}\phi_Y(u\sigma) = \exp\left[iu\mu - \frac{(u\sigma)^2}{2}\right], -\infty < u < \infty,\end{aligned} \tag{8.68}$$

which again could have been found, in a short cut, by formally substituting $t = iu$ into the MGF of $N(\mu, \sigma^2)$ given by (8.29), although it is not a mathematically allowable derivation. There is, however, a mathematically acceptable derivation of the above CF that does not involve complex integration (see Problem 8.13).

As is the case with the MGF discussed in Section 8.1, it is convenient to define the logarithm of the characteristic function, sometimes referred to as the **cumulant generating function (CGF)**, given by

$$\psi_X(u) \triangleq \ln \phi_X(u) = iu\mu - \frac{(u\sigma)^2}{2}. \tag{8.69}$$

By differentiating the above expression, we obtain

$$\mu_X = (-i)\psi_X'(0) = \mu \text{ and } \sigma_X^2 = (-i)^2\psi_X''(0) = \sigma^2. \tag{8.70}$$

□

8.2.2 Sum of independent random variables and convolution

In Example 5.3 of Section 5.2, the notion of **convolution** was introduced and discussed. We will derive here the same result by using CFs. Consider the sum of two independent RVs, say, $Y = X_1 + X_2$. The CF of Y is

$$\phi_Y(u) = E[e^{iuY}] = E[e^{iuX_1}]E[e^{iuX_2}] = \phi_{X_1}(u)\phi_{X_2}(u), -\infty < u < \infty, \tag{8.71}$$

where we use the property that the expectation of the product of functions of independent RVs is equal to the product of the expectations of the functions. Thus, by inverting the above result, we have

$$f_Y(y) = \frac{1}{2\pi}\int_{-\infty}^{\infty} \phi_{X_1}(u)\phi_{X_2}(u)e^{-iuy}\,du. \tag{8.72}$$

By substituting $\phi_{X_1}(u)$ of (8.56) into the above, and rearranging the order of integration, we obtain

$$f_Y(y) = \int_{-\infty}^{\infty} f_{X_1}(x)\left[\frac{1}{2\pi}\int_{-\infty}^{\infty} \phi_{X_2}(u)e^{-iu(y-x)}\,du\right]dx, \tag{8.73}$$

which, with (8.59), leads to

$$f_Y(y) = \int_{-\infty}^{\infty} f_{X_1}(x)f_{X_2}(y-x)\,dx. \tag{8.74}$$

This expression is the **convolution integral** (or simply **convolution**) of $f_{X_1}(\cdot)$ and $f_{X_2}(\cdot)$, as we defined in (5.27). Using the symbol $\circledast$ for convolution, we write

$$\boxed{f_Y(y) = f_{X_1}(y) \circledast f_{X_2}(y).} \tag{8.75}$$

The results (8.71) and (8.75) can be generalized by mathematical induction to many RVs. Let $\{X_k; k = 1, 2, \ldots, m\}$ be m independent RVs with CFs $\phi_k(u)$. Define the RV Y to be the sum of X_k:

$$Y = \sum_{k=1}^{m} X_k. \tag{8.76}$$

Then the CF of Y is given by the product of the $\phi_{X_k}(u)$:

$$\phi_Y(u) = \prod_{k=1}^{m} \phi_{X_k}(u). \tag{8.77}$$

Correspondingly, the PDF of Y is given by the ***m*-fold convolution**

$$f_Y(y) = f_{X_1}(y) \circledast f_{X_2}(y) \circledast \cdots \circledast f_{X_m}(y). \tag{8.78}$$

Example 8.6: Sum of independent normal variables. Let X_1 and X_2 be independent normal variables with distributions $N(\mu_1, \sigma_1^2)$ and $N(\mu_2, \sigma_2^2)$ respectively. Let $Y = X_1 + X_2$. Then, the CF of the RV Y is found from (8.68) as

$$\phi_Y(u) = \phi_{X_1}(u)\phi_{X_2}(u) = \exp\left[iu(\mu_1 + \mu_2) - \frac{u^2(\sigma_1^2 + \sigma_2^2)}{2}\right]. \tag{8.79}$$

Thus, we see that the RV Y is also a normal variable with distribution $N(\mu, \sigma^2)$, with

$$\mu = \mu_1 + \mu_2 \text{ and } \sigma^2 = \sigma_1^2 + \sigma_2^2.$$

The **reproductive property** of normal variables or distributions holds for the sum of any number of independent normal variables defined by (8.76). □

8.2.3 Moment generation from characteristic function

As we stated earlier, the CF is related to the MGF by $\phi_X(u) = M_X(iu)$. Thus, it is apparent that we should be able to generate moments $E[X^n]$ using the CF. By taking the nth derivative of the CF (8.55) with respect to u, we have

$$\phi_X^{(n)}(u) \triangleq \frac{d^n \phi_X(u)}{du^n} = \int_{-\infty}^{\infty} (ix)^n e^{iux} f_X(x)\, dx, \; -\infty < u < \infty. \tag{8.80}$$

Evaluating (8.80) at $u = 0$, we find

$$\boxed{E[X^n] = (-i)^n \phi_X^{(n)}(0), \text{ for } n = 1, 2, \ldots} \tag{8.81}$$

Suppose that the Taylor-series expansion of the CF exists throughout some interval in u that contains the origin. We may then write

$$\phi_X(u) = \sum_{n=0}^{\infty} \frac{\phi_X^{(n)}(0) u^n}{n!}. \tag{8.82}$$

It then follows, using (8.81) and the property that $i^2 = -1$, that

$$\boxed{\phi_X(u) = \sum_{n=0}^{\infty} E[X^n] \frac{(iu)^n}{n!}, \; -\infty < u < \infty.} \tag{8.83}$$

Therefore, the CF is uniquely determined in this interval by the moments of the RV.

The logarithm of the CF $\phi_X(u)$, as defined in (8.69), is the CGF and is denoted $\psi_X(u)$:

$$\psi_X(u) = \ln \phi_X(u). \tag{8.84}$$

We may also expand $\psi_X(u)$ as follows:

$$\psi_X(u) = \sum_{n=0}^{\infty} \kappa_n \frac{(iu)^n}{n!}. \tag{8.85}$$

The quantities κ_n are called **cumulants**. Note that any cumulant κ_n is a polynomial in the moments, and vice versa. In particular, the first two cumulants are the mean μ and the variance σ^2 respectively:

$$\kappa_1 = (-i)\psi_X'(0) = \mu \text{ and } \kappa_2 = (-i)^2 \psi_X''(0) = \sigma^2. \tag{8.86}$$

8.2.4 Joint characteristic function of multivariate random variables

Suppose that $\boldsymbol{X} = (X_1, X_2, \ldots, X_m)^\top$ is an m-dimensional random vector having the joint PDF $f_{\boldsymbol{X}}(\boldsymbol{x})$. Analogous to the joint MGF defined in Section 8.1.3, we define the **joint CF** $\phi_{\boldsymbol{x}}(\boldsymbol{u})$:

$$\begin{aligned}\phi_{\boldsymbol{X}}(\boldsymbol{u}) &= E\left[e^{i(u_1X_1+u_2X_2+\cdots+u_mX_m)}\right] \\ &= \int_{-\infty}^{\infty}\int_{-\infty}^{\infty}\cdots\int_{-\infty}^{\infty} \exp(i\langle\boldsymbol{u},\boldsymbol{x}\rangle) f_{\boldsymbol{x}}(\boldsymbol{x})\,dx_1\,dx_2\cdots dx_m, \end{aligned} \tag{8.87}$$

where $\langle\boldsymbol{u},\boldsymbol{x}\rangle$ represents the inner product of the two vectors. The joint CF always exists, and it assumes its greatest magnitude (which is unity) at the origin $\boldsymbol{u} = \boldsymbol{0}$. The joint moment $E\left[X_1^{n_1}\cdots X_m^{n_m}\right]$, if it exists, can be obtained by differentiating the CF n_1 times with respect to u_1, n_2 times with respect to u_2, …, n_m times with respect to u_m, and setting $u_1 = u_2 = \cdots = u_m = 0$:

$$E\left[X_1^{n_1}\cdots X_m^{n_m}\right] = (-i)^{n_1+\cdots+n_m}\left[\frac{\partial^{n_1+\cdots+n_m}\phi_{\boldsymbol{X}}(\boldsymbol{u})}{\partial u_1^{n_1}\cdots\partial u_m^{n_m}}\right]_{\boldsymbol{u}=\boldsymbol{0}}. \tag{8.88}$$

It should be observed that the joint CF of any subset of the components of the random vector $\boldsymbol{X}$ is obtained by setting equal to zero those u that correspond to the RVs not included in the subset. For example, the CF of (X_1, X_2) is $\phi_{\boldsymbol{X}}(u_1,\ u_2,\ 0,\ \ldots,\ 0)$.

The extension of the inverse transform (8.59) to the case of random vectors is straightforward and may be given as follows:

$$f_{\boldsymbol{X}}(\boldsymbol{x}) = \left(\frac{1}{2\pi}\right)^m \int_{-\infty}^{\infty}\int_{-\infty}^{\infty}\cdots\int_{-\infty}^{\infty} \exp(-i\langle\boldsymbol{u},\boldsymbol{x}\rangle)\phi_{\boldsymbol{x}}(\boldsymbol{x})\,du_1\,du_2\cdots du_m. \tag{8.89}$$

Note that the joint CF can be defined for discrete random vectors as well. The joint CF for the multinomial distribution given in Table 8.2 is such an example (see Problem 8.26).

Example 8.7: Bivariate normal distribution. Let us consider a vector of bivariate normal variables, $\boldsymbol{X} = (X_1, X_2)^\top$, whose PDF is given by (4.108) of Section 4.3.1, also referred to in Example 8.4. Because of the similarity of the CF to the MGF, the following argument exactly parallels that of Example 8.4. Define a scalar variable Y by

$$Y = \langle\boldsymbol{u},\boldsymbol{X}\rangle. \tag{8.90}$$

Since X_1 and X_2 are both normal RVs, so is Y, with mean

$$\mu_Y = \langle\boldsymbol{u},\boldsymbol{\mu}\rangle \tag{8.91}$$

and variance

$$\sigma_Y^2 = \boldsymbol{u}^\top\boldsymbol{C}\boldsymbol{u}, \tag{8.92}$$

where $\boldsymbol{C}$ is the covariance matrix of the bivariate normal distribution defined by (4.117) and (8.46), and $\boldsymbol{u}^\top$ is the transpose of the row vector $\boldsymbol{u}$. By writing $Y = \sigma_Y V + \mu_Y$, where V is the standard normal variable,[6] the CF of Y can be expressed as

$$\phi_Y(u) = E\left[e^{iuY}\right] = e^{iu\mu_Y} E\left[e^{iu\sigma_Y V}\right] = e^{iu\mu_Y} e^{-(u\sigma_Y)^2/2}, \tag{8.93}$$

where we used $\phi_V(v) = e^{-v^2/2}$. If we set $u = 1$ in the last equation, we have

$$\phi_Y(1) = E\left[e^{iY}\right] = e^{i\mu_Y - (\sigma_Y^2/2)}. \tag{8.94}$$

But

$$E\left[e^{iY}\right] = E\left[e^{i\langle \boldsymbol{u}, \boldsymbol{X}\rangle}\right] = \phi_{\boldsymbol{X}}(\boldsymbol{u}). \tag{8.95}$$

Thus, we have found the joint CF of the two-dimensional RV $\boldsymbol{X}$ as

$$\boxed{\phi_{\boldsymbol{X}}(\mathbf{u}) = \exp\left(i\langle \boldsymbol{u}, \boldsymbol{\mu}\rangle - \frac{\boldsymbol{u}^\top \boldsymbol{C}\boldsymbol{u}}{2}\right).} \tag{8.96}$$

□

The generalization of the bivariate normal distribution to a general multivariate normal distribution is straightforward. In Table 8.2 we show the **multivariate normal distribution** and its characteristic function.

Table 8.2. Some joint characteristic functions

1.	Multinominal $P[\boldsymbol{X} = \boldsymbol{k}] = \frac{n!}{k_1!k_2!\cdots k_m!} p_1^{k_1} p_2^{k_2} \cdots p_m^{k_m}$, where $k_i \geq 0$ for all i and $k_1 + k_2 + \cdots + k_m = n$. $\phi_{\boldsymbol{X}}(\boldsymbol{u}) = \left[1 + p_1(e^{iu_1} - 1) + p_2(e^{iu_2} - 1) + \cdots + p_m(e^{iu_m} - 1)\right]^n$
2.	Bivariate normal $f_{\boldsymbol{X}}(\boldsymbol{x}) = \frac{1}{2\pi\sigma_1\sigma_2\sqrt{1-\rho^2}} \exp\left[-\frac{1}{2}Q(x_1, x_2)\right]$, where $Q(x_1, x_2) = \frac{1}{1-\rho^2}\left[\frac{(x_1-\mu_1)^2}{\sigma_1^2} - 2\rho\frac{(x_1-\mu_1)(x_2-\mu_2)}{\sigma_1\sigma_2} + \frac{(x_2-\mu_2)^2}{\sigma_2^2}\right]$ $\phi_{\boldsymbol{X}}(\boldsymbol{u}) = \exp\left[i(\mu_1 u_1 + \mu_2 u_2) - \frac{1}{2}\left(\sigma_1^2 u_1^2 + 2\rho\sigma_1\sigma_2 u_1 u_2 + \sigma_2^2 u_2^2\right)\right]$
3.	Multivariate normal $f_{\boldsymbol{X}}(\boldsymbol{x}) = \frac{1}{(2\pi)^{m/2}\lvert\det \boldsymbol{C}\rvert^{1/2}} \exp\left[-\frac{1}{2}(\boldsymbol{x} - \boldsymbol{\mu})^\top \boldsymbol{C}^{-1}(\boldsymbol{x} - \boldsymbol{\mu})\right]$ $\phi_{\boldsymbol{X}}(\boldsymbol{u}) = \exp\left(i\boldsymbol{u}^\top \boldsymbol{\mu} - \frac{1}{2}\boldsymbol{u}^\top \boldsymbol{C}\boldsymbol{u}\right)$, where $\boldsymbol{C} = E[(\boldsymbol{X} - \boldsymbol{\mu})(\boldsymbol{X} - \boldsymbol{\mu})^\top]$: covariance matrix $\boldsymbol{\mu}^\top$ is the transpose of a column vector $\boldsymbol{\mu}$, etc.

[6] We use V here instead of U, since the symbol u is used as the variable of the CF.

8.2.5 Application of the characteristic function: the central limit theorem (CLT)

Now we are ready to discuss the **central limit throrem (CLT)**, which will explain why the normal distribution appears in many circumstances. Suppose that we have a population with an arbitrary distribution function $F(x)$, but with finite mean μ and variance σ^2. Let $\{X_i; 1 \leq i \leq n\}$ be n independent samples from the population. Then we can make the following statement about the manner in which the distribution of the sample average

$$\overline{X}_n = \frac{1}{n}\sum_{i=1}^{n} X_i \tag{8.97}$$

behaves as $n \to \infty$.

THEOREM 8.2 (The central limit theorem). *If $\overline{X}_n$ is the average of n* ***independent samples*** *from a distribution having finite variance σ^2 and mean μ, then*

$$\boxed{\lim_{n\to\infty} P\left[\frac{\sqrt{n}}{\sigma}(\overline{X}_n - \mu) \leq x\right] = \frac{1}{\sqrt{2\pi}}\int_{-\infty}^{x} e^{-u^2/2}\,du\,.} \tag{8.98}$$

Thus, $\overline{X}_n$ is ***asymptotically normally distributed*** *according to $N(\mu, \sigma^2/n)$.*

Proof. Let $\phi_n(u)$ be the CF of the RV $\sqrt{n}(\overline{X}_n - \mu)/\sigma$:

$$\phi_n(u) = E[e^{iu\sqrt{n}(\overline{X}_n-\mu)/\sigma}] = E\left[\exp\left[iu\sum_{i=1}^{n}\frac{(X_i-\mu)}{\sqrt{n}\sigma}\right]\right] = (\phi(u))^n\,, \tag{8.99}$$

where $\phi(u)$ (without the subscript n) is the CF of $(X_i - \mu)/\sqrt{n}\sigma$, common to all X_i, $i = 1, 2, \ldots, n$. By applying the Taylor expansion formula (8.83) to $\phi(u)$, we have

$$\phi(u) = 1 + iE\left[\frac{X-\mu}{\sqrt{n}\sigma}\right]u - \frac{1}{2}E\left[\left(\frac{X-\mu}{\sqrt{n}\sigma}\right)^2\right]u^2 + o\left(\frac{u^2}{n}\right), \tag{8.100}$$

where $o(u^2/n)$ represents the sum of all terms with higher order than u^2, which approaches zero at least as fast as u^2/n, as $u^2/n \to 0$. Therefore, for any given u, we have

$$\boxed{\lim_{n\to\infty}\phi_n(u) = \lim_{n\to\infty}\left[1 - \frac{u^2}{2n} + o\left(\frac{u^2}{n}\right)\right]^n = e^{-u^2/2},} \tag{8.101}$$

where we used the formula $\lim_{x\to 0}(1-x)^{1/x} = e^{-1}$, in which we set $x = u^2/2n$. We know from (8.62) that $e^{-u^2/2}$ is the CF associated with the unit normal distribution $N(0, 1)$. Therefore, the distribution function of $\sqrt{n}\sigma^{-1}(\overline{X}_n - \mu)$ converges to that of the distribution $N(0, 1)$ as $n \to \infty$; i.e.,

$$\lim_{n\to\infty} P\left[\frac{\sqrt{n}}{\sigma}(\overline{X}_n - \mu) \leq x\right] = \frac{1}{\sqrt{2\pi}}\int_{-\infty}^{x} e^{-u^2/2}\,du. \tag{8.102}$$

Thus, according to Lévy's continuity theorem [27], $\overline{X}_n$ is **asymptotically** normally distributed according to $N(\mu, \sigma^2/n)$.[7] □

Note that the CLT does not imply that the PDF of the sample average $\overline{X}_n$ (or its normalized quantity) approaches the normal PDF. If the original variables $\{X_i\}$ are discrete variables (e.g., binomial), then the PDF of $\overline{X}_n$ will be a collection of impulses (i.e., delta functions) no matter how large the sample size n is. In such a case the PDF of $\overline{X}_n$ never approaches the continuous function of a normal PDF. As is implied by (8.102), however, the integral of the PDF over any fixed interval approaches an integrated value of the normal density function.

The above stated theorem is known as the **equal component** case of the CLT that holds under weaker restrictions. For discussions of various forms of the CLT, the reader is directed to Feller [99, 100], Chung [53], and Gnedenko and Kolmogorov [123]. However, we will further discuss various versions of the CLT in Section 11.3.4.

While the limiting distribution of the sample mean (8.97) is normal, we sometimes find that the normal limit gives a relatively poor approximation for the tail of the actual distribution of $\overline{X}_n$ when n is finite. See Feller [99] (Chapter 7) and [100] (Chapter 8) for further details. The well-known example of a probability distribution to which the CLT does not apply is the **Cauchy–Lorentz distribution** (or simply **Cauchy distribution**) defined by

$$f(x) = \frac{1}{\pi\alpha\left[1 + \frac{(x-\mu)^2}{\alpha^2}\right]} = \frac{\alpha}{\pi\left[\alpha^2 + (x-\mu)^2\right]}, \quad -\infty < x < \infty. \tag{8.103}$$

It is not difficult to show that the variance of the Cauchy distribution is infinite (Problem 8.17(b)). In fact, the sum of any number of independent Cauchy RVs has the same distribution as any one of them; hence, the average of n independent observations is no better than a single observation in this case (Problem 8.27).

8.2.6 Characteristic function of multivariate complex-valued normal variables

Now let us find the CF of the multivariate normal variables discussed in Section 7.6.2. We use real-valued vector parameters $\boldsymbol{\alpha}$ and $\boldsymbol{\beta}$, instead of $\boldsymbol{u}$, in expressing the CF of the circular symmetric multivariate normal variables $\boldsymbol{X}$ and $\boldsymbol{Y}$:

$$\begin{aligned}\phi_{\boldsymbol{XY}}(\boldsymbol{\alpha}, \boldsymbol{\beta}) &= \left[e^{i(\langle\boldsymbol{\alpha},\boldsymbol{X}\rangle + \langle\boldsymbol{\beta},\boldsymbol{Y}\rangle)}\right] \\ &= \exp\left(-\frac{1}{2}[\boldsymbol{\alpha}^\top, \boldsymbol{\beta}^\top]\begin{bmatrix} \boldsymbol{A} & -\boldsymbol{B} \\ \boldsymbol{B} & \boldsymbol{A}\end{bmatrix}\begin{bmatrix}\boldsymbol{\alpha} \\ \boldsymbol{\beta}\end{bmatrix}\right) \\ &= \exp\left(-\frac{1}{2}\left[\boldsymbol{\alpha}^\top\boldsymbol{A}\boldsymbol{\alpha} + \boldsymbol{\beta}^\top\boldsymbol{A}\boldsymbol{\beta} + \boldsymbol{\alpha}^\top\boldsymbol{B}\boldsymbol{\beta} - \boldsymbol{\beta}^\top\boldsymbol{B}\boldsymbol{\alpha}\right]\right).\end{aligned} \tag{8.104}$$

[7] Lévy's continuity theorem states that if a sequence of characteristic functions converges pointwise to a function $\phi(u)$, which is continuous at $u = 0$, then the sequence of the corresponding CDFs converges to a CDF $F(x)$ and the characteristic function associated with $F(x)$ is $\phi(u)$.

By defining

$$\boxed{Z \triangleq X + iY \text{ and } \gamma \triangleq \alpha + i\beta,} \tag{8.105}$$

and defining the inner product[8]

$$\langle \gamma, Z \rangle \triangleq \gamma^\top Z^*, \tag{8.106}$$

we have

$$\langle \gamma, Z \rangle + \langle Z, \gamma \rangle = 2(\langle \alpha, X \rangle + \langle \beta, Y \rangle), \tag{8.107}$$

$$\gamma^\top C \gamma^* = \alpha^\top A \alpha + \beta^\top A \beta + \alpha^\top B \beta - \beta^\top B \alpha, \tag{8.108}$$

where we used $\alpha^\top B \alpha = \beta^\top B \beta = 0$ because B is skew symmetric. We define the CF of the complex-valued RV Z (not necessarily Gaussian) by

$$\boxed{\phi_Z(\gamma) \triangleq E\left[e^{(i/2)(\langle \gamma, Z \rangle + \langle Z, \gamma \rangle)}\right] = E\left[e^{i\Re\{\langle \gamma, Z \rangle\}}\right].}$$

Then, from (8.104), (8.107), and (8.108), we find that the CF of $Z = X + iY$ (where X and Y are circularly symmetric normal variables) is given by

$$\boxed{\phi_Z(\gamma) = \exp\left(-\frac{1}{2}\gamma^\top C \gamma^*\right).} \tag{8.109}$$

We write

$$\gamma = (\gamma_1, \ldots, \gamma_m, \ldots, \gamma_M), \tag{8.110}$$

$$\gamma^* = (\gamma_1^*, \ldots, \gamma_m^*, \ldots, \gamma_M^*). \tag{8.111}$$

Then, the CF defined by (8.109) satisfies the following moment generation properties:

$$\phi_Z(\mathbf{0}) = 1, \tag{8.112}$$

$$\left.\frac{\partial \phi_Z(\gamma)}{\partial \gamma_m}\right|_{\gamma=\mathbf{0}} = \frac{i}{2} E[Z_m^*], \; 1 \le m \le M, \tag{8.113}$$

$$\left.\frac{\partial \phi_Z(\gamma)}{\partial \gamma_m^*}\right|_{\gamma=\mathbf{0}} = \frac{i}{2} E[Z_m], \; 1 \le m \le M, \tag{8.114}$$

$$\left.\frac{\partial^2 \phi_Z(\gamma)}{\partial \gamma_m \gamma_n^*}\right|_{\gamma=\mathbf{0}} = -\frac{1}{4} E[Z_m Z_n^*], \; 1 \le m, n \le M, \tag{8.115}$$

$$\left.\frac{\partial^4 \phi_Z(\gamma)}{\partial \gamma_m \partial \gamma_n^* \partial \gamma_p \partial \gamma_q^*}\right|_{\gamma=\mathbf{0}} = \frac{1}{16} E[Z_m^* Z_n Z_p^* Z_q], \; 1 \le m, n, p, q \le M. \tag{8.116}$$

[8] We can alternatively define the inner product by $\langle \gamma, Z \rangle \triangleq \gamma^{*T} Z$, yet (8.107) and (8.108) remain unchanged.

In particular, for the circularly symmetric normal variables, we obtain the following results (Problem 8.28):

$$\boxed{\begin{aligned} &E[Z_m] = E[Z_m^*] = 0, \\ &E[Z_m Z_n^*] = 2C_{mn}, \\ &E[Z_m Z_n^* Z_p] = 0, \\ &E[Z_m Z_n^* Z_p Z_q^*] = 4[C_{mn}C_{pq} + C_{pn}C_{mq}]. \end{aligned}} \tag{8.117}$$

The derivation of the last property (8.117) makes use of the following property that holds for complex-valued normal variables:

$$\frac{\partial^4 \phi(\mathbf{0})}{\partial\gamma_m \partial\gamma_n^* \partial\gamma_p \partial\gamma_q^*} = \frac{\partial^2 Q(\mathbf{0})}{\partial\gamma_p \partial\gamma_q^*}\frac{\partial^2 Q(\mathbf{0})}{\partial\gamma_m \partial\gamma_n^*} + \frac{\partial^2 Q(\mathbf{0})}{\partial\gamma_p \partial\gamma_n^*}\frac{\partial^2 Q(\mathbf{0})}{\partial\gamma_m \partial\gamma_q^*}, \tag{8.118}$$

where

$$Q(\boldsymbol{\gamma}) = -\frac{1}{2}\boldsymbol{\gamma}^\top \boldsymbol{C}\boldsymbol{\gamma}^* \text{ and } \frac{\partial^2 Q(\mathbf{0})}{\partial\gamma_p \partial\gamma_q^*} \triangleq \left.\frac{\partial^2 Q(\boldsymbol{\gamma})}{\partial\gamma_p \partial\gamma_q^*}\right|_{\boldsymbol{\gamma}=\mathbf{0}},$$

etc., and we use the following properties in arriving at (8.118):

$$\boxed{Q(\mathbf{0})=0,\ \left.\frac{\partial Q(\boldsymbol{\gamma})}{\partial\gamma_m}\right|_{\boldsymbol{\gamma}=\mathbf{0}}=0,\ \frac{\partial^2 Q(\boldsymbol{\gamma})}{\partial\gamma_m \partial\gamma_n^*}=-\frac{1}{4}E[Z_m Z_n^*], \text{ and } \frac{\partial^3 Q(\boldsymbol{\gamma})}{\partial\gamma_m \partial\gamma_n^* \partial\gamma_p}=0.} \tag{8.119}$$

8.3 Summary of Chapter 8

MGF:	$M_X(t) = E[e^{tX}] = \int_{-\infty}^{\infty} e^{tx}\, dF_X(x)$	(8.1)
The nth moment:	$E[X^n] = M_X^{(n)}(0),\ n = 0, 1, 2, \ldots$	(8.5)
Logarithmic MGF:	$m_X(t) = \ln M_X(t),\ -\infty < t < \infty$	(8.6)
The expectation of X:	$E[X] = m'_X(0)$	(8.7)
The variance of X:	$\sigma_X^2 = m''_X(0)$	(8.8)
MGF of the binomial:	$M_X(t) = \sum_{k=0}^{n} e^{tk}\binom{n}{k}p^k q^{n-k} = (p\,e^t + q)^n$	(8.10)
log-MGF of Poisson:	$m_X(t) = \lambda(e^t - 1),\ -\infty < t < \infty$	(8.19)
log-MGF of $N(\mu, \sigma^2)$:	$m_X(t) = t\mu + \frac{(t\sigma)^2}{2},\ -\infty < t < \infty$	(8.30)
Central moments of $N(\mu, \sigma^2)$:	$E[(X-\mu)^n] = \begin{cases} 0, & n \text{ odd} \\ 1 \times 3 \cdots (n-1)\sigma^n, & n \text{ even} \end{cases}$	(8.35)
MGF of sum of independent RVs:	$M_Y(t) = \prod_{i=1}^{m} M_{X_i}(t)$	(8.39)
Definition of joint MGF:	$M_{\boldsymbol{X}}(\boldsymbol{t}) \triangleq E\left[e^{t_1X_1+t_2X_2+\cdots+t_mX_m}\right]$	(8.40)
Joint moments:	$E\left[X_1^{n_1}\cdots X_m^{n_m}\right] = \left.\frac{\partial^{n_1+\cdots+n_m} M_{\boldsymbol{X}}(\boldsymbol{t})}{\partial t_1^{n_1}\cdots\partial t_m^{n_m}}\right\|_{\boldsymbol{t}=\mathbf{0}}$	(8.41)

PDF of multivariate normal:	$f_{\boldsymbol{X}}(\boldsymbol{x})=\frac{1}{(2\pi)^{m/2}\lvert\det \boldsymbol{C}\rvert^{1/2}}e^{-[(\boldsymbol{x}-\boldsymbol{\mu})^\top \boldsymbol{C}^{-1}(\boldsymbol{x}-\boldsymbol{\mu})/2]}$	(8.52)
MGF of multivariate normal:	$M_{\boldsymbol{X}}(\boldsymbol{t}) = \exp\left(\boldsymbol{t}^\top\boldsymbol{\mu} + \frac{\boldsymbol{t}^\top\boldsymbol{C}\boldsymbol{t}}{2}\right)$	(8.54)
Characteristic function:	$\phi_X(u) = E[e^{iuX}] = \int_{-\infty}^{\infty} e^{iux}\, dF_X(x)$	(8.55)
CF of the unit normal distribution:	$\phi_X(u) = e^{-u^2/2}, -\infty < u < \infty$	(8.67)
Moment generation by CF:	$E[X^n] = (-i)^n\phi_X^{(n)}(0)$	(8.81)
Joint CF:	$\phi_{\boldsymbol{X}}(\boldsymbol{u}) = E\left[e^{i(u_1X_1+u_2X_2+\cdots+u_mX_m)}\right]$	(8.87)
Join moment generation:	$E\left[X_1^{n_1}\cdots X_m^{n_m}\right] = (-i)^{n_1+\cdots+n_m}\left[\frac{\partial^{n_1+\cdots+n_m}\phi_{\boldsymbol{X}}(\boldsymbol{u})}{\partial u_1^{n_1}\cdots\partial u_m^{n_m}}\right]_{\boldsymbol{u}=\boldsymbol{0}}$	(8.88)
CF of multivariate normal:	$\phi_{\boldsymbol{X}}(\mathbf{u}) = \exp\left(i\langle\boldsymbol{u},\boldsymbol{\mu}\rangle - \frac{\boldsymbol{u}^\top\boldsymbol{C}\boldsymbol{u}}{2}\right)$	(8.96)
CLT:	$\lim_{n\to\infty} P\left[\frac{\sqrt{n}}{\sigma}(\overline{X}_n-\mu) \le x\right] = \frac{1}{\sqrt{2\pi}}\int_{-\infty}^{x} e^{-u^2/2}\,du$	(8.98)
CF of complex RVs:	$\phi_{\boldsymbol{Z}}(\boldsymbol{\gamma}) \triangleq E\left[e^{(i/2)(\langle\boldsymbol{\gamma},\boldsymbol{Z}\rangle+\langle\boldsymbol{Z},\boldsymbol{\gamma}\rangle)}\right] = E\left[e^{i\Re\{\langle\boldsymbol{\gamma},\boldsymbol{Z}\rangle\}}\right]$	(8.109)
Circularly symmetric Gaussian:	$\phi_{\boldsymbol{Z}}(\boldsymbol{\gamma}) = \exp\left(-\frac{1}{2}\boldsymbol{\gamma}^\top\boldsymbol{C}\boldsymbol{\gamma}^*\right)$	(8.109)

8.4 Discussion and further reading

The CF is discussed in virtually all graduate-level textbooks on probability theory, since it is such an important concept and a useful mathematical device. Unfortunately, however, many authors incorrectly treat a complex integral as if it were an integration of a function defined over the real line $\mathbb{R}$. Even though this "malpractice" [131] yields correct answers in most cases, the right procedure is to integrate a given integrand along a contour appropriately defined in the x–y plane, as demonstrated in Example 8.5. There may sometimes be a way to avoid the use of complex analysis by an ad hoc method (e.g., Problem 8.13), but the orthodox method relies on the contour integral and Cauchy's residue theorem.

The MGF provides some advantages over the CF, especially for those who would prefer to avoid complex integration. But the MGF does not always exist, and does not allow us to recapture the PDF as the CF does: there is no inversion formula that corresponds to the Fourier inversion that applies to the CF. We make extensive use of the MGF in the discussion of Chernoff's bounds and large deviation theory in Chapter 10.

As we noted in Chapter 1, restrictive forms of what we now call the CLT were discussed by De Moivre and Laplace in 1718 and 1812 respectively. The task of perfecting the proof and relaxing its assumptions was taken up by the Russian probability theorists Chebyshev and his students Markov and Lyapunov. Chebyshev and Markov proved the

CLT under conditions weaker than Laplace's using moments, and Lyapunov later proved the same theorem using the CF.

Lindeberg proved the CLT under less restrictive assumptions in 1922; i.e., independent but not identically distributed X_i (as opposed to the **equal components case** assumed in Theorem 8.2 of the present chapter). His streamlined proof that uses the CF was discussed by Lévy in his 1925 book. We will further discuss the CLT in Chapter 11.

8.5 Problems

Section 8.1: Moment-generating function (MGF)

8.1* Properties of logarithmic MGF. Show that the logarithmic (or cumulant) MGF satisfies the following properties

$$m'_X(0) = E[X] = \mu_X$$

and

$$m''_X(0) = E[X^2] - (E[X])^2 = \sigma_X^2.$$

8.2 Uniform distributions.

(a) Show that the MGF of the uniform distribution

$$f_X(x) = \frac{1}{a}, 0 < x < a,$$

is given by

$$M_X(t) = \begin{cases} \frac{e^{at}-1}{at}, & t \neq 0, \\ 1, & t = 0. \end{cases}$$

(b) Derive the MGF of the uniform distribution

$$f_X(x) = \frac{1}{2a}, |x| < a.$$

8.3* Exponential distribution. Find the MGF of the exponential distribution

$$f_X(x) = \mu\, e^{-\mu x}, 0 \leq x < \infty.$$

8.4 Bilateral exponential distribution. Find the MGF of a **bilateral exponential** distribution defined by

$$f_X(x) = \frac{\mu}{2} e^{-\mu|x|}, -\infty < x < \infty,$$

Under what condition does the MGF exist?

8.5 Triangular distribution. Find the MGF of a **triangular distribution** defined by

$$f_X(x) = \frac{1}{a}\left(1 - \frac{|x|}{a}\right), |x| < a.$$

8.6 Negative binomial distribution. The **negative binomial distribution** with parameters (r, p) discussed in Section 3.3.4 takes three different forms:

$$\begin{aligned} P[X = i] &= P[r - 1 \text{ successes in } i - 1 \text{ trials and a success at the } i\text{th trial}] \\ &= \binom{i-1}{r-1} p^r (1-p)^{i-r} = \binom{i-1}{i-r} p^r q^{i-r} \\ &= \binom{-r}{i-r} p^r (-1)^i (1-p)^{i-r}, i = r, r+1, r+2, \ldots . \end{aligned} \tag{8.120}$$

Show that the MGF is given by

$$M_X(t) = \left[\frac{p\, e^t}{1-(1-p)e^t}\right]^r, -\infty < t < \ln\left(\frac{1}{1-p}\right). \tag{8.121}$$

8.7* Multivariate normal distribution. Show that the MGF of the multivariate normal distribution is given by (8.54).

8.8 Multinomial distributions. Consider the **multinomial distribution** defined in Problem 3.12:

$$P[(X_1, X_2, \ldots, X_m) = (k_1, k_2, \ldots, k_m)] = \frac{n!}{k_1!k_2!\cdots k_m!} p_1^{k_1} p_2^{k_2} \cdots p_m^{k_m},$$

where $\sum_{i=1}^m p_i = 1$ and $\sum_{i=1}^m k_i = n$. Show that the joint MGF of this distribution is given by

$$M_{\boldsymbol{X}}(\boldsymbol{t}) = \left(p_1 e^{t_1} + p_2 e^{t_2} + \cdots + p_m e^{t_m}\right)^n, \boldsymbol{t} \in \mathbb{R}^m. \tag{8.122}$$

8.9* Erlang distribution. Consider the r-stage Erlang distribution defined by (4.165), with PDF given by

$$f_{S_r}(x) = \frac{r\lambda(r\lambda x)^{r-1}}{(r-1)!} e^{-r\lambda x}, x \geq 0.$$

Show that the MGF is given by

$$M_{S_r}(t) = \left(\frac{r\lambda}{r\lambda - t}\right)^r, \text{ for } t < r\lambda. \tag{8.123}$$

8.10 Gamma distribution. Consider the Gamma distribution with parameter (λ, β) defined by (4.30):

$$f_{Y_{\lambda,\beta}}(y) = \frac{\lambda^\beta}{\Gamma(\beta)} y^{\beta-1} e^{-\lambda y}, y \geq 0. \tag{8.124}$$

Show that the MGF of this distribution is given by

$$M_{Y_{\lambda,\beta}}(t) = \left(\frac{\lambda}{\lambda - t}\right)^{\beta}, t < \lambda. \tag{8.125}$$

Section 8.2: Characteristic function (CF)

8.11* CF of the binomial distribution. Show that the CF of the binomial distribution $B(k; n, p), k = 0, 1, 2, \ldots, n$, is given by

$$\phi(u) = \left(p\, e^{iu} + 1 - p\right)^{n}, -\infty < u < \infty.$$

8.12 CF of the Poisson distribution. Show that the Poisson distribution with mean λ has the CF given by

$$\phi(u) = e^{\lambda(e^{iu}-1)}, -\infty < u < \infty.$$

8.13 Alternative derivation of (8.67). Obtain (8.67) by taking the following steps.

(a) Show that $\frac{d}{du}\phi_Y(u) = -u\phi_Y(u)$.
(b) Then show $\phi_Y(u)$ is given by (8.67).

8.14 Contour integration. Referring to Example 8.5, the integration in the complex plane must be done along the contour of Figure 8.1 (b) for $u < 0$. Obtain an equation for this case that is comparable to (8.66).

8.15* CF of the exponential distribution. Using contour integration in the complex domain, show that the CF of the exponential distribution with mean a is given by

$$\boxed{\phi(u) = \frac{1}{1 - iau}, -\infty < u < \infty.} \tag{8.126}$$

8.16 CF of the bilateral exponential distribution. The **bilateral exponential** density function is defined by

$$f(x) = \frac{e^{-|x|}}{2}, -\infty < x < \infty. \tag{8.127}$$

Find its CF.

8.17 CF of the Cauchy distribution. The Cauchy distribution (see Problem 5.4 and Section 7.5.1) is defined by

$$f_X(x) = \frac{1}{\pi(1 + x^2)}, -\infty < x < \infty. \tag{8.128}$$

(a) Show that the distribution function is given as

$$\boxed{F_X(x) = \frac{\tan^{-1} x}{\pi} + \frac{1}{2}.} \tag{8.129}$$

(b) Does the mean μ_X exist? What about the second moment $E[X^2]$ and higher moments? Does the Cauchy distribution possess an MGF?

(c) Show that the CF of the Cauchy distribution exists and is given by

$$\boxed{\phi_X(u) = e^{-|u|}, \; -\infty < u < \infty.} \tag{8.130}$$

8.18 Alternative derivation of the CF of the Cauchy distribution. Using the Fourier inverse formula of Theorem 8.1 (c) and the result of Problem 8.16, obtain the CF of the Cauchy distribution.

8.19 CF of the gamma and χ^2 distributions.

(a) Consider the gamma distribution

$$f_X(x) = \frac{x^{\beta-1}e^{-x}}{\Gamma(\beta)}.$$

Show that its CF is

$$\phi_X(u) = \frac{1}{(1-iu)^{\beta}}.$$

(b) Show that the CF of the χ_n^2-distribution is

$$\phi_{\chi_n^2}(u) = \frac{1}{(1-2iu)^{n/2}}.$$

8.20 CF of the noncentral χ^2 distribution. Show that the CF of the noncentral chi-squared variable $\chi_n^2(\mu^2)$ defined by (7.28) is given by (7.29):

$$\phi_{\chi_n^2(\mu^2)}(u) = \frac{1}{(1-2iu)^{n/2}} \exp\left(\frac{iu\mu^2}{1-2iu}\right).$$

8.21 Independent RVs. Let X and Y be independent RVs. Show that the CF of $Z = X + Y$ is given by (8.71).

8.22 CF of a symmetric distribution. If a PDF $f_X(x)$ is symmetric about $x = 0$, show that the CF $\phi_X(u)$ takes on only real values.

8.23 Sum of independent unit normal variables. Let $U_1, U_2, \ldots, U_n$ be independent RVs all having the unit (or standard) normal distribution. Find the PDF of the RV $(1/\sqrt{n})(U_1 + U_2 + \cdots + U_n)$.

8.24 Poisson distribution. The Poisson distribution with parameter λ is defined by

$$p_k = P(k; \lambda) = \frac{\lambda^k}{k!} e^{-\lambda}, k = 0, 1, 2, \ldots.$$

(a) Find the CF of the Poisson distribution.
(b) Compute the mean and variance.
(c) Let X_i be independent Poisson RVs with corresponding parameters λ_i, $i = 1, 2, \ldots, n$. Find the distribution of the RV $Y = X_1 + X_2 + \cdots + X_n$.

8.25 Bernoulli trials. Consider a sequence of Bernoulli trials with probability of success p, and that of failure $q = 1 - p$. The number of trials that precede the first success is a discrete RV, which we denote by X.

(a) Find the probability distribution of X.
(b) Find the mean and variance of X using the CF of this distribution.

8.26 The joint CF of a multinomial distribution. In Problem 3.12 we defined the following multinomial distribution as a generalization of the binomial distribution:

$$p_{\mathbf{k}} = \frac{n!}{k_1!k_2! \cdots k_m!} p_1^{k_1} p_2^{k_2} \cdots p_m^{k_m}. \tag{8.131}$$

(a) Find the joint CF of the above m-dimensional random vector $\mathbf{k} = [k_1, k_2, \ldots, k_m]$.
(b) Apply the moment-generation formula (8.88) and find the mean and variance of k_i and the **covariance** of k_i and k_j defined by

$$\text{Cov}[k_i, k_j] = E\left[(k_i - E[k_i])(k_j - E[k_j])\right].$$

8.27 Sample mean of the Cauchy variables. Show that the sample mean of independent Cauchy variables has the same Cauchy distribution as the component variables.

8.28 Moments of complex-valued multivariate normal variables. Derive the four equations in (8.117).

9 Generating functions and Laplace transform

In addition to the moment generating function (MGF) and characteristic function (CF) methods discussed in the preceding chapter, there are two other related methods that are frequently used in the study of probability theory. They are the **generating function** and the **Laplace transform** (LT).

Discrete RVs often assume integers or integral multiples of some unit, as is the case in counting applications and discrete-time systems. Then, the generating function method will be found to be a convenient device in probability analysis. When a random variable is continuous but nonnegative (e.g., waiting time and service time in a queueing system), we can make use of the rich theory of LTs in the analysis.

Since the CF exists for all distribution functions, both discrete and continuous, why should we study all these other transform methods that seem redundant? Certainly the CF should suffice in most situations, but generating functions and LTs are preferred whenever they are applicable, partly because their notation is somewhat simpler than that of the CF, and partly because there is a rich theory behind the generating function and LT methods, both of which have been widely used as operational methods in system theory that involves differential and integral equations. Thus, it is important for us to be sufficiently familiar with these transform methods to study the literature on probability theory and its applications.

9.1 Generating function

The notion of *generating function* can be more general than the **probability generating function (PGF)** that we will primarily discuss in this section. For a given sequence $\{f_k;\ k = 0, \pm 1, \pm 2, \dots\}$, the generating function is defined as a power series in z^k having as coefficients the values f_k. If the sequence $\{f_k\}$ is bounded, then its generating function converges at least for $|z| < 1$ (Problem 9.1):

$$F(z) = \sum_{k=0}^{\infty} f_k z^k, \ |z| < 1. \tag{9.1}$$

In the field of system analysis, the name "Z-transform" has gained wide acceptance in which usually a transformation based on power series in z^{-k} is used (e.g., see Freeman [112]):

$$\tilde{F}(z) = \sum_{k=0}^{\infty} f_k z^{-k}, |z| > 1. \tag{9.2}$$

If the summation is over a finite number of terms, say $0 \le k \le N$, then $F(z)$ is a *polynomial* in z of order N. This type of representation is used, for instance, for digital filters with finite impulse response (FIR). In coding theory, a *codeword* of length N, $c = (c_0, c_1, \ldots, c_{N-1})$ is compactly represented by a polynomial $C(D) = \sum_{k=0}^{N-1} c_k D^k$. Thus, the generating functions and polynomial representations are used in a variety of scientific and engineering applications, and the PGF to be discussed is just one example of these widely practiced mathematical techniques.

9.1.1 Probability-generating function (PGF)

Consider a random variable that takes on values from a countable (but possibly infinite) set, which we label by nonnegative integers 0, 1, 2, ... Let the probabilities associated with this nonnegative integer variable, denoted X, be

$$P[X = k] = p_k, \ k = 0, 1, 2, \ldots. \tag{9.3}$$

We define the PGF $P_X(z)$ by $E[z^X]$:

$$\boxed{P_X(z) \triangleq E\left[z^X\right] = \sum_{k=0}^{\infty} p_k z^k, |z| \le 1.} \tag{9.4}$$

$P_X(z)$ is clearly a function of the "parameter" z only, since it is obtained by summing over the index k. But it is also a single quantity that represents the entire probability distribution $\{p_0, p_1, p_2, \ldots\}$. We can recover (or generate) the values $p_0, p_1, p_2, \ldots$ from the function $P_X(z)$ assuming that the infinite sum in (9.4) exists for some values of z. The use of generating functions gives us an extremely powerful technique when we deal with certain operations involving RVs or their probabilities.

Example 9.1: Shifted geometric distribution. Consider a sequence of Bernoulli trials with probability of success p and that of failure $q = 1 - p$. Let X represent the number of failures until the first success occurs. Then X has the *shifted geometric distribution* defined in (3.123):

$$p_k = q^k p, k \ge 0. \tag{9.5}$$

The corresponding PGF is then

$$P(z) = p \sum_{k=0}^{\infty} q^k z^k = \frac{p}{1 - qz}, |z| < q^{-1}. \tag{9.6}$$

The region $|z| < q^{-1}$ is called the region of convergence of (9.6), and the number q^{-1} is called the **radius of convergence**. □

Example 9.2: PGF of Poisson distribution. The *Poisson distribution* with mean λ is defined by

$$p_k = \frac{\lambda^k}{k!}e^{-\lambda}, k \geq 0. \tag{9.7}$$

Using the formula $e^x = \sum_{k=0}^{\infty} \frac{x^k}{k!}$, we obtain

$$\boxed{P(z) = \sum_{k=0}^{\infty} \frac{(\lambda z)^k}{k!}e^{-\lambda} = e^{\lambda(z-1)}, |z| < \infty.} \tag{9.8}$$

Thus, the radius of convergence is infinite. □

Table 9.1 summarizes the distributions and the PGFs discussed in the above examples and some other distributions. The reader is suggested to derive these PGFs (Problem 9.2).

9.1.1.1 Generating function of the complementary distribution

Let X be a random variable that assumes integer k with probability p_k and let q_k be the distribution for its tails:

$$q_k \triangleq P[X > k] = p_{k+1} + p_{k+2} + \ldots. \tag{9.9}$$

Table 9.1. Some probability distributions and their PGFs

No.	Name	Probability distribution	Range	PGF
1.	Binomial	$\binom{n}{k}p^k q^{n-k}$	$k = 0, 1, 2, \ldots, n$	$(pz + q)^n$
2.	Poisson	$\frac{\lambda^k}{k!}e^{-\lambda}$	$k = 0, 1, 2, \ldots$	$e^{\lambda(z-1)}$
3.	Geometric[a]	$q^j p$	$j = 0, 1, 2, \ldots$	$\frac{p}{1-qz}$
4.	Geometric	$q^{k-1}p$	$k = 1, 2, 3, \ldots$	$\frac{pz}{1-qz}$
5.	Negative binomial (shifted)	$\binom{-r}{j}p^r(-q)^j = \binom{r+j-1}{j}p^r q^j$	$j = 0, 1, 2, \ldots$	$\left(\frac{p}{1-qz}\right)^r$

[a] j is the number of failures before the first success is attained in a sequence of Bernoulli trials; i.e. a *shifted* geometric distribution.

We denote the PGF of $\{p_k\}$ by $P(z)$ and the generating function of $\{q_k\}$ by $Q(z)$. Then it is not difficult to find the following simple relation (Problem 9.3):

$$\boxed{Q(z) = \frac{1 - P(z)}{1 - z}}. \tag{9.10}$$

9.1.1.2 Expectation and factorial moments

Recall that the MGF $M_X(t)$ and the CF $\phi_X(\theta)$ were used to generate the moments of X. Analogous results that we can obtain from the PGF are the expectation and the **factorial moments**, as shown below.

First we examine

$$P'(z) = \sum_{k=1}^{\infty} k p_k z^{k-1}, \; |z| < 1. \tag{9.11}$$

If we set $z = 1$, the right-hand side reduces to $\sum_k k p_k = E[X]$. Whenever the expectation exists, $P'(z)$ will be continuous in the closed interval $-1 \leq z \leq 1$ on the real line. If $\sum_k k p_k$ diverges, then $P'(z) \to \infty$ as $z \to 1$. In this case, we may write $P'(1) = E[X] = \infty$ and we say the expectation does not exist, or X has infinite expectation.

By applying the *mean value theorem* to the relation (9.10), we see that $Q(z) = P'(w)$, where w is a point lying between z and 1. The function $Q(z)$ increases monotonically as $z \to 1$, and $Q(z)$ approaches $P'(1) = E[X]$; thus, we have the following two different expressions for the expectation:

$$E[X] = \sum_{k=1}^{\infty} k p_k = \sum_{k=0}^{\infty} q_k \tag{9.12}$$

or

$$E[X] = P'(1) = Q(1). \tag{9.13}$$

We obtained an expression equivalent to $E[X] = Q(1)$ in (4.11).

Differentiate (9.11) once more and use the relation $P'(z) = Q(z) - (1 - z)Q'(z)$. Then you will find

$$E[X(X-1)] = \sum_k k(k-1)p_k = P''(1) = 2Q'(1). \tag{9.14}$$

The variance of X is thus expressed as

$$\mathrm{Var}[X] = P''(1) + P'(1) - P'^2(1) = 2Q'(1) + Q(1) - Q^2(1). \tag{9.15}$$

If $P''(1) = \infty$, we say that X has infinite variance, or the variance of X does not exist. The formulas (9.13) and (9.15) provide a quicker way to calculate the mean and variance of X.

Taking the nth derivative with respect to z of (8.89) and setting $z = 1$, we find

$$P_X^{(n)}(1) = \sum_{k=0}^{\infty} k(k-1)\cdots(k-n+1)p_k$$
$$= E[X(X-1)\cdots(X-n+1)]. \tag{9.16}$$

The right-hand side expression is referred to as the **nth factorial moment.**

Example 9.3: Moments of Poisson distribution. For the Poisson distribution of Example 9.2, substituting (9.8) into the above formulas yields

$$E[X] = \lambda, \; E[X^2] - E[X] = \lambda^2. \tag{9.17}$$

Therefore, the variance is given by

$$\sigma_X^2 = E[X^2] - E^2[X] = \lambda^2 + \lambda - \lambda^2 = \lambda. \tag{9.18}$$

□

9.1.2 Sum of independent variables and convolutions

In Section 8.2.2 we observed that the PDF of the sum of two statistically independent, continuous RVs is given by the convolution integral of the individual PDFs. We now consider the discrete analog of the convolution formula (8.74). Let X and Y be independent RVs with probability distributions $\boldsymbol{p} = \{p_k;\; 0 \leq k < \infty\}$ and $\boldsymbol{q} = \{q_k;\; 0 \leq k < \infty\}$ respectively, and let their PGFs be denoted by $P_X(z)$ and $P_Y(z)$ respectively. Then the sum

$$W = X + Y \tag{9.19}$$

has PGF $P_W(z)$, which is the product of the individual PGFs $P_X(z)$ and $P_Y(z)$:

$$P_W(z) = E[z^W] = E[z^X z^Y] = E[z^X]E[z^Y] = P_X(z)P_Y(z). \tag{9.20}$$

Let the probability distribution of the new variable W be denoted by $\boldsymbol{r} = \{r_k;\; 0 \leq k < \infty\}$. Then (9.20) can be written as

$$\sum_{k=0}^{\infty} r_k z^k = \left(\sum_{i=0}^{\infty} p_i z^i\right)\left(\sum_{j=0}^{\infty} q_j z^j\right). \tag{9.21}$$

By equating the coefficients of the terms z^k of both sides, we obtain

$$r_k = p_0 q_k + p_1 q_{k-1} + \cdots + p_k q_0 = \sum_{i=0}^{k} p_i q_{k-i}, \tag{9.22}$$

or, equivalently,

$$r_k = p_k q_0 + p_{k-1} q_1 + \cdots + p_0 q_{k-1} = \sum_{j=0}^{k} p_{k-j} q_j. \tag{9.23}$$

Equation (9.21) or (9.22) is known as the **convolution summation** or simply **convolution** of the distributions $\{p_k\}$ and $\{q_k\}$. We may simply write the above relation, analogously to (8.75), as

$$\boldsymbol{r} = \boldsymbol{p} \circledast \boldsymbol{q} \text{ or } \{r_k\} = \{p_k\} \circledast \{q_k\}. \tag{9.24}$$

We can generalize the foregoing results to the case of an arbitrary number of RVs. Let X_i, $i = 1, 2, \ldots, n$, be statistically independent nonnegative RVs with PGFs $P_i(z)$. A random variable W defined by

$$W = X_1 + X_2 + \cdots + X_n = \sum_{i=1}^{n} X_i \tag{9.25}$$

has a PGF that is the product of the $P_i(z)$ (see Problem 9.8):

$$P_W(z) = P_1(z)P_2(z)\cdots P_n(z) = \prod_{i=1}^{n} P_i(z). \tag{9.26}$$

The probability distribution $\boldsymbol{r}$ of W is given by the n-fold convolution:

$$\boldsymbol{r} = \boldsymbol{p}_1 \circledast \boldsymbol{p}_2 \circledast \cdots \circledast \boldsymbol{p}_n, \tag{9.27}$$

where $\boldsymbol{p}_i$ is the probability distribution of the variable X_i, $n = 1, 2, \ldots, n$.

The summation of RVs given by (9.25) and the product of their PGFs expressed by (9.26) are both **associative and commutative** operations. Thus, the convolution operation is also associative and commutative.

If the X_i have a common distribution $\boldsymbol{p} = \{p_k\}$, and hence a common PGF $P_X(z)$, then the distribution of W will be denoted by

$$\boldsymbol{r} = \boldsymbol{p}^{n\circledast} \text{ or } \{r_k\} = \{p_k\}^{n\circledast}. \tag{9.28}$$

It is apparent that

$$\boldsymbol{p}^{n\circledast} = \boldsymbol{p}^{(n-1)\circledast} \circledast \boldsymbol{p} \text{ or } \{p_k\}^{n\circledast} = \{p_k\}^{(n-1)\circledast} \circledast \{p_k\}. \tag{9.29}$$

Example 9.4: Convolution of binomial distributions. Consider n independent Bernoulli trials and let X be the number of success. We can write the binomial variable X as

$$X = B_1 + B_2 + \cdots + B_n, \tag{9.30}$$

where B_i is the result of ith trial, which takes on 1 with probability p and 0 with probability $q = 1 - p$. Thus, its PGF is given by $P_{B_i}(z) = E\left[z^{B_i}\right] = pz + q$ for all $i = 1, 2, \ldots, n$. Thus, we have

$$P_X(z) = \prod_{i=1}^{n} P_{B_i}(z) = (pz + q)^n, \tag{9.31}$$

as expected (see Table 9.1 and Problem 9.2 (a)). Thus,

$$\{B(k; n, p)\} = \{B(k; 1, p)\}^{n\circledast}. \tag{9.32}$$

The multiplicative property $(pz + q)^m(pz + q)^n = (pz + q)^{m+n}$ implies

$$\{B(k; m, p)\} \circledast \{B(k; n, p)\} = \{B(k; m + n, p)\}, \tag{9.33}$$

from which we find the following formula for binomial coefficients:

$$\binom{m}{0}\binom{n}{k} + \binom{m}{1}\binom{n}{k-1} + \cdots + \binom{m}{k}\binom{n}{0} = \binom{m+n}{k}. \tag{9.34}$$

□

9.1.3 Sum of a random number of random variables

We are often interested in the sum of i.i.d. discrete RVs X_j, $j = 1, \ldots, N$:

$$S_N = X_1 + X_2 + \ldots + X_N, \tag{9.35}$$

where the number N itself is also a random variable, independent of the X_j. Let $\{q_n\}$ be the probability distribution of N and let $P_N(z)$ be its PGF:

$$P[N = n] \triangleq q_n, n = 0, 1, 2, \ldots, \text{ and } P_N(z) \triangleq E\left[z^N\right] = \sum_{n=0}^{\infty} q_n z^n. \tag{9.36}$$

Suppose the probability distribution of X_j, denoted by $\{p_k\}$, is common to all j, and let its PGF be denoted by $P_X(z)$; i.e.,

$$P[X_j = k] \triangleq p_k,\ k \geq 0,\ \text{for all } j \geq 1,\ \text{and } P_X(z) \triangleq E\left[z^{X_j}\right] = \sum_{k=0}^{\infty} p_k z^k. \tag{9.37}$$

We are interested in the probability distribution $\{r_s\}$ of S_N, where $r_s = P[S_N = s]$, $s = 0, 1, 2, \ldots$. Towards this end we first seek its PGF $P_S(z)$, which can be expressed, using the law of iterated expectations (cf. (3.38) and (4.106)), as

$$P_S(z) = E\left[z^{S_N}\right] = E\left[E\left[z^{S_N}|N\right]\right], \tag{9.38}$$

where the *outer* expectation is taken with respect to the random variable N, whereas the *inner* conditional expectation is with respect to the RV S_N, given N. From the result of the previous section, it is apparent that

$$E\left[z^{S_N}|N = n\right] = [P_X(z)]^n, n = 1, 2, 3, \ldots. \tag{9.39}$$

By substituting this into (9.38), we find

$$\boxed{P_S(z) = E\left[(P_X(z))^N\right] = P_N(P_X(z)) \triangleq (P_N \circ P_X)(z).} \tag{9.40}$$

Thus, the PGF $P_S = P_N \circ P_X$ is the **compound function** of P_N and P_X.

Example 9.5: $N \sim$ Poisson and $X_j \sim$ Bernoulli. When the RV N is a Poisson variable with mean λ, i.e., $q_n = \frac{\lambda^n}{n!}e^{-\lambda}$, $n = 0, 1, 2, \ldots$, then $P_N(z) = e^{-\lambda(1-z)}$. If each X_j is a Bernoulli variable with $P[X_j = 1] = p$ and $P[X_j = 0] = 1 - p$, then $P_X(z) = pz + 1 - p$. Then by substituting these into (9.40), we find

$$P_S(z) = e^{-\lambda[1-(pz+1-p)]} = e^{-\lambda p(1-z)}. \tag{9.41}$$

Thus, the corresponding probability distribution is another Poisson distribution with mean λp. Hence,

$$r_s = P\left[S_N = s\right] = \frac{(\lambda p)^s}{s!}e^{-\lambda p}, \quad s = 0, 1, 2, \ldots. \tag{9.42}$$

□

9.1.4 Inverse transform of generating functions

A number of methods exist for finding the probability $\boldsymbol{p} = \{p_k\}$ for a given PGF $P(z)$. Table 9.2 lists five different methods. An obvious inversion method is to find by *inspection* the coefficient of each power term z^k of $P(z)$. The second approach is to obtain the *Taylor-series expansion* of $P(z)$ around $z = 0$:

$$p_k = \frac{P^{(k)}(0)}{k!}. \tag{9.43}$$

The transform pairs given in Table 9.3 will be useful in the inspection method for the inverse transform, as well as in computing the PGF of a given distribution. Note that the transform pairs in Table 9.3 are more broadly applicable to generating functions or polynomial methods in general, not just PGFs.

In some cases $P(z)$ is given in a rather complicated form, and the inspection method or Taylor expansion may not be practical. By treating z as a complex variable, we can apply **Cauchy's residue formula**:

$$p_k = \frac{1}{2\pi i}\oint \frac{P(z)}{z^{k+1}}dz, \tag{9.44}$$

where the integration to be performed is a **contour integral** in the complex plane. We used this technique in Section 8.2, Example 8.5, and Problem 8.17. This requires us to find all poles of $P(z)$, and the amount of computation involved will be comparable to the partial-fraction method when $P(z)$ is a rational function of z.

Table 9.2. Properties of PGF $P(z)$

1.	Definition: $P(z) = E[z^X] = \sum_{k=0}^{\infty} p_k z^k$
2.	Factorial moment: $E[X(X-1)\cdots(X-n+1)] = \lim_{z\to 1} \frac{d^n}{dz^n} P(z) = P^{(n)}(1)$
3.	Inversion of $P(z)$ to find $\{p_k\}$: (a) Inspection: p^k = coefficient of z^k term in $P(z)$ (b) Taylor series expansion around $z = 0$: $p_k = \frac{1}{k!} P^{(k)}(0)$ (c) Contour integral in the complex plane: $p_k = \frac{1}{2\pi i} \oint \frac{P(z)}{z^{k+1}} dz$ (use Cauchy's residue theorem) (d) Partial-fraction method: see Section 9.1.4.1 (e) Recursion method: see Section 9.1.4.3
4.	Use of complementary distribution (or survivor function) $\{q_k\}$, $q_k = P[X > k] = \sum_{j>k} p_j$. $Q(z) \triangleq \sum_{k=0}^{\infty} q_k z^k = \frac{1-P(z)}{1-z}$ (a) Mean: $E[X] = P'(1) = Q(1)$ (b) Second factorial moment: $E[X(X-1)] = P''(1) = 2Q'(1)$ (c) Variance: $\mathrm{Var}[X] = P''(1) + P'(1) - P'^2(1)$ $= 2Q^{(1)}(1) + Q(1) - Q^2(1)$

Table 9.3. Some important transform pairs of generating functions

No.	f_k	$F(z) = \sum_{k=0}^{\infty} f_k z^k$
1.	α^k	$\frac{1}{1-\alpha z}$
2.	$(k+1)\alpha^k$	$\frac{1}{(1-\alpha z)^2}$
3.	$\binom{k+n-1}{k}\alpha^k$	$\frac{1}{(1-\alpha z)^n}$
4.	1 (for all $k \geq 0$)	$\frac{1}{1-z}$
5.	$\delta_{k,0}$	1
6.	k	$\frac{z}{(1-z)^2}$
7.	k^2	$\frac{z(1+z)}{(1-z)^3}$
8.	k^3	$\frac{z(1+4z+z^2)}{(1-z)^4}$
9.	k^n	$\left(z\frac{d}{dz}\right)^n \frac{1}{1-z}$

In terms of numerical inversion by a computer program, the fifth method, i.e., the **recursion method** will probably be more practical than the partial-fraction method or the Cauchy integral method. In the remainder of this section we will discuss the latter two methods.

9.1.4.1 Partial-fraction expansion method

One method of finding $\{p_k;\ k = 0, 1, 2, \ldots\}$ for a given rational function $P(z)$ is to carry out a partial-fraction expansion of $P(z)$. The partial-fraction expansion is a purely algebraic operation for expressing a rational function of z as sum of simple recognizable terms.

Consider a PGF $P(z)$ that is given as a ratio of two polynomials in z:

$$P(z) = \frac{N(z)}{D(z)} = \frac{a_n z^n + a_{n-1} z^{n-1} + \cdots + a_1 z + a_0}{b_d z^d + b_{d-1} z^{d-1} + \cdots\ b_1 z + b_0}. \tag{9.45}$$

If $n \geq d$, we divide $N(z)$ by $D(z)$ until a remainder polynomial $\tilde{N}(z)$ of degree $d - 1$ or less is obtained:

$$P(z) = \sum_{k=0}^{n-d} c_k z^k + \frac{\tilde{N}(z)}{D(z)}. \tag{9.46}$$

First, we assume that $D(z) = 0$ has d **distinct** roots (real or complex), which we denote as $\{z_i;\ i = 1, 2, \ldots, d\}$:

$$D(z) = b_d (z - z_1)(z - z_2) \cdots (z - z_d). \tag{9.47}$$

Then it is known from algebra that $\tilde{N}(z)/D(z)$ can be expanded into **partial fractions**

$$\frac{\tilde{N}(z)}{D(z)} = \frac{f_1}{z_1 - z} + \frac{f_2}{z_2 - z} + \cdots + \frac{f_d}{z_d - z}, \tag{9.48}$$

where $f_1, f_2, \ldots, f_d$ are constants to be determined. To determine f_1, for instance, we multiply (9.48) by $(z_1 - z)$ and let $z \to z_1$. Then the product $(z_1 - z)\frac{\tilde{N}(z)}{D(z)}$ tends to[1] f_1. Thus,

$$f_1 = \lim_{z \to z_1} (z_1 - z) \frac{\tilde{N}(z)}{D(z)} = \frac{-\tilde{N}(z_1)}{b_d (z_1 - z_2)(z_1 - z_3) \cdots (z_1 - z_d)}. \tag{9.49}$$

Now we observe that the denominator, $(z_1 - z_2)(z_1 - z_3) \cdots (z_1 - z_d)$, is equal to $-D'(z_1)$. This can be shown by writing

$$D(z) = (z_1 - z)\tilde{D}(z);$$

differentiating, we have

$$D'(z) = -\tilde{D}(z) + (z_1 - z)\tilde{D}'(z),$$

which leads to $\tilde{D}(z_1) = -D'(z_1)$. Thus, we find $f_1 = -\frac{\tilde{N}(z_1)}{D'(z_1)}$. The same argument applies to the other roots, as well. Thus,

[1] Note that $(z_1 - z)P(z)$ also tends to f_1, as $z \to z_1$. Thus, we can write $f_1 = \lim_{z \to z_1}(z_1 - z)P(z)$.

$$f_i = -\frac{\tilde{N}(z_i)}{D'(z_i)}, \; 1 \le i \le d. \tag{9.50}$$

By writing

$$\frac{1}{z_i - z} = \frac{\alpha_i}{1 - \alpha_i z}, \text{ where } \alpha_i = z_i^{-1}, \tag{9.51}$$

we expand each partial-fraction term into a geometric series

$$\frac{1}{1 - \alpha_i z} = 1 + \alpha_i z + (\alpha_i z)^2 + (\alpha_i z)^3 + \ldots, \; |z| < |z_i|, \; 1 \le i \le d. \tag{9.52}$$

By substituting this into (9.46), we find the coefficient p_k of the z^k term in $P(z)$:

$$p_k = c_k + f_1 \alpha_1^{k+1} + f_2 \alpha_2^{k+1} + \cdots + f_d \alpha_d^{k+1}, \tag{9.53}$$

where $c_k = 0$ for $k \ge n - d + 1$.

Next, we consider the case where there are only $r(< d)$ distinct roots and the ith root $z_i = \alpha_i^{-1}$ has **multiplicity** $m_i (\ge 1)$ $(i = 1, 2, \ldots, r)$. The set of m_i must satisfy $\sum_{i=1}^{r} m_i = d$, which is the degree of $D(z)$. Then, we can write

$$D(z) = b_d (z - z_1)^{m_1} (z - z_2)^{m_2} \cdots (z - z_r)^{m_r}. \tag{9.54}$$

Then the partial-fraction expansion of $\frac{\tilde{N}(z)}{D(z)}$ takes the following form:

$$\begin{aligned} \frac{\tilde{N}(z)}{D(z)} &= \left[\frac{f_{1,1}}{z_1 - z} + \frac{f_{1,2}}{(z_1 - z)^2} + \cdots + \frac{f_{1,m_1}}{(z_1 - z)^{m_1}} \right] + \left[\frac{f_{2,1}}{z_2 - z} + \frac{f_{2,2}}{(z_2 - z)^2} + \cdots \right. \\ &\left. + \frac{f_{2,m_2}}{(z_2 - z)^{m_2}} \right] + \cdots + \left[\frac{f_{r,1}}{z_r - z} + \frac{f_{r,2}}{(z_r - z)^2} + \cdots + \frac{f_{r,m_r}}{(z_r - z)^{m_r}} \right]. \end{aligned} \tag{9.55}$$

To find f_{i,m_i}, first observe that we can write

$$\boxed{D(z) = (z_i - z)^{m_i} \tilde{D}(z),} \tag{9.56}$$

where the polynomial $\tilde{D}(z)$ does not have z_i as a root. We multiply $\frac{\tilde{N}(z)}{D(z)}$ by $(z_i - z)^{m_i}$ and let z tend to z_i. Then

$$f_{i,m_i} = \frac{\tilde{N}(z_i)}{\lim_{z \to z_i} \frac{D(z)}{(z_i - z)^{m_i}}} = \frac{(-1)^{m_i} m_i! \tilde{N}(z_i)}{D^{(m_i)}(z_i)}, \tag{9.57}$$

where we used l'Hôpital's rule,[2] whereby we differentiate both $D(z)$ and $(z_i - z)^{m_i}$ in the above expression m_i times and then set $z = z_i$.

To find f_{i,m_i-1}, after multiplying (9.55) by $(z_i - z)^{m_i}$, we differentiate both sides and let z tend to z_1, obtaining

$$(-1)^{m_i} m_i! \lim_{z\to z_i} \frac{d}{dz}\left[\frac{\tilde{N}(z)}{D^{(m_i)}(z)}\right] = -f_{i,m_i-1}, \tag{9.58}$$

where we observe that the denominator of $(z_i - z)^{m_i}\frac{\tilde{N}(z)}{D(z)} = \frac{\tilde{N}(z)}{\tilde{D}(z)}$ and $D^{(m_i)}(z)$ are related by

$$D^{(m_i)}(z) = (-1)^{m_i} m_i!\tilde{D}(z) + R_i(z), \tag{9.59}$$

where the polynomial $R_i(z)$ contains a factor $(z_i - z)$, and thus will vanish when we set $z = z_i$. Thus, we have

$$\begin{aligned} f_{i,m_i-1} &= (-1)^{m_i-1} m_i! \frac{d}{dz}\left[\frac{\tilde{N}(z)}{D^{(m_i)}(z)}\right]_{z=z_i} \\ &= (-1)^{m_i-1} m_i!\left[\frac{\tilde{N}'(z_i)}{D^{(m_i)}(z_i)} - \frac{\tilde{N}(z_i)D^{(m_i+1)}(z_i)}{\left(D^{(m_i)}(z_i)\right)^2}\right]. \end{aligned} \tag{9.60}$$

By extending the above argument, we can determine the partial expansion coefficients $f_{i,j}$ for $1 \le i \le r$ and $1 \le j \le m_i$ as follows:

$$\boxed{f_{i,j} = \frac{(-1)^{m_i-j}}{(m_i-j)!}\left[\frac{d^{m_i-j}}{dz^{m_i-j}}\left(\frac{\tilde{N}(z)}{\tilde{D}(z)}\right)\right]_{z=z_i} = \frac{(-1)^j m_i!}{(m_i-j)!}\left[\frac{d^{m_i-j}}{dz^{m_i-j}}\left(\frac{\tilde{N}(z)}{D^{(m_i)}(z)}\right)\right]_{z=z_i},} \tag{9.61}$$

where we use the relation (9.59) in the last step.

Since $z_i - z = \alpha_i^{-1} - z = \alpha_i^{-1}(1 - \alpha_i)$, we can write

$$\frac{f_{ij}}{(z_i - z)^j} = \frac{f_{ij}\alpha_i^j}{(1 - \alpha_i z)^j}, \tag{9.62}$$

and using the formula from Table 9.3, we find

$$\frac{1}{(1 - \alpha_i z)^j} = \sum_{k=0}^{\infty}\binom{k+j-1}{k}\alpha_i^k z^k. \tag{9.63}$$

[2] The rule is named after the seventeenth-century French mathematician Guillaume de l'Hôpital (1661–1704), who published the rule in his book (1696), but the rule is believed to be the work of his teacher, the Swiss mathematician Johann Bernoulli (1667–1748).

Thus, we obtain

$$\boxed{p_k = c_k z^k + \sum_{i=1}^{r} \sum_{j=1}^{m_i} f_{ij} \binom{k+j-1}{k} \alpha_i^{j+k},} \tag{9.64}$$

where $c_k = 0$ for $k \geq n - d + 1$.

If we define coefficients $g_{i,j}$ by

$$g_{0,k} \triangleq c_j,\ 1 \leq k \leq n - d, \tag{9.65}$$

$$g_{i,j} \triangleq f_{i,j}\, \alpha_i^j,\ 1 \leq i \leq r,\ 1 \leq j \leq m_i, \tag{9.66}$$

then we have the following alternative expression:

$$p_k = g_{0,k} z^k + \sum_{i=1}^{r} \sum_{j=1}^{m_i} g_{i,j} \binom{k+j-1}{k} \alpha_i^k,\ k = 0, 1, 2, \ldots, \tag{9.67}$$

where $g_{0,k} = 0$ for $k > n - d$, and

$$g_{ij} = \frac{1}{(m_i - j)!(-\alpha_i)^{m_i - j}} \frac{d^{m_i - j}}{dz^{m_i - j}} \left[(1 - \alpha_i z)^{m_i} \frac{\tilde{N}(z)}{D(z)} \right]_{z = \alpha_i^{-1}}, \tag{9.68}$$

where $i = 1, 2, \ldots, r$ and $j = 1, 2, \ldots, m_i$.

Example 9.6: Partial-fraction expansion. Consider a PGF $P(z)$ given by

$$P(z) = \frac{66 - 69z + 3z^2 + 16z^3 - 4z^4}{12(18 - 33z + 20z^2 - 4z^3)}.$$

The numerator has a higher degree than the denominator; hence we divide $D(z)$ into $N(z)$, obtaining

$$P(z) = \frac{1+z}{12} + \frac{24 - 27z + 8z^2}{6(18 - 33z + 20z^2 - 4z^3)}.$$

We find that $D(z) = 0$ has roots $z_1 = 2$ and $z_2 = \frac{3}{2}$ with $m_1 = 1$ and $m_2 = 2$. Thus,

$$\begin{aligned} &\tilde{N}(z) = 24 - 27z + 8z^2,\ \tilde{N}'(z) = 16z - 27, \\ &D(z) = -24(z - z_1)(z - z_2)^2,\ \tilde{D}(z) = -24(z - z_1),\ \tilde{D}'(z) = -24. \end{aligned} \tag{9.69}$$

The partial-fraction expansion of $P(z)$ is

$$P(z) = \frac{1+z}{12} + \frac{f_{1,1}}{z_1 - z} + \frac{f_{2,1}}{z_2 - z} + \frac{f_{2,2}}{(z_2 - z)^2}. \tag{9.70}$$

The expansion coefficients are given, from the above formulas, as

$$c_{00} = c_{01} = \frac{1}{12}$$

and

$$f_{1,1} = -\frac{\tilde{N}(z_1)}{D'(z_1)} = \frac{1}{3}, \quad f_{2,2} = \frac{\tilde{N}(z_2)}{\tilde{D}(z_2)} = \frac{1}{8}, \text{ and } f_{2,1} = \left.\frac{d}{dz}\frac{\tilde{N}(z)}{\tilde{D}(z)}\right]_{z=\frac{3}{2}} = \frac{1}{4} - \frac{1}{4} = 0,$$

which lead to the result

$$P(z) = \frac{1}{12} + \frac{z}{12} + \frac{\frac{1}{3}}{2-z} + \frac{\frac{1}{8}}{(\frac{3}{2}-z)^2} = \frac{1}{12} + \frac{z}{12} + \frac{\frac{1}{6}}{1-\frac{1}{2}z} + \frac{\frac{1}{18}}{(1-\frac{2}{3}z)^2}.$$

Thus, we obtain

$$p_k = \frac{\delta_{k,0}}{12} + \frac{\delta_{k,1}}{12} + \frac{1}{6}\left(\frac{1}{2}\right)^k + \frac{(k+1)}{18}\left(\frac{2}{3}\right)^k,$$

where $\delta_{i,j}$ is the Kronecker delta; that is, $\delta_{i,j} = 1$ for $i = j$ and $\delta_{i,j} = 0$ for $i \neq j$. The first few terms are calculated as

$$p_0 = \frac{1}{12} + \frac{1}{6} + \frac{1}{18} = \frac{11}{36},$$
$$p_1 = \frac{1}{12} + \frac{1}{12} + \frac{4}{54} = \frac{13}{54},$$
$$p_2 = \frac{1}{6}\left(\frac{1}{2}\right)^2 + \frac{3}{18}\left(\frac{2}{3}\right)2 = \frac{25}{216}, \text{ etc.}$$

□

9.1.4.2 Asymptotic formula in partial-fraction expansion

In (9.53) we have exact expressions for the probability p_k. The effort involved in calculating all d roots may be too laborious. In such a case, we may be content with an approximate solution for p_k. Suppose that z_1 is the smallest in absolute value among all the d distinct roots. Then,

$$\alpha_1 \geq \alpha_i, \text{ for } i = 2, 3, \ldots, d.$$

As k increases, the term α_1^k becomes dominant compared with the other terms in (9.53):

$$p_k \sim f_1\alpha_1^{k+1} = g_1\alpha_1^k, \text{ where } g_1 \triangleq f_1\alpha_1. \tag{9.71}$$

Here, the sign $\sim$ means that the ratio of the two sides tends to one, as $k \to \infty$. This **asymptotic** formula provides surprisingly good approximations even for relatively small k.

Now let us consider the case when the smallest root z_1 is a double root; i.e., $m_1 = 2$. Then the term α_1^k in (9.67) has the coefficient

$$g_{1,1} + g_{1,2}(k+1) = g_{1,1} + g_{1,2} + g_{1,2}k.$$

Thus, we have an asymptotic expansion

$$p_k \sim g_{1,2}\, k\alpha_1^k, \text{ for } k \gg 1. \tag{9.72}$$

By generalizing the above argument, we obtain the following asymptotic expression, when the smallest root z_1 is of multiplicity m_1:

$$p_k \sim g_{1,m_1} \binom{k+m_1-1}{k} \alpha_1^k, \text{ for } k \gg 1, \tag{9.73}$$

where $\alpha_1 = z_1^{-1}$ and the constant $g_{1,m_1} = f_{1,m_1}\alpha_1^{m_1}$, and

$$f_{1,m_1} = \frac{(-1)^{m_1} m_1! \tilde{N}(z_1)}{D^{(m)}(z_1)}. \tag{9.74}$$

9.1.4.3 Recursion method

In the partial-fraction method, it is necessary to find the zeros of the denominator $D(z)$. This is not a simple task when the degree d is not small. An alternative technique is to return to the original equation (9.45):

$$\sum_{k=0}^{\infty} p_k z^k = \frac{\sum_{i=0}^{n} a_i z^i}{\sum_{j=0}^{d} b_j z^j}. \tag{9.75}$$

On multiplying the denominator on both sides, we obtain

$$\sum_{k=0}^{\infty}\sum_{j=0}^{d} p_k b_j z^{k+j} = \sum_{i=0}^{n} a_i z^i. \tag{9.76}$$

Comparison of the terms z^i on both sides leads to the following set of **linear difference equations**:

$$\sum_{j=0}^{\min\{d,i\}} p_{i-j} b_j = \begin{cases} a_i, & \text{for } i = 0, 1, \ldots, n, \\ 0, & \text{for } i > n. \end{cases} \tag{9.77}$$

We can then solve for $\{p_i;\ i = 0, 1, 2, \ldots\}$ in a recursive manner:

$$\boxed{p_i = \frac{1}{b_0}\left[a_i - \sum_{j=1}^{\min\{d,i\}} b_j p_{i-j}\right], i = 0, 1, 2, \ldots,} \tag{9.78}$$

where $a_i = 0$ for $i > n$. The recursion method is useful if the numerical evaluation is to be performed on a computer, since the above formula is extremely simple to program.

Table 9.2 gives the various inversion methods discussed above. One method that we did not elaborate in this section is the "Contour integral in the complex plane" method (item 3(c) in Table 9.2). This is similar to what we discussed in Section 8.2 for the Fourier inversion of the CF. Also shown in Table 9.2 are some useful formulas that relate moments to $P(z)$ and $Q(z)$, the generating function of the complementary distribution defined in (9.10).

9.2 Laplace transform method

9.2.1 Laplace transform and moment generation

For continuous RVs, the Laplace transform (LT) method plays a role similar to what the PGF method does for discrete variables. Although the LT method can be applied to continuous RVs that take on values in $(-\infty, \infty)$,[3] we limit our discussion in this section to the case where the RV in question is defined only over $[0, \infty)$.

Let X be a random variable assuming values only on the nonnegative real line, with the PDF $f_X(x)$. We define the LT of $f_X(x)$ by

$$\Phi_X(s) = E[e^{-sX}] = \int_0^\infty f_X(x)e^{-sx}\,dx, \tag{9.79}$$

where s is a complex parameter. It is not difficult to show that

$$|\Phi_X(s)| \le \Phi_X(0) = 1 \text{ for } \Re(s) > 0, \tag{9.80}$$

where $\Re(s)$ means the real part of the complex-valued parameter s. Note the similarity between this transform and the CF defined earlier. The CF exists for any PDF, whereas the LT defined here applies only to nonnegative random variables. We can compute moments of the variable X by differentiating $\Phi_X(s)$ in much the same way as we generate the moments from the CF:

$$E[X^n] = (-1)^n \Phi_X^{(n)}(0), \tag{9.81}$$

which is quite similar to (8.81).

Example 9.7: Exponential random variable. Consider the PDF of the exponentially distributed RV X:

$$f_X(x) = \lambda e^{-\lambda x}. \tag{9.82}$$

The LT of f_X is thus evaluated as

$$\Phi_X(s) = \int_0^\infty \lambda e^{-\lambda x} e^{-sx}\,dx = \frac{\lambda}{s+\lambda}. \tag{9.83}$$

On taking the natural logarithm[4] of $\Phi_X(s)$ and differentiating it with respect to s, we obtain

$$\frac{\Phi'_X(s)}{\Phi_X(s)} = -\frac{1}{s+\lambda}, \tag{9.84}$$

[3] In such a case we often call the method the **bilateral** or **double-sided** LT.

[4] Direct differentiation of $\Phi_X(s)$ is straightforward in this case. If $\Phi_X(s)$ is a rational function of s, the logarithmic transformation significantly simplifies the computation.

which immediately leads to

$$E[X] = -\Phi'_X(0) = \frac{\Phi_X(0)}{\lambda} = \frac{1}{\lambda}. \tag{9.85}$$

By differentiating (9.84) again and setting $s = 0$, we find

$$\Phi''_X(0) - (\Phi'_X(0))^2 = \frac{1}{\lambda^2}, \tag{9.86}$$

which yields

$$E[X^2] = \Phi''_X(0) = \frac{2}{\lambda^2}. \tag{9.87}$$

□

Example 9.8: Consider RV Y, which is also exponentially distributed, but with parameter μ:

$$f_Y(y) = \mu\, e^{-\mu y}. \tag{9.88}$$

Hence, its LT is

$$\Phi_Y(s) = \frac{\mu}{s + \mu}. \tag{9.89}$$

Let us further assume that the variable X of Example 9.7 and Y are statistically independent, and consider their sum

$$W = X + Y. \tag{9.90}$$

The LT of the PDF of the new random variable W is then

$$\begin{aligned}\Phi_W(s) &= E[e^{-sW}] = E[e^{-sX}]E[e^{-sY}] \\ &= \Phi_X(s)\Phi_Y(s) = \frac{\lambda\mu}{(s+\lambda)(s+\mu)}.\end{aligned} \tag{9.91}$$

By a simple algebraic manipulation we can write $\Phi_W(s)$ as

$$\Phi_W(s) = \frac{\lambda\mu}{\mu - \lambda}\left(\frac{1}{s+\lambda} - \frac{1}{s+\mu}\right). \tag{9.92}$$

Then by applying (9.83), we can find the PDF:

$$f_W(x) = \frac{\lambda\mu}{\mu - \lambda}\left(e^{-\lambda x} - e^{-\mu x}\right) u(x), \tag{9.93}$$

where $u(x)$ is the unit step function. □

Table 9.4 lists some important PDFs and their LTs, together with means and variances of these distributions. The various LT pairs and important properties of the LT are tabulated in Tables 9.5 and 9.6.

Table 9.4. Some PDFs and their LTs

Name	PDF	LT	Mean	Variance
Exponential	$\mu e^{-\mu x}, x \geq 0$	$\frac{\mu}{\mu+s}$	$\frac{1}{\mu}$	$\frac{1}{\mu^2}$
k-stage Erlang	$\frac{(k\mu x)^k x^{-1}}{(k-1)!} e^{-k\mu x}, x \geq 0$	$(\frac{k\mu}{k\mu+s})^k$	$\frac{1}{\mu}$	$\frac{1}{k\mu^2}$
Deterministic	$\delta\left(x - \frac{1}{\mu}\right)$	$e^{-s/\mu}$	$\frac{1}{\mu}$	0
Uniform	$\frac{1}{a}, 0 < x < a$	$\frac{e^{as}-1}{as}$	$\frac{a}{2}$	$\frac{a^2}{12}$
Uniform	$\frac{1}{b-a}, a < x < b$	$\frac{e^{bs}-e^{as}}{(b-a)s}$	$\frac{a+b}{2}$	$\frac{(b-a)^2}{12}$
Hyperexponential	$\sum_{i=1}^{n} \pi_i \mu_i e^{-\mu_i x}, x \geq 0$	$\sum_{i=1}^{n} \frac{\pi_i \mu_i}{\mu_i+s}$	$\sum_{i=1}^{n} \frac{\pi_i}{\mu_i}$	$2\sum_{i=1}^{n} \frac{\pi_i}{\mu_i^2} - \left(\sum_{i=1}^{n} \frac{\pi_i}{\mu_i}\right)^2$

Table 9.5. Important LT pairs

No.	Name	Function	LT
1.	A function	$f(x)u(x)$	$\Phi(s) = \int_0^\infty f(x)e^{-sx}dx$
2.	Unit impulse	$\delta(x)$	1
3.	Shifted impulse	$\delta(x-a)$	e^{-as}
4.	Unit step	$u(x)$	$\frac{1}{s}$
5.	Shifted unit step	$u(x-a)$	$\frac{e^{-as}}{s}$
6.	Ramp	$xu(x) = \int_0^x u(t)\,dt$	$\frac{1}{s^2}$
7.	$(n-1)$st power function	$\frac{x^{n-1}}{(n-1)!}u(x)$	$\frac{1}{s^n}$
8.	αth power function	$x^\alpha u(x)$	$\begin{cases} \frac{\Gamma(\alpha+1)}{s^{\alpha+1}}, & \alpha > -1 \\ \frac{\alpha!}{s^{\alpha+1}}, & \alpha \text{ is a positive integer} \end{cases}$
9.	Negative exponential	$e^{-ax}u(x)$	$\frac{1}{s+a}$
10.		$x\,e^{-ax}u(x)$	$\frac{1}{(s+a)^2}$
11.		$\frac{x^{n-1}e^{-ax}}{(n-1)!}u(x)$	$\frac{1}{(s+a)^n}$
12.	Cosine	$(\cos bx)u(x)$	$\frac{s}{s^2+b^2}$
13.	Sine	$(\sin bx)u(x)$	$\frac{b}{s^2+b^2}$
14.	Exponential cosine	$(e^{-ax}\cos bx)u(x)$	$\frac{s+a}{(s+a)^2+b^2}$
15.	Exponential sine	$(e^{-ax}\sin bx)u(x)$	$\frac{b}{(s+a)^2+b^2}$

Table 9.6. Properties of the LT

No.	Name	Function	LT
1.	Shift	$f(x-a)u(x-a)$	$e^{-as}\Phi(s)$
2.	Truncation	$f(x)u(x-a)$	$\Phi(s)-\int_0^a f(x)e^{-sx}dx$
3.	Scaling	$f\left(\frac{x}{a}\right)u(x)$	$a\Phi(as)(a>0)$
4.	Exponential decay	$e^{-ax}f(x)u(x)$	$\Phi(s+a)(a>0)$
5.	Linear growth window	$xf(x)u(x)$	$-\frac{d\Phi(s)}{ds}$
6.	nth power window	$x^n f(x)u(x)$	$(-1)^n\frac{d^n\Phi(s)}{ds^n}$
7.	$1/x$ window	$\frac{f(x)}{x}u(x)$	$\int_s^\infty \Phi(s_1)ds_1$
8.	$1/x^n$ window	$\frac{f(x)}{x^n}u(x)$	$\int_s^\infty ds_1\int_{s_1}^\infty ds_2\cdots\int_{s_{n-1}}^\infty ds_n\Phi(s_n)$
9.	Differentiation	$\frac{df(x)}{dx}u(x)$	$s\Phi(s)-f(0^+)$
10.		$\frac{d\delta(x)}{dx}$	s
11.	Multiple differentiation	$\frac{d^n f(x)}{dx^n}u(x)$	$s^n\Phi(s)-s^{n-1}f(0^+)-s^{n-2}f^{(1)}(0^+)$ $\cdots-f^{(n-1)}(0^+)$
12.		$\frac{d^n\delta(x)}{dx^n}$	s^n
13.	Integration	$\int_a^x f(t)\,dt$	$\frac{\Phi(s)}{s}+\frac{1}{s}\int_a^{0^+}f(t)\,dt$
14.	Integration	$\int_{-\infty}^x f(t)\,dt$	$\frac{\Phi(s)}{s}+\frac{1}{s}\int_{-\infty}^{0^+}f(t)\,dt=\frac{\Phi(s)}{s}+\frac{f^{(-1)}(0^+)}{s}$
15.	Multiple integration	$f^{(-n)}(x)$	$\frac{\Phi(s)}{s^n}+\frac{f^{(-1)}(0^+)}{s^n}+\frac{f^{(-2)}(0^+)}{s^{n-1}}+$ $\cdots+\frac{f^{(-n)}(0^+)}{s}$

9.2.2 Inverse Laplace transform

The Laplace transform of a real-valued function $f(t)$ (not necessarily a PDF) is defined as

$$\Phi(s)=\int_0^\infty e^{-st}f(t)\,dt. \tag{9.94}$$

If $f(t)$ is a piecewise continuous function of **exponential order** α (i.e., $|f(t)|\le Me^{\alpha t}$), the transform function $\Phi(s)$ is defined for $\Re(s)>\alpha$. Here, the parameter α is often called the **abscissa of convergence**.

Conversely, for a given function $\Phi(s)$ of the Laplacian variable s, the inverse transformation to obtain the corresponding $f(t)$ is given by the formula

$$f(t)=\frac{1}{2\pi i}\int_{c-i\infty}^{c+i\infty}\Phi(s)e^{st}\,ds, \tag{9.95}$$

where $i=\sqrt{-1}$ and c can be any real number greater than α. This integral formula is analogous to the inverse formula (8.59) for the CF and the inverse formula (9.44) for

Table 9.7. Other properties of the LT

No.	Name	Property
1.	Linearity	$af_1(x) + bf_2(x) \Leftrightarrow a\Phi_1(s) + b\Phi_2(s)$
2.	Convolution	$f_1(x) \otimes f_2(x) \Leftrightarrow \Phi_1(s)\Phi_2(s)$
3.	Integral property	$\int_0^\infty f(x)dx = \Phi(0)$ If $f_X(x)$ is a PDF, $\Phi_X(0) = 1$
4.	Initial value theorem	$\lim_{x\to 0} f(x) = \lim_{s\to\infty} s\Phi(s)$
5.	Final value theorem	$\lim_{x\to\infty} f(x) = \lim_{s\to 0} s\Phi(s)$ if $s\Phi(s)$ is analytic for $\Re\{s\} \geq 0$
6.	Mean	$E[X] = -\frac{d\Phi_X(s)}{ds}\vert_{s=0} = -\Phi'_X(0)$
7.	nth moment	$E[X^n] = (-1)^n \Phi_X^{(n)}(0)$
8.		$\Gamma(s) = s\Phi(s) \Longrightarrow g(x) = \frac{d}{dx} f(x)$
9.		$\Gamma(s) = \frac{\Phi(s)}{s} \Longrightarrow g(x) = \int_0^x f(t)dt$
10.		$\Gamma(s) = \Phi(s+a) \Longrightarrow g(x) = e^{-ax} f(x)$
11.		$\Gamma(s) = e^{-as}\Phi(s) \Longrightarrow g(x) = f(x-a)u(x-a)$
12.		$\Gamma(s) = \frac{d\Phi(s)}{ds} \Longrightarrow g(x) = -xf(x)$
13.		$\Gamma(s) = \int_s^\infty \Phi(s)ds \Longrightarrow g(x) = \frac{f(x)}{x}$

the PGF. A straightforward evaluation based on this inversion formula would require a contour integral in the complex s-domain, and involves proper use of Cauchy's residue theorem as we remarked earlier in reference to the inverse transform of PGF (see Table 9.2, item 3(c)). In practice, however, one should attempt to represent a given $\Phi(s)$ in terms of the well-studied functions such as those listed in the right column of Table 9.5, together with useful properties summarized in Tables 9.6 and 9.7. If such an inspection method for the inversion cannot be successfully carried out, one needs to explore an alternative approach.

In this section, we discuss two different methods of carrying out the inverse LT: (1) the partial-fraction method and (2) the numerical-inversion method.

A remark is in order concerning the LT of a PDF of some nonnegative RV X. Although most of the results to be presented below apply to any piecewise continuous function of exponential order, we are primarily interested in the case where $f(\cdot)$ is the PDF of some nonnegative RV X. For this class of functions, the LT $\Phi_X(s)$ always exists for any positive value α (the exponential order defined in the sentence that followed (9.94)), since it is bounded according to

$$|\Phi_X(s)| \leq \int_0^\infty |e^{-sx} f_X(x)|\, dx \leq \int_0^\infty f_X(x)\, dx = 1 \text{ for } \Re(s) > 0. \tag{9.96}$$

In fact, the transform function could exist even for a negative value of α.

9.2.2.1 Partial-fraction expansion method

This method is a continuous counterpart of the partial-fraction expansion method discussed earlier for the z-transform or the PGF. Let us assume that $\Phi_X(s)$ is a rational function of s:

$$\Phi_X(s) = \frac{N(s)}{D(s)} = \frac{a_n s^n + a_{n-1}s^{n-1} + \cdots + a_1 s + a_0}{b_d s^d + b_{d-1}s^{d-1} + \cdots + b_1 s + b_0}. \tag{9.97}$$

Since $\Phi_X(s)$ is the LT of a PDF $f_X(x)$, the degree of $N(s)$ cannot exceed that of $D(s)$ (see Problem 9.21); that is,

$$n \leq d. \tag{9.98}$$

Furthermore, the property $\Phi_X(0) = 1$ immediately implies

$$a_0 = b_0. \tag{9.99}$$

If $n = d$, we divide $N(s)$ by $D(s)$ and obtain the expression

$$\Phi_X(s) = \frac{a_d}{b_d} + \frac{\tilde{N}(s)}{D(s)}, \tag{9.100}$$

where $\tilde{N}(s)$ is a polynomial of degree $d - 1$ or less. We then determine the zeros $\{-\lambda_i, i = 1, 2, \ldots, r\}$ of $D(s)$, obtaining an expression similar to (9.54),

$$D(s) = \prod_{i=1}^{r}(s + \lambda_i)^{m_i}, \tag{9.101}$$

which leads to the following partial-fraction expansion of $\Phi_X(s)$:

$$\Phi_X(s) = \frac{a_d}{b_d} + \sum_{i=1}^{r}\sum_{j=1}^{m_i}\frac{f_{i,j}}{(s + \lambda_i)^j}, \tag{9.102}$$

where the coefficients $\{f_{i,j}\}$ are given by

$$f_{i,j} = \frac{1}{(m_i - j)!}\frac{d^{m_i - j}}{ds^{m_i - j}}\left[(s + \lambda_i)^{m_i}\frac{N_1(s)}{D(s)}\right]_{s=-\lambda_i}. \tag{9.103}$$

Then by applying the formula (see Problem 9.17; also Table 9.4)

$$\int_0^\infty \frac{x^{j-1}}{(j-1)!}e^{-\lambda x}e^{-sx}\,dx = \frac{1}{(s + \lambda)^j}, \tag{9.104}$$

we obtain the PDF

$$f_X(x) = \frac{a_d}{b_d}\delta(x) + \sum_{i=1}^{r}\sum_{j=1}^{m_i}\frac{f_{i,j}x^{j-1}}{(j-1)!}e^{-\lambda_i x}, \tag{9.105}$$

where $\delta(x)$ is the unit-impulse function or the Dirac delta function. The corresponding distribution function is

$$F_X(x) = \frac{a_d}{b_d} + \sum_{i=1}^{r}\sum_{j=1}^{m_i} \frac{f_{i,j}}{\lambda_i}\left[1 - e^{-\lambda_i x}\sum_{k=0}^{j-1}\frac{(\lambda_i x)^k}{k!}\right]. \tag{9.106}$$

Example 9.9: Let $\Phi_X(s)$ be given by

$$\Phi_X(s) = \frac{s^3 + 8s^2 + 22s + 16}{4(s^3 + 5s^2 + 8s + 4)}.$$

By noting $D(s) = 4(s+1)(s+2)^2$, we readily find the zeros: $-\lambda_1 = -1$ with $m_1 = 1$ and $-\lambda_2 = -2$ with $m_2 = 2$. Thus, we can have the following partial-fraction expansion:

$$\Phi_X(s) = \frac{1}{4} + \frac{1}{4(s+1)} + \frac{1}{2(s+2)} + \frac{1}{(s+2)^2}.$$

The corresponding PDF and the distribution function can be readily found using the formulas in Table 9.4 (or No. 9 and No. 10 in Table 9.5) as

$$f_X(x) = \frac{\delta(x)}{4} + \frac{e^{-x}}{4} + \frac{e^{-2x}}{2} + x\,e^{-2x}, 0 \le x < \infty,$$

and

$$\begin{aligned} F_X(x) &= \frac{1}{4} + \frac{1 - e^{-x}}{4} + \frac{1 - e^{-2x}}{4} + \frac{1 - e^{-2x}(1+2x)}{4} \\ &= 1 - \frac{e^{-x} + e^{-2x} + e^{-2x}(1+2x)}{4}, 0 \le x < \infty. \end{aligned} \tag{9.107}$$

□

As discussed in the section on partial-fraction expansion of the PGF, an asymptotic expression can be very useful: it requires us to find the root whose real part is smallest in absolute value. Let $-\lambda_1$ be such a smallest root with multiplicity m_1. Then for large x, we have the asymptotic expression

$$f_X(x) \sim \frac{f_{1,m_1}}{(m_1 - 1)!} x^{m_1 - 1} e^{-\lambda_1 x}, x \gg 1, \tag{9.108}$$

because other terms in (9.105) decay faster as x increases. In the above example problem, an asymptotic expression is given by

$$f_X(x) \sim \frac{e^{-x}}{4}, x \gg 1.$$

9.2.2.2 Numerical-inversion method

Use of the partial-fraction method will become difficult when the degree of $D(s)$ becomes large. A computer program for finding the roots of a polynomial may be available in many scientific program libraries, but the computation of the coefficients c_{ij} is a rather cumbersome task even if we are given a set of λ_i. Another method frequently discussed in the literature is the evaluation of the integral by the **residue theorem**, as is well studied in the theory of complex variables. When calculating residues, however, one faces essentially the same type of difficulty as that pointed out for the partial-fraction method. An alternative approach is to adopt one of **numerical inversion methods**. The literature on numerical inversion of the LT abounds (e.g., see Valkó [335]).

The method to be outlined below is essentially to replace a contour integral by the finite Fourier transform, or to be more specific, by the discrete cosine transform. It is easy to understand and a fast Fourier transform (FFT) program can be used in numerical evaluations. By setting the complex variable s as

$$s = c + i\omega, \tag{9.109}$$

where $i = \sqrt{-1}$, and using the fact that $f_X(x)$ is a real-valued function and $f_X(x) = 0$ for $x < 0$, we can rewrite (9.95) in terms of the cosine transform (Problem 9.22):

$$f_X(x) = \frac{2e^{cx}}{\pi} \int_0^\infty \cos(\omega x)\, \Re\{\Phi(c + i\omega)\}\, d\omega, \ \ x \geq 0. \tag{9.110}$$

Here, the constant c, as discussed earlier, is a number greater than α, the abscissa of convergence. In other words, all poles of the complex function $\Phi_X(s)$ must have their real parts strictly less than c. In the numerical evaluation of $f_X(x)$, choice of a suitable parameter c is an important consideration.

Suppose we wish to evaluate $f_X(x)$ over a finite range $0 \leq x \leq T$, and at $(N+1)$ regularly spaced points $x = 0, \delta, 2\delta, \ldots, N\delta (= T)$. First we approximate the integration of (9.110) along the ω by the following summation:

$$f_X(x) \approx \frac{e^{cx}}{T}\left[\frac{1}{2}\Re\{\Phi(c)\} + \sum_{k=1}^{\infty} \Re\left\{\Phi\left(c + \frac{\pi i k}{2T}\right)\right\} \cos\left(\frac{\pi k x}{2T}\right)\right]. \tag{9.111}$$

The above summation formula is nothing more than a **trapezoidal rule** applied to the integral (Problem 9.23). Using the periodic property of the cosine function, the value of $f_X(x)$ at $x = j\delta = j(T/N)$ can be rewritten as

$$f_X(j\delta) \approx \frac{e^{jc\delta}}{T}\left[\frac{1}{2}g_0 + \sum_{k=1}^{2N-1} g_k \cos\frac{\pi j k}{2N} + \frac{(-1)^j}{2} g_{2N}\right], \ j = 0, 1, \ldots, N, \tag{9.112}$$

where

$$g_0 = \Re\{\Phi(c)\} + 2\sum_{m=0}^{\infty} \Re\left\{\Phi_X\left(c + \frac{2\pi i m}{\delta}\right)\right\} \tag{9.113}$$

and

$$g_k = \sum_{m=0}^{\infty}\left[\Re\left\{\Phi_X\left[c+\frac{2\pi i}{\delta}\left(\frac{k}{4N}+m\right)\right]\right\}\right.$$
$$\left.+\Re\left\{\Phi_X\left[c+\frac{2\pi i}{\delta}\left(1-\frac{k}{4N}+m\right)\right]\right\}\right], \quad k=1,2,\ldots,2N. \tag{9.114}$$

Since the above formula involves only $\Re\{\Phi_X(s)\}$, it is easy to program on a computer.

An appropriate choice of the parameters c, N, and δ is a rather involved question. The interested reader is directed to the literature: e.g., see Dubner and Abate [83], Cooley *et al.* [65], and IBM [156] for the error analysis and further details of this numerical approximation method. Crump [75] and Abate and Whitt [2] also discuss the numerical inversion method outlined above. It should be noted that the choice of $c = 0$ reduces the problem to the Fourier transform (or the CF if $f(t)$ is a PDF), and this selection may be quite acceptable provided the other parameters are appropriately specified.

For other work on the numerical methods for inversion of the LT, see the list of references in [335]. Jagerman [160, 161] presents error analysis of a numerical inversion technique for the LT. Abate *et al.* [1] discuss another numerical inversion method that involves the Laguerre polynomials.

9.3 Summary of Chapter 9

PGF:	$P_X(z) \triangleq E\left[z^X\right] = \sum_{k=0}^{\infty} p_k z^k,\ \lvert z\rvert \le 1$	(9.4)
PGF of Poisson distribution:	$P(z) = \sum_{k=0}^{\infty} \frac{(\lambda z)^k}{k!} e^{-\lambda} = e^{\lambda(z-1)},\ \lvert z\rvert < \infty$	(9.8)
Generating func. of $P[X > k]$:	$Q(z) = \frac{1-P(z)}{1-z}$	(9.10)
PGF of sum of RVs:	$P_W(z) = P_1(z)P_2(z)\cdots P_n(z) = \prod_{i=1}^{n} P_i(z)$	(9.26)
Inverse transform of PGF:	$p_i = \frac{1}{b_0}\left[a_i - \sum_{j=1}^{\min\{d,i\}} b_j p_{i-j}\right],\ i = 0, 1, \ldots$	(9.78)
LT of a nonnegative RV:	$\Phi_X(s) = E[e^{-sX}] = \int_0^{\infty} f_X(x)e^{-sx}\,dx$	(9.79)
LT of the exponential dist.:	$\Phi_X(s) = \int_0^{\infty} \lambda e^{-\lambda x} e^{-sx}\,dx = \frac{\lambda}{s+\lambda}$	(9.83)
Initial value theorem:	$\lim_{x\to 0^-} f(x) = \lim_{s\to\infty} s\Phi(s)$	Table 9.7
Final value theorem:	$\lim_{x\to\infty} f(x) = \lim_{s\to 0} s\Phi(s)$	Table 9.7
nth moment:	$E[X^n] = (-1)^n \Phi_X^{(n)}(0)$	Table 9.7
Inverse LT:	$f(t) = \frac{1}{2\pi i}\int_{c-i\infty}^{c+i\infty} \Phi(s)e^{st}\,ds$	(9.95)
Numerical inversion of LT:	$f_X(j\delta) \approx \frac{e^{jc\delta}}{T}\left[\frac{1}{2}g_0 + \sum_{k=1}^{2N-1} g_k \cos\frac{\pi jk}{2N} + \frac{(-1)^j}{2} g_{2N}\right]$	(9.112)

9.4 Discussion and further reading

Feller [99] devotes one chapter to generating functions, and we follow much of his discussion. Grimmett and Stirzaker [131] discuss the PGF, CF, and MGF under the general title "Generating functions and their applications." Our treatment of the inversion methods for the PGF and the LT, especially that of the partial-fraction expansion method and its asymptotic formula, is perhaps more comprehensive than found in other textbooks.

Queueing theory is a major branch of applied probability, and the random variables associated with queueing models are nonnegative integers (such as the number of customers in a queue) or nonnegative continuous random variables (e.g., waiting time), thus both generating functions and LTs are extensively used (e.g., see Kleinrock [189, 190], Kobayashi [197], and Kobayashi and Mark [203]), where a number of formulas and applications that involve generating functions and LTs will be found. See Feller [100] for comprehensive discussions on LTs and applications, many of which, however, are beyond the scope of this textbook.

9.5 Problems

Section 9.1: Generating function

9.1* Region of convergence for PGF, generating function, and Z-transform. Suppose a given sequence $\{f_k\}$ is bounded.

(a) Show that its generating function $F(z)$ defined by (9.1) converges at least for $|z| < 1$.
(b) Show that its z-transform $\tilde{F}(z)$ defined by (9.2) converges at least for $|z| > 1$.
(c) Show that if the sequence $\{f_k\}$ is a probability distribution $\{p_k\}$, then its PGF converges at least for $|z| \leq 1$.

9.2* Derivation of PGFs in Table 9.1. Derive the PGFs given in Table 9.1.

(a) Binomial distribution: show that the PGF of the binomial distribution is given by $P(z) = (pz + q)^n$.
(b) Geometric distribution: show that the PGF of the unshifted geometric distribution is given by $P(Z) = pz/(1 - qz)$.
(c) Shifted negative binomial distribution: as defined in Section 3.3.4, the shifted negative binomial distribution is defined as the distribution of the number of failures before the rth success is attained in a sequence of Bernoulli trials. Show that the PGF of this distribution is given by $P(z) = \left(\frac{p}{1-qz}\right)^r$.

9.3 Derivation of (9.10). Prove the formula (9.10).

9.4 Moments of binomial distribution. Compute the mean and variance of the binomial distribution using its PGF.

9.5 Examples of a generating function. Find the generating function $F(z)$ for the following sequence f_k:

(a) $f_k = 1$ for all $k = 0, 1, \ldots$.
(b) $f_k = 0$ for $k = 0, 1, \ldots, n$ and $f_k = 1$ for all $k \geq n+1$.
(c) $f_k = \frac{1}{k!}$ for all $k \geq 0$.
(d) $f_k = \binom{n}{k}$, for $k = 0, 1, \ldots, n$.

9.6 Some properties of the PGF. Let $P(z)$ be the PGF of the probability distribution $\{p_k;\ k = 0, 1, 2, \ldots\}$. Show the following properties:

(a) $P(1) = 1$,
(b) $P(0) = p_0$.

9.7 Generating function of a sequence. Let $F(z)$ be the generating function of a sequence or vector $\{f_k;\ k = 0, 1, 2, \ldots\}$ defined by

$$F(z) = \sum_{k=0}^{\infty} f_k z^k.$$

Find $\{f_k;\ k = 0, 1, 2, \ldots\}$ for the following $F(z)$:

(a) $F(z) = \frac{1}{1-\alpha z}$;
(b) $F(z) = \frac{1}{(1-\alpha z)^2}$;
(c) $F(z) = \frac{\alpha z}{(1-\alpha z)^2}$.

9.8 PGF of a sum of RVs. Show that the PGF of the sum of independent RVs is given by (9.26).

9.9 Convolution of Poisson distributions. Consider Poisson distributions of means λ_1 and λ_2. What is the convolution of the two distributions?

9.10* Shifted negative binomial distributions. Consider the Bernoulli trials.

(a) Let X be the number of failures until the first success; i.e., the *waiting time* for the first success. What is the mean and variance of X?
(b) Let Z_r denote the number of failures needed to achieve r successes as defined in (3.3.4.1). Find the PGF of Z_r and find its mean and variance.
(c) The PGF of Z_r obtained in (b) can also be written as

$$P_{Z_r}(z) = p^r \sum_{j=0}^{\infty} \binom{-r}{j} (-qz)^j,\ |z| < q^{-1}. \tag{9.115}$$

Hint: Use the following binomial expansion formula (also known as **Newton's generalized binomial formula**):

$$(1+t)^a = 1 + \binom{a}{1}t + \binom{a}{2}t^2 + \binom{a}{3}t^3 + \cdots = \sum_{j=0}^{\infty} \binom{a}{j} t^j, \tag{9.116}$$

where $|t| < 1$ and a is any number (real or complex).

(d) Show that the probability distribution of Z_r is

$$f(j; r, p) \triangleq \binom{-r}{j} p^r (-q)^j = \binom{r+j-1}{j} p^r q^j, \ j = 0, 1, 2, \ldots, \tag{9.117}$$

which equals the *shifted negative binomial distribution* obtained in (3.108).

(e) Show that the shifted negative binomial distribution possesses the following reproductive property:

$$\{f(k; r_1, p)\} \circledast \{f(k; r_2, p)\} = \{f(k; r_1 + r_2, p)\} \tag{9.118}$$

9.11 Formula (9.63). Prove the formula (9.63). What does this equation mean for the case $j = 1$?

9.12 Final value theorem. Refer to Problem 9.7. Show that

$$\lim_{z \to 1} (1 - z) F(z) = \lim_{k \to \infty} f_k.$$

9.13 Joint PGF. Suppose $\boldsymbol{X} = (X_1, X_2, \ldots, X_m)^\top$ is an m-dimensional random vector with probability distribution $p_{\boldsymbol{k}} = P[\boldsymbol{X} = \boldsymbol{k}]$, where $\boldsymbol{k} = (k_1, k_2, \ldots, k_m)$. The **joint PGF** is defined by

$$\boxed{P_{\boldsymbol{X}}(\boldsymbol{z}) \triangleq E\left[z_1^{X_1} z_2^{X_2} \cdots z_m^{X_m}\right] = \sum_{k_1} \sum_{k_2} \cdots \sum_{k_m} z_1^{k_1} z_2^{k_2} \cdots z_m^{k_m} p_{\boldsymbol{k}}.} \tag{9.119}$$

(a) Show that the inversion formula is given by

$$p_{\boldsymbol{k}} = \frac{1}{k_1! k_2! \cdots k_m!} \left. \frac{\partial^{k_1 + k_2 + \cdots k_m} P_{\boldsymbol{X}}(\boldsymbol{z})}{\partial z_1^{k_1} \partial z_2^{k_2} \cdots \partial z_m^{k_m}} \right|_{\boldsymbol{z}=\boldsymbol{0}},$$

where $\boldsymbol{0}$ is the vector whose components are all zeros.

(b) Find the joint PGF of the multinomial distribution defined in Problem 3.12.

9.14 Negative binomial (or Pascal) distribution. Recall the negative binomial distribution (or Pascal distribution):

$$P[Y_r = k] = \binom{k-1}{r-1} p^r q^{k-r}, \ k = r, r+1, r+2, \ldots, \tag{9.120}$$

which represents the probability that rth success occurs at the kth trial in a series of Bernoulli trials.

(a) In the Bernoulli trials, let $X^{(i)}$ be the number of additional trials necessary to achieve the ith success, counting from the trial just after the $(i-1)$th success. Let S_r be the sum of the variables

$$S_r = X^{(1)} + X^{(2)} + \cdots + X^{(r)},$$

Show that the probability $S_r = k$ is equivalent to $P[Y_r = k]$ given above.

(b) Find the PGF of the binomial distribution and calculate the mean and variance of the distribution.

9.15* Derivation of the binomial distribution via a two-dimensional generating function $C(z, w)$. Derive the binomial distribution $B(k; n, p)$ using the following steps:

(a) In order to obtain exactly k successes after n trials, either we must have already k successes after $(n-1)$ trials and then fail on the nth trial, or we must have $(k-1)$ successes after $(n-1)$ trials and then succeed on the nth trial. Based on this observation, find a linear difference equation for $B(k; n, p)$, $0 \leq k \leq n$.

(b) Define the PGF $G(z; n, p)$ by

$$G(z; n, p) = \sum_{k=0}^{n} B(k; n, p) z^k.$$

Find a recursive equation which $\{G(z; , n, p), n = 0, 1, \ldots\}$ must satisfy.

(c) Define a two-dimensional generating function $C(z, w)$ by

$$C(z, w; p) = \sum_{n=0}^{\infty} G(z; n, p) w^n.$$

Find closed-form expressions for $C(z, w; p)$ and $G(z; n, p)$. Then obtain the binomial distribution $B(k; n, p)$.

9.16 Example of the recursion method. Apply the recursion method to the $P(z)$ considered in Example 9.6 and obtain $\{p_0, p_1, p_2, \ldots\}$.

Section 9.2: Laplace transform method

9.17 Derivation of the Erlang distribution. Let $X_1, X_2, \ldots, X_k$ be i.i.d. RVs with the exponential distribution of parameter λ (see Section 4.2.3). Let Y be their sample mean defined by

$$Y = \frac{1}{k} \sum_{i=1}^{k} X_i.$$

(a) Find the LT of the PDF of Y.

(b) Show that the PDF of Y is given by

$$\boxed{f_Y(y) = \frac{k\lambda(k\lambda x)^{k-1}}{(k-1)!}e^{-k\lambda y},\ y \geq 0,} \tag{9.121}$$

which is the ***k*-stage Erlang distribution**, as defined in Problem 4.11.

(c) Find the mean and variance of the above distribution.

9.18* Convolution and the LT. Let $f_1(x)$ and $f_2(x)$ be two PDFs defined in the range $x \in [0, \infty)$ and let $g(x)$ be their convolution of the form

$$g(x) = \int_0^\infty f_1(x-y)f_2(y)\,dy.$$

Find the LT of $g(x)$.

9.19 LTs of the distribution function and survivor function. Given a PDF $f(x)$ and the corresponding distribution function $F(x)$, let $\Phi(s)$ be the LT of $f(x)$. Show that:

(a) the LT of the distribution function is

$$\boxed{\int_0^\infty e^{-sx}F(x)\,dx = \frac{\Phi(s)}{s};} \tag{9.122}$$

(b) the LT of the survivor function is

$$\boxed{\int_0^\infty e^{-sx}[1-F(x)]\,dx = \frac{1-\Phi(s)}{s}.} \tag{9.123}$$

9.20 The *n*-fold convolutions of the uniform distribution. Let X_i, $i = 1, 2, \ldots, n$, be i.i.d. RVs with the uniform distribution over the interval [0, 1]; that is,

$$F_X(x) = \begin{cases} 0, & x < 0, \\ x, & 0 \leq x \leq 1, \\ 1, & x > 1. \end{cases}$$

Let Y be their sum $Y = X_1 + X_2 + \cdots + X_n$. Show that the distribution of Y is given by

$$F_Y(y) = \frac{1}{n!}\sum_{k=0}^{n}(-1)^k\binom{n}{k}([y-k]_+)^n,$$

where

$$[x]_+ = \max\{x, 0\}.$$

Hint:

$$\int_0^\infty e^{-sx}\frac{([x-k]_+)^{n-1}}{(n-1)!}\,dx = s^{-n}e^{-ks}.$$

9.21* Discontinuities in a distribution function. Show that the magnitude of the discontinuity of $F_X(x)$ at the origin $x = 0$ is obtained by

$$\lim_{x\to 0+} F_X(x) = \lim_{s\to\infty} \Phi_X(s).$$

Generally, if $F_X(x)$ contains discontinuities of magnitudes p_k at points $x = x_k$, $\Phi_X(s)$ contains the corresponding terms $p_k e^{-sx_k}$.

9.22 Derivation of the cosine transform of (9.110). Show that the inverse LT formula (9.95) can be reduced to (9.110) of the cosine transform.
Hint: Use the fact that $f_X(x)$ is a real-valued function and that $f_X(x) = 0$ for $x < 0$. The latter condition leads to the following equation that the real and imaginary parts of $\Phi(c + i\omega)$ must satisfy:

$$\int_{-\infty}^{\infty} [\Re\{\Phi(c+i\omega)\}\cos\omega t + \Im\{\Phi(c+i\omega)\}\sin\omega t]\, d\omega = 0.$$

9.23 Trapezoidal approximation of an integral. Derive the trapezoidal approximation given by (9.112).
Hint: A trapezoidal approximation of a definite integral

$$\int_a^b f(x)\, dx$$

works as follows. Divide the area under the curve $y = f(x)$ into n strips, each of equal width $h = (b - a)/n$. Then the shape of each strip is approximated by a trapezoid. The kth strip is approximately

$$\frac{h}{2}[f(a + (k-1)) + f(a + kh)],\ k = 1, 2, \ldots, n.$$

Thus, the above definite integral is

$$\begin{aligned}\int_a^b f(x)\, dx &\approx \sum_{k=1}^{n} \frac{h}{2}[f(a + (k-1)) + f(a + kh)] \\ &= h\left[\frac{1}{2}f(a) + \sum_{k=1}^{n-1} f(a + kh) + \frac{1}{2}f(b)\right].\end{aligned}$$

10 Inequalities, bounds, and large deviation approximation

In this chapter we will discuss some important inequalities used in probability and statistics and their applications. They include the Cauchy–Schwarz inequality, Jensen's inequality, Markov and Chebyshev inequalities. We then discuss Chernoff's bounds, followed by an introduction to large deviation theory.

10.1 Inequalities frequently used in probability theory

10.1.1 Cauchy–Schwarz inequality

The Cauchy–Schwarz[1] inequality is perhaps the most frequently used inequality in many branches of mathematics, including linear algebra, analysis, and probability theory. In engineering applications, a matched filter and correlation receiver are derived from this inequality. Since the Cauchy–Schwarz inequality holds for a general *inner product space*, we briefly review its properties and in particular the notion of *orthogonality*. We assume that the reader is familiar with the notion of **field** and **vector space** (e.g., see Birkhoff and MacLane [28] and Hoffman and Kunze [153]). Briefly stated, a field is an algebraic structure with notions of addition, subtraction, multiplication, and division, satisfying certain axioms. The most commonly used fields are the field of real numbers, the field of complex numbers, and the field of rational numbers, but there is also a finite field, known as a Galois field. Any field may be used as the scalars for a vector space.

DEFINITION 10.1 (Inner product and the norm). *An* ***inner product*** *on a* vector space *V with a* field *F is a function that assigns to each ordered pair of vectors $\boldsymbol{x}, \boldsymbol{y} \in V$ a scalar $\langle \boldsymbol{x}, \boldsymbol{y} \rangle \in F$ such that*

(a) $\langle \boldsymbol{x} + \boldsymbol{y}, \boldsymbol{z} \rangle = \langle \boldsymbol{x}, \boldsymbol{z} \rangle + \langle \boldsymbol{y}, \boldsymbol{z} \rangle$,
(b) $\langle c\boldsymbol{x}, \boldsymbol{y} \rangle = c \langle \boldsymbol{x}, \boldsymbol{y} \rangle$,
(c) $\langle \boldsymbol{y}, \boldsymbol{x} \rangle = \overline{\langle \boldsymbol{x}, \boldsymbol{y} \rangle}$, *the bar denoting complex conjugation,*[2]

[1] Karl Hermann Amandus Schwarz (1843–1921) was a German mathematician.

[2] In this chapter we adopt the notation $\overline{x}$ instead of x^* for complex conjugate of x, since β^* is used for an optimal value of β. In other chapters, however, the symbol x^* may be used, because $\overline{X_n}$ denotes the average of variables $X_i, i = 1, 2, \ldots, n$ (e.g., see Chapter 11).

(d) $\langle \boldsymbol{x}, \boldsymbol{x} \rangle > 0$, *if* $\boldsymbol{x} \neq 0$.

The positive square root of $\langle \boldsymbol{x}, \boldsymbol{x} \rangle$ *denoted by*

$$\|\boldsymbol{x}\| = \sqrt{\langle \boldsymbol{x}, \boldsymbol{x} \rangle}$$

is called the ***norm*** *of* $\boldsymbol{x}$ *with respect to the inner product.* □

The conditions (a), (b), and (c) imply

(e) $\langle \boldsymbol{x}, c\boldsymbol{y} + \boldsymbol{z} \rangle = \overline{c}\langle \boldsymbol{x}, \boldsymbol{y} \rangle + \langle \boldsymbol{x}, \boldsymbol{z} \rangle$.

With (c), we have, for instance,

$$\langle i\boldsymbol{x}, i\boldsymbol{x} \rangle = -i\langle i\boldsymbol{x}, \boldsymbol{x} \rangle = -i^2\langle \boldsymbol{x}, \boldsymbol{x} \rangle = \langle \boldsymbol{x}, \boldsymbol{x} \rangle = \|\boldsymbol{x}\|^2.$$

Since we can define the distance or **metric** between any elements $\boldsymbol{x}$ and $\boldsymbol{y}$ by $d(\boldsymbol{x}, \boldsymbol{y}) = \|\boldsymbol{x} - \boldsymbol{y}\|$, the inner product space is a **metric space**. Note that when we are dealing with the real number field, the complex conjugates appearing in (c) and (e) are superfluous.

DEFINITION 10.2 (Orthogonality). *Let* $\boldsymbol{x}$ *and* $\boldsymbol{y}$ *be vectors in an inner product space* $\boldsymbol{V}$. *We say that* $\boldsymbol{x}$ *and* $\boldsymbol{y}$ *are* ***orthogonal*** *if* $\langle \boldsymbol{x}, \boldsymbol{y} \rangle = 0$. □

With these preliminaries, we are ready to state the Cauchy–Schwarz inequality.

THEOREM 10.1 (Cauchy–Schwarz inequality). *Let* $\boldsymbol{x}$ *and* $\boldsymbol{y}$ *be arbitrary vectors in an inner-product space* $\boldsymbol{V}$ *over a field* F. *Then we have*

$$|\langle \boldsymbol{x}, \boldsymbol{y} \rangle|^2 \leq \|\boldsymbol{x}\|^2 \|\boldsymbol{y}\|^2, \text{ or equivalently } |\langle \boldsymbol{x}, \boldsymbol{y} \rangle| \leq \|\boldsymbol{x}\| \|\boldsymbol{y}\|, \tag{10.1}$$

where the equality holds if and only if $\boldsymbol{x}$ *and* $\boldsymbol{y}$ *are linearly dependent; i.e.,* $\boldsymbol{x} = c\boldsymbol{y}$ *for some scalar constant* $c \in F$.

Proof. Consider $\|\boldsymbol{x} - c\boldsymbol{y}\| > 0$. This metric becomes minimum when $\boldsymbol{x} - c\boldsymbol{y}$ is orthogonal to $\boldsymbol{y}$ (see Figure 10.1); i.e.,

$$\langle \boldsymbol{x} - c\boldsymbol{y}, \boldsymbol{y} \rangle = 0,$$

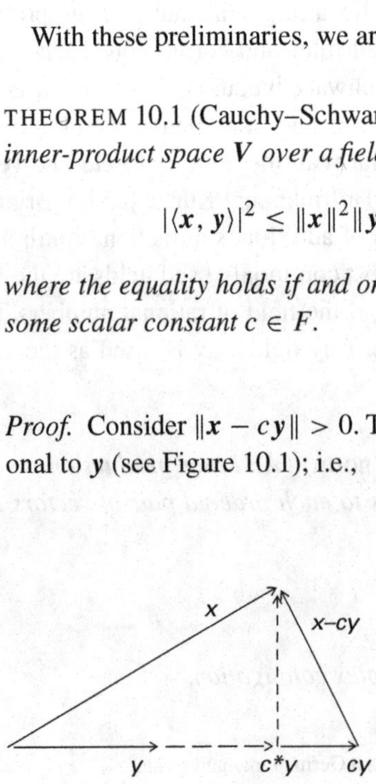

Figure 10.1 Geometric interpretation of the Cauchy–Schwarz inequality: the norm of vector $\boldsymbol{x} - c\boldsymbol{y}$ is minimal when it is orthogonal to vector $\boldsymbol{y}$.

which gives

$$c^* = \frac{\langle \boldsymbol{x}, \boldsymbol{y} \rangle}{\|\boldsymbol{y}\|^2}. \tag{10.2}$$

Then from **Pythagoras'[3] theorem,**

$$\|\boldsymbol{x} - c^*\boldsymbol{y}\|^2 = \|\boldsymbol{x}\|^2 - \|c^*\boldsymbol{y}\|^2 = \|\boldsymbol{x}\|^2 - |c^*|^2\|\boldsymbol{y}\|^2 \geq 0.$$

By substituting c^* of (10.2) into this, we readily find

$$\|\boldsymbol{x}\|^2 - \frac{|\langle \boldsymbol{x}, \boldsymbol{y} \rangle|^2}{\|\boldsymbol{y}\|^2} \geq 0,$$

which leads to (10.1). It should be clear that the equality holds if and only if $\boldsymbol{x} - c^*$ $\boldsymbol{y} = 0$; i.e., $\boldsymbol{x}$ and $\boldsymbol{y}$ are linearly dependent.

Note that both the notion of orthogonality and the Pythagorean theorem are applicable to **any inner-product space**, although our sketch is for the two-dimensional case. □

From the Cauchy–Schwarz inequality, we have for any pair of vectors

$$\frac{|\langle \boldsymbol{x}, \boldsymbol{y} \rangle|}{\|\boldsymbol{x}\| \|\boldsymbol{y}\|} \leq 1. \tag{10.3}$$

Thus, we define the angle ϕ between the two vectors by

$$\cos\phi = \frac{\langle \boldsymbol{x}, \boldsymbol{y} \rangle}{\|\boldsymbol{x}\| \|\boldsymbol{y}\|}. \tag{10.4}$$

Thus, we say that $\boldsymbol{x}$ and $\boldsymbol{y}$ are orthogonal if the angle between them is $\pi/2$. This definition is consistent with the definition we already gave in Definition 10.2.

10.1.1.1 Alternative proofs

1. The inequality can be derived immediately using the following **Lagrange identity** (Problem 10.1):

$$\|\boldsymbol{x}\|^2\|\boldsymbol{y}\|^2 - |\langle \boldsymbol{x}, \boldsymbol{y} \rangle|^2 = \sum_{i<j} |x_i y_j - x_j y_i|^2, \tag{10.5}$$

where x_i and y_i are the components of vectors $\boldsymbol{x}$ and $\boldsymbol{y}$ respectively. For n-dimensional vectors, we can write the above identity as

$$\left(\sum_{i=1}^{n} |x_i|^2\right)\left(\sum_{i=1}^{n} |y_i|^2\right) - \left|\sum_{i=1}^{n} x_i \overline{y}_i\right|^2 = \sum_{i=1}^{n-1} \sum_{j=i+1}^{n} \left|x_i y_j - x_j y_i\right|^2.$$

2. First, let us consider the case of **real vectors** $\boldsymbol{x}, \boldsymbol{y}$ and for a **real scalar** c:

$$f(c) = \|\boldsymbol{x} - c\boldsymbol{y}\|^2 = \|\boldsymbol{y}\|^2\left(c - \frac{\langle \boldsymbol{x}, \boldsymbol{y} \rangle}{\|\boldsymbol{y}\|^2}\right)^2 + \|\boldsymbol{x}\|^2 - \frac{\langle \boldsymbol{x}, \boldsymbol{y} \rangle^2}{\|\boldsymbol{y}\|^2}.$$

[3] Pythagoras of Samos (circa 570–circa 495 BC) was an Ionian Greek philosopher.

In order for the function $f(c)$ to be nonnegative for any real number c, it is necessary and sufficient that

$$\|\boldsymbol{x}\|^2 - \frac{\langle \boldsymbol{x}, \boldsymbol{y}\rangle^2}{\|\boldsymbol{y}\|^2} \geq 0,$$

from which (10.1) results. This proof and an equivalent argument given in many textbooks that the discriminant D of the quadratic equation $f(c) = 0$ must be nonnegative are valid only for the real number field.

The above algebraic proof can be generalized to the complex-number field as follows:

$$f(c) = \|\boldsymbol{x} - c\boldsymbol{y}\|^2 = \|\boldsymbol{y}\|^2 \left| c - \frac{\langle \boldsymbol{x}, \boldsymbol{y}\rangle}{\|\boldsymbol{y}\|^2} \right|^2 + \|\boldsymbol{x}\|^2 - \frac{|\langle \boldsymbol{x}, \boldsymbol{y}\rangle|^2}{\|\boldsymbol{y}\|^2} \geq 0,$$

where the equality holds if and only if $\|\boldsymbol{x} - c\boldsymbol{y}\| = 0$. This, together with the property (d) of the inner-product space, implies that $\boldsymbol{x} = c\boldsymbol{y}$.

10.1.1.2 The Cauchy–Schwarz inequality in an integral form

For the inner product space[4] of square-integrable complex-valued functions, we define the inner product of two functions by

$$\langle f, g\rangle = \int f(x)\overline{g(x)}\,dx.$$

Then the Cauchy–Schwarz inequality becomes

$$\left| \int f(x)\overline{g(x)}\,dx \right|^2 \leq \int |f(x)|^2\,dx \int |g(x)|^2\,dx. \tag{10.6}$$

Needless to say, the above inequality holds even if $\overline{g(x)}$ is replaced by $g(x)$, because $|\overline{g(x)}| = |g(x)|$.

10.1.1.3 The Cauchy–Schwarz inequality for random variables

Before we state the Cauchy–Schwarz inequality in the probabilistic context, we need to define the inner product space of random variables.

DEFINITION 10.3 (Inner product and orthogonality of random variables). *We define the* inner product *of complex-valued* ***random variables*** *X and Y by*

$$\boxed{\langle X, Y\rangle \triangleq E[X\overline{Y}].} \tag{10.7}$$

The norm *of the RV X is defined by*

$$\|X\| = \sqrt{\langle X, X\rangle} = \sqrt{E[|X|^2]}. \tag{10.8}$$

We say the RVs X and Y are ***orthogonal*** *if* $\langle X, Y\rangle = E[X\overline{Y}] = 0$. □

4 Such a space is often denoted as L_2 or L^2 space.

The reader is suggested to check that all the properties (a) through (e) of the inner product space specified in Definition 10.1 are satisfied by the inner product defined by (10.7).

The above definition can be generalized for random vectors $\boldsymbol{X}$ and $\boldsymbol{Y}$.

$$\langle \boldsymbol{X}, \boldsymbol{Y} \rangle \triangleq E[\boldsymbol{X}^\top \overline{\boldsymbol{Y}}], \tag{10.9}$$

where $\boldsymbol{X}^\top$ is a row vector and $\overline{\boldsymbol{Y}}$ is a column vector.

Then the Cauchy–Schwarz inequality can be stated for RVs

$$\left|E[X\overline{Y}]\right|^2 \leq E\left[|X|^2\right] E\left[|Y|^2\right] \tag{10.10}$$

and for random vectors

$$\boxed{\left|E[\boldsymbol{X}^\top \overline{\boldsymbol{Y}}]\right|^2 \leq E\left[\boldsymbol{X}^\top \overline{\boldsymbol{X}}\right] E\left[\boldsymbol{Y}^\top \overline{\boldsymbol{Y}}\right].} \tag{10.11}$$

From (10.10) we can derive (Problem 10.3)

$$|\text{Cov}[X, Y]|^2 \leq \text{Var}[X]\text{Var}[Y]. \tag{10.12}$$

Thus, the correlation coefficient satisfies the inequality

$$|\rho_{XY}| = \left| \frac{\text{Cov}[X, Y]}{\sqrt{\text{Var}[X]}\sqrt{\text{Var}[Y]}} \right| \leq 1. \tag{10.13}$$

The Cauchy–Schwarz inequality is generalized to **Hölder's inequality** and **Minkoswki's inequality** (see Problem 11.6 of Chapter 11).

10.1.2 Jensen's inequality

Jensen's inequality applies to a convex or concave function, which we define below for the one-dimensional case.

DEFINITION 10.4 (Convex and concave functions). *A real-valued function $g(x)$ is said to be* ***convex*** *if, for any $x_1, x_2 \in \mathbb{R}$ and for any p such that $0 \leq p \leq 1$,*

$$g(px_1 + (1-p)x_2) \leq p\,g(x_1) + (1-p)g(x_2). \tag{10.14}$$

A function $g(x)$ is said to be ***concave*** *if $-g(x)$ is convex.* □

It can be shown (Problem 10.5) that, for a convex function $g(x)$, the right and left derivatives exist and thus $g(x)$ is continuous at all points in the open interval where $g(x)$ is defined. We can also show (Problem 10.7) that if a continuous function has its second derivative $g''(x)$, then $g(x)$ is convex if and only if

$$g''(x) \geq 0, \text{ for all } x.$$

THEOREM 10.2 (Jensen's inequality). *For a convex function $g(x)$ and a finite set of points $x_1, x_2, \ldots, x_n$ in an open interval in which $g(x)$ is defined, the following inequality holds with nonnegative weights $p_1, p_2, \ldots, p_n$ such that $\sum_{i=1}^n p_i = 1$:*

$$g\left(\sum_i p_i x_i\right) \leq \sum_{i=1}^n p_i g(x_i). \tag{10.15}$$

Proof. For $n = 2$ the above equation reduces to the definition of convexity given in (10.14). So it is clear that the finite form of Jensen's inequality can be proved by the method of mathematical induction (Problem 10.2). The equality in (10.15) holds if and only if all the n points degenerate to a single point or $g(x)$ is constant. □

10.1.2.1 Jensen's inequality for random variables

Since the set of weighting coefficients $\{p_i\}$ defines the probability distribution, we can interpret the above inequality in terms of a discrete RV X whose probability distribution (or PMF) is $\{p_i\}$:

$$g(E[X]) \leq E\left[g(X)\right]. \tag{10.16}$$

The above inequality holds for a continuous RV X as well, if we express it in terms of the distribution function or PDF. Suppose that the function $g(x)$ is convex in the interval (a, b) and a random variable X has the PDF in the same interval, where a may be $-\infty$ and/or b may be ∞. Then we have

$$g\left(\int_a^b x f_X(x)\,\mathrm{d}x\right) \leq \int_a^b g(x) f_X(x)\,\mathrm{d}x. \tag{10.17}$$

Example 10.1: Simple examples of Jensen's inequality.

1. $g_1(x) = x^2$ is a convex function. Thus, we find
$$E[X^2] \geq (E[X])^2.$$
2. $g_2(x) = \log x$ is a concave function for $x > 0$. Thus, we have
$$E[\log X] \leq \log(E[X]),$$
3. $g_3(x) = |x|$ is a convex function. Thus,
$$E[|X|] \geq |E[X]|.$$

□

10.1.3 Shannon's lemma and log-sum inequality

In this section we discuss two inequalities that are often used in information theory and related fields. The first inequality is a lemma in Shannon's seminal paper [300]. This inequality is sometimes called **Gibbs' inequality**.

THEOREM 10.3 (Shannon's lemma). *Let* $\boldsymbol{f} = [f_1, f_2, \ldots, f_n]$ *and* $\boldsymbol{g} = [g_1, g_2, \ldots, g_n]$ *be two probability distributions. Then*

$$\boxed{\sum_{i=1}^{n} f_i \log g_i \le \sum_{i=1}^{n} f_i \log f_i, \text{ i.e.,} \sum_{i=1}^{n} f_i \log \frac{f_i}{g_i} \ge 0,} \tag{10.18}$$

where the equality holds if and only if $\boldsymbol{g} = \boldsymbol{f}$; *i.e.,* $g_i = f_i$, *for all* i.

Proof. Consider the following inequality,[5] frequently used in information theory:

$$\ln x \le x - 1, \text{ for } x > 0, \tag{10.19}$$

where the equality holds only at $x = 1$. Then for the logarithm of x with any base, we have

$$\log x \le (\log e)(x - 1).$$

Thus,

$$\begin{aligned} -\sum_{i=1}^{n} f_i \log g_i &= -\sum_{i=1}^{n} f_i \log \frac{g_i}{f_i} f_i \ge -\left[\log e \sum_{i=1}^{n} f_i \left(\frac{g_i}{f_i} - 1\right) + \sum_{i=1}^{n} f_i \log f_i\right] \\ &= -\sum_{i=1}^{n} f_i \log f_i \ge 0, \end{aligned} \tag{10.20}$$

where $-\sum_i f_i \log f_i$ is the **entropy** of a random variable whose probability distribution is $\boldsymbol{f}$, and denoted as $H(\boldsymbol{f})$. Equality holds when $g_1/f_i = 1$ for all i. The left-hand side of the second inequality in (10.18), $\sum_{i=1}^{n} f_i \log(f_i/g_1)$, is the **Kullback–Leibler divergence** (KLD), denoted as $D(\boldsymbol{f}\|\boldsymbol{g})$, which will be discussed in Chapter 19.

There are other ways to prove this lemma (Problem 10.9). □

Next, we will introduce the **log-sum inequality**, which is frequently used also in information theory (e.g., see Cover and Thomas [69]) and related fields. We will extensively use this inequality in the expectation-maximization (EM) algorithm and hidden Markov model to be discussed in Chapters 18 and 20.

THEOREM 10.4 (The log-sum inequality). *Let* $\boldsymbol{a} = (a_1, a_2, \ldots, a_n)$ *and* $\boldsymbol{b} = (b_1, b_2, \ldots, b_n)$ *be such that* $a_i, b_i \ge 0$, *for all* $i = 1, 2, \ldots, n$, $\sum_{i=1}^{n} a_i \triangleq a$ *and* $\sum_{i=1}^{n} b_i \triangleq b$. *Then*

[5] Apply Taylor's expansion to $\log x$ at $x = 1$, obtaining

$$\ln x = x - 1 - \frac{(x-1)^2}{2\xi^2},$$

where ξ is a point between 1 and x.

$$\sum_{i=1}^{n} a_i \log \frac{a_i}{b_i} \geq a \log \frac{a}{b}, \tag{10.21}$$

where the equality holds when $b_i/b = a_i/a$, for all i.

Proof. Set $f_i = a_i/a$ and $g_i = b_i/b$ in (10.18). Then

$$\sum_{i=1}^{n} a_i \log b_i \leq \sum_{i=1}^{n} a_i \log a_i + a \log\left(\frac{b}{a}\right), \tag{10.22}$$

which readily leads to (10.21). The condition for equality also readily follows from $f_i = g_i$ in Shannon's lemma. □

Note the similarity between this inequality and Shannon's lemma or Gibbs' inequality. As a matter of fact, the inequality formula in (10.18) holds even if $\boldsymbol{f}$ and $\boldsymbol{g}$ are not probability distributions: all that is required is $g_i > 0$, $f_i \geq 0$, and $\sum_i f_i = \sum_i g_i$ (Problem 10.10).

10.1.4 Markov's inequality

We start with Markov's inequality, which applies to the complementary distribution function (also called the survivor function) of any nonnegative RV:

THEOREM 10.5 (Markov's Inequality). *Let X be a nonnegative RV with finite mean $E[X]$. Then for any $a > 0$, the following inequality holds:*

$$\boxed{P[X \geq a] \leq \frac{E[X]}{a},} \tag{10.23}$$

*which is called **Markov's inequality**.*[6]

Proof. For a nonnegative RV, the expectation is equivalent to the integration of the complementary distribution $F_X^c(x) = 1 - F_X(x)$ from zero to infinity:

$$E[X] = \int_0^\infty [1 - F_X(x)]\,dx \geq \int_0^a [1 - F_X(x)]\,dx \geq a[1 - F_X(a)]. \tag{10.24}$$

Hence, we have

$$P[X \geq a] = 1 - F_X(a) \leq \frac{E[X]}{a}. \tag{10.25}$$

Figure 10.2a shows a simple pictorial explanation of the above inequality. The shaded area is equal to the expectation $E[X]$ and the rectangle has the area $aF_X^c(a)$, which is clearly smaller than the former.

[6] Some authors call this inequality Chebyshev's inequality. But we reserve the latter for the next inequality, which is a special case of Markov's inequality.

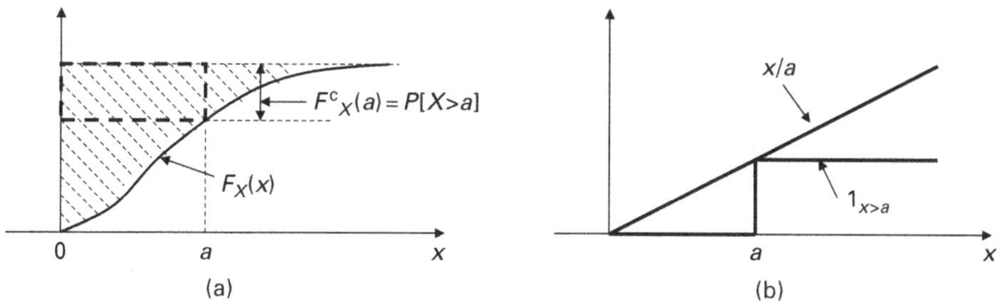

Figure 10.2 Illustration of proof of Markov's inequality: (a) the shaded region is equal to $E[X]$ and the rectangle's area is $aP[X \geq a]$; (b) second proof based on $x/a \geq \mathbf{1}_{x \geq a}$.

An alternative proof of Markov's inequality uses the following simple inequality as sketched in Figure 10.2b:

$$\mathbf{1}_{x \geq a} \leq \frac{x}{a}, \tag{10.26}$$

where $\mathbf{1}_E$ is equal to unity if the event E is true, and is zero, otherwise. This **indicator function** may alternatively be written as

$$\mathbf{1}_{x \geq a} = u(x - a), \tag{10.27}$$

where $u(x)$ is the unit step function. Then multiplication of $f_X(x)$ on both sides of (10.26) and integration over $x \in [0, \infty)$ yields

$$\int_a^\infty f_X(x)\,dx = P[X \geq a] \leq a^{-1} \int_0^\infty x f_X(x)\,dx, \tag{10.28}$$

from which (10.23) results. □

This upper bound by Markov is rather loose, since it assumes only knowledge of the mean of the distribution. However, it can be useful in quickly assessing the tail distribution of a random variable when a is relatively large compared with the mean. For instance, suppose that the expected response time of a web server is 2 s. Then what is the probability that one will experience a response time greater than a minute? The above inequality readily shows that at most 3.33% of the response times can be larger than a minute.

10.1.5 Chebyshev's inequality

Now let us proceed to the case where both the mean and variance of the distribution exist and are known. Qualitatively speaking, the smaller the variance is, the less likely it is that large deviations from the mean will occur. The following inequality due to Cheyshev makes a precise statement about this qualitative observation.

THEOREM 10.6 (Chebyshev's inequality). *Let X be a random variable. Then for any $b > 0$, we have*

$$\boxed{P[|X - E[X]| \geq b] \leq \frac{\sigma^2}{b^2}.} \tag{10.29}$$

Proof. Let $Y = (X - E[X])^2$ and $a = b^2$, and substitute them into Markov's inequality (10.23). The above inequality immediately follows. □

If we choose b to be n times the standard deviation, i.e., $b = n\sigma$, then the above inequality shows

$$P[|X - E[X]| \geq n\sigma] \leq \frac{1}{n^2}.$$

This suggests, for instance, that the probability that any RV deviates from its mean by more than three standard deviations is less than $\frac{1}{9} \approx 0.11$.

Example 10.2: A simple proof of Bernoulli's theorem. Consider n repeated Bernoulli trials, and let S_n be the number of successes, i.e.,

$$S_n = B_1 + B_2 + \cdots + B_n,$$

where B_i is a 1–0 variable, depending on the ith trial being a success or a failure. Let $P[B_i = 1] = p$, and $P[B_i = 0] = 1 - p = q$, for all $i = 1, 2, \ldots, n$. Define RV X_n by $X_n = S_n/n$. Then we readily see that

$$E[X_n] = \frac{E[S_n]}{n} = p \text{ and } \sigma^2_{X_n} = \frac{\sigma^2_{S_n}}{n^2} = \frac{npq}{n^2} = \frac{pq}{n}.$$

Hence, by applying Chebyshev's inequality to X_n, we obtain

$$P\left[\left|\frac{S_n}{n} - p\right| \geq \epsilon\right] \leq \frac{pq}{n\epsilon^2}, \quad (10.30)$$

which is (2.45) of Bernoulli's theorem discussed in Section 2.3. We will use Chebyshev's inequality subsequently to establish the **weak law of large numbers (WLLN)** that applies to the sample average or statistical average. □

10.1.6 Kolmogorov's inequalities for martingales and submartingales

Kolmogorov and Doob generalized Markov's inequality to martingales and submartingales. We first give definitions of martingale and submartingale.

DEFINITION 10.5 (Martingale). *A sequence of RVs $\{S_k\}$ is called a* ***martingale*** *if $E[|S_k|] < \infty$ for all k and if*

$$E[S_k|S_1, S_2, \ldots, S_{k-1}] = S_{k-1}, \textit{for all } k. \quad (10.31)$$

□

A martingale is a mathematical description of a fair game. If we interpret S_k as a player's (i.e., gambler's) fortune after the kth game, then (10.31) states that the player's expected fortune after the $(k+1)$st game is equal to his/her fortune after the kth game regardless of what happened before.

Example 10.3: A random walk is a martingale. Consider an i.i.d RV X_i with zero mean, i.e., $E[X_i] = 0$, and let $S_n = \sum_{i=1}^{n} X_i$. Then clearly $S_n = S_{n-1} + X_n$, and $E[S_n | S_1, S_2, \ldots, S_{n-1}] = S_{n-1}$. Thus, S_n is a martingale. If we interpret each X_i as the step size of a walk, then S_n is the position after n steps, staring from S_0. This discrete-time random process is called a (one-dimensional) **random walk**. We will discuss the random walk model and its mathematical limit (as the step size becomes infinitesimally small), called **Brownian motion**, in Chapter 17. □

DEFINITION 10.6 (Submartingale and supermartingale). *A sequence of RVs $\{Y_k\}$, with $E[|Y_k|] < \infty$ for all k, is called a* ***submartingale****, if*

$$E[Y_k | Y_1, Y_2, \ldots, Y_{k-1}] \geq Y_{k-1}, \textit{for all } k, \tag{10.32}$$

and is called a ***supermartingale****, if*

$$E[Y_k | Y_1, Y_2, \ldots, Y_{k-1}] \leq Y_{k-1}, \textit{for all } k. \tag{10.33}$$

□

So a submartingale embodies a game that is *superfair* to a player and *subfair* to the house, because his/her expected fortune after the next game is no less than his/her current fortune. A supermartingale is the opposite, i.e., more favorable to the house.

THEOREM 10.7 (Doob–Kolmogorov's inequality for a nonnegative submartingale). *Let $\{Y_k\}$ be a submartingale and $Y_k \geq 0$ for all k. Then for any $a > 0$,*

$$\boxed{P\left[\max\{Y_1, Y_2, \ldots, Y_n\} \geq a\right] \leq \frac{E[Y_n]}{a}.} \tag{10.34}$$

Proof. [100].
For given a, denote ξ the smallest subscript $1 \leq j \leq k$ such that $Y_j \geq a$ and set $\xi = 0$ if no such event occurs. Then ξ is a random variable with possible values $0, 1, \ldots, n$. Then the expectation of Y_n can be written as

$$E[Y_n] = \sum_{j=0}^{n} E[Y_n | \xi = j] P[\xi = j] \geq \sum_{j=1}^{n} E[Y_n | \xi = j] P[\xi = j]. \tag{10.35}$$

For $j \geq 1$ the event $\{\xi = j\}$ depends only on $Y_1, Y_2, \ldots, Y_j$, and therefore

$$\begin{aligned} E[Y_n | \xi = j] &= E[E[Y_n | Y_1, Y_2, \ldots, Y_n] | \xi = j] \\ &= E[E[Y_n | Y_1, Y_2, \ldots, Y_j] | \xi = j] \\ &\geq E[Y_j | \xi = j] \\ &\geq a. \end{aligned} \tag{10.36}$$

The first two equalities use properties of conditional expectation. The first inequality uses the definition of submartingale. The second inequality follows from the definition of ξ.

By substituting (10.36) into (10.35), it is evident that

$$\sum_{j=1}^{n} P[\xi = j] \leq \frac{E[Y_n]}{a}. \tag{10.37}$$

However, the left-hand side can be written as

$$\sum_{j=1}^{n} P[\xi = j] = P[Y_j \geq a \text{ for some } j = 1, 2, \ldots, n.]$$

$$= P[\max\{Y_1, Y_2, \ldots, Y_n\} \geq a]. \tag{10.38}$$

Thus, we have completed the proof. □

Now let us consider a random variable sequence X_k such that $E[X_k] = 0$ for all $k \geq 1$. If we define a new sequence

$$S_k = X_1 + X_2 + \cdots + X_k, \tag{10.39}$$

then clearly the sequence S_k is a martingale.

THEOREM 10.8 (Kolmogorov's inequality for a martingale). *Let the sequence S_k be a martingale. Then*

$$\boxed{P\left[\max\{|S_1|, |S_2|, \ldots, |S_n|\} \geq b\right] \leq \frac{E[S_n^2]}{b^2}.} \tag{10.40}$$

Proof. Set $Y_k = S_k^2$ and $a = b^2$ in the submartingale inequality (10.34). Then the martingale inequality (10.40) readily ensues. □

Note the almost identical structure between the above two inequalities. Indeed, we can show that $|S_k|$ is a submartingale when S_k is a martingale (Problem 10.15 (a)). If we set $|S_k| = Y_k$ and $b = a$ in the left-hand side of (10.40), the right-hand side of this inequality becomes Y_n^2/a^2 in contrast with the right-hand side of the inequality in (10.34). We can indeed generalize the inequality (10.34) (Problem 10.15 (b)) as follows:

$$\boxed{P\left[\max\{Y_1, Y_2, \ldots, Y_n\} \geq a\right] \leq \frac{E[Y_n^p]}{a^p}, \text{ for any } p \geq 1.} \tag{10.41}$$

When we paraphrase Kolmogorov's martingale inequality (10.40) in terms of the original independent sequence X_k that defines the martingale S_n in (10.39), we have

$$P\left[\max_k \left\{\left|\sum_{i=1}^{k} X_i\right|\right\} > b\right] \leq \frac{\sum_{k=1}^{n} \sigma_k^2}{b^2}, \tag{10.42}$$

where $E[X_k] = 0$ and $\sigma_k^2 = \text{Var}[X_k] = E[X_k^2]$.

10.2 Chernoff's bounds

We have shown the use of the MGF $M_X(t)$ in computing various moments of X. Now we will show that by applying the MGF to Markov's inequality we can obtain a much tighter upper bound on the complementary distribution function than the ones given by the Markov or Chebyshev inequalities. This improved bound, called Chernoff's bound,[7] is achieved when the MGF of the distribution is made available. Knowledge of the MGF is equivalent to having information about all moments of the RV, which is a lot more than what Chebyshev's inequality requires. As will be shown, Chernoff's bound is an especially powerful technique when we deal with a sum of i.i.d. RVs. But we start with a single RV.

10.2.1 Chernoff's bound for a single random variable

For given X, define a new RV $Y = e^{\xi X}$, where ξ is a real-valued parameter.[8] Then the RV Y is nonnegative. Furthermore,

$$X \geq b \iff Y \geq e^{\xi b}, \text{ for } \xi \geq 0, \tag{10.43}$$

and

$$X \leq b \iff Y \geq e^{\xi b}, \text{ for } \xi \leq 0, \tag{10.44}$$

where $\iff$ stands for "if and only if." Then by applying Markov's inequality to Y in (10.43), we obtain

$$P[X \geq b] = P[Y \geq e^{\xi b}] \leq \frac{E[Y]}{e^{\xi b}}, \xi \geq 0. \tag{10.45}$$

From the definition of Y we can write

$$E[Y] = E[e^{\xi X}] = M_X(\xi), \tag{10.46}$$

where $M_X(\xi)$ is the MGF of X. Thus, we have

$$P[X \geq b] = P[Y \geq e^{\xi b}] \leq e^{-\xi b} M_X(\xi), \xi \geq 0. \tag{10.47}$$

Since the inequality (10.47) holds for all $\xi \geq 0$, we can obtain the best (i.e., tightest) upper bound by selecting ξ^* that gives the infimum[9] of the upper bounds. Thus, we find the following theorem.

[7] Herman Chernoff (1923–present) is an American mathematician and statistician.

[8] In this chapter we use the symbol ξ, instead of t, as the parameter of the MGF.

[9] The infimum of a given set $\mathcal{S}$ is the **greatest lower bound** of $\mathcal{S}$ and is denoted as $\inf\{\mathcal{S}\}$. It is not necessarily the case that $\inf\{\mathcal{S}\} \in \mathcal{S}$. If $\mathcal{S} = (a, b)$, $\inf\{\mathcal{S}\} = a$. The "infimum" is equivalent to the "minimum" when $\inf\{\mathcal{S}\} \in \mathcal{S}$. In this chapter, inf is replaced by min since we are almost always dealing with cases where the greatest lower bound is achievable within the set $\mathcal{S}$. Similarly, sup, which stands for "supremum," lowest upper bound, is replaced by max in this chapter.

THEOREM 10.9 (Chernoff's bound). *Let X be a real-valued RV whose moments all exist. Then*

$$P[X \geq b] \leq \min_{\xi \geq 0} \left\{ e^{-\xi b} M_X(\xi) \right\}, \text{ where } M_X(\xi) = E[e^{\xi X}]. \quad (10.48)$$

If there exists the "best" ξ^* *such that*

$$m'(\xi^*) = \left. \frac{d(\ln M_X(\xi))}{d\xi} \right|_{\xi=\xi^*} = b, \quad (10.49)$$

then

$$\boxed{P[X \geq b] \leq e^{-\xi^* b} M_X(\xi^*).} \quad (10.50)$$

Proof. The proof is rather straightforward. Consider a function

$$U(\xi) = e^{-\xi b} M_X(\xi), \quad (10.51)$$

which should be minimized. By taking its first derivative and setting it to zero we obtain (10.49), which the optimum $\xi^\star$ must satisfy. In order to show that this value gives the minimum value of $U(\xi)$ within the admissible interval I_ξ, it suffices to show that $U(\xi)$ is a convex function by showing that

$$U''(\xi) = e^{-\xi b}[b^2 M_X(\xi) - 2bM'_X(\xi) + M''_X(\xi)] \quad (10.52)$$

is nonnegative. Details are left to the reader as an exercise (Problems 10.20 and 10.21). □

Note that the above Chernoff bound was derived by considering $\xi > 0$. If we choose $\xi < 0$ instead, then $e^{\xi X} \geq e^{\xi b}$ if and only if $X \leq b$, thus yielding

$$P[X \leq b] \leq e^{-\xi b} M_X(\xi), \xi < 0. \quad (10.53)$$

Equation (10.47) can be used in a meaningful way only in bounding the "right-end tail" (i.e., $X \geq b(> \mu_X)$) of the distribution (Problem 10.22). Similarly, the expression (10.53) can be used in bounding the "left-end tail" ($X \leq b(< \mu_X)$) of the distribution. These bounds are especially useful when a random variable is a sum of independent RVs, as shown below.

10.2.2 Chernoff's bound for a sum of i.i.d. random variables

Let $X_1, X_2, \ldots, X_n$ be a sequence of i.i.d. RVs, and form the sum

$$S_n = \sum_{i=1}^{n} X_i. \quad (10.54)$$

Then the MGF of S_n is given by

$$M_{S_n}(\xi) = (M_X(\xi))^n, \quad (10.55)$$

where $M_X(\xi)$ is the common MGF of the variables X_i. Then by applying the above argument to the RV S_n, we obtain

$$P[S_n \geq b] \leq \min_{\xi \geq 0} \left\{ e^{-\xi b + n m_X(\xi)} \right\} \tag{10.56}$$

and

$$P[S_n \leq b] \leq \min_{\xi \leq 0} \left\{ e^{-\xi b + n m_X(\xi)} \right\}, \tag{10.57}$$

where $m_X(\xi)$ is the logarithmic MGF (or cumulant MGF) defined in (8.6):

$$m_X(\xi) = \ln M_X(\xi). \tag{10.58}$$

The value ξ^* that achieves a minimum is a root of

$$m'_X(\xi) = \frac{M'_X(\xi)}{M_X(\xi)} = \frac{b}{n}. \tag{10.59}$$

Thus, we obtain

$$\boxed{P[S_n \geq b] \leq e^{-n[\xi^* m'_X(\xi^*) - m_X(\xi^*)]}, \ \xi^* \geq 0,} \tag{10.60}$$

and

$$\boxed{P[S_n \leq b] \leq e^{-n[\xi^* m'_X(\xi^*) - m_X(\xi^*)]}, \ \xi^* \leq 0.} \tag{10.61}$$

The last two expressions are the **Chernoff bounds for the sum variable**.

If we define the "normalized" threshold β

$$\beta \triangleq \frac{b}{n}, \tag{10.62}$$

the last two equations can be rewritten as

$$P[S_n \geq n\beta] \leq e^{-n[\xi^* \beta - m_X(\xi^*)]}, \ \xi^* \geq 0, \tag{10.63}$$

and

$$P[S_n \leq n\beta] \leq e^{-n[\xi^* \beta - m_X(\xi^*)]}, \ \xi^* \leq 0. \tag{10.64}$$

Example 10.4: Coin tossing. Consider the experiment of tossing a fair coin n times discussed in Problem 10.11, where

$$S_n = B_1 + B_2 + \cdots + B_n, \tag{10.65}$$

with $P[B_i = 1] = P[B_i = 0] = \frac{1}{2}$ for all $i = 1, 2, \ldots, n$.

The MGFs of the individual B_i and S_n are

$$M_{B_i}(\xi) = \frac{1}{2}e^{0\cdot\xi} + \frac{1}{2}e^{1\cdot\xi} = \frac{1+e^{\xi}}{2} \text{ and } M_{S_n}(\xi) = \left(\frac{1+e^{\xi}}{2}\right)^n. \tag{10.66}$$

Thus, the Chernoff bound is given as

$$P[S_n \geq n\beta] \leq \min_{\xi \geq 0} \left\{ e^{-\xi n\beta} \left(\frac{1+e^{\xi}}{2} \right)^n \right\}. \tag{10.67}$$

The optimum parameter ξ^* should be the solution of

$$\frac{d}{d\xi} \ln \left(\frac{1+e^{\xi}}{2} \right)^n = n\beta, \tag{10.68}$$

which leads to

$$\frac{e^{\xi^*}}{1+e^{\xi^*}} = \beta \text{ or } \xi^* = \ln \left(\frac{\beta}{1-\beta} \right). \tag{10.69}$$

Therefore, the bound (10.67) becomes

$$P[S_n \geq n\beta] \leq \left(\frac{\beta}{1-\beta} \right)^{-n\beta} \left(\frac{1+\frac{\beta}{1-\beta}}{2} \right)^n = \left[2\beta^{\beta}(1-\beta)^{1-\beta} \right]^{-n}. \tag{10.70}$$

By using the **entropy function** [300]

$$\mathcal{H}(\beta) \triangleq -\beta \log_2 \beta - (1-\beta)\log_2(1-\beta), \tag{10.71}$$

(10.70) can be more compactly expressed as

$$\boxed{P[S_n \geq n\beta] \leq 2^{-n(1-\mathcal{H}(\beta))}.} \tag{10.72}$$

Let us find the numerical value of the upper bound, for the same case discussed in Problem 10.11; i.e., $\beta = 0.8$ and $n = 100$ and $n = 1000$. The entropy function for $\beta = 0.8$ is $\mathcal{H}(0.8) = 0.7219$. Hence, $n(1 - \mathcal{H}(0.8)) = 0.2781n$, and Chernoff's bound is $2^{-0.2781n}$, which yields

$$P[S_{100} \geq 80] \leq 2^{-27.81} \approx 4.298 \times 10^{-9} \text{ and}$$
$$P[S_{1000} \geq 800] \leq 2^{-278.1} \approx 1.9589 \times 10^{-84}. \tag{10.73}$$

Thus, the upper bounds computed in Problem 10.11 using Chebyshev's inequality (i.e., 2.78×10^{-2} for $n = 100$ and 2.78×10^{-3} for $n = 1000$) are found too loose to be useful. The reader is suggested to compute the Chernoff bound for a case where β is much closer to the mean, say $\beta = 0.51$ (Problem 10.25).

10.2.2.1 The role of entropy function

Before we close this example, it will be instructive to discuss the role of the entropy function $\mathcal{H}(\beta)$ in this biased coin tossing problem. The term $2^{-n(1-\mathcal{H}(\beta))}$ that appeared in (10.72), and later in (10.159) of Problem 10.28, can be written as

$$2^{-n[1-\mathcal{H}(\beta)]} = \frac{2^{n\mathcal{H}(\beta)}}{2^n}, \tag{10.74}$$

where the denominator 2^n is the total number of distinct outcomes in the experiment of tossing any coin (fair or unfair) n times, whereas the numerator $2^{n\mathcal{H}(\beta)}$ can be interpreted, when n is sufficiently large, as the number of **typical sequences** [300] that will be obtained when a **biased** coin (that lands on head with probability β) is tossed n times. In a typical sequence, there are on average $n\beta$ heads and $n(1-\beta)$ tails.

When the coin is fair, and n is large, any outcome that shows "head" $n\beta$ times or more in n tossings will be extremely improbable when the threshold parameter β largely deviates from 0.5. In other words, the event $\{S_n \geq n\beta\}$ will be a **rare event**, making the tail end distribution $P[S_n \geq n\beta]$ extremely small.

When we change the experiment of fair coin tossing to that of an unfair coin that lands on head with probability β, then a sequence that contains $n\beta$ heads becomes a typical sequence, and the number of such typical sequences approaches $2^{n\mathcal{H}(\beta)}$ for large n. Each typical sequence will occur with probability $2^{-n\mathcal{H}(\beta)}$. There are 2^n possible distinct sequences. The difference $2^n - 2^{n\mathcal{H}(\beta)}$ is the number of nontypical sequences, and the total probability of such sequences becomes negligibly small for sufficiently large n.

In essence, the large deviation approximation method transforms the computation of the probability of rare events to that of typical events, by exponentially twisting the underlying probability measure.

10.3 Large deviation theory

As discussed in the previous section, Chernoff's bound is very useful when we have the sum of independent RVs as given in (10.54). We will now further improve Chernoff's bound, based on the **theory of large deviations**.

10.3.1 Large deviation approximation

We assume that the variables X_i are independent as before, but they are **not necessarily identically** distributed. Before we consider the sum of independent RVs, let us start with a single RV X with the distribution function $F_X(x)$. We define a random variable Y whose distribution $F_Y(y)$ is an **exponentially tilted** (or **exponentially twisted**) version of $F_X(x)$, as defined by

$$dF_Y(x) = k\, e^{\xi x} dF_X(x), \tag{10.75}$$

where ξ is a positive real parameter, and k is a normalization constant. Here, $dF_X(x)$ should be interpreted as $dF_X(x) = P[x < X \leq x + dx] = F_X(x+dx) - F_X(x)$; similarly, $dF_Y(x) = P[x < Y \leq x + dx] = F_Y(x+dx) - F_Y(x)$.[10] By integrating both sides over $-\infty < x < \infty$, we have

[10] Some authors write $F_X(dx)$ instead of $dF_X(x)$, etc.

$$1 = k \int_{-\infty}^{\infty} e^{\xi x}\, dF_X(x) = kE[e^{\xi X}] = kM_X(\xi), \tag{10.76}$$

where $M_X(t)$ is the MGF of X as defined earlier:

$$M_X(t) = E[e^{tX}] = \int_{-\infty}^{\infty} e^{tx}\, dF_X(x). \tag{10.77}$$

Hence, the constant k is determined as $k = M_X(\xi)^{-1}$. Thus,

$$\boxed{dF_Y(x) = \frac{e^{\xi x}}{M_X(\xi)} dF_X(x).} \tag{10.78}$$

If X is a continuous RV and its PDF $f_X(x)$ exists, the above relation reduces to

$$f_Y(x) = \frac{e^{\xi x}}{M_X(\xi)} f_X(x), \tag{10.79}$$

where

$$M_X(t) = \int_{-\infty}^{\infty} e^{tx} f_X(x)\, dx. \tag{10.80}$$

Equation (10.78) is preferred since it holds regardless of whether the RV X is discrete or continuous. Of course, (10.79) can be properly interpreted for a discrete RV as well, by using delta functions at those values x_i where $F_X(x)$ is discontinuous.

The parameter ξ used in (10.78) must be chosen within the **interval of convergence** of the MGF $M_X(t)$. The CDF $F_Y(x)$ (or the PDF $f_Y(x)$) thus defined is called an *exponentially twisted* CDF (or PDF) of the CDF $F_X(x)$ (or the PDF $f_X(x)$). Sometimes the twisted CDF or PDF is referred to as the **exponential change of measure (ECM)**. By defining the logarithmic MGF $m_X(t)$ by

$$m_X(t) = \ln M_X(t), \tag{10.81}$$

we find the following important relations between the mean and variance of the RV Y and $m_X(t)$ of the original RV X (Problem 10.27):

$$\boxed{E[Y] = \left.\frac{dm_X(t)}{dt}\right|_{t=\xi} = m'_X(\xi)} \tag{10.82}$$

and

$$\boxed{\mathrm{Var}[Y] = \left.\frac{d^2 m_X(t)}{dt^2}\right|_{t=\xi} = m''_X(\xi).} \tag{10.83}$$

With the above preparation, let us now turn our attention to the sum variable

$$S_n = \sum_{i=1}^{n} X_i.$$

Let $F_{Y_i}(x)$ be an exponentially tilted CDF of $F_{X_i}(x)$; i.e.,

$$dF_{Y_i}(x) = \frac{e^{\xi x}}{M_{X_i}(\xi)} dF_{X_i}(x), i = 1, 2, \ldots, n, \tag{10.84}$$

where the **tilting parameter** ξ is common to all $i = 1, 2, \ldots, n$. Then consider the sum of the twisted RVs:

$$T_n = \sum_{i=1}^{n} Y_i. \tag{10.85}$$

Then, the relationship between the distribution functions $F_{T_n}(x)$ and $F_{S_n}(x)$ should be similar to that between $F_{Y_i}(x)$ and $F_{X_i}(x)$ given by (10.84). Thus,

$$dF_{T_n}(x) = \frac{e^{\xi x}}{M_{S_n}(\xi)} dF_{S_n}(x), \tag{10.86}$$

where $M_{S_n}(t)$ is the MGF of $S_n = \sum_{i=1}^{n} X_i$:

$$M_{S_n}(t) = E[e^{tS_n}] = \prod_{i=1}^{n} M_{X_i}(t). \tag{10.87}$$

The MGF of T_n can be derived from (10.86) as follows:

$$\begin{aligned} M_{T_n}(t) = E[e^{tT_n}] &= \int_{-\infty}^{\infty} e^{tx} \frac{e^{\xi x}}{M_{S_n}(\xi)} \, dF_{S_n}(x) \\ &= \frac{M_{S_n}(\xi + t)}{M_{S_n}(\xi)}. \end{aligned} \tag{10.88}$$

The same relation holds between the MGFs of the individual element RVs X_i and Y_i; i.e.,

$$\boxed{M_{Y_i}(t) = \frac{M_{X_i}(\xi + t)}{M_{X_i}(\xi)}, i = 1, 2, \ldots, n.} \tag{10.89}$$

The probability that the sum variable S_n exceeds b is expressed as

$$\begin{aligned} P[S_n \geq b] &= \int_b^{\infty} dF_{S_n}(x) = \int_b^{\infty} M_{S_n}(\xi) e^{-\xi x} \, dF_{T_n}(x) \\ &= e^{-\xi b} M_{S_n}(\xi) \int_b^{\infty} e^{-\xi(x-b)} \, dF_{T_n}(x) = e^{-\xi b} M_{S_n}(\xi) A(\xi), \end{aligned} \tag{10.90}$$

where

$$\boxed{A(\xi) = \int_b^\infty e^{-\xi(x-b)}\, dF_{T_n}(x).} \tag{10.91}$$

Recall that S_n is a sum of independent RVs. Then by virtue of the CLT it can be shown that the distribution function will approach the normal distribution as n becomes large, provided that any individual component is not big enough to appreciably influence the sum S_n. Even if such conditions for the CLT are satisfied, however, the tail of the distribution of S_n approaches the normal distribution much more slowly than the distributional form around the mean. For instance, if the $X_i's$ are all 1–0 variables, then $0 \le S_n \le n$, whereas the normal distribution ranges from $-\infty$ to ∞. Thus, for $b \gg E[S_n]$, $P[S_n \ge b]$ cannot be accurately approximated by the tail end of the normal distribution, unless n becomes really large.

Now we make use of the property that the exponential twisting, with proper choice of the parameter ξ, can translate the tail end of the distribution $F_{S_n}(x)$ to the main part of the distribution $F_{T_n}(x)$. Then, even for a moderately large n, the main part of the tilted distribution function $F_{T_n}(x)$ can be well approximated by the normal distribution with mean $m'_{S_n}(\xi)$ and variance $m''_{S_n}(\xi)$ (see (10.82) and (10.83) and Problem 10.27); i.e.,

$$dF_{T_n}(x) \approx \frac{1}{\sqrt{2\pi m''_{S_n}(\xi)}} \exp\left[-\frac{(x - m'_{S_n}(\xi))^2}{2m''_{S_n}(\xi)}\right] dx. \tag{10.92}$$

By following an argument similar to the derivation of Chernoff's bound, we choose an "optimum" tilting parameter $\xi = \xi^*$ that satisfies

$$m'_{S_n}(\xi^*) = b = n\beta, \tag{10.93}$$

where b is the threshold parameter as defined in (10.90).

Therefore, for the tilting parameter ξ^*, $A(\xi^*)$ corresponds to the integration of the upper half of an approximately normal distribution $N(b, m''_{S_n}(\xi^*))$ multiplied by the exponentially decaying function $e^{-\xi^*(x-b)}$, which takes on unity at $x = b$, the center of the tilted distribution $f_{T_n}(x)$. In Figure 10.3 (a) we show an example of the tilted PDF $f_{T_n}(x)$. Since the multiplication by $e^{-\xi^*(x-b)} \propto e^{-\xi^* x}$ is the reverse operation of the tilting operation defined in (10.78), $e^{-\xi^*(x-b)} dF_{T_n}(x)$ results in a Gaussian-shape function which is approximately proportional to the original distribution $dF_{S_n}(x)$ as sketched in Figure 10.3 (b). The shaded area corresponds to the integration $A(\xi)$ of (10.91) evaluated at $\xi = \xi^*$. Since $f_{T_n}(x)$ with $\xi = \xi^*$ has its center at $x = b$, and $e^{-\xi(x-b)} \le 1$ for $x \ge b$, we readily see from Figure 10.3 (b) that $A(\xi^*)$ is typically much smaller than $\frac{1}{2}$. Note that Chernoff's bound (10.72) corresponds to an approximation where $A(\xi^*)$ is replaced by unity.

An approximate evaluation of $A(\xi^*)$ can be obtained by carrying out the integration in (10.91) by

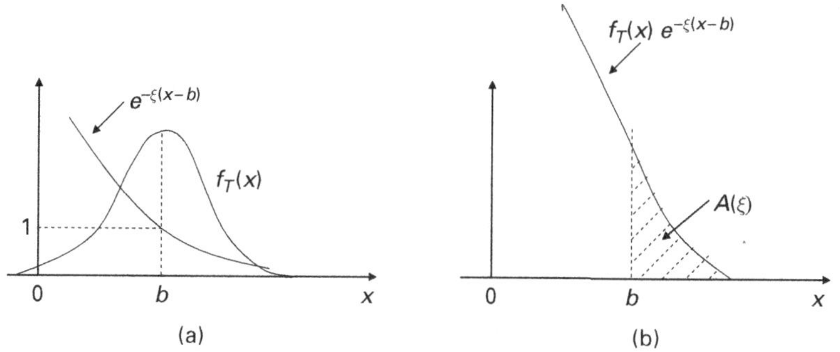

Figure 10.3 (a) The tilted PDF $f_{T_n}(x)$ and $e^{-\xi(x-b)}$; (b) the relation between the function $e^{-\xi(x-b)} f_{T_n}(x)$ and $A(\xi)$.

$$A(\xi^*) \approx \int_b^\infty \frac{1}{\sqrt{2\pi m''_{S_n}(\xi^*)}} \exp\left[-\frac{(x-b)^2}{2m''_{S_n}(\xi^*)} - \xi^*(x-b)\right] dx$$
$$= e^{c^2/2} \int_c^\infty \frac{1}{\sqrt{2\pi}} e^{-y^2/2}\, dy = e^{c^2/2} Q(c). \qquad (10.94)$$

Here, we changed the integration variable by $y = (x-b)/\sqrt{m''_{S_n}(\xi^*)} + c$, where

$$c = \xi^* \sqrt{m''_{S_n}(\xi^*)} \qquad (10.95)$$

and $Q(c)$ is the upper tail end of the unit normal distribution,

$$Q(c) = \int_c^\infty \frac{1}{\sqrt{2\pi}} \exp\left(-\frac{t^2}{2}\right) dt, \qquad (10.96)$$

and is called the ***Q*-function**[11] in digital communications (e.g., see [361]). It is related to the **complementary error function** (e.g., see [3]) according to

$$\boxed{Q(c) = \frac{1}{2}\text{erfc}\left(\frac{c}{\sqrt{2}}\right), \text{ where } \text{erfc}(c) = \frac{2}{\pi}\int_c^\infty \exp(-t^2)\, dt.} \qquad (10.97)$$

The Q-function is bounded by

$$\frac{\exp\left(-\frac{c^2}{2}\right)}{c\sqrt{2\pi}}\left(1 - \frac{1}{c^2}\right) < Q(c) < \frac{\exp\left(-\frac{c^2}{2}\right)}{c\sqrt{2\pi}}. \qquad (10.98)$$

Thus, for large $c \gg 1$, we can approximate it by

$$Q(c) \approx \frac{1}{c\sqrt{2\pi}} e^{-c^2/2}. \qquad (10.99)$$

[11] This Q-function has nothing to do with the auxiliary function Q, used in maximum-likelihood estimation by the EM algorithm discussed in Chapter 18.

Hence, an approximate value of the integration is given by

$$A(\xi^*) \approx \frac{1}{c\sqrt{2\pi}} = \frac{1}{\xi^*\sqrt{2\pi m''_{S_n}(\xi^*)}}. \tag{10.100}$$

Thus, we finally obtain the following approximation to evaluate the small probability of the tail end of the distribution:

$$\boxed{P[S_n \geq b] \approx A(\xi^*)e^{-\xi^* b}M_{S_n}(\xi^*) = \frac{1}{\xi^*\sqrt{2\pi m''_{S_n}(\xi^*)}}e^{-[\xi^* b - m_{S_n}(\xi^*)]},} \tag{10.101}$$

where in the last step we used $M(t) = e^{m(t)}$.

Example 10.5: Large deviation approximation of a sum of normal RVs. Let us consider a random variable X that has the $N(0, 1)$ distribution. From Example 8.3 of Section 8.1 we have $M_X(t) = e^{t^2/2}$ and $m_X(t) = t^2/2$, which yield $m'_X(t) = t$ and $m''_X(t) = 1$. Then the associated RV Y has the exponentially twisted PDF

$$f_Y(x) = \frac{e^{\xi x}}{e^{\xi^2/2}}\frac{1}{\sqrt{2\pi}}e^{-x^2/2} = \frac{1}{\sqrt{2\pi}}e^{-(x-\xi)^2/2}, \tag{10.102}$$

which is again the normal distribution with unit variance, but its mean gets shifted to ξ. This confirms the formulas (10.82) and (10.83) that relate the mean and variance of the twisted variable Y to the MGF of the original variable X; i.e., $E[Y] = \xi = m'_X(\xi)$ and $\mathrm{Var}[Y] = 1 = m''_X(\xi)$.

Now consider the sum of n independent unit normal variables $S_n = X_1 + X_2 + \cdots + X_n$. Take for instance $n = 3$ and $b = 15$. Then the exact value of $P[S_3 \geq 15]$ can be found, by using, for example, a MATLAB function, as

$$P[S_3 \geq 15] = P[\overline{X}_3 \geq 5] = 2.35 \times 10^{-18}, \tag{10.103}$$

where $\overline{X}_3$ is the sample mean; i.e., $\overline{X}_3 = (X_1 + X_2 + X_3)/3 = S_3/3$.

Since the logarithmic MGF of the sum variable S_n is $m_{S_n}(t) = \sum_{i=1}^n m_{X_i}(t) = nt^2/2$, the optimum tilting parameter is found from $m'_{S_n}(\xi^*) = b$ as $\xi^* = b/n = \beta = 5$. Thus, c of (10.95) is $c = \xi^*\sqrt{m''_{S_n}(\xi^*)} = \beta\sqrt{n} = 5\sqrt{3}$. Hence, $A(\xi^*) \approx 1/c\sqrt{2\pi} = 0.133$, and we find from (10.101)

$$P[S_n \geq b] = P[\overline{X}_n \geq \beta] \approx \frac{1}{\beta\sqrt{2\pi n}}e^{-n\beta^2 + n\beta^2/2} = \frac{1}{\beta\sqrt{2\pi n}}e^{-n\beta^2/2} \tag{10.104}$$

$$= \frac{1}{5\sqrt{6\pi}}e^{-3\times 5^2/2} = 0.0461 \times 5.17 \times 10^{-17} = 2.38 \times 10^{-18}. \tag{10.105}$$

which is very close to the exact value in (10.103). This small discrepancy between 2.35×10^{-18} versus 2.38×10^{-18} obtained here, however, is not due to the large deviation approximation, but due to the approximation we adopted in computing $A(\xi^*)$; i.e., approximating $Q(c)$ by $Q(c) \approx 1/(c\sqrt{2\pi})e^{-c^2/2}$, assuming $c \gg 1$.

Since $f_{T_n}(x)$ is exactly a normal distribution for any n in this particular case, $A(\xi)$ of (10.91) has an exact expression:

$$A(\xi) = \int_b^\infty e^{-\xi(x-b)} \frac{1}{\sqrt{2\pi n}} e^{-(x-n\xi)^2/2n}. \tag{10.106}$$

Hence, for $\xi = \xi^* = b/n = \beta$, we find $A(\xi^*) = \frac{e^{b^2/2n}}{\sqrt{2\pi n}} \int_b^\infty e^{-x^2/2n}\,dx = e^{n\beta^2/2} Q(\sqrt{n}\beta)$. Substituting the above result into (10.90), we have

$$A(\xi^*)e^{-\xi^* b} M_{S_n}(\xi^*) = e^{n\beta^2/2} Q(\sqrt{n}\beta) e^{-n\beta^2} e^{n\beta^2/2} = Q(\sqrt{n}\beta), \tag{10.107}$$

which is, not surprisingly, equal to $P[S_n \geq b]$. □

10.3.2 Large deviation rate function

In this section we will further investigate large deviations theory by introducing the notion of **rate function**. Let us consider again the sum S_n of i.i.d. RVs X_i:

$$S_n = X_1 + X_2 + \cdots + X_n.$$

In Section 10.2 we found the probability that S_n exceeds some threshold $b = n\beta$ satisfies Chernoff's bound given by (10.56):

$$P[S_n \geq n\beta] \leq \min_{\xi \geq 0} \left\{ e^{-n[\xi\beta - m_X(\xi)]} \right\}. \tag{10.108}$$

Taking the natural logarithm of both sides and dividing them by n yields

$$\frac{1}{n} \ln P[S_n \geq n\beta] \leq \min_{\xi \geq 0} \{-[\xi\beta - m_X(\xi)]\} = -\max_{\xi \geq 0} \{\xi\beta - m_X(\xi)\}. \tag{10.109}$$

Using the optimum tilting parameter ξ^* that satisfies

$$m'_X(\xi^*) = \beta, \tag{10.110}$$

we have an alternative expression for the Chernoff bound (10.60):

$$\frac{1}{n} \ln P[S_n \geq b] \leq -[\xi^*\beta - m_X(\xi^*)],\ \xi^* \geq 0. \tag{10.111}$$

In the last section we improved upon Chernoff's bound and obtained the large deviation approximation (10.101), which can be rewritten for the case of the sum of **identically distributed** RVs as

$$\frac{1}{n}\ln P[S_n \geq b] \approx -[\xi^*\beta - m_X(\xi^*)] - \frac{1}{n}\ln\left[\xi^*\sqrt{2\pi n m''_X(\xi^*)}\right]$$
$$= -[\xi^*\beta - m_X(\xi^*)] + O\left(\frac{\ln n}{n}\right), \tag{10.112}$$

where $O(\frac{\ln n}{n})$ is the term that converges to zero as n grows at the same rate as $\frac{\ln n}{n}$.

These results motivate us to define

$$\boxed{I(\beta) \triangleq \max_{\xi \geq 0}\{\xi\beta - m_X(\xi)\} = \xi^*\beta - m_X(\xi^*),} \tag{10.113}$$

which is referred to as the **large deviation rate function**. We can also express the rate function as

$$I(\beta) = -\ln\left(\min_{\xi \geq 0}\left\{e^{-\xi\beta}M_X(\xi)\right\}\right). \tag{10.114}$$

It is not difficult to show that

1. $I(\beta)$ is a convex function.
2. It takes its minimal value at $\beta = E[X]$ and $I(E[X]) = 0$; hence, $I(\beta) \geq 0$ for all β.

Chernoff's bound can be expressed in terms of the rate function $I(\beta)$ as

$$\frac{1}{n}\ln P[S_n \geq n\beta] \leq -I(\beta); \text{ that is, } P[S_n \geq n\beta]^{1/n} \leq e^{-I(\beta)}. \tag{10.115}$$

Taking the limit of (10.115), we have

$$\lim_{n\to\infty} \max \frac{1}{n}\ln P[S_n \geq n\beta] \leq -I(\beta). \tag{10.116}$$

Large deviation theory essentially shows that $-I(\beta)$ is **also a lower bound**; i.e.,

$$\lim_{n\to\infty} \min \frac{1}{n}\ln P[S_n \geq n\beta] \geq -I(\beta). \tag{10.117}$$

The last equation, combined with (10.116), yields

$$\boxed{\lim_{n\to\infty} \frac{1}{n}\log P[S_n \geq n\beta] = -I(\beta).} \tag{10.118}$$

In order to show (10.117) we write

$$P[S_n \geq n\beta] = \int_{x_1+\cdots+x_n \geq n\beta} f_X(x_1)\cdots f_X(x_n)\,dx_1\cdots dx_n. \tag{10.119}$$

Since the tilted RV Y_i has the PDF given by

$$f_Y(y) = \frac{e^{\xi^* y} f_X(y)}{M_X(\xi^*)}, \tag{10.120}$$

we find

$$P[S_n \geq n\beta] = M_X(\xi^*)^n \int_{y_1+\cdots+y_n \geq n\beta} e^{-\xi^*(y_1+\cdots+y_n)} f_Y(y_1)\cdots f_Y(y_n)\, dy_1\cdots dy_n. \tag{10.121}$$

Let ϵ be an arbitrary positive constant, and let us restrict the integration range from $[n\beta, \infty)$ to $[n\beta, n(\beta+\epsilon)]$. Then we have the following lower bound expression:

$$\begin{aligned} P[S_n \geq n\beta] &\geq M_X(\xi^*)^n \int_{n\beta \leq y_1+\cdots+y_n \leq n(\beta+\epsilon)} e^{-\xi^*(y_1+\cdots+y_n)} f_Y(y_1)\cdots f_Y(y_n)\, dy_1\cdots dy_n \\ &\geq M_X(\xi^*)^n e^{-\xi^* n(\beta+\epsilon)} \int_{n\beta \leq y_1+\cdots+y_n \leq n(\beta+\epsilon)} f_Y(y_1)\cdots f_Y(y_n)\, dy_1\cdots dy_n. \end{aligned} \tag{10.122}$$

Noting that the RVs Y_i have the common mean β (Problem 10.29), i.e.,

$$E[Y_i] = \beta, \tag{10.123}$$

we see that the distribution of $Y_1 + \cdots + Y_n$ will become concentrated around its mean $n\beta$ as n increases. This property, called the **strong law of large numbers**, will be discussed Section 11.3.3. In other words, the integral in (10.122) converges to unity as $n \to \infty$:

$$\lim_{n\to\infty} \int_{\beta \leq (y_1+\cdots+y_n)/n \leq \beta+\epsilon} f_Y(y_1)\cdots f_Y(y_n)\, dy_1 \cdots dy_n = 1. \tag{10.124}$$

Taking the logarithm of both sides of (10.122), dividing them by n, and letting $n \to \infty$, we obtain (10.117).

A more rigorous, but somewhat more involved, derivation of the lower bound (10.122) is found in, for example, Grimmett and Stirzaker [131].

Example 10.6: Large deviation approximation for Erlang distribution. Let X_i be i.i.d. RVs with exponential distribution

$$F_X(x) = 1 - e^{-\mu x},\ f_X(x) = \mu e^{-\mu x}, x \geq 0.$$

Then their sum

$$S_n = X_1 + X_2 + \cdots + X_n$$

is distributed according to the n-stage Erlang distribution with mean n/μ (see Problem 4.11). We can readily find the following functions related to the large deviation approximation:

$$
\begin{aligned}
M_X(t) &= \int_0^\infty \mu e^{-\mu x} e^{tx}\, dx = \frac{\mu}{\mu - t},\\
m_X(t) &= \ln \mu - \ln(\mu - t),\\
I(\beta) &= \xi^* \beta - \ln \mu + \ln(\mu - \xi^*).
\end{aligned}
$$

Noting that

$$\frac{dI(\beta)}{d\xi^*} = \beta - \frac{1}{\mu - \xi^*} = 0,$$

we find

$$\xi^* = \mu - \frac{1}{\beta} = \frac{\mu\beta - 1}{\beta},$$

and thus

$$I(\beta) = \mu\beta - 1 - \ln(\mu\beta). \tag{10.125}$$

Hence, the large deviation approximation for the tail of the n-stage Erlang distribution is

$$P[S_n \geq n\beta] = e^{-nI(\beta)+o(n)} = e^{-n(\mu\beta-1)+n\ln(\mu\beta)+o(n)} = (\mu\beta)^n e^{-n(\mu\beta-1)+o(n)}. \tag{10.126}$$

It will be of interest to derive the following approximate expression for the distribution function of the n-stage Erlang variable S_n. By substituting

$$\beta = \frac{x}{n}$$

into the result obtained above we have

$$
\begin{aligned}
F_{S_n}(x) &= 1 - P[S_n \geq x] = 1 - \left(\frac{\mu x}{n}\right)^n e^{-\mu x + n + o(n)}\\
&= 1 - \frac{(\mu x)^n}{(n/e)^n} e^{-\mu x + o(n)} \approx 1 - \frac{(\mu x)^n}{n!} e^{-\mu x} \sqrt{2\pi n}\, e^{o(n)},
\end{aligned}
$$

where we used Stirling's formula $n! \approx \sqrt{2\pi n}\,(n/e)^n$.

Then, in the limit $n \to \infty$, we find

$$F_{S_n}(x) \approx 1 - \frac{(\mu x)^n}{n!} e^{-\mu x} \sqrt{2\pi n}. \tag{10.127}$$

□

10.4 Summary of Chapter 10

Cauchy–Schwarz inequality:	$\lvert\langle \boldsymbol{x}, \boldsymbol{y}\rangle\rvert \le \lVert\boldsymbol{x}\rVert\,\lVert\boldsymbol{y}\rVert$	(10.1)
Lagrange identity:	$\lVert\boldsymbol{x}\rVert^2\lVert\boldsymbol{y}\rVert^2 - \lvert\langle \boldsymbol{x}, \boldsymbol{y}\rangle\rvert^2 = \sum_{i<j} \lvert x_i y_j - x_j y_i\rvert^2$	(10.5)
Cauchy–Schwarz inequality for random vectors:	$\left\lvert E[\boldsymbol{X}^\top \overline{\boldsymbol{Y}}]\right\rvert^2 \le E\left[\lVert\boldsymbol{X}\rVert^2\right] E\left[\lVert\boldsymbol{Y}\rVert^2\right]$	(10.11)
Convex function:	$g(px_1 + (1-p)x_2) \le p\,g(x_1) + (1-p)g(x_2)$	(10.14)
Jensen's inequality:	$g\left(\sum_i p_i x_i\right) \le \sum_{i=1}^n p_i g(x_i)$	(10.15)
Jensen's inequality for a random variable:	$g(E[X]) \le E\left[g(X)\right]$	(10.16)
Shannon's lemma:	$\sum_{i=1}^n f_i \log(f_i/g_i) \ge 0$	(10.18)
Markov inequality:	$P[X \ge a] \le E[X]/a$	(10.23)
Chebyshev's inequality:	$P[\lvert X - E[X]\rvert \ge b] \le \sigma^2/b^2$	(10.29)
Martingale:	$E[S_k \mid S_1, S_2, \ldots, S_{k-1}] = S_{k-1}$	(10.31)
Submartingale:	$E[Y_k \mid Y_1, Y_2, \ldots, Y_{k-1}] \ge Y_{k-1}$	(10.33)
Supermartingale:	$E[Y_k \mid Y_1, Y_2, \ldots, Y_{k-1}] \le Y_{k-1}$	(10.33)
Doob–Kolmogorov's ineq. for a nonnegative submartingale:	$P\left[\max\{Y_1, Y_2, \ldots, Y_n\} \ge a\right] \le E[Y_n]/a$	(10.34)
Kolmogorov's inequality for a martingale:	$P\left[\max\{\lvert S_1\rvert, \lvert S_2\rvert, \ldots, \lvert S_n\rvert\} \ge b\right] \le E[S_n^2]/b^2$	(10.40)
Chernoff's bound:	$P[X \ge b] \le e^{-\xi^* b} M_X(\xi^*) = e^{-\xi^* b} M_X(\xi^*)$	(10.50)
where	$m'(\xi^*)$=b	(10.49)
Chernoff's bound for sum of i.i.d. RVs:	$e^{-n[\xi^* m'_X(\xi^*) - m_X(\xi^*)]}$	(10.60)
Chernoff's bound for coin tossing:	$P[S_n \ge n\beta] \le 2^{-n(1-\mathcal{H}(\beta))}$	(10.72)
Stirling's approx. formula:	$n! \approx \sqrt{2\pi n}\,(n/e)^n$	(10.150)
Exponentially tilted distribution:	$dF_Y(x) = \frac{e^{\xi x}}{M_X(\xi)} dF_X(x)$	(10.78)
MGF of Y:	$M_Y(t) = \frac{M_X(\xi + t)}{M_X(\xi)}$	(10.89)
Q-function:	$Q(c) = \frac{1}{2}\text{erfc}\left(c/\sqrt{2}\right)$,	(10.97)
where	$\text{erfc}(c) = \frac{2}{\pi}\int_c^\infty \exp(-t^2)\,dt$	(10.97)
Large deviation approximation:	$P[S_n \ge b] \approx= \frac{1}{\xi^*\sqrt{2\pi m''_{S_n}(\xi^*)}} e^{-[\xi^* b - m_{S_n}(\xi^*)]}$	(10.101)
Large deviation rate function:	$I(\beta) \triangleq \max_{\xi \ge 0}\{\xi\beta - m_X(\xi)\} = \xi^*\beta - m_X(\xi^*)$	(10.113)

10.5 Discussion and further reading

In engineering or other applications, inequalities or bounds may be useful, for instance, when we want to be conservative in performance analysis of a given system, or to provide a minimum performance guarantee in system design.

The proof we provided for the Cauchy–Schwarz inequality is more general than the proof discussed in many textbooks that is applicable only to inner product spaces over the real number field. Markov's inequality is the basis for many other inequalities, bounds, and approximations presented in this chapter. Chebyshev's inequality will be used in a streamlined proof of the weak law of large numbers to be discussed in Section 11.3.2.

We introduced the concepts of martingale, submartingale, and supermartingale, together with Kolmogorov's inequalities. Although not discussed in most textbooks written for engineering students, **martingale theory**, pioneered by Doob [82], is an active research topic in applied probability theory. Its applications are diverse, including random walks, Brownian motion, limit theorems, game theory, queueing theory (e.g., see Asmussen [6]) and mathematical finance (e.g., see Shafer and Vovk [299]). See Ross [289], Rogers and Williams [282], and Williams [354] for further study of martingales.

Chernoff's bound is a powerful technique when we deal with the computation of the tail end of the distribution. For instance, the computation of the probability of decoding error in communication systems (e.g., see Gallager [113] and Wozencraft and Jacobs [361]) can be facilitated by proper use of this bounding method. In communication networks, a simple expression for the probability of overflow (due to insufficient buffer allocation) or the probability of call blocking (due to insufficient bandwidth allocation) can be derived from Chernoff's bound (e.g., see Hui [155]).

An approximation technique, such as the large deviations approximation, may be preferred when an accurate evaluation of a performance metric is more important than a conservative evaluation. Its application domains are essentially the same as those for Chernoff's bound; i.e., evaluation of the *bit error rate* in a digital communication system or that of the *packet loss rate* at routers or switches in a packet-switched network.

The large deviation approximation is also useful when we must resort to a *simulation experiment* in order to evaluate the probability of some **rare event** in a system to be studied. If the probability of interest is as small as 10^{-8}–10^{-10}, the run time required to obtain an accurate estimate of such a small probability in a brute-force simulation would be too excessive to be practical. Use of the *exponential change of measure* based on the large deviation theory leads to a **fast simulation** technique called **importance sampling**. The interested reader is directed to, for example, Jeruchim *et al.* [167], Ross [289], and Kobayashi and Mark [203]. For a further mathematical study of large deviations theory, see, for example, Bucklew [42] and Shwartz and Weiss [304]. Ellis [90] discusses the large deviations theory developed in the field of statistical mechanics.

10.6 Problems

Section 10.1: Inequalities frequently used in probability theory

10.1 Lagrange identity. Verify the Lagrange identity given by (10.5).

10.2 Proof of Jensen's inequality. Prove Jensen's inequality (10.15) by using mathematical induction.

10.3 Inequality for covariance. Derive the inequality (10.12).

10.4 Arithmetic mean and geometric mean. Prove that the **arithmetic mean** of x_i (all nonnegative) is not smaller than their **geometric mean**, i.e.,

$$\frac{\sum_{i=1}^{n} x_i}{n} \geq (x_1 x_2 \cdots x_n)^{1/n}, \; x_i \geq 0, \text{ for all } i. \tag{10.128}$$

Hint: $g(x) = \log x$ is a concave function in $(0, \infty)$.

10.5 Convex function is continuous. Let $x_{-1} < x_0 < x_1 < x_2$.

(a) If we write

$$x_1 = p x_0 + (1 - p) x_2,$$

what is p?

(b) Show that for a convex function $g(x)$,

$$\frac{g(x_{-1}) - g(x_0)}{x_{-1} - x_0} < \frac{g(x_1) - g(x_0)}{x_1 - x_0} \leq \frac{g(x_2) - g(x_0)}{x_2 - x_0}, \text{ for } x_0 < x_1 < x_2.$$

Show that the right-hand derivative at x_0, denoted $f'_+(x_0)$, exists; i.e.,

$$\lim_{x \downarrow x_0} \frac{g(x) - g(x_0)}{x - x_0} = g'_+(x_0).$$

Similarly show that the left-hand derivative at x_0, $g'_-(x_0)$, exists and that $g'_-(x_0) \leq g'_+(x_0)$.

(c) Show that $g(x)$ is **continuous at all points**.

10.6 A convex function is above its tangent. Show that if $g(x)$ is a convex function, it is **above its tangent** at every point x; i.e.,

$$f(x) \geq f(x_0) + a(x - x_0), \text{ for some constant } a. \tag{10.129}$$

10.7 A twice-differentiable function and a convex function. Show that if a continuous function $g(x)$ has a second derivative $g''(x) \geq 0$ at all points, then it is a convex function.

10.8 Another derivation of Jensen's inequality. Show that if a function $g(x)$ is continuous and twice-differentiable, then Jensen's inequality holds.

10.9 Alternative derivation of Shannon's lemma.

(a) Derive (10.18) from Jensen's inequality.
Hint: $-\log x$ is a convex funtion.
(b) Derive (10.18) using the Lagrangian multiplier method.

10.10 Inequalities in information theory [278].

(a) Show that if $a_i, b_i > 0$ for all i and $\sum_i a_i = \sum_i b_i$, then

$$\sum_i \log \frac{a_i}{b_i} \geq 0, \qquad (10.130)$$

where the equality is attained if and only if $a_i = b_i$.
(b) In addition to the assumptions in (a), further assume that $a_i, b_i \leq 1$ for all i. Then show

$$\sum_i a_i \log \frac{a_i}{b_i} \geq \frac{1}{2} \sum_i a_i (a_i - b_i)^2. \qquad (10.131)$$

Hint: Use the Taylor expansion of $\log x$ at $x = 1$.

10.11 Coin tossing and Markov and Chebyshev inequalities. Consider the experiment of tossing a fair coin n times. Let S_n be the total number of "head $(=1)$":

$$S_n = B_1 + B_2 + \cdots + B_n, \qquad (10.132)$$

where B_i are independent binary variables, with $P[B_i = 1] = P[B_i = 0] = \frac{1}{2}$ for all $i = 1, 2, \ldots, n$. We assume the threshold value $b > E[S_n] = n/2$; i.e., $\frac{1}{2} < \beta < 1$, where $\beta = b/n$.

(a) Apply Markov's inequality and find an upper bound on the probability that S_n exceeds $b = \beta n$. Compute the upper bound for cases where $n = 100$ and $n = 1000$ with $\beta = 0.8$ for both cases.
(b) Apply Chebyshev's inequality and find an upper bound on the probability that S_n exceeds $b = \beta n$. Compute the upper bound for cases where $n = 100$ and $n = 1000$ with $\beta = 0.8$ for both cases.

10.12 Bienaymé's[12] inequality.

(a) Let X be a random variable with $E[|X|^r] < \infty$ for $r > 0$, where r is not necessarily an integer. Show that for any $b > 0$ we have

$$\boxed{P[|X| \geq b] \leq \frac{E[|X|^r]}{b^r}.} \qquad (10.133)$$

[12] Irénée-Jules Bienaymé (1796–1878) was a French mathematician. Among his contributions was a translation of Chebyshev's (1821–1894) work, written in Russian, into French.

(b) Show that if $E[|X - E[X]|^r] < \infty$, then

$$P[|X - E[X]| \geq b] \leq \frac{E[|X - E[X]|^r]}{b^r}. \tag{10.134}$$

10.13 Markov–Chebyshev–Bienaymé's inequality. Let $g(x)$ be an increasing non-negative function defined on $[0, \infty)$. Show that for any $b > 0$,

$$\boxed{P[|X| \geq b] \leq \frac{E[g(|X|)]}{g(b)},} \tag{10.135}$$

whenever the right side exists.

10.14 One-sided Chebyshev's inequality [289].

(a) Let X be a random variable with $E[X] = 0$ and $\text{Var}[X] = \sigma^2$. Show that for any $a > 0$

$$P[X \geq a] \leq \frac{\sigma^2}{\sigma^2 + a^2}.$$

Hint: For any $b > 0$,

$$X \geq a \text{ if and only if } X + b \geq a + b (> 0).$$

(b) If $E[X] = \mu$ and $\text{Var}[X] = \sigma^2$. Show that

$$P[X \geq \mu + a] \leq \frac{\sigma^2}{\sigma^2 + a^2},$$
$$P[X \leq \mu - a] \leq \frac{\sigma^2}{\sigma^2 + a^2}.$$

10.15 Submartingale derived from a martingale.

(a) Show that $|S_k|$ is a submartingale when S_k is a martingale.
Hint: Use Jensen's inequality.
(b) Prove the inequality (10.41).

10.16* Bernstein's[13] inequality [21, 131]. Let $B_i, 1 \leq i \leq n$, be a sequence of Bernoulli trials; i.e., i.i.d. RVs with $P[B_i = 1] = p$ and $P[B_i] = 1 - p = q$. Let $S_n = \sum_{i=1}^{n} B_i$; that is, the number of successes in the Bernoulli trials. Then show the following inequality, called **Bernstein's inequality**:

$$P\left[\left|\frac{S_n}{n} - p\right| \geq \epsilon\right] \leq 2\exp\left(-\frac{n\epsilon^2}{4}\right), \text{ for } \epsilon > 0. \tag{10.136}$$

Take the following steps to derive the above inequality.

[13] Sergei Natanovich Bernstein (1880–1968) was a Soviet mathematician.

(a) Show

$$P\left[\frac{S_n}{n} - p \geq \epsilon\right] \leq \exp(-\lambda n\epsilon)\left(p\,e\lambda q + q^{-\lambda p}\right)^n,$$

where $m = \lceil n(p+\epsilon)\rceil$ and $\lambda > 0$.
Hint: Use $1 \leq e^{\lambda k - m}$ for $k \geq m$.

(b) Show

$$P\left[\frac{S_n}{n} - p \geq \epsilon\right] \leq \exp(\lambda^2 n - \lambda n\epsilon).$$

Hint: Use the following inequality:$e^x \leq x + e^{x^2}$ for any real number x.

(c) Find λ that gives a tightest upper bound.

10.17* Hoeffding's[14] inequality for a martingale [152, 288]. Let $Y_i : i = 1, 2, \ldots$ be a martingale, with mean $E[Y_i] = \mu$ and let $Y_0 = \mu$. Suppose that Y_i have bounded differences in the sense that

$$-a_i \leq Y_i - Y_{i-1} \leq b_i, \text{ where } a_i, b_i \geq 0.$$

Show that, for all positive integers n and $t > 0$, the following **Hoeffding inequalities**[15] hold:

$$P[Y_n - \mu \geq t] \leq \exp\left(-\frac{2t^2}{\sum_{i=1}^n (a_i + b_i)^2}\right), \tag{10.137}$$

$$P[Y_n - \mu \leq -t] \leq \exp\left(-\frac{2t^2}{\sum_{i=1}^n (a_i + b_i)^2}\right). \tag{10.138}$$

(a) Assume first that $\mu = 0$. Let $W_i = e^{\lambda Y_i}$ with some $\lambda > 0$. Show that

$$P[Y_n \geq t] \leq e^{-\lambda t} E[W_n] \tag{10.139}$$

and

$$E[W_n | Y_{n-1}] \leq W_{n-1} \frac{b_n e^{-\lambda a_n} + a_n e^{\lambda b_n}}{a_n + b_n}. \tag{10.140}$$

Hint: Let X be such that $E[X] = 0$ and $P[-a \leq X \leq b] = 1$. Then, for a convex function f,

$$E[f(X)] \leq \frac{bf(-a) + af(b)}{a+b}. \tag{10.141}$$

(b) Obtain

$$P[Y_n \geq t] \leq \exp\left(-\lambda t + \lambda^2 \frac{\sum_{i=1}^n (a_i + b_i)^2}{8}\right). \tag{10.142}$$

[14] Wassily Hoeffding (1914–1991) was an American statistician and probabilist.
[15] Ross [288] credits this inequality to **Kazuoki Azuma** (1939–) [8], a Japanese mathematician.

Hint: For $0 \le \theta \le 1$, the following inequality holds for any real x:

$$\theta\, e^{(1-\theta)x} + (1-\theta)e^{-\theta x} \le e^{x^2/8}.$$

The derivation of this inequality is rather involved (see Ross [288] Lemma 6.3.2).

(c) Find an optimal value of λ and obtain the Azuma–Hoeffding inequality.

10.18* Upper bound on the waiting time in a G/G/1 queueing system [196]. Consider a single server queue in which the service time of the nth customer is denoted as S_n and the interarrival time between the nth and $n+1$st customer is denoted as T_n. Assume that the S_n are i.i.d. and so are the T_n. Also S_n and T_n are mutually independent.

If we define a new RV by

$$X_n = S_n - T_n,$$

then $E[X_n] < 0$ for the queueing system to be stable. Furthermore, it is known (and not difficult to show) that the sequence $\{W_n\}$ of waiting times in the queue is given recursively as

$$W_0 = 0 \text{ and } W_{n+1} = \max\{0, W_n + X_n\}, \tag{10.143}$$

where we assume that the queue is initially empty.

(a) Define a sequence $\{Y_j\}$ by

$$Y_0 = 1 \text{ and } Y_j = e^{\theta(X_{n-1}+X_{n-2}+\cdots+X_{n-j})},\ 1 \le j \le n, \tag{10.144}$$

where θ is a real-valued parameter to be determined. Show that if $\theta > 0$, then

$$e^{\theta W_n} = \max\{Y_0, Y_1, \ldots, Y_n\}. \tag{10.145}$$

(b) Show that Y_n forms a **submartingale** when θ is suitably chosen.
Hint: Consider $M_X(\theta)$, the MGF of the i.i.d. RVs X_n, and use its properties.

(c) Show the complementary distribution function of W_n has the exponential upper bound

$$F^c_{W_n}(t) = P[W_n > t] \le e^{-\theta t + n m_X(\theta)},$$

where $m_X(\theta)$ is the logarithmic MGF or semi-invariant function, as defined in (8.6):

$$m_X(\theta) \triangleq \ln M_X(\theta).$$

(d) For given n and t, find the value $\theta^\star$ that gives the tightest upper bound.

(e) Show that in the limit $n \to \infty$,

$$\lim_{n\to\infty} F^c_{W_n}(t) = F^c_W(t) \le e^{-\theta_0}, \tag{10.146}$$

where $\theta_0 \in I_\theta$ such that

$$m_X(\theta_0) = 0, \text{ or equivalently } M_X(\theta_0) = 1. \tag{10.147}$$

The inequality (10.146) for the equilibrium distribution is **Kingman's upper bound** [184, 185] which he obtained by using a martingale.

Section 10.2: Chernoff's bounds

10.19 Altenative derivation of (10.47). Obtain an alternative derivation of the inequality (10.47) by using the indicator function as we did for the alternative derivation of Markov's inequality.

10.20 Derivation of the Chernoff bound (10.50). Derive the Chernoff bound (10.50). *Hint:* Show that $U(\xi) = e^{-\xi b} M_X(\xi)$ is a convex function.

10.21 Alternative proof of Chernoff's bound. Prove the convexity of $U(\xi) = e^{-b\xi} M_X(\xi)$ using the following steps. Define a function $f(b, \xi) = M_X(\xi) b^2 - 2bM'_X(\xi) + M''_X(\xi)$. Then (10.52) can be written as

$$U''(\xi) = e^{-\xi b} f(b, \xi). \tag{10.148}$$

(a) What are the necessary and sufficient conditions for $f(b, \xi)$ to be nonnegative for all b?

(b) Show that the above conditions are satisfied.
Hint: Use the version of the Cauchy–Schwarz inequality for RVs.

10.22 When is the Chernoff bound meaningful? Sketch the upper bound function $U(\xi) = e^{-\xi b} M_X(\xi)$ versus ξ for the case (a) $b > \mu_X = E[X]$ and for the case (b) $b < \mu_X$, and show that the Chernoff bound (10.47) can be meaningful only in bounding the "right-end tail" (i.e., $X \geq b > \mu_X$) of the distribution.

10.23 Chernoff's bound for sum of normal RVs. Let S be a sum of n independent unit (or standard) normal variables U_i; i.e., $U_i \sim N(0, 1)$: $S = U_1 + U_2 + \ldots + U_n$. Apply Chernoff's bound and show that $P[S \geq n\beta] \leq e^{-n\beta^2/2}$.

10.24 Chernoff's bound for the sum of Poisson variables. Let $S = X_1 + X_2 + \cdots + X_n$, where X_i are i.i.d. Poisson variables with mean $1/\lambda$. Apply Chernoff's bound and show $P[S \geq n\beta] \leq e^{-nB}$, where $B = \beta \ln(\beta/\lambda) + \lambda(1 - \beta)$.

10.25 Numerical evaluation of the Chernoff bound. In the fair coin tossing experiment discussed in Example 10.4, set $\beta = 0.51$. Then numerically evaluate the upper bounds for $P[S_n \geq n\beta]$ for $n = 100, 1000, 10^4, 10^5, 10^6$.

10.26 Assessment of Chernoff's bound.
In order to assess the accuracy of the Chernoff bound, consider the coin tossing experiment in Example 10.4.

(a) Find an exact expression for $P[S_n \geq b]$.

(b) Let C_b be the sum of the binomial coefficient from $k = b$ to $k = n$. For $\beta < 1/2$, show that C_b has the following upper and lower bounds:

$$\binom{n}{b} < C_b < (n - b + 1)\binom{n}{b}. \tag{10.149}$$

(c) The well-known **Stirling approximation formula**[16] [98, 99] gives

$$n! \approx \sqrt{2\pi n}\left(\frac{n}{e}\right)^n. \tag{10.150}$$

Show that the binomial coefficient can be approximated as follows:

$$\binom{n}{k} \approx \sqrt{\frac{n}{2\pi k(n-k)}}\frac{n^n}{(n-k)^{n-k}k^k}. \tag{10.151}$$

(d) The following simple upper and lower bounds are known to hold for all n (see Feller [99] and Nelson [254]):

$$\sqrt{2\pi n}\left(\frac{n}{e}\right)^n B_{\mathrm{L}}(n) < n! < \sqrt{2\pi n}\left(\frac{n}{e}\right)^n B_{\mathrm{U}}(n), \tag{10.152}$$

where

$$B_{\mathrm{L}}(n) = \exp\left(\frac{1}{12n+1}\right) \text{ and } B_{\mathrm{U}}(n) = \exp\left(\frac{1}{12n}\right). \tag{10.153}$$

Show that upper and lower bounds for the binomial coefficient are obtainable from (10.152) as

$$\sqrt{\frac{n}{2\pi k(n-k)}}\frac{n^n\theta_{\mathrm{L}}(n,k)}{(n-k)^{n-k}k^k} < \binom{n}{k} < \sqrt{\frac{n}{2\pi k(n-k)}}\frac{n^n\theta_{\mathrm{U}}(n,k)}{(n-k)^{n-k}k^k}, \tag{10.154}$$

where

$$\theta_{\mathrm{L}}(n,k) = \frac{B_{\mathrm{L}}(n)}{B_U(k)B_{\mathrm{U}}(n-k)} \text{ and } \theta_{\mathrm{U}}(n,k) = \frac{B_{\mathrm{U}}(n)}{B_L(k)B_{\mathrm{L}}(n-k)}. \tag{10.155}$$

(d) Since Stirling's approximation formula and the above bounds hold only for integer k, we set $b = \lceil n\beta \rceil$ if $n\beta$ is not an integer. Then replace β by

$$\tilde{\beta} = \frac{b}{n} = \frac{\lceil n\beta \rceil}{n}. \tag{10.156}$$

$$\binom{n}{\lceil n\beta \rceil} \approx \frac{(1-\tilde{\beta})^{-n(1-\tilde{\beta})}\tilde{\beta}^{-n\tilde{\beta}}}{\sqrt{2\pi n\tilde{\beta}(1-\tilde{\beta})}} = \frac{1}{\sqrt{2\pi n\tilde{\beta}(1-\tilde{\beta})}}2^{n\mathcal{H}(\tilde{\beta})}. \tag{10.157}$$

Show that the following are upper and lower bounds of $P[S_n \geq n\beta]$ for all n:

$$\frac{\theta_{\mathrm{L}}(n,b)}{\sqrt{2\pi n\tilde{\beta}(1-\tilde{\beta})}}2^{-n(1-\mathcal{H}(\tilde{\beta}))} < P[S_n \geq n\beta] < \frac{(n-b+1)\theta_{\mathrm{U}}(n,b)}{\sqrt{2\pi n\tilde{\beta}(1-\tilde{\beta})}}2^{-n(1-\mathcal{H}(\tilde{\beta}))}. \tag{10.158}$$

(e) The term $2^{-n(1-\mathcal{H}(\tilde{\beta}))}$ of (10.72) now appears in both upper and lower bounds. Thus, Chernoff's bound is off from the exact value only by a certain factor, which is insignificant in the sense that the exponential term $2^{-n(1-\mathcal{H}(\tilde{\beta}))}$ largely determines the behavior of $P[S_n \geq n\tilde{\beta}]$ when n gets larger.

[16] James Stirling (1692–1770) was a Scottish mathematician and surveyor.

Numerically evaluate the upper and lower bounds given by (10.158) for the cases of $\beta = 0.8$ and $\beta = 0.51$ with different values of n, and assess the Chernoff bounds obtained in Example 10.4 (for $\beta = 0.8$) and in Problem 10.25 (for $\beta = 0.51$).

Section 10.3: Large deviations theory

10.27 Derivation of (10.82) and (10.83). Show that the expectation and variance of the exponentially twisted variable Y are given by (10.82) and (10.83); i.e., the first and second derivatives of the logarithmic MGF $m_X(t)$, evaluated at $t = \xi$.

10.28 Application of large deviation approximation to coin tossing. Consider the coin tossing problem discussed in Example 10.4.

(a) Find the distribution function $F_{B_i}(x)$ and its tilted counterpart F_{Y_i}.
(b) We are again interested in estimating the probability $P[S_n \geq n\beta]$. For given β, what is the optimum tilting parameter ξ^*?
(c) What is the the mean and variance of the sum variable T_n?
(d) Show that the large deviation approximation of $P[S_n \geq n\beta]$ is given by

$$P[S_n \geq n\beta] \approx \frac{1}{\ln\left(\frac{\beta}{1-\beta}\right)\sqrt{2\pi n\beta(1-\beta)}} 2^{-n(1-\mathcal{H}(\beta))}. \qquad (10.159)$$

Compare this with the Chernoff bound.

10.29 Derivation of (10.123). Show that (10.123) holds.

11 Convergence of a sequence of random variables and the limit theorems

So far we have been somewhat imprecise or vague when we state that a sequence of RVs X_n converges to some limit X, as n tends to infinity. In this chapter we discuss various types (or modes) of convergence for a sequence of RVs. The types of convergence we discuss in this chapter are:

1. *Convergence in distribution.*
2. *Convergence in probability.*
3. *Almost sure convergence* (also known as *convergence with probability one*, or *convergence almost everywhere*).
4. *Convergence in mean square.*

We will begin with a brief review of convergence of a sequence of numbers and a sequence of (nonrandom) functions.

11.1 Preliminaries: convergence of a sequence of numbers or functions

11.1.1 Sequence of numbers

Let $\{a_n\} = \{a_1, a_2, \ldots, a_n, \ldots\}$ be a sequence of numbers.

DEFINITION 11.1 (Convergence of a sequence of numbers). *We say that a sequence $\{a_n\}$ converges to a, written*

$$\lim_{n\to\infty} a_n = a, \quad \text{or } a_n \to a,$$

if, for each $\epsilon > 0$, there exists a number $N(\epsilon)$ such that

$$|a_n - a| < \epsilon, \textit{ for all } n \geq N(\epsilon).$$

The above definition of convergence requires knowledge of the limit a. It is sometimes convenient to use a criterion that does not rely on such knowledge.

DEFINITION 11.2 (Cauchy[1] convergence). *We say that a sequence $\{a_n\}$ is Cauchy convergent if, for any $\epsilon > 0$, there exists a number $N(\epsilon)$ such that*

[1] Augustin-Louis Cauchy (1789–1857) was a French mathematician.

$$|a_m - a_n| < \epsilon, \text{ for all } m, n \geq N(\epsilon).$$

We state the following theorem without proof, since the reader may well be familiar with this.

THEOREM 11.1 (Cauchy criterion for convergence). *A sequence $\{a_n\}$ converges if and only if it is Cauchy convergent.*

11.1.2 Sequence of functions

We now consider a sequence of (real-valued) functions $\{g_n(x)\}$ defined over some interval $[a, b]$. If x is fixed to some point $x_0 \in [a, b]$, then the sequence $\{g_n(x_0)\}$ is just a sequence of numbers.

DEFINITION 11.3 (Pointwise convergence). *We say that a sequence $\{g_n(x_0)\}$ converges to $g(x_0)$ **pointwise** at point $x = x_0$ if, for given $\epsilon > 0$, there exists a number $N(\epsilon, x_0)$ such that*

$$|g_n(x_0) - g(x_0)| < \epsilon \text{ for all } n \geq N(\epsilon, x_0).$$

DEFINITION 11.4 (Convergence everywhere). *We say that a sequence $\{g_n(x)\}$ converges to $g(x)$ **everywhere** in $[a, b]$ if it converges pointwise at every point; i.e., if for given $\epsilon > 0$ and x there exists a number $N(\epsilon, x)$ such that*

$$|g_n(x) - g(x)| < \epsilon \text{ for all } n \geq N(\epsilon, x) \text{ and } x \in [a, b].$$

A stronger condition on the sequence $\{g_n(x)\}$ is the notion of *uniform convergence.*

DEFINITION 11.5 (Uniform convergence). *We say that a sequence $\{g_n(x)\}$ converges to $g(x)$ **uniformly** in $[a, b]$ if, for given $\epsilon > 0$, there exists a number $N(\epsilon)$ such that*

$$|g_n(x) - g(x)| < \epsilon \text{ for all } n \geq N(\epsilon) \text{ and for all } x \in [a, b].$$

It is clear that uniform convergence in $[a, b]$ implies pointwise convergence at every point in $[a, b]$. However, the converse is not true.

Example 11.1: Convergence everywhere, but not uniformly [319]. Consider the following sequence of functions (see Figure 11.1)

$$g_n(x) = n^2 x\, e^{-nx}, \quad 0 \leq x < \infty, \quad n = 1, 2, \ldots$$

For given $x \in [0, \infty)$, the term e^{-nx} decays geometrically as n increases. Thus, despite the increasing factor n^2, $g_n(x)$ converges to zero everywhere in the interval $x \in [0, \infty)$ as $n \to \infty$. However, this sequence does not converge uniformly in $[0, \infty)$.

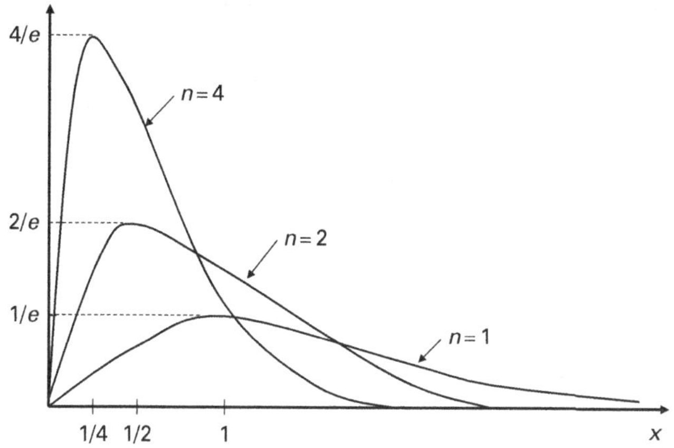

Figure 11.1 A sequence of functions $g_n(x) = n^2 x\, e^{-nx}$ for $n = 1, 2, 4, \ldots$ in Example 11.1.

In order to show this, we observe that, for given n, the function $g_n(x)$ is an increasing function in the interval $x \in [0, n^{-1}]$, and its maximum value is $g_n(n^{-1}) = n\, e^{-1}$. If the sequence were to converge uniformly, we would have to show that for arbitrary $\epsilon > 0$,

$$|g_n(x) - 0| = n^2 x\, e^{-nx} < \epsilon$$

for all $x \in [0, \infty)$ when n is larger than some number $N(\epsilon)$. Let us choose specifically the value of x that maximizes $g_n(x)$, i.e., $x = n^{-1}$, so that

$$|g_n(n^{-1}) - 0| = n\, e^{-1} < \epsilon.$$

Obviously there is no number $N(\epsilon)$ such that $N(\epsilon)e^{-1} < \epsilon$ for arbitrary $\epsilon > 0$. Thus, the sequence does not converge uniformly in $[0, \infty)$.

It will be instructive to note that the above function is a special case of the PDF of the gamma distribution

$$f(x) = \frac{e^{-\alpha x}}{\Gamma(\beta)} \alpha(\alpha x)^{\beta-1}, \quad x \geq 0,$$

in which $\alpha = n$ and $\beta = 2$. The mean is $\beta/\alpha = 2n^{-1}$ and the variance is $\beta/\alpha^2 = 2n^{-2}$. The mode of the distribution is n^{-1}, as shown above. As n increases, the PDF $g_n(x)$ approaches the impulse function at $x = 0^+$. When β is an integer k, the gamma distribution reduces to the k-stage Erlang distribution with the exponential rate parameter $\mu = \alpha$. Thus, the function $g_n(x)$ defined above is the PDF of the two-stage Erlang distribution with the rate $\mu = n$ (i.e., the mean $\mu^{-1} = n^{-1}$) of the exponential distribution.

Note also that we can replace the interval $[0, \infty)$ with any interval $[a, b]$ $(0 \leq a < b < \infty)$ in the above argument. □

11.2 Types of convergence for sequences of random variables

A sequence of RVs $X_1, X_2, \ldots$ can be viewed as a sequence of **functions** $X_1(\omega), X_2(\omega), \ldots$ that map $\omega \in \Omega$ into the real line $(-\infty, \infty)$. Therefore, a direct application of the concept given in Definition 11.4 for a sequence of nonrandom function $\{f_n(x)\}$ to a sequence of RVs $\{X_n(\omega)\}$ is as follows.

DEFINITION 11.6 (Convergence everywhere). *We say that a sequence of RVs $\{X_n\}$ converges to X* ***everywhere*** *if*

$$X_n(\omega) \to X(\omega) \text{ as } n \to \infty \textit{ for every } \omega \in \Omega;$$

that is, if for any $\epsilon > 0$ and for every $\omega \in \Omega$, there exists a number $N(\epsilon, \omega)$ such that

$$|X_n(\omega) - X(\omega)| < \epsilon \textit{ for all } n \geq N(\epsilon, \omega).$$

If we define set A

$$A = \left\{\omega : \lim_{n\to\infty} X_n(\omega) = X(\omega)\right\}, \tag{11.1}$$

Then

$$A = \Omega, \text{ if } X_n \text{ converges to } X \text{ everywhere.}$$

Similarly, the concept of uniform convergence can be applied to X_n. But such modes of convergence are not of much interest to us, since they contain no reference to probability. RVs are associated with some probability space $(\Omega, \mathcal{F}, P)$; therefore, we will be interested in interpreting, from probabilistic points of view, the statement $X_n \to X$ as $n \to \infty$.

11.2.1 Convergence in distribution

DEFINITION 11.7 (Convergence in distribution). *We say that a sequence of RVs $\{X_n\}$ converges to X* ***in distribution****, written*

$$X_n \xrightarrow{\mathrm{D}} X, \text{ or } X_n \xrightarrow{\mathrm{d}} X,$$

if the distribution function $F_n(x) = P[X_n \leq x]$ converges pointwise to $F_X(x)$ at all continuity points of $F_X(x) = P[X \leq x]$; that is, if

$$\boxed{\lim_{n\to\infty} F_n(x) = F_X(x)} \tag{11.2}$$

at all points x where $F_X(x)$ is continuous. □

Example 11.2: A sequence of $N(0, \sigma^2/n)$. Consider a sequence of RVs $\{X_n\}$, where X_n is distributed according to $N(0, \sigma^2/n)$:

$$F_n(x) = \int_{-\infty}^{x} \frac{\sqrt{n}}{\sqrt{2\pi}\,\sigma} e^{-nu^2/2\sigma^2}\, du.$$

Then, in the limit $n \to \infty$, we have

$$\lim_{n\to\infty} F_n(x) = \begin{cases} 0, & x < 0, \\ \frac{1}{2}, & x = 0, \\ 1, & x > 0. \end{cases}$$

Thus, $\{X_n\}$ converges in distribution to the RV X with distribution function

$$F_X(x) = \begin{cases} 0, & x < 0, \\ 1, & x \geq 0. \end{cases} \tag{11.3}$$

Note that $F_X(x)$ is a distribution function, so it must be right-continuous. Therefore, $F_X(0) = 1$, not $\frac{1}{2}$. Consequently, $\{F_n(x)\}$ does not converge to $F_X(x)$ at this discontinuity point $x = 0$. □

Example 11.3: Poisson distribution as a limit of binomial distributions. Let X_n be a discrete RV that is binomially distributed according to $B(x; n, p)$; i.e.,

$$F_n(x) = \sum_{k=0}^{n} \binom{n}{k} p^k (1-p)^{n-k} u(x-k), \quad x \geq 0,$$

where $u(x)$ is the unit step function. We take the limit $n \to \infty$ in such a way that

$$\lim_{n\to\infty} np = \lambda,$$

where λ is a fixed nonzero but finite number. Then, as discussed in Section 3.3.3, we have

$$\lim_{n\to\infty} F_n(x) = \sum_{k=0}^{\infty} \frac{\lambda}{k!} e^{-\lambda} u(x-k), \quad x \geq 0,$$

which is the Poisson distribution with mean λ. Thus, the binomial RV converges in distribution to a Poisson RV. □

Note that convergence in distribution is a condition on the probability distributions $F_n(x)$, not on the RVs $X_n(\omega)$. Thus, the sequence $X_n(\omega)$ may not converge to any fixed point for given ω, although we write, somewhat confusingly, $X_n \xrightarrow{D} X$. For example, consider a set $\{X_n(\omega),\ \omega \in \Omega\}$ at some n. Apply an arbitrary permutation to the set $\{X_n(\omega)\}$. Then $F_n(x)$ should remain unchanged if all sample functions are equally likely. If such shuffling is applied at every point n in sequence, $X_n(\omega)$ will not converge to any fixed point $X(\omega)$.

11.2.2 Convergence in probability

DEFINITION 11.8 (Convergence in probability). *We say that X_n converges to X **in probability**, written*

$$X_n \xrightarrow{\mathrm{P}} X, \text{ or } X_n \xrightarrow{\mathrm{p}} X,$$

if

$$\boxed{\lim_{n\to\infty} P[|X_n - X| > \epsilon\,] = 0 \text{ for any } \epsilon > 0,} \tag{11.4}$$

i.e., if for arbitrary $\epsilon > 0$ and $\delta > 0$ there exists a number $N(\epsilon, \delta)$ such that

$$P[\,|X_n - X| > \epsilon\,] < \delta, \quad \text{for all } n \geq N(\epsilon, \delta). \tag{11.5}$$

□

Let us define the set

$$A_n(\epsilon) = \{\omega : |X_n(\omega) - X(\omega)| \leq \epsilon\}; \tag{11.6}$$

hence,

$$A_n^{\mathrm{c}}(\epsilon) = \{\omega : |X_n(\omega) - X(\omega)| > \epsilon\}. \tag{11.7}$$

Then convergence in probability is equivalent to claiming

$$\lim_{n\to\infty} P[A_n^{\mathrm{c}}(\epsilon)] = 0 \text{ for any } \epsilon > 0; \tag{11.8}$$

i.e.,

$$\lim_{n\to\infty} P[A_n(\epsilon)] = 1 \text{ for any } \epsilon > 0. \tag{11.9}$$

From (11.5), we may state that X_n converges to X in probability if there exists $N(\epsilon, \delta)$ such that

$$P[A_n(\epsilon)] \geq 1 - \delta \text{ for all } n \geq N(\epsilon, \delta).$$

Convergence in probability is sometimes called **stochastic convergence** and we say that $\{X_n\}$ converges stochastically to X.

In Section 2.3 we discussed Bernoulli's theorem (Theorem 2.1 of Chapter 2), in which the sequence $X_n = k/n$ converges to p in probability (see (2.46)). This is a degenerate case where the limit RV X is a constant p, i.e., $P[X = p] = 1$, or equivalently,

$$F_X(x) = u(x - p) = \begin{cases} 0, & x < p, \\ 1, & x \geq p, \end{cases}$$

where $u(x)$ is the unit step function.

Example 11.4: A sequence of Cauchy distributions. Recall the Cauchy distribution discussed in Problem 5.4 of Section 5.1. Let the RV X_n have the PDF given by

$$f_n(x) = \frac{n}{\pi}\frac{1}{1+n^2x^2}, \quad -\infty < x < \infty,$$

and the distribution function

$$F_n(x) = \frac{n}{\pi}\int_{-\infty}^{x} \frac{1}{1+n^2u^2}\,du.$$

In the limit, the PDF $f_n(x)$ approaches Dirac's delta-function $\delta(x)$:

$$\lim_{n\to\infty} f_n(x) = \delta(x),$$

and $F_n(x)$ approaches the unit step function $u(x)$:

$$\lim_{n\to\infty} F_n(x) = u(x) = \begin{cases} 0, & x < 0, \\ 1, & x \geq 1. \end{cases}$$

Thus, the limit RV X is a constant; i.e., $X = 0$ with probability one. We see that X_n indeed converges in probability to zero, because

$$\lim_{n\to\infty} P[\,|X_n - 0| > \epsilon\,] = \int_{-\infty}^{-\epsilon} \delta(x)\,dx + \int_{\epsilon}^{\infty} \delta(x)\,dx = 0.$$

□

We will further discuss the notion of stochastic convergence in Section 11.3.2, where we introduce various **weak laws of large numbers**, one of which is Bernoulli's theorem stated earlier. We now state an important relationship between convergence in probability and convergence in distribution.

THEOREM 11.2 (Stochastic convergence versus convergence in distribution). *Convergence in probability implies convergence in distribution; i.e.,*

$$\boxed{X_n \xrightarrow{\text{P}} X \implies X_n \xrightarrow{\text{D}} X.} \tag{11.10}$$

But the converse is not true.

Proof. Consider first the probability $F_{X_n}(x) = P[X_n \leq x]$. The event $\{X_n \leq x\}$ may occur either when $X \leq x + \epsilon$ or when $X > x + \epsilon$. Since the latter events are mutually exclusive,

$$F_{X_n}(x) = P[X_n \leq x, X \leq x + \epsilon] + P[X_n \leq x, X > x + \epsilon]. \tag{11.11}$$

Similarly, we find

$$F_X(x+\epsilon) = P[X \leq x + \epsilon, X_n \leq x] + P[X \leq x + \epsilon, X_n > x]. \tag{11.12}$$

Subtraction yields

$$F_{X_n}(x) - F_X(x+\epsilon) = P[X_n \leq x, X > x + \epsilon] - P[X \leq x + \epsilon, X_n > x]. \tag{11.13}$$

The joint event $\{X_n \leq x\} \cap \{X > x + \epsilon\}$ implies $|X_n - X| > \epsilon$. But this is one way in which the event $\{|X_n - X| > \epsilon\}$ can occur. Therefore, on defining

$$\delta_n \triangleq P[|X_n - X| > \epsilon], \tag{11.14}$$

we have

$$F_{X_n}(x) \leq F_X(x + \epsilon) + \delta_n. \tag{11.15}$$

Similarly,

$$\begin{aligned} F_X(x - \epsilon) &= P[X \leq x - \epsilon, X_n \leq x] + P[X \leq x - \epsilon, X_n > x] \\ &\leq P[X_n \leq x] + P[|X_n - X| > \epsilon]. \end{aligned} \tag{11.16}$$

Thus, from the last two equations,

$$F_X(x - \epsilon) - \delta_n \leq F_{X_n}(x) \leq F_X(x + \epsilon) + \delta_n. \tag{11.17}$$

It is apparent that $\delta_n \to 0$ as $n \to \infty$, whenever $X_n \stackrel{\text{P}}{\to} X$. Therefore,

$$F_X(x - \epsilon) \leq \lim_{n \to \infty} F_{X_n}(x) \leq F_X(x + \epsilon) \text{ for every } \epsilon > 0. \tag{11.18}$$

At every point of continuity of $F_X(x)$, we have

$$F_X(x - \epsilon) \uparrow F_X(x) \text{ and } F_X(x + \epsilon) \downarrow F_X(x), \text{ as } \epsilon \downarrow 0. \tag{11.19}$$

Thus, we have shown that stochastic convergence

$$X_n \stackrel{\text{P}}{\longrightarrow} X$$

implies

$$\lim_{n \to \infty} F_{X_n}(x) = F_X(x)$$

at all continuity points of $F_X(x)$. Thus, $X_n \stackrel{\text{D}}{\longrightarrow} X$. □

Example 11.5: $X_n \stackrel{\text{D}}{\to} X$ versus $X_n \stackrel{\text{P}}{\to} X$. Here, we provide a simple example [131] that shows that $X_n \stackrel{\text{D}}{\to} X$ does not imply $X_n \stackrel{\text{P}}{\to} X$. Let X be a Bernoulli variable such that

$$P[X = 0] = P[X = 1] = \frac{1}{2}.$$

Let $X_1, X_2, \ldots, X_n, \ldots$ be identical RVs such that

$$X_n = X \text{ for all } n.$$

The X_n are certainly not independent, but $X_n \stackrel{\text{D}}{\to} X$.

Let us define variable Y by $Y = 1 - X$. Clearly, Y is also a 1–0 variable, having the same distribution as X; i.e., $P[Y = 0] = P[Y = 1] = \frac{1}{2}$. Hence $X_n \stackrel{\text{D}}{\to} Y$. But X_n

cannot converge to Y in any other mode (than convergence in distribution), because $|X_n - Y| = 1$ for all n. □

Example 11.6: $X_n \xrightarrow{\text{D}} X$ versus $X_n \xrightarrow{\text{P}} X$ (continued). Here is another example that will show that convergence in distribution does not imply convergence in probability. Consider the case where the X_n are i.i.d. RVs with a common distribution function $F_X(x)$. Then clearly $X_n \xrightarrow{\text{D}} X$, since $F_n(x) = F_X(x)$ for all n. However, $P[|X_n - X| > \epsilon]$ cannot be made arbitrarily small. In order to show this, define a random variable Z:

$$Z = X_n - X = X_n + Y,$$

where $Y = -X$. Since X_n and Y are independent,

$$\begin{aligned} F_Z(z) &= \int_{-\infty}^{\infty} F_{X_n}(z - y) f_Y(y)\, dy \\ &= \int_{-\infty}^{\infty} F_X(z - y) f_X(-y)\, dy \\ &= \int_{-\infty}^{\infty} F_X(z + x) f_X(x)\, dx. \end{aligned} \tag{11.20}$$

Thus,

$$\begin{aligned} P[\,|X_n - X| \le \epsilon] &= F_Z(\epsilon) - F_Z(-\epsilon) \\ &= \int_{-\epsilon}^{\epsilon} \int_{-\infty}^{\infty} f_X(z + x) f_X(x)\, dx\, dz. \end{aligned} \tag{11.21}$$

For small $\epsilon > 0$, we can approximate the last equation by

$$P[\,|X_n - X| \le \epsilon\,] \approx 2\epsilon \int_{-\infty}^{\infty} f_X^2(x)\, dx. \tag{11.22}$$

Therefore,

$$P[\,|X_n - X| > \epsilon\,] \approx 1 - 2\epsilon \int_{-\infty}^{\infty} f_X^2(x)\, dx. \tag{11.23}$$

This last expression is independent of n and cannot be zero for arbitrary ϵ. Thus, we have shown that X_n does not converge to X in probability. □

11.2.3 Almost sure convergence

DEFINITION 11.9 (Convergence almost surely). *We say that $\{X_n\}$ converges **almost surely** (or **almost everywhere**, or **with probability one**) to X, written*

$$X_n \xrightarrow{\text{a.s.}} X, \quad X_n \xrightarrow{\text{a.e.}} X, \text{ or } X_n \xrightarrow{\text{w.p.1.}} X,$$

if the sequence of numbers $\{X_n(\omega)\}$ converges to $X(\omega)$ for all sample points $\omega \in \Omega$ with probability one; that is, if

$$\boxed{P[\lim_{n\to\infty} X_n = X] = 1.} \tag{11.24}$$

□

Almost sure (a.s.) convergence is a rather strong type of convergence and implies other modes of convergence. Before we show that a.s. convergence leads to convergence in probability, we state the following condition for a.s. convergence that is equivalent to (11.24).

LEMMA 11.1 (Conditions for a.s. convergence). $X_n \xrightarrow{\text{a.s.}} X$, *if and only if, for arbitrary $\epsilon > 0$ and $\delta > 0$, there exists a number $M(\epsilon, \delta)$ such that*

$$P\left[\bigcap_{n=m}^{\infty}\{\omega : |X_n(\omega) - X(\omega)| < \epsilon\}\right] \geq 1 - \delta \tag{11.25}$$

for all $m \geq M(\epsilon, \delta)$.

Proof. The proof is somewhat long and involved, so, in the interest of space, it will be provided in the material that will be made available in "Supplementary Materials" on the book's website. □

Given arbitrary $\epsilon > 0$ and $\delta > 0$, convergence in probability as defined in Definition 11.8 means that, for each $n \geq N(\epsilon, \delta)$, $|X_n(\omega) - X(\omega)| > \epsilon$ for less than $\delta \times 100\%$ of the sample points $\omega \in \Omega$. It is possible, however, that there is not even one $\omega^* \in \Omega$ such that $|X_n(\omega^*) - X(\omega^*)| \leq \epsilon$ for *all* $n \geq N(\epsilon, \delta)$. Convergence almost surely, on the other hand, requires that, for each $m \geq M(\epsilon, \delta)$, more than $(1 - \delta) \times 100\%$ of the sample points $\omega \in \Omega$ satisfy $|X_n(\omega) - X(\omega)| < \epsilon$ for *all* $n \geq m$. Figure 11.2 illustrates the difference between these two types of convergence.

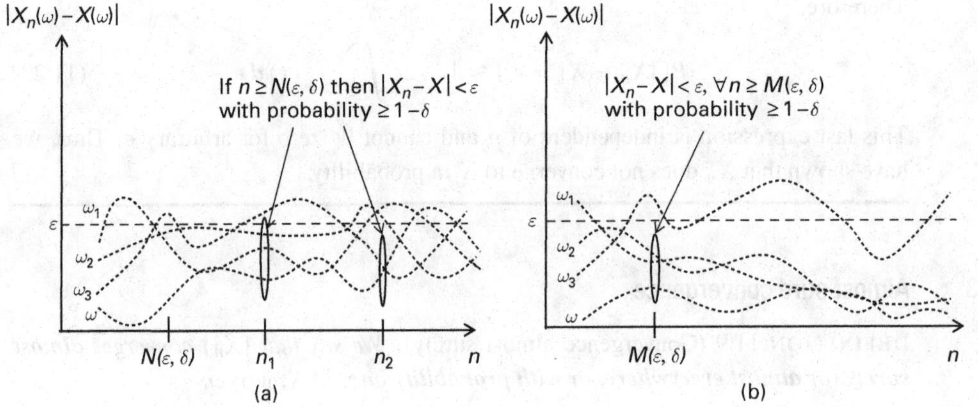

Figure 11.2 Comparison of two types of convergence: (a) convergence in probability; (b) almost sure convergence.

It will be convenient to define, for $\epsilon > 0$ and m a positive integer, the following event:

$$B_m(\epsilon) \triangleq \bigcap_{n=m}^{\infty} \{\omega : |X_n(\omega) - X(\omega)| < \epsilon\} = \bigcap_{n=m}^{\infty} A_n(\epsilon), \tag{11.26}$$

where $A_n(\epsilon)$ is as defined in (11.6).

THEOREM 11.3 (Almost sure convergence and convergence in probability). *Almost sure convergence implies convergence in probability; i.e.,*

$$\boxed{X_n \stackrel{\text{a.s.}}{\to} X \implies X_n \stackrel{\text{P}}{\to} X.} \tag{11.27}$$

Proof. The criterion (11.25) in the lemma can be restated, using (11.26), as

$$P[B_m(\epsilon)] \geq 1 - \delta. \tag{11.28}$$

Thus, $X_n \stackrel{\text{a.s}}{\to} X$ if and only if

$$\lim_{m\to\infty} P[B_m(\epsilon)] = 1 \text{ for any } \epsilon > 0. \tag{11.29}$$

Since $A_m(\epsilon) \supseteq B_m(\epsilon)$, it readily follows that

$$\lim_{m\to\infty} P[A_m(\epsilon)] = 1, \text{ for any } \epsilon > 0,$$

or

$$X_n \stackrel{\text{P}}{\to} X.$$

□

Although a.s. convergence implies convergence in probability as established in the above theorem, the converse is not true, as the following example shows.

Example 11.7: P convergence does not imply almost sure convergence [131]. Let us define an independent sequence $\{X_n\}$ by

$$X_n = \begin{cases} 1, & \text{with probability } \frac{1}{n}, \\ 0, & \text{with probability } 1 - \frac{1}{n}. \end{cases}$$

Clearly $X_n \stackrel{\text{P}}{\to} 0$. However, $X_n(\omega) \stackrel{\text{a.s.}}{=} 0$ does not hold, because $P[B_m(\epsilon)]$ does not converge to one, as would be required due to (11.29). To prove this, first, we write $P[B_m(\epsilon)]$, from (11.26), as

$$P[B_m(\epsilon)] = P\left[\bigcap_{n=m}^{\infty} A_n(\epsilon)\right].$$

Since the X_n are assumed to be independent,

$$P\left[\bigcap_{n=m}^{\infty} A_n(\epsilon)\right] = \prod_{n=m}^{\infty} P[A_n(\epsilon)].$$

Since X_n is a discrete RV and takes on only zero or one, we find that for $0 < \epsilon < 1$ the event $A_n(\epsilon)$ is given by

$$A_n(\epsilon) = \{\omega : |X_n(\omega) - 0| < \epsilon\} = \{\omega : X_n(\omega) = 0\}$$

and its probability is

$$P[A_n(\epsilon)] = 1 - \frac{1}{n} = \frac{n-1}{n}.$$

Clearly

$$\lim_{n\to\infty} P[A_n(\epsilon)] = 1,$$

as it should, but

$$\lim_{m\to\infty} P[B_m(\epsilon)] = 0,$$

because

$$\begin{aligned} P[B_m(\epsilon)] &= \frac{m-1}{m}\frac{m}{m+1}\frac{m+1}{m+2}\cdots \\ &= \lim_{M\to\infty} \frac{m-1}{m}\frac{m}{m+1}\frac{m+1}{m+2}\cdots\frac{M}{M+1} \\ &= \lim_{M\to\infty} \frac{m-1}{M+1} = 0. \end{aligned}$$

Therefore, (11.29) cannot be met; hence X_n does not converge a.s. □

Before we conclude this section, we restate the condition for almost sure convergence (see (11.24)) and that for convergence in probability (see (11.4)):

$$P[\lim_{n\to\infty} |X_n - X| < \epsilon] = 1 \text{ for any } \epsilon > 0 \text{ (almost sure convergence)},$$

and

$$\lim_{n\to\infty} P[|X_n - X| < \epsilon] = 1 \text{ for any } \epsilon > 0 \text{ (convergence in probability)}.$$

11.2.4 Convergence in the rth mean

DEFINITION 11.10 (Convergence in the **r**th mean). *We say that X_n converges to X in **the rth mean** (or in the mean of order r) ($r \geq 1$), written*

$$X_n \xrightarrow{r} X, \text{ or } X_n \to X \text{ (mean } r\text{)},$$

if $E[|X_n|^r] < \infty$, $E[|X|^r] < \infty$ *and*

$$\boxed{\lim_{n\to\infty} E[|X_n - X|^r] = 0.} \tag{11.30}$$

When $r = 1$, $\{X_n\}$ *is said to* ***converge in the mean*** *to* X. *When* $r = 2$, $\{X_n\}$ *is said to* ***converge in mean square*** *or* ***converge in the mean square sense.*** □

Mean square convergence is often written as

$$\boxed{X_n \overset{\text{m.s.}}{\to} X, \quad \text{or } \underset{n\to\infty}{\text{l.i.m.}}\ X_n = X.} \tag{11.31}$$

The following two theorems are important to the notion of **ergodicity** that relates the sample mean (or statistical average) to the expectation (or ensemble average).

THEOREM 11.4 (Markov's theorem). *Given a sequence of RVs* $\{X_n\}$, *denote its nth arithmetic average by*

$$\overline{X}_n = \frac{X_1 + X_2 + \cdots + X_n}{n}.$$

If the expectation of $\overline{X}_n$ *converges to a constant* μ, *and its variance converges to zero,*[2] *i.e.,*

$$\boxed{\overline{\mu}_n \triangleq E[\overline{X}_n] \longrightarrow \mu \text{ and } \overline{\sigma}_n^2 \triangleq E[(\overline{X}_n - \overline{\mu}_n)^2] \longrightarrow 0, \text{ as } n \to \infty,} \tag{11.32}$$

then $\overline{X}_n$ *converges to* μ *in mean square:*

$$\boxed{\lim_{n\to\infty} E[(\overline{X}_n - \mu)^2] = 0.} \tag{11.33}$$

Proof. We write

$$\begin{aligned}(\overline{X}_n-\mu)^2 &= (\overline{X}_n - \overline{\mu}_n+\overline{\mu}_n-\mu)^2=(\overline{X}_n-\overline{\mu}_n)^2+(\overline{\mu}_n-\mu)^2+2(\overline{X}_n-\overline{\mu}_n)(\overline{\mu}_n-\mu)\\ &\le 2(\overline{X}_n-\overline{\mu}_n)^2+2(\overline{\mu}_n-\mu)^2,\end{aligned} \tag{11.34}$$

where the last expression was obtained by using a simple inequality $2ab \le a^2 + b^2$. Taking the expectation of both sides of this inequality, we obtain

$$E[(\overline{X}_n - \mu)^2] \le 2E[(\overline{X}_n - \overline{\mu}_n)^2] + 2E[(\overline{\mu}_n - \mu)^2] = 2\overline{\sigma}_n^2 + 2(\overline{\mu}_n - \mu)^2. \tag{11.35}$$

Using the assumptions (11.32), the desired result (11.33) follows. □

[2] Note that both sequences $\{\overline{\mu}_n\}$ and $\{\overline{\sigma}_n^2\}$ are sequences of numbers, not RVs. Thus, the notion of convergence and limits discussed in Section 11.1.1 apply to these sequences.

THEOREM 11.5 (Chebyshev's condition for mean square convergence). *If the RVs X_k with finite variances σ_k^2, $k = 1, 2, \ldots, n$ are uncorrelated, and if*

$$\frac{\sigma_1^2 + \sigma_2^2 + \cdots \sigma_n^2}{n^2} \longrightarrow 0 \text{ as } n \to \infty, \tag{11.36}$$

then

$$\boxed{\overline{X}_n \overset{\text{m.s.}}{\to} \mu,} \tag{11.37}$$

where

$$\mu = \lim_{n\to\infty} \overline{\mu}_n = \lim_{n\to\infty} \frac{1}{n} \sum_{k=1}^{n} E[X_k]. \tag{11.38}$$

Proof. For uncorrelated RVs, the left side of (11.36) is equal to $\overline{\sigma}_n^2$ defined by (11.32). Therefore, Markov's theorem yields (11.37). □

Now by returning to convergence in the rth mean for any integer r, we state the following important theorem.

THEOREM 11.6 (Convergence in the **r**th mean and convergence in probability). *Convergence in the rth mean implies convergence in probability, i.e.,*

$$\boxed{X_n \overset{\text{m.s.}}{\to} X \implies X_n \overset{\text{P}}{\to} X.} \tag{11.39}$$

Proof. We make use of Bienaymè's inequality (see Problem 10.12):

$$P[|X_n - X| > \epsilon] \le \frac{E[|X_n - X|^r]}{\epsilon^r}$$

for arbitrary $\epsilon > 0$. Then if $\{X_n\}$ converges in the rth mean to X, it immediately follows that $\{X_n\}$ converges to X in probability, since

$$\lim_{n\to\infty} P[|X_n - X| > \epsilon] = 0.$$

□

The converse of the above theorem is not true, as shown in the next example.

Example 11.8: A sequence of Cauchy RVs, does not converge in m.s. Consider the sequence of RVs with Cauchy's distribution discussed in Example 11.4. Then

$$E[|X_n - 0|^2] = E[X_n^2] = \frac{n}{\pi} \int_{-\infty}^{\infty} \frac{x^2}{1 + n^2 x^2}\, dx.$$

This integral does not exist even for finite n. Thus, the sequence X_n does not converge to zero in mean square, even though it converges in probability as shown in Example 11.4. □

Now we wish to find the relationship between convergence in the mean of different orders. Let us define a **norm** (Problem 11.6) of RV Y by

$$\|Y\|_r = \left(E[|Y|^r]\right)^{1/r}. \tag{11.40}$$

Then the following inequality applies to the norm $\|Y\|_r$ with different values of r.

THEOREM 11.7 (Lyapunov's[3] inequality). *The norm $\|Y\|_r$ is a nondecreasing function of r, i.e.,*

$$\|Y\|_s \leq \|Y\|_r, \quad \text{for } 0 < s < r, \tag{11.41}$$

which is known as Lyapunov's inequality.

Proof. The proof is rather involved. The interested reader is suggested to follow the steps provided in Problem 11.5. □

Now we are ready to state the following theorem.

THEOREM 11.8 (Convergence in the mean of a lower order). *Convergence in the mean of order r implies convergence in the mean of a lower order; i.e.,*

$$X_n \xrightarrow{r} X \implies X_n \xrightarrow{s} X, \quad 1 \leq s < r. \tag{11.42}$$

Proof. We substitute

$$Y = X_n - X$$

into Lyapunov's inequality (11.41), and take logarithms, obtaining

$$\frac{\log E[|X_n - X|^s]}{s} \leq \frac{\log E[|X_n - X|^r]}{r}.$$

If $\{X_n\}$ converges in the rth mean to X, then the right side of the above equation approaches $-\infty$ in the limit $n \to \infty$. Thus, the left side also approaches $-\infty$, which implies that $X_n \xrightarrow{s} X$. □

Convergence in the rth mean does not imply almost sure convergence (see Problem 11.7). Neither does almost sure convergence imply convergence in the rth mean (see Problem 11.8).

[3] Aleksandr Mikhailovich Lyapunov (1857–1918) was a Russian mathematician.

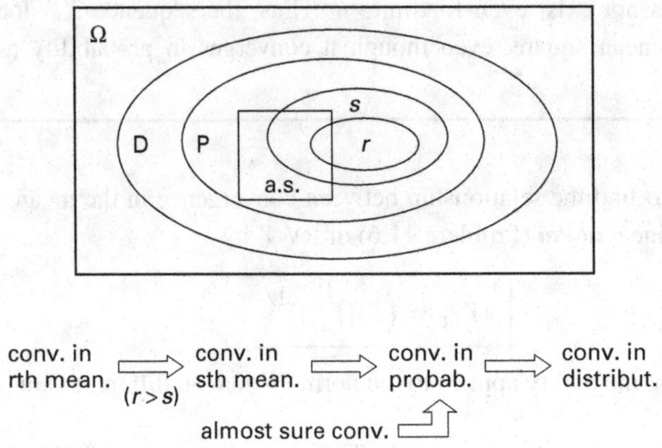

Figure 11.3 Relationships among various modes of convergence: D, in distribution; P, in probability; s, in sth mean ($s > 1$); r, in rth mean ($r > s$); and a.s., almost surely.

11.2.5 Relations between the modes of convergence

The schematic diagram in Figure 11.3 summarizes the relations concerning the various modes of convergence discussed above.

It is clear that convergence in distribution is the weakest, followed by convergence in probability. Although convergence in distribution does not, in general, imply convergence in probability, we can state the following theorem.

THEOREM 11.9 (Convergence in distribution to a constant). *Convergence in distribution to a constant implies convergence in probability; i.e.,*

$$\boxed{X_n \xrightarrow{\text{D}} c \Longrightarrow X_n \xrightarrow{\text{P}} c.} \tag{11.43}$$

Proof. Convergence in distribution means

$$\lim_{n\to\infty} F_n(x) = F_X(x),$$

where $F_X(x)$ is in this case

$$F_X(x) = \begin{cases} 0, & x < c, \\ 1, & x \geq c. \end{cases}$$

Then, for any $\epsilon > 0$

$$P[|X_n - c| \leq \epsilon] = P[c - \epsilon \leq X_n \leq c + \epsilon] = 1 - 0 = 1.$$

Thus, $\{X_n\}$ converges in probability to the constant c. □

As shown in Figure 11.3, both a.s. convergence and convergence in the rth mean are strong modes of convergence. In order to investigate the direct relation between these two modes of convergence, we need the following theorem, which we state without proof. Interested readers are referred, for example, to Munroe [252], Royden [290], Williams [354], and Billingsley [26].

THEOREM 11.10 (Lebesgue's[4] dominated convergence theorem). *Let $\{Y_n\}$ be a sequence of RVs that converges almost surely to RV Y, and let RV Z exist such that $|Y_n| \le Z$ for all n. If $E[Z] < \infty$, then*

$$\boxed{\lim_{n\to\infty} E[Y_n] = E[Y],} \tag{11.44}$$

or equivalently

$$\lim_{n\to\infty} E[|Y_n - Y|] = 0. \tag{11.45}$$

□

This theorem says, in essence, that the expectation and limit *commute* (i.e., can be changed in order), when all RVs $|Y_n|$ are all dominated by a common RV Z.

Then using Lebesgue's dominated convergence theorem, we can derive the following theorem asserting that, when some bounded condition is met on $|X_n - X|^r$ for all n, a.s. convergence implies convergence in the mean square of order r.

THEOREM 11.11 (Conditions under which convergence a.s. implies convergence in rth mean). *Suppose that there exists a random variable Z such that $E[Z] < \infty$ and $|X_n - X|^r \le Z$ for all n. Then a.s. convergence of the sequence X_n to X implies its convergence in the rth mean; i.e.,*

$$\boxed{X_n \stackrel{\text{a.s.}}{\to} X \implies X_n \stackrel{r}{\to} X.} \tag{11.46}$$

Proof. Let $Y_n = |X_n - X|^r$ and let $Y = 0$ with probability one. Then by using the above Lebesgue dominated convergence theorem,

$$\lim_{n\to\infty} E[|X_n - X|^r] = 0;$$

that is, $\{X_n\}$ converges in the rth mean to X. □

11.3 Limit theorems

In Section 2.3 a limit theorem known as Bernuoulli's theorem (Theorem 2.1) was derived, and we remarked there that this theorem is a special case of a limit theorem known as the weak law of large numbers (WLLN). In this section we derive several limit theorems of general nature, including:

[4] Henri Lon Lebesgue (1875–1941) was a French mathematician, most famous for his theory of integration.

1. the weak law of large numbers (WLLN);
2. the strong law of large numbers (SLLN);
3. the central limit theorem (CLT).

11.3.1 Infinite sequence of events

Before developing these limit theorems, let us discuss some important results involving an **infinite sequence of events**. Let $A_1, A_2, \ldots, A_k, \ldots$ be an infinite sequence of events from some probability space $(\Omega, \mathcal{F}, P)$. We start with a special case where $\{A_k\}$ is an "increasing" sequence of events.

THEOREM 11.12 (Limit of an increasing sequence of events). *If* $A_1 \subset A_2 \subset \cdots$, *then*

$$\boxed{P\left[\bigcup_{k=1}^{\infty} A_k\right] = \lim_{m\to\infty} P[A_m].} \tag{11.47}$$

Proof. Let $B_1 = A_1,\ B_2 = A_2 \cap A_1^c = A_2 \setminus B_1,\ B_3 = A_3 \setminus (B_1 \cup B_2),\ \ldots,\ B_m = A_m \setminus \bigcup_{k=1}^{m-1} B_k,\ \ldots$ Then the B_k are mutually exclusive and their union is equal to $\bigcup_{k=1}^{\infty} A_k$. Furthermore,

$$\bigcup_{k=1}^{m} B_k = A_m. \tag{11.48}$$

Hence,

$$\begin{aligned} P\left[\bigcup_{k=1}^{\infty} A_k\right] &= P\left[\bigcup_{k=1}^{\infty} B_k\right] = \sum_{k=1}^{\infty} P[B_k] = \lim_{m\to\infty} \sum_{k=1}^{m} P[B_k] \\ &= \lim_{m\to\infty} P\left[\bigcup_{k=1}^{m} B_k\right] = \lim_{m\to\infty} P[A_m]. \end{aligned} \tag{11.49}$$

□

Note that a key step in (11.49) is to make use of Axiom 4 (2.29) to derive

$$P\left[\lim_{m\to\infty} \bigcup_{k=1}^{m} B_k\right] = \lim_{m\to\infty} P\left[\bigcup_{k=1}^{m} B_k\right].$$

Now consider the opposite case; i.e., the sequence of events $\{A_k\}$ is a "decreasing" sequence.

THEOREM 11.13 (Limit of a decreasing sequence of events). *If* $A_1 \supset A_2 \supset \cdots$, *then*

$$\boxed{P\left[\bigcap_{k=1}^{\infty} A_k\right] = \lim_{m\to\infty} P[A_m].} \tag{11.50}$$

Proof. Consider $A_1^c \subset A_2^c \subset \cdots$ and apply Theorem 11.12. Then

$$\begin{aligned} P\left[\bigcap_{k=1}^{\infty} A_k\right] &= 1 - P\left[\bigcup_{k=1}^{\infty} A_k^c\right] = 1 - \lim_{m\to\infty} P[A_m^c] \\ &= \lim_{m\to\infty}(1 - P[A_m^c]) = \lim_{m\to\infty} P[A_m]. \end{aligned} \tag{11.51}$$

□

Theorems 11.12 and 11.13 are summarized as follows.

THEOREM 11.14. *If A_n is either an increasing or decreasing sequence of events, then*

$$\boxed{P\left[\lim_{n\to\infty} A_n\right] = \lim_{n\to\infty} P[A_n].} \tag{11.52}$$

□

Now we derive the following important upper bound on the probability of countably infinite union of events $\{A_k\}$, when the sequence is formed from arbitrary events.

THEOREM 11.15 (Boole's[5] inequality or the union bound). *For arbitrary events $A_1, A_2, \ldots,$*

$$\boxed{P\left[\bigcup_{k=1}^{\infty} A_k\right] \le \sum_{k=1}^{\infty} P[A_k].} \tag{11.53}$$

Proof. As in the proof of Theorem 11.12, we express $\bigcup_{k=1}^{\infty} A_k$ as the union of mutually exclusive events $B_1, B_2, \ldots,$ where $B_k \subset A_k$. Therefore, $P[B_k] \le P[A_k]$ for all k. Taking the infinite sum of this inequality and using the relation $P\left[\bigcup_{k=1}^{\infty} A_k\right] = P\left[\bigcup_{k=1}^{\infty} B_k\right] = \sum_{k=1}^{\infty} P[B_k]$, we arrive at the above inequality. □

Now let us consider an infinite sequence of events denoted as $E_1, E_2, \ldots, E_k, \ldots$ We are interested in finding how many of the E_n occur. Let A_n represent the event that at least one of $E_n, E_{n+1}, E_{n+2}, \ldots$ occurs:

$$A_n = \bigcup_{k=n}^{\infty} E_k. \tag{11.54}$$

Then $\{A_n\}$ is a decreasing sequence.

[5] George Boole (1815–1864) was an English mathematician and philosopher, well known as the inventor of Boolean logic.

Let A represent the event "infinitely many of events $E_1, E_2, \ldots$ occur." A occurs if and only if A_n occurs for every n. This is because:

1. If an infinite number of the E_k occur, then A_n occurs for each n; thus $\cap_{n=1}^{\infty} A_n$ occurs.
2. Conversely, if $\cap_{n=1}^{\infty} A_n$ occurs, then A_n occurs for each n. Thus, for each n at least one of the events $E_k, k \geq n$ occurs; hence, an infinite number of the E_k occur.

Thus,

$$A = \bigcap_{n=1}^{\infty} A_n = \bigcap_{n=1}^{\infty} \left(\bigcup_{k=n}^{\infty} E_k \right). \tag{11.55}$$

For the event A thus defined, we have

$P[A] = 0 \iff$ With probability 1, only finitely many of the events E_n occur and

$P[A] = 1 \iff$ With probability 1, infinitely many of the events E_n occur.

Furthermore, $A_1 \supset A_2 \supset \cdots$ Hence, from Theorem 11.13 we have

$$P[A] = \lim_{n \to \infty} P[A_n], \tag{11.56}$$

where $P[A_n]$ is bounded from above due to Theorem 11.15:

$$P[A_n] \leq \sum_{k=n}^{\infty} P[E_k]. \tag{11.57}$$

With these preparations, we are now in a position to state one of the most important theorems in probability theory, usually referred to as **Borel**[6]**–Cantelli**[7] lemmas.

THEOREM 11.16 (Borel–Cantelli lemmas). *Let $\{E_k\}$ be an infinite sequence of events, and let A be the event that infinitely many of the events E_k occur, as defined by (11.55). Then:*

- ***First lemma.*** *Regardless of the events E_k being independent or not,*

$$\boxed{\text{if } \sum_{k=1}^{\infty} P[E_k] < \infty, \text{ then } P[A] = 0;} \tag{11.58}$$

that is, with probability 1 only finitely many of the events $E_1, E_2, \ldots$ occur.

- ***Second lemma.*** *Suppose that $E_1, E_2, \ldots$ are independent events. Then:*

$$\boxed{\text{if } \sum_{k=1}^{\infty} P[E_k] = \infty, \text{ then } P[A] = 1;} \tag{11.59}$$

that is, infinitely many of the events $E_1, E_2, \ldots$ occur with probability 1.

[6] Félix Édouard Justin Émile Borel (1871–1956) was a French mathematician and politician.
[7] Francesco Paolo Cantelli (1875–1966) was an Italian mathematician.

Proof. If $\sum_k P[E_k]$ converges, then (11.57) shows

$$\lim_{n\to\infty} P[A_n] \le \lim_{n\to\infty} \sum_{k=n}^{\infty} P[E_k] = 0. \tag{11.60}$$

By applying (11.56) to this decreasing sequence $\{A_n\}$, we find

$$P[A] = P\left[\lim_{n\to\infty} A_n\right] = \lim_{n\to\infty} P[A_n] = 0.$$

This completes the proof of the first lemma.

Now we proceed to prove the second lemma. Take the complement of A_n of (11.54):

$$A_n^c = \bigcap_{k=n}^{\infty} E_k^c. \tag{11.61}$$

Then using the relation

$$A_n^c \subset \bigcap_{k=n}^{n+m} E_k^c, \quad \text{for every } m = 0, 1, 2, \ldots, \tag{11.62}$$

and the assumption of the second lemma that the E_k are mutually independent and hence so are the E_k^c (Problem 11.10), we find

$$\begin{aligned} P[A_n^c] &\le P\left[\bigcap_{k=n}^{n+m} E_k^c\right] = P[E_n^c]\cdots P[E_{n+m}^c] = (1 - P[E_n])\cdots(1 - P[E_{n+m}]) \\ &\le \exp\left(-\sum_{k=n}^{n+m} P[E_k]\right), \quad \text{for every } m = 0, 1, 2, \ldots, \end{aligned} \tag{11.63}$$

where we used the inequality $1 - x \le e^{-x}$, $x \ge 0$. If $\sum_{k=1}^{\infty} P[E_k] = \infty$, then $\sum_{k=n}^{n+m} P[E_k] \to \infty$ as $m \to \infty$. Hence, by taking the limit $m \to \infty$ in the last equation, we have

$$P[A_n^c] = 0 \ \text{ for every } n = 1, 2, \ldots.$$

Take the complement of A of (11.55):

$$A^c = \bigcup_{n=1}^{\infty} A_n^c.$$

Thus,

$$\boxed{P[A^c] \le \sum_{n=1}^{\infty} P[A_n^c] = 0.} \tag{11.64}$$

Therefore, we finally have $P[A] = 1 - P[A^c] = 1$. This completes the proof of the second lemma. Needless to say, when the E_n are independent, the first lemma still applies;

that is,

$$\sum_{k=1}^{\infty} P[E_k] < \infty \Longrightarrow P[A] = 0.$$

□

Example 11.9: Sequences of RVs and constants. Consider a sequence of RVs $\{X_k\}$ and a sequence of constants $\{c_k\}$. Define the *event* E_k by

$$E_k = \{|X_k| > c_k\}.$$

Then the Borel–Cantelli lemmas imply that if $\sum_{k=1}^{\infty} P[|X_k| > c_k] < \infty$, then with probability one, only finitely many of the events $E_k = \{|X_k| > c_k\}$ occur. If the events X_k are independent and $\sum_{k=1}^{\infty} P[|X_k| > c_k] = \infty$, then, with probability one, infinitely many of the events E_k occur. □

11.3.2 Weak law of large numbers (WLLN)

Let $\{X_k\} = (X_1, X_2, \ldots, X_k, \ldots)$ be a sequence of RVs with finite mean and variance:

$$E[X_k] = \mu_k \text{ and } \mathrm{Var}[X_k] = \sigma_k^2, \quad k = 1, 2, \ldots.$$

Define a new sequence of RVs $\{S_n\}$ by the nth partial sum

$$S_n = \sum_{k=1}^{n} X_k, \quad n = 1, 2, \ldots. \tag{11.65}$$

The mean and variance of the RV S_n are given by

$$m_n = E[S_n] = \sum_{k=1}^{n} \mu_k \tag{11.66}$$

and

$$s_n^2 = \mathrm{Var}[S_n] = E[(S_n - m_n)^2]. \tag{11.67}$$

Define the nth **arithmetic average**[8] $\overline{X}_n$ by

$$\overline{X}_n = \frac{S_n}{n}, \tag{11.68}$$

[8] This quantity is equivalent to the **sample average** or **sample mean**, if we interpret X_k as the kth **sample** of a certain RV X. Here, we are assuming, up to this point, that the X_k are neither independent nor identically distributed; thus, we avoid use of the term "sample."

whose mean and variance are given by

$$E[\overline{X}_n] = \frac{m_n}{n} \text{ and } \mathrm{Var}[\overline{X}_n] = E\left[\left(\overline{X}_n - \frac{m_n}{n}\right)^2\right] = \frac{s_n^2}{n}. \quad (11.69)$$

We are interested in the asymptotic behavior of S_n and $\overline{X}_n$ as $n \to \infty$. A number of convergence statements can be made about the asymptotic behavior of $\{S_n\}$ and $\{\overline{X}_n\}$, depending on the properties of the original sequence $\{X_k\}$. The *weak law of large numbers* (*WLLN*) is concerned about such convergence statements.

Suppose the variance of $\{X_k\}$ approaches zero in the limit; i.e.,

$$\lim_{k\to\infty} \sigma_k^2 = 0. \quad (11.70)$$

By applying Chebyshev's inequality to X_k, we have for any $\epsilon > 0$

$$P[|X_k - \mu_k| \geq \epsilon] \leq \frac{\sigma_k^2}{\epsilon^2}, \quad (11.71)$$

which suggests, together with (11.70), that the sequence $\{X_k - \mu_k\}$ converges in probability to zero:

$$\lim_{k\to\infty} P[|X_k - \mu_k| \geq \epsilon] = 0. \quad (11.72)$$

By applying a similar argument to the averaged sequence $\{\overline{X}_n\}$, we find that

$$\boxed{\text{if } \lim_{n\to\infty} \frac{s_n^2}{n^2} = 0, \text{ then } \lim_{n\to\infty} P\left[\left|\overline{X}_n - \frac{m_n}{n}\right| \geq \epsilon\right] = 0;} \quad (11.73)$$

thus, the sequence $\{\overline{X}_n - m_n/n\}$ converges in probability to zero.

If the X_k are *independent* and their variances are bounded – that is, if there exists a positive number M such that $\sigma_k^2 \leq M$ for all k – then (11.73) is satisfied because

$$\lim_{n\to\infty} \frac{s_n^2}{n^2} = \lim_{n\to\infty} \frac{1}{n^2} \sum_{k=1}^{n} \sigma_k^2 \leq \lim_{n\to\infty} \frac{M}{n} = 0. \quad (11.74)$$

If the X_k are, in addition, i.i.d. with common mean μ, then the condition in (11.73) obviously holds and the result takes the form

$$\lim_{n\to\infty} P[|\overline{X}_n - \mu| \geq \epsilon] = 0. \quad (11.75)$$

This last result involving i.i.d. RVs is often called the **weak law of large numbers (WLLN)**.

THEOREM 11.17 (Weak law of large numbers). *Let $X_1, X_2, \ldots, X_k, \ldots$ be independent and also identically distributed RVs with finite mean $E[X_k] = \mu$ and finite variance. Let $\overline{X}_n = \frac{1}{n}\sum_{k=1}^{n} X_k$. Then we have*

$$\overline{X}_n \xrightarrow{\text{P}} \mu,$$

which is equivalently stated as

$$\boxed{\lim_{n\to\infty} P[|\overline{X}_n - \mu| \geq \epsilon] = 0 \text{ for any } \epsilon > 0.} \quad (11.76)$$

□

As discussed above, the WLLN is easily generalizable to cases where the X_k are not identically distributed or even not independent, although the i.i.d. assumption is commonly associated with the statement of the law.

Example 11.10: Bernoulli's theorem. Recall Bernoulli's theorem (2.45) discussed in Section 2.3. It should be clear that this theorem is a special case of the WLLN. Let RV B_k take on the value "1" or "0", depending on whether the kth Bernoulli trial is a "success" or a "failure." Let p be the probability of success; i.e.,

$$B_k = \begin{cases} 1, & \text{with probability } p, \\ 0, & \text{with probability } 1-p. \end{cases} \quad (11.77)$$

Then $\mu_k = p$ and $\sigma_k^2 = p(1-p)$ for all k, and the partial sum $S_n = \sum_{k=1}^n B_k$ is the number of successes in n trials and has mean $m_n = np$ and variance $s_n^2 = np(1-p)$. The nth arithmetic average

$$\overline{B}_n = \frac{S_n}{n} = \frac{1}{n}\sum_{k=1}^{n} B_k$$

is the relative frequency of successes in n trials. Its mean and variance are $m_n/n = p$ and $s_n^2/n^2 = p(1-p)/n$ respectively. Thus, the condition in (11.73) is satisfied and the relative frequency $\overline{B}_n$ converges in probability to p; i.e., for any $\epsilon > 0$ we have

$$\boxed{\lim_{n\to\infty} P[|\overline{B}_n - p| < \epsilon] = 1.} \quad (11.78)$$

□

11.3.3 Strong laws of large numbers (SLLN)

The WLLN discussed in the previous section stated the conditions under which the nth arithmetic average $\overline{X}_n$ of a sequence $\{X_k\}$ converges *in probability* to the average of the means; i.e., $\overline{X}_n \stackrel{P}{\to} \frac{1}{n}\sum_{k=1}^n \mu_k$. The **strong law of large numbers (SLLN)**, to be discussed below, is concerned with the **almost sure convergence** of the sequence $\{\overline{X}_n\}$. Since **almost sure** convergence implies convergence in probability, any sequence that obeys an SLLN also obeys the corresponding WLLN. The SLLN was first formulated and proved by Borel [34] in 1909.

THEOREM 11.18 (Borel's strong law of large numbers). *Let $\{B_k\}$ be a sequence of Bernoulli trials with the probability of success p. Then the sequence $\{B_k\}$ obeys the strong law of large numbers; i.e.,*

$$\overline{B}_n \overset{\text{a.s.}}{\to} p, \tag{11.79}$$

or equivalently,

$$\boxed{P\left[\lim_{n\to\infty} |\overline{B}_n - p| < \epsilon\right] = 1.} \tag{11.80}$$

Proof. The proof provided by Borel is based on the number-theoretic interpretation. First, consider the case $p = 1/2$ (fair coin tossing). Any real number ω taken at random with uniform distribution in the interval (0, 1) can be converted into an infinite sequence $\{B_k(\omega)\}$ by using the binary expansion

$$\omega = \sum_{k=1}^{\infty} B_k(\omega)2^{-k},$$

where the $B_k(\omega)$ assume zero and one with probability 1/2 each and are independent RVs. The sum $S_n(\omega) = \sum_{k=1}^{n} B_k(\omega)$ is equal to the number of ones among the first n digits in the binary expansion of ω and can be also interpreted as the number of successes in n independent Bernoulli trials in which the probability of success is 1/2. Borel showed that the portion of ones, $\overline{B}_n(\omega) = S_n(\omega)/n$, tends to 1/2 for almost all ω in (0, 1). In a similar manner, if we expand ω to the base 10, any one of the digits $0, 1, 2, \ldots, 9$ appears with probability 1/10, and any group of r digits appears with probability $r/10$. The above argument extends to any rational number $p \in (0, 1)$. □

Another proof of Borel's SLLN follows as a special case of Kolmogorov's sufficient criterion for the SLLN given below (Theorem 11.19).

Note that, compared with the weaker version, i.e., Bernoulli's theorem (11.78), the "$\lim_{n\to\infty}$" moves inside the expression $P[\ ]$. The SLLN makes a statement regarding individual **sample sequences** or **sample paths** $B_k(\omega)$ that correspond to each sample point $\omega \in \Omega$ of this Bernoulli experiment. That is, for large n, the $\overline{B}_n(\omega)$ computed from any (except for those belonging to a set of probability measure zero) sample path $\{B_k(\omega)\}$ approaches arbitrarily close to p. Thus, the SLLN suggests that we can estimate the probability p with sufficient accuracy by conducting a single stream of Bernoulli experiments of sufficient length n. In contrast, the WLLN (i.e., Bernoulli's theorem) makes a statement regarding the **entire ensemble** of such sample paths. That is, when we consider all possible sample paths $\{B_k(\omega)\}$, $\omega \in \Omega$, then, probabilistically speaking, the RV $\overline{B}_n(\omega)$ becomes arbitrarily close to the constant p as n is made sufficiently large.

In 1917 Cantelli derived sufficient conditions for the SLLN for independent RVs X_k in terms of the second and fourth moments of the summands. A further extension of the condition for the SLLN was made by **Khinchin**,[9] who introduced the term "strong law of large numbers (la loi forte des grands nombres)" and derived a sufficient condition applicable to **correlated** summands X_k.

For the case of independent (but not necessarily identical) summands, the best known condition for the applicability of the SLLN was established in 1930 by Kolmogorov [207], which is often referred to as Kolmogorov's first theorem for the SLLN (cf. Rao [278], Thomas [319]).

THEOREM 11.19 (Kolmogorov's sufficient criterion for the SLLN when the X_k are independent). *Let $\{X_k\}$ be a sequence of independent RVs such that $E[X_k] = \mu_k$ and $\text{Var}[X_k] = \sigma_k^2$. Define*

$$\overline{X}_n = \frac{\sum_{k=1}^n X_k}{n} \text{ and } \overline{\mu}_n = \frac{\sum_{k=1}^n \mu_k}{n}.$$

Then,

$$\boxed{\text{if } \sum_{k=1}^{\infty} \frac{\sigma_k^2}{k^2} < \infty, \text{ then } \overline{X}_n - \overline{\mu}_n \stackrel{\text{a.s.}}{\to} 0;} \tag{11.81}$$

that is, the sequence $X_1, X_2, \ldots$ obeys the SLLN.

Proof. See Feller [99] (pp. 243–244), in which he uses Kolmogorov's inequality, a generalization of Chebyshev's inequality. Rao [278] uses the Hajek–Renyi inequality, which is a generalization of Kolmogorov's inequality. □

If the summands X_k are i.i.d., the Kolmogorov criterion (11.81) is replaced by $\sigma^2 < \infty$ (or equivalently $E[X_i^2] = \sigma^2 + \mu^2 < \infty$). Grimmett and Stirzaker [131] (pp. 294–296) provide the proof for this case. They also show that $\overline{X}_n$ converges to μ in mean square as well as almost surely:

THEOREM 11.20 (Sufficient condition for the SLLN when the X_k are i.i.d.). *Let $\{X_k\}$ be a sequence of i.i.d. RVs with common mean $\mu = E[X_k]$. Then,*

$$\boxed{\text{if } E[X_k^2] < \infty, \text{ then } \overline{X}_n \stackrel{\text{a.s.}}{\to} \mu \text{ and } \overline{X}_n \stackrel{\text{m.s.}}{\longrightarrow} \mu.} \tag{11.82}$$

□

Subsequently, in 1933, Kolmogorov showed the necessary and sufficient condition for the SLLN to hold in a sequence of i.i.d. RVs. The following theorem is sometimes referred to as Kolmogorov's second theorem for the SLLN.

[9] Aleksandr Yakovlevich Khinchin (1894–1959) was a Russian mathematician who contributed to number theory, probability theory, queueing theory, statistical mechanics, and information theory.

THEOREM 11.21 (Strong law of large numbers when the X_k are i.i.d.). *Let $\{X_k\}$ be a sequence of i.i.d. RVs. Then*

$$\boxed{\overline{X}_n \overset{\text{a.s.}}{\to} \mu \text{ if and only if } E[|X_k|] < \infty,} \tag{11.83}$$

where $\mu = E[X_1]$.

Proof. This theorem is also referred to as "the necessary and sufficient condition for the strong law" [319]. The proof involves use of the first lemma of the Borel–Cantelli lemmas stated in Theorem 11.16. For example, see Feller [99] (pp. 244–245) and [100] (p. 233), Rao [278] (pp. 94–96), and Grimmett and Stirzaker [131]). □

11.3.4 The central limit theorem (CLT) revisited

The SLLN states that the sample average $\overline{X}_n = S_n/n$ converges in a strong sense (i.e., for individual sample paths) to $\mu = E[X]$. But the law does not provide any information about the distribution of $\overline{X}_n$ other than its mean. The central limit theorem (CLT) stated in Theorem 8.2 of Section 8.2.5 is concerned about this question. Specifically, Eq. (8.98) can now be paraphrased as

$$\frac{\sqrt{n}}{\sigma}(\overline{X}_n - \mu) \overset{\text{D}}{\to} U, \tag{11.84}$$

where U is the standard normal variable. In this section we will discuss several variations of the CLT.

Let us consider the case where $\{X_k\}$ is a sequence of independent (but not necessarily identical) RVs with means $E[X_k] = \mu_k$ and variance $\text{Var}[X_k] = \sigma_k^2$. Define the **normalized average** Z_n by

$$Z_n = \frac{S_n - m_n}{s_n}, \tag{11.85}$$

where S_n, $m_n = E[S_n]$, and $s_n^2 = \text{Var}[S_n]$ are given by (11.65), (11.66), and (11.67) respectively. The new RV Z_n has zero mean and variance of unity. We now discuss several forms of the CLT under which Z_n converges **in distribution** to the unit normal variable.

Let us consider the simplest case where the X_k are not only **independent** but also **identically distributed** with common mean μ and variance σ^2. Then (11.85) reduces to

$$Z_n = \frac{S_n - n\mu}{\sqrt{n}\sigma} = \sum_{k=1}^{n} \frac{X_k - \mu}{\sqrt{n}\sigma} = \frac{\sqrt{n}}{\sigma}(\overline{X}_n - \mu). \tag{11.86}$$

We restate Theorem 8.2 as follows.

THEOREM 11.22 (Lindeberg–Lévy's CLT for i.i.d. RVs).[10] *Let $\{X_k\}$ be a sequence of i.i.d. RVs. Then,*

$$\boxed{\text{if } E[X_1] = \mu < \infty,\ \ \text{Var}[X_1] = \sigma^2 < \infty, \text{ then } Z_n \xrightarrow{D} U,} \tag{11.87}$$

where Z_n is defined in (11.86) and U is the unit normal variable.

Proof. Note that we write $E[X_1]$ instead of $E[X_k]$ for all k, since that would be redundant, given the i.i.d. assumption. The proof has already been given, when we discussed Theorem 8.2, where we used the characteristic function (CF) of the normalized average Z_n. The CF of Z_n tends to $e^{-u^2/2}$, the CF of the unit normal distribution. To justify the transition from the convergence of the CFs to convergence of the corresponding distributions, we need the continuity theorem (often referred to as the Lévy continuity theorem), which states that a sequence $\{F_n(x)\}$ of probability distributions converges to a probability distribution $F(x)$ if and only if the sequence $\{\phi_n(x)\}$ of their CFs converges to a continuous limit $\phi(u)$. In this case $\phi(u)$ is the CF of $F(x)$, and the sequence $\{\phi_n(u)\}$ converges to $\phi(u)$ uniformly. See Feller [100] (pp. 481, 487–491). □

We could have used the moment generating function (MGF) instead of the CF (Problem 11.12) in the proof. This i.i.d. RV version of the CLT is referred to as the Lindeberg–Lévy theorem [278, 299, 319].

Now we return to a more general case where the X_k are independent but not necessarily identically distributed. In order for the CLT to hold in this case, it is necessary to add some condition to insure that no single term of X_k dominates. There are several variations for the CLT to apply for nonidentical X_k.

Lyapunov [231, 232] provides a sufficient condition for the CLT in terms of the *third absolute central moments* m_k^3, as well as the means μ_k and variances σ_k^2. Let us define

$$m_k^3 \triangleq E\left[|X_k - \mu_k|^3\right]. \tag{11.88}$$

THEOREM 11.23 (Lyapunov's CLT for independent but nonidentical RVs). *Let X_k be independent RVs with $E[X_k] = \mu_k$ and $\text{Var}[X_k] = \sigma_k^2$, and let s_n^2 be as defined in (11.67). Then,*

$$\boxed{\text{if } m_k < \infty \textit{ for all } k \text{ and } \lim_{n\to\infty} \frac{\sum_{k=1}^n m_k}{s_n} = 0, \text{ then } Z_n \xrightarrow{D} U,} \tag{11.89}$$

where Z_n is defined in (11.86) and U is the unit normal variable.

Proof. Note that the term in the limit can be written as

$$\frac{\sum_{k=1}^n m_k}{s_n} = \frac{\sum_{k=1}^n \left(E\left[|X_k - \mu_k|^3\right]\right)^{1/3}}{\left(\sum_{k=1}^n E\left[|X_k - \mu_k|^2\right]\right)}.$$

[10] Jarl Waldemar Lindeberg (1876–1932) was a Finnish mathematician known for his work on the CLT. Paul Pierre Lévy (1886–1971) was a French mathematician, known for introducing martingales and the Lévy process.

The proof of this theorem involves expanding the CF of Z_n and taking the limit as $n \to \infty$, similar to the proof of the Lindeberg–Lévy theorem (Theorem 11.22; i.e., Theorem 8.2). The reader is referred to, for example, Cramér [74] and Billingsley [26] for the proof of this theorem. □

Lyapunov generalized the sufficient condition (11.89) and showed that if

$$\lim_{n\to\infty} \frac{\sum_{k=1}^{n} E\left[(X_k - m_k)^{2+\delta}\right]}{s_n^{2+\delta}} = 0 \tag{11.90}$$

for some $\delta > 0$, then the Z_n converges in distribution to the unit normal variable U. The condition (11.89) corresponds to the case $\delta = 1$.

A stronger result of sufficient condition of the CLT was obtained by Lindeberg in 1922 [226]. Subsequently Feller [97] proved the necessity of Lindeberg's condition. Hence, the following version of the CLT is often referred to as Lindeberg–Feller theorem [278, 353]. This is a stronger result, since it requires knowledge of the distribution functions, not just the moments. They define the following quantities:

$$\sigma_k^{\star 2}(\epsilon) \triangleq \int_{|x-\mu_k|\le\epsilon s_n} (x-\mu_k)^2 dF_{X_k}(x), \quad k = 1, 2, \ldots, n, \tag{11.91}$$

$$s_n^{\star 2}(\epsilon) \triangleq \sum_{k=1}^{n} \sigma_k^{\star 2}(\epsilon). \tag{11.92}$$

THEOREM 11.24 (Lindeberg–Feller's CLT for independent but nonidentical RVs). *Let X_k be independent RVs with $E[X_k] = \mu_k$, $\mathrm{Var}[X_k] = \sigma_k^2$, and $s_n^{\star 2}(\epsilon)$ defined in (11.92). Then,*

$$\boxed{Z_n \xrightarrow{\mathrm{D}} D, \text{ if and only if } \lim_{n\to\infty} s_n^2 = \infty, \text{ and } \lim_{n\to\infty} \frac{s_n^{\star 2}(\epsilon)}{s_n^2} = 1 \text{ for every } \epsilon > 0,} \tag{11.93}$$

where Z_n is defined in (11.86) and U is the unit normal variable.

Proof. See Feller [100] pp. 256–257, 491–493. See also Billingsley [26]. □

The condition (11.93), called the **Lindeberg condition**, guarantees that the individual variances σ_k^2 are all small in comparison to their sum s_n^2. It can be shown that the **Lyapunov condition** (11.90) satisfies Lindeberg's condition (11.93), because the sum in (11.90) is bounded by

$$\frac{1}{s_n^2}\sum_{k=1}^{n}\int_{|x-\mu_k|>\epsilon s_n} \frac{(x-\mu_k)^{2+\delta}}{\epsilon^\delta s_n^\delta} dF_{X_k}(x) \le \frac{1}{\epsilon^\delta}\frac{\sum_{k=1}^{n} E\left[(X_k-m_k)^{2+\delta}\right]}{s_n^{2+\delta}}.$$

See Billingsley [26] for details.

Before closing this section, we should reiterate that the preceding results on the CLT are concerned with convergence in distribution. In other words, the distribution of the normalized average Z_n converges to that of the unit normal variable U. This does not necessarily imply that the PDF of Z_n converges to that of U. If the X_k are continuous RVs, then, under some regularity conditions, the PDFs of Z_n will converge to that of U. But for a finite value of n, the normal distribution may well give a *poor approximation* to the tails of the PDF, as we discussed in Chapter 10. In fact, the large deviations approximation discussed in that chapter was motivated by this poor approximation of tail end of the distribution by the normal distribution. If the X_k are discrete RVs, the situation is more complicated. For example, the sum of independent Poisson RVs is also a Poisson variable for any n, although the *envelope* of the density function of Z_n may converge to that of U.

11.4 Summary of Chapter 11

Convergence types:		
$X_n \overset{\text{D}}{\to} X$	$\lim_{n\to\infty} F_n(x) = F_X(x)$, x a continuity point	(11.2)
$X_n \overset{\text{P}}{\to} X$	$\lim_{n\to\infty} P[\lvert X_n - X\rvert < \epsilon] = 1,\ \epsilon > 0$	(11.4)
$X_n \overset{\text{a.s.}}{\to} X$	$P[\lim_{n\to\infty} X_n = X] = 1$	(11.24)
$X_n \overset{\text{r}}{\to} X$	$E[\lvert X_n\rvert^r] < \infty,\ E[\lvert X\rvert^r] < \infty$	
	$\Rightarrow \lim_{n\to\infty} E[\lvert X_n - X\rvert^r] = 0$	(11.30)
$X_n \overset{\text{m.s.}}{\to} X$	$X_n \overset{\text{r=2}}{\to} X$	(11.31)
Convergence to RV X:	$X_n \overset{\text{P}}{\to} X \Rightarrow X_n \overset{\text{D}}{\to} X$	(11.10)
	$X_n \overset{\text{a.s.}}{\to} X \Rightarrow X_n \overset{\text{P}}{\to} X$	(11.27)
	$X_n \overset{\text{r}}{\to} X \Rightarrow X_n \overset{\text{P}}{\to} X$	(11.39)
Markov's theorem:	$E[\overline{X}_n] \to \mu,\ E[(\overline{X}_n - \mu)^2] \to 0$	
	$\Rightarrow X_n \overset{\text{m.s.}}{\to} \mu$	(11.33)
Chebyshev's condition:	X_k uncorrelated, $\sigma_k^2 < \infty$, $\frac{\sigma_1^2+\sigma_2^2+\cdots+\sigma_n^2}{n^2} \to 0$	
	$\Rightarrow \overline{X}_n \overset{\text{m.s.}}{\to} \lim_{n\to\infty} \frac{1}{n}\sum_{k=1}^{n} E[X_k]$	(11.37)
rth norm	$\lVert Y\rVert_r = (E[\lvert Y\rvert^r])^{1/r}$	(11.40)
Lyapunov's Inequality:	$\lVert Y\rVert_s \le \lVert Y\rVert_r,\ 0 < s < r$	(11.41)
Convergence in rth versus sth mean:	$X_n \overset{\text{r}}{\to} X \Rightarrow X_n \overset{\text{s}}{\to} X$, if $1 \le s < r$	(11.42)
Convergence in D to constant:	$X_n \overset{\text{D}}{\to} c \Rightarrow X_n \overset{\text{P}}{\to} c$	(11.43)
Convergence a.s. versus in rth mean:	$E[Z] < \infty,\ \lvert X_n - X\rvert^r \le Z$	
	$X_n \overset{\text{a.s.}}{\to} X \Rightarrow X_n \overset{\text{r}}{\to} X$	(11.46)
Lebesgue dominated convergence theorem:	$\lvert Y_n\rvert \le Z \Rightarrow \lim_{n\to\infty} E[Y_n] = E[Y]$	(11.44)
Limit of increasing sequence:	$P\left[\bigcup_{k=1}^{\infty} A_k\right] = \lim_{m\to\infty} P[A_m]$	(11.47)

Limit of decreasing sequence:	$P\left[\bigcap_{k=1}^{\infty} A_k\right] = \lim_{m\to\infty} P[A_m]$	(11.50)
Limit of monotone sequence:	$P\left[\lim_{m\to\infty} A_m\right] = \lim_{m\to\infty} P[A_m]$	(11.52)
Boole's inequality:	$P\left[\bigcup_{k=1}^{\infty} A_k\right] \le \sum_{k=1}^{\infty} P[A_k]$	(11.53)
Borel–Cantelli	$A = \bigcap_{n=1}^{\infty}\left(\bigcup_{k=n}^{\infty} E_k\right)$	
Lemma 1:	$\sum_{k=1}^{\infty} P[E_k] < \infty \Rightarrow P[A] = 0$	(11.58)
Lemma 2:	$\sum_{k=1}^{\infty} P[E_k] = \infty \Rightarrow P[A] = 1$	(11.64)
WLLN:	$\{X_n\}$ i.i.d., $E[X_k] = \mu$, $\text{Var}\,[X_k] < \infty$ $\Rightarrow \overline{X}_n \overset{\text{P}}{\to} X$	(11.76)
Borel's SLLN:	$\{B_k\}$ Bernoulli sequence, parameter p $\Rightarrow \overline{B}_n \overset{\text{a.s.}}{\to} p$	(11.80)
Kolmogorov criterion for SLLN:	$\sum_{k=1}^{\infty} \frac{\sigma_k^2}{k^2} < \infty$	(11.81)
Sufficient condition for SLLN for i.i.d. RVs:	$E[X_k^2] < \infty$	(11.82)
SLLN for i.i.d. RVs:	$\overline{X}_n \overset{\text{a.s.}}{\to} \mu$, if and only if $E[\lvert X_k\rvert] < \infty$	(11.83)
Lindeberg–Lévy's CLT for i.i.d. RVs:	if $E[X_k] = \mu < \infty$, $\text{Var}\,[X_k] = \sigma^2 < \infty$ for all k then $Z_n \overset{\text{D}}{\to} U$,	(11.87)
Lyapunov's CLT for independent RVs:	if $m_k < \infty$ for all k and $\lim_{n\to\infty} \frac{\sum_{k=1}^{n} m_k}{s_n} = 0$, then $Z_n \overset{\text{D}}{\to} U$	(11.89)
Lindeberg–Feller's CLT for independent RVs:	$Z_n \overset{\text{D}}{\to} D$, if and only if $\lim_{n\to\infty} s_n^2 = \infty$, and $\lim_{n\to\infty} \frac{s_n^{\star 2}(\epsilon)}{s_n^2} = 1$ for every $\epsilon > 0$	(11.93)

11.5 Discussion and further reading

The limiting behavior of a sequence of RVs $\{X_n\}$ plays a central role in probability theory and statistical analysis. Indeed, in Chapter 2 the axiomatic definition of probability was motivated by the empirical limit of a sequence of relative frequency values as the number of repetitions of an experiment increases without limit (cf. (2.1), (2.2)). In this chapter, we considered several types of convergence of a sequence of RVs $\{X_n\}$ and studied the relationships among them.

The laws of large numbers state various conditions under which the sample average $\overline{X}_n = \frac{1}{n}\sum_{k=1}^{n} X_k$ converges to the average of the means $\overline{\mu}_n = \frac{1}{n}\sum_{k=1}^{n} \mu_k$ of the sequence $\{X_k; k = 1, 2, \ldots, n\}$, as $n \to \infty$. These laws provide important justification for the use of the sample average in statistical analysis (see Chapter 6). The WLLN states that if $\{X_n\}$ is i.i.d. with finite mean μ and finite variance, the sample average converges in probability to μ. The SLLN establishes that a stronger type of convergence, i.e., convergence a.s., holds if and only if the stronger condition $E[\lvert X_k\rvert] < \infty$ for all k is satisfied.

The different forms of the CLT give conditions for which a sequence $\{X_n\}$ converges in distribution to a normally distributed random variable. In particular, if $\{X_n\}$ is an i.i.d. sequence with finite mean and variance, it converges in distribution to a normal RV with the same mean and variance. The CLT provides justification for the prevalent use of the normal distribution in error analysis and for characterizing noise in physical systems.

Our presentation largely follows Thomas [319], including several examples and exercise problems. For further study of the laws of large numbers and the CLT, the reader is directed to advanced books on probability theory, e.g., Billingsley [26], Breiman [35], Chung [54], Cramér [74], Doob [82], Feller [99, 100], Gnedenko [122], Grimmett and Stirzaker [131], Loève [230], Rao [278], and Williams [354]. Shafer and Vovk [299] provide useful historical notes on the subjects.

11.6 Problems

Section 11.2: Types of convergence for sequences of random variables

11.1* Example of D convergence [319]. Let $\{X_n\}$ be a sequence of i.i.d. RVs with common distribution function $F_X(x)$ given by

$$F_X(x) = \begin{cases} 0, & x \le 0, \\ x, & 0 < x \le 1, \\ 1, & x > 1. \end{cases}$$

Define two sequences $\{Y_n\}$ and $\{Z_n\}$ by

$$Y_n = \max\{X_1, X_2, \dots, X_n\}$$

and

$$Z_n = n(1 - Y_n).$$

Show that $\{Z_n\}$ converges in distribution to a random variable Z with distribution

$$F_Z(z) = 1 - e^{-z}.$$

11.2 Bernstein's lemma. Let X_n, Y_n and Z_n be RVs such that

$$Z_n = X_n + Y_n.$$

Assume that X_n and Y_n have a joint distribution function $F_n(x, y)$ and X_n has a distribution function $F_n(x)$. Let the variance of Y_n

$$\mathrm{Var}[Y_n] \to 0, \text{ as } n \to \infty,$$

and

$$F_n(x) \xrightarrow{D} F_X(x).$$

Show that the distribution of Z_n converges to $F_X(x)$; that is,

$$Z_n \xrightarrow{\text{D}} X.$$

Hint: Show that, for any $\epsilon > 0$,

$$\lim_{n\to\infty} P[Z_n \leq z] \leq F_X(z+\epsilon)$$

and

$$\lim_{n\to\infty} P[Z_n \geq z] \geq F_X(z-\epsilon).$$

11.3* Convergence of sample average [319]. Define a random variable X_k by

$$X_k = c + N_k, \quad k = 1, 2, \ldots,$$

where c is a constant and the N_k are i.i.d. RVs with zero mean.

Show that the sequence of sample averages $\{\overline{X}_n\}$

$$\overline{X}_n = \frac{1}{n}\sum_{k=1}^{n} X_k, \quad n = 1, 2, \ldots$$

converges in probability to c.

11.4 P convergence and m.s. convergence [319]. Let $\{X_n\}$ be a sequence of RVs that converges in probability to a random variable X; that is,

$$X_n \xrightarrow{\text{P}} X.$$

Assume that the PDFs $f_n(x)$ of X_n are such that, for some $N > 0$,

$$f_n(x) = 0, \quad \text{for } |x| > x_0 \text{ for all } n > N.$$

Show that $\{X_n\}$ converges in mean square to X.

11.5 Proof of Lyapunov's inequality. Prove Lyapunov's inequality (11.41) by following the steps suggested below.

(a) Denote the rth absolute moment of Y by μ_r:

$$\mu_r = E[|Y|^r].$$

Then show

$$\mu_r^2 \leq \mu_{r-s}\mu_{r+s}. \tag{11.94}$$

Hint: Consider a new RV X defined by

$$X = a|Y|^{(r-s/2)} + |Y|^{(r+s/2)},$$

where a is a real parameter.

(b) Define a function $g(r)$ by

$$g(r) = \log \mu_r.$$

Show that $g(r)$ is a convex function and is a nondecreasing function of r for $r > 0$.

(c) Define a function $h(r)$ by

$$h(r) = \frac{g(r)}{r}. \tag{11.95}$$

Show that $h(r)$ is also a nondecreasing function.

(d) Using the result of part (c), show that (11.41) holds.

11.6* Properties of $\|Y\|_r$ [131]. Consider $\|Y\|_r$ defined by (11.40):

$$\|Y\|_r = \left(E[|Y|^r]\right)^{1/r}. \tag{11.96}$$

Show the following properties:

(a) **Hölder's[11] inequality.** If $r, s > 1$ and $r^{-1} + s^{-1} = 1$, then

$$\|XY\|_1 \leq \|X\|_r \|Y\|_s, \quad \text{or } E[|XY|] \leq \left(E[|X|^r]\right)^{1/r} \left(E[|Y|^s]\right)^{1/s}, \tag{11.97}$$

with equality if and only if $X \propto Y^{r-1}$. The case $r = s = 2$ reduces to the Cauchy–Schwarz inequality (see Section 10.1.1).

Hint: For any real numbers u and v, and $\frac{1}{r} + \frac{1}{s} = 1$,

$$\exp\left(\frac{u}{r} + \frac{v}{s}\right) \leq \frac{e^u}{r} + \frac{e^v}{s}. \tag{11.98}$$

(b) **Hölder's inequality for nonrandom vectors and functions.** Show that for $x_i, y_i \geq 0,\ i = 1, 2, \ldots, n$, and $\frac{1}{r} + \frac{1}{s} = 1$,

$$\sum_{i=1}^{n} x_i y_i \leq \left(\sum_{i=1}^{n} x_i^r\right)^{1/r} \left(\sum_{i=1}^{n} y_i^s\right)^{1/s}, \tag{11.99}$$

with equality if and only if $y_i \propto x_i^{r-1}$.
Similarly, show that for real-valued functions $f(u), g(u) \leq 0$, and $\frac{1}{r} + \frac{1}{s} = 1$,

$$\int f(u)g(u)\,du \leq \left(\int f(u)^r\,du\right)^{1/r} \left(\int g(u)^s\,du\right)^{1/s}, \tag{11.100}$$

with equality if and only if $g(u) \propto f(u)$ for all $-\infty < u < \infty$.

[11] Otto Ludwig Hölder (1859–1937) was a German mathematician.

A different hint: Try the following hint that is different from that in (a). Consider $F(x) \triangleq \frac{x^r}{r} + \frac{x^{-s}}{s}$, $x > 0$. Find x that minimize $F(x)$ and prove

$$uv \leq \frac{u^r}{r} + \frac{v^s}{s},$$

with equality when $v = u^{r-1}$.

(c) **Minkowski's[12] inequality.** If $r \geq 1$,

$$\|X + Y\|_r \leq \|X\|_r + \|Y\|_r;$$

or equivalently,

$$E[|X + Y|^r]^{1/r} \leq E[|X|^r]^{1/r} + E[|Y|^r]^{1/r}.$$

Note: Because of the above "triangular property," the quantity $\|Y\|_r$ qualifies as a **norm**.

11.7 Convergence in the rth mean versus almost sure convergence [319]. Consider the following example to illustrate that convergence in the rth mean does not imply almost sure convergence.

Let $\{X_n\}$ be a sequence of independent RVs defined by

$$X_n = \begin{cases} n^{1/2r}, & \text{with probability } \frac{1}{n}, \\ 0, & \text{with probability } 1 - \frac{1}{n}. \end{cases}$$

Let X be the degenerate RV

$$X = 0, \quad \text{with probability one.}$$

(a) Show that $\{X_n\}$ converges in the rth mean to $X = 0$.
(b) Show that for $\epsilon > 0$ and for arbitrary integer $m > 0$,

$$P\left[\bigcap_{n=m}^{\infty} \{\omega : X_n(\omega) < \epsilon\}\right] = 0.$$

Hint: Follow the argument in Example 11.7.

(c) Show that $\{X_n\}$ does not converge almost surely to $X = 0$.

11.8 Convergence in the rth mean versus almost sure convergence – continued [319]. Consider the following example to prove that almost sure convergence does not imply convergence in the rth mean, either.

Let $\{X_n\}$ be a sequence of independent RVs defined by

$$X_n = \begin{cases} e^n, & \text{with probability } \frac{1}{n^2}, \\ 0, & \text{with probability } 1 - \frac{1}{n^2}. \end{cases}$$

Let X be again the degenerate RV that is always zero.

[12] Hermann Minkowski (1864–1909) was a German mathematician.

(a) Show that $\{X_n\}$ does not converge to zero in the rth mean for any $r > 0$.
(b) Show that for arbitrary $\epsilon > 0$

$$P[\bigcap_{n=m}^{\infty}\{\omega : X_n(\omega) < \epsilon\}] = 1.$$

Hint: Use the inequality

$$\frac{1}{2} \leq \prod_{n=m}^{\infty}\left(1 - \frac{1}{n^2}\right) \leq 1, \quad \text{for } m \geq 2,$$

and

$$\lim_{m\to\infty} \prod_{n=m}^{\infty}\left(1 - \frac{1}{n^2}\right) = 1.$$

(c) Show that the sequence $\{X_n\}$ converges almost surely to X.

Section 11.3: Limit theorems

11.9 Limits in Bernoulli trials. Consider an infinite series of independent Bernoulli trials, where the probability of success in each trial is p $(0 < p < 1)$.

(a) Let E_k represent a success in the kth trial. What is the meaning of A_n defined by (11.55)? Find $P[A_n]$ for $n \geq 1$.
(b) Let E_k be the event that all trials up to the kth one are successful. Find $P[A]$, where A represents the event that infinitely many of the E_k occur.

11.10 Independence of complements of independent events. Show that if $E_1, E_2, \ldots, E_n$ are independent, their complementary events $E_1^c, E_2^c, \ldots, E_n^c$ are also independent.

11.11 Borel–Cantelli lemmas and Bernoulli trials. Apply the Borel–Cantelli lemmas to the two cases formulated in Problem 11.9.

11.12 Proof of the CLT. Prove the CLT for i.i.d. RVs (Theorem 11.22) by using the MGF of the normalized average Z_n.

11.13 Product of independent RVs. Let $Y_k; k \geq 1$ be independent positive RVs. We form their product.

$$R_n = Y_1 Y_2 \cdots Y_n. \tag{11.101}$$

(a) Can you find an approximate PDF of $f_{R_n}(r)$ for sufficiently large n?
(b) What are the conditions for the above approximation to be valid? Can you state these conditions in terms of the means and variances of the Y_k?

Part III

Random processes

12 Random processes

In this and following chapters, we will discuss random processes. After a brief introduction to this subject in Section 12.1, we will give an overview of various random processes in Section 12.2 and then discuss (strictly) stationary and wide-sense stationary random processes and introduce the notion of ergodicity. The last section focuses on complex-valued Gaussian processes, which will be useful in the study of communication systems and other applications.

12.1 Random process

There are many situations in which the *time dependency* of a set of probability functions is important. One example is a *noise process* that accompanies a signal process and should be suppressed or filtered out so that we can recover the signal reliably and accurately. Another example is the amount of outstanding packets yet to be processed at a network router or switch, which may lead to undesirable packet loss due to buffer overflows if not properly attended to in time.

Such a process can be conveniently characterized probabilistically by extending the notion of a random variable (RV) as follows: we assign to each sample point $\omega \in \Omega$ a **real-valued function** $X(\omega, t)$, where t is the **time parameter** or *index parameter* in some range $\mathcal{T}$, which may be, for instance, $\mathcal{T} = (-\infty, \infty)$ or $\mathcal{T} = \{0, 1, 2, \dots\}$ (see Figure 12.1). Imagine that we can observe this set of time functions $\{X(\omega, t); \omega \in \Omega, t \in \mathcal{T}\}$ at some instant $t = t_1$. Since each point $\omega \in \Omega$ has associated with it both the number $X(\omega, t_1)$ and its probability, the collection of numbers $\{X(\omega, t_1); \omega \in \Omega\}$ forms a random variable. By observing the time functions at another time, say at $t = t_2$, we will have another collection of numbers with a possibly different probability measure. Indeed, this set of time-indexed functions defines a separate RV for each choice of the time.

A probability system, which is composed of a sample space, a set of real-valued time-indexed functions, and a probability measure, is called a **random process** or a **stochastic process** and is usually denoted by a notation such as $X(t); t \in \mathcal{T}$, or simply as $X(t)$, if $\mathcal{T}$ is implicitly understood. The individual time functions of the random process $X(t)$ are called **sample functions**, and the particular sample function associated with a sample point $\omega \in \Omega$ is denoted as $X(\omega, t); t \in \mathcal{T}$. The set of all possible sample functions, together with a probability law, is called the **ensemble**. Naming a random

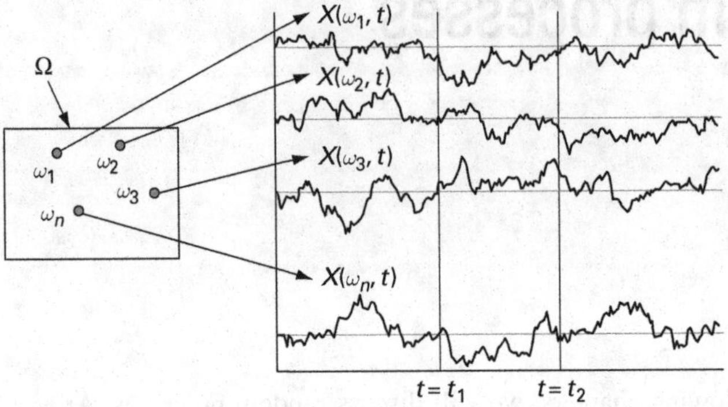

Figure 12.1 A random process $X(\omega, t)$ as a mapping from a sample point $\omega \in \Omega$ to a real-valued function.

process $X(t)$ and denoting the sample function associated with the sample point ω as $X(\omega, t)$ is consistent with our previous practice of naming a random variable X and denoting the sample value associated with the sample point ω as $X(\omega)$.

By definition, a random process implies the existence of an infinite number of RVs, one for each t in some range. Thus, we may speak of the PDF $f_{X(t_1)}(\cdot)$ of the random variable $X(t_1)$ obtained by observing $X(t)$ at time t_1. Generally, for N time instants $\{t_i : i = 1, 2, \ldots, N\}$, we define the N random variables $X_i = X(t_i)$; $i = 1, 2, \ldots, N$. Then we can speak of the joint PDF of $X_1, X_2, \ldots, X_N$.

12.2 Classification of random processes

We saw in Chapter 3 that RVs can be classified into two types. For random processes we have the time-index parameter t, so we can further classify them depending on whether t takes on continuous or discrete values. In addition, we have a class of random processes, **point processes** or **counting processes**, where the intervals between points of events are RVs. In this section we give several dichotomies of random processes and at the same time provide a brief and *informal* preview of some of the most frequently encountered random processes, many of which will be more formally discussed in later sections in the present and subsequent chapters.

12.2.1 Discrete-time versus continuous-time processes

When the time index takes on values from a set of discrete of time instants, say $\mathcal{T} = \{0, 1, 2, \ldots\}$, the process is said to be a **discrete-time random process** or a **random sequence**, and is often denoted as X_t; $t \in \mathcal{T}$ or X_k; $k \in \mathcal{T}$ instead of $X(t)$, which is usually used for a continuous-time process, where the interval $\mathcal{T}$ for the time-index is typically a *real line* $\mathcal{T} = (-\infty, \infty)$ or $\mathcal{T} = [0, \infty)$. A sequence of random variables $\{X_k\}$, discussed in Chapter 11, is indeed a discrete-time random process if the index k can be interpreted as a time index. A random process may intrinsically be of the discrete-time nature, but in many cases it is the result of observing or sampling a continuous-time process at discrete points in time, either at regular intervals or at some suitably defined moments.

In statistics, signal processing, econometrics, and social sciences the term **time series** is often used to represent a sequence of data points, measured typically at successive times spaced at uniform time intervals. This term is also used often almost synonymously with discrete-time random process, when it is discussed in the context of certain types of statistical analysis, such as spectral analysis, estimation, or prediction. In this book we use capital letters, say $\{X_t\}$ or $\{X_n\}$, to represent random variables and lower cases for observed data, e.g., $\{x_t\}$ or $\{x_n\}$, where the time index t or n is typically natural numbers, i.e., $\mathbb{Z}_+ = \{1, 2, 3, \ldots\}$. In most cases, underlying assumptions of a time series are that it is both **stationary** and **ergodic** in the sense explained later in this section.

12.2.2 Discrete-state versus continuous-state processes

The set of possible values that $X(t)$ may take on is called its **state space**, often denoted $\mathcal{S}$. For a Bernoulli trial sequence, $\mathcal{S} = \{\text{s}, \text{f}\}$ or $\mathcal{S} = \{0, 1\}$, where s and f stand for "success" and "failure." Thus, the Bernoulli trial sequence is a discrete-time, discrete-state random process. A simple **random walk**, which moves to the right or to the left by a unit step (i.e., $X_k - X_{k-1} = \pm 1$), is another example of a discrete-time, discrete-state process.

If the set $\mathcal{S}$ is continuous, such as $\mathcal{S} = (-\infty, \infty)$ or $\mathcal{S} = [0, \infty)$, the process $X(t)$ is called a continuous-state process. If $X(t)$ represents the temperature (of a certain place or object) at time t, it is a continuous-state process. If $X(t)$, however, is the price of a stock at time t, it is a discrete-state process. A Gaussian (or normal) process that we will discuss in Section 12.3.2 is a continuous-time, continuous-state process. **Brownian motion** or the **Wiener process**, which can be obtained as a limit of the random walk – by making both time interval $h = t_k - t_{k-1}$ and the step size δ infinitesimally small, while keeping $\delta^2/h = \sigma^2$ (constant) – is another example of a continuous-time, continuous-state process. We will discuss Brownian motion and its generalization, **diffusion processes**, in Chapter 17.

Quantization or digitization used in modern signal processing converts a continuous-state process into a discrete-state process. For instance, digitized information stored on an audio CD represents a discrete-time, discrete-state process, although the original acoustic signal and the replayed signal from the speaker are continuous-time, continuous-state processes.

12.2.3 Stationary versus nonstationary processes

As we will discuss in detail in Section 12.3, a **stationary** process is a process such that its distribution function

$$F_X(x; t) \triangleq P[X(t) \leq x], \quad t \in \mathcal{T}, \tag{12.1}$$

is invariant to shifts in time for all values of its arguments. In other words, the distribution functions $F_X(x; t)$ are independent of t. Otherwise, the process is called **nonstationary**.

For the case of discrete-time, discrete-state processes, we can easily think of a stationary random process. One example is an infinite series of Bernoulli trials that defines a stationary random sequence $X_k; k = 0, 1, 2, \ldots$. **Gaussian processes**, which are often used to represent a noise process or a stochastic signal like a multi-path fading signal, are usually modeled as stationary Gaussian processes. A random walk and Brownian motion are examples of nonstationary processes: their variance increases in proportion to the steps n or time t (hence, the standard deviation grows in proportion to $\sqrt{n}$ or $\sqrt{t}$ – often referred to as the **square-root law**).

Many random processes of interest in real life are nonstationary. For instance, the price of a stock or the Dow Jones' index and the packet traffic in a network are both nonstationary processes. But if we limit ourselves to a relatively short interval $\mathcal{T}$, then some of these processes may be well approximated as stationary processes. Somewhat paradoxically, however, we often write a stationary process as $X(t); -\infty < t < \infty$, because a stationary process must have begun in the infinite past and will continue into the infinite future: a process that has started only in a finite past must still be in its transient state and require an infinite amount of time in order to reach its steady state or equilibrium state. Thus, a stationary process assumption is, at best, a mathematical idealization in modeling a real system.

12.2.4 Independent versus dependent processes

Suppose we arbitrarily choose n time instants and consider the joint distribution function $F_{\boldsymbol{X}}(\boldsymbol{x}, \boldsymbol{t})$ of the set of random variables $\boldsymbol{X} = (X_1, X_2, \ldots, X_n)$, where $X_i = X(t_i); i = 1, 2, \ldots, n$. If this distribution function factors into the product

$$F_{\boldsymbol{X}}(\boldsymbol{x}; \boldsymbol{t}) \triangleq F_{X_1 X_2 \cdots X_n}(x_1, x_2, \ldots, x_n; t_1, t_2, \ldots, t_n) = F_{X_1}(x_1; t_1) F_{X_2}(x_2; t_2) \cdots F_{X_n}(x_n; t_n), \tag{12.2}$$

for any finite n and for any choice of the instants $\boldsymbol{t}$, we say $X(t)$ is an **independent process**.

In the case of a discrete-time process, the independent random sequence X_k discussed in Chapter 11 provides such an example.

In the case of a continuous-time, continuous-state process, a random process is commonly called **white noise** if its power spectral is flat for all frequencies $f \in (-\infty, \infty)$. Brownian motion, which can be viewed as an integration of white noise, is a **dependent process**, and so are the random walk and its generalized versions. The random step process $Y_k \triangleq X_k - X_{k-1}$ of the random walk X_k, however, is usually treated as an independent sequence.

12.2.5 Markov chains and Markov processes

12.2.5.1 Discrete-time Markov chain (DTMC)

In 1906 A. A. Markov [237] defined a **simple chain** as "an infinite sequence

$$X_1, X_2, \ldots, X_k, X_{k+1}, \ldots$$

of variables connected in such a way that X_{k+1} for any k is independent of

$$X_1, X_2, \ldots, X_{k-1}$$

in case X_k is known." (see Basharin *et al.* [14]). In his subsequent paper in 1908 (later translated into German [239] and into English [240]) he extended the notion to *complex chains* in which "every member is directly connected not with single but with several preceding numbers." A more common term used today is a **high-order** Markov chain. A **Markov chain of order *h*** is defined as a sequence in which X_k depends on its past only through its h previous values, $X_{k-1}, X_{k-2}, \ldots, X_{k-h}$:

$$p(x_k|x_{k-i};\ i \geq 1) = p(x_k|x_{k-1}, x_{k-2}, \ldots, x_{k-h}). \tag{12.3}$$

The case $h = 1$ reduces to a simple Markov chain and $h = 0$ to an independent sequence. Any higher-order Markov chain with finite h defined over state space $\mathcal{S}$ can be transformed into a simple Markov chain by defining the state space $\mathcal{S}^h = \mathcal{S} \times \mathcal{S} \cdots \times \mathcal{S}$, the h-times Cartesian product of $\mathcal{S}$ with itself. We apply this observation in Section 13.4.3 where we discuss an autoregressive process of order p, denoted as AR(p). We shall discuss Markov chains, also called **Markov proccess**, and their properties in Chapters 15, 16, and 20.

Markov [237] introduced the notion of Markov chain in order to extend the **law of large numbers** and the **central limit theorem** to dependent sequences, but he also applied the Markov chain model to the sequence of 20 000 letters in A. S. Pushkin's poem *Eugene Onegin*, computing the probability of a vowel, the probability of a vowel following a vowel, the probability of a vowel following a consonant, etc. [14]. Today, Markov chain models are used in a variety of fields, including linguistic models for speech recognition, DNA and protein sequences, and network traffic, as we remarked in Chapter 1 and shall discuss in Chapter 20.

The simple Markov chain $\{X_k\}$ defined above is referred to as a **discrete-time Markov chain** (DTMC). If there are M different states that the Makov chain can take on, we can label them, without loss of generality, by integers, $0, 1, 2, \ldots, M-1$. We denote this set of states by $\mathcal{S}$:

$$\mathcal{S} = \{0, 1, 2, \ldots, M-1\}, \tag{12.4}$$

where M may be infinite. We write $X_n = i$, when the DTMC $\{X_k\}$ assumes state $i \in \mathcal{S}$ at time n. We illustrate in Figure 12.2 (a) a sample path of a DTMC.

12.2.5.2 Continuous-time Markov chain (CTMC)

For a given DTMC $\{X_k\}$, we can construct a **continuous-time Markov chain** (CTMC) $X(t)$ as follows. Let $X(t)$ make the kth state transition at time $t = t_k$ and enter state $i \in \mathcal{S}$, and stay in this state until the next epoch of transition at time t_{k+1}; i.e.,

$$X(t) = i, \quad \text{for } t_k \leq t < t_{k+1}, \quad \text{where } i = X(t_k) \text{ and } X(t_{k+1}) = j (\neq i), \tag{12.5}$$

and let the interval $\tau_k \triangleq t_{k+1} - t_k$ be **exponentially distributed** with mean λ_k^{-1}. Figure 12.2 (b) shows a sample path of a CTMC.

Given the current time t_n, the future behavior of the process $X(t); t \geq t_n$ depends on its past $X(s); -\infty < s < t_n$ only through its current state $X(t_n) = i \in \mathcal{S}$. How long

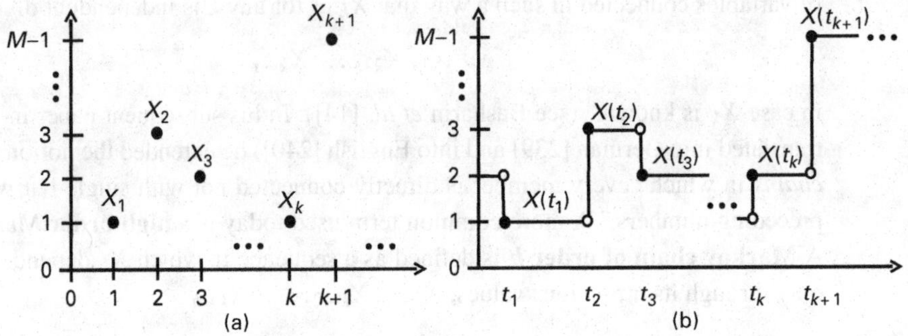

Figure 12.2 (a) A discrete-time Markov chain (DTMC); (b) a continuous-time Markov chain (CTMC).

$X(t)$ has been in its current state i is immaterial because of the **memoryless property** of the exponential distribution, as discussed in Section 4.2.2. In a DTMC, the interval that the chain remains in state $i \in \mathcal{S}$ before it moves to one of the other states in $\mathcal{S}$ is geometrically distributed with parameter $p = 1 - P_{i,i} = \sum_{j \neq i} P_{i,j}$, where $P_{i,j} = P[X_{k+1} = j | X_k = i]$ is the transition probability from state i to state $j \neq i$ $(i, j \in \mathcal{S})$. The reader may recall that the geometric distribution also possesses the memoryless property.

A **Poisson process** process is an example of CTMC, with $\lambda_k = \lambda$ for all k, and the number of states $M = \infty$; i.e., $\mathcal{S} = \{0, 1, 2, \ldots\}$. Here, $X(t)$ represents the cumulative number of arrivals (or births) up to time t. In this definition, the Poisson process is a monotone nondecreasing function; thus, it is a **counting process**.

12.2.5.3 Continuous-time, continuous-space (CTCS) Markov process

A continuous-time, continuous-state (CTCS) Markov process is also known as a **diffusion process**, and its state transition probability (or conditional probability) distribution function $F_{X(t_1)|X(t_0)}(x_0, x_1; t_0, t_1)$ satisfies a partial differential equation known as a *diffusion equation*. We denote the corresponding transition PDF by $f_{X(t_1)|X(t_0)}(x_0, x_1; t_0, t_1)$. We will further discuss these diffusion processes in Chapter 17.

12.2.5.4 Discrete-Time, Continuous-Space (DTCS) Markov Process

Another class of Markov process, which is seldom discussed in the literature on Markov chains and Markov processes, is what we term a discrete-time, continuous-state (DTCS) Markov process. An autoregressive (AR) time series, discussed in Section 13.4.1, is such an example. Similarly, the state sequence associated with an autoregressive moving average (ARMA) time series, defined in Section 13.4.3, forms a multidimensional DTCS Markov process (see (13.236)). If we observe a diffusion process $X(t)$ at discrete-time moments, $t_0, t_1, t_2, \ldots$, then the time series $X_0, X_1, X_2, \ldots$ (where $X_i = X(t_i)$) defines a DTCS Markov process. In Section 21.7 we will introduce a simulation technique called Markov chain Monte Carlo (MCMC), whereby we generate (dependent) random variates $x_0, x_1, x_2, \ldots$ sampled from a *target distribution* $\pi(\boldsymbol{x})$. Such a sequence can be viewed as an instance of a DTCS Markov process $\{X_t\}$.

12.2.5.5 Semi-Markov process and embedded Markov chain

If the interval $\tau_k = t_{k+1} - t_k$ in (12.5) is not exponentially distributed, the process $X(t)$ does not possess the Markovian or memoryless property. Such a process is called a **semi-Markov process**. Semi-Markov processes will be further discussed in Section 16.1.

Conversely, for a given CTMC or semi-Markov process $X(t)$, a DTMC $\{X_k\}$ defined by observing at the epochs of state transitions in $X(t)$ (and the state $X_k = i \in \mathcal{S}$ is entered at the kth transition) is said to be *embedded* (or *imbedded*) in the original process $X(t)$. In queueing theory, we often find ourselves in a situation where the **embedded Markov chain** (EMC) $\{X_k\}$ is amenable to analysis, even when the original process $X(t)$ is not.

12.2.6 Point processes and renewal processes

A **point process** is a random process that consists of a sequence of epochs

$$t_1, t_2, t_3, \ldots, t_n, \ldots,$$

where "point events" occur. The point process is thought of as a sequence of events in which only the times of their occurrence are of interest.

The one-dimensional point process can be represented by $N(t)$,

$$N(t) = \max\{n : t_n \le t\}, \tag{12.6}$$

which is the cumulative count of events in the time interval $(0, t]$. Thus, the process $\{N(t)\}$ is called a **counting process**. It is clearly an integer-valued process and is a nondecreasing function of t. The difference of the event points

$$X_n = t_n - t_{n-1} \tag{12.7}$$

represents the interval between the $(n-1)$st and the nth point events.

If we assume that the X_n are independent and identically distributed (i.i.d.) RVs with a common distribution function $F_X(x)$, the corresponding point process $N(t)$ is called a **renewal process**. The event points t_n are called **renewal points** and the intervals X_n are the **lifetimes**.

The **Poisson process** is a point process, where the $X_n(= t_n - t_{n-1})$ are independent. Thus, it is a renewal process. It is also a CTMC, since the X_n are exponentially distributed.

Both point processes and renewal processes are extensively used in queueing theory, reliability theory, risk analysis, and mathematical finance. Chapter 14 is devoted to these processes.

12.2.7 Real-valued versus complex-valued processes

Although random processes we encounter are usually real-valued, it is often mathematically convenient to deal with a complex-valued random process associated with a real-valued random process in much the same way as we often deal with $e^{i\omega t}$ instead of the sinusoidal function $\cos \omega t$. In communication systems, for instance, the class of carrier-modulated data transmission systems that adopt **linear modulation** (e.g., see [193]) which include amplitude shift keying (ASK), phase shift keying (PSK), amplitude–phase shift keying (APSK), quadrature amplitude modulation (QAM), as well as conventional analog modulation schemes such as AM, PM, SSB (single-sideband) and VSB (vestigial-sideband) modulation, can be concisely represented in terms of complex-valued processes, known as **analytic signals**. Furthermore, Gaussian noise that goes through a bandpass filter at the receiver can be compactly represented in terms of a **complex-valued Gaussian process**.

In a radio channel with multi-path propagation, the amplitude and phase of the received signal vary randomly. This phenomenon, called **fading**, introduces a multiplicative factor which can be compactly expressed by a complex-valued Gaussian random variable (as discussed in Section 7.6.2) or complex-valued Gaussian process. We will further discuss this topic in Section 12.4.

12.2.8 One-dimensional versus vector processes

Vector representation of a random process has two types. The first type occurs when a scalar or one-dimensional random process is observed at multiple instants $\boldsymbol{t} = (t_1, t_2, \ldots, t_n)$; the corresponding observations may be represented as a random vector, i.e., $\boldsymbol{X} = (X(t_1), X(t_2), \ldots, X(t_n))^\top$. The vector processes that we specifically refer to in this section are the second type, where we deal with a **multidimensional** process instead of a one-dimensional scalar-valued process. Such a situation occurs, for instance, in **diversity reception** [329] where one attempts to detect or estimate signals using multiple antennas placed at different positions, as is the case in phased array radars and large-aperture seismic array systems.

In statistics and econometrics, *vector-valued* time series, or simply, **vector time series**, or **multivariate time series** are extensively used [141, 145]. Such vector-valued processes will be denoted in this book as $\boldsymbol{X}_t$, where $\boldsymbol{X}_t = (X_{1,t}, X_{2,t}, \ldots, X_{n,t})$; $t \in \mathcal{T} = (1, 2, \ldots, T)$. Then an observed sample path $\{\boldsymbol{x}_t;\ t \in \mathcal{T}\}$ of this time series $\boldsymbol{X}_t$ can be presented as a two-dimensional array and is often referred to as **panel data**. In communication engineering, there has been an increasing level of research and development activities lately concerning **multiple-input, multiple-output (MIMO)** systems that are designed to improve signal recovery in the presence of multipath fading, interference, and noise. Such multiple signals can be compactly represented as complex-valued vector processes.

We start with scalar real-valued random processes.

12.3 Stationary random process

In dealing with random processes in the real world, we often notice that statistical properties of interest are relatively independent of the time at which observation of the random process is begun. A **stationary random process** is defined as one for which all the distribution functions are *invariant* under a shift of the time origin.

12.3.1 Strict stationarity versus wide-sense stationarity

DEFINITION 12.1 (Strictly stationary random process). *A real-valued random process $X(t)$ is said to be* **strictly stationary, strict-sense stationary (SSS),** *or* **strongly stationary** *if for* every *finite set of time instants $\{t_i; i = 1, 2, \ldots, n\}$ and for every constant h the joint distribution functions of $X_i = X(t_i)$ and those of $X'_i = X(t_i + h)$; $i = 1, 2, \ldots, n$ are the same:*

$$\boxed{F_{X_1 X_2 \cdots X_n}(x_1, x_2, \ldots, x_n) = F_{X'_1 X'_2 \cdots X'_n}(x_1, x_2, \ldots, x_n).} \tag{12.8}$$

□

The above condition for stationarity is often unnecessarily restrictive, and we define below a weaker form of stationarity. We define the mean of a random process $X(t)$ by

$$\mu_X(t) = E[X(t)] \tag{12.9}$$

and the **autocorrelation function** of $X(t)$ by

$$\boxed{R_X(t_1, t_2) \triangleq E[X(t_1)X(t_2)].} \tag{12.10}$$

If the joint density function exists, we can write

$$R_X(t_1, t_2) = \int_{-\infty}^{\infty} \int_{-\infty}^{\infty} x_1 x_2 f_{X_1 X_2}(x_1, x_2)\, dx_1\, dx_2 \tag{12.11}$$

Now we are ready to give a second type of stationarity.[1]

DEFINITION 12.2 (Wide-sense stationary random process). *A random process $X(t)$ is said to be* **wide-sense stationary** *(WSS),* **weakly stationary**, **covariance stationary**, *or* **second-order stationary** *if it has*

1. *a constant mean: $E[X(t)] = \mu_X$ for all $-\infty < t < \infty$;*
2. *finite second moments $E[X(t)^2] < \infty$, for all $-\infty < t < \infty$; and*
3. *a covariance*

$$E[(X(s) - \mu)(X(t) - \mu)]$$

that depends only on the time difference $|s - t|$.

□

If $X(t)$ is WSS, then the autocorrelation function $R_X(s, t)$ satisfies

$$R_X(s, t) = R_X(s - t, 0) = R_X(0, t - s),$$

[1] Some authors (e.g. [82, 263, 319]) drop the condition 1. $E[X(t)] = \mu$, constant, of the definition below.

indicating that it is also a function of the time difference $|s-t|$ only.[2]

It is not difficult to see that if a random process is stationary in the strict sense and has finite means and second moments, then it must also be stationary in the wide sense. However, the converse does not hold true.

Example 12.1: An i.i.d. sequence. Let $\{X_n\}$ be a sequence of i.i.d. real-valued RVs with a common mean μ_X and variance σ_X^2. Then it is strictly stationary, and its autocorrelation function is given by

$$R_X(k) = E[X_{n+k}X_n] = \begin{cases} \sigma_X^2 + \mu_X^2, & k = 0, \\ \mu_X^2, & k \neq 0. \end{cases} \tag{12.12}$$

Thus, $\{X_n\}$ is strictly stationary as well as WSS.

From the strong law of large numbers (SLLN), we can assert

$$\frac{\sum_{i=1}^{n} X_i}{n} \xrightarrow{\text{a.s.}} \mu_X. \tag{12.13}$$

□

Example 12.2: An uncorrelated sequence with identical mean and variance. Let $\{Y_n\}$ be a sequence of uncorrelated real-valued RVs, with a common mean μ_Y and variance σ_Y^2. Then, the process $\{Y_n\}$ is WSS, since $E[Y_n] = \mu_Y$ for all n and the autocorrelation function takes the same form as (12.12). However, it is not strictly stationary, unless all of the Y_n are mutually independent and have a common distribution. Note, however, that stationarity does not require independence or uncorrelatedness of the Y_n. If we pass the above stationary sequence into a time-invariant filter, the output sequence is correlated, yet strictly stationary. □

We now consider a stationary process quite different from the two previous examples.

Example 12.3: Identical sequences [131, 175]. Let Z be a single RV with known distribution, and set $Z_1 = Z_2 = \cdots = Z$; i.e.,

$$Z_n = Z, \quad \text{for all } n.$$

Then the process $\{Z_n\}$ is easily seen to be strictly stationary as well as being WSS. The auto-covariance function is

$$R_Z[k] \triangleq E[Z_n Z_{n+k}] = \sigma^2, \quad \text{for all } k. \tag{12.14}$$

[2] If $X(t)$ is a complex-valued process, the autocorrelation function is defined by $R_X(s,t) = E[X(s)X^*(t)]$. If $R_X(s,t)$ is a function of the difference $(s-t)$, $X(t)$ is said to be WSS (see Section 13.1.3). The autocorrelation function is no longer a symmetric function, but $R_X(-t) = R_X^*(t)$.

Similar to (12.13), we have

$$\frac{\sum_{i=1}^{n} Z_i}{n} \xrightarrow{\text{a.s.}} Z, \tag{12.15}$$

since each term Z_i in the sum is the same as the limit Z. □

Example 12.4: Sinusoidal function with random amplitudes [131, 175, 262]. Consider a random process defined by

$$X(t) = A\cos\omega_0 t + B\sin\omega_0 t, \tag{12.16}$$

where A and B are real-valued RVs. If A and B are time functions, representing some information-carrying waveforms, then the process $X(t)$ is what is known as a **quadrature-amplitude modulated** (QAM) signal in communication systems.

Taking the expectation

$$E[X(t)] = E[A]\cos\omega_0 t + E[B]\sin\omega_0 t,$$

we find that, for the mean $E[X(t)]$ to be constant, it is necessary that

$$\boxed{E[A] = E[B] = 0.} \tag{12.17}$$

In order for the process $X(t)$ to be WSS,

$$\begin{aligned} &E[X(t)X(u)]\\ &= E[A^2\cos\omega_0 t\cos\omega_0 u + B^2\sin\omega_0 t\sin\omega_0 u + AB(\cos\omega_0 t\sin\omega_0 U + \sin\omega_0 t\cos\omega_0 u)]\\ &= \frac{E[A^2]}{2}[\cos\omega_0(t+u) + \cos\omega_0(t-u)] - \frac{E[B^2]}{2}[\cos\omega_0(t+u) - \cos\omega_0(t-u)]\\ &\quad + E[AB]\sin\omega_0(t+u) \end{aligned} \tag{12.18}$$

must be a function of $t - u$ only. Suppose that

$$\boxed{E[A^2] = E[B^2] = \sigma^2} \tag{12.19}$$

and

$$\boxed{E[AB] = 0} \tag{12.20}$$

hold. Then it readily follows that

$$E[X(t)X(u)] = \sigma^2\cos\omega_0(t-u) \triangleq R_X(t-u). \tag{12.21}$$

Hence, $X(t)$ becomes WSS.

Conversely, suppose that $X(t)$ is WSS. Then

$$E[X(t_1)X(t_1)] = E[X(t_2)X(t_2)] = R_X(0) \tag{12.22}$$

for any t_1 and t_2. Letting $t_1 = 0$ and $t_2 = \pi/2\omega_0$, we have

$$X(t_1) = X(0) = A \text{ and } X(t_2) = X\left(\frac{\pi}{2\omega_0}\right) = B.$$

Then (12.22) becomes

$$E[A^2] = E[B^2] = R_X(0) = \sigma^2,$$

which is (12.19). Furthermore, we have

$$E[X(t)X(u)] = \sigma^2 \cos\omega_0(t-u) + E[AB]\sin\omega_0(t+u).$$

The last expression is a function of $t - u$ only if (12.20) is met. Thus, we have shown that the above $X(t)$ is WSS if and only if the conditions (12.17), (12.19), and (12.20) are met. □

12.3.2 Gaussian process

An important class of random processes is that of Gaussian processes.

DEFINITION 12.3 (Gaussian process). *A real-valued continuous-time process $X(t)$; $-\infty < t < \infty$ is called a* Gaussian process *if, for every finite set of time instants*

$$\mathcal{T} = \{t_1, t_2, \ldots, t_n\},$$

the vector

$$\boldsymbol{X} = (X_1, X_2, \ldots, X_n)^\top, \text{ where } X_i = X(t_i),\ i = 1, 2, \ldots, n,$$

has the multivariate normal distribution with some mean vector $\boldsymbol{\mu}$ and covariance matrix $\boldsymbol{C}$, both of which may depend on $\mathcal{T}$:

$$f_X(\boldsymbol{x}) = \frac{1}{(2\pi)^{n/2}|\det\boldsymbol{C}|^{1/2}} \exp\left[-\frac{1}{2}(\boldsymbol{x}-\boldsymbol{\mu})^\top \boldsymbol{C}^{-1}(\boldsymbol{x}-\boldsymbol{\mu})\right]. \tag{12.23}$$

□

The covariance matrix

$$\boldsymbol{C} = \left[C_{ij}\right]$$

is related to the autocorrelation function $R_X(t_i, t_j)$ as follows:

$$\begin{aligned} C_{ij} &= E\left[(X_i - \mu_i)(X_j - \mu_j)\right] \\ &= E\left[(X(t_i) - \mu(t_i))(X(t_j) - \mu(t_j))\right] \\ &= R_X(t_i, t_j) - \mu(t_i)\mu(t_j). \end{aligned} \tag{12.24}$$

Thus, a Gaussian process $X(t)$ can be characterized by providing its mean function $\mu_X(t)$; $-\infty < t < \infty$ and autocorrelation function $R_X(t_1, t_2)$; $-\infty < t_1, t_2 < \infty$, which then yield the required mean vectors $\boldsymbol{\mu}$ and covariance matrices $\boldsymbol{C}$ for any finite set of time instants $\mathcal{T}$.

12.3.2.1 Stationary Gaussian process

If $X(t)$ is a WSS Gaussian process, then $\mu(t) = \mu$ and

$$R_X(t_i, t_j) = R_X(0, t_i - t_j) \triangleq R_X(t_i - t_j). \tag{12.25}$$

Hence,

$$C_{ij} = R_X(t_i - t_j) - \mu^2, \tag{12.26}$$

which is a function of $t_i - t_j$ only. Since the mean vector and covariance matrix completely specify the probability distribution of multivariate normal variables, a WSS Gaussian process is also SSS, which we state as the following theorem.

THEOREM 12.1 (Stationary Gaussian process). *A real-valued continuous-time process* $X(t)$; $-\infty < t < \infty$ *is a* stationary Gaussian process *if*

$$E[X(t)] = \mu \text{ and } E[X(t)X(s)] = R_X(t - s), \text{ for } -\infty < t, s < \infty, \tag{12.27}$$

with some constant μ *and autocorrelation function* $R_X(\cdot)$ *and if for every finite set of time instants*

$$\mathcal{T} = \{t_1, t_2, \ldots, t_n\}$$

the vector

$$\boldsymbol{X} = (X_1, X_2, \ldots, X_n)^\top, \text{ where } X_i = X(t_i),\ i = 1, 2, \ldots, n,$$

has the multivariate normal distribution with mean $\boldsymbol{\mu} = (\mu, \mu, \ldots, \mu)$ *and covariance matrix* $\boldsymbol{C}$*, whose* (i, j) *entry* C_{ij} *is given by (12.26).* □

Both stationary and nonstationary Gaussian processes will appear repeatedly in the rest of this volume. In Section 12.4 we will extend the Gaussian process to a multivariate complex-valued Gaussian process.

12.3.3 Ergodic processes and ergodic theorems

Recall that the sequence $\{X_n\}$ in Example 12.1 and the sequence $\{Z_n\}$ in Example 12.3 are both strictly stationary, but their behaviors are entirely different. Observing $X_1, X_2, \ldots, X_n$ provides no information that could be used to predict X_{n+1}, whereas observing only Z_1 allows us to predict $Z_2, Z_3, \ldots$ exactly. Although the convergence modes in (12.13) and (12.15) are both almost sure convergence, the sample average $\overline{X}_n (= \frac{\sum_{i=1}^n X_i}{n})$ converges to the population mean μ_X, whereas in the nth sample average $\overline{Z}_n$ there is just as much randomness as in the first observation Z_1.

Most sequences of interest lie somewhere between the above two extreme cases, in that the sequences are not uncorrelated but the covariance between X_n and X_{n+k} approaches zero as the lag k increases. In such circumstances, the time average and ensemble average can be shown to be equivalent under some regularity conditions. We will address this problem in this subsection.

If sample averages formed from a single realization of a stationary process converge to some underlying parameter of the process, the process X_i is said to be **ergodic**. In other words, an ergodic process is a stationary process such that its **time average** (or **statistical average**) is equivalent to the **ensemble average** (or **population mean**). In order to make inference about the underlying laws that govern an ergodic process, one need only observe a single sample path, but over a sufficiently long span of time. Thus, it is important to determine conditions that lead to a stationary process being ergodic. The **ergodic theory** addresses this question and an **ergodic theorem** specifies conditions for a stationary process to be ergodic. Ergodic theorems, although mathematically advanced concepts, have important practical implications. For instance, when we perform a simulation experiment of a stochastic system, the underlying assumption is that the process to be observed is an ergodic process.

Laws of large numbers for sequences that are *not uncorrelated* are called *ergodic theorems*. Just as there are strong and weak laws of large numbers, there are a variety of ergodic theorems, depending on their assumptions and on modes of convergence. Recall that the SLLN for an i.i.d. sequence states

$$P[\lim_{n\to\infty} \overline{X}_n = \mu_X] = 1.$$

Stationary processes, strong or weak, provide a natural setting to generalize the law of large numbers, since for such processes the mean value is a constant independent of time; i.e., $\mu_X(n) = \mu_X$ for all n.

Let us digress for the moment and assume that we do not know whether the process $\{X_n\}$ we are concerned with is stationary or nonstationary, let alone about its possible ergodicity. In order to estimate an unknown mean sequence $\mu_X(n)$, we must in general observe a large number M of separate realizations of the process, say

$$\{X_n^{(1)}; n = 1, 2, \ldots, N\}, \{X_n^{(2)}; n = 1, 2, \ldots, N\}, \ldots, \{X_n^{(M)}; n = 1, 2, \ldots, N\}.$$

We then compute the ensemble average at each time n:

$$\overline{X}(n) \triangleq \frac{X_n^{(1)} + X_n^{(2)} + \cdots + X_n^{(M)}}{M}, \quad n = 1, 2, \ldots, N,$$

which we will use as an estimate for $\mu_X(n)$.

If it were known (or judged on the observations) that $\mu_X(n)$ is constant, which would be the case for a stationary process, we might obtain the grand average by computing

$$\begin{aligned}\overline{X} &= \frac{\overline{X}(1) + \overline{X}(2) + \cdots + \overline{X}(N)}{N} \\ &= \frac{\sum_{n=1}^{N}\sum_{m=1}^{M} X_n^{(m)}}{MN}.\end{aligned}$$

If, in addition, we knew that the process $X(t)$ was ergodic, the same estimation problem could be simplified so that we would need to observe, as stated earlier, only a single realization (or sample path) of the process over a sufficiently long period.

12.4 Complex-valued Gaussian process

The RVs and random processes we have discussed so far have been mostly real-valued, except for the discussion on **circularly symmetric** Gaussian RVs given in Section 7.6.1. But many of the definitions and properties discussed in the present chapter can be generalized to the complex-valued RVs and random processes. The notion of complex-valued Gaussian processes is very useful in analyzing communication systems. Many signal processes that we deal with in communication systems, including radar, sonar, and wireless systems, have a spectrum of finite bandwidth around some carrier frequency ω_0. The amplitude of a signal process $S(t)$, at the receiving end, will often exhibit a Gaussian distribution because of the multi-path fading effect. Then such a signal can be represented as

$$S(t) = \Re\{(X(t) + iY(t))e^{i\omega_0 t}\} = X(t)\cos\omega_0 t - Y(t)\sin\omega_0 t,$$

where $X(t)$ and $Y(t)$ are jointly Gaussian and have their spectrum around $\omega = 0$. The process $Z(t) \triangleq X(t) + iY(t)$ is what we term a **complex-valued Gaussian process** or **complex Gaussian process**, for short. The real component $X(t)$ is often referred to as the in-phase component, whereas $Y(t)$ is called the quadrature component.

Even in the absence of such a fading effect, an additive noise process $N(t)$, after passing through a front-end amplifier, is often a Gaussian process with finite bandwidth (typically a bandwidth comparable to that of the signal process). Such a bandpass noise can be represented as

$$N(t) = \Re\{(N_x(t) + iN_y(t))e^{i\omega_0 t}\} = N_x(t)\cos\omega_0 t - N_y(t)\sin\omega_0 t,$$

and $N_z(t) = N_x(t) + iN_y(t)$ is a complex Gaussian process.

12.4.1 Complex-valued Gaussian random variables

We define an M-dimensional complex Gaussian variable as a vector $\boldsymbol{Z} = (Z_1, Z_2, \dots, Z_M)^\top$, which is an M-tuple of complex Gaussian variables. We define the $2M$-dimensional Gaussian variable $\boldsymbol{W}$ by (7.97); i.e.,

$$\boldsymbol{W} \triangleq \begin{bmatrix} \boldsymbol{X} \\ \boldsymbol{Y} \end{bmatrix} = (X_1, X_2, \dots, X_M, Y_1, Y_2, \dots, Y_M)^\top. \tag{12.28}$$

Assuming that both $\boldsymbol{X}$ and $\boldsymbol{Y}$ have mean zero, we have $E[\boldsymbol{W}] = \boldsymbol{0}$, and its covariance matrix is given by

$$\boldsymbol{K} \triangleq E\left[\boldsymbol{W}\boldsymbol{W}^\top\right] = \begin{bmatrix} E[\boldsymbol{X}\boldsymbol{X}^\top] & E[\boldsymbol{X}\boldsymbol{Y}^\top] \\ E[\boldsymbol{Y}\boldsymbol{X}^\top] & E[\boldsymbol{Y}\boldsymbol{Y}^\top] \end{bmatrix} = \begin{bmatrix} \boldsymbol{A} & \boldsymbol{B}^\top \\ \boldsymbol{B} & \boldsymbol{D} \end{bmatrix}, \tag{12.29}$$

where $\boldsymbol{A}$ and $\boldsymbol{D}$ are symmetric; i.e., $\boldsymbol{A}^\top = \boldsymbol{A}$ and $\boldsymbol{D}^\top = \boldsymbol{D}$. The joint PDF can be written as

$$f_{XY}(\boldsymbol{x}, \boldsymbol{y}) = \frac{1}{(2\pi)^M |\det \boldsymbol{K}|^{1/2}} \exp\left(-\frac{1}{2}[\boldsymbol{x}^\top \boldsymbol{y}^\top]\boldsymbol{K}^{-1}\begin{bmatrix}\boldsymbol{x}\\ \boldsymbol{y}\end{bmatrix}\right). \tag{12.30}$$

If, in particular, $\boldsymbol{B}^\top = -\boldsymbol{B}$ and $\boldsymbol{D} = \boldsymbol{A}$, then the complex Gaussian variable $\boldsymbol{Z}$ is **circularly symmetric**, and the matrix $\boldsymbol{K}$ of (12.29) reduces to the matrix Σ of (7.98) defined in Section 7.6.2. There, we showed that the complex RVs $\boldsymbol{Z}$ and $\boldsymbol{Z}^*$ have the PDF (7.109):

$$\boxed{f_{ZZ^*}(\boldsymbol{z}, \boldsymbol{z}^*) = \frac{1}{(4\pi)^M |\det \boldsymbol{C}|} \exp\left(-\frac{1}{2}\boldsymbol{z}^\top \boldsymbol{C}^{-1}\boldsymbol{z}^*\right),} \tag{12.31}$$

where

$$\boxed{\boldsymbol{C} = \boldsymbol{A} + i\boldsymbol{B} = \frac{1}{2}E\left[\boldsymbol{Z}\boldsymbol{Z}^{\mathrm{H}}\right],} \tag{12.32}$$

as defined in (7.100). We now state the following theorem due to Wooding [360].

THEOREM 12.2 (Complex-valued multivariate normal variables). *The probability distribution of (12.30) has the equivalent representation of (12.31) if and only if*

$$\begin{aligned} E\left[\boldsymbol{X}\boldsymbol{X}^\top\right] &= E\left[\boldsymbol{Y}\boldsymbol{Y}^\top\right] = \boldsymbol{A} \\ E\left[\boldsymbol{Y}\boldsymbol{X}^\top\right] &= -E\left[\boldsymbol{X}\boldsymbol{Y}^\top\right] = \boldsymbol{B}, \end{aligned} \tag{12.33}$$

or equivalently,

$$\boldsymbol{Q}_{ZZ} \triangleq E\left[\boldsymbol{Z}\boldsymbol{Z}^\top\right] = \boldsymbol{0}. \tag{12.34}$$

Proof. The "if part" is clear on using (7.102), (7.105), and (12.32) with the following relation:

$$d\boldsymbol{x}\, d\boldsymbol{y} = 2^M\, d\boldsymbol{z}\, d\boldsymbol{z}^*, \tag{12.35}$$

which is obtained from (7.108). The proof of the "only if part" is straightforward and therefore is not given here. □

Note that the matrix $\boldsymbol{Q}_{ZZ}$ is different from the covariance matrix of $\boldsymbol{Z}$, which should be defined as $\boldsymbol{R}_{ZZ} = E\left[\boldsymbol{Z}\boldsymbol{Z}^{\mathrm{H}}\right]$. The matrix $\boldsymbol{Q}_{ZZ}$ is sometimes called the **pseudo-covariance matrix**. As we have seen, this matrix plays an important role in determining circular symmetry of a complex Gaussian variable.

12.4.2 Complex-valued Gaussian process

Now we turn our attention to a complex-valued Gaussian process $Z(t) = Z(t) + iY(t)$, in which the real and imaginary parts $(X(t), Y(t))$ are jointly Gaussian processes.

For a given random process, we often find a random variable that is a **linear observable** Z of the random process $Z(t)$ defined by

$$Z = \int_{\mathcal{T}} h(t)Z(t)\,dt, \tag{12.36}$$

where $h(t)$ is a nonrandom function and may be complex-valued. Typical examples of a linear observable Z are:

1. Evaluation of $Z(t)$ at some $t = t_0 \in \mathcal{T}$; i.e.,

   $$Z(t_0) = \int_{\mathcal{T}} \delta(t - t_0)Z(t)\,dt.$$

   Similarly,

   $$Z'(t_0) = \int_{\mathcal{T}} \delta'(t - t_0)Z(t)\,dt.$$

2. The coefficient of an eigenfunction expansion as discussed in Section 13.2.2:

   $$z_k = \int_{\mathcal{T}} v_k^*(t)Z(t)\,dt, \quad k = 1, 2, \ldots,$$

   where $v_k(t)$ is an eigenfunction.
3. The value of a linear filter output at some instant $t = t_0$ having $Z(t)$ as the input.

If the integral of (12.36) exists in the Riemann sense for every sample function $Z(\omega, t)$ of the random process $Z(t)$, it defines a number $z(\omega)$, where ω is a simple event over which a probability measure is defined. Thus, z is a random variable. It has been shown (see Doob [82]) that the integral (12.36) exists for each sample function of $Z(t)$ if

$$\int_{\mathcal{T}} E[|h(t)Z(t)|]\,dt = \int_{\mathcal{T}} |h(t)|E[|X(t)|]\,dt < \infty. \tag{12.37}$$

Even if the integral of (12.36) does not exist for each sample function $Z(\omega, t)$, it will be sufficient for our purposes if the integrals can be defined as **limits in the mean-square sense**. Let the interval $\mathcal{T} = [a, b]$ be partitioned by the set of points

$$a = t_1 < t_2 < \cdots < t_{n+1} = b.$$

Let S_n be the approximating sum

$$S_n = \sum_{i=1}^{n} h(t_i)Z(t_i)(t_{i+1} - t_i). \tag{12.38}$$

Clearly, S_n is a random variable, dependent on the particular partition. Now we consider the limit of the above sum, by letting $n \to \infty$ and $\max\{t_{i+1} - t_i\} \to 0$. We say that the integral (12.36) converges in the mean-square sense (or in quadratic mean) to Z if

$$\lim_{n\to\infty} E\left[|Z - S_n|^2\right] = 0. \tag{12.39}$$

It can be shown (Problem 12.3) that if the ordinary Riemann integral

$$Q = \int_a^b \int_a^b h(t) R_{ZZ}(t,s) h^*(s)\, dt\, ds, \quad \text{where } R_{ZZ}(t,s) = E[Z(t)Z^*(s)], \tag{12.40}$$

exists, then mean-square convergence is insured.

12.4.2.1 Criteria for circular symmetry of a linear observable

We now consider the necessary and sufficient condition which a complex Gaussian process $Z(t)$ must satisfy so that any finite collection of linear observables generated from the process are circularly symmetric and allow the representation of (12.31). Let $\mathbf{Z}$ be an N-variate complex Gaussian variable consisting of linear observables of $Z(t)$ such that $\mathbf{Z} = (Z_1, Z_2, \ldots, Z_N)^\top$, where

$$Z_k = \int_{\mathcal{T}} f_k(t) Z(t)\, dt \tag{12.41}$$

for some function $f_k(t),\ k = 1, 2, \ldots, N$. Then it is not difficult to see that $\mathbf{Z}$ has the representation of (12.31) if and only if

$$E[X(s)X(t)] = E[Y(s)Y(t)], \tag{12.42}$$

$$E[X(s)Y(t)] = -E[Y(s)X(t)], \tag{12.43}$$

or, equivalently,

$$\boxed{Q_{ZZ}(s,t) \triangleq E[Z(s)Z(t)] = 0,} \tag{12.44}$$

for all $s, t \in \mathcal{T}$.

The function $Q_{ZZ}(s,t)$ defined for the complex Gaussian process, which may be aptly called the **pseudo-covariance function**, plays a role similar to the pseudo-covariance matrix $\boldsymbol{Q}_{ZZ}$ defined for a complex Gaussian variable. We have seen $\boldsymbol{Q}_{ZZ} = \mathbf{0}$ and $Q_{ZZ}(s,t) = 0$ are respectively the conditions for circular symmetry of the complex Gaussian variable $\mathbf{Z}$ and the underlying complex Gaussian process $Z(t)$. Grettenberg [130] showed that the condition (12.44) holds if and only if the probability distribution of $Z(t)$ and $Z(t)e^{i\theta}$ are the same; i.e, the distribution is invariant under rotation (Problem 12.4).

The above results do not assume stationarity of the Gaussian processes involved. Now let us focus on stationary Gaussian processes.

12.4.2.2 Wide-sense stationarity and strict-sense stationarity of complex Gaussian processes

We have shown earlier that, for real-valued Gaussian processes, wide-sense stationarity and strict-sense stationarity are equivalent. Such is not the case, however, for complex Gaussian processes.

THEOREM 12.3 (Strict-sense stationarity of a complex Gaussian process). *For a complex Gaussian WSS process $Z(t)$ with $E[Z(t)] = 0$ to be strictly stationary, it is necessary and sufficient that*

$$\boxed{Q_{ZZ}(s,t) = E[Z(s)Z(t)] \text{ is a function of } s - t.} \tag{12.45}$$

Proof. Let $Z(t) = X(t) + iY(t)$. If $Q(s,t)$ is a function of $s-t$, it follows that $E[X(s)X(t)]$, $E[Y(s)Y(t)]$, and $E[X(s)Y(t)]$ are all functions of $s-t$, since

$$
\begin{aligned}
E[X(s)X(t)] &= \frac{1}{4}E[(Z(s) + Z^*(s))(Z(t) + Z^*(t))] \\
&= \frac{1}{4}[Q_{ZZ}(s-t) + Q^*_{ZZ}(s-t) + R_{ZZ}(s-t) + R^*_{ZZ}(s-t)],
\end{aligned}
$$

etc. Thus, the covariance matrix of the $2M$ Gaussian RVs

$$X(t_1), \ldots, X(t_M), Y(t_1), \ldots, Y(t_M)$$

depends only on the time difference $t_i - t_j$, so that the joint distribution of these variables is invariant under a time translation, and the process $Z(t)$ is strictly stationary. The proof of the converse statement is straightforward. □

Example 12.5: Wide-sense stationarity and strict-sense stationarity. Let $X(t)$ be a real-valued stationary Gaussian process with

$$E[X(t)] = 0 \text{ and } E[X(s)X(t)] = R_X(s-t).$$

Then, $Z(t) = X(t)e^{i\omega_0 t}$ is WSS with

$$E[Z(t)] = 0 \text{ and } E[Z(s)Z^*(t)] = R_X(s-t)e^{i\omega_0(s-t)}.$$

However, $Z(t)$ is not strictly stationary, since

$$E[Z(s)Z(t)] = R_X(s-t)e^{i\omega_0(s+t)},$$

which is not a function of $s-t$. □

It should be noted that (12.44) $Q(s,t) = 0$ is a sufficient condition for *strict stationarity*. It will be shown in the next two sections that the **analytic signal** and the **complex envelope** associated with a stationary Gaussian process $X(t)$ are strictly stationary. In fact, these processes satisfy (12.44).

12.4.3 Hilbert transform and analytic signal

Before we define the *Hilbert transform* we introduce the notion of the **Cauchy principal value**.

DEFINITION 12.4 (Principal value integral). *Let $[a,b]$ be a real interval and $g(t)$ be a (real or complex) function defined over $[a,b]$. If $g(t)$ is unbounded near $t = u$, where u is an interior point of $[a,b]$, and the limit*

$$\lim_{\epsilon \to 0^+} \left(\int_a^{u-\epsilon} g(t)\,dt + \int_{u+\epsilon}^{b} g(t)\,dt \right)$$

exists, it is called the **principal value integral** *of $g(t)$ from a to b and is denoted as*

$$P\int_a^b g(t)\,dt.$$

Example 12.6: Integration of a real-valued function $g(t) = 1/t$ from $-a$ to a can be written as

$$\int_{-a}^{a}\frac{dt}{t} = \lim_{\epsilon_1\to 0+}\int_{-a}^{-\epsilon_1}\frac{dt}{t} + \lim_{\epsilon_2\to 0+}\int_{\epsilon_2}^{a}\frac{dt}{t},$$

but these integrals cannot be evaluated separately because of the pole at $t = 0$. But if we let $\epsilon_1 = \epsilon_2 = \epsilon$, then

$$P\int_{-a}^{a}\frac{dt}{t} = \lim_{\epsilon\to 0+}\left(\int_{-a}^{-\epsilon}\frac{dt}{t} + \int_{\epsilon}^{a}\frac{dt}{t}\right).$$

The first integral can be rewritten as follows by setting $t = -u$:

$$\int_{-a}^{-\epsilon}\frac{dt}{t} = \int_{a}^{\epsilon}\frac{-du}{-u} = -\int_{\epsilon}^{a}\frac{du}{u}.$$

Thus, it is clear that the principal value integral exists and

$$P\int_{-a}^{a}\frac{dt}{t} = 0.$$

□

DEFINITION 12.5 (Hilbert transform). *The Hilbert transform $\hat{f}(t)$ of a (real or complex) function $f(t)$ is defined by*

$$\hat{f}(t) = \frac{1}{\pi}P\int_{-\infty}^{\infty}\frac{f(u)}{t-u}\,du, \tag{12.46}$$

if the principal value *of the integral exists.* □

Because of the pole at $t = u$, it is generally not possible to obtain the Hilbert transform as an ordinary improper integral, so we assign the Cauchy principal value to the integral.

Since the Hilbert transform $\hat{f}(t)$, if it exists, can be viewed as the convolution integral of $f(t)$ and

$$h(t) \triangleq \frac{1}{\pi t},$$

its Fourier transform is given by

$$\mathcal{F}\{\hat{f}(t)\} = \mathcal{F}\{f(t)\}\mathcal{F}\{h(t)\}.$$

If we denote the Fourier transform of $h(t)$ by $H(\omega)$, we find

$$\begin{aligned} H(\omega) &= \int_{-\infty}^{\infty} \frac{e^{-i\omega t}}{\pi t}\,dt = \int_{-\infty}^{0} \frac{e^{-i\omega t}}{\pi t}\,dt + \int_{0}^{\infty} \frac{e^{-i\omega t}}{\pi t}\,dt \\ &= \int_{0}^{\infty} \frac{e^{-i\omega t} - e^{i\omega t}}{\pi t}\,dt = -2i \int_{0}^{\infty} \frac{\sin \omega t}{\pi t}\,dt \\ &= -i\,\mathrm{sgn}(\omega) = \begin{cases} -i, & \omega > 0, \\ 0, & \omega = 0, \\ i, & \omega < 0. \end{cases} \end{aligned} \tag{12.47}$$

In deriving the last result, we used the formula

$$\int_{0}^{\infty} \frac{\sin \pi x}{\pi x}\,dx = \int_{0}^{\infty} \mathrm{sinc}(x)\,dx = \frac{1}{2}, \tag{12.48}$$

where the function

$$\mathrm{sinc}(x) \triangleq \frac{\sin \pi x}{\pi x} \tag{12.49}$$

is known as the **sampling function** or the **sinc function** (an abbreviation of "sine cardinal" function), and is the Fourier transform of the ideal **brick-wall** low-pass filter (Problem 12.5).

As seen from (12.47) the system function $H(\omega)$ of **Hilbert-transform filter**, also called the **quadrature filter**, is an all-pass filter with $-90°$ phase shift. Therefore, its response to $\cos \omega t$ equals $\cos(\omega t - \pi/2) = \sin \omega t$ and its response to $\sin \omega t$ is $\sin(\omega t - \pi/2) = -\cos \omega t$. It readily follows from (12.47) that

$$H(\omega)^2 = \begin{cases} -1, & \omega \neq 0, \\ 0, & \omega = 0. \end{cases} \tag{12.50}$$

So if we apply the Hilbert transform to $\hat{f}(t)$, we should obtain $-f(t)$, but its zero-frequency component will be lost.

DEFINITION 12.6 (Analytic signal). *For a given (real or complex) process $X(t)$, the complex-valued process $X_a(t)$, defined by*

$$\boxed{X_a(t) = X(t) + i\hat{X}(t),} \tag{12.51}$$

is called the ***analytic signal*** *(also called the* ***pre-envelope****) associated with $X(t)$, where $X(t)$ and $\hat{X}(t)$ form a Hilbert transform pair; i.e.,*

$$\hat{X}(t) = \frac{1}{\pi} \int_{-\infty}^{\infty} \frac{X(u)}{t-u}\,du = \frac{1}{\pi} \int_{-\infty}^{\infty} \frac{X(t-u)}{u}\,du, \tag{12.52}$$

$$X(t) = -\frac{1}{\pi} \int_{-\infty}^{\infty} \frac{\hat{X}(u)}{t-u}\,du = \frac{1}{\pi} \int_{-\infty}^{\infty} \frac{\hat{X}(t+u)}{u}\,du. \tag{12.53}$$

□

If we denote the Fourier transform of $X(t)$ and $X_a(t)$ by $\tilde{X}(\omega)$ and $\tilde{X}_a(\omega)$ respectively, we have from (12.51) that

$$\tilde{X}_a(\omega) = (1 + iH(\omega))\tilde{X}(\omega) = (1 + \operatorname{sgn}(\omega))\tilde{X}(\omega) = \begin{cases} 2\tilde{X}(\omega), & \omega > 0, \\ \tilde{X}(\omega), & \omega = 0, \\ 0, & \omega < 0. \end{cases} \tag{12.54}$$

Thus, the analytic signal suppresses the negative frequency components of $\tilde{X}(\omega)$. The main idea behind this representation is that the negative frequency components of the Fourier transform of a *real-valued* function $X(t)$ are superfluous and thus can be discarded, with no loss of information, because of the property $\tilde{X}(-\omega) = \tilde{X}(\omega)^*$. The analytic signal representation makes the derivation and analysis of some modulation and demodulation in communication systems (e.g., single-sideband modulation) extremely simple (e.g., see Schwartz *et al.* [295]). Note that the Hilbert transform and the composition of the analytic signal do not require $X(t)$ to be real valued, so we assume throughout this section that $X(t)$ is complex valued, unless stated otherwise.

Example 12.7: Let $X(t) = \cos\omega_0 t$ for some $\omega_0 > 0$. Then $\hat{X}(t) = \sin\omega_0 t$. Thus, $X_a(t) = \cos\omega_0 t + i\sin\omega_0 t = e^{i\omega_0 t}$, often referred to as Euler's formula. $X(t) = \cos\omega_0 t = \frac{1}{2}\left(e^{i\omega_0 t} + e^{-i\omega_0 t}\right)$ contains both positive and negative frequency components, but $X_a(t) = e^{i\omega_0 t}$ has just the positive frequency. □.

Suppose that the real-valued process $X(t)$ is **WSS** with zero mean and covariance function

$$R_{XX}(\tau) = E[X(t+\tau)X(t)]. \tag{12.55}$$

Then it is easily seen (Problem 12.6) that $\hat{X}(t)$ is also WSS with

$$R_{\hat{X}\hat{X}}(\tau) = R_{XX}(\tau), \tag{12.56}$$

$$R_{\hat{X}X}(\tau) = -R_{X\hat{X}}(\tau). \tag{12.57}$$

Furthermore,

$$R_{\hat{X}X}(\tau) = \frac{1}{\pi}\int_{-\infty}^{\infty} \frac{R_{XX}(u)}{\tau - u}\,du = \hat{R}_{XX}(\tau), \tag{12.58}$$

and the analytic signal $X_a(t)$ is WSS with

$$R_{X_aX_a}(\tau) = E[X_a(t+\tau)X_a^*(t)] = 2[R_{XX}(\tau) + i\hat{R}_{XX}(\tau)]. \tag{12.59}$$

As is shown below, Equations (12.52) and (12.53) imply (12.44) with $Z(t) = X_a(t)$, for a WSS process $X(t)$:

$$\boxed{Q_{X_a X_a}(s,t) = E[X_a(t)X_a(s)] = 0.} \tag{12.60}$$

Conversely, if (12.60) holds, then $X(t)$ must be WSS. Noting that a WSS real-valued Gaussian process is strictly stationary, we have the following useful theorem.

THEOREM 12.4 (Criterion for an analytic signal to be circularly symmetric). *Let $X_a(t) = X(t) + i\hat{X}(t)$ be the analytic signal associated with a real-valued Gaussian process $X(t)$ with zero mean, and let $\mathbf{Z}$ be an N-variate complex Gaussian variable consisting of linear observables of $X_a(t)$. Then $\mathbf{Z}$ is circularly symmetric if and only if $X(t)$ is **stationary**.*

Proof. If $X(t)$ is stationary, then $X(t)$ and $\hat{X}(t)$ are jointly stationary. From (12.52), it follows that

$$E[\hat{X}(s)\hat{X}^*(t)] = \frac{1}{\pi}\int_{-\infty}^{\infty} \frac{E[\hat{X}(s)X^*(t-u)]}{u}\,du = \frac{1}{\pi}\int_{-\infty}^{\infty} \frac{R_{\hat{X}X}(s-t+u)}{u}\,du.$$

Similarly, from (12.53),

$$E[X(s)X^*(t)] = \frac{1}{\pi}\int_{-\infty}^{\infty} \frac{E[\hat{X}(s+u)X^*(t)]}{u}\,du = \frac{1}{\pi}\int_{-\infty}^{\infty} \frac{R_{\hat{X}X}(s-t+u)}{u}\,du.$$

Therefore,

$$E[\hat{X}(s)\hat{X}^*(t)] = E[X(s)X^*(t)]. \tag{12.61}$$

Similarly,

$$E[\hat{X}(s)X^*(t)] = -E[X(s)\hat{X}^*(t)]. \tag{12.62}$$

Then, from our earlier result on the criterion for circular symmetry for a complex Gaussian process, we see that $\mathbf{Z}$ is circularly symmetric. The converse statement is proved as follows. From (12.52) and (12.53),

$$\begin{aligned} E[\hat{X}(s)X^*(t)] &= \frac{1}{\pi}\int_{-\infty}^{\infty} \frac{E[X(s)X^*(t)]}{u}\,du \\ -E[X(s)\hat{X}^*(t)] &= -\frac{1}{\pi}\int_{-\infty}^{\infty} \frac{E[X(s)X^*(t-u)]}{u}\,du \\ &= \frac{1}{\pi}\int_{-\infty}^{\infty} \frac{E[X(s)X^*(t+u)]}{u}\,du. \end{aligned}$$

In order to satisfy (12.62), the following relation must hold:

$$E[X(s-u)X^*(u)] = E[X(s)X^*(t+u)], \quad s,t \in \mathcal{T}.$$

This last relation implies that the Gaussian process $X(t)$ is WSS and hence SSS. □

12.4.4 Complex envelope

It is well known that, for *bandpass* (sometimes called *passband*) signals $X(t)$ and $\hat{X}(t)$, the function

$$A(t) = |X_a(t)| = \sqrt{X^2(t) + \hat{X}^2(t)} \tag{12.63}$$

may be identified approximately with the *amplitude envelope* of the signal $X(t)$ and the function

$$\phi(t) = \angle X_a(t) = \tan^{-1}\frac{\hat{X}(t)}{X(t)} \tag{12.64}$$

with the *instantaneous phase*. Let ω_0 be the center frequency or carrier frequency of the narrowband process. Thus, we have the polar coordinate representation of the analytic signal:

$$X_a(t) = A(t)e^{i\phi(t)}. \tag{12.65}$$

If we shift, in the frequency domain, $X_a(t)$ towards 0 Hz by ω_0, we will have

$$E(t) = X_a(t)e^{-i\omega_0 t} = A(t)e^{i\phi(t)-i\omega_0 t}, \tag{12.66}$$

which is called the **complex envelope** of $X(t)$. Thus,

$$X(t) = \Re\{X_a(t)\} = \Re\left\{E(t)e^{i\omega_0 t}\right\}. \tag{12.67}$$

It is easy to see that

$$|E(t)| = |X_a(t)|, \tag{12.68}$$

$$\angle E(t) = \angle X_a(t). \tag{12.69}$$

Furthermore, it is seen from (12.66) that the following relations hold for the covariance functions and pseudo-covariance functions between $E(t)$ and $X_a(t)$:

$$R_{EE}(s,t) = R_{X_aX_a}(s,t)e^{-i\omega_0(s-t)}, \tag{12.70}$$

$$Q_{EE}(s,t) = Q_{X_aX_a}(s,t)e^{-i\omega_0(s+t)}. \tag{12.71}$$

From (12.70), the complex envelope $E(t)$ is WSS if and only if $X_a(t)$ is WSS. From (12.71) we see that $Q_{EE}(s,t) = 0$ if and only if $Q_{X_AX_a}(s,t) = 0$. We also see that $X_a(t)$ and $E(t)$ are strictly stationary, when $X(t)$ is a stationary Gaussian process.

Example 12.8: In many communication systems, signals are modulated in both amplitude and phase.

$$S(t) = A_S(t)\cos[\omega_c t + \phi_S(t)].$$

Then by defining the in-phase (I) and quadrature (Q) components

$$S_{\rm I}(t) = A_S(t)\cos\phi_S(t),$$
$$S_{\rm Q}(t) = A_S(t)\sin\phi_S(t),$$

we have an equivalent representation

$$S(t) = S_{\rm I}(t)\cos\omega_{\rm c}t - S_{\rm Q}(t)\sin\omega_c t. \qquad (12.72)$$

Thus, we can represent the bandpass signal $S(t)$ as

$$\begin{aligned} S(t) &= \Re\left\{\left(A_S(t)e^{i\phi_S(t)}\right)e^{i\omega_{\rm c}t}\right\} \\ &= \Re\left\{[S_{\rm I}(t) + iS_{\rm Q}(t)]e^{i\omega_{\rm c}t}\right\} \\ &= \Re\left\{E_S(t)e^{i\omega_{\rm c}t}\right\}. \end{aligned}$$

Thus, $E_S(t) = S_{\rm I}(t) + S_{\rm Q}(t)$ is the complex envelope of the bandpass signal $S(t)$, or the **baseband equivalent signal** of $S(t)$. □

12.5 Summary of Chapter 12

Random process $X(t)$:	$X(\omega, t), \quad \omega \in \Omega$	Fig. 12.1
Discrete- or continuous-time:	$\mathcal{T} = \{0, 1, 2, \ldots\}$ or $\mathcal{T} = (-\infty, \infty)$	Sec. 12.2.1
Discrete- or continuous-state:	$\mathcal{S} = \{0, 1, \ldots, M-1\}$ or $\mathcal{S} = \{-\infty, \infty\}$	Sec. 12.2.2
Stationary process:	$F_X(x; t)$ is independent of t	Sec. 12.2.3
Independent process:	$F_X(\boldsymbol{x}; \boldsymbol{t}) = \prod_{i=1}^n F_{X_i}(x_i; t_i)$	(12.2)
Markov chain of order h:	$p(x_k \mid x_{k-i};\ i \geq 1) = p(x_k \mid x_{k-1}, \ldots, x_{k-h})$	(12.3)
DTMC:	Markov chain of order 1	Fig. 12.2
CTMC:	$\tau_k = t_{k+1} - t_k$, exponentially distributed	(12.5)
Semi-Markov process:	$\tau_k = t_{k+1} - t_k$, not exponentially distributed	(12.5)
Strictly stationary:	$F_{X_1 \cdots X_n}(x_1, \ldots, x_n) = F_{X'_1 \cdots X'_n}(x_1, \ldots, x_n)$	(12.8)
Autocorrelation func.:	$R_X(t_1, t_2) = E[X(t_1)X(t_2)]$	(12.10)
WSS:	$R_X(t_1, t_2) = R_X(t_1 - t_2)$	Def. 12.2
Gaussian process:	$(X(t_1) \ldots X(t_n))$ are Gaussian distributed	Def. 12.3
Stationary Gaussian process:	$E[X(t)] = \mu$ and $E[X(t)X(s)] = R_X(t-s)4$	Thm. 12.1
Circularly symmetric $\boldsymbol{Z}$:	$\boldsymbol{Q}_{\boldsymbol{ZZ}} = E[\boldsymbol{ZZ}^\top] = 0$	(12.34)
Circularly symmetric $Z(t)$:	$Q_{ZZ}(s, t) = E[Z(s)Z(t)] = 0$	(12.44)

Strictly stationary $Z(t)$:	$Q_{ZZ}(s,t)=Q_{ZZ}(s-t)$, $R_{ZZ}(s,t)=R_{ZZ}(s-t)$	Thm. 12.3
Hilbert transform:	$\hat{f}(t)=\frac{1}{\pi}P\int_{-\infty}^{\infty}\frac{f(u)}{t-u}\,du$	(12.46)
Hilbert transform filter:	$H(\omega)=-i\,\mathrm{sgn}(\omega)$	(12.47)
Analytic signal:	$X_a(t)=X(t)+i\hat{X}(t)$	(12.51)
Circularly symmetric $X_a(t)$:	Stationary $X(t)$	Thm. 12.4
Complex envelope:	$E(t)=X_a(t)e^{-i\omega_0 t}=A(t)e^{i\phi(t)-i\omega_0 t}$	(12.66)

12.6 Discussion and further reading

In this chapter we first gave an overview of a variety of random processes, some of which will be further studied in subsequent chapters, especially Markov chains and Markov processes, point processes, and diffusion processes.

We introduced the notions of (strict) stationarity and wide-sense stationarity and ergodicity of random processes. In the last section we focused on complex-valued Gaussian processes and discussed various criteria for linear observables to be circularly symmetric complex Gaussian variables. The presentation of these materials draws from Kobayashi [192]. The Hilbert transform is used in many physical science and engineering disciplines. The recent two volumes by King [181, 182] are entirely devoted to this subject and its applications. Among the earliest works that discuss analytic signals are Arens [5], Dugundji [84], and Zakai [368]. Analytic signals are discussed in many textbooks on communication systems, including e.g., Helstrom [150] and Schwartz *et al.* [295]

12.7 Problems

Section 12.1: Random process

12.1 Sinusoidal functions with different frequencies and random amplitudes [175]. Consider the following extension of Example 12.4. Let $A_0, A_1, \ldots, A_m$ and $B_0, B_1, \ldots, B_m$ be uncorrelated RVs having zero mean. Assume that A_i and B_i have a common variance σ_i^2 and let $\sigma^2 = \sum_{k=0}^{m}\sigma_i^2$. Let $\omega_0, \omega_1, \ldots, \omega_m$ be independent (angular) frequencies in $[0, \pi]$ and set

$$X(t) = \sum_{i=0}^{m}(A_i \cos\omega_i + B_i \sin\omega_i t)\,.$$

(a) Show that $X(t)$ is WSS and its covariance function is given by

$$R_X(\tau) = \sigma^2 \sum_{i=0}^{m} f_i \cos\omega_i\tau, \tag{12.73}$$

where $f_i = \sigma_i^2/\sigma^2$; thus, $\sum_{i=0}^{m} f_i = 1$.

(b) In part (a) ω was a discrete variable. Now let ω be a continuous RV having possible values in $[0, \pi]$ with the CDF $F(\omega)$. How do you suggest (without a rigorous mathematical argument) to extend the expression (12.73) to this continuous case?

(c) Consider the special case where $F(\omega)$ is a uniform distribution in $[0, \pi]$. Find the expression for the covariance function $R_X(\tau)$.

12.2 Moving average process [175]. Let $\{X_n; -\infty < n < \infty\}$ be uncorrelated RVs having a common mean μ and variance σ^2. Let $a_0, a_1, \ldots, a_{m-1}$ be arbitrary real numbers. Consider a *moving average process* $\{Y_n\}$ defined by

$$Y_n = a_0 X_n + a_1 X_{n-1} + \cdots + a_{m-1} X_{n-m+1} = \sum_{i=0}^{m-1} a_i X_{n-i}. \tag{12.74}$$

(a) Find $E[Y_n]$ and $\text{Var}[Y_n]$.

(b) Find the auto-covariance function of $\{Y_n\}$ and show that $\{Y_n\}$ is a WSS sequence.

(c) Let $a_i = 1/\sqrt{m}$ for $i = 0, 1, \ldots, m-1$. Obtain the autocovariance function $R_Y(k)$ of $\{Y_n\}$ and sketch the curve.

(d) Consider the two extreme cases $m = 1$ and $m = \infty$ and identify similar sequences among the examples given in the text.

Section 12.4: Complex-valued Gaussian process

12.3 Condition for integration in mean-square. Show that if the integral (12.40) exists, mean-square convergence (12.39) is insured.

12.4 Circular symmetry criterion for a complex Gaussian process. Show that the distribution of the complex Gaussian process $Z(t)e^{i\theta}$ is invariant for all rotations $0 \leq \theta \leq 2\pi$ if and only if the pseudo-covariance function $Q_{ZZ}(t, s) = E[Z(t)Z(s)] = 0$.

12.5 Brick-wall filter and the sampling function. Consider the system function $H(f)$ of an ideal brick-wall low-pass filter defined by

$$H(f) = \begin{cases} 1, & |f| \leq W, \\ 0, & |f| > W. \end{cases} \tag{12.75}$$

Find the impulse response function $h(t)$. Also derive (12.48).

12.6 Properties of the Hilbert transform. Prove the following properties of the Hilbert transform.

(a) **Hermitian symmetry.** Let the Fourier transform of a real-valued function $f(t)$ and its Hilbert transform $\hat{f}(t)$ be denoted as $F(\omega)$ and $\hat{F}(\omega)$. Show that both $F(\omega)$ and $\hat{F}(\omega)$ exhibit Hermitian symmetry; i.e., $F(-f) = F^*(f)$ and $\hat{F}(-\omega) = \hat{F}^*(\omega)$.

(b) **Orthogonality of $\hat{g}(t)$ and $g(t)$.** Show that $g(t)$ (not necessarily real valued) and its Hilbert transform are orthogonal; i.e.,

$$\langle g(t), \hat{g}(t) \rangle \triangleq \int_{-\infty}^{\infty} g(t)\hat{g}^*(t)\,dt = 0.$$

(c) **Hilbert transform of a symmetric function.** If $g(t)$ is symmetric around $t = 0$, i.e., $g(-t) = g(t)$, then $\hat{g}(t)$ is skew-symmetric, i.e., $\hat{g}(-t) = -\hat{g}(t)$.

(d) **Hilbert transform of a WSS process.** Prove that $\hat{X}(t)$ is WSS when $X(t)$ is WSS, and derive (12.56) through (12.58).

13 Spectral representation of random processes and time series

In this chapter we discuss spectral representations and eigenvector-based time-series analysis. We begin our discussion with a review of the Fourier series and Fourier transform of nonrandom functions, followed by the Fourier analysis of periodic WSS processes. Then we introduce the power spectrums of non-periodic WSS random processes, the Wiener–Khinchin formula, and the peoriodogram analysis of time-series data. The eigenvector-based orthogonal expansion of random vectors and its continuous-time analog, known as the Karhuenen–Loéve expansion, are discussed in detail. Principal component analysis (PCA) and singular-value decomposition (SVD) are two commonly used statistical techniques applicable to any data presentable in matrix form, where correlation exists across its rows and/or columns. We also briefly discuss algorithms being developed for Web information retrieval, and they can be viewed as instances of general *spectral expansion*, the common theme of the present chapter.

The chapter ends with discussion of an important class of time series known as autoregressive moving average (ARMA), which is widely used in statistics and econometrics. Its spectral representation and state space formulation are also discussed.

13.1 Spectral representation of random processes and time series

In this section we consider the problem of representing a random process in terms of a series or integral with respect to some system of deterministic functions, such that the coefficients in this expansion are uncorrelated RVs. Such a representation is referred to as **spectral representation** or **spectral expansion**. Before we pursue this subject, let us briefly review the Fourier series expansion.[1]

13.1.1 Fourier series

Let $g(t)$ be a real or complex-valued **periodic** function of a *real* variable t, which is typically time in our applications.

[1] Jean Baptiste Joseph Fourier (1768–1830) was a French mathematician and physicist who is best known for his work on Fourier series and their application to problems of heat flow.

If $g(t)$ is **absolutely integrable** over a period T,[2] i.e.,

$$\int_0^T |g(t)|\,dt < \infty, \tag{13.1}$$

then $g(t)$ has associated with it the following Fourier series expansion

$$\tilde{g}_N(t) = \sum_{k=-N}^{N} g_k e^{i2\pi n f_0 t}, \quad \text{where } i = \sqrt{-1},\ f_0 \triangleq \frac{1}{T}, \tag{13.2}$$

and g_k is the kth Fourier coefficient

$$g_k = \frac{1}{T}\int_0^T g(t)e^{-i2\pi k f_0 t}\,dt. \tag{13.3}$$

If $g(t)$ is of *bounded variation*[3] in the interval $0 \le t \le T$, then

$$\boxed{\lim_{N\to\infty} \tilde{g}_N(t) = g(t)} \tag{13.4}$$

at all t where $g(t)$ is continuous.

If $g(t)$ is **square integrable** over the period T,[4] i.e.,

$$\int_0^T |g(t)|^2\,dt < \infty, \tag{13.5}$$

then the sum (13.2) converges to $g(t)$ in the mean-square sense:

$$\boxed{\lim_{N\to\infty} \int_0^T |\tilde{g}_N(t) - g(t)|^2\,dt = 0,} \tag{13.6}$$

which we write

$$g(t) = \underset{N\to\infty}{\text{l.i.m.}} \sum_{n=-N}^{N} g_n e^{-i2\pi n f_0 t}. \tag{13.7}$$

One of the most important properties of the Fourier series expansion is the identity

$$\boxed{\sum_{n=-\infty}^{\infty} |g_n|^2 = \frac{1}{T}\int_0^T |g(t)|^2\,dt,} \tag{13.8}$$

which is known as **Parseval's formula** or **Parseval's identity**.[5] Note that the condition (13.5) is equivalent to function $g(t)$ having finite power, and the left and right sides of (13.8) represent the power defined in the spectral and time domains respectively.

[2] In functional analysis, we say such $g(t)$ belongs to the $L_1(T)$ space.

[3] A function of *bounded variation* is a real-valued function whose total variation is bounded; i.e., finite. See the online Supplementary Materials for details.

[4] In functional analysis we say that $g(t) \in L_2(T)$.

[5] Marc-Antoine Parseval des Chênes (1755–1836) was a French mathematician.

13.1.2 Fourier transform

Now let us assume $g(t)$ is a real- or complex-valued **nonperiodic** function $g(t)$. Then it is known from the theory of Fourier transforms (e.g., Titchmarsh [324]) that:

1. If $g(t)$ is **square integrable** on the whole line, i.e.,

$$\int_{-\infty}^{\infty} |g(t)|^2\, dt < \infty, \tag{13.9}$$

then its Fourier transform $G(f)$ exists, which is also square integrable on the whole line:

$$\boxed{G(f) = \underset{A\to\infty}{\text{l.i.m.}} \int_{-A}^{A} g(t)e^{-i2\pi ft}\, dt} \tag{13.10}$$

and

$$\boxed{g(t) = \underset{B\to\infty}{\text{l.i.m.}} \int_{-B}^{B} G(f)e^{i2\pi ft}\, df,} \tag{13.11}$$

where the notation l.i.m. means "limit in the mean," and (13.10) should be interpreted as

$$\lim_{A\to\infty} \int_{-\infty}^{\infty} \left| G(f) - \int_{-A}^{A} g(t)e^{-i2\pi ft}\, dt \right|^2 df = 0. \tag{13.12}$$

Using the notation for the *Cauchy principal value* of an integral (see Definition 12.4), the above equations may be expressed as

$$G(f) \overset{\text{m.s.}}{=} P \int_{-\infty}^{\infty} g(t)e^{-i2\pi ft}\, dt \tag{13.13}$$

and

$$g(t) \overset{\text{m.s.}}{=} P \int_{-\infty}^{\infty} G(f)e^{i2\pi ft}\, df. \tag{13.14}$$

2. If $g(t)$ is **absolutely integrable**, i.e., $\int_{-\infty}^{\infty} |g(t)|\, dt < \infty$, then (13.13) should be replaced by

$$G(f) = \int_{-\infty}^{\infty} g(t)e^{-i2\pi ft}\, dt, \tag{13.15}$$

where the integral is the ordinary **Riemann**[6] integral. Similarly, if the transform $G(f)$ is absolutely integrable, then (13.11) becomes

$$g(t) = \int_{-\infty}^{\infty} G(f)e^{i2\pi ft}\, df. \tag{13.16}$$

[6] Georg Friedrich Bernhard Riemann (1826–1866) was a German mathematician.

The above results are due to **Plancherel's theorem**,[7] which states that if a function $g(t); t \in \mathbb{R} \triangleq (-\infty, \infty)$ is in both $L_1(\mathbb{R})$ and $L_2(\mathbb{R})$, then its Fourier transform is also in $L_2(\mathbb{R})$; moreover, the Fourier transform mapping is **isometric** (i.e., distance preserving). Under the condition (13.9) we can show (Problem 13.1)

$$\boxed{\int_{-\infty}^{\infty} |G(f)|^2\, df = \int_{-\infty}^{\infty} |g(t)|^2\, dt,} \tag{13.17}$$

which is another form of Parseval's identity. The condition (13.9) means that $g(t)$ is a function with finite energy, and the left and right sides of (13.17) represent its energy in the frequency and time domains respectively.

13.1.3 Analysis of periodic wide-sense stationary random process

By generalizing the definition we gave for a real-valued WSS process, we say that a complex-valued process $X(t)$ is WSS if it has a constant mean and its autocorrelation function $E[X(s)X^*(t)]$ is a function of $s - t$ only; i.e., if we can write it as

$$E[X(s)X^*(t)] = R_X(s-t). \tag{13.18}$$

Note that unlike in the real-valued process case, the argument is not $|s-t|$, since $R_X(\cdot)$ is not a symmetric function. Rather, it is self-adjoint: take the complex conjugate of (13.18); then

$$R_X^*(s-t) = [X(t)X^*(s)] = R_X(t-s); \text{ i.e., } R_X(-\tau) = R_X^*(\tau).$$

Now suppose that $R_X(\tau)$ is a **periodic function** with some period T:

$$R_X(\tau + T) = R_X(\tau), \text{ for all } -\infty < \tau < \infty. \tag{13.19}$$

Then, the WSS random process is said to be periodic with period T. Then we can show (Problem 13.2) that for any t

$$E\left[|X(t) - X(t+T)|^2\right] = 0; \tag{13.20}$$

hence, the RVs $X(t)$ and $X(t+T)$ are said to be **equal in mean square** or *equivalent in mean square*. As we shall show at the end of this section, equivalence in mean square implies equivalence with probability one (i.e., equivalence almost surely) and vice versa. Thus, (13.20) implies that RVs $X(t)$ and $X(t+T)$ are equivalent with probability one. If all the sample functions $X(\omega, t); \omega \in \Omega$ (possibly except for a set which occurs with probability zero) are periodic, the process $X(t)$ is periodic in the sense defined above. Then if a sample function $x(t) = X(\omega, t)$ for some ω is periodic and can be expanded in Fourier series, we will have

$$x(t) = \sum_{n=-\infty}^{\infty} x_n e^{i2\pi f_0 t},\ f_0 = \frac{1}{T}, \tag{13.21}$$

[7] Michel Plancherel (1885–1967) was a Swiss mathematician.

where

$$x_n \triangleq \frac{1}{T}\int_0^T x(t)e^{-i2\pi n f_0 t}\,dt. \tag{13.22}$$

For different sample functions, (13.22) yields different values for x_n. It can be shown[8] that under the condition

$$\int_0^T E[|X(t)|]\,dt < \infty, \tag{13.23}$$

all the sample functions, except a set of probability zero, are absolutely integrable and, in addition,

$$E\left[\int_0^T X(t)\,dt\right] = \int_0^T E[X(t)]\,dt. \tag{13.24}$$

The condition (13.23), which is referred to as the *measurability condition*, can be assumed to hold in practice. Thus, we are free to consider integrals of the sample functions of a random process $X(t)$ whenever the mean value of the process $|X(t)|$ is integrable. Further, we can calculate means involving these integrals by using (13.24). Thus, the integral

$$X_n \triangleq \frac{1}{T}\int_0^T X(t)e^{-i2\pi n f_0 t}\,dt \tag{13.25}$$

exists with probability one and it can be shown below that an expansion similar to (13.21) exists, namely

$$\boxed{X(t) = \underset{N\to\infty}{\text{l.i.m.}} \sum_{n=-N}^{N} X_n e^{i2\pi f_0 t},\ 0 \le t \le T,} \tag{13.26}$$

where l.i.m. stands for "limit in the mean" as defined in (11.31) for mean square convergence.

The Fourier series expansion of a periodic WSS process, as claimed in (13.26) with (13.25), can be derived using the following **double orthogonality** (i.e., (13.27) and (13.28)). Namely:

1. The set of trigonometric functions $\{e^{i2\pi n f_0 t};\ n = 0, \pm1, \pm2, \ldots\}$, where $f_0 = 1/T$, forms an **orthogonal** basis for $L_2(T)$, the space of **square integrable** functions over the interval $[0, T]$:

$$\int_0^T e^{-i2\pi m f_0 t} e^{i2\pi n f_0 t}\,dt = \delta_{m,n} T. \tag{13.27}$$

2. The Fourier expansion coefficients $\{X_n\}$ defined by (13.25) are RVs, which are **orthogonal** to each other in the sense that any pair of RVs X_m and $X_n (m \neq n)$ are

[8] See Doob [82], Chapter 2, Theorem 2.7.

uncorrelated. Recall that we defined the orthogonality of RVs in Section 10.1.1 (cf. Definition 10.3):

$$\boxed{\langle X_m, X_n \rangle \triangleq E[X_m X_n^*] = r_n \delta_{m,n},} \tag{13.28}$$

where $\langle X, Y \rangle \triangleq E[XY^*]$ is the *inner product* of the RVs X and Y, as defined in Definition 10.3, where Y^* is the complex conjugate of Y.[9] The reader is suggested to prove the second orthogonality (Problem 13.4). Then, we can show that the Fourier series expansion converges to the process $X(t)$ in the mean square sense, as claimed in (13.26), or equivalently

$$\lim_{N\to\infty} E\left[\left|\sum_{n=-N}^{N} X_n e^{i2\pi n f_0 t} - X(t)\right|^2\right] = 0. \tag{13.29}$$

When the process is nonstationary we cannot use the trigonometric functions as a basis, and even when the process is WSS the second orthogonality (13.28) does not hold for finite T unless $X(t)$ is periodic in the sense of (13.19). Therefore, we need to investigate a way to generalize the Fourier series expansion method. Before we investigate the general case in Section 13.2, we will study the power spectrum in Sections 13.1.4 and 13.1.5, which can be defined when the process is WSS and the observation period is "infinite."

Remark on the equivalence of random variables

The concept of equivalence of two RVs X and Y almost surely, or with probability one, was first introduced in Section 3.1: $X \overset{\text{a.s.}}{=} Y$ if $P[X = Y] = 1$. Earlier in this section, the notion of equivalence in mean square arose in (13.20): two RVs X and Y are defined to be equivalent in mean square, $X \overset{\text{m.s.}}{=} Y$, if $E[|X - Y|^2] = 0$. It is not difficult to show that these two notions of equivalence of RVs are actually the same; i.e., $X \overset{\text{a.s.}}{=} Y$ if and only if $X \overset{\text{m.s.}}{=} Y$ (Problem 13.3). Thus, we may write simply $X = Y$ in most contexts without ambiguity. We remark that equivalence in mean square and equivalence almost surely are analogous to certain notions of convergence discussed in Chapter 11. However, the corresponding notions of convergence are *not* the same. In particular, we saw in Chapter 11 that $X_n \overset{\text{a.s.}}{\to} Y$ does not imply $X_n \overset{\text{m.s.}}{\to} Y$ and $X_n \overset{\text{m.s.}}{\to} Y$ does not imply $X_n \overset{\text{a.s.}}{\to} Y$.

13.1.4 Power spectrum

With these preparations, we are in a position to discuss spectral analysis of WSS random processes, when the observation period is $(-\infty, \infty)$.

[9] In Section 10.1.1 we used the notation $\overline{Y}$ instead.

DEFINITION 13.1 (Power spectral density). *The* **power spectral density** *(or simply* **power spectrum***) of a WSS random process $X(t)$ is defined by*

$$\boxed{S_X(f) \triangleq \mathcal{F}\{R_X(\tau)\} = \int_{-\infty}^{\infty} R_X(\tau)e^{-i2\pi f\tau}\,d\tau,} \tag{13.30}$$

where $R_X(\tau)$ is the autocorrelation function *of $X(t)$:*

$$R_X(\tau) = E[X(t)X(t-\tau)]. \tag{13.31}$$

□

Equation (13.30) implies that the inverse transform gives

$$\boxed{R_X(\tau) = \mathcal{F}^{-1}\left[S_X(f)\right] = \int_{-\infty}^{\infty} S_X(f)e^{i2\pi f\tau}\,df.} \tag{13.32}$$

This important property that the autocorrelation function and power spectrum form a Fourier transform pair is often referred to as the **Wiener–Khinchin**[10] **formula** or *Wiener–Khinchin theorem.* We often use the the *correlation function* (or, more aptly, the *normalized autocovariance function*) $\rho(\tau)$:

$$\rho_X(\tau) \triangleq \frac{\mathrm{Cov}[X(t)X(t-\tau)]}{(\mathrm{Var}[X(t)]\mathrm{Var}[X(t-\tau)])^{1/2}} = \frac{R_X(\tau) - \mu_X^2}{\sigma_X^2}, \tag{13.33}$$

which is related to the the autocorrelation function according to

$$R_X(\tau) = \sigma_X^2\rho_X(\tau) + \mu_X^2. \tag{13.34}$$

Some authors (e.g., Grimmett and Stirzaker [131]) define the **spectrum distribution function** $\tilde{F}_X(\omega)$ in terms of the correlation function:

$$\rho_X(\tau) = \int_{-\infty}^{\infty} e^{i\omega\tau}\,d\tilde{F}_X(\omega), \tag{13.35}$$

where $\omega = 2\pi f$ is called, in electrical engineers' terms, the **angular frequency**, as opposed to the *frequency* f. In terms of the **spectrum density function** defined by $\tilde{f}_X(\omega) \triangleq dF(\omega)/d\omega$, the Wiener–Khinchin formula is expressed as

$$\rho_X(\tau) = \mathcal{F}^{-1}\{\tilde{f}_X(\omega)\} = \int_{-\infty}^{\infty} \tilde{f}_X(\omega)e^{i\omega\tau}\,d\omega \tag{13.36}$$

and

$$\tilde{f}_X(\omega) = \mathcal{F}\{\rho_X(\tau)\} = \frac{1}{2\pi}\int_{-\infty}^{\infty} \rho_X(\tau)e^{-i\omega\tau}\,d\tau. \tag{13.37}$$

[10] Aleksandr Yakovlevich Khinchin (1894–1959) was a Soviet mathematician. His name is often spelled as Alexandre Khintchine (French spelling).

The density function $\tilde{f}_X(\omega)$ is similar to, but not quite the same as, the power spectral density function $S_X(f)$ defined above. Since

$$\rho_X(\tau) \triangleq \frac{\mathrm{Cov}[X(t)X(t-\tau)]}{(\mathrm{Var}[X(t)]\mathrm{Var}[X(t-\tau)])^{1/2}} = \frac{R_X(\tau) - \mu_X^2}{\sigma_X^2}, \tag{13.38}$$

the autocorrelation function can be written as

$$R_X(\tau) = \sigma_X^2 \rho_X(\tau) + \mu_X^2, \tag{13.39}$$

which leads to the following relation between $S_X(f)$ and $\tilde{f}_X(\omega)$:

$$S_X(f) = 2\pi\sigma_X^2 \tilde{f}_X(2\pi f) + \mu_X^2 \delta(f). \tag{13.40}$$

Before we close this section we should note that the power spectrum method applies only when a WSS process is observed over $(-\infty, \infty)$. The assumption of infinite observation period may be dropped in practice, if the correlation function $\rho(\tau)$ decays to practically zero beyond $|\tau| > T$ for some finite T. However, if the autocorrelation function decays according to a power law, i.e.,

$$\boxed{\rho(\tau) \propto \frac{1}{|\tau|^\beta}, \text{ as} |\tau| \to \infty,} \tag{13.41}$$

with $0 < \beta < 1$, then the autocorrelation function is not integrable; i.e.,

$$\int_0^\infty \rho(\tau)\, d\tau = \infty. \tag{13.42}$$

A process with the property (13.42) is called a long-range dependent (LRD) process. It can be shown that, for the LRD process with (13.42), the power spectrum takes the form

$$\boxed{S(f) \propto |f|^\gamma \text{ for} f \approx 0,} \tag{13.43}$$

where $\gamma = 1 - \beta$. Therefore, the spectral density of an LRD process has a singularity at $f = 0$; i.e., it diverges at zero frequency. For a discussion of LRD processes, see, for example, [203] and references therein.

13.1.5 Power spectrum and periodogram of time series

In practice, we often deal with cases in which the time index is discrete. This situation occurs when either (i) the random process itself is a discrete-time process by its own nature or (ii) the random process is a continuous-time process but is observed at only discrete points in time. Thus, instead of the process $\{X(t); -\infty < t < \infty\}$, we must deal with a **discrete-time random process** $\{X_n; -\infty < n < \infty\}$. The theory of discrete-time processes is well developed in statistics and econometrics. Various statistical analysis techniques such as correlation analysis, spectral analysis, smoothing and prediction are practiced in these applications. Such analysis is generally called "time-series analysis." A **time series** is a sequence of data points, measured typically at successive times, spaced at (often uniform) time intervals. Thus, we use the term time series synonymously with discrete-time random process. When we discuss data, i.e., an instance of the time series $\{X_n\}$, we use the lower case symbols $\{x_n\}$, and sometimes a vector notation $\boldsymbol{x} = (x_1, x_2, \ldots, x_T)$.

Similar to Definition 12.2, we say that a time series $\{X_n\}$ is WSS if

$$E[X_n] = \mu_X \text{ for all } n \tag{13.44}$$

and

$$E[X_n X_{n+k}] = R_X[k] \text{ for all } n \text{ and } k. \tag{13.45}$$

The function $R_X[k]$[11] is called the **autocorrelation function** of $\{X_n\}$. Similarly, the **autocovariance function** of the WSS time series $\{X_n\}$ is defined as

$$C_X[k] \triangleq E[(X_n - \mu_X)(X_{n+k} - \mu_X)] = R_X[k] - \mu_X^2 \text{ for all } n \text{ and } k. \tag{13.46}$$

For a given WSS $\{X_n\}$ we define the **serial correlation coefficient** of *lag* k (or *order* k) $\{\rho_X[k]\}$ by

$$\rho_X[k] \triangleq \frac{C_X[k]}{R_X[0]} = \frac{R_X[k] - \mu_X^2}{\sigma_X^2}, \; -\infty < k < \infty. \tag{13.47}$$

It can be shown (see (10.13)) that

$$-1 \le \rho_X[k] \le 1 \text{ and } \rho_X[0] = 1. \tag{13.48}$$

The serial correlation coefficient of a time series corresponds to the correlation function $\rho(\tau)$ defined in (13.33) for the continuous-time process.

13.1.5.1 Power spectrum and the Wiener–Khinchin formula

We define the **Fourier transform** of the autocovariance function $C_X[k]$[12] by

$$\begin{aligned} P_X(\omega) &\triangleq \frac{1}{2\pi} \sum_{k=-\infty}^{\infty} C_X[k] e^{-ik\omega} \\ &= \frac{1}{2\pi} \left[C_X[0] + 2\sum_{k=1}^{\infty} C_X[k] \cos(k\omega) \right], \; -\pi \le \omega \le \pi. \end{aligned} \tag{13.49}$$

[11] We write $R_X[k]$, instead of $R_X(t)$, to emphasize that the time index k is discrete. Similarly, we write the covariance function as $C_X[k]$.

[12] For a continuous-time process, we normally define the power spectrum $S_X(f)$, as given by (13.30), as the Fourier transform of the autocorrelation function $R_X(\tau)$ instead of the autocovariance function $C_X(\tau)$. This definition cannot be carried over to the discrete-time case when the mean μ_X is nonzero, because the summation does not converge. For the continuous-time case, however, the nonzero mean results in a term $(\mu_X^2/2\pi)\delta(\omega)$ in the power spectrum.

The inverse relationship is given by

$$\boxed{C_X[k] = \int_{-\pi}^{\pi} P_X(\omega)e^{ik\omega}\,d\omega = \int_{-\pi}^{\pi} P_X(\omega)\cos k\omega\,d\omega.} \tag{13.50}$$

The transform pair (13.49) and (13.50) is a discrete-time version of the **Wiener–Khinchin** formula defined by (13.30) and (13.32).

Let us set $k = 0$ in the last equation. By noting $C_X[0] = E[(X_k - \mu_X)^2] = \sigma_X^2$, we find

$$\sigma_X^2 = \int_{-\pi}^{\pi} P_X(\omega)\,d\omega. \tag{13.51}$$

Thus, the variance σ_X^2 is made up of infinitesimal contributions $P_X(\omega)\,d\omega$ in small bands around each *frequency* ω. Therefore, the function $P_X(\omega)$ is aptly called the **(power) spectrum**. Its normalized version

$$\boxed{\frac{P_X(\omega)}{\sigma_X^2} = \frac{1}{2\pi}\sum_{k=-\infty}^{\infty} \rho_X[k]e^{-ik\omega}} \tag{13.52}$$

is called the **(power) spectral density function**.

13.1.5.2 Periodogram of a time series

For the observed sequence of finite length denoted as $\{x_n; 0 \le n \le N-1\}$,[13] let us consider the Fourier transform:

$$\tilde{X}^{(N)}(\omega) \triangleq \frac{1}{\sqrt{N}}\sum_{n=0}^{N-1}(x_n - \overline{x}^{(N)})e^{in\omega}, -\pi < \omega < \pi, \tag{13.53}$$

where

$$\overline{x}^{(N)} = \frac{1}{N}\sum_{n=0}^{N-1} x_n. \tag{13.54}$$

We restrict our attention to values of ω of the form $\omega = 2\pi m/N$, $m = 0, 2, \ldots, N-1$. Although other values of ω may be considered, no additional information is obtained, since $C_X(\omega)$ can be interpolated with weighting functions of the form

$$\frac{\sin[\omega - (2\pi m/N)]}{\omega - (2\pi m/N)}.$$

This property is known as the **sampling theorem** in communication theory.[14] The theorem states that a continuous-time *bandlimited* signal $x(t)$; $-\infty < t < \infty$, can be

[13] Here the time label n starts with zero, instead of one, so that we can use the convention of the discrete Fourier transform notation.

[14] This theorem is attributed to Kotelnikov [212], Shannon [300, 301], Someya [308], and others, but its mathematical foundation on interpolation functions by Whittaker [348] and Ogura [258] predates the discovery by communication theorists. Although Nyquist is often mentioned in the literature as one of the

reconstructed from sampled data $x_n = x(n\Delta t)$; $n = 0, \pm 1, \pm 2, \ldots$, if the sample time spacing Δt [s] satisfies $\Delta t \leq 1/2W$, where W [Hz] is the maximum frequency component in $s(t)$ (e.g., see [301]). Observing the duality that exists between the time domain and the frequency domain and between the Fourier transform and its inverse transform, it is apparent that the sampling theorem also suggests that all information for the spectrum $S(f)$ of a *time-limited signal* (continuous time or discrete time) of duration T [s] can be contained in the uniformly spaced samples of the spectrum $S_n = S(n\Delta f)$, if the spectral spacing Δf satisfies $\Delta f \leq 1/T$.

By denoting the Fourier transform $\tilde{X}^{(N)}(\omega)$ evaluated at $\omega = 2\pi m/N$ as $\tilde{X}_m^{(N)}$, i.e.,

$$\tilde{X}_m^{(N)} = \tilde{X}^{(N)}\left(\frac{2\pi m}{N}\right), \tag{13.55}$$

we have

$$\tilde{X}_m^{(N)} = \frac{1}{\sqrt{N}} \sum_{n=0}^{N-1} (x_n - \overline{x}^{(N)}) e^{in2\pi m/N} = \begin{cases} \frac{1}{\sqrt{N}} \sum_{n=0}^{N-1} x_n W^{mn}, & m = 1, 2, \ldots, N-1, \\ 0, & m = 0, \end{cases} \tag{13.56}$$

where $W = e^{i2\pi/N}$. Note the constant term $\overline{x}^{(N)}$ disappears in $\tilde{X}_m^{(N)}$ for all $m = 0, 1, \ldots, N-1$, because $\sum_{n=0}^{N-1} e^{in2\pi m/N} = \delta_{0,m}$. The relation (13.56) is what is commonly called the **discrete Fourier transform** (DFT) of $\{x_n - \overline{x}^{(N)}\}$, and its **inverse discrete Fourier transform** (IDFT) is

$$x_n - \overline{x}^{(N)} = \frac{1}{\sqrt{N}} \sum_{m=1}^{N-1} \tilde{X}_m^{(N)} W^{-mn}, n = 0, 2, \ldots, N-1. \tag{13.57}$$

The transformed sequence $\tilde{X}_m^{(N)}$ is generally complex valued. Take the absolute square of the DFT:

$$\boxed{P_m^{(N)} = \left|\tilde{X}_m^{(N)}\right|^2 = \frac{1}{N} \left| \sum_{n=0}^{N-1} (x_n - \overline{x}^{(N)}) W^{mn} \right|^2, m = 0, 1, 2, \ldots, N-1,} \tag{13.58}$$

which is called the **periodogram** . If we rearrange the double sum involved in (13.58), we can show (Problem 13.5) that

$$P_m^{(N)} = \sum_{k=-N+1}^{N-1} \left(1 - \frac{|k|}{N}\right) \hat{C}_X^{(N)}[k] W^{mk}, m = 0, 1, 2, \ldots, N-1, \tag{13.59}$$

contributors to this theorem, his work [257] was concerned about the maximum transmission rate, i.e., the so-called *Nyquist rate* of telegraph signals, but not a direct proof of the minimum sampling rate required for the reconstruction of a bandlimited continuous-time waveform. For example, see Butzer *et al.* [44].

where $\hat{C}_X^{(N)}[k]$ is an estimate of the autocovariance function based on N samples:

$$\hat{C}_X^{(N)}[k] \triangleq \frac{1}{N-|k|} \sum_{n=\max(1,k+1)}^{\min(N-k,N)} (x_n - \overline{x}^{(N)})(x_{n+k} - \overline{x}^{(N)}),\ k = 0, \pm 1, \ldots, \pm(N-1). \tag{13.60}$$

We can show that $\hat{C}_X^{(N)}[k]$ is an *asymptotically unbiased estimate* of $C_X[k]$; i.e.,

$$\lim_{N\to\infty} E[\hat{C}_X^{(N)}[k]] = C_X[k]. \tag{13.61}$$

See Definition 18.1 of Chapter 18 for the definition of unbiasedness and other desired properties of an estimator.

By taking the expectation of both sides in (13.59), we have

$$E[P_m^{(N)}] = \sigma_X^2 \sum_{k=-N+1}^{N-1} \left(1 - \frac{|k|}{N}\right) \rho_X[k] \cos\left(\frac{2\pi mk}{N}\right), \tag{13.62}$$

where we used the property

$$E\left[\hat{C}_X[k]\right] = \sigma_X^2 \rho_X[k]$$

and the identity

$$W^{mk} = e^{i2\pi mk/N} = \cos\left(\frac{2\pi mk}{N}\right) + i \sin\left(\frac{2\pi mk}{N}\right).$$

By letting $N \to \infty$ (13.62), we find, for sufficiently large N,

$$\boxed{E[P_m^{(N)}] \approx \sum_{k=-\infty}^{\infty} C_X[k] \cos\left(\frac{2\pi km}{N}\right) = 2\pi P_X\left(\frac{2\pi m}{N}\right),\ m = 0, 1, \ldots, N-1.} \tag{13.63}$$

Thus, use of the periodogram has been practiced as an intuitively appealing way to estimate spectral density $P_X(\omega)$ of a time series defined by (13.49). For example, at the top of Figure 13.1 we plot a signal $\{S_n; n = 0, 1, \ldots, N-1\}$ generated from white noise $\{W_n; n = 0, 1, \ldots, N-1\}$ as follows:

$$S_n = \alpha S_{n-1} + W_n,\ n = 0, 2, \ldots, N-1, \tag{13.64}$$

with $S_{-1} = 0$. This random signal S_n is Gaussian, because W_n is Gaussian, and $\{S_n\}$ forms a simple Markov chain, because S_n depends on its past only through S_{n-1}, as implied by (13.64). Thus, S_n is called a **Gauss–Markov process** (GMP). We chose the parameters $\alpha = 0.7$ and $N = 1024$. The second curve is the periodogram of this GMP signal. Note that $m = 0$ and $m = N$ in (13.58) correspond to $\omega = 0$ and $\omega = 2\pi$, although we define the spectral density function $P_S(\omega)$ over $-\pi \leq \omega \leq \pi$.

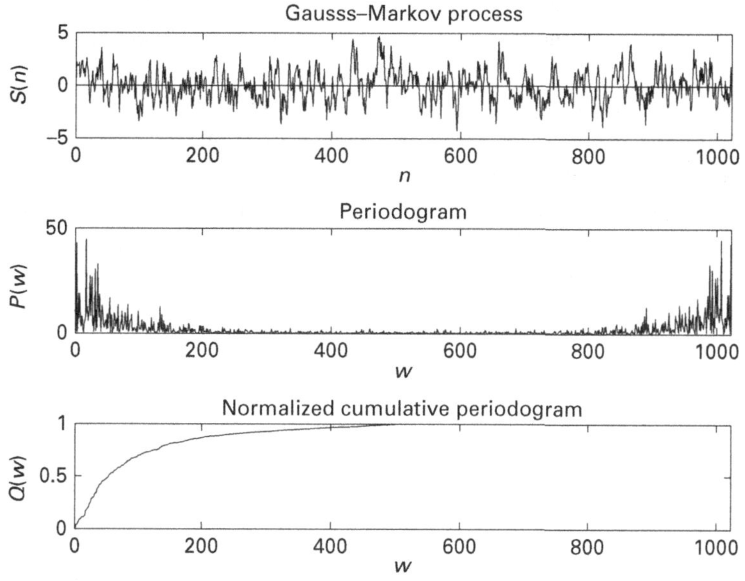

Figure 13.1 A Gauss–Markov signal S_n, its peoriodogram, and its normalized cumulative periodogram.

As you see from the middle plot, the periodogram is not a smooth curve. So it seems a rather questionable and unreliable estimate of the spectral density function. This apparent shortcoming of the periodogram can be explained as follows. Let us write the term $\tilde{X}_m^{(N)}$ of (13.55) as

$$\tilde{X}_m^{(N)} = A_m^{(N)} + iB_m^{(N)}, m = 1, 2, \ldots, N-1, \tag{13.65}$$

where

$$A_m^{(N)} = \frac{1}{\sqrt{N}} \sum_{n=0}^{N-1} x_n \cos\left(\frac{2\pi nm}{N}\right) \text{ and } B_m^{(N)} = \frac{1}{\sqrt{N}} \sum_{n=0}^{N-1} x_n \sin\left(\frac{2\pi nm}{N}\right). \tag{13.66}$$

Note we exclude the term $m = 0$, since $\tilde{X}_0^{(N)} = 0$, as shown in (13.56). It is not difficult to show that both $A_m^{(N)}$ and $B_m^{(N)}$ have zero mean and variance $\sigma_X^2/2$. Furthermore, they are uncorrelated because of the orthogonality between $\sin(2\pi nm/N)$ and $\cos(2\pi nm/N)$. Also, by virtue of the CLT, they are both asymptotically (that is, as $N \to \infty$) normally distributed; consequently, they are asymptotically independent. The individual components $\{P_m^{(N)}\}$ for different values of m are asymptotically independent (note that if the X_n are independent and normally distributed, so are $\{A_m^{(N)}\}$ and $\{B_m^{(N)}\}$ for any sample size N). Therefore, for a given index m, the variable

$$P_m^{(N)} = A_m^{(N)2} + B_m^{(N)2}, m = 1, 2, \ldots, N-1, \tag{13.67}$$

has asymptotically (and exactly for the case of normal X_n) a distribution proportional to the *chi-squared distribution* (Section 7.1) with two degrees of freedom; i.e., an **exponential distribution**. Thus, the coefficient of variation of the variable $P_m^{(N)}$

remains unity, no matter how large the sample size N becomes. In other words, $P_m^{(N)}$ does *not* converge to $2\pi P_X(2\pi m/N)$ in the mean-square sense. This property, together with the asymptotic independence of $P_m^{(N)}$ and $P_{m'}^{(N)} (m \neq m')$, implies that the periodogram fluctuates highly erratically, when it is plotted as a function of m. Therefore, the periodogram based on a single observation sequence will not provide a reliable estimate of the spectrum, no matter how large the sample size N may be. Multiple observation sequences $\{X_n\}$ ought to be obtained and the corresponding multiple periodograms must be properly processed in order to obtain a smoothed estimate of the spectral density.

However, the periodogram $\{P_m^{(N)}\}$ is a convenient statistic to use in **testing for independence** of the process $\{X_n\}$. If the variables $\{X_n\}$ are independent, then the variables $\{P_m^{(N)}; 1 \leq m \leq \lceil N/2 \rceil\}$ are i.i.d. with an exponential distribution. Hence, if we form a point process with interarrival times $\{P_m^{(N)}\}$, then it is a Poisson process! A convenient way of testing for the uniformity of this associated Poisson process (see Section 14.1.2) is done graphically through the **normalized cumulative periodogram**:

$$\boxed{Q_m^{(N)} = \frac{\sum_{n=1}^{m} P_n^{(N)}}{\sum_{n=1}^{M} P_n^{(N)}}, m = 1, 2, \ldots, M(= \lceil N/2 \rceil).} \tag{13.68}$$

In Figure 13.2 we plot an example of white noise, its periodogram, and the normalized cumulative periodogram $Q_m^{(N)}$. We see clearly that this curve is a straight line, whereas the corresponding curve in Figure 13.1 reveals no independence of the GMP.

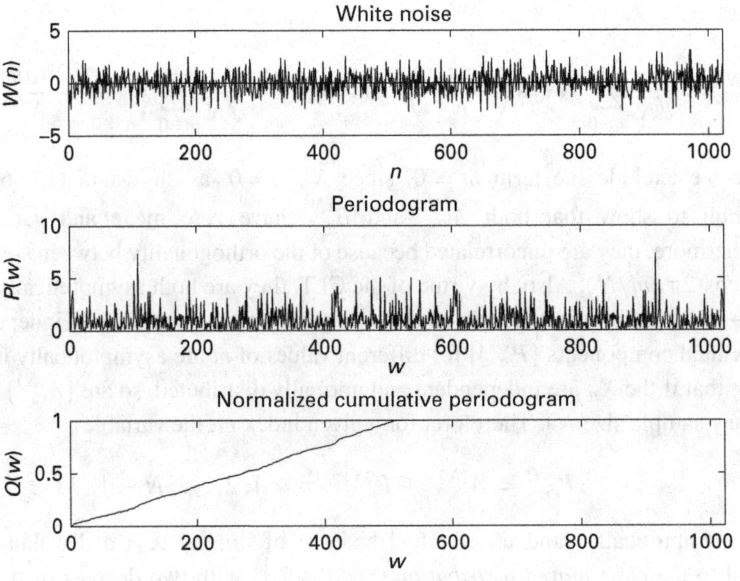

Figure 13.2 White noise W_n, its peoriodogram, and normalized cumulative periodogram.

It will be worth noting that the periodogram was originally proposed to detect hidden periodic tendencies in time series, as discussed by Schuster [294][15] in 1897. In the 1950s and the early 1960s, however, a new power spectrum estimation method proposed by Blackman and Tukey [30] became prevalent: the autocorrelation is first computed and then the Fourier transformation of the autocorrelation is taken to produce an estimate of the spectrum, as suggested by the Wiener–Khinchin formula. In order to obtain a "smoother" estimate of the power spectrum, the autocorrelation function may be multiplied by a "weight function" (or **window function**) before the transformation is taken (e.g., see Hannan [144]).

With the advent of the FFT discovered by Cooley and Tukey [66] to greatly speed up calculation of DFT (as in (13.56)), the smoothed version of periodiogram has become popular again. See Welch [347] and Chatfield [51].

13.2 Generalized Fourier series expansions

In the preceding section we extended the classical Fourier series to spectral representation of a *periodic* WSS random processes, and then the Fourier integral of nonperiodic WSS random processes, where we assumed that the observation period is $(-\infty, \infty)$. In this section we will generalize these spectral representation methods and discuss decomposition of a wide class of random processes and time series using eigenfunction-based expansion, called the Karhunen–Loève expansion method. The set of eigenvalues λ_i associated with the autocorrelation function $R_X(s, t)$ can be viewed as the generalized "spectrum" of a (possibly nonstationary) random process $X(t)$. We can assume nonstationary complex-valued processes, and the observation period may be finite. We begin with a brief review of matrix theory that is relevant to the rest of this and succeeding sections.

13.2.1 Review of matrix analysis

Let $\boldsymbol{X}$ be a complex-valued *column vector* of dimension n,

$$\boldsymbol{X} = \begin{bmatrix} X_1 \\ X_2 \\ \vdots \\ X_n \end{bmatrix}, \tag{13.69}$$

and

$$\boldsymbol{X}^{\mathrm{H}} \triangleq \boldsymbol{X}^{*\top} = (X_1^*, X_2^*, \ldots, X_n^*), \tag{13.70}$$

[15] Arthur Friedrich Schuster (1851–1934) was a German physicist known for his work in spectroscopy, and the application of harmonic analysis to physics.

where the superscript H, called **Hermitian**, is a compact notation for $*\mathsf{T}$, "complex-conjugate and transpose."

The correlation matrix $\boldsymbol{R}$ is defined by

$$\boldsymbol{R} = E[\boldsymbol{X}\boldsymbol{X}^{\mathrm{H}}] = [R_{ij}],\ R_{ij} = E[X_i X_j^*],\ 1 \le i, j \le n. \tag{13.71}$$

We list below the important properties of $\boldsymbol{R}$ and related quantities.[16]

1. $\boldsymbol{R}$ is a **self-adjoint** matrix (or a *Hermitian* matrix):[17]

$$\boldsymbol{R}^{\mathrm{H}} = \boldsymbol{R}. \tag{13.72}$$

2. $\boldsymbol{R}$ is **nonnegative definite**, i.e.,

$$\boldsymbol{a}^{\mathrm{H}}\boldsymbol{R}\boldsymbol{a} = \langle \boldsymbol{a}, \boldsymbol{R}\boldsymbol{a}\rangle \ge 0, \text{ for any vector } \boldsymbol{a}, \tag{13.73}$$

where $\langle \boldsymbol{a}, \boldsymbol{b}\rangle$ is the *inner product* (or *scalar product*) of complex-valued vectors $\boldsymbol{a}$ and $\boldsymbol{b}$:

$$\langle \boldsymbol{a}, \boldsymbol{b}\rangle \triangleq \boldsymbol{a}^{\mathrm{H}}\boldsymbol{b} = \sum_{j=1}^{n} a_j^* b_j. \tag{13.74}$$

3. Let λ be an **eigenvalue** and $\boldsymbol{u}(\neq \boldsymbol{0})$ be its associated **right-eigenvector**:

$$\boldsymbol{R}\boldsymbol{u} = \lambda\boldsymbol{u} \quad \text{or} \quad (\boldsymbol{R} - \lambda\boldsymbol{I})\,\boldsymbol{u} = \boldsymbol{0}. \tag{13.75}$$

Since a vector $c\boldsymbol{u}$, where c is any nonzero scalar, is also an eigenvector associated with λ, we can assume without loss of generality that $\boldsymbol{u}$ is **normalized**:

$$\|\boldsymbol{u}\| \triangleq \sqrt{\langle \boldsymbol{u}, \boldsymbol{u}\rangle} = 1. \tag{13.76}$$

From the second equation in (13.75), we see that an eigenvalue is a solution of the following **characteristic polynomial** in λ of order N:

$$\det[\boldsymbol{R} - \lambda\boldsymbol{I}] = 0, \tag{13.77}$$

where $\boldsymbol{I}$ is the $n \times n$ identity matrix.

4. Let λ_j $(j = 1, 2, \ldots, n)$ be the set of eigenvalues and $\boldsymbol{u}_j$ be their associated eigenvectors:

$$\boldsymbol{u}_j = \begin{bmatrix} u_{j,1} \\ u_{j,2} \\ \vdots \\ u_{j,n} \end{bmatrix},\ j = 1, 2, \ldots, n. \tag{13.78}$$

[16] For a more comprehensive discussion of matrix theory, see the online Supplementary Materials.
[17] If it is real, a self-adjoint matrix is called a **symmetric** matrix.

The eigenvalues λ_j are all real and nonnegative (Problem 13.7), which we order in the decreasing order:

$$\lambda_1 \geq \lambda_2 \geq \ldots \geq \lambda_n \geq 0. \tag{13.79}$$

The set of eigenvalues is also called the **spectrum** of $\boldsymbol{R}$. Since we can show (Problem 13.9) that eigenvectors associated with different eigenvalues are **orthogonal** to each other, we find for the *normalized* eigenvectors $\boldsymbol{u}_j$

$$\langle \boldsymbol{u}_i, \boldsymbol{u}_j \rangle = \boldsymbol{u}_i^{\mathrm{H}} \boldsymbol{u}_j = \delta_{i,j} \triangleq \begin{cases} 1, & i = j, \\ 0, & i \neq j, \end{cases} \tag{13.80}$$

where $\delta_{i,j}$ is called the **Kronecker**[18] delta. The set of eigenvectors $\boldsymbol{u}_j$ are said to be **orthonormal**.

5. $\boldsymbol{u}_j^{\mathrm{H}}$ is the **left-eigenvector** associated with λ_j:

$$\boldsymbol{u}_j^{\mathrm{H}} \boldsymbol{R} = \lambda_j \boldsymbol{u}_j^{\mathrm{H}}. \tag{13.81}$$

6. $\boldsymbol{R}$ can be expanded as

$$\boxed{\boldsymbol{R} = \sum_{j=1}^{n} \lambda_j \boldsymbol{u}_j \boldsymbol{u}_j^{\mathrm{H}} = \sum_{j=1}^{n} \lambda_j \boldsymbol{E}_j,} \tag{13.82}$$

where

$$\boldsymbol{E}_j = \boldsymbol{u}_j \boldsymbol{u}_j^{\mathrm{H}} \tag{13.83}$$

is called a **projection** matrix in the sense $\boldsymbol{E}_j^2 = \boldsymbol{E}_j$. Furthermore, they are orthogonal to each other:

$$\boldsymbol{E}_i \boldsymbol{E}_j = \delta_{i,j} \boldsymbol{E}_j. \tag{13.84}$$

7. Define an $n \times n$ matrix $\boldsymbol{U}$ by

$$\boldsymbol{U} = [\boldsymbol{u}_1, \boldsymbol{u}_2, \cdots, \boldsymbol{u}_n]; \text{ hence, } \boldsymbol{U}^{\mathrm{H}} = \begin{bmatrix} \boldsymbol{u}_1^{\mathrm{H}} \\ \boldsymbol{u}_2^{\mathrm{H}} \\ \vdots \\ \boldsymbol{u}_n^{\mathrm{H}} \end{bmatrix}. \tag{13.85}$$

Then

$$\boldsymbol{U}^{\mathrm{H}} \boldsymbol{U} = \boldsymbol{I} \Longrightarrow \boldsymbol{U}^{\mathrm{H}} = \boldsymbol{U}^{-1}. \tag{13.86}$$

A matrix $\boldsymbol{U}$ with this property is called a **unitary matrix**.

[18] Leopold Kronecker (1823–1891) was a German mathematician and logician.

8. We can write $\boldsymbol{R}$ as

$$\boxed{\boldsymbol{R} = \boldsymbol{U}\boldsymbol{\Lambda}\boldsymbol{U}^{\mathrm{H}} = \boldsymbol{U}\boldsymbol{\Lambda}\boldsymbol{U}^{-1},} \tag{13.87}$$

where

$$\boldsymbol{\Lambda} = \begin{bmatrix} \lambda_1 & 0 & \cdots & 0 \\ 0 & \lambda_2 & \cdots & 0 \\ \vdots & \vdots & \ddots & \vdots \\ 0 & 0 & \cdots & \lambda_n \end{bmatrix} \tag{13.88}$$

and

$$\boldsymbol{\Lambda} = \boldsymbol{U}^{-1}\boldsymbol{R}\boldsymbol{U} = \boldsymbol{U}^{\mathrm{H}}\boldsymbol{R}\boldsymbol{U}. \tag{13.89}$$

This relation between $\boldsymbol{R}$ and $\boldsymbol{\Lambda}$ is called a **similarity transformation.**

9. Now we are ready to discuss the main topic of this section, namely, an eigenvector expansion of a random variable vector $\boldsymbol{X}$. Recall the Fourier series expansion of a periodic WSS random process defined by (13.26) with the expansion coefficients (13.25). Using the same line of argument, it can be shown that a random variable vector $\boldsymbol{X}$ with the correlation matrix $\boldsymbol{R}$ can be expanded using the set of eigenvectors of the matrix $\boldsymbol{R}$ as follows:

$$\boxed{\boldsymbol{X} \overset{\text{m.s.}}{=} \sum_{j=1}^{n} \chi_j \boldsymbol{u}_j,} \tag{13.90}$$

where $\overset{\text{m.s.}}{=}$ is *equivalence in mean square*, as defined in the previous section and is analogous to the notion of convergence in mean square discussed in Section 11.2.4; i.e.,

$$E[\|\boldsymbol{X} - \sum_{j=1}^{n} \chi_j \boldsymbol{u}_j\|^2] = 0, \tag{13.91}$$

where $\|\boldsymbol{Y}\|^2 = \langle \boldsymbol{Y}, \boldsymbol{Y}\rangle$ is the norm square of a random vector $\boldsymbol{Y}$ as defined by (10.9). The expansion coefficient

$$\boxed{\chi_j = \langle \boldsymbol{u}_j, \boldsymbol{X}\rangle = \sum_{i=1}^{n} u_{j,i}^* X_i} \tag{13.92}$$

can be geometrically interpreted as the *projection* of vector $\boldsymbol{X}$ onto the basis vector $\boldsymbol{u}_j$. The expansion coefficients χ_j are RVs and are orthogonal to each other (Problem 13.10):

$$\langle \chi_i, \chi_j \rangle = E[\chi_i \chi_j^*] = \lambda_j \delta_{i,j}, \tag{13.93}$$

which is analogous to the orthogonality (13.28).

When n is large or infinite, we may choose to approximate $\boldsymbol{X}$ by using only the first $k(< n)$ terms in (13.90):

$$\hat{\boldsymbol{X}}_k = \sum_{j=1}^{k} \chi_j \boldsymbol{u}_j,\ k < n. \tag{13.94}$$

Then the mean-squared error is given (Problem 13.11) by

$$\boxed{\mathcal{E}^2 \triangleq E\left[\left|\hat{\boldsymbol{X}}_k - \boldsymbol{X}\right|^2\right] = \sum_{j=k+1}^{n} \lambda_j,} \tag{13.95}$$

which may be made small enough with an appropriate choice of k because of the ordering (13.79) of the spectrum $\{\lambda_i\}$. Equation (13.94) is the basis of the **principal component analysis** (**PCA**), which is widely practiced in data compression, pattern classification, and clustering analysis. We will discuss PCA in Section 13.3.1.

10. If the sequence $\{X_j\}$ is WSS, then its autocorrelation matrix $\boldsymbol{R} = [R_{i,j}]$ satisfies the following property:

$$R_{i,j} = E[X_i X_j^*] = R_{i-j},\ 1 \le i, j \le n. \tag{13.96}$$

Such a matrix is called a **Toeplitz** [124, 125, 129][19] matrix, which is defined by the following structure:

$$\boldsymbol{T} = \begin{bmatrix} a_0 & a_1 & a_2 & \cdots & a_{n-1} \\ a_{-1} & a_0 & a_1 & \cdots & a_{n-2} \\ a_{-2} & a_{-1} & a_0 & \cdots & a_{n-3} \\ \vdots & \vdots & \vdots & \ddots & \vdots \\ a_{-n+1} & a_{-n+2} & a_{-n+3} & \cdots & a_0 \end{bmatrix}. \tag{13.97}$$

In the correlation matrix of a WSS process, $a_i = R_i,\ -n+1 \le i \le n-1$, and the matrix-vector equation (13.75) becomes a set of convolution sum expressions:

$$\sum_{k=0}^{n-1} R_{i-k} u_k = \lambda u_k,\ i = 0, 1, 2, \ldots, n-1, \tag{13.98}$$

where

$$\boldsymbol{u} = (\boldsymbol{u}_0, \boldsymbol{u}_1, \ldots, \boldsymbol{u}_{n-1})^\top. \tag{13.99}$$

[19] Otto Toeplitz (1881–1940) was a German-born mathematician and worked on algebraic geometry and spectral theory.

11. A special class of Toeplitz matrices are the **circulant** matrices [124, 125] in which each row vector is a cyclic shift of the row vector above:

$$C = \begin{bmatrix} c_0 & c_1 & c_2 & \cdots & c_{n-1} \\ c_{n-1} & c_0 & c_1 & \cdots & c_{n-2} \\ c_{n-2} & c_{n-1} & c_0 & \cdots & c_{n-3} \\ \vdots & \vdots & \vdots & \ddots & \vdots \\ c_1 & c_2 & c_3 & \cdots & c_0 \end{bmatrix}. \quad (13.100)$$

Thus, a circular matrix is completely specified by one vector $\boldsymbol{c} = (c_0, c_1, c_2, \cdots, c_{n-1})^\top$, which appears as the first row of $\boldsymbol{C}$. Note that the last column of $\boldsymbol{C}$ is the vector $\boldsymbol{c}$ in the reverse order, and the remaining columns are each a cyclic shift of the last column. The jth eigenvalue is given by (Problem 13.12)

$$\lambda_j = \sum_{k=0}^{n-1} c_k W^{jk}, \; j = 0, 1, 2, \ldots, n-1, \quad (13.101)$$

where

$$W = e^{i2\pi/n}, i = \sqrt{-1}.$$

The corresponding eigenvector is given by

$$\boldsymbol{u}_j = (1, W^j, W^{2j}, \ldots, W^{(n-1)j})^\top, \; j = 0, 1, 2, \ldots, n-1. \quad (13.102)$$

Equation (13.101) shows that the set of eigenvalues $(\lambda_0, \lambda_1, \ldots, \lambda_{n-1})$ is the DFT of $(c_0, c_1, \ldots, c_{n-1})$. Note that the correlation matrix $\boldsymbol{R}$ of a WSS discrete-time process has the circulant structure of (13.100) if and only if the process is periodic with a period that divides n (i.e., period n/k with some integer k).[20]

12. If the Toeplitz matrix is of infinite dimension, i.e., in the limit $n \to \infty$, we can use the **generating function** method (i.e., a *polynomial representation*) instead of vectors (Problem 13.13):

$$R(z) = \sum_{i=-\infty}^{\infty} R_i z^i, \; u(z) = \sum_{i=0}^{\infty} u_i z^i. \quad (13.103)$$

Then (13.98) becomes

$$\boxed{(R(z) - \lambda)\, u(z) = 0.} \quad (13.104)$$

By setting $z = e^{i\omega}$, we find

$$\lambda = R(e^{i\omega}) = 2\pi P(\omega), \quad (13.105)$$

[20] See Section 13.1.3 for the definition of a periodic WSS process.

where

$$P(\omega) \triangleq \frac{1}{2\pi} \sum_{k=-\infty}^{\infty} R_k e^{-ik\omega} \tag{13.106}$$

is the power spectrum of the WSS discrete-time process $\boldsymbol{X} = (X_n; 0 \le n < \infty)$, as discussed in Section 13.1.5. The set of eigenvalues becomes the power spectrum $P(\lambda)$. In this case the number of distinct eigenvalues is infinite, and not even countable. This is because the observation period is infinite.

13.2.2 Karhunen–Loève expansion and its applications

Now we are in a position to generalize the Fourier series expansion to a random process $X(t)$ which is not periodic. It may not be even a WSS process. Even if the process is WSS, the power spectrum discussed in Section 13.1.4 will not apply if the observation period is finite.

The theory to be developed below applies to nonstationary processes as well as stationary processes, and the observation period can be finite or infinite. The method can be viewed as a continuous-time analog of the eigenvalue and eigenvector analysis of a correlation matrix reviewed in the preceding section.

Our objective is to find a set of (possibly, infinitely many) orthogonal functions $\{u_i(t); i = 1, 2, \ldots\}$ that allows us to expand a random process $X(t)$ when we are given its autocorrelation (matrix) function $R_X(t, s)$:

$$X(t) \overset{\text{m.s.}}{=} \sum_i X_i u_i(t), 0 \le t \le T. \tag{13.107}$$

We want to have **double orthogonality** in the sense that the functions are *orthonormal*; i.e.,

$$\boxed{\langle u_m, u_n \rangle \triangleq \int_0^T u_m^*(t) u_n(t)\, dt = \delta_{m,n}} \tag{13.108}$$

and

$$\boxed{\langle X_m, X_n \rangle \triangleq E[X_m^* X_n] = \lambda_m \delta_{m,n},} \tag{13.109}$$

where λ_n are some constants yet to be determined. Suppose that the above three requirements (13.107), (13.108), and (13.109) are satisfied. Then the autocorrelation function should be represented as[21]

[21] We write $R(t, s)$ instead of $R_X(s, t)$ for notational simplicity.

$$R(t,s) = E[X(t)X^*(s)] = E\left[\sum_n X_n u_n(t) \sum_m X_m^* u_m^*(s)\right]$$
$$= \sum_n \sum_m \lambda_n \delta_{n,m} u_n(t) u_m^*(s) = \sum_n \lambda_n u_n(t) u_n^*(s), 0 \le t, s \le T. \quad (13.110)$$

Then we find

$$\int_0^T R(t,s) u_k(s)\, ds = \sum_n \lambda_n u_n(t) \int u_n^*(s) u_k(s)\, ds$$
$$= \sum_n \lambda_n u_n(t) \delta_{n,k} = \lambda_k u_k(t), 0 \le t \le T. \quad (13.111)$$

This result implies that, in the language of integral equations, the λ_k must be **eigenvalues** (or *characteristic values*) and the functions $u_k(t)$ must be the **eigenfunctions** (or *characteristic functions*) of the integral equation:

$$\boxed{\int_0^T R(t,s) u(s)\, ds = \lambda u(t), 0 \le t \le T.} \quad (13.112)$$

In the theory of integral equations, $R(t,s)$ is called the **integral kernel** (or simply *kernel*). Thus, we have shown that when $X(t)$ satisfies the properties (13.107), (13.108), and (13.109), then its autocorrelation (13.110) satisfies the integral equation (13.112) for $\lambda = \lambda_n, n = 1, 2, \ldots$.

Conversely, for a random process with continuous $R(t,s)$, we can construct an orthogonal expansion method for $X(t)$ by finding eigenvalues λ_n and the corresponding eigenfunctions $u_n(t)$ of the integral equation (13.111). We scale $u_n(t)$ so that they form an orthonormal set. Let

$$\boxed{X_n = \int_0^T X(t) u_n^*(t)\, dt = \langle u_n(t), X(t) \rangle.} \quad (13.113)$$

Then

$$E[X_n X_m^*] = \int_0^T \int_0^T R(t,s) u_n^*(t) u_m(s)\, dt\, ds$$
$$= \int_0^T \lambda_m u_m(t) u_n^*(t)\, dt = \lambda_m \delta_{m,n}. \quad (13.114)$$

Thus, the orthogonality of the expansion coefficients given by (13.109) is satisfied. The orthogonality condition (13.108) for the functions $u_n(t)$ is satisfied, because they are constructed from the eigenfunctions of the integral equation (Problem 13.14). In order to show that the property (13.107) is met, let

$$X_N(t) = \sum_{n=1}^{N} X_n u_n(t). \quad (13.115)$$

Then it readily follows that

$$E[X(t)X_N^*(t)] = E[X_N(t)X_N^*(t)] = \sum_{n=1}^{N} \lambda_n u_n(t) u_n^*(t) \tag{13.116}$$

and

$$E[\|X(t) - X_N(t)\|^2] = R(t,t) - \sum_{n=1}^{N} \lambda_n u_n(t) u_n^*(t). \tag{13.117}$$

From (13.110) we have

$$\lim_{N\to\infty} \sum_{n=1}^{N} \lambda_n u_n(t) u_n^*(t) = R(t,t). \tag{13.118}$$

Hence,

$$\underset{N\to\infty}{\text{l.i.m.}} X_N(t) = X(t). \tag{13.119}$$

If we consider a linear vector space (of infinite dimension) spanned by the orthonormal functions $\{u_n(t)\}$, the expansion coefficient X_n is a **projection** of $X(t)$ onto the nth coordinate $u_n(t)$ of this linear space. Thus, the continuous-time random process $X(t)$ is mapped to a vector of infinite dimension:

$$\boldsymbol{X} = (X_1, X_2, \ldots, X_n, \ldots). \tag{13.120}$$

Conversely, for a given $\boldsymbol{X}$ in the vector space, we construct the corresponding continuous-time process according to (13.107), or more precisely

$$\boxed{X(t) = \underset{N\to\infty}{\text{l.i.m.}} \sum_{n=1}^{N} X_n u_n(t), \; 0 \le t \le T.} \tag{13.121}$$

The **energy** of the square integrable function $X(t)$ is defined as

$$\begin{aligned} E\left[\int_0^T |X(t)|^2\, dt\right] &= E\left[\int_0^T \sum_{n=1}^{\infty} X_n u_n(t) \sum_{m=1}^{\infty} X_m^* u_m^*(t)\, dt\right] \\ &= E\left[\sum_{n=1}^{\infty} |X_n|^2\right] = \sum_{n=1}^{\infty} \lambda_n. \end{aligned} \tag{13.122}$$

The expansion (13.107) or (13.121) is called the **Karhunen–Loève expansion**, because this representation was proposed by **Loève** [229][22] and **Karhunen** [174].[23] It can be viewed as a generalization of the Fourier series expansion. By dividing equation (13.122) by T we have

[22] Michel Loève (1907–1979) was a French–American probabilist and mathematical statistician who taught at UC Berkeley.

[23] Kari Karhunen (1915–1992) was a Finnish mathematician.

$$E\left[\frac{1}{T}\int_0^T |X(t)|^2\,dt\right] = \frac{1}{T}\sum_{n=1}^{\infty}\lambda_n, \tag{13.123}$$

which may be referred to as the **spectral expansion** of power with respect to the functions $u_n(t)$.

Example 13.1: Signal detection problem. Consider two hypotheses:[24] under the first hypothesis, denoted H_0, the received signal $X(t)$ is noise alone, and under the alternative hypothesis, denoted H_1, it includes the signal $S(t)$; i.e.,

$$\boxed{\begin{aligned} &H_0 : X(t) = N(t), 0 \le t \le T;\\ &H_1 : X(t) = S(t) + N(t) 0 \le t \le T.\end{aligned}} \tag{13.124}$$

We assume that the signal $S(t)$ is a known deterministic function, whereas the noise $N(t)$ is a complex-valued Gaussian process (see Section 12.4) with mean zero and autocovariance function $R_N(s,t), 0 \le s,t \le T$. Let λ_k and $u_k(t)$ $(k = 1, 2, \ldots)$ be the eigenvalues and eigenvectors of the **integral kernel** $R_N(s,t)$; i.e.,

$$\int_0^T R_N(s,t)u_k(t)\,dt = \lambda_k u_k(s), 0 \le s \le T, k = 1, 2, \ldots \tag{13.125}$$

We then convert the continuous-time functions $X(t)$, $S(t)$, and $N(t)$ into a set of the first M coefficients, $\boldsymbol{X} = [X_1, X_2, \ldots, X_M]$, $\boldsymbol{S} = [S_1, S_2, \ldots, S_M]$, and $\boldsymbol{N} = [N_1, N_2, \ldots, N_M]$, where

$$X_k = \int_0^T u_k^*(t)X(t)\,dt,\; S_k = \int_0^T u_k^*(t)S(t)\,dt, \text{ and } N_k = \int_0^T u_k^*(t)N(t)\,dt.$$

Then, the hypothesis testing (13.124) can be represented in terms of the M-dimensional vectors:

$$\boxed{\begin{aligned} &H_0 : \boldsymbol{X} = \boldsymbol{N},\\ &H_1 : \boldsymbol{X} = \boldsymbol{S} + \boldsymbol{N}.\end{aligned}} \tag{13.126}$$

The expanded noise coefficients N_k are the complex representation of circularly symmetric Gaussian variables discussed in Sections 7.6.1 and 12.4. They have zero mean and covariances

$$E[N_j N_k^*] = \lambda_j \delta_{j,k} \text{ or } E[\boldsymbol{N}\boldsymbol{N}^{\mathrm{H}}] = \begin{bmatrix} \lambda_1 & 0 & \cdots & 0\\ 0 & \lambda_2 & \cdots & 0\\ \vdots & \vdots & \ddots & \vdots\\ 0 & 0 & \cdots & \lambda_M \end{bmatrix} \triangleq \boldsymbol{\Lambda}. \tag{13.127}$$

[24] Hypothesis testing is discussed in Section 18.2.1.

Then the conditional PDFs of the complex-valued random vector $\boldsymbol{X}$ under the hypotheses H_0 and H_1 are given, respectively, by

$$f_X(\boldsymbol{x}|H_0) = \prod_{k=1}^{M} \frac{1}{2\pi\lambda_k} \exp\left(-\sum_{k=1}^{M} \frac{|x_k|^2}{\lambda_k}\right) = \frac{1}{(2\pi)^M |\det \boldsymbol{\Lambda}|} \exp\left(-\boldsymbol{x}^{\mathrm{H}} \boldsymbol{\Lambda}^{-1} \boldsymbol{x}\right) \tag{13.128}$$

$$f_X(\boldsymbol{x}|H_1) = \frac{1}{(2\pi)^M |\det \boldsymbol{\Lambda}|} \exp\left[-(\boldsymbol{x}-\boldsymbol{s})^{\mathrm{H}} \boldsymbol{\Lambda}^{-1} (\boldsymbol{x}-\boldsymbol{s})\right]. \tag{13.129}$$

As we will fully discuss in Chapter 18, an optimal choice between the two hypotheses can be made on the basis of the **likelihood ratio**

$$L(\boldsymbol{x}) = \frac{f_X(\boldsymbol{x}|H_1)}{f_X(\boldsymbol{x}|H_0)} = \exp\left(\sum_{k=1}^{M} \frac{s_k^* x_k + x_k^* s_k - |s_k|^2}{\lambda_k}\right), \tag{13.130}$$

which is to be compared with some fixed critical value λ_α (see (18.65)).[25] We declare that there is no signal present if $L(\boldsymbol{x}) < \lambda_\alpha$, or equivalently if

$$T_M(\boldsymbol{x}) \triangleq \sum_{k=1}^{M} \frac{s_k^* x_k + x_k^* s_k}{\lambda_k} < \ln \lambda_\alpha + \sum_{k=1}^{M} \frac{|s_k|^2}{\lambda_k}. \tag{13.131}$$

A larger M, the number of expansion terms, implies that more information is used about the input process $X(t)$. Therefore, we consider the limit $M \to \infty$:

$$T(\boldsymbol{x}) \triangleq \sum_{k=1}^{\infty} \frac{s_k^* x_k + x_k^* s_k}{\lambda_k} = 2\Re\left\{\sum_{k=1}^{\infty} \frac{s_k^* x_k}{\lambda_k}\right\}. \tag{13.132}$$

If we set $q_k = s_k/\lambda_k$ and define the function $Q(t)$ by

$$Q(t) = \sum_{k=1}^{\infty} q_k u_k(t), \tag{13.133}$$

we can write the above **test statistic** $T(\boldsymbol{x})$ as

$$\boxed{T(\boldsymbol{x}) = 2\Re\left\{\sum_{k=1}^{\infty} q_k^* x_k\right\} = 2\Re\left\{\int_0^T Q^*(t) X(t)\, dt\right\},} \tag{13.134}$$

which is a manifestation of Parseval's identity. Using representation (13.110) of $R_N(s,t)$, i.e.,

[25] The symbol λ of λ_α has nothing to do with the eigenvalues λ_k.

$$R_N(s,t) = \sum_{k=1}^{\infty} \lambda_k u_k(s) u_k^*(t), \tag{13.135}$$

and (13.133) of $Q(t)$, we can show (Problem 13.15) that $Q(t)$ is the solution to the following **Fredholm integral equation of the first kind**:

$$\boxed{\int_0^T R_N(t,u) Q(u)\, du = S(t).} \tag{13.136}$$

The integration (13.134) is a linear operation on the random process $X(t)$ and can be realized by passing $X(t)$ into a linear filter whose impulse response is

$$h(t) \triangleq Q^*(T-t), 0 \le t \le T, \tag{13.137}$$

because the integration (13.134) can be represented as the convolution integral of $X(t)$ with $h(t)$ evaluated at $t = T$ (Problem 13.16):

$$\boxed{\int_0^T Q^*(t) X(t)\, dt = \left. \int_0^t h(s) X(t-s)\, ds \right|_{t=T}.} \tag{13.138}$$

The left side is the cross-correlation between $Q(t)$ and $X(t)$ and is referred to as the output of the **correlator** or **correlation receiver** in communications engineering. The filter $h(t)$ is often referred to as a **matched filter** in the sense it is an optimum filter that receives the signal $S(t)$ in the presence of noise with autocorrelation function $R_N(t)$ (Problem 13.17: see also Problems 18.11 and 18.12 of Section 18.2.4).

Thus, the signal detection problem can be summarized as follows:

1. Solve the integral equation (13.136) to obtain $Q(t)$.
2. Cross-correlate the input process $X(t)$ with $Q(t)$ and obtain the **test statistic** $T(\boldsymbol{x})$ of (13.134). Alternatively, pass the input $X(t)$ into the matched filter $h(t)$ defined by (13.137) and sample the output at $t = T$ to obtain $T(\boldsymbol{x})$.
3. Compare $T(\boldsymbol{x})$ against some preset threshold T_0. If $T(\boldsymbol{x}) \ge T_0$, accept the hypothesis H_1 (i.e., the signal is present). Otherwise, accept the hypothesis H_0 (i.e., the signal is absent).

The lower the threshold T_0 is set, the higher the false alarm probability; i.e., the probability of accepting H_1 when H_0 is indeed true. Conversely, the higher T_0 is set, the higher the signal miss probability, i.e., the probability of accepting H_0 when H_1 is true. An optimum value of the threshold T_0 is determined by proper balance between the above two types of error. Both the false alarm probability and signal miss probability can be expressed in terms of the parameter d defined by

$$d^2 \triangleq \sum_{k=0}^{\infty} \frac{|s_k|^2}{\lambda_k} = \int_0^T Q^*(t) S(t)\, dt. \tag{13.139}$$

The quantity d^2 represents the **signal-to-noise ratio** (SNR) at the output of the correlator or matched filter. See Section 18.2.3 for a further discussion of this signal detection problem, including the receiver operating characteristic (ROC). □

As far as the practicality of the above procedure is concerned, the most difficult part will be to solve the Fredholm integral equation (13.136) in the first step.

Example 13.2: The Karhunen–Loève expansion when noise is white. Now let us consider the case where the noise $N(t)$ in the previous example is white; i.e.,

$$R_N(s,t) = \sigma^2 \delta(s-t). \tag{13.140}$$

Then the integral equation (13.125) reduces to

$$\sigma^2 \int_0^T \delta(s-t)u(t)\,dt = \lambda u(s), 0 \le s \le T, \tag{13.141}$$

which is simply

$$\sigma^2 u(s) = \lambda u(s), 0 \le s \le T. \tag{13.142}$$

Then, the set of orthonormal basis functions $u_k(t)$ is any set of orthonormal set of functions and their corresponding eigenvalues are all σ^2; i.e., $\lambda_k = \sigma^2$ for $k = 1, 2, \ldots$. For the binary hypothesis testing problem in the previous example, the most convenient choice is

$$u_1(t) = \frac{S(t)}{\|S\|}, \text{ where } \|S\|^2 = \int_0^T |S(t)|^2\,dt. \tag{13.143}$$

The expansion coefficients of an observation $x(t)$, an instance of the random process $X(t)$, and the deterministic signal $S(t)$ in terms of $u_1(t) = S(t)/\|S\|$ are

$$x_1 = \frac{1}{\|S\|}\int_0^T S^*(t)x(t)\,dt, s_1 = \frac{1}{\|S\|}\int_0^T S^*(t)S(t)\,dt = \|S\|. \tag{13.144}$$

We could find the other orthonormal basis functions $u_k(t), k \ge 2$, but the expansion coefficients s_k are all zero for $k \ge 2$. Thus, as far as the signal detection problem is concerned, the expansion coefficients x_k for $k \ge 2$ are irrelevant, because they do not contribute to the likelihood ratio statistic, since the numerators and denominators for the terms $k \ge 2$ in (13.130) cancel with each other. The test statistic $T_1(\boldsymbol{x})$ is a **sufficient statistic**[26] for this binary hypothesis testing problem:

[26] The notion of *sufficient statistic* will be formally defined and discussed in Section 18.1. Informally speaking, when any information contained in data $\boldsymbol{x}$, other than that summarized in statistic $T(\boldsymbol{x})$, does not provide additional information useful for a given decision or estimation problem, the statistic $T(\boldsymbol{x})$ is called a sufficient statistic.

$$T_1(\boldsymbol{x}) = \frac{s_1^* x_1 + s_1 x_1^*}{\sigma^2} = \frac{2\|S\|\Re\{x_1\}}{\sigma^2}. \tag{13.145}$$

The solution to the integral equation (13.136) becomes

$$Q(t) = \frac{S(t)}{\sigma^2} \propto S(t) \tag{13.146}$$

and the SNR is given by

$$d^2 = \int_0^T Q^*(t)S(t)\,dt = \frac{\|S\|^2}{\sigma^2}. \tag{13.147}$$

□

Example 13.3: Signal space method for digital communications. Now we generalize the binary hypothesis testing problem. In a digital communication systems, there are multiple signals. Let $S_1(t), S_2(t), \ldots S_M(t)$ be M possible signals. Then the hypothesis test is to select among the following multiple hypotheses:

$$\boxed{\begin{aligned} H_0 &: X(t) = N(t), \\ H_i &: X(t) = S_i(t) + N(t), i = 1, 2, \ldots, M, \end{aligned}} \tag{13.148}$$

where the noise $N(t)$ is assumed again to be white noise. Unlike in the signal detection problem of the previous example, the null hypothesis H_0 is usually not considered in the communication problem, because the observer at the receiving end normally assumes one of the M signals is transmitted by the sender. Inclusion of H_0, however, is mathematically convenient, because the likelihood ratio of $f(\boldsymbol{x}|H_i)$ to $f(\boldsymbol{x}|H_0)$ is well defined, even when the conditional PDFs $f(\boldsymbol{x}|H_i)$ are not.

Note that not all the M signals may be linearly independent of each other. So let $m(\leq M)$ be the dimensionality of the functional space spanned by the signal set. For instance, bipolar signals with plus and minus pulses have only $m = 1$; the quadrature amplitude modulation (QAM) and PSK (phase-shift keying) signals both have a signal space of $m = 2$; CDMA (code division multiple access) signals of n-chip code have the dimension $m = n$, and so forth.

Then, it suffices to find a set of m orthonormal functions using the **Gram–Schmidt orthogonalization process**;[27] i.e.,

[27] See the online Supplementary Materials of this textbook, or books on linear algebra or matrix theory; e.g., [124].

$$u_1(t) \triangleq \frac{S_1(t)}{\|S_1\|},$$

$$r_2(t) = S_2(t) - s_{2,1}u_1(t), \quad \text{where } s_{2,1} = \int_0^T u_1^*(t)S_2(t)\,dt,$$

$$u_2(t) \triangleq \frac{r_2(t)}{\|r_2\|},$$

$$r_i(t) = S_i(t) - s_{i,1}u_1(t) - s_{i,2}u_2(t) - \cdots - s_{i,i-1}u_{i-1}(t),$$

$$\text{where } s_{i,j} = \int_0^T u_j^*(t)S_i(t)\,dt,$$

$$u_i(t) \triangleq \frac{r_i(t)}{\|r_i\|}, i \geq 2.$$

In the above step, $r_i(t)$ represents the difference (or residue) of $S_i(t)$ and its projection to the subspace spanned by all the preceding basis functions $u_1(t), \ldots, u_{i-1}(t)$. Thus, $r_i(t)$ is orthogonal to all of $u_1(t), \ldots u_{i-1}(t)$, and should be added as a new basis, and $u_i(t)$ is the normalized version of $r_i(t)$; i.e., $\|u_i\| = 1$.

This special case of the Karhunen–Loève expansion method for $R(t) \propto \delta(t)$ is applicable to communication systems with additive white noise, and is known as the **signal space method** (e.g., see Wozencraft and Jacobs [361]). The input signal $X(t)$ is **projected** onto the m-dimensional subspace spanned by the set of signals $S_i(t)$; $1 \leq i \leq M$, and $m \leq M$. Then, as we have done in the above signal detection problem, the observed process $x(t)$, an instance of the random process $X(t)$, is converted into the m-dimensional vector $\boldsymbol{x} = [x_1, x_2, \ldots, x_m]$ and the conditional PDFs and the likelihood ratio can be defined in the same manner.

Assuming that all the signals are equally likely, the optimum decision rule is the **maximum-likelihood decision** rule. This decides on the signal $S_j(t)$ for which the likelihood ratio $L(\boldsymbol{x}|H_j)$ is the largest among all $L(\boldsymbol{x}|H_i)$, where

$$L(\boldsymbol{x}|H_i) \triangleq \frac{f(\boldsymbol{x}|H_i)}{f(\boldsymbol{x}|H_0)} = \exp\left(\sum_{k=1}^{m} \frac{s_{i,k}^* x_k + x_k^* s_{i,k} - |s_{i,k}|^2}{\sigma^2}\right), i = 1, 2, \ldots, M, \tag{13.149}$$

with the vector elements defined by

$$x_k = \int_0^T u_k^*(t)X(t)\,dt, k = 1, 2, \ldots, m; \tag{13.150}$$

$$s_{i,k} = \int_0^T u_k^*(t)S_i(t)\,dt, k = 1, 2, \ldots, m, \quad \text{and } i = 1, 2, \ldots, M. \tag{13.151}$$

□

Solving the Karhunen–Loève integral equation is reduced to the problem of solving the corresponding differential equation. Problem 13.18 on the Karhunen–Loève expansion of the Wiener process, which is a nonstationary process, also deduces a differential equation from the integral equation.

13.3 Principal component analysis and singular value decomposition

In this section we present two important statistical analysis techniques that can be applied to data which contain some correlation among them, and we wish to capitalize on it for the purpose of data compression, pattern classification, retrieval of information, etc., depending on the nature of data and our objective. They are principal component analysis (PCA) and Singular value decomposition (SVD).

We assume the data are presented in a two-dimensional array; i.e., as a matrix of size $m \times n$. We denote this *data set* (or *panel data*, as it is called in econometrics) by $\mathbf{X} = [x_{ij}]$; namely,

$$\mathbf{X} = \begin{bmatrix} x_{11} & x_{12} & \cdots & x_{1n} \\ x_{21} & x_{22} & \cdots & x_{2n} \\ \vdots & \vdots & \ddots & \vdots \\ x_{m1} & x_{m2} & \cdots & x_{mn} \end{bmatrix}. \tag{13.152}$$

In a *multivariate time series*, for instance, x_{ij} may represent the value of the ith variate at discrete-time j. In a *gene expression data* [344], x_{ij} may represent the expression level of the ith gene in the jth assay. In a *pattern classification* problem, x_{ij} may represent the value of the m feature of the jth object. In *latent semantic analysis* (LSA) used in information retrieval (IR) [23], the data set $\mathbf{X}$ is called the *keyword-document matrix*, in which x_{ij} is the number of occurrences of the ith keyword in the jth document. In *Web information retrieval*, such as Google's search engine, $\mathbf{X}$ is an estimate of the state transition probability matrix $\boldsymbol{P}$ or the adjacency matrix $\boldsymbol{A}$ defined over the web graph. We will discuss the web information retrieval application in Section 13.3.3.

13.3.1 Principal component analysis (PCA)

Principal component analysis is a statistical procedure that transforms a set of correlated data into a smaller number of uncorrelated variates called *principal components.* The conception of PCA may be traced back to the 1901 paper by K. Pearson [267]. Its mathematical basis is that the correlation matrix $\boldsymbol{R}$ can be expanded as in (13.82).

13.3.1.1 Correlation across the m variates

Suppose we interpret the data $\mathbf{X}$ as an m-variate time-series data. An *estimate of the correlation matrix* across the m variates, based on n samples, is given by the following $m \times m$ matrix:

$$R = [R_{ii'}] \triangleq \frac{1}{n}\mathbf{X}\mathbf{X}^{\mathrm{H}} = \frac{1}{n}\sum_{j=1}^{n} \boldsymbol{x}_{\cdot j}\boldsymbol{x}_{\cdot j}^{\mathrm{H}}, \text{ i.e.,} R_{ii'} = \frac{1}{n}\sum_{j=1}^{n} x_{ij}x_{i'j},\ 1 \le i, i' \le m, \tag{13.153}$$

where

$$\boldsymbol{x}_{\cdot j} = \begin{bmatrix} x_{1j} \\ x_{2j} \\ \vdots \\ x_{mj} \end{bmatrix}.$$

We find the eigenvalues and eigenvectors of $\boldsymbol{R}$ as discussed in Section 13.2.1:

$$\boldsymbol{R}\boldsymbol{u}_i = \lambda_i \boldsymbol{u}_i,\ i = 1, 2, \ldots, m. \tag{13.154}$$

By writing

$$\boldsymbol{\Lambda} = \begin{bmatrix} \lambda_1 & 0 & \cdots & 0 \\ 0 & \lambda_2 & \cdots & 0 \\ \vdots & \vdots & \ddots & \vdots \\ 0 & 0 & \cdots & \lambda_m \end{bmatrix},\ \boldsymbol{U} = [\boldsymbol{u}_1\boldsymbol{u}_2\cdots\boldsymbol{u}_m],\ \boldsymbol{u}_i = \begin{bmatrix} u_{i1} \\ u_{i2} \\ \vdots \\ u_{im} \end{bmatrix},\ i = 1, 2, \ldots, m, \tag{13.155}$$

we can expand $\boldsymbol{R}$ as follows:

$$\boldsymbol{R} = \boldsymbol{U}\boldsymbol{\Lambda}\boldsymbol{U}^{\mathrm{H}} = \sum_{i=1}^{m} \lambda_i \boldsymbol{u}_i\boldsymbol{u}_i^{\mathrm{H}} = \sum_{i=1}^{m} \lambda_i \boldsymbol{E}_i, \tag{13.156}$$

where

$$\boldsymbol{E}_i = \boldsymbol{u}_i\boldsymbol{u}_i^{\mathrm{H}},\ i = 1, 2, \ldots, m,$$

are the *projection matrices* and are orthogonal to each other; namely,

$$\boldsymbol{E}_i\boldsymbol{E}_{i'} = \boldsymbol{E}_i\delta_{i,i'} \text{ and } \boldsymbol{R}\boldsymbol{E}_i = \lambda_i\boldsymbol{E}_i. \tag{13.157}$$

We then decompose data $\mathbf{X}$ by projecting it to the new coordinates whose unit vectors are given by $\boldsymbol{u}_1, \boldsymbol{u}_2, \ldots, \boldsymbol{u}_m$:

$$\boxed{\boldsymbol{\chi}_i^{\top} = \boldsymbol{u}_i^{\mathrm{H}}\mathbf{X},\ i = 1, 2, \ldots, m,} \tag{13.158}$$

where each $\boldsymbol{\chi}_i^{\top}$ is an n-dimensional row vector. By forming an $m \times n$ matrix $\boldsymbol{\chi}$,

$$\boldsymbol{\chi} = \begin{bmatrix} \boldsymbol{\chi}_1^{\top} \\ \boldsymbol{\chi}_2^{\top} \\ \vdots \\ \boldsymbol{\chi}_m^{\top} \end{bmatrix},$$

we can write

$$\boldsymbol{\chi} = \boldsymbol{U}^{\mathrm{H}}\mathbf{X} = \boldsymbol{U}^{-1}\mathbf{X}.$$

Then, we can reconstruct $\mathbf{X}$ from the $\boldsymbol{\chi}_i$ as

$$\boxed{\mathbf{X} = U\boldsymbol{\chi} = \sum_{i=1}^{m} \boldsymbol{u}_i \boldsymbol{\chi}_i^{\top}.} \tag{13.159}$$

In other words, we transform the $m \times n$ array $\mathbf{X}$ into another $m \times n$ array $\boldsymbol{\chi}$, whose m row vectors $\boldsymbol{\chi}_i$ are *orthogonal* to each other (Problem 13.19):

$$\langle \boldsymbol{\chi}_i, \boldsymbol{\chi}_{i'} \rangle \triangleq \boldsymbol{\chi}_i^{\top} \boldsymbol{\chi}_{i'}^{*} = \mu_i \delta_{i,i'}, \tag{13.160}$$

where $\mu_i = n\lambda_i$ are eigenvalues of the matrix $\mathbf{X}\mathbf{X}^{\mathrm{H}}$. We order the spectrums μ_i in decreasing order; i.e., $\mu_1 \geq \mu_2 \geq \cdots \geq \mu_m$.

Because of the relation $\|\boldsymbol{\chi}_i\|^2 = \langle \boldsymbol{\chi}_i, \boldsymbol{\chi}_i \rangle = \mu_i$ established in (13.160), we see that the vector variable $\boldsymbol{\chi}_1$ carries the largest amount of "information," and $\boldsymbol{\chi}_2$ is the next largest, and so forth. Instead of keeping all the $\boldsymbol{\chi}_i$, we may select only the k largest spectral components, where k is a suitably chosen number. In other words, we use the k **principal components** to approximate the original variable $\boldsymbol{X}$ by

$$\hat{\mathbf{X}} = \sum_{i=1}^{k} \boldsymbol{u}_i \boldsymbol{\chi}_i^{\top}. \tag{13.161}$$

In order to quantify the quality of this approximation, we need to define the inner product and norm of matrices.

DEFINITION 13.2 (Inner product and the norm of matrices). *For $m \times n$ matrices $\boldsymbol{A}$ and $\boldsymbol{B}$, we define their inner product by*

$$\langle \boldsymbol{A}, \boldsymbol{B} \rangle = \mathrm{trace}(\boldsymbol{A}\boldsymbol{B}^{\mathrm{H}}). \tag{13.162}$$

Then the norm of $\boldsymbol{A}$ is defined as

$$\|\boldsymbol{A}\| = \sqrt{\mathrm{trace}(\boldsymbol{A}^{\mathrm{H}}\boldsymbol{A})} = \sqrt{\sum_{i=1}^{m}\sum_{j=1}^{n} |a_{ij}|^2} \tag{13.163}$$

*which is called the **Frobenius**[28] **norm.*** □

The reader is suggested to check that the above definition of inner product satisfies the five properties of an inner product space defined in Definition 10.1.

Then we can show (Problem 13.20) that the approximation error given in terms of the **sum of squares of the differences** can be expressed by using the Frobenius norm

$$\boxed{Q = \|\mathbf{X} - \hat{\mathbf{X}}\|^2 = \sum_{i=1}^{m}\sum_{j=1}^{n} |x_{ij} - \hat{x}_{ij}|^2 = \sum_{j>k} \mu_j,} \tag{13.164}$$

[28] Ferdinand Georg Frobenius (1849–1917) was a German mathematician.

where μ_j are the eigenvalues of the matrix $\mathbf{XX}^{\mathrm{H}}$, as defined earlier. Equation (13.161) forms the basis of *data compression* or *data reduction* based on the principal components. Since the principal components χ_i, $1 \le i \le k$, contain much of the information of $\mathbf{X}$, they can be also used for classification of data, and for feature extraction for pattern recognition.

Example 13.4: Let us assume $m = 2$ and $n = 3$, and the data are given by

$$\mathbf{X} = \begin{bmatrix} 2 & 1 & 0 \\ 4 & 3 & 0 \end{bmatrix}; \text{ hence, } \mathbf{X}^{\top} = \begin{bmatrix} 2 & 4 \\ 1 & 3 \\ 0 & 0 \end{bmatrix}.$$

Then

$$\mathbf{XX}^{\top} = \begin{bmatrix} 5 & 11 \\ 11 & 25 \end{bmatrix}; \text{ hence, } \boldsymbol{R} = \frac{1}{n}\mathbf{XX}^{\top} = \frac{1}{3}\begin{bmatrix} 5 & 11 \\ 11 & 25 \end{bmatrix}.$$

Then we have the characteristic equation

$$\det\left|\frac{1}{n}\mathbf{XX}^{\top} - \lambda \boldsymbol{I}\right| = 0 \text{ or } \det\begin{vmatrix} 5-\mu & 11 \\ 11 & 25-\mu \end{vmatrix} = 0, \text{ where } \mu = n\lambda = 3\lambda,$$

which gives

$$\mu^2 - 30\mu + 4 = 0,$$

from which we find

$$\mu_1 = 15 + \sqrt{221} \approx 29.8661,\ \mu_2 = 15 - \sqrt{221} \approx 0.1339.$$

Thus

$$\lambda_1 = \frac{\mu_1}{3} = 9.9537,\ \lambda_2 = \frac{\mu_2}{3} = 0.0446; \text{ hence, } \boldsymbol{\Lambda} = \begin{bmatrix} 9.9537 & 0 \\ 0 & 0.0446 \end{bmatrix}.$$

Then, we find the orthonormal vectors $\boldsymbol{u}_1$ and $\boldsymbol{u}_2$ as

$$\boldsymbol{u}_1 = \begin{bmatrix} 0.4045 \\ 0.9145 \end{bmatrix},\ \boldsymbol{u}_2 = \begin{bmatrix} -0.9145 \\ 0.4045 \end{bmatrix}; \text{ hence, } \boldsymbol{U} = \begin{bmatrix} 0.4045 & -0.9145 \\ 0.9145 & 0.4045 \end{bmatrix}.$$

Thus, from (13.158), we find

$$\chi_1 = \begin{bmatrix} 2 & 4 \\ 1 & 3 \\ 0 & 0 \end{bmatrix}\begin{bmatrix} 0.4045 \\ 0.9145 \end{bmatrix} = \begin{bmatrix} 4.4670 \\ 3.1480 \\ 0 \end{bmatrix} \text{ and}$$

$$\chi_2 = \begin{bmatrix} 2 & 4 \\ 1 & 3 \\ 0 & 0 \end{bmatrix}\begin{bmatrix} -0.9145 \\ 0.4045 \end{bmatrix} = \begin{bmatrix} -0.2110 \\ 0.2990 \\ 0 \end{bmatrix}.$$

Thus, we can expand the data $\mathbf{X}$ using the eigenvectors as follows:

$$\mathbf{X} = \boldsymbol{u}_1 \boldsymbol{\chi}_1^\top + \boldsymbol{u}_2 \boldsymbol{\chi}_2^\top .$$

If we set $k = 1$ in the approximation formula (13.161), we obtain

$$\hat{\mathbf{X}} = \boldsymbol{u}_1 \boldsymbol{\chi}_1^\top = \begin{bmatrix} 1.8069 & 1.2734 & 0 \\ 4.0851 & 2.8788 & 0 \end{bmatrix}.$$

Thus, the difference is

$$\mathbf{X} - \hat{\mathbf{X}} = \begin{bmatrix} 0.1931 & -0.2734 & 0 \\ -0.0851 & 0.1212 & 0 \end{bmatrix}. \tag{13.165}$$

Alternatively, we can find

$$\mathbf{X} - \hat{\mathbf{X}} = \boldsymbol{u}_2 \boldsymbol{\chi}_2^\top = \begin{bmatrix} 0.1930 & -0.2734 & 0 \\ -0.0853 & 0.1209 & 0 \end{bmatrix}. \tag{13.166}$$

If we use (13.165), the sum of the squares of the differences is

$$Q = \sum_{i=1}^{m} \sum_{j=1}^{n} |x_{ij} - \hat{x}_{ij}|^2 = 0.1339,$$

which is equal to the eigenvalue μ_2, confirming the formula (13.165). If we use (13.166) instead, we obtain $Q = 0.1338$, which is slightly off, but this is due to the truncation error in numerical computation.

Figure 13.3a shows the data as three points (x_{1j}, x_{2j}); $j = 1, 2, 3$ and their transformations in the new coordinate axes $\boldsymbol{u}_1$ and $\boldsymbol{u}_2$, which are represented as (χ_{1j}, χ_{2j});

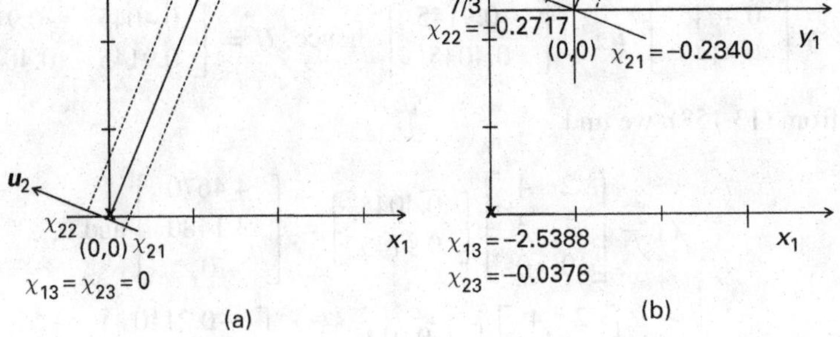

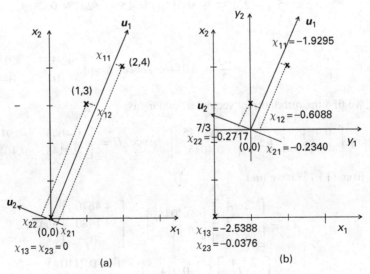

Figure 13.3 Geometric interpretations of (a) PCA directly applied to the data $\mathbf{X}$; (b) PCA applied to the mean adjusted data $\mathbf{Y} = \mathbf{X} - \bar{\boldsymbol{x}}\,\mathbf{1}_n^\top$.

$j = 1, 2, 3$. It is clear that the variation of the three points in the direction of $\boldsymbol{u}_1$ is much larger than that along the $\boldsymbol{u}_2$ axis. This is reflected in the enormous difference in eigenvalues $\lambda_1 = 9.9537$ versus $\lambda_2 = 0.0446$. We can interpret λ_i as the second moment of the three data points in the direction of $\boldsymbol{u}_i$; $i = 1, 2$.

□

13.3.1.2 Adjustment by the empirical means

When the variance (i.e., the central second moments) of data is believed to be more informative than the second moments, we should subtract the sample means (or empirical means) before we apply PCA. We denote the translated data by $\mathbf{Y}$:

$$\mathbf{Y} = \mathbf{X} - \overline{\boldsymbol{x}}\,\mathbf{1}_n^\top,$$

where

$$\overline{\boldsymbol{x}} = \begin{bmatrix} \overline{x}_{1\cdot} \\ \overline{x}_{2\cdot} \\ \vdots \\ \overline{x}_{m\cdot} \end{bmatrix} \text{ and } \overline{x}_{i\cdot} = \frac{\sum_{j=1}^{n} x_{ij}}{n},$$

and $\mathbf{1}_n$ is an n-dimensional column vector whose elements are all one. Then we apply PCA to $\mathbf{Y}$ using the eigenvectors, which we denote as $\tilde{\boldsymbol{u}}_i$, of the correlation matrix of $\mathbf{Y}$, which corresponds to the *covariance matrix* of $\mathbf{X}$. Then the PCA expansion of the original data $\mathbf{X}$ can be presented as follows:

$$\mathbf{X} = \mathbf{Y} + \overline{\boldsymbol{x}}\,\mathbf{1}_n^\top = \sum_{i=1}^{m} \tilde{\boldsymbol{u}}_i \tilde{\chi}_i^{\mathrm{H}} + \overline{\boldsymbol{x}}\,\mathbf{1}_n^\top, \tag{13.167}$$

where

$$\tilde{\chi}_i = \mathbf{Y}^{\mathrm{H}} \tilde{\boldsymbol{u}}_i = (\mathbf{X}^{\mathrm{H}} - \mathbf{1}\,\overline{\boldsymbol{x}}^{\mathrm{H}})\tilde{\boldsymbol{u}}_i,\ i = 1, 2, \ldots, m. \tag{13.168}$$

The reader is suggested to work on the data $\mathbf{X}$ used in Example 13.4 (Problem 13.21). Generally, the eigenvalues will decrease, since they are the central second moment of the expansion coefficients χ_j which have zero mean. In Figure 13.3 (b) we give a geometric illustration of PCA applied to the mean adjusted data. The directions of the eigenvectors are slightly different from the original case presented in Figure 13.3 (a).

13.3.1.3 Correlation across n data points

If we expect that the correlation across the n data points exists, we should form the following $n \times n$ correlation matrix:

$$\tilde{\boldsymbol{R}} = \frac{1}{m}\mathbf{X}^{\mathrm{H}}\mathbf{X}. \tag{13.169}$$

If $n > m$, as in Example 13.4, this matrix is singular, since the *rank*[29] r of a matrix cannot be larger than $\min(m, n)$. As will be proved below, the first m eigenvalues of $\mathbf{X}^{\mathrm{H}}\mathbf{X}$ are the same as those of $\mathbf{X}\mathbf{X}^{\mathrm{H}}$. Thus, the first m eigenvalues of $\tilde{\boldsymbol{R}}$ are the eigenvalues of $\boldsymbol{R}$ multiplied by n/m, and the remaining $(n-m)$ eigenvalues are all zeros. We denote the eigenvector associated with the jth eigenvalue λ_j as $\boldsymbol{v}_j$ and the collection of eigenvectors by $\boldsymbol{V}$:

$$\boldsymbol{V} = [\boldsymbol{v}_1 \boldsymbol{v}_2 \cdots \boldsymbol{v}_m]. \tag{13.170}$$

Then, similar to (13.89) and (13.156), we have the following similarity transformation:

$$\tilde{\boldsymbol{R}} = \boldsymbol{V}\tilde{\boldsymbol{\Lambda}}\boldsymbol{V}^{\mathrm{H}} = \boldsymbol{V}\tilde{\boldsymbol{\Lambda}}\boldsymbol{V}^{-1}, \tag{13.171}$$

where $\tilde{\boldsymbol{\Lambda}}$ is a diagonal matrix, whose left upper corner is $\boldsymbol{\Lambda}$ and the rest of diagonal terms are all zeros.

Example 13.5: Example 13.4 continued. Let us illustrate the above procedure by using the same data of Example 13.4. Then we have

$$\mathbf{X}^{\top}\mathbf{X} = \begin{bmatrix} 20 & 14 & 0 \\ 14 & 10 & 0 \\ 0 & 0 & 0 \end{bmatrix}.$$

The characteristic equation for eigenvalues is given by

$$\det|\mathbf{X}^{\top}\mathbf{X} - \mu \boldsymbol{I}| = 0; \text{ i.e., } \det \begin{vmatrix} 20-\mu & 14 & 0 \\ 14 & 10-\mu & 0 \\ 0 & 0 & -\mu \end{vmatrix} = 0,$$

where the eigenvalue μ is related to the eigenvalue $\tilde{\lambda}$ of $\tilde{\boldsymbol{R}}$ simply by $\mu = m\tilde{\lambda} = 2\tilde{\lambda}$. Eigenvectors of $\mathbf{X}^{\top}\mathbf{X}$ and those of $\tilde{\boldsymbol{R}}$ are the same as we already pointed out. The last expression is reduced to

$$\mu(\mu^2 - 30\mu + 4) = 0.$$

Thus, μ_1 and μ_2 are the same as those found in Example 13.4 and $\mu_3 = 0$:

$$\mu_1 = 29.8661, \mu_2 = 0.1339, \text{ and } \mu_3 = 0.$$

Thus,

$$\tilde{\lambda}_1 = 14.9331, \tilde{\lambda}_2 = 0.0669, \lambda_3 = 0; \text{ hence, } \tilde{\boldsymbol{\Lambda}} = \begin{bmatrix} 14.9331 & 0 & 0 \\ 0 & 0.0669 & 0 \\ 0 & 0 & 0 \end{bmatrix}.$$

[29] The rank of a matrix is the number of linearly independent columns or rows.

The corresponding orthonormal eigenvectors are

$$\boldsymbol{v}_1 = \begin{bmatrix} 0.8174 \\ 0.5760 \\ 0 \end{bmatrix}, \boldsymbol{v}_2 = \begin{bmatrix} -0.5760 \\ 0.8174 \\ 0 \end{bmatrix}, \boldsymbol{v}_3 = \begin{bmatrix} 0 \\ 0 \\ 1 \end{bmatrix}.$$

Thus,

$$V = \begin{bmatrix} 0.8174 & -0.5760 & 0 \\ 0.5760 & 0.8174 & 0 \\ 0 & 0 & 1 \end{bmatrix}.$$

Then, by projecting $\mathbf{X}$ to the orthonormal eigenvectors $\boldsymbol{v}_1, \boldsymbol{v}_2, \boldsymbol{v}_3$, we can obtain the new coordinates χ_1, χ_2, and χ_3, which are now two-dimensional vectors. They are found as

$$\chi_1 = \mathbf{X}\boldsymbol{v}_1 = \begin{bmatrix} 2.2108 \\ 4.9976 \end{bmatrix}, \chi_2 = \mathbf{X}\boldsymbol{v}_2 = \begin{bmatrix} -0.3346 \\ 0.1482 \end{bmatrix}, \chi_3 = \mathbf{X}\boldsymbol{v}_3 = \begin{bmatrix} 0 \\ 0 \end{bmatrix}.$$

Thus, the data set can be written as

$$\mathbf{X} = \chi_1 \boldsymbol{v}_1^\top + \chi_2 \boldsymbol{u}^\top.$$

If we choose $k = 1$, then the PCA approximation is

$$\hat{\mathbf{X}} = \chi_1 \boldsymbol{v}_1^\top = \begin{bmatrix} 1.8071 & 1.2734 & 0 \\ 4.0850 & 2.8786 & 0 \end{bmatrix}.$$

Then the difference between the original data and the PCA approximation is

$$\mathbf{X} - \hat{\mathbf{X}} = \begin{bmatrix} 0.1929 & -0.2734 & 0 \\ -0.0850 & 0.1214 & 0 \end{bmatrix}, \tag{13.172}$$

which can be alternatively computed as

$$\mathbf{X} - \hat{\mathbf{X}} = \chi_2 \boldsymbol{u}^\top = \begin{bmatrix} 0.1927 & -0.2735 & 0 \\ -0.0854 & 0.1211 & 0 \end{bmatrix}. \tag{13.173}$$

If we use the latter approximation, the sum of squares of the difference between the data and its approximation is

$$Q = \sum_{i=1}^{m} \sum_{j=1}^{n} |x_{ij} - \hat{x}_{ij}|^2 = 0.1339,$$

which is again equal to the eigenvalue μ_2, as expected. If we use the error (13.172), we find $Q = 0.1338 \approx \mu_2$. □

13.3.2 Singular value decomposition (SVD)

In the previous section we established the following two similarity transformations for the $m \times n$ data set $\mathbf{X}$:

$$\mathbf{X}\mathbf{X}^{\mathrm{H}} = \boldsymbol{U}\boldsymbol{M}\boldsymbol{U}^{\mathrm{H}}, \tag{13.174}$$

$$\mathbf{X}^{\mathrm{H}}\mathbf{X} = \boldsymbol{V}\tilde{\boldsymbol{M}}\boldsymbol{V}^{\mathrm{H}}, \tag{13.175}$$

where $\boldsymbol{M}$ and $\tilde{\boldsymbol{M}}$ are diagonal matrices of size m and n respectively. In the rest of this section we assume $m \le n$, unless stated otherwise. Then we can write these matrices as

$$\boldsymbol{M} = \begin{bmatrix} \mu_1 & 0 & \cdots & 0 \\ 0 & \mu_2 & \cdots & 0 \\ \vdots & \vdots & \ddots & \vdots \\ 0 & 0 & \cdots & \mu_m \end{bmatrix}, \quad \tilde{\boldsymbol{M}} = \begin{bmatrix} \boldsymbol{M} & 0 & \cdots & 0 \\ 0 & 0 & \cdots & 0 \\ \vdots & \vdots & \ddots & \vdots \\ 0 & 0 & \cdots & 0 \end{bmatrix}.$$

If $m \ge n$, the structure of $\boldsymbol{M}$ and $\tilde{\boldsymbol{M}}$ should be reversed, and the eigenvalues run from μ_1 through μ_n, and the rest of the arguments should remain unchanged.

From (13.174) we can write

$$\mathbf{X}\mathbf{X}^{\mathrm{H}} = \boldsymbol{U}\sqrt{\boldsymbol{M}}\sqrt{\boldsymbol{M}}\boldsymbol{U}^{\mathrm{H}} = \boldsymbol{U}\Sigma\Sigma^{\top}\boldsymbol{U}^{\mathrm{H}}, \tag{13.176}$$

where $\sqrt{\boldsymbol{M}}$ is an $m \times m$ diagonal matrix with $\sqrt{\mu_i}$ as its diagonal elements:

$$\sqrt{\boldsymbol{M}} = \begin{bmatrix} \sqrt{\mu_1} & 0 & \cdots & 0 \\ 0 & \sqrt{\mu_2} & \cdots & 0 \\ \vdots & \vdots & \ddots & \vdots \\ 0 & 0 & \cdots & \sqrt{\mu_m} \end{bmatrix} = \begin{bmatrix} \sigma_1 & 0 & \cdots & 0 \\ 0 & \sigma_2 & \cdots & 0 \\ \vdots & \vdots & \ddots & \vdots \\ 0 & 0 & \cdots & \sigma_m \end{bmatrix} \tag{13.177}$$

and

$$\Sigma = \left[\sqrt{\boldsymbol{M}}\mathbf{O}_{(n-m)\times m}\right], \text{ or } \Sigma = \begin{bmatrix} \sigma_1 & 0 & \cdots & 0 & 0 & \cdots & 0 \\ 0 & \sigma_2 & \cdots & 0 & 0 & \cdots & 0 \\ \vdots & \vdots & \ddots & \vdots & \vdots & \ddots & \vdots \\ 0 & 0 & \cdots & \sigma_m & 0 & \cdots & 0 \end{bmatrix}, \text{ with } \sigma_i = \sqrt{\mu_i}, \tag{13.178}$$

and we readily find the following simple relations among Σ, $\boldsymbol{M}$, and $\tilde{\boldsymbol{M}}$:

$$\Sigma\Sigma^{\top} = \boldsymbol{M} \text{ and } \Sigma^{\top}\Sigma = \tilde{\boldsymbol{M}}. \tag{13.179}$$

Now for any $n \times n$ unitary matrix $\boldsymbol{A}$ (i.e., a matrix such that $\boldsymbol{A}\boldsymbol{A}^{\mathrm{H}} = \boldsymbol{I}$), we have from (13.176)

$$\mathbf{X}\mathbf{X}^{\mathrm{H}} = \boldsymbol{U}\Sigma\boldsymbol{A}\boldsymbol{A}^{\mathrm{H}}\Sigma^{\top}\boldsymbol{U}^{\mathrm{H}} = (\boldsymbol{U}\Sigma\boldsymbol{A})(\boldsymbol{U}\Sigma\boldsymbol{A})^{\mathrm{H}}. \tag{13.180}$$

Then we readily find that $\mathbf{X}$ can be written as

$$\mathbf{X} = \boldsymbol{U}\Sigma\boldsymbol{A}; \text{ hence, } \mathbf{X}^{\mathrm{H}}\mathbf{X} = \boldsymbol{A}^{\mathrm{H}}\Sigma^{\top}\boldsymbol{U}^{\mathrm{H}}\boldsymbol{U}\Sigma\boldsymbol{A} = \boldsymbol{A}^{\mathrm{H}}\tilde{\boldsymbol{M}}\boldsymbol{A}.$$

By comparing the last expression with (13.175), we identify $\boldsymbol{A}^{\mathrm{H}} = \boldsymbol{V}$. Thus, we finally obtain the following decomposition for $\mathbf{X}$:

$$\boxed{\mathbf{X} = \boldsymbol{U}\Sigma\boldsymbol{V}^{\mathrm{H}} = \sum_{i=1}^{\min(m,n)} \sigma_i \boldsymbol{u}_i \boldsymbol{v}_i^{\mathrm{H}},} \tag{13.181}$$

which is called the **SVD** of **X** [124]. The SVD expression (13.181) can be also written as

$$\mathbf{X} = \sum_{i=1}^{\min(m,n)} \sigma_i \boldsymbol{E}_i, \text{ where } \boldsymbol{E}_i \triangleq \boldsymbol{u}_i \boldsymbol{v}_i^{\mathrm{H}}. \tag{13.182}$$

Here, $\boldsymbol{E}_i$ are $m \times n$ matrices, satisfying the following orthogonality conditions:

$$\boldsymbol{E}_i \boldsymbol{E}_j^{\mathrm{H}} = \boldsymbol{u}_i \boldsymbol{u}_i^{\mathrm{H}} \delta_{i,j} \text{ and } \boldsymbol{E}_j^{\mathrm{H}} \boldsymbol{E}_i = \boldsymbol{v}_i \boldsymbol{v}_i^{\mathrm{H}} \delta_{i,j}, \tag{13.183}$$

which are similar to (13.157). We also find (Problem 13.22)

$$\boxed{\mathbf{X}\boldsymbol{v}_j = \sigma_j \boldsymbol{u}_j, \mathbf{X}^{\mathrm{H}}\boldsymbol{u}_i = \sigma_i \boldsymbol{v}_i,} \tag{13.184}$$

for $1 \le i \le m, 1 \le j \le n$, where$\sigma_j = 0, \min(m,n) < j \le \max(m,n)$. This last expression resembles (13.154).

We should note that, when $m \le n$, the $m \times m$ matrix $\boldsymbol{U}$ can be obtained once the $n \times n$ matrix $\boldsymbol{V}$ and the $m \times m$ eigenvalue matrix $\boldsymbol{M}$ are found (Problem 13.23) as

$$\boldsymbol{U} = [\mathbf{X}\boldsymbol{V}]_{m \times m} \boldsymbol{M}^{-1/2}, \tag{13.185}$$

where $[\mathbf{X}\boldsymbol{V}]_{m \times m}$ is the first m columns of the $m \times n$ matrix $\mathbf{X}\boldsymbol{V}$. Needless to say, when $m \ge n$, $\boldsymbol{V}$ can be found once $\boldsymbol{U}$ is obtained.

Suppose that we select the $k(< \min(m,n))$ largest singular values $\sigma_1 \ge \sigma_2 \ge \cdots \ge \sigma_k$ and ignore the terms for $i > k$ in the decomposition (13.181) or (13.181). Then we have the following approximate reconstruction of X:

$$\hat{\mathrm{X}} = \sum_{i=1}^{k} \sigma_i \boldsymbol{u}_i \boldsymbol{v}_i^{\mathrm{H}} = \sum_{i=1}^{k} \sigma_i \boldsymbol{E}_i. \tag{13.186}$$

THEOREM 13.1 (Frobenius norm and singular values). *The Frobenius norm of an $m \times n$ matrix $\boldsymbol{A}$, $\|\boldsymbol{A}\|$ defined in Definition 13.2, is given by*

$$\|A\|^2 = \text{trace}(AA^{\mathrm{H}}) = \sum_{i=1}^{\min(m,n)} \sigma_i^2, \tag{13.187}$$

where σ_i are the singular values of **A**.

Proof. The proof is left to the reader as an exercise problem (Problem 13.24). □

Then the sum of squares of differences between **X** and its approximation can be expressed as

$$Q = \|\mathbf{X} - \hat{\mathbf{X}}\|^2 = \sum_{i=1}^{m}\sum_{j=1}^{n} |x_{ij} - \hat{x}_{ij}|^2 = \sum_{i=k+1}^{m} \sigma_i^2. \tag{13.188}$$

Therefore, the SVD decomposition, similar to the PCA decomposition, may allow us to substantially reduce or compress the data and yet retain much of information, if k is appropriately chosen.

Example 13.6: Consider the same 2×3 data **X** used in Examples 13.4 and 13.5. The entries of the singular matrix are computed as $\sigma_1 = \sqrt{\mu_1} = 5.4650$ and $\sigma_2 = \sqrt{\mu_2} = 0.3660$. Then the SVD decomposition

$$\mathbf{X} = U\Sigma V^{\mathrm{H}} = \sum_{i=1}^{m} \sigma_i \boldsymbol{u}_i \boldsymbol{v}_i^{\mathrm{H}}$$

is found for this data **X** as

$$\begin{aligned}
\begin{bmatrix} 2 & 1 & 0 \\ 4 & 3 & 0 \end{bmatrix} &= \begin{bmatrix} 0.4045 & -0.9145 \\ 0.9145 & 0.4045 \end{bmatrix} \begin{bmatrix} 5.4650 & 0 & 0 \\ 0 & 0.3660 & 0 \end{bmatrix} \begin{bmatrix} 0.8174 & 0.5760 & 0 \\ -0.5760 & 0.8174 & 0 \\ 0 & 0 & 1 \end{bmatrix} \\
&= 5.4650 \begin{bmatrix} 0.4045 \\ 0.9145 \end{bmatrix} \begin{bmatrix} 0.8174 & 0.5760 & 0 \end{bmatrix} \\
&\quad + 0.3660 \begin{bmatrix} -0.9145 \\ 0.4045 \end{bmatrix} \begin{bmatrix} 0.5760 & 0.8174 & 0 \end{bmatrix}.
\end{aligned} \tag{13.189}$$

We can confirm the formula (13.185) by computing

$$\mathbf{X}V = \begin{bmatrix} 2 & 1 & 0 \\ 4 & 3 & 0 \end{bmatrix} \begin{bmatrix} 0.8174 & -0.5760 & 0 \\ 0.5760 & 0.8174 & 0 \\ 0 & 0 & 1 \end{bmatrix} = \begin{bmatrix} 2.2108 & -0.3346 & 0 \\ 4.9976 & 0.1482 & 0 \end{bmatrix}.$$

Hence,

$$[\mathbf{X}V]_{2\times 2}\,M^{-1/2} = \begin{bmatrix} 2.2108 & -0.3346 \\ 4.9976 & 0.1482 \end{bmatrix} \begin{bmatrix} 5.465^{-1} & 0 \\ 0 & 0.3660^{-1} \end{bmatrix}$$
$$= \begin{bmatrix} 0.4045 & -0.9142 \\ 0.9145 & 0.4049 \end{bmatrix} = V.$$

If we set $k = 1$ in (13.186) and discard the second term in the SVD decomposition, we have an approximation of $\mathbf{X}$:

$$\hat{\mathbf{X}} = \sigma_1 \boldsymbol{u}_1 \boldsymbol{v}_1^{\mathrm{H}} = \begin{bmatrix} 1.8069 & 1.2733 & 0 \\ 4.0852 & 2.8787 & 0 \end{bmatrix}.$$

Thus, the sum of the squares of differences is

$$Q = \|\mathbf{X} - \hat{\mathbf{X}}\|^2 = \left\| \begin{matrix} 0.1931 & -0.2733 & 0 \\ -0.0852 & 0.1213 & 0 \end{matrix} \right\|^2$$
$$= 0.1931^2 + 0.2733^2 + 0.0852^2 + 0.1213^2 = 0.1347,$$

which is practically equal to $\sigma_2^2 = 0.1340$. Thus, we confirmed the formula (13.188).

□

In the above example, we readily notice that the third row of V^{H} (i.e., the third column of V) and the third row of Σ^{H} are superfluous. Note that this is not because the third column of $\mathbf{X}$ happens to be all zeros, but because the rank of the matrix $\mathbf{X}$ is two. We can simplify the above decomposition to

$$\mathbf{X} = \begin{bmatrix} 2 & 1 & 0 \\ 4 & 3 & 0 \end{bmatrix} = \begin{bmatrix} 0.4045 & -0.9145 \\ 0.9145 & 0.4045 \end{bmatrix} \begin{bmatrix} 5.4650 & 0 \\ 0 & 0.3660 \end{bmatrix} \begin{bmatrix} 0.8174 & 0.5760 & 0 \\ -0.5760 & 0.8174 & 0 \end{bmatrix}.$$

By generalizing this observation, we can have a *reduced form* of singular value decomposition of an $m \times n$ matrix $\mathbf{X}$ with $m \le n$:

$$\mathbf{X} = U\sqrt{M}\tilde{V}^{\mathrm{H}}, \tag{13.190}$$
$$\mathbf{X}^{\mathrm{H}} = \tilde{V}\sqrt{M}U^{\mathrm{H}}, \tag{13.191}$$

where $\tilde{V}$ is an $n \times m$ matrix, obtained by deleting the $(n - m)$ columns of the right end of V.

The formula (13.185) to derive U from V can be rewritten as

$$U = \mathbf{X}\tilde{V}M^{-\frac{1}{2}}.$$

13.3.2.1 Computation of pseudo-inverse

The SVD has a broad range of application in matrix algebra, such as computation of the pseudo-inverse, in addition to the least-squares fitting of data, matrix approximation, and determining the rank, range, and null space of a matrix. Here, we briefly discuss the **pseudo-inverse**.

The *pseudo-inverse* $A^{\dagger}$ of an $m \times n$ matrix A (whose entries can be real or complex numbers) is defined as the unique $n \times m$ matrix satisfying all of the following four criteria:

1. $\boldsymbol{A}\boldsymbol{A}^{\dagger}\boldsymbol{A} = \boldsymbol{A}$ (here $\boldsymbol{A}\boldsymbol{A}^{\dagger}$ need not be the identity matrix, but it maps all column vectors of $\boldsymbol{A}$ to themselves);
2. $\boldsymbol{A}^{\dagger}\boldsymbol{A}\boldsymbol{A}^{\dagger} = \boldsymbol{A}^{\dagger}$;
3. $(\boldsymbol{A}\boldsymbol{A}^{\dagger})^{\mathrm{H}} = \boldsymbol{A}\boldsymbol{A}^{\dagger}$ (i.e., $\boldsymbol{A}\boldsymbol{A}^{\dagger}$ is Hermitian); and
4. $(\boldsymbol{A}^{\dagger}\boldsymbol{A})^{\mathrm{H}} = \boldsymbol{A}^{\dagger}\boldsymbol{A}$ (i.e., $\boldsymbol{A}^{\dagger}\boldsymbol{A}$ is also Hermitian).

The SVD provides a way of computing the *pseudo-inverse* of $\mathbf{X}$, denoted $\mathbf{X}^{\dagger}$, as follows

$$\mathbf{X}^{\dagger} = \boldsymbol{V}\Sigma^{\dagger}\boldsymbol{U}^{\mathrm{H}},$$

where $\Sigma^{\dagger}$, the pseudo-inverse of Σ, is obtained by taking the reciprocal of each nonzero element on the diagonal in $\Sigma^{\top}$.

13.3.3 Matrix decomposition methods for Web information retrieval

Web information retrieval is of increasing importance and poses technical challenges and opportunities different from traditional information retrieval because of its huge size and its rapidly changing nature. Google's PageRank algorithm (e.g., see Del Corso *et al.* [79] and Langville and Meyer [218]) is a computationally efficient procedure to find the *stationary distribution vector* $\boldsymbol{\pi}$ of the transition probability matrix (TPM) $\boldsymbol{P} = [p_{ij}; i, j = 1, 2, \ldots, n]$ of the Markov chain defined over a Web graph, where each of n nodes (or states in the chain) represents a web page. The stationary distribution vector is given as the solution of $\boldsymbol{\pi}^{\top} = \boldsymbol{\pi}^{\top}\boldsymbol{P}$, as discussed in Section 15.2.2; so the PageRank statistic $\boldsymbol{\pi} = (\pi_i; i = 1, 2, \ldots, n)$ is equivalent to the *left eigenvector* associated with the eigenvalue $\lambda_1 = 1$ of $\boldsymbol{P}$.

Another search engine algorithm, known as HITS (hypertext induced topics search) (see Kleinberg [187]) is based on PCA applied to the $n \times n$ *adjacency matrix* $\boldsymbol{A} = [a_{ij}]$ of the Web graph, which is a 1–0 matrix defined by $a_{ij} = 1$ if there is a link from node i to node j and $a_{ij} = 0$ otherwise. In terms of the Markov chain defined above, it is simply given as $a_{ij} = \operatorname{sgn} p_{ij}$.

Computational procedures for such *spectral expansion* of the matrix should take advantage of the *sparsity* of the matrix $\boldsymbol{A}$ or $\boldsymbol{P}$; namely, the number of links (i, j) with $a_{ij} = 1$ (i.e., $p_{ij} > 0$) is much smaller than n^2. An algorithmic study for a Web graph as large as $n = 24$ million nodes with more than 100 million links is reported in the aforementioned 2005 publication [79], and these numbers will continue to grow rapidly.

13.4 Autoregressive moving average time series and its spectrum[30]

In econometrics, signal processing (e.g., video data stream), computer network traffic modeling, and other numerous applications, autoregressive moving average (ARMA) time-series models are very frequently used to represent autocorrelated discrete-time data.

13.4.1 Autoregressive (AR) time series

An autoregressive time series of order p, denoted AR(p), is defined by

$$\boxed{X_n = \sum_{i=1}^{p} a_i X_{n-i} + e_n, n = 1, 2, 3, \ldots,} \tag{13.192}$$

where a_i, $1 \leq i \leq p$, are constants and e_n[31] is the error variable and is usually assumed to be **white noise**, meaning

$$E[e_n] = 0 \text{ and } E[e_n e_m] = \sigma^2 \delta_{n,m}. \tag{13.193}$$

Although the time-series data $\{x_n\}$ may be first observed at time $n = 1$, we consider that the underlying random process $\{X_n\}$ has started at some point in the past. In order for the process to be stationary or WSS, we must define X_n for $-\infty < n < \infty$. We assume the *error* or *disturbance* e_n is independent of $X_{n-1}, \ldots, X_{n-p}$. You may want to add a constant a_0 to the right-hand side of (13.192), but it does not change the essential structure of the model. It simply changes the expectation of X_n from zero to $E[X_n] = a_0/(1 - \sum_{i=1}^{p} a_i) \neq 0$, provided $\sum_{i=1}^{p} a_i \neq 1$ (Problem 13.25).

Example 13.7: AR(1) process. Let us first examine the simplest model, i.e., AR(1). Then, by denoting $a_1 = a$, we have

$$X_n = aX_{n-1} + e_n, -\infty < n < \infty. \tag{13.194}$$

Since e_n are uncorrelated and the current disturbance e_n is independent of X_{n-1}, we see that the AR(1) process is a **simple Markov chain**. Furthermore, if the white noise is Gaussian, X_n is also Gaussian, and thus it becomes a Gauss-Markov process (GMP). We plot a sample path of a GMP in Figure 13.1 (see also Figure 22.8 of Chapter 22).

[30] The reader may skip this section at first reading.

[31] Although we use the lower case e, this is a random variable. In order to be consistent with the rest of the book, we should use some capital letter, but e_n or ϵ_n seems conventionally used in the time series literature. So we follow this abused notation in this section.

By substituting the above recursive expression successively, we find

$$X_n = aX_{n-1} + e_n = a(aX_{n-2} + e_{n-1}) + e_n = \cdots$$
$$= \sum_{j=0}^{k-1} a^j e_{n-j} + a^k X_{n-k}, k = 1, 2, 3, \ldots \quad (13.195)$$

The first term consists of the lagged values of the white noise e_{n-j} that drives the time series X_n, and the effect of past noise diminishes geometrically, provided $|a| < 1$. If $|a| > 1$, the process is nonstationary, because the weighting coefficient a^j will grow infinity in its magnitude as $j \to \infty$. If $a = 1$, $\{X_n\}$ is a **random walk** (see Section 17.1) and is also a **martingale** (see Definition 10.5).

If X_{n-k} is the starting value of this time series, its influence becomes negligibly small as $k \to \infty$. So, if the time series is considered as having started at some time in the distant past, we can write

$$X_n = \lim_{k\to\infty} \sum_{j=0}^{k-1} a^j X_{n-j} = \sum_{j=0}^{\infty} a^j X_{n-j}, \quad (13.196)$$

where the limit should be interpreted as the limit in *mean square* or l.i.m. (see Section 11.2.4, and also Problem 13.26). The above result can be also found by taking the *double-sided* Z-transform of (13.194), yielding

$$X(z) = az^{-1}X(z) + E(z), \quad (13.197)$$

where

$$X(z) = \sum_{n=-\infty}^{\infty} X_n z^{-n} \text{ and } E(z) = \sum_{n=-\infty}^{\infty} e_n z^{-n}. \quad (13.198)$$

From (13.197), we have

$$\boxed{X(z) = \frac{E(z)}{1 - az^{-1}} = \left(\sum_{j=0}^{\infty} a^j z^{-j}\right) E(z),} \quad (13.199)$$

where the region of convergence is $|az^{-1}| < 1$; i.e., $|z| > |a|$. By equating the coefficients of the z^{-n} terms in both sides of the above equation, we have

$$X_n = \sum_{j=0}^{\infty} a^j e_{n-j}, \quad (13.200)$$

which is (13.196). Note that the last equation expresses X_n as a **moving average** (of infinite order) of the white noise e_n, where the weighting coefficients a^j are exponentially decaying.

Since $E[X_n] = 0$, we have

$$\text{Var}[X_n] = E[X_n^2] = R_X[0],$$

where $R_X[k] = E[X_n X_{n+k}]$ is the autocorrelation function of the time series X_n. From (13.196), we can obtain (Problem 13.27 (a))

$$R_X[0] = \frac{\sigma^2}{1-a^2}. \tag{13.201}$$

Multiplying (13.196) by X_{n+k} and taking the expectation we have (Problem 13.27 (b))

$$R_X[k] = aR_X[|k|-1], k = 0, \pm 1, \pm 2, \ldots, \tag{13.202}$$

which leads to

$$\boxed{R_X[k] = \frac{a^{|k|}\sigma^2}{1-a^2}, k = 0, \pm 1, \pm 2, \ldots.} \tag{13.203}$$

Using the *Wiener–Khinchin formula* (13.49), we find the power spectrum

$$\boxed{P_X(\omega) = \frac{1}{2\pi}\sum_{k=0}^{\infty} R_X[k]e^{-ik\omega} = \frac{\sigma^2}{2\pi(1+a^2-2a\cos\omega)}, -\pi \le \omega \le \pi.} \tag{13.204}$$

This result could have been directly obtained by using the relation between the input and output spectrum and the transfer function of a linear system. If we write $H(z) = 1/(1-az^{-1})$, then (13.199) can be written as

$$X(z) = H(z)E(z), |z| > |a|, \text{where} H(z) = \frac{1}{1-az^{-1}}. \tag{13.205}$$

As is well known in the linear system theory, the *power spectrum* of the output process is found by multiplying the power spectrum by the magnitude square of the system transfer function $|H(e^{i\omega})|^2$:

$$P_X(\omega) = P_e(\omega)|H(e^{i\omega})|^2. \tag{13.206}$$

By substituting the power spectrum of the white noise

$$P_e(\omega) = \frac{\sigma^2}{2\pi}, -\pi \le \omega \le \pi, \tag{13.207}$$

we obtain

$$\boxed{P_X(\omega) = \frac{\sigma^2}{2\pi}\frac{1}{|1-a\,e^{-i\omega}|^2},} \tag{13.208}$$

which equals (13.204). □

Now we return to the AR(p) time series defined by (13.192). By taking the double-sided Z-transform, we obtain as in the case of AR(1) the following expression:

$$X(z) = \frac{E(z)}{1 - A(z)}, \text{ where} A(z) = \sum_{i=1}^{p} a_i z^{-i}. \tag{13.209}$$

Let $\alpha_1, \alpha_2, \ldots, \alpha_p$ be the roots of the **characteristic equation:**

$$1 - A(z) = 1 - a_1 z^{-1} - a_2 z^{-2} - \cdots - a_p z^{-p} = 0. \tag{13.210}$$

Then, we can write

$$1 - A(z) = (1 - \alpha_1 z^{-1})(1 - \alpha_2 z^{-1}) \cdots (1 - \alpha_p z^{-1}). \tag{13.211}$$

Then it is not difficult to see that (13.200) can be generalized to

$$X_n = c_1 \sum_{j=0}^{\infty} \alpha_1^j e_{n-j} + c_2 \sum_{j=0}^{\infty} \alpha_2^j e_{n-j} + \cdots + c_p \sum_{j=0}^{\infty} \alpha_p^j e_{n-j}, \tag{13.212}$$

where $c_1, c_2, \ldots, c_p$ are constants that can be uniquely determined when we find the partial fraction expansion of $1/(1 - A(z))$, namely,

$$\begin{aligned} c_1^{-1} &= (1 - \alpha_2 \alpha_1^{-1}) \cdots (1 - \alpha_p \alpha_1^{-1}),\ c_2^{-1} = (1 - \alpha_1 \alpha_2^{-1}) \cdots (1 - \alpha_p \alpha_2^{-1}), \cdots, \\ c_p^{-1} &= (1 - \alpha_1 \alpha_p^{-1}) \cdots (1 - \alpha_{p-1} \alpha_p^{-1}). \end{aligned} \tag{13.213}$$

This procedure is similar to the steps required in the inversion of the PGF discussed in Section 9.1.4. Note, however, that we defined the PGF as a polynomial in z, whereas here we represent a time series as a polynomial in z^{-1}, as done in the conventional Z-transform. The power spectrum of the AR(p) time series $\{X_n\}$ can be easily obtained from the formula (13.206):

$$P_X(\omega) = \frac{\sigma^2}{2\pi} \frac{1}{|1 - \sum_{i=1}^{p} a_i e^{-i\omega}|^2}. \tag{13.214}$$

We observed earlier that an AR(1) process is a simple Markov chain. Equation (13.192) defines that an AR(p) process is a Markov chain of order p. We remarked in Section 12.2.5 that a higher order Markov chain can be reduced to a simple Markov chain by defining the state space that is a p-fold Cartesian product of the original state space. An AR(3) process, for instance, can be expressed as

$$\begin{bmatrix} X_n \\ X_{n-1} \\ X_{n-2} \end{bmatrix} = \begin{bmatrix} a_1 & a_2 & a_3 \\ 1 & 0 & 0 \\ 0 & 1 & 0 \end{bmatrix} \begin{bmatrix} X_{n-1} \\ X_{n-2} \\ X_{n-3} \end{bmatrix} + \begin{bmatrix} e_n \\ 0 \\ 0 \end{bmatrix}. \tag{13.215}$$

By generalizing this observation, we write an AR(p) as a simple Markov chain in terms of the p-dimensional time series $\{\boldsymbol{X}_n\}$, where

$$\boldsymbol{X}_n = (X_n, X_{n-1}, \ldots, X_{n-p+2}, X_{n-p+1})^{\top}, \tag{13.216}$$

which satisfies

$$\boxed{\boldsymbol{X}_n = \boldsymbol{A}\boldsymbol{X}_{n-1} + \boldsymbol{e}_n,} \tag{13.217}$$

where

$$\boldsymbol{A} = \begin{bmatrix} a_1 & a_2 & \cdots & a_{p-1} & a_p \\ 1 & 0 & \cdots & 0 & 0 \\ 0 & 1 & \cdots & 0 & 0 \\ \vdots & \vdots & \ddots & \vdots & \vdots \\ 0 & 0 & \cdots & 1 & 0 \end{bmatrix} \tag{13.218}$$

and

$$\boldsymbol{e}_n = (e_n, 0, \ldots, 0, 0)^\top. \tag{13.219}$$

This vector representation of a univariate AR(p) will be convenient when we later discuss the problem of choosing optimal AR coefficients $a_1, a_2, \ldots, a_p$ in the model construction or the problem of predicting the value of X_n given the AR model. We can readily apply results from the **Kalman filter theory**, when we represent the AR(p) process in this form.

The **state-space representation** (13.217) suggests that we can formulate a **multivariate autoregressive process** or a **vector autoregressive process** of first-order by an $m \times 1$ vector

$$\boldsymbol{X}_n = (X_{1,n}, X_{2,n}, \ldots, X_{m,n}), -\infty < n < \infty, \tag{13.220}$$

and the matrix $\boldsymbol{A}$ is an arbitrary $m \times m$ matrix of autoregressive parameters, not restricted to the structure of (13.218), and $\boldsymbol{e}_n$ is an $m \times 1$ vector of serially uncorrelated RVs with mean $\mathbf{0}$ and covariance matrix $\boldsymbol{C}$. Such vector-valued autoregressive processes are used in econometrics (e.g., see Hamilton [141] and Harvey [145]).

13.4.2 Moving average (MA) time series

We defined a moving-average process in Problem 12.2 and showed that the AR(1) process can be represented as a moving average of infinite order as shown in (13.200). We define a **moving-average** process of order q, denoted MA(q), as

$$\boxed{X_n = \sum_{j=0}^{q} b_j e_{n-j}, -\infty < n < \infty,} \tag{13.221}$$

where $b_j; 0 \le j \le q$ are constants and e_n are white noise, as defined in (13.193). The MA(q) process X_n has zero mean and the autocorrelation function

$$R_X[k] = \sigma^2 \sum_{i=0}^{q-|k|} b_i b_{i+|k|}. \tag{13.222}$$

By defining $B(z)$ by

$$B(z) = \sum_{j=0}^{q} b_j z^{-j}, \tag{13.223}$$

the power spectrum of the MA(q) time series is given by

$$P_X(\omega) = \frac{\sigma^2}{2\pi}|B(e^{i\omega})|^2 = \frac{\sigma^2}{2\pi}\left|\sum_{j=0}^{q} b_j e^{-i\omega}\right|^2. \tag{13.224}$$

13.4.3 Autoregressive moving average (ARMA) time series

By combining the AR(p) and MA(q), we define the **ARMA** process $\{X_n\}$ of order (p, q), denoted ARMA(p, q), by

$$\boxed{X_n = \sum_{i=1}^{p} a_i X_{n-i} + \sum_{j=0}^{q} b_j e_{n-j}, -\infty < n < \infty.} \tag{13.225}$$

By taking the double-sided Z-transform, we find

$$X(z) = A(z)X(z) + B(z)E(z), \tag{13.226}$$

from which we have

$$\boxed{X(z) = \frac{B(z)E(z)}{1 - A(z)},} \tag{13.227}$$

where $X(z)$ and $E(z)$ are defined in (13.198) and $A(z)$ and $B(z)$ are as defined in (13.209) and (13.223) respectively.

The power spectrum of ARMA(p, q) is readily obtained by generalizing the argument that led to (13.214) and (13.224):

$$P_X(\omega) = \frac{\sigma^2}{2\pi} \frac{\left|\sum_{j=0}^{q} b_j e^{-i\omega}\right|^2}{\left|1 - \sum_{i=1}^{p} a_i e^{-i\omega}\right|^2}. \tag{13.228}$$

The representation of ARMA(p, q) suggests that we can write X_n as the output of a **feedback** and **feed-forward filter** driven by white noise, as shown in Figure 13.4, where the case $q > p$ is shown. The case $p \geq q$ should be self-explanatory. There exists a large body of literature of digital filters of this type. The feed-forward component $B(D)$ corresponds to finite impulse response (FIR) and the feedback component $(1 - A(D))^{-1}$ provides infinite impulse response (IIR). The problem of how to choose the best coefficients $a_i, 1 \leq i \leq p$, and $b_j, 0 \leq j \leq q$, to allow the ARMA model to fit the observed data is mathematically equivalent to an optimal filter design problem when the mean square error (MSE) is a criterion for optimality.

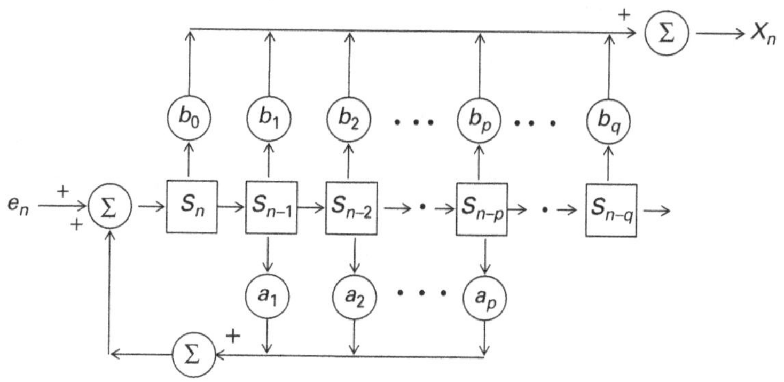

Figure 13.4 Representation of the ARMA(p, q) as a feedback/feed-forward filter driven by white noise.

We rewrite (13.227) in two steps:

$$X(z) = B(z)S(z), \tag{13.229}$$

$$S(z) = \frac{E(z)}{1 - A(z)}, \tag{13.230}$$

The second equation can be written as

$$S(z) = A(z)S(z) + E(z). \tag{13.231}$$

In terms of time series, (13.229) and (13.231) can be written as

$$X_n = \sum_{j=0}^{q} b_j S_{n-j}, \tag{13.232}$$

$$S_n = \sum_{i=1}^{p} a_i S_{n-1} + e_n. \tag{13.233}$$

In order to obtain a vector-matrix representation similar to (13.217), let $r = \max\{p, q\}$ and define the r-dimensional state vector

$$\boldsymbol{S}_n = (S_n, S_{n-1}, \ldots, S_{n-r})^\top \tag{13.234}$$

and an r-dimensional parameter vector $\boldsymbol{b}$

$$\boldsymbol{b} = \begin{cases} (b_1, b_2, \ldots, b_q)^\top, & \text{if } r = q \geq p, \\ (b_1, b_2, \ldots, b_q, 0, \ldots, 0), & \text{if } q < p = r. \end{cases} \tag{13.235}$$

Then we find the following *state-space representation* that is equivalent to (13.232) and (13.233):

$$\boxed{\begin{aligned} X_n &= \boldsymbol{b}^\top \boldsymbol{S}_n, \\ \boldsymbol{S}_n &= \boldsymbol{A}\boldsymbol{S}_{n-1} + \boldsymbol{e}_n, \end{aligned}} \tag{13.236}$$

where $\boldsymbol{A}$ is an $r \times r$ matrix. If $r = p \geq q$, then its structure is exactly the same as that of (13.218). If $r = q > p$, then we add $(r - p)$ columns to the right end of the matrix $\boldsymbol{A}$ of (13.218) and $(r - p)$ rows to the bottom of $\boldsymbol{A}$. Obviously $a_{p+1} = \cdots = a_r = 0$, but otherwise the structure of the expanded matrix $\boldsymbol{A}$ is the same as (13.218), having unity in all $(i + 1, i)$ entries. Note that the state sequence $\boldsymbol{S}_n$ is unobservable, or hidden. If the driving noise is Gaussian, as is often assumed in the ARMA process, this **hidden Markov process (HMM)**, a subject fully discussed in Chapter 20, becomes an *r-dimensional GMP*. Then the Kalman filter theory, which will be discussed in Section 22.3), is directly applicable to estimation and prediction of the ARMA time series.

13.4.4 Autoregressive integrated moving average (ARIMA) time series

Related to ARMA(p, q) is an *autoregressive integrated moving average (ARIMA)* process, denoted ARIMA(p, d, q), when the characteristic equation (13.210) has d multiple roots of $z = 1$; i.e.,

$$1 - \sum_{i=1}^{p} a_i z^{-1} = (1 - z^{-1})^d \left(1 - \sum_{i=1}^{p-d} a'_i z^{-i} \right), \tag{13.237}$$

where a'_i; $1 \leq i \leq p - d$ are a set of parameters that can be uniquely determined by the factorization. Thus, we can write

$$\boxed{X(z) = \frac{B(z)E(z)}{(1 - z^{-1})^d \left(1 - \sum_{i=1}^{p-d} a'_i z^{-i}\right)}.} \tag{13.238}$$

If we define a new process $\{Y_n\}$ by

$$Y(z) = (1 - z^{-1})^d X(z), \tag{13.239}$$

then

$$Y(z) = \frac{\sum_{j=0}^{q} b_j z^{-j}}{1 - \sum_{i=1}^{p-d} a'_i z^{-i}}, \tag{13.240}$$

which implies that $\{Y_n\}$ is an ARMA$(p - d, q)$. If $d = 1$, we have $Y(z) = (1 - z^{-1})X(z)$, i.e.,

$$Y_n = X_n - X_{n-1}, -\infty < n < \infty; \tag{13.241}$$

hence,

$$X_n = Y_n + X_{n-1} = Y_n + Y_{n-1} + X_{n-2} = \cdots = \sum_{j=0}^{\infty} Y_{n-j}. \tag{13.242}$$

Thus, X_n is an integration of the ARMA process Y_n; hence, it is named an ARIMA process. The ARIMA$(p, 1, q)$ process can be viewed as a *generalized random walk*, where the step sizes Y_j are not independent, but are correlated to according to ARMA$(p-1, q)$. The ARIMA$(p, 1, q)$ is a **martingale** because $E[Y_n] = 0$.

If $p - 1 = q = 0$, then $\{Y_n\}$ reduces to white noise $\{e_n\}$ and X_n is indeed a **random walk**. Thus, ARIMA(1,1,0) is a random walk and ARIMA$(p, 0, q)$ is equivalent to ARMA(p, q). When d is a noninteger, it is called **fractional ARIMA** (FARIMA) or autoregressive fractionally integrated moving average (ARFINA), and this model is found to be flexible enough to represent a LRD process (i.e., *self-similar* process) as well as a *short-range-dependent* (SRD) process. If $d \in (-0.5, 0.5)$, FARIMA(p, d, q) is a stationary process. For example, see [162] and references therein.

13.5 Summary of Chapter 13

Fourier series expansion:	$g(t) = \sum_{k=-\infty}^{\infty} g_k e^{i2\pi n f_0 t}$, $f_0 = 1/T$	(13.4)
where	$g_k = \frac{1}{T}\int_0^T g(t)e^{-i2\pi k f_0 t}\,dt$	(13.3)
Parseval's formula:	$\sum_{n=-\infty}^{\infty} \lvert g_n\rvert^2 = \frac{1}{T}\int_0^T \lvert g(t)\rvert^2\,dt$	(13.8)
Fourier transform:	$G(f) = \int_{-\infty}^{\infty} g(t)e^{-i2\pi f t}\,dt$	(13.15)
Fourier inverse transform:	$g(t) = \int_{-\infty}^{\infty} G(f)e^{i2\pi f t}\,df$	(13.16)
Parseval's formula:	$\int_{-\infty}^{\infty} \lvert G(f)\rvert^2\,df = \int_{-\infty}^{\infty} \lvert g(t)\rvert^2\,dt$	(13.17)
Periodic WSS process:	$R_X(\tau + T) = R_X(\tau)$	(13.19)
Fourier series expansion:	$X(t) = \text{l.i.m.}_{N\to\infty} \sum_{n=-N}^{N} X_n e^{i2\pi n f_0 t}$	(13.26)
where	$X_n = \frac{1}{T}\int_0^T X(t)e^{-i2\pi n f_0 t}\,dt$	(13.25)
Orthogonality of coefficients:	$\langle X_m, X_n\rangle = E[X_m^* X_n] = r_n\delta_{m,n}$	(13.28)
Power spectral density:	$S_X(f) = \int_{-\infty}^{\infty} R_X(\tau)e^{-i2\pi f\tau}\,d\tau$	(13.30)
WSS time series:	$E[X_n] = \mu_X$ and $E[X_n X_{n+k}] = R_X[k]$	(13.45)
Power spectrum:	$P_X(\omega) = \frac{1}{2\pi}\sum_{k=-\infty}^{\infty} C_X[k]e^{-ik\omega}$, $-\pi \le \omega \le \pi$	(13.49)
where	$C_X[k] = E[(X_n - \mu_X)(X_{n+k} - \mu_X)]$	(13.46)
Inverse relation:	$C_X[k] = \int_{-\pi}^{\pi} P_X(\omega)e^{ik\omega}\,d\omega$	(13.50)
Periodogram:	$P_m^{(N)} = \frac{1}{N}\left\lvert\sum_{n=1}^{N}(x_n - \overline{x}^{(N)})W^{mn}\right\rvert^2$	(13.58)
Limit of $E[P_m^{(N)}]$:	$\lim_{N\to\infty} E[P_m^{(N)}] = 2\pi P_X(2\pi m/N)$	(13.63)
Expansion of $\boldsymbol{R}$:	$\boldsymbol{R} = \sum_{n=1}^{N} \lambda_n \boldsymbol{u}_n \boldsymbol{u}_n^H = \sum_{n=1}^{N} \lambda_n \boldsymbol{E}_n$	(13.82)
Similarity transformation:	$\boldsymbol{R} = \boldsymbol{U}\boldsymbol{\Lambda}\boldsymbol{U}^{\mathrm{H}} = \boldsymbol{U}\boldsymbol{\Lambda}\boldsymbol{U}^{-1}$	(13.87)
PCA:	$\boldsymbol{X} = \sum_{n=1}^{N} \chi_n \boldsymbol{u}_n$	(13.90)
where	$\chi_n = \langle \boldsymbol{u}_n, \boldsymbol{X}\rangle = \sum_{i=1}^{N} u_{n,i}^* X_i$	(13.92)
MSE of PCA:	$\mathcal{E}^2 \triangleq E[\lvert\hat{\boldsymbol{X}}_M - \boldsymbol{X}\rvert^2] = \sum_{n=M+1}^{N} \lambda_n$	(13.95)
Karhunen–Loève expansion:	$X(t) = l.i.m._{N\to\infty} \sum_{n=1}^{N} X_n u_n(t)$	(13.121)
where	$X_n = \int_0^T X(t)u_n^*(t)\,dt = \langle u_n(t), X(t)\rangle$	(13.113)

Double orthogonality:	$\langle u_m, u_n\rangle \triangleq \int_0^T u_m^*(t)u_n(t)\,dt = \delta_{m,n}$	(13.108)
	$\langle X_m, X_n\rangle \triangleq E[X_m^* X_n] = \lambda_m \delta_{m,n}$	(13.109)
Characteristic equation:	$\int_0^T R_X(t,s)u(s)\,ds = \lambda u(t)$	(13.112)
PCA expansion:	$\mathbf{X} = \sum_{i=1}^m \boldsymbol{u}_i \boldsymbol{\chi}_i^\top$	(13.159)
where	$\boldsymbol{\chi}_i = \mathbf{X}^\top \boldsymbol{u}_i^*$	(13.158)
Frobenius norm of a matrix:	$\|\boldsymbol{A}\| = \sqrt{\mathrm{trace}(\boldsymbol{A}^{\mathrm{H}}\boldsymbol{A})}$ $= \sqrt{\sum_{i=1}^m \sum_{j=1}^n \lvert a_{ij}\rvert^2}$	(13.163)
Sum of squared PCA errors:	$Q = \|\mathbf{X} - \hat{\mathbf{X}}\|^2 = \sum_{j>k} \mu_j$	(13.164)
SVD expansion:	$\mathbf{X} = \boldsymbol{U}\Sigma\boldsymbol{V}^{\mathrm{H}} = \sum_{i=1}^{\min(m,n)} \sigma_i \boldsymbol{u}_i \boldsymbol{v}_i^{\mathrm{H}}$	(13.181)
where	$\Sigma = [\sqrt{\boldsymbol{M}}\boldsymbol{O}_{(n-m)\times m}]$	(13.178)
Frobenius norm and SVD:	$\|\boldsymbol{A}\|^2 = \mathrm{trace}(\boldsymbol{A}\boldsymbol{A}^{\mathrm{H}}) = \sum_{i=1}^{\min(m,n)} \sigma_i^2$	(13.187)
Sum of squared SVD errors:	$Q = \|\mathbf{X} - \hat{\mathbf{X}}\|^2 = \sum_{i=k+1}^m \sigma_i^2$	(13.188)
AR(p) process:	$X_n = \sum_{i=1}^p a_i X_{n-i} + e_n$	(13.192)
Z-transform of AR(1):	$X(z) = E(z)/(1 - az^{-1})$ $= \left(\sum_{j=0}^\infty a^j z^{-j}\right) E(z)$	(13.199)
Autocorrelation of AR(1):	$R_X[k] = a^{\lvert k\rvert}\sigma^2/(1-a^2)$	(13.203)
Spectrum of AR(1):	$P_X(\omega) = \sigma^2/(1 + a^2 - 2a\cos\omega)$ $-\pi \le \omega \le \pi$	(13.204)
State-space rep. of AR(1):	$\boldsymbol{X}_n = \boldsymbol{A}\boldsymbol{X}_{n-1} + \boldsymbol{e}_n$	(13.217)
MA(q):	$X_n = \sum_{j=0}^q b_j e_{n-j}$	(13.221)
ARMA(p,q):	$X_n = \sum_{i=1}^p a_i X_{n-i} + \sum_{j=0}^q b_j e_{n-j}$	(13.225)
Z-transform of ARMA(p,q):	$X(z) = B(z)E(z)/(1 - A(z))$	(13.227)
State-space representation of ARMA(p,q):	$X_n = \boldsymbol{b}^\top \boldsymbol{S}_n$ $\boldsymbol{S}_n = \boldsymbol{A}\boldsymbol{S}_{n-1} + \boldsymbol{e}_n$	(13.236)
ARIMA(p,d,q):	$X(z) = \dfrac{B(z)E(z)}{(1-z^{-1})^d\left(1-\sum_{i=1}^{p-d} a_i' z^{-i}\right)}$	(13.238)

13.6 Discussion and further reading

The Karhunen–Loève representation is a general theory that extends the Fourier analysis to continuous-time nonstationary random processes. We discussed its application to signal detection problems and showed that the signal space method widely practiced in communication systems analysis can be viewed as a special case of the Karhunen–Loève expansion. The Karhunen–Loève expansion is discussed in many textbooks on random processes written for engineering students, including Davenport and Root [77], Gubner [133], Papoulis and Pillai [262], and Thomas [319].

Although we interpreted data **X** in Section 13.3 primarily as a multivariate time series, both PCA and SVD should be applicable to any data presentable in a matrix form.

They are applied, for instance, to gene expression analysis in bioinformatics [344], latent semantic analysis (LSA) in natural language processing in such applications as information retrieval [23]. SVD has been shown effective in extracting signal from noisy data, especially in space-diversity reception such as multiple-input, multiple-output (MIMO) systems. For example, see Kung *et al.* [217], van der Veen [336], and Zhang [370].

Jolliffe [169] devotes an entire volume to PCA, and Golub and Van Loan [124] is a good source for matrix theory, including SVD. Skillicorn [306] discusses applications of several matrix decompositions, including SVD and the PageRank algorithms, to data mining. As remarked in Section 13.3.3, Web information is an active area of research for algorithmic development. The matrix decomposition method applied to the Markov transition matrix (as in the PageRank algorithm) or to the adjacency matrix (as in the HITS algorithm) can be viewed as instances of general spectral expansion, the common theme of this chapter. For example, see Kleinberg [187], Del Corso *et al.* [79], and Langville and Meyer [218] for reading of Web information retrieval algorithms.

Econometricians extensively use ARMA and ARIMA models and their multivariate versions. For example, see Hamilton [141] and Harvey [145]. Jagerman *et al.* [162] report that autoregessive-type traffic models have been applied to model variable bit rate (VBR) video traffic. They also review the literature of fractional ARIMA(p, d, q).

13.7 Problems

Section 13.1: Spectral representations of random processes and time series

13.1* Parseval's identity. Prove the identity (13.17).
Hint: Use the following identity.

$$\int_{-\infty}^{\infty} e^{i2\pi f t}\, df = \delta(t).$$

13.2 Periodic WSS random process. Show that if the autocorrelation function $R_X(\tau)$ is periodic with period T, $X(t)$ satisfies (13.20).

13.3 Equivalence of random variables. Show that $X \overset{\text{m.s.}}{=} Y$ if and only if $X \overset{\text{a.s.}}{=} Y$.

13.4* Orthogonality of Fourier expansion coefficients of a periodic WSS process. Show that (13.28) holds among the Fourier series expansion coefficients X_n's.

13.5 Derivation of (13.59). Derive the expression for the periodogram given by (13.59).

Section 13.2: Generalized Fourier series expansions

13.6 Nonnegative definite matrix. Show that the correlation matrix $\boldsymbol{R}$ satisfies the property (13.73).

13.7 Nonnegativity of spectrum. Prove (13.79); that is, show that the spectrum (i.e., set of eigenvalues) is nonnegative.

13.8 Left and right eigenvectors. Show that when $\boldsymbol{u}$ is a right eigenvector of $\boldsymbol{R}$ associated with eigenvalue λ, $\boldsymbol{u}^{\mathrm{H}}$ is the left eigenvector. In other words, prove (13.81).

13.9* Orthogonality of eigenvectors. Prove that eigenvectors associated with different eigenvalues are orthogonal; i.e.,

$$\langle \boldsymbol{u}_i, \boldsymbol{u}_j \rangle = \boldsymbol{u}_i^{\mathrm{H}} \boldsymbol{u}_j = 0, \quad \text{for } i \neq j.$$

13.10 Variance of expansion coefficients in the eigenvector expansion. Prove (13.93); that is, the expansion coefficients obtained in the eigenvector expansion are orthogonal to each other.

13.11 Mean square error of eigenvector expansion approximation. Show that the mean square error of the approximation (13.94) is given by (13.95).

13.12* Eigenvectors and eigenvalues of a circulant matrix. Show that the eigenvalues and eigenvectors of the circulant matrix (13.100) are given by (13.101) and (13.102), by taking the following steps.

(a) Let $\boldsymbol{u} = (u_0, u_1, \ldots, u_j, \ldots, u_{n-1})^{\top}$ be an eigenvector and λ be the corresponding eigenvalue; i.e., $\boldsymbol{C}\boldsymbol{u} = \lambda \boldsymbol{u}$. Show that the following equations must be satsified:

$$\sum_{k=0}^{n-k-1} c_k u_{k+j} + \sum_{k=n-j}^{n-1} u_{k-n+j} = \lambda u_j,\ j = 0, 1, 2, \ldots, n-1. \tag{13.243}$$

(b) Assume the solution form $u_j = \alpha^j$ with some constant α. Show that the above difference equations are reduced to

$$\sum_{k=0}^{n-j-1} c_k \alpha^k + \alpha^{-n} \sum_{k=n-j}^{n-1} c_k \alpha^k = \lambda,\ j = 0, 1, 2, \ldots, n-1. \tag{13.244}$$

(c) Let $\alpha^{-n} = 1$; i.e., α is one of the n distinct complex nth roots of unity. Then derive (13.101) and (13.102).

13.13 Generating function method. Derive (13.103).

13.14 Orthogonality of eigenfunctions. Show that eigenfunctions $u_n(t)$ are orthogonal to each other.

13.15 Derivation of the integral equation (13.136). Show that the function $Q(t)$ defined by (13.133) satisfies the integral equation (13.136).

13.16 Matched filter equivalent to a correlation receiver. Show that the output of a linear filter matched to $Q^*(T-t)$ sampled at time $t = T$, as shown in the right side of (13.138), is equal to cross-correlation between the input $X(t)$ and $Q(t)$ given by the left side of (13.138).

13.17* Matched filter and SNR. Let $h(t)$ be the impulse response of an arbitrary linear filter. Let $S_{\mathrm{O}}(t)$ be the filter output when the signal alone $S(t)$ is added to the filter. Similarly, let $N_O(t)$ be the filter output when the WSS noise $N(t)$ alone is added to the filter.

(a) Write down the expression for the signal power at time $t = T$; i.e., $P_S = |S_O(t)|^2_{t=T}$. Similarly, write down the expression for the noise power $P_N = E[|N_O(t)|^2]$, by assuming that the noise is white; i.e., $R_N(s,t) = \sigma^2\delta(s-t)$.
(b) Find a linear filter $h(t)$ that maximizes the signal-to-noise-ratio SNR $= P_S/P_N$. *Hint*: Use the Cauchy–Schwarz inequality.
(c) Find a linear filter $h(t)$ that maximizes SNR when the the noise process $N(t)$ is not white, not even stationary, and the autocovariance function is given by $R_N(t,s)$.

13.18* Orthogonal expansion of Wiener process. The Wiener process[32] $W(t)$ has the autocorrelation function

$$R_W(t,s) = \sigma^2 \min(t,s) = \begin{cases} \sigma^2 t, & t < s, \\ \sigma^2 s, & t > s. \end{cases} \tag{13.245}$$

Follow the following steps and find eigenfunctions $\psi(t)$ and eigenvalues λ that satisfy

$$\int_0^T R_W(t,s)\psi(s)\,ds = \lambda\psi(t). \tag{13.246}$$

(a) Show that $\psi(t)$ should satisfy a differential equation

$$\lambda\psi''(t) + \sigma^2\psi(t) = 0. \tag{13.247}$$

(b) *Case 1.* $\lambda < 0$*:* Show that no solution exists in this case.
(c) *Case 2.* $\lambda > 0$*:* Let $\sigma^2/\lambda = \omega^2$. Show that the solutions are given by

$$\omega_n = \frac{(2n+1)\pi}{2T}; \text{ thus, } \lambda_n = \frac{\sigma^2}{\omega_n^2}, \; n = 0, \pm 1, \pm 2, \ldots,$$

and

$$\psi_n(t) = \sqrt{\frac{2}{T}} \sin \omega_n t.$$

(d) Hence the Wiener–Levy process can be expanded as

$$W(t) = \sum_{n=-\infty}^{\infty} W_n \psi_n(t) = \sqrt{\frac{2}{T}} \sum_{n=-\infty}^{\infty} W_n \sin \omega_n t. \tag{13.248}$$

Show that the coefficients $\{W_n\}$ are uncorrelated and

$$E[W_n^2] = 4\left[\frac{\sigma T}{(2n+1)\pi}\right]^2, \; n = 0, \pm 1, \pm 2, \ldots. \tag{13.249}$$

[32] In an earlier section we used $W(t)$ to represent a white noise process, so the reader should not confuse the two processes. The Wiener process is an integration of white noise.

(e) An alternative expansion of $W(t)$ is

$$W(t) = \sqrt{\frac{2}{T}} \sum_{n=0}^{\infty} U_n \sin \omega_n t, \tag{13.250}$$

where

$$E[U_n^2] = 8 \left[\frac{\sigma T}{(2n+1)\pi} \right]^2, n = 0, 1, 2, 3, \ldots.$$

Section 13.3: Principal component analysis and singular value decomposition

13.19 Orthogonality of the expansion coefficient vectors χ_i. Show the orthogonality given in (13.160); i.e.,

$$\langle \chi_i, \chi_{i'} \rangle \triangleq \chi_i^\top \chi_{i'}^* = \mu_i \delta_{i,i'}.$$

13.20* Sum of squares of the difference.

(a) Let A be an $m \times n$ matrix; i.e., $A = [a_{ij}]$. We define $\|A\|^2 \triangleq \sum_{i=1}^m \sum_{j=1}^n |a_{ij}|^2$. Suppose A has the following structure:

$$A = bc^\top, \text{ i.e., } a_{ij} = b_i c_j,$$

where b is an m-dimensional column vector and $c^\top$ is an n-dimensional row vector. Show that

$$\|A\| = \|b\|^2 \|c\|^2,$$

where the norm squares of vectors b and c are

$$\|b\|^2 = \sum_{i=1}^m |b_i|^2 \text{and} \|c\|^2 = \sum_{j=1}^n |c_j|^2.$$

(b) Suppose that A is given as

$$A = b^{(1)} c^{(1)^\top} + b^{(2)} c^{(2)^\top},$$

where the n-dimensional vectors $c^{(1)}$ and $c^{(2)}$ are orthogonal. Find the expression for $\|A\|^2$.

(c) Prove (13.164) for the *sum of squares of the differences*; i.e.,

$$Q = \sum_{i=1}^m \sum_{j=1}^n |x_{ij} - \hat{x}_{ij}|^2 = \sum_{j>k} \mu_j.$$

13.21 The covariance matrix-based PCA. Consider the 3×2 data $\mathbf{X}$ discussed in Example 13.4. Apply the PCA method after adjusting the data by subtracting the empirical average.

13.22 Derivation of SVD (13.184). Show that the singular values σ_i and singular vectors $\boldsymbol{u}_i$ and $\boldsymbol{v}_j$ satisfy the equations in (13.184).

13.23 Derivation of U from V. Derive (13.185).

13.24 Frobenius norm and singular values. Prove the formula (13.187).

Section 13.4: Autoregressive moving average time series and its spectrum

13.25 Inclusion of a constant a_0 in AR(p) of (13.192). Consider modifying (13.192) by adding a constant term a_0:

$$Y_n = a_0 + \sum_{i=1}^{p} a_i Y_{n-i} + e_n, n = 1, 2, 3, \ldots. \tag{13.251}$$

Show that the effect is simply to have Y_n shifted from X_n by some constant, provided $\sum_{i=1}^{p} a_i \neq 1$.

13.26* Mean square convergence of (13.196). Show that the first term in the final expression of (13.195) converges to X_n of (13.196) in mean square; i.e.,

$$X_n = \underset{k\to\infty}{\text{l.i.m.}} \sum_{j=0}^{k-1} a^j X_{n-j}. \tag{13.252}$$

13.27 Variance and autocorrelation function of the AR(1) time series.

(a) Show that the AR(1) sequence X_n has the variance given by (13.201).
(b) Show that the autocorrelation function satisfies the recursion (13.202) and is given by (13.203).

14 Poisson process, birth–death process, and renewal process

In Section 12.2.6 we briefly described point processes and renewal processes. The Poisson process and birth-and-death (BD) process are the simplest examples of a point process. A renewal process is a generalization of the Poisson process and is used in a variety of applications, including queueing theory and reliability theory. Some of the results discussed in this chapter will be extended in Chapter 23, which discusses various queueing and loss system models.

14.1 The Poisson process

14.1.1 Derivation of the Poisson process

We have already informally referred to the Poisson process on several occasions in previous chapters without really providing its mathematical definition. One way to define the Poisson process is to assume that the interarrival times X_n (cf. (12.7)) of a point process $N(t)$ (cf. (12.6)) are independent and identically distributed (i.i.d.) RVs with a common exponential distribution (Problem 14.1).

$$F_X(x) = 1 - e^{-\lambda x}, x \geq 0. \tag{14.1}$$

Alternatively, the Poisson process can be derived as a limiting case of Bernoulli trials (Problem 14.2). But a more formal definition of the Poisson process is given as follows.

DEFINITION 14.1 (Poisson process). *A Poisson process of rate λ is a counting process $N(t)$, $t \geq 0$, taking values in the set $\mathbb{Z}^+ = \{0, 1, 2, \ldots\}$ such that*

(a) $N(0) = 0$ and $N(t)$ is nondecreasing; i.e., if $t_2 > t_1$, then $N(t_2) \geq N(t_1)$.
(b) Its transition probabilities

$$P_{mn}(h) \triangleq P[N(t+h) = n | N(t) = m]$$

are stationary (i.e., independent of t) and satisfy

$$\boxed{P_{mn}(h) = \begin{cases} 1 - \lambda h + o(h), & \text{if } n = m, \\ \lambda h + o(h), & \text{if } n = m+1, \\ o(h), & \text{if } n \geq m+2, \end{cases}} \tag{14.2}$$

where $o(h)$ represents a quantity that approaches zero faster than h as $h \to 0$; i.e., $\lim_{h\to 0} o(h)/h = 0$.

(c) If $t > s$, then $N(t) - N(s)$, the number of arrivals in $(s, t]$, is independent of $N(s)$, the number of arrivals in $(0, s]$. □

The Poisson process has been shown to be a plausible representation of a number of physical phenomena; e.g., the occurrence of telephone calls, order arrivals at a service facility, and random failures of equipment. The first two examples are typical of problems encountered in *queueing theory*. Simply put, the Poisson process is a mathematical model of **completely random** arrival patterns.

THEOREM 14.1. *The Poisson counting process $N(t)$ is Poisson distributed with mean λt; i.e.,*

$$P[N(t) = n] = \frac{(\lambda t)^n}{n!} e^{-\lambda t}, n = 0, 1, 2, \ldots \tag{14.3}$$

Proof. Let

$$p_n(t) \triangleq P[N(t) = n]. \tag{14.4}$$

To calculate $p_n(t + h)$ we note that $N(t + h) = n$ occurs if one of the following two situations occurs:

(i) $N(t) = n$ and no arrival in $(t, t + h]$;
(ii) $N(t) = n - 1$ and one arrival in $(t, t + h]$.

Then we have

$$\begin{aligned} p_n(t + h) &= (1 - \lambda h + o(h))p_n(t) + (\lambda h + o(h))p_{n-1}(t) + o(h) \\ &= (1 - \lambda h)p_n(t) + \lambda h p_{n-1}(t) + o(h), \quad n \geq 1, \end{aligned} \tag{14.5}$$

where the last term $o(h)$ reflects the fact that $P_{mn}(h)p_{n-m}(t)$ is negligibly small for $m \leq n - 2$ as seen from (14.2). For $n = 0$ the above equation reduces to

$$p_0(t + h) = (1 - \lambda h)p_0(t) + o(h). \tag{14.6}$$

By subtracting $p_n(t)$ from both sides in (14.5), dividing them by h, and letting $h \to 0$ we obtain

$$\boxed{\begin{aligned} \frac{dp_n(t)}{dt} &= \lambda p_{n-1}(t) - \lambda p_n(t), n \geq 1, \\ \frac{dp_0(t)}{dt} &= -\lambda p_0(t). \end{aligned}} \tag{14.7}$$

The above equations are ordinary homogeneous differential-difference equations. The initial condition $N(0) = 0$ gives the following boundary condition:

$$p_n(0) = \delta_{n,0}. \tag{14.8}$$

There are two methods to solve (14.7): (a) the induction method and (b) the PGF method.

(a) Induction method: From (14.7) we readily find

$$p_0(t) = C_0 e^{-\lambda t}, \tag{14.9}$$

and the constant C_0 can be determined, from the initial condition $p_0(0) = 1$, as $C_0 = 1$:

$$p_0(t) = e^{-\lambda t}. \tag{14.10}$$

Substituting this result into (14.7) with $n = 1$ yields

$$p_1(t) = \lambda t\, e^{-\lambda t} + C_1 e^{-\lambda t}, \tag{14.11}$$

where $C_1 = p_1(0) = 0$ so that

$$p_1(t) = \lambda t\, e^{-\lambda t}. \tag{14.12}$$

We repeat the same procedure and find

$$\boxed{p_n(t) = \frac{(\lambda t)^n}{n!} e^{-\lambda t}, n = 0, 1, 2, \ldots} \tag{14.13}$$

This result can be easily verified by direct substitution into (14.7).

(b) Probability generating function (PGF) method: Alternatively, let us consider the PGF of $N(t)$:

$$G(z, t) \triangleq E[z^{N(t)}] = \sum_{n=0}^{\infty} p_n(t) z^n. \tag{14.14}$$

Multiplying (14.7) by z^n and summing it over $n = 0, 1, 2, \ldots$, we have

$$\boxed{\frac{\partial}{\partial t} G(z, t) = \lambda z G(z, t) - \lambda G(z, t) = \lambda(z - 1) G(z, t).} \tag{14.15}$$

The boundary condition (14.8) becomes

$$G(z, 0) = 1. \tag{14.16}$$

Then from the last two equations we find

$$G(z, t) = e^{\lambda(z-1)t} = e^{-\lambda t} \sum_{n=0}^{\infty} \frac{(\lambda t)^n}{n!} z^n. \tag{14.17}$$

By comparing the coefficients of z^n in (14.14) and (14.17), we arrive at (14.13). □

In Problem 14.1 the reader is asked to show that independent interarrival times of exponential distribution lead to the Poisson processes. Now we show the argument of the reverse direction. Namely, from Definition 14.1 and Theorem 14.1 we can show that the interarrival times of the Poisson process are indeed i.i.d. exponential RVs.

THEOREM 14.2 (Interarrival times of the Poisson process). *The interarrival times* $X_1, X_2, \ldots, X_n, \ldots$ *of the Poisson process are independent exponential RVs with mean* $1/\lambda$*; i.e.,*

$$P[X_n \leq x] = 1 - e^{-\lambda x}, x \geq 0, \text{ for all } n. \tag{14.18}$$

Proof. Let us consider first X_1, the interval from $t = 0$. Noting that

$$X_1 > x, \quad \text{if and only if no arrival in } (0, x],$$

we find, with $n = 0$ in (14.3), that

$$\boxed{P[X_1 > x] = P[N(x) = 0] = p_0(x) = e^{-\lambda x}.} \tag{14.19}$$

Now, conditioned on $X_1 = t_1$,

$$X_2 > x, \quad \text{if and only if no arrival in } (t_1, t_1 + x].$$

Thus, we find

$$P[X_2 > x|X_1 = t_1] = P[\text{No arrival in } (t_1, t_1 + x]|X_1 = t_1]. \tag{14.20}$$

The event "No arrival in $(t_1, t_1 + x]$" is independent of the event "$X_1 = t_1$" because of property (c) in Definition 14.1. Thus,

$$P[X_2 > x|X_1 = t_1] = P[\text{No arrival in } (t_1, t_1 + x]] = e^{-\lambda x}. \tag{14.21}$$

Thus, X_2 is independent of X_1 and has the same exponential distribution. Similarly,

$$\begin{aligned} &P[X_n > x|X_1 = t_1, X_2 = t_2 - t_1, \ldots, X_{n-1} = t_{n-1} - t_{t-2}] \\ &= P[\text{No arrival in } (t_{n-1}, t_{n-1} + x]] = e^{-\lambda x}. \end{aligned} \tag{14.22}$$

Hence, X_n is independent of $X_1, X_2, \ldots, X_{n-1}$ and has the distribution function

$$\boxed{F_X(x) = P[X_n \leq x] = 1 - e^{-\lambda x}, \text{ for all } n.} \tag{14.23}$$

□

14.1.2 Properties of the Poisson process

In this section we discuss several important properties of the Poisson process.

14.1.2.1 Memoryless property of Poisson process

The first property of the Poisson process has to do with the memoryless property of the exponential variable as discussed in Section 4.2.2. Assume that Y time units have elapsed since the last arrival. Let us define a random variable

$$R = X - Y, \tag{14.24}$$

which is the **remaining time** until the next arrival. By applying the definition of conditional probability, we can write

$$\begin{aligned} P[R \le r | X \ge Y] &= \frac{P[R \le r, X \ge Y]}{P[X \ge Y]} = \frac{P[Y \le X \le Y + r]}{P[X \ge Y]} \\ &= \frac{F_X(Y + r) - F_X(Y)}{1 - F_X(Y)} = \frac{e^{-\lambda Y} - e^{-\lambda(Y+r)}}{e^{-\lambda Y}} = 1 - e^{-\lambda r}. \end{aligned} \tag{14.25}$$

Thus,

$$\boxed{P[R \le r | X \ge Y] = 1 - e^{-\lambda r} = F_X(r).} \tag{14.26}$$

Therefore, the conditional distribution of R is independent of Y and has the same distribution as the interarrival time. Thus, the Poisson process is said to have the **memoryless** property such that, in calculating the probability of the remaining time before the next arrival, we do not need to consider when the last arrival took place.

14.1.2.2 Superposition of Poisson processes

Consider m independent sources that generate streams of arrivals and assume that each stream is a Poisson process of rate λ_k, $k = 1, 2, \ldots, m$. If we merge these streams into a single stream, then we again have a Poisson process of rate $\lambda = \lambda_1 + \lambda_2 + \cdots + \lambda_m$. This **reproductive additivity** of Poisson processes is analogous to the sum of independent Gaussian RVs. This property can be proved by using the PGF method as follows.

Let $N_k(t)$ be the number of arrivals from the kth source in the interval $(0, t]$. Since it is Poisson distributed with mean $\lambda_k t$, its PGF is, as found in (14.17), given by

$$G_k(z, t) = E\left[z^{N_k(t)}\right] = e^{-\lambda_k t(1-z)}. \tag{14.27}$$

Therefore, the total number of arrivals $N(t) = \sum_{k=1}^{m} N_k(t)$ in the interval $(0, t]$ from all m sources has the PGF

$$\boxed{G(z, t) = E\left[z^{\sum_{k=1}^{m} N_k(t)}\right] = \prod_{k=1}^{m} G_k(z, t) = e^{-\lambda t(1-z)},} \tag{14.28}$$

with $\lambda = \sum_{k=1}^{m} \lambda_k$, where the product form for $G(z, t)$ is due to statistical independence of the RVs $N_1(t), \ldots, N_k(t)$ for any t.

Thus, the total number of arrivals in the merged stream is Poisson distributed with mean λt. This in turn implies that the combined stream forms a Poisson process with rate λ. (See Problem 14.3 for different ways to derive this result.)

14.1.2.3 Decomposition of a Poisson process

Let us now consider the opposite situation. Given a Poisson arrival stream of rate λ, let us split the stream into m output substreams. For each arrival, we place it into the kth output substream with probability r_k, and such assignments are done independently for all arrivals. Then, we will show that the kth output stream is a Poisson process with rate $r_k\lambda$, $k = 1, 2, \ldots, m$. Furthermore, these k streams are statistically independent.

Let $N(t)$ denote, as before, the number of input arrivals in $(0, t]$ and let $N_k(t)$ be the number of arrivals that are sent into the kth substream. Then, the conditional joint distribution of $N_k(t)$ $(k = 1, 2, \ldots, m)$, given $N(t) = n$, is the following **multinomial distribution**:

$$P[n_1, n_2, \ldots, n_m | n] = \frac{n!}{n_1! n_2! \cdots n_m!} r_1^{n_1} r_2^{n_2} \cdots r_m^{n_m}. \tag{14.29}$$

By multiplying the probability distribution (14.13), we obtain

$$\begin{aligned} P[n_1, n_2, \ldots, n_m] &= \frac{n!}{n_1! n_2! \cdots n_m!} r_1^{n_1} r_2^{n_2} \cdots r_m^{n_m} \frac{(\lambda t)^n}{n!} e^{-\lambda t} \\ &= \prod_{k=1}^{m} \frac{(r_k \lambda t)^{n_k}}{n_k!} e^{-r_k \lambda t} = \prod_{k=1}^{m} P(n_k; r_k \lambda t), \end{aligned} \tag{14.30}$$

where $P(i; a)$ is the Poisson distribution defined by (3.77). Since the joint probability factors into m Poisson distributions, the random variables $N_1(t), N_2(t), \ldots, N_m(t)$ are statistically independent for an arbitrarily chosen interval $(0, t]$. Therefore, the m output substreams are independent of each other for an arbitrarily chosen interval $(0, t]$; hence, we have proved that the substreams form independent Poisson processes.

14.1.2.4 Uniformity of Poisson arrivals

Another important property of the Poisson process is what is called its **uniformity**: given that there is an arrival in the interval $(0, T]$, its arrival epoch τ is uniformly distributed over the interval; i.e.,

$$P[t \le \tau \le t + h | \text{ an arrival in } (0, T]] = \frac{h}{T} + o(h). \tag{14.31}$$

We can prove (14.31) as follows.

Suppose there are n arrivals in the interval $(0, T]$. The joint probability that there are i arrivals in a subinterval $(0, t]$, one arrival in $(t, t + h]$, and $n - i - 1$ arrivals in $(t + h, T]$ is the product of three Poisson distributions:

$$\begin{aligned} &\left[\frac{(\lambda t)^i}{i!} e^{-\lambda t}\right] \left(\lambda h\, e^{-\lambda h}\right) \left\{\frac{[\lambda(T - t - h)]^{n-i-1}}{(n - i - 1)!} e^{-\lambda(T-t-h)}\right\} \\ &= \frac{\lambda^n h\, e^{-\lambda T}}{(n - 1)!} \binom{n-1}{i} t^i (T - t - h)^{n-i-1}. \end{aligned} \tag{14.32}$$

Summing the above expression over the possible values of i, we find that the joint probability that there are n arrivals in $(0, T]$ with an arrival in $(t, t + h]$ is

$$\frac{\lambda^n h\, e^{-\lambda T}}{(n-1)!}\sum_{i=0}^{n-1}\binom{n-1}{i}t^i(T-t-h)^{n-i-1} = \frac{[\lambda(T-h)]^{n-1}}{(n-1)!}\lambda h\, e^{-\lambda T}, \tag{14.33}$$

where we used the binomial formula $\sum_{i=0}^{k}\binom{k}{i}x^i y^{k-i} = (x+y)^k$, with $k = n-1$.

Since h is an infinitesimal interval, we rewrite (14.33) as

$$P[n \text{ arrivals in } (0, T] \text{ with an arrival in } (t, t+h]] = \frac{(\lambda T)^{n-1}}{(n-1)!}\lambda h\, e^{-\lambda T} + o(h), \tag{14.34}$$

obtaining the conditional probability

$$\begin{aligned} P[\text{an arrival in } (t, t+h] | n \text{ arrivals in } (0, T]] &= \frac{\frac{(\lambda T)^{n-1}}{(n-1)!}\lambda h\, e^{-\lambda T} + o(h)}{\frac{(\lambda T)^n}{n!}e^{-\lambda T}} \\ &= \frac{nh}{T} + o(h). \end{aligned} \tag{14.35}$$

Since the n arrivals are independent, any one of them will fall into the interval $(t, t+h]$ with equal chance, with probability h/T. This final result is independent of n. Thus, we have proved (14.31).

14.1.2.5 Infinitesimal generator of Poisson process

The Poisson counting process $N(t)$ is a Markov process defined over a countably infinite set of states

$$\mathcal{S} = \{0, 1, 2, \ldots, n, \ldots\}. \tag{14.36}$$

The process $N(t)$ in state $n-1$ at time t makes a transition to state n in $(t, t+h]$ with probability λh irrespective of how the process has reached the current state, and how long the sojourn time in this state has been. These are the properties of Markov processes, but $N(t)$ is a special Markov process in that transitions to states other than to the immediate right in an infinitesimal interval $(t, t+h]$ can be ignored. In other words, the probability of transition from $n-1$ to $n+1$ or larger is $o(h)$.

Let $p_n(t) = P[N(t) = n]$ and let $\boldsymbol{p}(t)$ be a column vector of infinite dimension defined as

$$\boldsymbol{p}(t) = (p_0(t), p_1(t), \ldots, p_n(t), \ldots)^\top. \tag{14.37}$$

The **infinitesimal generator** or **transition rate matrix** (see (16.23)) of $N(t)$ is given by

$$\boldsymbol{Q} = \begin{bmatrix} -\lambda & \lambda & 0 & 0 & \cdots \\ 0 & -\lambda & \lambda & 0 & \cdots \\ 0 & 0 & -\lambda & \lambda & \cdots \\ 0 & 0 & 0 & -\lambda & \cdots \\ \vdots & \vdots & \vdots & \vdots & \ddots \end{bmatrix}. \tag{14.38}$$

Then, the set of differential-difference equations (14.7) can be concisely represented by a single equation:

$$\boxed{\frac{d\boldsymbol{p}(t)^\top}{dt} = \boldsymbol{p}(t)^\top \boldsymbol{Q},} \tag{14.39}$$

which is a variant of Kolmogorov's forward equation defined in (16.34).

14.2 Birth–death (BD) process

In this section we study properties of the random process $N(t)$ that takes the form

$$N(t) = A(t) - D(t), \tag{14.40}$$

where $A(t)$ is the arrival counting process and $D(t)$ is the departure counting process. If we view $N(t)$ as the the number of alive individuals in some *population*, then $A(t)$ is the number of **births** in the interval $(0, t]$ and $D(t)$ is the number of **deaths** in $(0, t]$. Therefore, $N(t)$ represents the cumulative effect of births and deaths, and is aptly called a **birth–death (BD) process**. This process has a number of applications, including population biology and queueing and loss system models (see Chapter 23).

The formal definition of a **BD process** applies to the following special class of point processes.

DEFINITION 14.2 (BD process). *A random process $N(t)$ is called a BD process if its transition probabilities*

$$P_{m,n}(h) = P[N(t+h) = n | N(t) = m] \tag{14.41}$$

are independent of t and satisfy

$$\boxed{P_{m,n}(h) = \begin{cases} \lambda_m h + o(h), & n = m+1, \\ \mu_m h + o(h), & n = m-1, \\ 1 - \{\lambda_m + \mu_m\}h + o(h), & n = m, \\ o(h), & |n-m| \geq 2. \end{cases}} \tag{14.42}$$

□

The values $m = 0, 1, 2, \ldots$ that $N(t)$ assumes are called the **states** of the process, and the coefficients λ_m and μ_m represent, respectively, the **birth rate** and **death rate** when the process is in state m. If λ_m and μ_m are independent of m, then $N(t)$ is called a homogeneous BD process. The Poisson process is a special case of the BD process, and is referred to as a **pure birth process**, where the birth rate is homogeneous, i.e., $\lambda_m = \lambda$ for all $m \geq 0$, and no deaths occur, i.e., $\mu_m = 0$ for all $m \geq 0$. Like the Poisson process, the BD process is a special class of Markov process, in which transitions within an infinitesimal interval from a given state m are limited only to either $m-1$ or $m+1$.

Let $p_n(t)$ be, as before, the probability that the population size at time t is n:

$$p_n(t) = P[N(t) = n]. \tag{14.43}$$

To calculate $p_n(t+h)$ we note that the population size at time $t+h$ is n only if one of the following events occurs:

1. $N(t) = n$ and neither birth nor death occurs in $(t, t+h]$;
2. $N(t) = n-1$ and one birth in $(t, t+h]$;
3. $N(t) = n+1$ and one death in $(t, t+h]$;
4. $N(t)$ is other than $n-1, n, n+1$, but two or more births/deaths occur in $(t, t+h]$, making $N(t+h) = n$.

From the definition of a BD process, $P_{m,n}(h) = o(h)$ for $|m-n| \geq 2$, and thus the probability of event 4 is $o(h)$. The first three events are mutually exclusive, so that their probabilities add. Therefore,

$$\begin{aligned} p_n(t+h) = {} & p_n(t)\{1 - \lambda_n h - \mu_n h + o(h)\} \\ & + p_{n-1}(t)\{\lambda_{n-1}h + o(h)\} \\ & + p_{n+1}(t)\{\mu_{n+1}h + o(h)\} + o(h), \end{aligned} \tag{14.44}$$

from which we obtain the following differential-difference equations:

$$\begin{aligned} \frac{dp_0(t)}{dt} &= -\lambda_0 p_0(t) + \mu_1 p_1(t), \\ \frac{dp_n(t)}{dt} &= -(\lambda_n + \mu_n)p_n(t) + \lambda_{n-1}p_{n-1}(t) + \mu_{n+1}p_{n+1}(t), \; n = 1, 2, \ldots, \end{aligned} \tag{14.45}$$

or in matrix notation

$$\frac{d\boldsymbol{p}(t)^\top}{dt} = \boldsymbol{p}(t)^\top \boldsymbol{Q}, \tag{14.46}$$

where the matrix $\boldsymbol{Q}$ is the **transition rate matrix** or **infinitesimal generator matrix** (see Definition 16.3) given by

$$\boldsymbol{Q} = \begin{bmatrix} -\lambda_0 & \lambda_0 & 0 & 0 & \cdots \\ \mu_1 & -\lambda_1 - \mu_1 & \lambda_1 & 0 & \cdots \\ 0 & \mu_2 & -\lambda_2 - \mu_2 & \lambda_2 & \cdots \\ 0 & 0 & \mu_3 & -\lambda_3 - \mu_3 & \cdots \\ \vdots & \vdots & \vdots & \vdots & \ddots \end{bmatrix}. \tag{14.47}$$

The solution of the above equations is manageable with certain restrictions imposed on the BD rates (see Problems 14.9–14.11). But the general time-dependent solution of

(14.45) is difficult to come by, so we should content ourselves with the equilibrium-state solution. Indeed, in many applications, the equilibrium-state (or steady-state) solution may be all we want.

We define an **equilibrium-state** distribution as π_n such that $p_n(t) = \pi_n$ specifies a (constant) solution to (14.45). If such a distribution exists, it is unique, and for each state n

$$\lim_{t\to\infty} p_n(t) = \pi_n. \tag{14.48}$$

Since we are interested only in the statistical equilibrium properties of the system, we take limits as $t \to \infty$ in (14.45), and set $\lim_{t\to\infty} dp_n(t)/dt = 0$, thus obtaining the following linear difference equations:

$$\mu_1\pi_1 - \lambda_0\pi_0 = 0, \tag{14.49}$$

$$\mu_{n+1}\pi_{n+1} - \lambda_n\pi_n = \mu_n\pi_n - \lambda_{n-1}\pi_{n-1},\ n = 1, 2, 3, \ldots \tag{14.50}$$

From the above equations,

$$\boxed{\mu_n\pi_n = \lambda_{n-1}\pi_{n-1}, \text{ for all } n = 1, 2, 3, \ldots.} \tag{14.51}$$

The left-hand side of the above equation represents the rate of transition from state n to state $n-1$ and the right-hand side is the transition rate from state $n-1$ to state n, and these two quantities should balance out each other in the steady state. The equations (14.51) are called **detailed balance equations**, which hold between every pair of adjacent states $n-1$ and n of the BD process, $n = 1, 2, \ldots$ They are shown schematically in Figure 14.1. Detailed balance equations for general Markov processes will be discussed in Section 16.3.

From the detailed balance equations we obtain the following recurrence relation:

$$\pi_n = \frac{\lambda_{n-1}}{\mu_n}\pi_{n-1}, \text{ for all } n = 1, 2, 3, \ldots. \tag{14.52}$$

The above balance equations can be derived from (14.46) as well. Setting $\lim_{t\to\infty} d\boldsymbol{p}(t)^\top/dt = \mathbf{0}^\top$,

$$\boldsymbol{\pi}^\top \boldsymbol{Q} = \mathbf{0}^\top, \tag{14.53}$$

where $\boldsymbol{\pi}^\top = (\pi_0, \pi_1, \ldots, \pi_n, \ldots)$ and $\mathbf{0}^\top$ is a row vector of size infinity whose elements are all zeros. By extracting the nth element in the above vector expression,

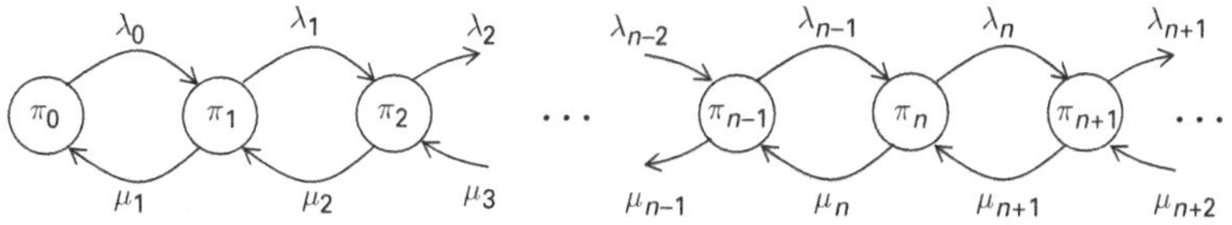

Figure 14.1 A schematic diagram of the balance equation in the BD process.

$$\pi_{n-1}\lambda_{n-1} - \pi_n(\lambda_n + \mu_n) + \pi_{n+1}\mu_{n+1} = 0, \tag{14.54}$$

which is, as expected, equivalent to (14.50).

By a simple recursion, (14.52) leads to

$$\boxed{\pi_n = \frac{\lambda_{n-1}}{\mu_n}\pi_{n-1} = \pi_0 \prod_{i=0}^{n-1} \frac{\lambda_i}{\mu_{i+1}},} \tag{14.55}$$

where π_0 is determined by the condition $\sum_{n=0}^{\infty} \pi_n = 1$ or

$$\pi_0 \sum_{n=0}^{\infty} \prod_{i=0}^{n-1} \frac{\lambda_i}{\mu_{i+1}} = 1. \tag{14.56}$$

Therefore, if the series

$$\boxed{G = \sum_{n=0}^{\infty} \prod_{i=0}^{n-1} \frac{\lambda_i}{\mu_{i+1}}} \tag{14.57}$$

converges, then

$$\pi_0 = G^{-1} = \left[\sum_{n=0}^{\infty} \prod_{i=0}^{n-1} \frac{\lambda_i}{\mu_{i+1}} \right]^{-1}. \tag{14.58}$$

This constant G of (14.57) is called the **normalization constant** or **partition function**.

14.3 Renewal process

As introduced in Section 12.2.6, a **point process** in one-dimensional space (i.e., the time axis) is a random process that consists of a sequence of epochs $\{t_1, t_2, \ldots, t_n, \ldots\}$ where "point events" occur. Examples of point events are births of babies, occurrences of traffic accidents, and transmissions of data packets. The Poisson process discussed in the preceding section is one of the simplest examples of a point process. A point process is a sequence of events in which only the times of their occurrence are of interest.

The point process, just like the Poisson process, can be represented by $N(t)$, the cumulative count of events in the time interval $(0, t]$:

$$N(t) = \max\{n : t_n \le t\}, \tag{14.59}$$

which is called a **counting process**. The difference of the event points

$$X_n = t_n - t_{n-1} \tag{14.60}$$

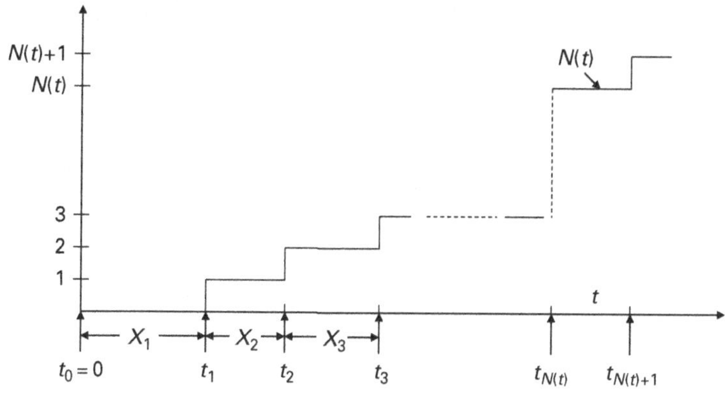

Figure 14.2 Renewal process $N(t)$ with renewal points $\{t_n\}$.

represents the interval between the $(n-1)$st and the nth point events. Conversely, we can write

$$t_n = t_{n-1} + X_n = \sum_{i=1}^{n} X_i, \tag{14.61}$$

where $t_0 = 0$.

If we assume that the X_i are i.i.d. RVs with a common distribution function $F_X(x)$, the point process $N(t)$ formed is called a **renewal process**. The event points t_i are called **renewal points** and the intervals X_i are the **lifetimes**. The distribution function $F_X(x)$ is thus called the **lifetime distribution function** of this renewal process.

It should be instructive to note that the renewal process $N(t)$ can be interpreted as a special **random walk** in which the t-axis corresponds to the one-dimensional real line, and all steps are positive (i.e., steps to the "right") with variable sizes (see Section 15.1 for the simple random walk model and Section 17.1 for general random walk models). The RV X_i represents the step size of the ith step and t_n is the position after the nth step.

14.3.1 Renewal function and renewal equation

The expected number of renewal points in $(0, t]$ is called the **renewal function** $M(t)$:

$$\boxed{M(t) = E[N(t)], t \geq 0,} \tag{14.62}$$

and

$$\boxed{m(t) = \frac{dM(t)}{dt}, t \geq 0,} \tag{14.63}$$

is called the **renewal density**.

Let us denote by $F_{t_n}(x)$ the distribution function of the nth renewal point t_n:

$$F_{t_n}(x) = P[t_n \leq x], x \geq 0. \tag{14.64}$$

Since t_{n-1} and X_n in (14.61) are independent, we have (cf. (5.29))

$$F_{t_n}(x) = \int_0^x F_{t_{n-1}}(x-y)\, dF_X(y),\ x \geq 0. \tag{14.65}$$

If the PDF $f_X(x)$ exists (and hence $f_{t_n}(x)$, which is the nth convolution of $f_X(x)$, exists as well), the above equation becomes

$$f_{t_n}(x) = \int_0^x f_{t_{n-1}}(x-y) f_X(y)\, dy = f_{t_{n-1}}(x) \circledast f_X(x),\ x \geq 0, \tag{14.66}$$

where $\circledast$ denotes convolution (cf. (5.27) and (8.75)).

By noting that $N(t)$ is at least k, if and only if the kth renewal point occurs before time t, i.e.,

$$N(t) \geq k \Longleftrightarrow t_k \leq t, \tag{14.67}$$

we find

$$P[N(t) \geq k] = P[t_k \leq t] = F_{t_k}(t) < t \geq 0. \tag{14.68}$$

Consequently, it follows that

$$\begin{aligned} P[N(t) = k] &= P[N(t) \geq k] - P[N(t) \geq k+1] \\ &= F_{t_k}(t) - F_{t_{k+1}}(t),\ t \geq 0,\ k = 1, 2, \ldots. \end{aligned} \tag{14.69}$$

Recall that the expectation of a nonnegative RV is the integral of its complementary distribution function (see (4.11)), which reduces to a sum for discrete RVs (see (3.120)). Applying this result, we have

$$\begin{aligned} M(t) = E[N(t)] &= \sum_{n=0}^{\infty} P[N(t) > n] = \sum_{k=1}^{\infty} P[N(t) \geq k] \\ &= \sum_{k=1}^{\infty} P[t_k \leq t] = \sum_{k=1}^{\infty} F_{t_k}(t),\ t \geq 0. \end{aligned} \tag{14.70}$$

The mean lifetime,

$$m_X = E[X_i] = \int_0^{\infty} x\, dF_X(x) \text{ for all } i, \tag{14.71}$$

is strictly positive. If m_X is finite, then by applying the strong law of large numbers (see Section 11.3.3), we can show that (Problem 14.14)

$$\frac{N(t)}{t} \xrightarrow{\text{a.s.}} \frac{1}{m_X}. \tag{14.72}$$

We remark that, in general, almost sure convergence does not imply convergence in the mean (see Section 11.2.5). Nevertheless, the expected value version of (14.72) holds indeed; i.e.,

$$\lim_{t\to\infty} \frac{M(t)}{t} \longrightarrow \frac{1}{m_X}. \tag{14.73}$$

A proof of (14.73) can be found in, for example, Wolff [357, Section 2-14].

Consider the first renewal point $t_1 = X_1$. If the first lifetime X_1 exceeds t, there are no renewals in $(0, t]$. If $X_1 < t$, there is clearly one renewal at $t_1 = X_1$, plus, on average, additional $M(t - X_1)$ renewals occurring in the remaining interval $(X_1, t]$. Thus, we have the following conditional expectation of $N(t)$, given X_1:

$$E[N(t)|X_1 = x] = \begin{cases} 0, & \text{if } x > t, \\ 1 + M(t - x), & \text{if } x \leq t. \end{cases} \tag{14.74}$$

Then, the unconditional expectation is obtained as

$$M(t) = E[N(t)] = \int_0^\infty E[N(t)|X_1 = x]\, dF_X(x) = \int_0^t (1 + M(t - x))\, dF_X(x). \tag{14.75}$$

Thus, the renewal function $M(t)$ satisfies

$$\boxed{M(t) = F_X(t) + \int_0^t M(t - x)\, dF_X(x), t \geq 0,} \tag{14.76}$$

which is called the **renewal equation**. The corresponding equation for the renewal density function is given by

$$m(t) = f_X(t) + m(t) \circledast f_X(t), t \geq 0. \tag{14.77}$$

By taking the Laplace transform of (14.77), we have

$$m^*(s) = f_X^*(s) + m^*(s) f_X^*(s), \tag{14.78}$$

from which we obtain

$$\boxed{m^*(s) = \frac{f_X^*(s)}{1 - f_X^*(s)}.} \tag{14.79}$$

14.3.2 Residual life in a renewal process

As we discussed earlier, a renewal process is a counting process in which the **lifetimes** $\{X_i; i = 1, 2, \ldots\}$ between successive *renewal points* are i.i.d. with a common distribution function $F_X(x)$. Consider a sufficiently long interval $[0, T)$ and randomly choose an epoch τ; that is, τ is uniformly distributed in $[0, T)$. Suppose that the point τ falls between two renewal points t_{k-1} and t_k. Let us denote by X^* the lifetime between these two renewal points:

$$X^* = t_k - t_{k-1}.$$

What is the distribution function of this RV?

Since $X^* = X_k$ and the distribution function of X_k is $F_X(x)$, we may hasten to conclude that the distribution function of X^* is also $F_X(x)$. But this answer is incorrect! This is because, although the observation epoch τ is distributed uniformly in $[0, T)$, the probability that a given subinterval $[t_i, t_{i+1})$ happens to include the epoch τ is not uniform. The longer the lifetime $X_i = t_{i+1} - t_i$ is, the more likely this lifetime contains the point τ.

The expected number of lifetimes in the interval $[0, T)$ is $\overline{N} = T/m_X$, where $m_X = E[X]$. Out of $\overline{N}$ such lifetimes, the number of lifetimes of length $(x, x + dx]$ should be on average $\overline{N} f_X(x)\,dx = T f_X(x)\,dx/m_X$, and the sum of these lifetimes is

$$I(x, x+dx) = \frac{T x f_X(x)\,dx}{m_X}.$$

Thus, the ratio $I(x, x+dx)/T$ is the probability that the RV X^* falls in $(x, x+dx]$:

$$f_{X^*}(x)\,dx = \frac{I(x, x+dx)}{T} = \frac{x f_X(x)}{m_X}\,dx. \tag{14.80}$$

Therefore,

$$\boxed{f_{X^*}(x) = \frac{x f_X(x)}{m_X},} \tag{14.81}$$

which leads to

$$F_{X^*}(x) = \frac{1}{m_X}\left\{\int_0^x [1 - F_X(u)]\,du - x\,[1 - F_X(x)]\right\}. \tag{14.82}$$

For given $X^* = x$, the **current lifetime** or **age** $Y = \tau - t_{k-1}$, as shown in Figure 14.3, is uniformly distributed in $[0, x)$. Then the **residual lifetime** (also called the **excess lifetime** or **forward recurrence time**) $R = x - Y$ is also uniformly distributed in $[0, x)$. Thus,

$$P[r < R \le r + dr | X^* = x] = \begin{cases} \frac{dr}{x}, & r < x, \\ 0, & r \ge x. \end{cases} \tag{14.83}$$

Then the unconditional probability that R falls in $(r, r + dr]$ is

$$\begin{aligned} P[r < R \le r + dr] &= \int_r^\infty \frac{dr}{x} f_{X^*}(x)\,dx = \frac{\int_r^\infty f_X(x)\,dx}{m_X}\,dr \\ &= \frac{1 - F_X(r)}{m_X}\,dr,\ r \ge 0. \end{aligned} \tag{14.84}$$

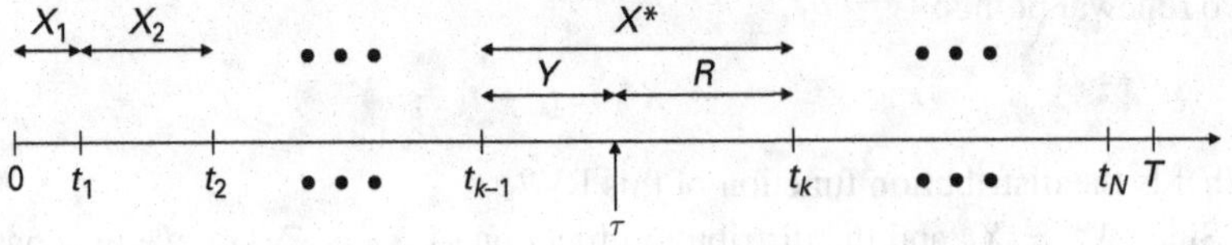

Figure 14.3 A renewal process with renewal points $\{t_i\}$. The point $t = \tau$ is an arbitrary observation point, R is the residual (or excess) lifetime, and Y is the current lifetime or age.

The left-hand side of the above expression is equal to $f_R(r)\,dr$; thus, the PDF of the residual life is given as

$$\boxed{f_R(r) = \frac{1 - F_X(r)}{m_X}, r \geq 0.} \tag{14.85}$$

By integrating this PDF, we obtain the distribution function

$$F_R(r) = \frac{1}{m_X} \int_0^r [1 - F_X(x)]\,dx. \tag{14.86}$$

It is not difficult to see that the distribution of the age Y is the same as that of the residual life R. For a given renewal process $N(t)$, $-\infty < t < \infty$, its time-reversed process $N(-t)$ is also a renewal process that is statistically identical to the original process; then the current lifetime Y in the original process becomes the residual lifetime R in its time-reversed version.

The Laplace transform (LT) of the PDF $f_R(x)$ is given by

$$f_R^*(s) = \frac{1}{m_X} \int_0^\infty [1 - F_X(r)]\,e^{-sr}\,dr = \frac{1}{m_X} \frac{1 - f_X^*(s)}{s}, \tag{14.87}$$

where we used the formula

$$\mathcal{L}\{F_X(x)\} = \mathcal{L}\left\{\int_0^x f_X(u)\,du\right\} = \frac{f_X^*(s)}{s}.$$

Then the **mean residual life**, $m_R = E[R]$, is obtained by

$$\begin{aligned} m_R &= -\lim_{s\to 0} \frac{df_R^*(s)}{ds} = -\frac{1}{m_X} \lim_{s\to 0} \frac{-sf_X^{*(1)}(s) - (1 - f_X^*(s))}{s^2} \\ &= -\frac{1}{m_X} \lim_{s\to 0} \frac{-f_X^{*(1)}(s) - sf_X^{*(2)}(s) + f_X^{*(1)}(s)}{2s} = \frac{f_X^{*(2)}(0)}{2m_X}, \end{aligned} \tag{14.88}$$

where $f_X^{*(1)}(s)$ and $f_X^{*(2)}(s)$ are the first and second derivatives of $f_X^*(s)$. Thus,

$$\boxed{m_R = \frac{E[X^2]}{2E[X]}.} \tag{14.89}$$

Example 14.1: Hyperexponential lifetime. Consider the k-phase hyperexponential (or mixed exponential) distribution described in Problem 4.12:

$$F_X(x) = \sum_{i=1}^{k} \pi_i(1 - e^{-\mu_i x}) = 1 - \sum_{i=1}^{k} \pi_i e^{-\mu_i x}, \tag{14.90}$$

which gives $m_X = \sum_{i=1}^{k} \pi_i/\mu_i$. Then the distribution of the residual lifetime is

$$\begin{aligned} F_R(r) &= \frac{1}{m_X}\int_0^r [1 - F_X(x)]\,dx = \frac{1}{m_X}\int_0^r \sum_{i=1}^{k} \pi_i e^{-\mu_i x}\,dx \\ &= \frac{1}{m_X}\sum_{i=1}^{k} \frac{\pi_i}{\mu_i}(1 - e^{-\mu_i r}). \end{aligned} \tag{14.91}$$

By defining the probability $p_i = \pi_i/\mu_i m_X$, we find

$$F_R(r) = \sum_{i=1}^{k} p_i(1 - e^{-\mu_i r}), \tag{14.92}$$

which is a k-phase hyperexponential distribution with the same set of exponential distributions, but with different weights.

Thus, the mean residual life is

$$m_R = E[R] = \sum_{i=1}^{k} \frac{p_i}{\mu_i} = \frac{1}{m_X}\sum_{i=1}^{k} \frac{\pi_i}{\mu_i^2}. \tag{14.93}$$

A more informative expression of m_R can be found by using the formula (14.89):

$$m_R = \frac{E[X^2]}{2m_X} = \frac{m_X^2 + \sigma_X^2}{2m_X}, \tag{14.94}$$

where $\sigma_X^2 = \text{Var}[X]$ and is given, for this hyperexponential distribution, by

$$\sigma_X^2 = m_X^2 + \sum_{i=1}^{k}\sum_{j=1}^{k} \pi_i\pi_j\left(\frac{1}{\mu_i} - \frac{1}{\mu_j}\right)^2. \tag{14.95}$$

Therefore, we find

$$m_R = m_X + \frac{1}{2m_X}\sum_i\sum_j \pi_i\pi_j\left(\frac{1}{\mu_i} - \frac{1}{\mu_j}\right)^2 \geq m_X. \tag{14.96}$$

Thus, the mean residual lifetime for a hyperexponential distribution is **longer** than the expected lifetime! This is owing to the fact that the coefficient of variation ($c_X = \sigma_X/m_X$) of this distribution is greater than one, as shown in the above. For a k-stage Erlang distribution we have a different result (see Problem 14.16). □

As we discussed in Section 6.3.3, if the RV X represents service time, then

$$h(x) = \frac{f_X(x)}{1 - F_X(x)} \tag{14.97}$$

is called the **completion rate function** or **hazard function** (see (6.36) and (6.37)). In the language of renewal theory, this is the age-dependent **failure rate**: $h(x)\,dx$ represents the probability that the life will terminate in $(x, x + dx]$ given that it has already attained the age of x.

We also discussed in Section 6.3.3 the **mean residual life curve** defined by

$$R_X(t) = E[R|X > t] = \frac{\int_t^{\infty}[1 - F_X(u)]du}{1 - F_X(t)}. \tag{14.98}$$

Plotting these curves is useful as a first step in determining whether an observed lifetime distribution should be modeled by an exponential or other "short-tailed" distribution or whether it warrants the use of a heavy-tailed distribution such as a Weibull, Pareto, or log-normal distribution.

14.4 Summary of Chapter 14

Poisson process $N(t)$:	(a) nondecreasing	
	(b) $P_{mn}(h) = \begin{cases} 1-\lambda h+o(h), & n = m, \\ \lambda h+o(h), & n=m+1, \\ o(h), & n\geq m+2, \end{cases}$	Def. 14.1
	(c) for $t > s$, $N(t) - N(s)$ is independent of $N(s)$	
Distribution of $N(t)$:	$P[N(t) = n] = \frac{(\lambda t)^n}{n!}e^{-\lambda t}, n \geq 0$	(14.3)
Interarrival time dist.:	$F_X(x) = P[X_n \leq x] = 1 - e^{-\lambda x}$	(14.23)
Properties of PP:	1. memoryless, 2. reproductive additivity, 3. reproductive decomposition, 4. uniformity	
Infini. generator:	$\boldsymbol{Q} = \begin{bmatrix} -\lambda & \lambda & 0 & 0 & \cdots \\ 0 & -\lambda & \lambda & 0 & \cdots \\ 0 & 0 & -\lambda & \lambda & \cdots \\ 0 & 0 & 0 & -\lambda & \cdots \\ \vdots & \vdots & \vdots & \vdots & \ddots \end{bmatrix}.$	(14.38)
BD process $N(t)$:	$P_{m,n}(h) = \begin{cases} \lambda_m h+o(h), & n=m+1, \\ \mu_m h+o(h), & n=m-1, \\ 1-\{\lambda_m+\mu_m\}h+o(h), & n=m, \\ o(h), & \lvert n-m\rvert\geq 2, \end{cases}$	(14.42)

Equations for $p_n(t)$:	$$\begin{aligned} \frac{dp_0(t)}{dt} &= -\lambda_0 p_0(t) + \mu_1 p_1(t), \\ \frac{dp_n(t)}{dt} &= -(\lambda_n + \mu_n) p_n(t) + \lambda_{n-1} p_{n-1}(t) \\ &+ \mu_{n+1} p_{n+1}(t), n = 1, 2, \ldots \end{aligned}$$	(14.45)
Equation for $\boldsymbol{p}(t)$:	$\frac{d\boldsymbol{p}(t)^\top}{dt} = \boldsymbol{p}(t)^\top \boldsymbol{Q}$	(14.46)
where	$$\boldsymbol{Q} = \begin{bmatrix} -\lambda_0 & \lambda_0 & 0 & 0 & \cdots \\ \mu_1 & -\lambda_1 - \mu_1 & \lambda_1 & 0 & \cdots \\ 0 & \mu_2 & -\lambda_2 - \mu_2 & \lambda_2 & \cdots \\ 0 & 0 & \mu_3 & -\lambda_3 - \mu_3 & \cdots \\ \vdots & \vdots & \vdots & \vdots & \ddots \end{bmatrix}$$	(14.47)
Equilibrium distribution:	$\pi_n = \lim_{t\to\infty} p_n(t)$	(14.48)
Balance equations:	$\mu_n \pi_n = \lambda_{n-1}\pi_{n-1}$, for all $n = 1, 2, 3, \ldots$	(14.51)
Equation for $\boldsymbol{\pi}$:	$\boldsymbol{\pi}^\top \boldsymbol{Q} = \boldsymbol{0}^\top$	(14.53)
Solution:	$\pi_n = \pi_0 \prod_{i=0}^{n-1} \frac{\lambda_i}{\mu_{i+1}}$	(14.55)
where	$\pi_0 = \left[\sum_{n=0}^{\infty} \prod_{i=0}^{n-1} \frac{\lambda_i}{\mu_{i+1}}\right]^{-1}$	(14.58)
Counting process $N(t)$:	$N(t) = \max\{n : t_n \le t\}$	(14.59)
Renewal process:	X_n are i.i.d., where $X_n = t_n - t_{n-1}$	(14.60)
Renewal function:	$M(t) = E[N(t)]$	(14.62)
Renewal density:	$m(t) = dM(t)/dt$	(14.63)
Laplace transform of $m(t)$:	$m^*(s) = \frac{f_X^*(s)}{1-f_X^*(s)}$	(14.79)
Renewal equation:	$M(t) = F_X(t) + \int_0^t M(t-x)\,dF_X(x)$	(14.76)
Residual life PDF:	$f_R(r) = \frac{1-F_X(r)}{m_X}, r \ge 0$	(14.85)
Mean residual life:	$m_R = \frac{E[X^2]}{2E[X]}$	(14.89)
Hazard function:	$h(x) = \frac{f_X(x)}{1-F_X(x)}$	(14.97)
Mean residual life curve:	$R_X(t) = E[R \vert X > t] = \frac{\int_t^\infty [1-F_X(u)]du}{1-F_X(t)}$	(14.98)

14.5 Discussion and further reading

Renewal processes may be viewed as a generalization of Poisson processes. Most textbooks on probability and random processes written for electrical engineers and computer scientists, however, discuss very little of the renewal theory, with a possible exception of Trivedi [327]. For a further study of renewal processes, the reader is referred to Çinlar [57], Cox [70], Cox and Lewis [71], Feller [100], Grimmett and Stirzaker [131], Kao [173], Karlin and Taylor [175], Nelson [254], Ross [288], and Wolff [358].

Birth–death processes have many applications in demography, queueing theory, and stochastic models in biology (e.g., the evolution of bacteria). All books on queueing theory discuss the BD process and its applications to various classes of queueing system models. However, queueing processes are often non-Markovian and are more difficult to analyze than the simple BD process models. For further reading on queueing theory, the reader is referred to, for example, Kleinrock [189], Kobayashi [197], Kobayashi and Mark [203], Nelson [254], Trivedi [327], and Wolff [358].

14.6 Problems

Section 14.1: The Poisson process

14.1* Alternative derivation of the Poisson process. Consider a point process $N(t)$, in which point events occur at $t_1, t_2, \ldots,$ and let each interarrival time (or lifetime) $X_n = t_n - t_{n-1}$ have the exponential distribution

$$F_X(x) = 1 - e^{-\lambda x}, x \geq 0, \tag{14.99}$$

and the corresponding PDF

$$f_X(x) = \frac{d}{dx} F_X(x) = \lambda\, e^{-\lambda x}, x \geq 0. \tag{14.100}$$

(a) A point event does not necessarily occur at time $t = 0$. Find $F_{t_1}(t) = P[t_1 \leq t]$.
(b) Show

$$f_{t_{n+1}}(t) = \int_0^t f_{t_n}(t-u) f_X(u)\, du, t \geq 0,\ n = 0, 1, 2, \ldots \tag{14.101}$$

(c) Derive

$$P[N(t) = n] = \frac{(\lambda t)^n}{n!} e^{-\lambda t}, n = 0, 1, 2, \ldots \tag{14.102}$$

Hint: Use (14.69).

14.2 Bernoulli process to Poisson process. Recall that we derived the Poisson distribution $P(i; a)$ of (3.77) as the limit of the binomial distribution in Section 3.3.3. Consider a semi-interval $(0, t]$ and divide it into m disjoint and contiguous subintervals of small length h (Figure 14.4), where

$$m = \lfloor t/h \rfloor.$$

Figure 14.4 Partitioning of the interval $(0, t]$ into m subintervals.

Let the arrival rate be a constant λ. For any subinterval,

1. The probability of one arrival is $\lambda h + o(h)$.
2. The probability of two or more arrivals is $o(h)$.
3. The probability of no arrival is $1 - \lambda h + o(h)$.

$o(h)$ represents a quantity that approaches to zero faster than h, as $h \to 0$.

(a) Treat an arrival as a "success" in a Bernoulli trial. Find an expression for the probability of i arrivals in the m subintervals.
(b) Take the limits $h \to 0$ and $m \to \infty$ while keeping $mh = t$ and find the probability distribution of the RV $N(t)$, the number of arrivals in the period $(0, t]$.

14.3 Superposition of Poisson processes.

(a) Let $\{X_j;\, j = 1, 2, \ldots, m\}$ be a set of independent RVs, exponentially distributed with parameters λ_j, $j = 1, 2, \ldots, m$, respectively. Find the distribution of the RV

$$Y = \min\{X_1, X_2, \ldots, X_m\}.$$

(b) Using the result of (a), show that the superposition of m independent Poisson processes with rates $\lambda_j (1 \leq m)$ generates a Poisson process with rate $\lambda = \sum_{i=1}^{m} \lambda_i$.

14.4 Consistency check of the Poisson process. Demonstrate the consistency of Definition 14.1 for the Poisson process as follows.

(a) Show from (14.13) that the probabilities of no arrival, one arrival, and multiple arrivals in a small interval h are given by $1 - \lambda h + o(h)$, $\lambda h + o(h)$, and $o(h)$ respectively.
(b) Derive the same set of probabilities as that given in part (a) from (14.99) alone.

14.5 Decomposition of a Poisson process.

(a) Let $\{X_j\}$ be a sequence of i.i.d. variables with the exponential distribution (14.99). Let $S_N = X_1 + X_2 + \cdots + X_N$, where N has the following geometric distribution:

$$P[N = n] = (1 - r)^{n-1} r, \quad n = 1, 2, 3, \ldots,$$

where $0 < r < 1$. Show that S_N has an exponential distribution.
(b) Consider the problem of decomposing a Poisson stream into m substreams, as discussed in the text. Using the result of part (a), show that the kth substream is a Poisson process with rate λr_k, $k = 1, 2, \ldots, m$.

14.6 Alternate decomposition of a Poisson stream. Refer again to the problem of decomposing a Poisson stream into the m substreams: rather than the independent selection, each substream receives every mth arrival; i.e., the first arrival, $(m + 1)$ arrival, $(2m + 1)$ arrival, …, go to substream 1, the second, $(m + 2)$, $(2m + 2)$, …, arrivals go to substream 2, etc. Find the interarrival time distribution of the individual substreams.

14.7 Derivation of the Poisson distribution. Obtain the Poisson distribution (3.77) through the following steps.

(a) Solve first for $p_0(t)$ from the differential equation (14.7). Insert this result into the case $n = 1$ in (14.7) and solve for $p_1(t)$. Continuing by induction, obtain $p_n(t)$ for all $n \geq 0$.

(b) Alternatively, define the Laplace transform

$$p_n^*(s) = \int_0^\infty p_n(t)e^{-st}\,dt.$$

Derive algebraic equations for $p_n^*(s)$ from the differential equations of part (a) and solve for $p_n^*(s)$ and then apply the inverse Laplace transform to obtain $p_n(t)$.

14.8* Uniformity and statistical independence of Poisson arrivals. Suppose that n (unordered) arrivals, $U_1, \ldots, U_n$, of a Poisson process occur in the interval $(0, T]$. Equation (14.31) establishes that the PDF of U_i conditioned on $\{N(T) = n\}$ is given by $f_{U_i}(u|N(T) = n) = 1/T$, $u \in (0, T]$, for each $i = 1, \ldots, n$.

(a) Show formally that the joint PDF of $U_1, \ldots, U_n$ conditioned on $\{N(T) = n\}$ is given by

$$f_{U_1 \cdots U_n}(u_1, \ldots, u_n|N(T) = n) = \frac{1}{T^n}, \tag{14.103}$$

for all $u_1, \ldots, u_n \in (0, T]$. Hence, $U_1, \ldots, U_n$ are independent, conditioned on $\{N(T) = n\}$.

(b) Based on (14.103), describe a way to generate a sequence of n Poisson arrivals on a given interval $(0, T]$.

Section 14.2: Birth–death (BD) process

14.9 Pure birth process. A BD process is called a **pure birth process** if $\lambda_n = \lambda$ for all $n \geq 0$ and $\mu_n = 0$ for all $n \geq 0$. Assuming $N(0) = 0$, solve the differential-difference equations (14.45) for $p_n(t)$ of the pure birth process.

14.10 Time-dependent solution. Set $\mu_n = 0$ for all $n \geq 0$ as in the above problem but permit state-dependent birth rates λ_n. Show that the time-dependent solution for $p_n(t)$ satisfies

$$p_n(t) = e^{-\lambda_n t}\left[\lambda_{n-1}\int_0^t p_{n-1}(x)e^{\lambda_n x}\,dx + p_n(0)\right].$$

14.11 Pure death process. Let us consider a pure death process; that is, $\lambda_n = 0$ for all $n \geq 0$ and $\mu_n = \mu$ for all $n \geq 1$. Let the initial population be $N(0) = N_0$. Show that the time-dependent solution for $p_n(t)$ is given by

$$p_n(t) = \frac{(\mu t)^{N_0-n}}{(N_0-n)!}e^{-\mu t}, \text{ for } 1 \le n \le N_0,$$

and

$$p_0(t) = 1 - \sum_{i=0}^{N_0-1} \frac{(\mu t)^i}{i!}e^{-\mu t}.$$

14.12 The time-dependent PGF. We extend the notion of PGF to the time-dependent probability distribution

$$G(z,t) = \sum_{n=0}^{\infty} p_n(t)z^n.$$

Solve the pure birth problem of Problem 14.9 using this generating function.

14.13* Time-dependent solution for a certain BD process. Let $\lambda(n) = \lambda$ for all $n \ge 0$ and $\mu(n) = n\mu$ for all $n \ge 1$. Find the partial differential equation that $G(z,t)$ must satisfy. Show that the solution to this equation is

$$G(z,t) = \exp\left[\frac{\lambda}{\mu}(1-e^{-\mu t})(z-1)\right].$$

Show that the solution for $p_n(t)$ is given as

$$p_n(t) = \frac{\left[\frac{\lambda}{\mu}(1-e^{-\mu t})\right]^j}{j!}\exp\left[-\frac{\lambda}{\mu}(1-e^{-\mu t})\right], \quad 0 \le n < \infty. \tag{14.104}$$

This particular BD process represents the $M/M/\infty$ queue, which we will discuss in Chapter 23.

Section 14.3: Renewal process

14.14* Derivation of (14.72). Derive (14.72) as follows:

(a) Apply the SLLN to show that

$$\frac{N(t)}{t_{N(t)}} \xrightarrow{\text{a.s.}} \frac{1}{m_X}. \tag{14.105}$$

(b) Show that $t/t_{N(t)} \xrightarrow{\text{a.s.}} 1$ as $t \to \infty$.
(c) Hence, establish (14.72).

14.15 The Poisson process as a renewal process. Consider the Poisson process $N(t)$ with rate λ.

(a) Find the renewal function $M(t)$ and the renewal density $m(t)$.
(b) Find $f_{t_n}(x)$; i.e., the PDF of the kth arrival time t_n.
(c) Show that $M(t)$ obtained in (a) satisfies the renewal equation (14.76).

(d) Find the Laplace transform $m^*(s)$ of $m(t)$ using (14.79) and then obtain $m(t)$.
(e) Obtain the residual life PDF $f_R(r)$.
(f) Obtain the mean residual lifetime m_R.

14.16 Residual lifetime of an Erlang distributed lifetime. Consider the following k-stage Erlang distribution for the lifetime (or the interval between consecutive renewal points) X:

$$F_X(x) = 1 - e^{-\lambda x} \sum_{j=0}^{k-1} \frac{(\lambda x)^j}{j!}.$$

(a) Find an expression for $f_R(r)$, the PDF of the residual lifetime.
(b) Find the Laplace transform $f_R^*(s)$ of the above PDF.
(c) Find the mean residual life m_R and compare it with m_X, the mean lifetime.

14.17 Residual lifetime of a uniformly distributed lifetime. Consider a renewal process where the lifetime is uniformly distributed over $[0, L]$ for some constant L.

(a) Find the distribution function and the expectation of the age Y seen by a random observer.
(b) Do the same for the residual lifetime R.

14.18 Moments of residual lifetime and age. Show that the nth moment of the age Y and the residual lifetime R is given by

$$E[Y^n] = E[R^n] = \frac{E[X^{n+1}]}{(n+1)E[X]}. \tag{14.106}$$

Hint: Generalize the following formula for a nonnegative RV X:

$$E[X] = \int_0^\infty (1 - F_X(x))\, dx.$$

15 Discrete-time Markov chains

In Section 12.2.5 we introduced the notions of Markov chains, both discrete-time Markov chains (DTMCs) and continuous-time Markov chains (CTMCs). In both the DTMC and CTMC, a *discrete state space* is implicitly assumed, although the number of states can be countably infinite. The term "Markov process" is more general than Markov chain, in that the state space can be continuous, although some authors use the two terms synonymously. Markov processes are extensively used in many stochastic modeling and statistical analyses, because many real systems can be adequately represented in terms of a Markov process, and because of the Markov property (i.e., the memoryless property) a Markov process is simple to analyze. Markov processes have been successfully used in queueing system modeling and traffic modeling, as discussed in Chapter 23. A special type of Markov process model, called a hidden Markov model (HMM), has been successfully used in a variety of scientific and engineering applications, including coding and decoding in communications, natural language processing such as speech recognition, and computational biology (e.g., DNA sequencing). Chapter 20 will be devoted to this important topic.

15.1 Markov processes and Markov chains

Let t_0 be the present time. Then a Markov process $X(t)$ is a process such that its evolution in the future, i.e., $\{X(t); t_0 < t\}$, depends on its past $\{X(s); s \leq t_0\}$ only through the present value $X(t_0)$. This property is called the **Markovian property** or **Markov property**. It is also sometimes referred to as the **memoryless property**, since the future behavior of $X(t)$ does not depend on how the current value $X(t_0)$ has been reached. Let us begin with a more precise definition.

DEFINITION 15.1 (Markov process). *A random process $X(t)$ is called a **Markov process** if, for any $t_1 < t_2$,*

$$P[X(t_2) \leq x | X(t); -\infty < t \leq t_1] = P[X(t_2) \leq x | X(t_1)]. \tag{15.1}$$

□

Consider a set of n arbitrarily chosen instants in time, denoted as $t_1 < t_2 < \cdots < t_{n-1} < t_n$. Then, the above Markov property implies

$$P[X(t_n) \leq x | X(t_i); i = 1, 2, \ldots, n-1] = P[X(t_n) \leq x | X(t_{n-1})]. \quad (15.2)$$

Markov processes can be either discrete-time or continuous-time. Often, the time index runs through the set of nonnegative integers, and in such a case we write $\{X_n; n = 0, 1, 2, \ldots\}$. A Markov process $\{X_n\}$ of this type is often called a **Markov sequence**. The points in time may be equally spaced, or their spacing may depend on some events in the physical system with which the random process is associated – for example, occurrences of customer arrivals or departures in a queueing system. When the *states* that X_n takes on are finite or, at most, countably infinite, we may label the states as $0, 1, 2, \ldots$; i.e.,

$$\mathcal{S} \triangleq \{0, 1, 2, \ldots\}. \quad (15.3)$$

The states may be nonquantitative or quantitative values.

DEFINITION 15.2 (*h*th order Markov chain). *A random sequence*[1] X_n *is called a* ***Markov chain*** *of* ***order h*** *(or an hth order Markov chain) when* X_n *takes a finite or countably infinite number of states* $\mathcal{S} = \{0, 1, 2, \ldots\}$ *and if the state of* X_n *depends only on the last h states; i.e., if*

$$P[X_n | X_k; -\infty < k \leq n-1] = P[X_n | X_{n-h}, \ldots, X_{n-1}], \textit{ for all } n. \quad (15.4)$$

A first-order Markov chain is called a ***simple*** *Markov chain.* □

An hth order Markov chain, with appropriate choice of h, has been found to be a useful statistical model to represent a variety of processes or phenomena we encounter in the real world; for instance, a syntactic structure of an English sentence (Shannon [300]), a sequence of phonemes of human speech (Jelinek [164]), a DNA sequence, the price changes in a stock market, etc.

If X_n is a second-order Markov chain, we define a new sequence $\{Y_n\}$ by

$$Y_n = (X_{n-1}, X_n). \quad (15.5)$$

Then $\{Y_n\}$ is a first-order Markov chain over the state space $\mathcal{S} \times \mathcal{S} (= \mathcal{S}^2)$. In general, if X_n is an hth order chain, then

$$\boxed{Y_n = (X_{n-h+1}, X_{n-h+2}, \ldots, X_{n-1}, X_n),} \quad (15.6)$$

where $Y_n \in \mathcal{S}^h$ reduces to a simple Markov chain. Thus, we will discuss only simple Markov chains in the rest of this chapter. The term DTMC usually means a first-order Markov chain or simple Markov chain as defined above.

The term Markov chain is often defined more broadly to include a "continuous-time" (as well as discrete-time) Markov process with a finite or countably infinite number of states. Thus, a continuous-time, discrete-state Markov process is often referred to as a CTMC.

[1] One may write $\{X_n\}$, but for notational brevity we simply write X_n, just like we write $W(t)$ for a continuous-type process.

DEFINITION 15.3 (Transition probabilities). *The conditional probabilities of a (simple) Markov chain*

$$P[X_{n+1} = j | X_n = i] = P_{ij}(n), i, j \in \mathcal{S}, n = 0, 1, 2, \ldots \tag{15.7}$$

are called the ***(one-step) transition probabilities****. The matrix*

$$\boldsymbol{P}(n) = [P_{ij}(n)] \tag{15.8}$$

is called the ***transition probability matrix*** *(TPM).* □

Example 15.1: Random walk. Let $X_1, X_2, \ldots, X_n, \ldots$ be an i.i.d. sequence of $+1$s and -1s:

$$P[X_n = +1] = p, P[X_n = -1] = 1 - p \triangleq q.$$

Let S_n denote the running sum:

$$S_n = X_1 + X_2 + \cdots + X_n = S_{n-1} + X_n.$$

Clearly, $\{S_n\}$ is a Markov chain with the transition probabilities

$$P_{ij}(n) = \begin{cases} p, & j = i + 1, \\ q, & j = i - 1, \\ 0, & \text{otherwise}, \end{cases}$$

for all integers $i, j \in \mathbb{Z} = \{0, \pm 1, \pm 2, \ldots\}$ and time index n. Such a Markov chain $\{S_n\}$ is often referred to as a **random walk**. □

We can write a simple recursion equation for the state probabilities of a Markov chain. Consider the probability $p_j(n+1)$ that X_{n+1} assumes state j. Now at time n the system may be in any state i with probability $p_i(n)$, $i \in \mathcal{S}$. For each state $i \in \mathcal{S}$, there is a transition probability $P_{ij}(n)$ that the system will make the transition from this state to state j. Hence,

$$p_j(n+1) = \sum_{i \in \mathcal{S}} p_i(n) P_{ij}(n), \; j \in \mathcal{S}, \tag{15.9}$$

or in matrix and vector representation

$$\boldsymbol{p}^\top(n+1) = \boldsymbol{p}^\top(n)\boldsymbol{P}(n), \tag{15.10}$$

where $\boldsymbol{p}^\top(n)$ is a row vector of dimension[2] $|\mathcal{S}|$ (possibly infinite) with elements $p_j(n)$, $j \in \mathcal{S}$, and $\boldsymbol{P}(n)$ is an $|\mathcal{S}| \times |\mathcal{S}|$ matrix as defined in (15.8). If all $P_{ij}(n)$ are independent of n, we have

[2] $|\mathcal{S}|$ denotes the cardinality of the set $\mathcal{S}$; i.e., the total number of states.

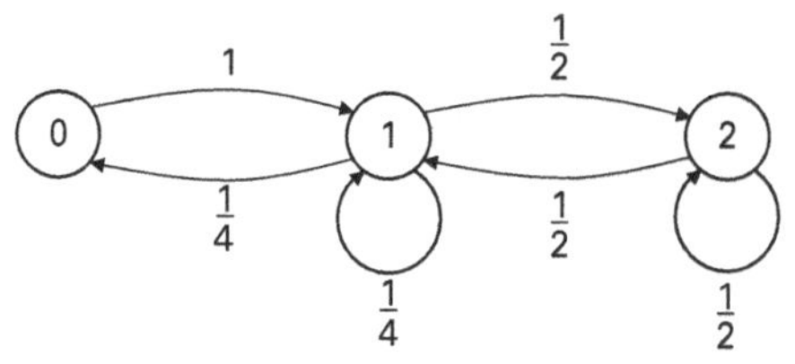

Figure 15.1 An example of a state transition diagram.

$$\boxed{\boldsymbol{p}^\top(n+1) = \boldsymbol{p}^\top(n)\boldsymbol{P} \text{ for all } n = 0, 1, 2, \ldots,} \tag{15.11}$$

and such a Markov chain is called a **homogeneous** or **stationary** Markov chain.

A Markov chain can be represented graphically by the corresponding **state transition diagrams**. It is a signal flow graph in which the nodes represent the states i and the directed arcs represent the transition probabilities P_{ij}. Figure 15.1 shows the state transition diagram of a homegeneous three-state Markov chain, whose set of states is $\mathcal{S} = \{0, 1, 2\}$ and its TPM is given by

$$\boldsymbol{P} = \begin{bmatrix} 0 & 1 & 0 \\ \frac{1}{4} & \frac{1}{4} & \frac{1}{2} \\ 0 & \frac{1}{2} & \frac{1}{2} \end{bmatrix}. \tag{15.12}$$

It is clear from the definition of conditional probabilities that the elements of the TPM $\boldsymbol{P}$ must satisfy the following properties, whether the process is stationary or non-stationary:

$$P_{ij}(n) \geq 0 \quad \text{for all} \quad i, j, \text{and } n = 0, 1, 2, \ldots \tag{15.13}$$

and

$$\sum_{j \in \mathcal{S}} P_{ij}(n) = 1 \quad \text{for all} \quad i \in \mathcal{S}, \text{and } n = 0, 1, 2, \ldots. \tag{15.14}$$

Matrices satisfying (15.13) and (15.14) are called **stochastic**. Any stochastic square matrix may serve as a TPM.

Example 15.2: Discrete memoryless channel and its stochastic matrix. Examples of a stochastic matrix that are **not** a Markov chain TPM are provided by **discrete memoryless channels** (DMCs), which play an important role in information theory [300]. Let $\{r_1, r_2, \ldots, r_m\}$ be the set of symbols that the channel input X takes on and let $\{s_1, s_2, \ldots, s_n\}$ be the set of discrete values that Y may take. Then the channel matrix is defined by

$$
C = \begin{bmatrix} P_{11} & P_{12} & \cdots & P_{1n} \\ P_{21} & P_{22} & \cdots & P_{2n} \\ \vdots & \vdots & \ddots & \vdots \\ P_{m1} & P_{m2} & \cdots & P_{mn} \end{bmatrix},
$$

where P_{ij} is the probability that the input symbol r_i is delivered by the channel as the output symbol s_j, where $1 \leq i \leq m$ and $1 \leq j \leq n$. The matrix $\boldsymbol{C}$ is not even a square matrix (unless $m = n$), but is always *stochastic*.

A simple case is where $m = n = 2$, $P_{12} = P_{21} = p$, and $P_{11} = P_{22} = q = 1 - p$. Both channel input and output symbols are binary symbols: "1" and "0." Such a DMC is commonly referred to as a **binary symmetric channel** (BSC):

$$
C = \begin{bmatrix} q & p \\ p & q \end{bmatrix} \text{ (BSC).}
$$

Another often used channel model is a **binary erasure channel** (BEC) in which $r = 2$ and $s = 3$. The input symbols are "1" and "0," but the output symbols are "1", "ϵ," and "0," where the intermediate symbol is called an **erasure**. The channel output may take this value with probability $p(< 1)$ when the channel is noisy. The channel matrix in this case is a 2×3 matrix given by

$$
C = \begin{bmatrix} q & p & 0 \\ 0 & p & q \end{bmatrix} \text{ (BEC).}
$$

□

From (15.11) we find

$$
\boldsymbol{p}^\top(n) = \boldsymbol{p}^\top(0)\boldsymbol{P}(0)\boldsymbol{P}(1)\cdots\boldsymbol{P}(n-1), \tag{15.15}
$$

if the Markov chain is nonhomogeneous, and

$$
\boldsymbol{p}^\top(n) = \boldsymbol{p}^\top(0)\boldsymbol{P}^n, \tag{15.16}
$$

if it is homogeneous. The latter equation shows that the state probabilities of a Markov chain are completely determined for all $n \geq 1$, if we know the TPM $\boldsymbol{P}$ and the initial state probability vector $\boldsymbol{p}(0)$.

If we set $n \leftarrow m + n$ in (15.16), we have

$$
\boldsymbol{p}^\top(m+n) = \boldsymbol{p}^\top(0)\boldsymbol{P}^m\boldsymbol{P}^n = \boldsymbol{p}^\top(m)\boldsymbol{P}^n. \tag{15.17}
$$

By letting the (i, j) element of the matrix $\boldsymbol{P}^n$ be denoted $P_{ij}^{(n)}$, i.e.,

$$
\boldsymbol{P}^n = \left[P_{ij}^{(n)}\right], i, j \in \mathcal{S}, \tag{15.18}
$$

we have

$$
p_j(m+n) = \sum_{i\in\mathcal{S}} p_i(m)P_{ij}^{(n)}, \text{ where } i, j \in \mathcal{S}, m, n = 0, 1, 2, \ldots. \tag{15.19}
$$

We interpret $P_{ij}^{(n)}$ as the conditional probability that the Markov chain, which is in state i at a given step, will be in state j after exactly n steps. Therefore, we call $\boldsymbol{P}^n$ the n-step TPM. Then we have the following fundamental equation, commonly called the **Chapman–Kolmogorov equations**.

THEOREM 15.1 (Chapman–Kolmogorov equations). *In a homogeneous Markov chain the following equations hold for all states $i, j, k \in \mathcal{S}$ and time (or step) indices $m, n = 0, 1, 2, \ldots$:*

$$\boxed{P_{ik}^{(m+n)} = \sum_{j\in\mathcal{S}} P_{ij}^{(m)} P_{jk}^{(n)}.} \tag{15.20}$$

Proof. The above result is immediately obtainable from the matrix multiplication formula applied to $\boldsymbol{P}$:

$$\boldsymbol{P}^{m+n} = \boldsymbol{P}^m \boldsymbol{P}^n. \tag{15.21}$$

But let us derive the formula (15.20) from the original definition of the n-step transition probabilities. Since the Markov chain is homogeneous, we can write

$$\begin{aligned}
P_{ik}^{(m+n)} &= P[X_{m+n} = k|X_0 = i] \\
&= \sum_{j\in\mathcal{S}} P[X_{m+n} = k, X_m = j|X_0 = i] \\
&\overset{\text{(a)}}{=} \sum_{j\in\mathcal{S}} P[X_{m+n} = k|X_m = j, X_0 = i]P[X_m = j|X_0 = i] \\
&\overset{\text{(b)}}{=} \sum_{j\in\mathcal{S}} P[X_{m+n} = k|X_m = j]P[X_m = j|X_0 = i] \\
&= \sum_{j\in\mathcal{S}} P_{jk}^{(n)} P_{ij}^{(m)},
\end{aligned} \tag{15.22}$$

where (a) is obtained using the formula $P[A \cap B|C] = P[A|B \cap C]P[B|C]$ and (b) makes use of the Markov property; i.e., $P[X_{m+n} = k|X_m = j, X_0 = i] = P[X_{m+n} = k|X_m = j]$ for any $i, j, k \in \mathcal{S}$ and $m, n = 0, 1, 2, \ldots$ □

15.2 Computation of state probabilities

The evaluation of the state probability vector $\boldsymbol{p}(n)$ of (15.16) for large n is most conveniently done by using one of the following two methods: (a) the generating function method or (b) the spectral expansion method. We will first discuss the generating function method.

15.2.1 Generating function method

If we let $\boldsymbol{g}(z)$ denote the generating function of the vector sequence $\{\boldsymbol{p}(n); n = 0, 1, 2, \ldots\}$, i.e.,

$$\boxed{\boldsymbol{g}(z) = \sum_{n=0}^{\infty} \boldsymbol{p}(n) z^n,} \tag{15.23}$$

then we have from (15.11)

$$\begin{aligned} \boldsymbol{g}^\top(z)\boldsymbol{P} &= \sum_{n=0}^{\infty} \boldsymbol{p}^\top(n+1)z^n = z^{-1}\sum_{n=0}^{\infty} \boldsymbol{p}^\top(n+1)z^{n+1} \\ &= z^{-1}\boldsymbol{g}^\top(z) - z^{-1}\boldsymbol{p}^\top(0), \end{aligned} \tag{15.24}$$

and hence,

$$\boxed{\boldsymbol{g}^\top(z) = \boldsymbol{p}^\top(0)[\boldsymbol{I} - \boldsymbol{P}z]^{-1},} \tag{15.25}$$

where $\boldsymbol{I}$ is the $M \times M$ identity matrix, where $M = |\mathcal{S}|$.

Example 15.3: Consider the Markov chain whose TPM is defined by (15.12). The two-step TPM is given by

$$\boldsymbol{P}^2 = \begin{bmatrix} 0 & 1 & 0 \\ \frac{1}{4} & \frac{1}{4} & \frac{1}{2} \\ 0 & \frac{1}{2} & \frac{1}{2} \end{bmatrix} \cdot \begin{bmatrix} 0 & 1 & 0 \\ \frac{1}{4} & \frac{1}{4} & \frac{1}{2} \\ 0 & \frac{1}{2} & \frac{1}{2} \end{bmatrix} = \begin{bmatrix} \frac{1}{4} & \frac{1}{4} & \frac{1}{2} \\ \frac{1}{16} & \frac{9}{16} & \frac{3}{8} \\ \frac{1}{8} & \frac{3}{8} & \frac{1}{2} \end{bmatrix}.$$

Similarly, the three-step TPM is calculated as

$$\boldsymbol{P}^3 = \begin{bmatrix} \frac{1}{4} & \frac{1}{4} & \frac{1}{2} \\ \frac{1}{16} & \frac{9}{16} & \frac{3}{8} \\ \frac{1}{8} & \frac{3}{8} & \frac{1}{2} \end{bmatrix} \cdot \begin{bmatrix} 0 & 1 & 0 \\ \frac{1}{4} & \frac{1}{4} & \frac{1}{2} \\ 0 & \frac{1}{2} & \frac{1}{2} \end{bmatrix} = \begin{bmatrix} \frac{1}{16} & \frac{9}{16} & \frac{3}{8} \\ \frac{9}{64} & \frac{25}{64} & \frac{15}{32} \\ \frac{3}{32} & \frac{15}{32} & \frac{7}{16} \end{bmatrix}.$$

It can be shown, as discussed below, that if $\boldsymbol{P}^n$ is calculated successively for $n = 4, 5, 6, \ldots$, it converges to the following limit:

$$\lim_{n\to\infty} \boldsymbol{P}^n = \boldsymbol{P}^\infty = \begin{bmatrix} \frac{1}{9} & \frac{4}{9} & \frac{4}{9} \\ \frac{1}{9} & \frac{4}{9} & \frac{4}{9} \\ \frac{1}{9} & \frac{4}{9} & \frac{4}{9} \end{bmatrix}.$$

Namely, all rows become identical. By substituting the above result into (15.16), we obtain

$$\lim_{n\to\infty} \boldsymbol{p}^\top(n) = (p_0(0)\,p_1(0)\,p_2(0))\,\boldsymbol{P}^\infty = \left(\frac{1}{9}\;\frac{4}{9}\;\frac{4}{9}\right).$$

That is, the probability of being in state j after a large number of steps (i.e., $n \to \infty$) approaches a certain definite value, independent of the initial state.

The same result can be obtained by using (15.25). From linear algebra, the inverse of a matrix $\boldsymbol{A}$ can be expressed as

$$\boldsymbol{A}^{-1} = \frac{\text{adj}(\boldsymbol{A})}{\det|\boldsymbol{A}|},$$

where adj $(\boldsymbol{A})$ is the *adjugate* (sometimes called *classical adjoint*) matrix of $\boldsymbol{A}$ and det $|\boldsymbol{A}|$ is the *determinant* of $\boldsymbol{A}$. Applying this result to (15.25), we have

$$[\boldsymbol{I} - \boldsymbol{P}z]^{-1} = \frac{\boldsymbol{B}(z)}{\Delta(z)}, \tag{15.26}$$

where

$$\boldsymbol{B}(z) \triangleq \text{adj}\,(\boldsymbol{I} - \boldsymbol{P}z) = \begin{bmatrix} 1 - \frac{3z}{4} - \frac{z^2}{8} & z\left(1 - \frac{z}{2}\right) & \frac{z^2}{2} \\ \frac{z}{4}\left(1 - \frac{z}{2}\right) & 1 - \frac{z}{2} & \frac{z}{2} \\ \frac{z^2}{8} & \frac{z}{2} & 1 - \frac{z}{4} - \frac{z^2}{4} \end{bmatrix}$$

and

$$\Delta(z) \triangleq \det|\boldsymbol{I} - \boldsymbol{P}z| = (1 - z)\left(1 + \frac{z}{4} - \frac{z^2}{8}\right). \tag{15.27}$$

Hence,

$$[\boldsymbol{I} - \boldsymbol{P}z]^{-1} = \frac{1}{\Delta(z)} \begin{bmatrix} 1 - \frac{3z}{4} - \frac{z^2}{8} & z\left(1 - \frac{z}{2}\right) & \frac{z^2}{2} \\ \frac{z}{4}\left(1 - \frac{z}{2}\right) & 1 - \frac{z}{2} & \frac{z}{2} \\ \frac{z^2}{8} & \frac{z}{2} & 1 - \frac{z}{4} - \frac{z^2}{4} \end{bmatrix}.$$

Then by substituting this expression and

$$\boldsymbol{p}^\top(0) = (p_0(0)p_1(0)p_2(0))$$

into (15.25), we obtain

$$\boldsymbol{g}^\top(z) = (g_0(z)g_1(z)g_2(z)),$$

where

$$g_0(z) = \frac{1}{\Delta(z)}\left[p_1(0)\left(1 - \frac{3z}{4} - \frac{z^2}{8}\right) + p_2(0)\frac{z}{4}\left(1 - \frac{z}{2}\right) + p_3(0)\frac{z^2}{8}\right],$$

$$g_1(z) = \frac{1}{\Delta(z)}\left[p_1(0)z\left(1 - \frac{z}{2}\right) + p_2(0)\left(1 - \frac{z}{2}\right) + p_3(0)\frac{z}{2}\right],$$

$$g_2(z) = \frac{1}{\Delta(z)}\left[p_1(0)\frac{z^2}{2} + p_2(0)\frac{z}{2} + p_3(0)\left(1 - \frac{z}{4} - \frac{z^2}{4}\right)\right].$$

By applying one of the PGF inversion techniques discussed in Section 9.1.4, we can obtain the sequence of probabilities $\{p_i(n); i = 0, 1, 2, n = 0, 1, 2, \ldots\}$. The limiting probabilities $\lim_{n\to\infty} p_i(n)$ can be obtained, however, without going through the inversion of the PGF. By applying the **final value theorem** (see Problem 9.12), we have

$$\lim_{n\to\infty} p_0(n) = \lim_{z\to 1}(1-z)g_0(z) = \frac{1}{9}.$$

Similarly, we find

$$\lim_{n\to\infty} p_1(n) = \lim_{z\to 1}(1-z)g_1(z) = \frac{4}{9}$$

and

$$\lim_{n\to\infty} p_2(n) = \lim_{z\to 1}(1-z)g_2(z) = \frac{4}{9}.$$

□

We now generalize the approach of Example 15.3 (see [330, p. 391]). Setting the determinant $\Delta(z)$ of (15.27) equal to zero, we denote its distinct roots as λ_i^{-1}; $i = 0, 1, \ldots, k-1$. The λ_i are the nonzero eigenvalues (see the next section) of $\boldsymbol{P}$, since they satisfy the characteristic equation $\det|\boldsymbol{P} - \lambda\boldsymbol{I}| = |\lambda|^M \Delta(\lambda^{-1}) = 0$. By noting $\Delta(0) = 1$, we can write

$$\Delta(z) = \prod_{i=0}^{k-1}(1-\lambda_i z)^{m_i}, \tag{15.28}$$

where m_i is the multiplicity of the eigenvalue λ_i. The degree $b(\leq M-1)$ of numerator (matrix) polynomial $\boldsymbol{B}(z)$ can be equal to or greater than the degree $d = \sum_{i=0}^{k-1} m_i (\leq M)$ of the polynomial $\Delta(z)$ when some of the eigenvalues of $\boldsymbol{P}$ are zero, as can be seen from (15.28). Then, we divide $\boldsymbol{B}(z)$ by $\Delta(z)$, resulting in a quotient polynomial $\boldsymbol{B}_0(z)$ of degree $b-d$ or less and a remainder polynomial $\boldsymbol{B}_1(z)$ of degree $d-1$ or less:

$$\frac{\boldsymbol{B}(z)}{\Delta(z)} = \boldsymbol{B}_0(z) + \frac{\boldsymbol{B}_1(z)}{\Delta(z)}.$$

By applying the partial-fraction expansion method discussed in Section 9.1.4 to (15.26) (see also [196], pp. 64–65), we can write

$$[\boldsymbol{I} - \boldsymbol{P}z]^{-1} = \sum_{n=0}^{b-d} \boldsymbol{B}_{0n} z^n + \sum_{i=0}^{k-1}\sum_{j=1}^{m_i} \frac{\boldsymbol{C}_{ij}}{(1-\lambda_i z)^j}, \tag{15.29}$$

where (cf. (9.61))

$$\boldsymbol{C}_{ij} = \frac{(-1)^{m_i-j}}{(m_i-j)!\lambda_i^{m_i-j}} \frac{d^{m_i-j}}{dz^{m_i-j}} \left[\frac{\boldsymbol{B}_1(z)(1-\lambda_i z)^{m_i}}{\Delta(z)}\right]_{z=\lambda_i^{-1}}. \tag{15.30}$$

Using the formal power series expansion

$$[\boldsymbol{I} - \boldsymbol{P}z]^{-1} = \sum_{n=0}^{\infty} \boldsymbol{P}^n z^n, \tag{15.31}$$

the matrix $\boldsymbol{P}^n$ is seen to be coefficient of z^n in the partial-fraction expansion (15.29). Using the Taylor series expansion or the formula (9.63), we find

$$\frac{1}{(1-\lambda_i z)^j} = \sum_{n=0}^{\infty} \binom{n+j-1}{n} \lambda_i^n z^n.$$

Applying this in (15.29) together with (15.31), we obtain (cf. (9.64))

$$\boldsymbol{P}^n = \boldsymbol{B}_{0n} + \sum_{i=0}^{k-1} \sum_{j=1}^{m_i} \boldsymbol{C}_{ij} \binom{n+j-1}{j-1} \lambda_i^n, \tag{15.32}$$

where $\boldsymbol{B}_{0n} = \boldsymbol{0}$ for $n > b - d$. The reader is referred to [320], p. 391 for a slightly different approach to this problem, where the formula $[\boldsymbol{I}z - \boldsymbol{P}]^{-1} = \sum_{n=0}^{\infty} \boldsymbol{P}^n z^{-n-1}$ is used.

15.2.2 Spectral expansion method

An alternative method to evaluate the state probability vector $\boldsymbol{p}(n)$ is to use the eigenvalues and eigenvectors of the TPM $\boldsymbol{P}$. This is often referred to as the **spectral expansion** method, since the set of eigenvalues of a matrix is also called its **spectrum**. The spectral expansion method is similar to the generalized Fourier series expansion or Karhunen–Loève expansion method discussed in Chapter 12. In this case, however, $\boldsymbol{P}$ is not a symmetric matrix.

Let λ_i be the ith **eigenvalue** and $\boldsymbol{u}_i$ be the associated **right-eigenvector** of the Markov TPM:

$$\boldsymbol{P}\boldsymbol{u}_i = \lambda_i \boldsymbol{u}_i, i \in \mathcal{S} = \{0, 1, 2, \ldots, M-1\}, \tag{15.33}$$

where $M = |\mathcal{S}|$ is the number of states and $\boldsymbol{u}_i$ is a column vector, making its transpose $\boldsymbol{u}_i^\top$ a row vector. We assume that all eigenvalues are distinct; i.e., there is no multiplicity of any of the eigenvalues. Let us form an $M \times M$ matrix $\boldsymbol{U}$ by

$$\boldsymbol{U} = \left[\boldsymbol{u}_0 \boldsymbol{u}_1 \cdots \boldsymbol{u}_{M-1}\right] \tag{15.34}$$

and a diagonal matrix $\boldsymbol{\Lambda}$ by

$$\boldsymbol{\Lambda} = \begin{bmatrix} \lambda_0 & 0 & \cdots & 0 \\ 0 & \lambda_1 & \cdots & 0 \\ \vdots & \vdots & \ddots & \vdots \\ 0 & 0 & \cdots & \lambda_{M-1} \end{bmatrix}. \tag{15.35}$$

Then the TPM $\boldsymbol{P}$ can be expanded as

$$\boxed{\boldsymbol{P} = \boldsymbol{U}\boldsymbol{\Lambda}\boldsymbol{U}^{-1} = \boldsymbol{U}\boldsymbol{\Lambda}\boldsymbol{V},} \tag{15.36}$$

where

$$\boldsymbol{V} = \boldsymbol{U}^{-1}. \tag{15.37}$$

By multiplying $\boldsymbol{V}$ on the left of (15.36), we have

$$\boldsymbol{V}\boldsymbol{P} = \boldsymbol{\Lambda}\boldsymbol{V}. \tag{15.38}$$

Defining a set of row vectors $\boldsymbol{v}_i^\top$ by

$$\boldsymbol{V} = \begin{bmatrix} \boldsymbol{v}_0^\top \\ \boldsymbol{v}_1^\top \\ \vdots \\ \boldsymbol{v}_{M-1}^\top \end{bmatrix}, \tag{15.39}$$

we find from (15.38)

$$\boldsymbol{v}_i^\top \boldsymbol{P} = \lambda_i \boldsymbol{v}_i^\top,\ i \in \mathcal{S}. \tag{15.40}$$

Therefore, $\boldsymbol{v}_i^\top$ is the **left-eigenvector** associated with the eigenvalue λ_i. From (15.2.2) we have

$$\boldsymbol{V}\boldsymbol{U} = \boldsymbol{I}, \tag{15.41}$$

which implies that $\boldsymbol{v}_i$ and $\boldsymbol{u}_j$ are **bi-orthonormal**; i.e.,

$$\boxed{\boldsymbol{v}_i^\top \boldsymbol{u}_j = \delta_{ij},\ i, j \in \mathcal{S}.} \tag{15.42}$$

From (15.36) we find

$$\boldsymbol{P}^2 = \boldsymbol{U}\boldsymbol{\Lambda}\boldsymbol{U}^{-1}\boldsymbol{U}\boldsymbol{\Lambda}\boldsymbol{U}^{-1} = \boldsymbol{U}\boldsymbol{\Lambda}^2\boldsymbol{U}^{-1}. \tag{15.43}$$

By repeating the same procedure $(n-1)$ times, we have

$$\boxed{\boldsymbol{P}^n = \boldsymbol{U}\boldsymbol{\Lambda}^n\boldsymbol{U}^{-1} = \sum_{i\in\mathcal{S}} \lambda_i^n \boldsymbol{u}_i \boldsymbol{v}_i^\top = \sum_{i\in\mathcal{S}} \lambda_i^n \boldsymbol{E}_i,} \tag{15.44}$$

where the matrices

$$\boldsymbol{E}_i = \boldsymbol{u}_i \boldsymbol{v}_i^\top,\ i \in \mathcal{S}, \tag{15.45}$$

are the **projection matrices**. By taking the transpose of (15.41) and expanding $\boldsymbol{I} = \boldsymbol{U}\boldsymbol{V}^\top = \sum_{i\in\mathcal{S}} \boldsymbol{v}_i \boldsymbol{u}_i^\top = \sum_{i\in\mathcal{S}} \boldsymbol{E}_i^\top$, we find the following identity:

$$\sum_{i\in\mathcal{S}} \boldsymbol{E}_i = \sum_{i\in\mathcal{S}} \boldsymbol{E}_i^\top = \boldsymbol{I}, \tag{15.46}$$

which corresponds to the case $n = 0$ in the expansion formula (15.44).

In reference to (15.18), we write the n-step transition probability from state i to state j as

$$P_{ij}^{(n)} = \sum_{k\in\mathcal{S}} \lambda_k^n u_{ki} v_{kj},\ i, j \in \mathcal{S},\ n = 0, 1, 2, \ldots. \tag{15.47}$$

The state probability vector at step n is given, from (15.16), as

$$\boldsymbol{p}^\top(n) = \boldsymbol{p}^\top(0)\boldsymbol{P}^n = \sum_{k\in\mathcal{S}} \lambda_k^n \boldsymbol{p}^\top(0)\boldsymbol{u}_k\boldsymbol{v}_k^\top. \tag{15.48}$$

Thus, the probability that the Markov chain is in state i at time n is given by

$$p_i(n) = \sum_{k\in\mathcal{S}} \lambda_k^n \left(\boldsymbol{p}^\top(0)\boldsymbol{u}_k\right) v_{ki},\ i \in \mathcal{S}. \tag{15.49}$$

Example 15.4: Let us discuss the same three-state Markov chain used in Example 15.3. The characteristic equation that determines the eigenvalues is given by

$$\det|\boldsymbol{P} - \lambda\boldsymbol{I}| = 0, \tag{15.50}$$

which can be rearranged as

$$\det|\boldsymbol{I} - \lambda^{-1}\boldsymbol{P}| = \Delta(\lambda^{-1}) = 0, \tag{15.51}$$

where $\Delta(z)$ was defined in (15.27). Since we find that there are three roots for $\Delta(z) = 0$, $z_0 = 1$, $z_1 = -2$, and $z_2 = 4$, we readily find the three eigenvalues of $\boldsymbol{P}$: $\lambda_0 = z_0^{-1} = 1$, $\lambda_1 = z_1^{-1} = -\frac{1}{2}$, and $\lambda_2 = z_2^{-1} = \frac{1}{4}$. Then, the corresponding right-eigenvectors are readily found as

$$\boldsymbol{u}_0 = \begin{bmatrix} 1 \\ 1 \\ 1 \end{bmatrix},\ \boldsymbol{u}_1 = \begin{bmatrix} 4 \\ -2 \\ 1 \end{bmatrix},\ \text{and}\ \boldsymbol{u}_2 = \begin{bmatrix} 4 \\ 1 \\ -2 \end{bmatrix}.$$

Thus, we find

$$\boldsymbol{U} = [\boldsymbol{u}_0\boldsymbol{u}_1\boldsymbol{u}_2] = \begin{bmatrix} 1 & 4 & 4 \\ 1 & -2 & 1 \\ 1 & 1 & -2 \end{bmatrix}$$

and its inverse

$$\boldsymbol{V} = \boldsymbol{U}^{-1} = \frac{1}{9}\begin{bmatrix} 1 & 4 & 4 \\ 1 & -2 & 1 \\ 1 & 1 & -2 \end{bmatrix} = \frac{1}{9}\boldsymbol{U},$$

from which we find

$$\boldsymbol{v}_0^\top = \frac{1}{9}(1, 4, 4),\ \boldsymbol{v}_1^\top = \frac{1}{9}(1, -2, 1),\ \boldsymbol{v}_2^\top = \frac{1}{9}(1, 1, -2).$$

The projection matrices are

$$E_0 = u_0 v_0^\top = \frac{1}{9}\begin{bmatrix} 1 & 4 & 4 \\ 1 & 4 & 4 \\ 1 & 4 & 4 \end{bmatrix},$$

$$E_1 = u_1 v_1^\top = \frac{1}{9}\begin{bmatrix} 4 & -8 & 4 \\ -2 & 4 & -2 \\ 1 & -2 & 1 \end{bmatrix},$$

$$E_2 = u_2 v_2^\top = \frac{1}{9}\begin{bmatrix} 4 & 4 & -8 \\ 1 & 1 & -2 \\ 2 & 2 & -4 \end{bmatrix}.$$

Then, the n-step TPM is readily computed as

$$\begin{aligned} P^n &= \sum_{i=0}^{2} \lambda_i^n E_i \\ &= \frac{1}{9}\begin{bmatrix} 1 & 4 & 4 \\ 1 & 4 & 4 \\ 1 & 4 & 4 \end{bmatrix} + \frac{1}{9}\left(-\frac{1}{2}\right)^n \begin{bmatrix} 4 & -8 & 4 \\ -2 & 4 & -2 \\ 1 & -2 & 1 \end{bmatrix} + \frac{1}{9}\left(\frac{1}{4}\right)^n \begin{bmatrix} 4 & 4 & -8 \\ 1 & 1 & -2 \\ 2 & 2 & -4 \end{bmatrix}. \end{aligned}$$

Suppose that the Markov chain is initially at state 1; i.e., $p^\top(0) = (1, 0, 0)$. Then,

$$\begin{aligned} p^\top(n) = p^\top(0)P^n &= \frac{1}{9}(1, 4, 4) + \frac{1}{9}\left(-\frac{1}{2}\right)^n (4, -8, 4) + \frac{1}{9}\left(\frac{1}{4}\right)^n (4, 4, -8) \\ &= (p_0^{(n)} p_1^{(n)} p_2^{(n)}), \end{aligned}$$

where

$$\begin{aligned} p_0^{(n)} &= \frac{1}{9} + \frac{4}{9}\left(-\frac{1}{2}\right)^n + \frac{4}{9}\left(\frac{1}{4}\right)^n, \\ p_1^{(n)} &= \frac{4}{9} - \frac{8}{9}\left(-\frac{1}{2}\right)^n + \frac{4}{9}\left(\frac{1}{4}\right)^n, \\ p_2^{(n)} &= \frac{4}{9} + \frac{4}{9}\left(-\frac{1}{2}\right)^n - \frac{8}{9}\left(\frac{1}{4}\right)^n. \end{aligned}$$

We can also verify that $p_0^{(n)} + p_1^{(n)} + p_2^{(n)} = 1$ for all n as expected, because the n-step TPM P^n is a stochastic matrix.

We also find that, in the limit $n \to \infty$, $p_0^{(n)} \to \frac{1}{9}$, $p_1^{(n)} \to \frac{4}{9}$, and $p_2^{(n)} \to \frac{4}{9}$. These steady-state probabilities are independent of the initial probability $p(0)$. The left-eigenvector v_0, associated with the eigenvalue $\lambda_0 = 1$, determines the steady-state solution. □

15.2.2.1 Multiplicity of eigenvalues

In the analysis and example given above we assumed that the eigenvalues of P were all distinct. Suppose that there are only k distinct eigenvalues $\lambda_i, i = 0, 1, \ldots, k-1$, and the degree of multiplicity of λ_i is $m_i (\geq 1)$. Clearly,

$$\sum_{i=0}^{k-1} m_i = M,$$

where $M = |\mathcal{S}|$, the number of states in the Markov chain. When the set of right-eigenvectors, $\{\boldsymbol{u}_0, \ldots, \boldsymbol{u}_{M-1}\}$, is linearly independent, the spectral decomposition of $\boldsymbol{P}$ given by (15.36) still holds and the matrix $\boldsymbol{P}$ is said to be *diagonalizable*. In general, a matrix is diagonalizable if and only if it has a set of M linearly independent eigenvectors. However, if a matrix is *not* diagonalizable, i.e., if it has $J(< M)$ linearly independent eigenvectors, we can still express $\boldsymbol{P}$ using what is known as the **Jordan canonical form**:

$$\boldsymbol{P} = \boldsymbol{U}\boldsymbol{\Lambda}\boldsymbol{U}^{-1} = \boldsymbol{U}\boldsymbol{\Lambda}\boldsymbol{V}, \tag{15.52}$$

where the matrix $\boldsymbol{\Lambda}$ is now a block diagonal matrix with the following structure:

$$\boldsymbol{\Lambda} = \begin{bmatrix} \boldsymbol{\Lambda}_1 & \boldsymbol{0} & \cdots & \boldsymbol{0} \\ \boldsymbol{0} & \boldsymbol{\Lambda}_2 & \cdots & \boldsymbol{0} \\ \vdots & \vdots & \ddots & \vdots \\ \boldsymbol{0} & \boldsymbol{0} & \cdots & \boldsymbol{\Lambda}_J \end{bmatrix}, \tag{15.53}$$

where the submatrix $\boldsymbol{\Lambda}_j$ $(j = 1, 2, \ldots, J)$ is an $n_j \times n_j$ **super-diagonal** matrix:

$$\boldsymbol{\Lambda}_j = \begin{bmatrix} \lambda_i & 1 & 0 & \cdots & 0 \\ 0 & \lambda_i & 1 & \cdots & 0 \\ \vdots & \vdots & \vdots & \ddots & \vdots \\ 0 & 0 & 0 & \cdots & \lambda_i \end{bmatrix}. \tag{15.54}$$

Here we assume that $\boldsymbol{\Lambda}_j$ is associated with one of $g_i(\leq m_i)$ linearly independent eigenvectors associated with eigenvalue λ_i. If $g_i = m_i$, then all g_i submatrices $\boldsymbol{\Lambda}_j$ are simply a scalar λ_i, which is a degenerate case of the structure (15.54). If $g_i < m_i$, there exists at least one super-diagonal matrix of the form (15.54) of size $n_j \times n_j$ such that $1 < n_j \leq m_i - g_i + 1$. The reader is referred to online Supplementary Materials for a full discussion on how such super-diagonal matrices can be found.

Then, the n-step TPM is given as

$$\boldsymbol{P}^n = \boldsymbol{U}\boldsymbol{\Lambda}^n\boldsymbol{U}^{-1}, \tag{15.55}$$

where

$$\boldsymbol{\Lambda}^n = \begin{bmatrix} \boldsymbol{\Lambda}_1^n & \boldsymbol{0} & \cdots & \boldsymbol{0} \\ \boldsymbol{0} & \boldsymbol{\Lambda}_2^n & \cdots & \boldsymbol{0} \\ \vdots & \vdots & \ddots & \vdots \\ \boldsymbol{0} & \boldsymbol{0} & \cdots & \boldsymbol{\Lambda}_J^n \end{bmatrix} \tag{15.56}$$

and the submatrices $\mathbf{\Lambda}_j^n$, $j = 1, 2, \ldots, J$, have the following structure:

$$\mathbf{\Lambda}_j^n = \begin{bmatrix} \lambda_i^n & \binom{n}{1}\lambda_i^{n-1} & \cdots & \binom{n}{n_j-1}\lambda_i^{n-n_j+1} \\ 0 & \lambda_i^n & \cdots & \binom{n}{n_j-2}\lambda_i^{n-n_j+2} \\ \vdots & \vdots & \ddots & \vdots \\ 0 & 0 & \cdots & \lambda_i^n \end{bmatrix} \tag{15.57}$$

With this modification, the rest of the analysis techniques presented in this section can be applied to find the state probability vector $\boldsymbol{p}(n)$ for any time index n.

15.3 Classification of states

In a Markov chain, each state can be classified into one of several categories (see Figures 15.2 and 15.3), and these categories partition the Markov states. The notion of *first passage time* plays an important role in the state classification.

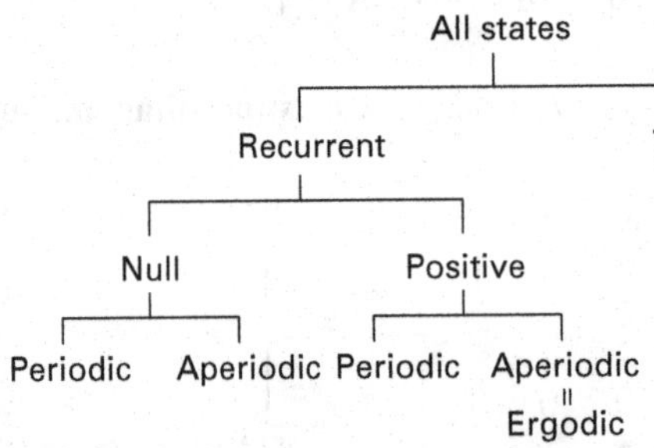

Figure 15.2 Classification of states in a Markov chain.

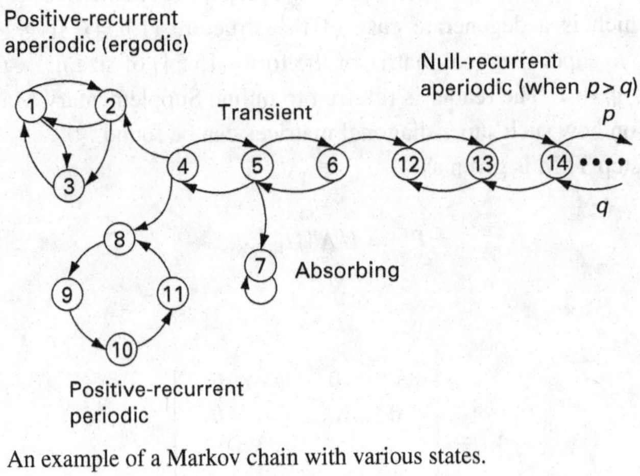

Figure 15.3 An example of a Markov chain with various states.

15.3.1 Recurrent and transient states

Consider a pair of states (i, j) in the state space $\mathcal{S}$ of a given Markov chain X_n, and suppose the chain is initially in state i; i.e., $X_0 = i$. The number of transitions T_{ij} that X_n takes in going from state i to reach state j for the first time is called the **first-passage time** from i to j. In other words, $X_{T_{ij}} = j$ but $X_n \neq j$ for all $1 \leq n \leq T_{ij} - 1$. When $j = i$, the RV T_{ii} is the number of transitions required until X_n returns to the same state i for the first time, and is called the **recurrence time** of state i.

Let $f_{ij}^{(n)}$ be the probability that the first-passage time T_{ij} is equal to n:

$$\boxed{f_{ij}^{(n)} \triangleq P[T_{ij} = n], i, j \in \mathcal{S}, n = 1, 2, \ldots.} \tag{15.58}$$

The sum

$$f_{ij} = \sum_{n=1}^{\infty} f_{ij}^{(n)}, i, j \in \mathcal{S}, \tag{15.59}$$

is the probability that X_n, starting from state i, ever reaches state j. Thus, if $f_{ij} < 1$, the process initially in state i may never reach state j. If $f_{ij} = 1$, the set $\{f_{ij}^{(n)} : n = 1, 2, \ldots\}$ represents the bona fide probability distribution of the first-passage time. In particular, for $j = i$, in (15.59) the sum

$$f_{ii} = \sum_{n=1}^{\infty} f_{ii}^{(n)} \tag{15.60}$$

is the probability that the system ever returns to state i.

DEFINITION 15.4 (Recurrent state and transient state). *State $i \in \mathcal{S}$ is called a* ***recurrent state*** *(or a* ***persistent*** *state) if $f_{ii} = 1$. It is called a* ***transient state*** *if $f_{ii} < 1$.* □

If the state i is recurrent, then the Markov chain returns infinitely often to state i as $t \to \infty$. If state i is transient, the chain returns to the state i only finitely often. For a recurrent state i, the recurrence time T_{ii} is a well-defined RV, and $\{f_{ii}^{(n)}; n = 1, 2, \ldots\}$ represents the probability distribution of the recurrence time.

Calculation of $f_{ij}^{(n)}$ for all n may be generally difficult, but it is relatively simple to obtain the **mean first-passage time** (or **expected first-passage time**). We define the mean first-passage time μ_{ij} by

$$\boxed{\mu_{ij} = E[T_{ij}] = \begin{cases} \sum_{n=1}^{\infty} n f_{ij}^{(n)}, & \text{if} \quad f_{ij} = 1, \\ \infty, & \text{if} \quad f_{ij} < 1, \end{cases} \quad i, j \in \mathcal{S}.} \tag{15.61}$$

When $f_{ij} = 1$ for a given pair of states i and j, the mean first-passage time μ_{ij} satisfies the equation

$$\mu_{ij} = P_{ij} + \sum_{k \neq j} P_{ik}(\mu_{kj} + 1), \tag{15.62}$$

since the system in state i either goes to j in one step or else goes to intermediate step k and then eventually to j.

When $j = i$, the mean first-passage time is called the **mean recurrence time** (or **expected recurrence time**).

DEFINITION 15.5 (Null-recurrent state and positive-recurrent state). *A recurrent state $i \in \mathcal{S}$ is called a **null-recurrent** state if $\mu_{ii} = \infty$ and is called a **positive-recurrent** (or **regular-recurrent**) state if $\mu_{ii} < \infty$.* □

It should be clear that in a finite Markov chain there are no null-recurrent states; that is, there are only positive-recurrent states and transient states.

If $P_{ii}^{(n)} = 0$ for $n \neq d_i, 2d_i, \ldots$, then state i is called **periodic** with period d_i. In other words, a return to state i is impossible except perhaps in $n = d_i, 2d_i, \ldots$ steps. We can write d_i as

$$d_i = \gcd\left\{n : P_{ii}^{(n)} > 0\right\}; \tag{15.63}$$

i.e., the greatest common divisor of the epochs at which the return to state i is possible. Thus, $P_{ii}^{(n)} = 0$, whenever n is not divisible by d_i.

DEFINITION 15.6 (Periodic state and aperiodic state). *A state $i \in \mathcal{S}$ is called **periodic** with period d_i if $d_i > 1$ and is called **aperiodic** if $d_i = 1$.* □

In the random walk model of Example 15.1, all states of the Markov chain $\{S_n\}$ are periodic states, having a period $d_i = 2$ for all $i \in \mathcal{S} = \mathbb{Z}$, the set of all integers.

DEFINITION 15.7 (Ergodic state). *State $i \in \mathcal{S}$ is called an **ergodic** state if it is positive-recurrent and aperiodic; i.e., $f_{ii} = 1$, $\mu_{ii} < \infty$, and $d_i = 1$.* □

DEFINITION 15.8 (Absorbing state). *State $i \in \mathcal{S}$ is called an **absorbing state** if the transition probability satisfies $P_{ii} = 1$.* □

An absorbing state is a special case of a positive-recurrent state since $P_{ii} = 1$ implies that $f_{ii} = f_{ii}^{(1)} = 1$. If a state is an absorbing state, the process will never leave the state, once entered.

Figure 15.2 shows a summary of state classifications discussed above.

15.3.2 Criteria for state classification

In order to further investigate the relation between the n-step transition probabilities $\boldsymbol{P}^n = [P_{ij}^{(n)}]$ and the first-passage probabilities $[f_{ij}^{(n)}]$, let us define generating functions $G_{ij}(z)$ and $F_{ij}(z)$ by

$$G_{ij}(z) \triangleq \sum_{n=0}^{\infty} P_{ij}^{(n)} z^n, i, j \in \mathcal{S}, \tag{15.64}$$

and

$$F_{ij}(z) \triangleq \sum_{n=0}^{\infty} f_{ij}^{(n)} z^n, i, j \in \mathcal{S}. \tag{15.65}$$

Then we have the following theorem.

THEOREM 15.2 (Generating functions of $\boldsymbol{n}$-step transition and first-passage probabilities). *The generating functions $G_{ij}(z)$ and $F_{ij}(z)$ defined above are related according to*

$$\boxed{G_{ij}(z) = \delta_{ij} + F_{ij}(z)G_{jj}(z), i, j \in \mathcal{S}.} \tag{15.66}$$

Proof. Suppose that $X_0 = i \in \mathcal{S}$. Let $A_n(j)$ represent the event $X_n = j \in \mathcal{S}$ and let $B_m(j)$ be the event that the first visit to state j occurs at time $m(\leq n)$; i.e.,

$$B_m(j) = \{X_t \neq j; 1 \leq t \leq m-1, X_m = j\}.$$

Then the $B_m(j)$ $(m \leq n)$ are disjoint events and $B_1(j) \cup B_2(j) \cup \cdots \cup B_n(j) = \Omega$, i.e., the sure event, so

$$P[A_n(j)|X_0 = i] = \sum_{m=1}^{n} P[A_n(j), B_m(j)|X_0 = i]. \tag{15.67}$$

Using the relation

$$\begin{aligned} P[A_n(j), B_m(j)|X_0 = i] &= P[A_n(j)|B_m(j), X_0 = i]P[B_m(j)|X_0 = i] \\ &= P[A_n(j)|X_m = j]P[B_m(j)|X_0 = i], \end{aligned}$$

(15.67) can be interpreted as

$$P_{ij}^{(n)} = \sum_{m=1}^{n} f_{ij}^{(m)} P_{jj}^{(n-m)}, n \geq 1. \tag{15.68}$$

By substituting the last equation into (15.64) and writing $n - m = k$, we obtain

$$\begin{aligned} G_{ij}(z) = P_{ij}^{(0)} + \sum_{n=1}^{\infty}\sum_{m=1}^{n} f_{ij}^{(m)} P_{jj}^{(n-m)} z^n &= \delta_{ij} + \sum_{m=1}^{\infty} f_{ij}^{(m)} z^m \sum_{k}^{\infty} P_{jj}^{(k)} z^k \\ = \delta_{ij} + F_{ij}(z)G_{jj}(z), \end{aligned} \tag{15.69}$$

where we use the convention $\boldsymbol{P}^0 = \boldsymbol{I}$, or $P_{ij}^{(0)} = \delta_{ij}$. The last equation makes use of the well-known property of the generating function (or z-transform) that convolution in the discrete-time domain (represented in the superscripts $(n-m)$ and (m)) corresponds to multiplication in the z-domain. □

For $i = j$, we obtain from (15.69)

$$G_{ii}(z) = \frac{1}{1 - F_{ii}(z)}. \tag{15.70}$$

The first-passage probability defined earlier is

$$f_{ij} = \sum_{n=1}^{\infty} f_{ij}^{(n)} = F_{ij}(1). \tag{15.71}$$

The mean first-passage time is

$$\mu_{ij} = \sum_{n=1}^{\infty} n f_{ij}^{(n)} = F'_{ij}(1). \tag{15.72}$$

Based on the above theorem, we find the following conditions that help us classify states.

THEOREM 15.3 (Conditions for recurrent and transient states). *The following criteria apply to classify whether a given state is recurrent or transient:*

1. *State $j \in \mathcal{S}$ is **recurrent** if and only if*

$$\sum_{n=0}^{\infty} P_{jj}^{(n)} = \infty. \tag{15.73}$$

If this holds, then for all $i \in \mathcal{S}$ such that $f_{ij} > 0$,

$$\sum_{n=0}^{\infty} P_{ij}^{(n)} = \infty. \tag{15.74}$$

2. *State $j \in \mathcal{S}$ is **transient** if and only if*

$$\sum_{n=0}^{\infty} P_{jj}^{(n)} < \infty. \tag{15.75}$$

If this holds, then, for all $i \in \mathcal{S}$,

$$\sum_{n=0}^{\infty} P_{ij}^{(n)} < \infty. \tag{15.76}$$

Proof. It suffices to show that state j is recurrent if and only if (15.73) holds. Let $z \uparrow 1$ in (15.70):

$$\lim_{z \uparrow 1} G_{jj}(z) = \infty, \text{ if and only if } f_{jj} = F_{jj}(1) = 1.$$

Then we use Abel's limit theorem[3] to obtain

$$\lim_{z\uparrow 1} G_{jj}(z) = \sum_{n=0}^{\infty} P_{jj}^{(n)},$$

and we have shown that (15.73) is a necessary and sufficient condition for the state to be recurrent. Any state that does not satisfy this condition must be transient, hence the condition (15.75) must hold for a transient state. The remaining part of the theorem can be proved by applying Abel's theorem to (15.69) for $i \neq j$. □

THEOREM 15.5 (Conditions for null-recurrent, ergodic, or periodic states). *The following criteria and properties apply, depending on whether a recurrent state is null, ergodic, or periodic:*

1. *A recurrent state j is* ***null-recurrent****, if and only if*

$$\lim_{n\to\infty} P_{jj}^{(n)} = 0. \tag{15.79}$$

If this holds, then, for all states $i \in \mathcal{S}$,

$$\lim_{n\to\infty} P_{ij}^{(n)} = 0. \tag{15.80}$$

2. *A recurrent state j is* ***ergodic*** *(i.e., positive-recurrent and aperiodic) if and only if*

$$\boxed{\lim_{n\to\infty} P_{jj}^{(n)} = \frac{1}{\mu_{jj}} > 0.} \tag{15.81}$$

If this holds, then, for all states $i \in \mathcal{S}$,

$$\boxed{\lim_{n\to\infty} P_{ij}^{(n)} = \frac{f_{ij}}{\mu_{jj}}.} \tag{15.82}$$

[3] Abel's limit theorem, to be stated below, allows us to find the limit of a generating function as z approaches one from below, even when the radius of convergence R of the generating function is equal to one and we do not know a priori whether the limit is finite or not.

THEOREM 15.4 (Abel's limit theorem). *Let $A(z) = \sum_{n=0}^{\infty} a_n z^n$.*

1. *If $\sum_{n=0}^{\infty} a_n = S < \infty$, then*

$$\lim_{z\uparrow 1} A(z) = S. \tag{15.77}$$

2. *If $a_n \geq 0$ for all n* and $\lim_{z\uparrow 1} A(z) = S \leq \infty$, *then*

$$\sum_{n=0}^{\infty} a_n = \lim_{N\to\infty} \sum_{n=0}^{N} a_n = S, \tag{15.78}$$

whether the sum S is finite or infinite.

Proof. See, for example, Karlin and Taylor [175]. □

3. *If state j is* **periodic** *with period d, then*

$$\boxed{\lim_{n\to\infty} P_{jj}^{(nd)} = \frac{d}{\mu_{jj}}.} \tag{15.83}$$

Proof. Let us define

$$q_n = P_{jj}^{(n)} - P_{jj}^{(n-1)}, n \geq 1, q_0 = P_{jj}^{(0)} = 1.$$

Then,

$$P_{jj}^{(n)} = \sum_{k=0}^{n} q_k.$$

Defining $Q(z) = \sum_{k=0}^{\infty} q_k z^k$, we have

$$Q(z) = (1-z)G_{jj}(z) = \frac{1-z}{1-F_{jj}(z)}.$$

Therefore,

$$\lim_{z\to 1} Q(z) = \lim_{z\to 1} \frac{1}{(F_{jj}(z)-1)/(z-1)} = \frac{1}{F'_{jj}(1)} = \frac{1}{\mu_{jj}}. \tag{15.84}$$

But from Abel's limit theorem, we have

$$\lim_{z\uparrow 1} Q(z) = \lim_{n\to\infty} \sum_{k=0}^{n} q_k = \lim_{n\to\infty} P_{jj}^{(n)}, \tag{15.85}$$

which, together with (15.84), implies

$$\lim_{n\to\infty} P_{jj}^{(n)} = \frac{1}{\mu_{jj}}. \tag{15.86}$$

Thus, state j is null-recurrent ($\mu_{jj} = \infty$) if and only if $P_{jj}^{(n)} \to 0$. If j is ergodic, then by definition $\mu_{jj} < \infty$; hence, (15.81) must hold.

If we take the limit $n \to \infty$ in formula (15.68):

$$\lim_{n\to\infty} P_{ij}^{(n)} = \lim_{n\to\infty} \sum_{m=1}^{n} f_{ij}^{(m)} P_{jj}^{(n-m)} = \sum_{m=1}^{\infty} \frac{f_{ij}^{(m)}}{\mu_{jj}} = \frac{f_{ij}}{\mu_{jj}}, \tag{15.87}$$

which is (15.82). If state j is null-persistent, then $\mu_{jj} = \infty$ so that (15.80) holds.

If state j is periodic with period d, then we can write

$$F_{jj}(z) = F(z^d), \tag{15.88}$$

for some function $F(\cdot)$. Then from (15.70)

$$G_{jj}(z) = \frac{1}{1-F_{jj}(z)} = \frac{1}{1-F(z^d)} = \sum_{k=0}^{\infty} P_{ij}^{(kd)} z^{kd}. \tag{15.89}$$

Then,

$$G_{jj}(z^{1/d}) = \frac{1}{1 - F(z)} = \sum_{k=0}^{\infty} P_{ij}^{(kd)} z^k. \tag{15.90}$$

By repeating the argument that led to (15.86), and setting $z = 1$ in the relation $F'_{jj}(z) = dz^{d-1}F'(z^d)$, we obtain

$$\lim_{k\to\infty} P_{jj}^{(kd)} = \frac{1}{F'(1)} = \frac{d}{\mu_{jj}}. \tag{15.91}$$

This completes the proof. □

15.3.3 Communicating states and an irreducible Markov chain

In the previous section we have classified states of a Markov chain into different categories, and the notion of **communication** among the states is closely related to the classification of the states.

DEFINITION 15.9 (Reachable states and communicating states). *We say that state $j \in \mathcal{S}$ is* ***reachable*** *(or* ***accessible****) from state $i \in \mathcal{S}$ if there is an integer $n \geq 1$ such that $P_{ij}^{(n)} > 0$. If state i is reachable from state j and state j is reachable from state i, then the states i and j are said to* ***communicate,***[4] *written as $i \leftrightarrow j$.* □

THEOREM 15.6 (Reachable states from a recurrent state). *Suppose state j is reachable from a recurrent state i (i.e., $j \leftarrow i$). Then state i is also reachable from state j ($i \leftarrow j$); hence, states i and j communicate ($i \leftrightarrow j$). Moreover, state j is also recurrent.*

Proof. Since $i \to j$, there exists some number m of steps such that

$$P_{ij}^{(m)} = p > 0. \tag{15.92}$$

After this, the system would not return to i, if i were not reachable from j. Thus, starting from state i, the probability that the system does not return to i would be at least p. Thus,

$$f_{ii} \leq 1 - p < 1.$$

This, however, contradicts the assumption that state i is recurrent. Thus, state i must be reachable from j; i.e., there should exist some ℓ such that

$$P_{ji}^{(\ell)} = q > 0. \tag{15.93}$$

Because of (15.20), we have

$$P_{ij}^{(m+n)} = \sum_{k\in\mathcal{S}} P_{ik}^{(m)} P_{kj}^{(n)} \geq P_{ik}^{(m)} P_{kj}^{(n)} \text{ for any } k \in \mathcal{S}. \tag{15.94}$$

[4] Some authors use the term "intercommunicate" to emphasize the communication is bidirectional.

Then,

$$P_{ii}^{(m+n+\ell)} \geq P_{ij}^{(m)} P_{ji}^{(n+\ell)} \geq P_{ij}^{(m)} P_{jj}^{(n)} P_{ji}^{(\ell)} = pq P_{jj}^{(n)}. \tag{15.95}$$

Similarly,

$$P_{jj}^{(m+n+\ell)} \geq P_{ji}^{(\ell)} P_{ij}^{(m+n)} \geq P_{ji}^{(\ell)} P_{ii}^{(n)} P_{ij}^{(m)} = pq P_{ii}^{(n)}. \tag{15.96}$$

Thus, the two series $\sum_{n=0}^{\infty} P_{ii}^{(n)}$ and $\sum_{j=0}^{\infty} P_{jj}^{(n)}$ converge or diverge together. Since state i is recurrent, Theorem 15.3 implies that $\sum_{n=0}^{\infty} P_{ii}^{(n)} = \infty$. Thus, $\sum_{n=0}^{\infty} P_{jj}^{(n)} = \infty$; hence state j must be recurrent. □

If two states i and j do not communicate, then either

$$P_{ij}^{(n)} = 0 \text{ for all } n \geq 1,$$

or

$$P_{ji}^{(n)} = 0 \text{ for all } n \geq 1,$$

or both relations are true. The relation of communications "$\leftrightarrow$" is an **equivalence relation**; that is, the following three properties hold:

$$\tag{15.97}$$

1. **Reflexive property**: $i \leftrightarrow i$ for all $i \in \mathcal{S}$.
2. **Symmetric property**: if $i \leftrightarrow j$, then $j \leftrightarrow i$.
3. **Transitive property**: if $i \leftrightarrow j$ and $j \leftrightarrow k$, then $i \leftrightarrow k$.

Property 1 is a consequence of $P_{ij}^{(0)} = \delta_{ij}$; Property 2 is apparent, and Property 3 is easy to verify (Problem 15.4).

DEFINITION 15.10 (Closed or open set and an irreducible set of states). *A set $\mathcal{C}$ of states is called **closed** if*

$$P_{ij} = 0 \textit{ for all } i \in \mathcal{C},\ j \notin \mathcal{C},$$

*and is called **open** otherwise.*

*A set $\mathcal{C}$ is called **irreducible** if*

$$i \leftrightarrow j \textit{ for all } i, j \in \mathcal{C}.$$

*If all the states $\mathcal{S}$ of a Markov chain are irreducible, the chain is called **irreducible**.* □

Once the chain assumes a state in a closed set $\mathcal{C}$ of states, it never leaves $\mathcal{C}$ afterwards. A closed set containing exactly one state is called an **absorbing** state.

We show below that the states of an irreducible Markov chain are either all recurrent or all transient. Furthermore, if one state in an irreducible chain is periodic with period d, all the states are periodic with period d. The following theorem summarizes the above discussion.

THEOREM 15.7 (Property of an irreducible Markov chain). *In an irreducible Markov chain all states belong to the same class: they are all transient, all null-recurrent, or all positive-recurrent. Furthermore, they are either all aperiodic or all periodic with the same period.*

Proof. If the chain is irreducible, every state is reachable from any other state. As discussed in the proof of Theorem 15.6 (see (15.95) and (15.96)), $\sum_{n=0}^{\infty} P_{ii}^{(n)}$ and $\sum_{n=0}^{\infty} P_{jj}^{(n)}$ converge or diverge together. Thus, all states are either transient or recurrent. If state i is null-recurrent, then $P_{ii}^{(n)} \to 0$ as $n \to \infty$. Then, from (15.95),

$$P_{jj}^{(n)} \to 0, \text{ as } n \to \infty.$$

Hence, state j is also null-recurrent.

If state i is positive-recurrent and periodic with period d, then $P_{ii}^{(n)} > 0$ only when n is a multiple of d. From (15.92), (15.93), and (15.94),

$$P_{ii}^{(m+\ell)} \geq P_{ij}^{(m)} P_{ji}^{(\ell)} = pq > 0,$$

since states i and j communicate. This implies that $(m + \ell)$ must be a multiple of d. From (15.96) we have

$$P_{jj}^{(m+n+\ell)} \geq pq P_{ii}^{(n)} > 0,$$

only when n and hence $(m + n + \ell)$ are multiples of d. Thus, state j also has the period d. □

THEOREM 15.8 (Decomposition of a Markov chain). *If a Markov chain is not an irreducible chain, its state space $\mathcal{S}$ can be uniquely partitioned as*

$$\mathcal{S} = \mathcal{T} \cup \mathcal{C}_1 \cup \mathcal{C}_2 \cup \cdots,$$

where $\mathcal{T}$ is the set of transient states and the $\mathcal{C}_r$ $(r = 1, 2, \ldots)$ are irreducible closed sets of recurrent states.

Proof. Let the $\mathcal{C}_r$ be the recurrent equivalence classes induced by the relation $\leftrightarrow$. It is clear that equivalence classes defined by the relation $\leftrightarrow$ are irreducible. Thus, it suffices to show that each $\mathcal{C}_r$ is closed. Suppose $\mathcal{C}_r$ is not closed, then there should exist $i \in \mathcal{C}_r$ and $j \notin \mathcal{C}_r$ such that $P_{ij} > 0$ (i.e., $i \to j$). The assumption $j \notin \mathcal{C}_r$ means that i and j do not communicate, or i is not reachable from j. Then, we have

$$P[X_n \neq i \text{ for all } n \geq 1 | X_0 = i] \geq P[X_1 = j | X_0 = i] = P_{ij} > 0,$$

because once the system assumes state j, it can never return to i, and there might be other events that prevent the system from returning to state i. But this conclusion that X_n will never assume state i contradicts the assumption that i is a recurrent state. Hence, $\mathcal{C}_r$ must be a closed (and irreducible) set. □

Figure 15.3 illustrates an example of a **reducible** Markov chain, which can be decomposed into $\mathcal{C}_1$, a set of positive-recurrent and aperiodic (i.e., ergodic) states (States 1 through 3); $\mathcal{T}$, a set of transient states (States 4 through 6); $\mathcal{C}_2$, a closed set of a singleton, i.e., an absorbing state (State 7); $\mathcal{C}_3$, a set of positive-recurrent and periodic states

(States 8 through 11); and C_4, a set of null-recurrent and aperiodic states (States 12, 13, 14, ...), assuming $p > q$, where p and q are the transition probabilities of shifting to the right and to the left respectively.

15.3.4 Stationary distribution of an aperiodic irreducible chain

In this section we restrict our discussion to aperiodic irreducible chains. Thus, all states of the chain must be either all positive-recurrent (hence, *ergodic*), all null-recurrent, or all transient. A null-recurrent chain occurs only when the state space $\mathcal{S}$ is infinite in its cardinality. An important question we want to address is that of stability. Does the system, regardless of its initial state, converge to some limiting distribution?

Assume that the system is stable and let the limit of the probability distribution vector be $\boldsymbol{\pi}$:

$$\lim_{n\to\infty} \mathbf{p}(n) = \boldsymbol{\pi}. \tag{15.98}$$

Then, by applying this limit to (15.11), we find that $\boldsymbol{\pi}$ must satisfy the equation

$$\boldsymbol{\pi}^\top = \boldsymbol{\pi}^\top \boldsymbol{P}. \tag{15.99}$$

DEFINITION 15.11 (Stationary distribution). *A probability distribution $\boldsymbol{\pi}$ satisfying (15.99) is called a* ***stationary distribution*** *or an* ***invariant distribution****: if the initial probability $\mathbf{p}(0)$ is set to $\boldsymbol{\pi}$, then $\mathbf{p}(n) = \boldsymbol{\pi}$ for all $n \geq 0$.* □

The next question we may ask will be "Does a solution of (15.99) always exist?"

THEOREM 15.9 (Stationary distribution of an irreducible aperiodic Markov chain). *In an irreducible aperiodic Markov chain, one of the following two alternatives holds:*

1. *The states are all transient or all null-recurrent; in this case there exists no stationary distribution and*

$$\lim_{n\to\infty} P_{ij}^{(n)} = 0, \textit{ for all } i, j \in \mathcal{S}. \tag{15.100}$$

2. *All states are ergodic; in this case there exists a* ***unique stationary distribution*** *and*

$$\lim_{n\to\infty} P_{ij}^{(n)} = \pi_j, \textit{ for all } i, j \in \mathcal{S}. \tag{15.101}$$

Furthermore, π_i is equal to the reciprocal of the mean recurrence time for state i; i.e.,

$$\boxed{\pi_i = \frac{1}{\mu_{ii}}, i \in \mathcal{S}.} \tag{15.102}$$

Proof. In Theorem 15.3 it was shown that, for a transient state i, $\sum_{i=0}^{\infty} P_{ii}^{(n)} < \infty$. Thus, it is apparent that (15.100) must hold. From Theorem 15.5 we have shown (15.100) is a necessary and sufficient condition for a recurrent state to be null. So we need to prove the remaining half; i.e, the case where the chain is ergodic. Equation (15.102) results from (15.81) or (15.86) of Theorem 15.5. As we show below, this choice of $\boldsymbol{\pi}$ satisfies (15.99): let $n \to \infty$ in the one-step transition formula

$$P_{ij}^{(n+1)} = \sum_k P_{ik}^{(n)} P_{kj},$$

leading to

$$\frac{f_{ij}}{\mu_{jj}} = \sum_k \frac{f_{ik}}{\mu_{kk}} P_{kj}.$$

In an ergodic chain, all states communicate with each other, hence $f_{ij} = f_{ik} = 1$, and thus

$$\frac{1}{\mu_{jj}} = \sum_k \frac{1}{\mu_{kk}} P_{kj}.$$

Therefore, $\{\pi_i = 1/\mu_{ii}, i \in \mathcal{S}\}$ satisfies (15.99). This completes the proof. □

The theorem assures us that (15.101) holds whenever the states are ergodic. Therefore, for any initial probability assignment $\{p_i(0); i \in \mathcal{S}\}$, we have the following asymptotic result:

$$\begin{aligned} \lim_{n\to\infty} p_j(n) &= \lim_{n\to\infty} \sum_{i\in\mathcal{S}} p_i(0) P_{ij}^{(n)} = \sum_{i\in\mathcal{S}} p_i(0)\pi_j \\ &= \pi_j,\ j \in \mathcal{S}. \end{aligned} \tag{15.103}$$

Thus, the stationary distribution becomes the **steady-state distribution**. The stationary distribution can be computed by solving a set of M (where $M = |\mathcal{S}|$) linear equations of (15.99) with the following linear constraint condition:

$$\boldsymbol{\pi}^\top \mathbf{1} = 1, \tag{15.104}$$

where $\mathbf{1}$ is an M-dimensional column vector of all unity elements.

Alternatively, a simple but useful computational formula can be found as follows: by repeating (15.104) column-wise M times, we form the matrix equation

$$\boldsymbol{\pi}^\top \boldsymbol{E} = \mathbf{1}^\top, \tag{15.105}$$

where $\boldsymbol{E}$ is an $M \times M$ matrix with all entries unity; i.e., $\boldsymbol{E} = \mathbf{1} \cdot \mathbf{1}^\top$. From (15.99) and (15.105) we obtain

$$\boldsymbol{\pi}^\top (\boldsymbol{P} + \boldsymbol{E} - \boldsymbol{I}) = \mathbf{1}^\top, \tag{15.106}$$

where $\boldsymbol{I}$ is the $M \times M$ identity matrix. Note that matrix $(\boldsymbol{P} + \boldsymbol{E} - \boldsymbol{I})$ is nonsingular, whereas $(\boldsymbol{P} - \boldsymbol{I})$ is singular. Thus, the stationary distribution vector is readily obtained as

$$\boxed{\boldsymbol{\pi}^\top = \mathbf{1}^\top (\boldsymbol{P} + \boldsymbol{E} - \boldsymbol{I})^{-1}.} \tag{15.107}$$

This formula can be further generalized (Problem 15.10).

When all states of a given Markov chain X_n are ergodic, the chain is called an **ergodic Markov chain**. As discussed in Section 12.3.3, a random process, not necessarily a Markov chain, is said to be *ergodic* if the **time average** of any function $f(\cdot)$ of any sample function $\{X_n(\omega)\}$ of the process converges *almost surely*, i.e., *with probability one*, to the **ensemble average** or **expectation**; i.e.,

$$\boxed{\lim_{N\to\infty} \frac{1}{N}\sum_{n=1}^{N} f(X_n(\omega)) \xrightarrow{\text{a.s.}} E[f(X_n)] = \sum_{i\in\mathcal{S}} \pi_i f(i),} \tag{15.108}$$

that has $\mathcal{S}$ as its domain and $[-\infty, \infty)$ as its range such that $\sum_{i\in\mathcal{S}} \pi_i\,|f(i)| < \infty$.

In most physical applications, it is assumed that stationary processes are ergodic and that time averages and expectations can be used interchangeably. A simulation study of a stochastic system is based on the assumption that the process of our interest is ergodic. By analyzing one realization $\{X(\omega_0, t)\}$ of a random process for a sufficiently long period, we can study properties of the ensemble $\{X(\omega, t);\ \omega \in \Omega\}$, if the process is ergodic.

15.4 Summary of Chapter 15

Markov process $X(t)$:	$P[X(t_2)\le x\mid X(t);\, t\le t_1] = P[X(t_2)\le x\mid X(t_1)]$	(15.1)
Markov chain of order h:	$P[X_n\mid X_k;\, k\le n-1] = P[X_n\mid X_{n-h},\ldots, X_{n-1}]$	(15.4)
Transition prob. matrix:	$\boldsymbol{P} = [P_{ij}] = P[X_{n+1} = j\mid X_n = i]$	(15.7)
Homogeneous MC:	$\boldsymbol{p}^\top(n+1) = \boldsymbol{p}^\top(n)\boldsymbol{P}$ for all $n\ge 0$	(15.11)
Chapman–Kolmogorov eq.:	$P_{ik}^{(m+n)} = \sum_{j\in\mathcal{S}} P_{ij}^{(m)} P_{jk}^{(n)}$	(15.20)
Generating function method:	$g^\top(z) = \boldsymbol{p}^\top(0)[\boldsymbol{I} - \boldsymbol{P}z]^{-1}$	(15.25)
Spectral expansion method:	$p_i(n) = \sum_{k\in\mathcal{S}} \lambda_k^n \left(\boldsymbol{p}^\top(0)\boldsymbol{u}_k\right) v_{ki},\ i\in \mathcal{S}$	(15.49)
First-passage time distr.:	$f_{ij}^{(n)} = P[T_{ij} = n],\ n\ge 1$	(15.58)
Recurrent state:	$f_{ii} = P[T_{ii}\ge 1] = 1$	Def. 15.4
	iff $\sum_{n=0}^{\infty} P_{ii}^{(n)} = \infty$	(15.73)
Transient state:	$f_{ii} < 1$	Def. 15.4
	iff $\sum_{n=0}^{\infty} P_{ii}^{(n)} < \infty$	(15.75)
Mean first-passage time:	$\mu_{ij} = E[T_{ij}] = \begin{cases}\sum_{n=1}^{\infty} n f_{ij}^{(n)}, & \text{if } f_{ij} = 1\\ \infty, & \text{if } f_{ij} < 1\end{cases}$	(15.61)

Null-recurrent state:	$\mu_{ii} = \infty$	Def. 15.5
	iff $\lim_{n\to\infty} P_{ii}^{(n)} = 0$	(15.79)
Positive-recurrent state:	$\mu_{ii} < \infty$	Def. 15.5
Period of state i:	$d_i = \gcd\left\{n : P_{ii}^{(n)} > 0\right\}$	(15.63)
Periodic state:	$d_i > 1$	Def. 15.6
Aperiodic state:	$d_i = 1$	Def. 15.6
Ergodic state:	$f_{ii} = 1, \mu_{ii} < \infty,$ and $d_i = 1$	Def. 15.7
	iff $\lim_{n\to\infty} P_{ii}^{(n)} = 1/\mu_{ii}$	(15.81)
Absorbing state:	$P_{ii} = 1 (\Longrightarrow f_{ii} = f_{ii}^{(1)} = 1)$	Def. 15.8
Communication relations:	Reflexive, symmetric, and transitive	(15.97)
Prop. of irreducible chains:	All transient, null-recurrent, or positive-recurrent	Thm. 15.7
Stationary distribution:	$\boldsymbol{\pi}^\top = \boldsymbol{\pi}^\top \boldsymbol{P}$	(15.99)
$\boldsymbol{\pi}$ of an ergodic chain:	$\pi_i = 1/\mu_{ii}, 1 \in \mathcal{S}$	(15.102)
Formula for $\boldsymbol{\pi}$:	$\boldsymbol{\pi}^\top = \mathbf{1}^\top(\boldsymbol{P} + \boldsymbol{E} - \boldsymbol{I})^{-1}$	(15.107)
Ergodic chain:	$\lim_{N\to\infty} \frac{1}{N} \sum_{n=1}^{N} f(X_n(\omega)) \xrightarrow{\text{a.s.}} E[f(X_n)]$	(15.108)

15.5 Discussion and further reading

Markov chains and Markov processes are among the core subjects of virtually all books on probability and random processes (e.g., Çinlar [57], Doob [82], Feller [99, 100], Nelson [254], Rosenblatt [286], Ross [288]), whereas they are often not adequately discussed, in comparison with Gaussian processes, in traditional textbooks written for electrical engineers.

Shannon in his 1948 seminal paper [300] introduced the concept of entropy through Markov modeling of the English language that formed a foundation for HMM (cf. Chapter 20), which has been successfully applied to modeling of natural languages, algorithms for speech recognition, identification of DNA and protein sequences, etc. Stochastic models, such as queueing and loss system models, make extensive use of Markov processes (e.g., see Kleinrock [189], Kobayashi and Mark [203]). Markov chains and Markov processes are increasingly important topics to modeling and analysis of communication networks, including the Internet and its applications. The PageRank of a web page as used by Google is defined by a Markov chain [218]. Hidden Markov or semi-Markov models may be used to analyze web navigation behavior of users (e.g., [366]). A user's web link transition on a particular website can be modeled using Markov models of some order.

Whereas this chapter focused on discrete-time Markov chains, the next chapter will discuss continuous-time Markov chains and semi-Markov processes.

15.6 Problems

Section 15.2: Computation of state probabilities

15.1* **Homogeneous Markov chain.** Consider a homogeneous Markov chain whose transition matrix is given by

$$\boldsymbol{P} = \begin{bmatrix} \frac{1}{2} & \frac{1}{2} & 0 \\ \frac{1}{3} & 0 & \frac{2}{3} \\ 0 & \frac{1}{5} & \frac{4}{5} \end{bmatrix}.$$

(a) Draw the state transition diagram.
(b) The system is initially at state 1; i.e.,

$$\boldsymbol{p}^{\top}(0) = (1\,0\,0).$$

Find $\boldsymbol{p}(1), \boldsymbol{p}(2), \boldsymbol{p}(3), \ldots$.
(c) Evaluate $\boldsymbol{p}(n)$ for an arbitrary positive integer $n \geq 0$.

15.2 **State probabilities.** Consider a Markov chain whose transition matrix is given by

$$\boldsymbol{P} = \begin{bmatrix} \frac{2}{3} & \frac{1}{3} & 0 \\ \frac{1}{2} & 0 & \frac{1}{2} \\ 0 & 0 & 1 \end{bmatrix}.$$

(a) Draw the state transition diagram of $\boldsymbol{P}$ and classify the states.
(b) Find the roots of the characteristic equation

$$\det|\boldsymbol{I} - z\boldsymbol{P}| = 0.$$

(c) Suppose that the system is in state 1 at $n = 0$. Find the probability vector

$$p(n) = [p_1(n)p_2(n)p_3(n)] \text{ for } n = 0, 1, 2, \ldots.$$

15.3 **Simple queueing problem.** Consider the following simple queueing problem. Let X_n be the number of customers awaiting service or being served at time n. We make the following assumptions: (1) if the server is servicing a customer at time n, this customer's service will be completed before time $n+1$ with probability β; (2) between times n and $n+1$, one customer will arrive with probability α, and with probability $1-\alpha$ no customer will arrive. Show that X_n is a Markov chain. Find the transition probabilities.

Section 15.3: Classification of states

15.4* **Transitive property.** Prove the transitive property of the communications relation "$\leftrightarrow$."

15.5* Stationary distribution. Find the stationary distributions of the Markov chains determined by the following TPMs using (15.107):

(a)

$$\boldsymbol{P} = \begin{bmatrix} 0 & 1 & 0 \\ \frac{1}{4} & \frac{1}{4} & \frac{1}{2} \\ 0 & \frac{1}{2} & \frac{1}{2} \end{bmatrix};$$

(b)

$$\boldsymbol{P} = \begin{bmatrix} \frac{1}{2} & \frac{1}{2} & 0 \\ \frac{1}{3} & 0 & \frac{2}{3} \\ 0 & \frac{1}{5} & \frac{4}{5} \end{bmatrix}.$$

15.6 Characteristic equation. Consider a Markov chain with the TPM

$$\boldsymbol{P} = \begin{bmatrix} 0 & 1 & 0 & 0 \\ \frac{1}{2} & 0 & \frac{1}{2} & 0 \\ 0 & \frac{1}{2} & 0 & \frac{1}{2} \\ 0 & 0 & 1 & 0 \end{bmatrix}.$$

(a) Draw the state transition diagram corresponding to $\boldsymbol{P}$.
(b) Discuss the properties of this chain.
(c) Find the roots of the characteristic equation

$$\det |\boldsymbol{I} - z\boldsymbol{P}| = 0.$$

15.7 Roots of a characteristic equation. Answer the following questions regarding the roots of the characteristic equation

$$\det |\boldsymbol{I} - z\boldsymbol{P}| = 0$$

of a Markov chain with TPM $\boldsymbol{P}$.

(a) Show that none of the roots may have a magnitude less than unity.
(b) Show that at least one root is equal to unity. If there is more than one root equal to unity, what does this imply?
(c) If the characteristic equation contains a factor $(z^k - 1)$, what does this mean?

15.8 Transition probability matrix of a certain structure. Consider a Markov chain with the transition probability matrix

$$\boldsymbol{P} = \begin{bmatrix} 1 & 0 & 0 & 0 \\ 0 & 1 & 0 & 0 \\ \frac{1}{2} & 0 & 0 & \frac{1}{2} \\ 0 & \frac{1}{2} & \frac{1}{2} & 0 \end{bmatrix}.$$

(a) Draw the state transition diagram and classify the states.

(b) Show that, when n is an odd integer,

$$\boldsymbol{P}^n = \begin{bmatrix} 1 & 0 & 0 & 0 \\ 0 & 1 & 0 & 0 \\ * & * & 0 & \frac{1}{2^n} \\ * & * & \frac{1}{2^n} & 0 \end{bmatrix}.$$

What will be the structure of matrix when n is even?

(c) Suppose that the matrix $\boldsymbol{P}$ is partitioned into the following form:

$$\boldsymbol{P} = \begin{bmatrix} \boldsymbol{Q}_1 & \boldsymbol{0} & \boldsymbol{0} \\ \boldsymbol{0} & \boldsymbol{Q}_2 & \boldsymbol{0} \\ \boldsymbol{A} & \boldsymbol{B} & \boldsymbol{C} \end{bmatrix}.$$

What form does $\boldsymbol{P}^n$ take?

15.9 First-passage time matrix. Consider an N-state Markov chain $\boldsymbol{P}$. Let $\boldsymbol{M}$ be the matrix whose (i, j) component is the expected first-passage time μ_{ij}, $1 \le i, j \le N$. Let $\boldsymbol{M}_{\text{dg}}$ be a matrix that has the same diagonal entries as $\boldsymbol{M}$ and zeros elsewhere.

(a) Show that $\boldsymbol{M}$ satisfies the following matrix equation:

$$\boldsymbol{M} = \boldsymbol{E} + \boldsymbol{P}(\boldsymbol{M} - \boldsymbol{M}_{\text{dg}}),$$

where $\boldsymbol{E}$ is an $N \times N$ matrix whose entries are all unity.
Hint: Start with (15.62).

(b) Derive (15.102), i.e., $\pi_i = 1/\mu_{ii}$, for all $i = 1, 2, \ldots, N$.
Hint: Multiply the equation of part (a) by $\boldsymbol{\pi}'$.

(c) Find the first-passage time matrix of the two Markov chains defined in Problem 15.5.

15.10 Computation formula for the stationary distribution.[5] Let $\boldsymbol{P}$ be the TPM of an M-state ergodic Markov chain, $\boldsymbol{E}$ an $M \times M$ matrix of all ones, and $\boldsymbol{1}$ an M-dimensional column vector of all ones. Show that the stationary distribution can be found from

$$\boldsymbol{\pi}^\top = k\boldsymbol{1}^\top(\boldsymbol{P} + k\boldsymbol{E} - \boldsymbol{I})^{-1}, \tag{15.109}$$

where k is any nonzero constant. For $k = 1$, we obtain (15.107).

15.11 Number of returns. Prove the following theorem

THEOREM 15.10 (Number of returns). *Suppose a Markov chain X_n assumes state i at some n_0. If the state i is recurrent, then the Markov chain returns infinitely often to state i as $n \to \infty$. If state i is transient, the chain returns to the state i only finitely often.*

Hint: Use the Borel–Cantelli lemmas (Theorem 11.16).

[5] This generalization of (15.107) is due to Yihong Wu.

16 Semi-Markov processes and continuous-time Markov chains

In this chapter we focus on a class of continuous-time processes that may be considered generalizations of the discrete-time Markov chain (DTMC) discussed in the previous chapter. We begin with a semi-Markov process (SMP) $X(t)$, which can be characterized by a DTMC $\{X_n\}$ together with a sojourn time process $\{\tau_n\}$ which has a distribution function that depends on the current and next states of the DTMC, X_n and X_{n+1} respectively. While $X(t)$ is not itself a Markov process, the embedded process $\{X_n\}$ is a Markov process, hence the "semi-Markov" designation for $X(t)$. By considering an SMP where the sojourn times τ_n are exponentially distributed, we obtain a continuous-time Markov chain (CTMC) $X(t)$, which is indeed a Markov process.

The concept of reversibility is particularly important in the study of Markov chains, as reversible DTMCs and CTMCs have special structures that often can be exploited in applications. In this chapter we also discuss some applications of the CTMC and SMP to modeling ion currents in biology and to the study of evolutionary or phylogenetic trees.

16.1 Semi-Markov process

Before we extend Markov chains or discrete-time Markov processes to their continuous counterparts, we discuss in this section a class of random processes called a semi-Markov process (SMP). As its name indicates, an SMP is similar to a Markov process but, in general, does not satisfy the Markovian or memoryless property of the Markov process.

DEFINITION 16.1 (Semi-Markov process). *A semi-Markov process $\{X(t),\ t \geq 0\}$ is a right-continuous, piecewise constant process, which takes values in a finite or countably infinite set of states $\mathcal{S}$ and transitions at times $t_1, t_2, \ldots$ We shall assume that $t_0 = 0$ and define the nth* ***sojourn time*** *by $\tau_n = t_n - t_{n-1}$, $n \geq 1$. An SMP is characterized by*

(i) its TPM

$$\boldsymbol{P} = [P_{ij}], \quad i, j \in \mathcal{S},$$

where $P_{ij} = P[X(t_n) = j \mid X(t_{n-1}) = i]$ is the probability with which the process makes a transition from state i to state j at time t_n, $n \geq 1$;

(ii) the set of distribution functions

$$F_{ij}(t) = P[\tau_n \leq t \mid X(t_{n+1}) = j,\ X(t_n) = i], \quad i, j \in \mathcal{S},$$

where $F_{ij}(t)$ is the distribution function of the nth sojourn time τ_n given that the states of the SMP at times t_n and t_{n-1} $(n \geq 1)$ are i and j respectively.

□

An SMP $X(t)$ may have the property that, upon reaching a state i, it remains in this state with probability one. In this case, we call state i an **absorbing state**. For an SMP $X(t)$, we shall assume that each time instant t_n $(n \geq 1)$ corresponds to an observable jump from a state i to a *different* state $j \neq i$, unless state i is an absorbing state. Between any two time instants t_n and t_{n+1}, $X(t)$ has a constant value; i.e.,

$$X(t) = X(t_n) \text{ for } t_n \leq t < t_{n+1}, \quad n = 0, 1, \ldots. \tag{16.1}$$

This assumption implies that the TPM $\boldsymbol{P}$ in Definition 16.1 satisfies $P_{ii} = 0$ for all states $i \in \mathcal{S}$, unless i is an absorbing state, in which case $P_{ii} = 1$. We shall refer to the DTMC corresponding to this TPM $\boldsymbol{P}$ as the **embedded Markov chain** (EMC) associated with the SMP $X(t)$.[1] Thus, the EMC, which we denote by $\{X_n\}$, is obtained by sampling $X(t)$ at its *jump* times $t_1, t_2, \ldots$ as follows:

$$X_n \triangleq X(t_n), \quad n = 1, 2, \ldots. \tag{16.2}$$

Figure 16.1 shows an example sample path of an SMP $X(t)$. Note that the sample path is piecewise constant and right-continuous, with jump discontinuities at $t_1, t_2, \ldots$.

Although $X(t)$ is not itself a Markov process, the pair process $\{(X_n, \tau_n)\}$ is a Markov process. It is apparent from Figure 16.1 that the sample path of the SMP $X(t)$ can be completely specified by the sequence of pairs,

$$\{(X_0, t_0), (X_1, t_1), \ldots, (X_n, t_n), (X_{n+1}, t_{n+1}), \ldots\}.$$

The process $X(t)$ stays in state X_n during the sojourn time $[t_n, t_{n+1})$, $n = 1, 2, \ldots$. This pair sequence $\{(X_n, t_n)\}$ is called a **Markov renewal process** (MRP). More generally,

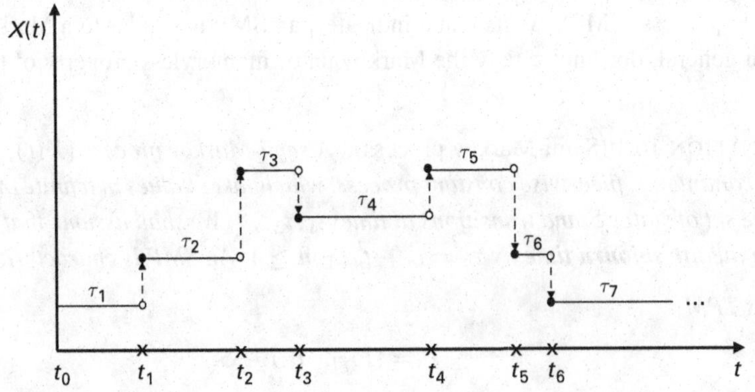

Figure 16.1 A sample path of a semi-Markov process.

[1] Some authors refer to this specification of the EMC as the canonical representation of an SMP (e.g., see [332]).

an MRP $\{(X_n, t_n)\}$ consists of an arbitrary DTMC $\{X_n\}$ together with a sequence of *renewal points*, $\{t_n\}$, such that the sojourn time $\tau_n = t_n - t_{n-1}$ depends only X_{n-1} and X_n, $n \geq 1$ (e.g., see [56, 58]). Given an MRP, $\{(X_n, t_n)\}$, we can define an associated continuous-time process $X(t)$ as follows:

$$X(t) = X_n \text{ for } t_n \leq t < t_{n+1}, \quad n = 0, 1, \ldots. \tag{16.3}$$

This associated process $X(t)$ is an SMP in the sense of Definition 16.1.

For an MRP $\{(X_n, t_n)\}$, we can define a **counting process** $N(t)$ as the cumulative count of state transitions up to time t; i.e.,

$$N(t) = \max\{n : t_n \leq t\}. \tag{16.4}$$

Then the SMP $X(t)$ specified in (16.3) can equivalently be expressed as follows:

$$\boxed{X(t) = X_{N(t)}, \quad t \geq 0.} \tag{16.5}$$

Equations (16.3) and (16.5) show the close relationship between SMPs and MRPs.

Given an SMP $X(t)$, we define the **sojourn time variable** S_{ij}, between states i and j, as the time between two successive jump points, say t_{n-1} and t_n, given that $X_{n-1} = i$ and $X_n = j$, $n \geq 1$. Clearly, the distribution function of S_{ij} is given by

$$F_{ij}(t) = P[\tau_n \leq t \mid X_n = j, X_{n-1} = i] = P[S_{ij} \leq t]. \tag{16.6}$$

Similarly, we define the sojourn time in state i, S_i, as the time between two successive jump points, t_{n-1} and t_n, such that $X_{n-1} = i$, $n \geq 1$. The variables S_i and S_{ij} are related by

$$S_i = S_{ij} \text{ with probability } P_{ij}, \quad i, j \in \mathcal{S}. \tag{16.7}$$

The distribution function of S_i is given by

$$F_i(t) = P[S_i \leq t] = P[\tau_{n-1} \leq t \mid X_{n-1} = i] = \sum_{j \in \mathcal{S}} F_{ij}(t) P_{ij}. \tag{16.8}$$

Suppose that the EMC $\{X_n\}$ associated with $X(t)$ possesses a stationary distribution $\tilde{\boldsymbol{\pi}} = (\tilde{\pi}_i, i \in \mathcal{S})^\top$, given as the solution of

$$\tilde{\boldsymbol{\pi}}^\top = \tilde{\boldsymbol{\pi}}^\top \tilde{\boldsymbol{P}} \text{ and } \tilde{\boldsymbol{\pi}}^\top \mathbf{1} = 1, \tag{16.9}$$

where $\mathbf{1}$ denotes a column vector all ones. Let $\boldsymbol{\pi} = (\pi_i, i \in \mathcal{S})^\top$ denote the stationary distribution of the SMP $X(t)$. Then the stationary distributions of $X(t)$ and its associated EMC can be related as follows.

THEOREM 16.1 (Stationary distributions of SMP and EMC). *Let $X(t)$ be an SMP with state sojourn times S_i and let $\tilde{\boldsymbol{P}}$ be the TPM of its associated EMC $\{X_n\}$. We further assume that $0 < E[S_i] < \infty$ for all states $i \in \mathcal{S}$. Then the stationary distribution of $X(t)$, $\boldsymbol{\pi}$, is related to the stationary distribution of $\{X_n\}$, $\tilde{\boldsymbol{\pi}}$, by*

$$\pi_i = \frac{\tilde{\pi}_i E[S_i]}{\sum_{j\in\mathcal{S}} \tilde{\pi}_j E[S_j]}, \quad i \in \mathcal{S}, \tag{16.10}$$

provided the sum on the right-hand side converges. Conversely, $\tilde{\pi}$ is given by

$$\tilde{\pi}_i = \frac{\pi_i / E[S_i]}{\sum_{j\in\mathcal{S}} \pi_j / E[S_j]}, \quad i \in \mathcal{S}, \tag{16.11}$$

provided the sum on the right-hand side converges.

Proof. Let us consider a sample path of $X(t)$ over the period $[0, t_{N+1})$ in which $X(t)$ makes N jumps (excluding the one possibly at $t_0 = 0$). Let n_i be the number of times state i is visited during this period. Clearly,

$$\sum_{i\in\mathcal{S}} n_i = N.$$

The assumption that the EMC is positive recurrent implies that $n_i \overset{\text{a.s.}}{\to} \infty$ as $N \to \infty$.

Let $s_i(k)$ be the sojourn time in state i in its kth visit by $X(t)$. Then the fraction of time the process spends in state i is

$$f_i(N) = \frac{\sum_{k=1}^{n_i} s_i(k)}{\sum_{j\in\mathcal{S}} \sum_{k=1}^{n_j} s_j(k)} = \frac{\frac{n_i}{N} \cdot \frac{1}{n_i} \sum_{k=1}^{n_i} s_i(k)}{\sum_{j\in\mathcal{S}} \frac{n_j}{N} \cdot \frac{1}{n_j} \sum_{k=1}^{n_j} s_j(k)}. \tag{16.12}$$

The strong law of large numbers implies that

$$\frac{1}{n_i} \sum_{k=1}^{n_i} s_i(k) \overset{\text{a.s.}}{\longrightarrow} E[S_i], \quad \text{as } N \longrightarrow \infty. \tag{16.13}$$

Similarly,

$$\frac{n_i}{N} \overset{\text{a.s.}}{\longrightarrow} \tilde{\pi}_i, \quad \text{as } N \longrightarrow \infty. \tag{16.14}$$

Thus, it follows that

$$f_i(N) \overset{\text{a.s.}}{\longrightarrow} \frac{\tilde{\pi}_i E[S_i]}{\sum_{j\in\mathcal{S}} \tilde{\pi}_j E[S_j]}, \quad i \in \mathcal{S}. \tag{16.15}$$

From the definition of $f_i(N)$, it is apparent that

$$f_i(N) \overset{\text{a.s.}}{\longrightarrow} \pi_i, \quad \text{as } N \to \infty. \tag{16.16}$$

Thus, (16.10) has been proved. Equation (16.11) then follows by rearranging (16.10).□

Similar to the classification of states of a DTMC (see Section 15.3), the states of an SMP can be classified as being **transient**, **null-recurrent**, and **positive-recurrent** by introducing analogous notions of first-passage time and mean recurrence time in terms of the SMP. However, owing to the underlying role of the EMC in the definition of an SMP, the states of an SMP can be classified more simply in terms of its associated EMC. In particular, a state i of an SMP $X(t)$ is transient, null-recurrent, or positive-recurrent according to whether state i is transient, null-recurrent, or positive-recurrent with respect to the EMC $\{X_n\}$.

Communication of states in an SMP also corresponds to communication of states with respect to the EMC (cf. Section 15.3.3). For example, state i is reachable from state j with respect to an SMP $X(t)$ if i is reachable from j in the associated EMC. Thus, states i and j of an SMP communicate (written as $i \leftrightarrow j$) if they communicate in the associated EMC. Hence, an SMP is **irreducible** if its associated EMC is irreducible. As mentioned earlier, state i of an SMP is an **absorbing state** if state i is absorbing with respect to the EMC; i.e., $P_{ii} = 1$. Owing to the random sojourn times between states of an SMP, the notion of a periodic state in a DTMC does not generally have a counterpart with respect to an SMP.

16.2 Continuous-time Markov chain (CTMC)

Now we are in a position to discuss the CTMC. Consider an SMP $X(t)$ for which the sojourn time distribution $F_{ij}(t)$ is exponential with parameter $\nu_i > 0$:

$$F_{ij}(t) = 1 - e^{\nu_i t}, \quad t \geq 0, \quad i, j \in \mathcal{S}. \tag{16.17}$$

Note that $F_{ij}(t)$ depends only on i, and not on j. It then follows from (16.8) that the distribution function of the sojourn time in state i is given by $F_{S_i}(t) = F_{ij}(t)$ for all $i, j \in \mathcal{S}$. Then $X(t)$ can be shown to be a Markov process in the sense of Definition 15.1 and hence is called a CTMC. The Markovian property of $X(t)$ is essentially due to the memoryless property of the exponentially distributed sojourn times as specified in (16.17) (see Problem 16.5).

16.2.1 Infinitesimal generator and embedded Markov chain of a continuous-time Markov chain

DEFINITION 16.2 (Transition probability matrix function (TPMF)). *For a CTMC $X(t)$, the conditional probabilities*

$$P_{ij}(t) = P[X(s+t) = j \mid X(s) = i], \quad i, j \in \mathcal{S}, \quad 0 \leq t < \infty, \tag{16.18}$$

are called ***transition probability functions****, and*

$$\boldsymbol{P}(t) = [P_{ij}(t)] \tag{16.19}$$

is called the ***transition probability matrix function*** *(TPMF). It is clear that (16.18) and (16.19) are continuous-time counterparts of (15.7) and (15.8) respectively.* □

The Chapman–Kolmogorov equation (15.20) of Theorem 15.1 takes the following form:

$$P_{ik}(s+t) = \sum_{j \in \mathcal{S}} P_{ij}(s) P_{jk}(t), \quad s, t \geq 0, \quad i, k \in \mathcal{S}, \tag{16.20}$$

or in matrix form

$$\boxed{\boldsymbol{P}(s+t) = \boldsymbol{P}(s)\boldsymbol{P}(t).} \tag{16.21}$$

Thus, many results that were obtained for the DTMC can be carried over to the CTMC, by replacing the TPM $\boldsymbol{P}$ of the discrete chain by the TPM function $\boldsymbol{P}(t)$ of the continuous chain. The Chapman–Kolmogorov equation is such an example. However, there is one important difference between discrete-time and continuous-time chains. In the continuous-time case we cannot have a meaningful definition of **transition probabilities per unit time**. Instead, we consider the **transition rates** and their matrix. The **infinitesimal generator matrix** or **transition rate matrix** defined below plays a role to fill this gap.

DEFINITION 16.3 (Infinitesimal generator matrix). *The infinitesimal generator (matrix) of a stationary Markov process with TPMF* $\boldsymbol{P}(t)$ *is defined by*

$$\boxed{\boldsymbol{Q} = \left.\frac{d\boldsymbol{P}(t)}{dt}\right|_{t=0}.} \tag{16.22}$$

□

Alternatively, we can define the matrix $\boldsymbol{Q}$ by

$$\boldsymbol{Q} = \left[Q_{ij}\right],$$

where

$$Q_{ij} = \lim_{h \to 0} \frac{P[X(t+h) = j \mid X(t) = i]}{h} = \lim_{h \to 0} \frac{P_{ij}(h)}{h}, \quad \text{for } i \neq j, \tag{16.23}$$

and

$$Q_{ii} = -\sum_{j \neq i} Q_{ij} = \lim_{h \to 0} \frac{P_{ii}(h) - 1}{h}. \tag{16.24}$$

A CTMC $X(t)$ is completely characterized by its infinitesimal generator $\boldsymbol{Q}$. On the other hand, $X(t)$ is also an SMP with sojourn time distributions satisfying (16.17). Thus, given an infinitesimal generator $\boldsymbol{Q}$, the CTMC $X(t)$ can be represented in terms of the TPM of its EMC, $\boldsymbol{P}$, together with a set of transition rates $\{\nu_i, i \in \mathcal{S}\}$ (see Problem 16.6). This representation is especially convenient for computer simulation of a CTMC.

The states of a CTMC $X(t)$ can be classified from its representation as an SMP; i.e., the states of $X(t)$ are transient, positive-recurrent, and null-recurrent if the corresponding states are transient, positive-recurrent, and null recurrent with respect to the EMC

$\{X_n\}$. Further, state i is reachable from state j in $X(t)$ if state i is reachable j in $\{X_n\}$. Thus, i communicates with j in $X(t)$ if $i \leftrightarrow j$ in $\{X_n\}$ and $X(t)$ is called irreducible if $\{X_n\}$ is irreducible. We shall call a CTMC **ergodic** if it is irreducible and all states are positive-recurrent. Note that the concept of ergodicity for a CTMC does not require the states to be aperiodic as in the definition of an ergodic DTMC. Since the sojourn times between state transitions are exponentially distributed, periodicity of the EMC does not imply that the recurrence time of a state in $X(t)$ has a periodic structure.

Suppose that the EMC $\{X_n\}$ possesses a unique stationary distribution $\tilde{\boldsymbol{\pi}} = (\tilde{\pi}_i; i \in \mathcal{S})^\top$; i.e., $\tilde{\pi}$ satisfies (16.9). Let $\boldsymbol{\pi} = (\pi_i; i \in \mathcal{S})^\top$ denote the stationary distribution of $X(t)$. Then, by applying Theorem 16.1, we find that the stationary distribution of $X(t)$ is given by

$$\pi_i = \frac{\tilde{\pi}_i / \nu_i}{\sum_{j \in \mathcal{S}} \tilde{\pi}_j / \nu_j}, \quad i \in \mathcal{S}. \tag{16.25}$$

THEOREM 16.2 (Stationary distribution of an ergodic CTMC in terms of its EMC). *Suppose $X(t)$ is an ergodic CTMC with infinitesimal generator matrix $\boldsymbol{Q}$. Let $\tilde{\boldsymbol{\pi}}$ denote the stationary distribution of the associated EMC $\{X_n\}$; i.e., $\tilde{\boldsymbol{\pi}}$ is the solution to* (16.9). *Then $X(t)$ has a unique stationary distribution $\boldsymbol{\pi}$ given by* (16.25).

Example 16.1: Ion channel model [62]. The endplate membrane of muscle fibers contains ion-permeable channels that can open and close under the influence of certain drugs. Opening of an ion channel increases the membrane conductance and hence increases the current through the channel. The signal produced by a single ion channel is too small to be observed using conventional techniques. However, the variation in the number of open channels over time causes fluctuations about the mean current that are amenable to stochastic modeling.

The law of mass action in the context of drug action mechanisms implies that the lifetime of each chemical species present in a system is a memoryless random variable that does not depend on the age of the species. This suggests that the state of the system can be modeled as a CTMC $X(t)$ defined on the set $\mathcal{S}$ of kinetically distinguishable states of the ion channel. Under the assumption of a constant drug concentration, the CTMC will be time-homogeneous.

As a simple example (cf. [62]), consider the case of two states, $\mathcal{S} = \{0, 1\}$, where state 0 represents the ion channel being *open* and state 1 represents the ion channel being *closed*. The classical theory of drug action states that, when a drug is combined with a closed channel (state 1), an open channel (state 0) is formed at rate $\eta\mu_1$, where $\mu_1 \geq 0$ is a constant and η is the *concentration* of the drug in a closed channel. Given an open channel, the rate of relaxation to the closed state is a constant $\mu_0 \geq 0$. In this case, the infinitesimal generator matrix of the CTMC is given by

$$\boldsymbol{Q} = \begin{bmatrix} -\mu_0 & \mu_0 \\ \eta\mu_1 & -\eta\mu_1 \end{bmatrix}. \tag{16.26}$$

The TPMF $\boldsymbol{P}(t) = [P_{ij}(t)]$ can be expressed as (see (16.35))

$$\boldsymbol{P}(t) = e^{\boldsymbol{Q}t} = \boldsymbol{I} + \boldsymbol{Q}t + \frac{\boldsymbol{Q}^2 t^2}{2} + \cdots . \tag{16.27}$$

Thus, for small t, we have

$$\boldsymbol{P}(t) = e^{\boldsymbol{Q}t} = \boldsymbol{I} + \boldsymbol{Q}t + o(t)\boldsymbol{E} = \begin{bmatrix} 1 - \mu_0 t + o(t) & \mu_0 t + o(t) \\ \eta\mu_1 + o(t) & 1 - \eta\mu_1 t + o(t) \end{bmatrix}, \tag{16.28}$$

where $\boldsymbol{E}$ denotes the matrix of all ones.

The EMC $\{X_n\}$ associated with $X(t)$ has TPM given by

$$\boldsymbol{P} = \begin{bmatrix} 0 & 1 \\ 1 & 0 \end{bmatrix} \tag{16.29}$$

and exponential sojourn time distributions with parameters $\nu_0 = \mu_0$ and $\nu_1 = \eta\mu_1$, corresponding to states 0 and 1 respectively (see Problem 16.6). It is apparent that states 0 and 1 are positive-recurrent with respect to $\{X_n\}$ and that $\{X_n\}$ is irreducible. The CTMC $X(t)$ is ergodic because it is irreducible and its states are positive-recurrent. However, the EMC $\{X_n\}$ is not ergodic, since states 0 and 1 have period two.

By solving (16.9), the stationary distribution of $\{X_n\}$ is obtained as $\tilde{\boldsymbol{\pi}} = (1/2, 1/2)^\top$. The transition rates of $X(t)$ are given by $\boldsymbol{D} = \text{diag}\{\mu_0, \eta\mu_1\}$. Applying (16.25), we obtain the stationary distribution of $X(t)$ as

$$\boldsymbol{\pi} = \left(\frac{\eta\mu_1}{\mu_0 + \eta\mu_1}, \frac{\mu_0}{\mu_0 + \eta\mu_1} \right)^\top . \tag{16.30}$$

□

16.2.2 Kolmogorov's forward and backward equations

In the Chapman–Kolmogorov equation (16.20), let s be replaced by an infinitesimal interval h:

$$P_{ik}(t+h) = \sum_{j \neq k} P_{ij}(t) P_{jk}(h) + P_{ik}(t) P_{kk}(h). \tag{16.31}$$

By subtracting $P_{ik}(t)$ from both sides and dividing them by h, we have

$$\frac{P_{ik}(t+h) - P_{ik}(t)}{h} = \sum_{j \neq k} P_{ij}(t) \frac{P_{jk}(h)}{h} + \frac{P_{kk}(h) - 1}{h} P_{ik}(t). \tag{16.32}$$

By letting $h \to 0$, we find

$$\frac{dP_{ik}(t)}{dt} = \sum_{j \neq k} P_{ij}(t) Q_{jk} + P_{ik}(t) Q_{kk} = \sum_{j \in \mathcal{S}} P_{ij}(t) Q_{jk}, \tag{16.33}$$

or in matrix form

$$\boxed{\frac{d\boldsymbol{P}(t)}{dt} = \boldsymbol{P}(t)\boldsymbol{Q},} \tag{16.34}$$

which is known as **Kolmogorov's forward (differential) equation**. The solution to this differential matrix equation is found as

$$\boxed{\boldsymbol{P}(t) = e^{\boldsymbol{Q}t},} \tag{16.35}$$

using the identity $P_{ij}(0) = \delta_{ij}$; i.e., $\boldsymbol{P}(0) = \boldsymbol{I}$. The exponent of matrix $\boldsymbol{A}$ is defined by

$$e^{\boldsymbol{A}} \triangleq \boldsymbol{I} + \boldsymbol{A} + \frac{\boldsymbol{A}^2}{2} + \cdots + \frac{\boldsymbol{A}^n}{n!} + \cdots \tag{16.36}$$

If we set $t = h$ in (16.20), we have

$$P_{ik}(h+u) = \sum_{j \neq i} P_{ij}(h)P_{jk}(u) + P_{ii}(h)P_{ik}(u). \tag{16.37}$$

A similar manipulation leads to

$$\boxed{\frac{d\boldsymbol{P}(t)}{dt} = \boldsymbol{Q}\boldsymbol{P}(t),} \tag{16.38}$$

which is known as **Kolmogorov's backward (differential) equation**. This differential equation also leads to the same solution (16.35). If all entries of $\boldsymbol{Q}$ are bounded, then $\boldsymbol{Q}$ is said to be **uniform** and

$$\boldsymbol{P}(t) = e^{\boldsymbol{Q}t} = \sum_{n=0}^{\infty} \frac{\boldsymbol{Q}^n t^n}{n!}, \quad t \geq 0. \tag{16.39}$$

If $\boldsymbol{Q}$ is not uniform, then it is possible that the series in (16.39) may not converge (see Problem 16.8).

The probability distribution $\boldsymbol{p}(t)^\top = (p_i(t),\ i \in \mathcal{S})$ of the CTMC $X(t)$ is given by

$$\boldsymbol{p}^\top(t) = \boldsymbol{p}^\top(0)\boldsymbol{P}(t), \quad t \geq 0. \tag{16.40}$$

The distribution $\boldsymbol{p}(t)$ does not depend on t if and only if $\boldsymbol{p}(0) = \boldsymbol{\pi}$, where $\boldsymbol{\pi}$ is the **invariant** or **stationary** distribution that satisfies (Problem 16.9)

$$\boxed{\boldsymbol{\pi}^\top \boldsymbol{Q} = \boldsymbol{0}^\top,} \tag{16.41}$$

which is called the **global balance equation**,[2] or simply the **balance equation**. By writing the components of the above matrix equation separately, we have

$$\sum_{j \in \mathcal{S}} \pi_j Q_{ji} = 0, \quad \text{for all } i \in \mathcal{S}, \tag{16.42}$$

which, using property (16.24), can be rewritten (Problem 16.10) as

$$\boxed{\sum_{j \neq i} \pi_j Q_{ji} = \pi_i \left(\sum_{j \neq i} Q_{ij} \right), \quad \text{for all } i \in \mathcal{S}.} \tag{16.43}$$

The above set of balance equations states that the rate of transitions into any state i and the rate of transitions out of this state are equal. In that case the process $X(t)$ is stationary; i.e., the distribution of $X(t)$ does not depend on $t \geq 0$ once we set $\boldsymbol{p}(0) = \boldsymbol{\pi}$.

Theorem 16.2 provides an expression for the stationary distribution of an ergodic CTMC $X(t)$, $\boldsymbol{\pi}$, in terms of the stationary distribution, $\tilde{\boldsymbol{\pi}}$, of its associated EMC $\{X_n\}$. On the other hand, from the above developments, we know that $\boldsymbol{\pi}$ must satisfy the global balance equation (16.41). Thus, we have the following result.

THEOREM 16.3 (Stationary distribution of an ergodic CTMC). *An ergodic CTMC with infinitesimal generator matrix* $\boldsymbol{Q}$ *has a unique stationary distribution* $\boldsymbol{\pi}$ *satisfying the global balance equation* (16.41).

In Example 16.1, the stationary distribution of the CTMC $X(t)$ was obtained from the stationary distribution of the associated EMC $\{X_n\}$. Alternatively, the stationary distribution of $X(t)$ can be obtained without consideration of the EMC by solving the global balance equation (16.41) together with the normalization constraint $\boldsymbol{\pi}^\top \mathbf{1} = 1$.

16.2.3 Spectral expansion of the infinitesimal generator

Similar to the expansion of the TPM given in (15.33), we can expand the infinitesimal generator $\boldsymbol{Q}$ as follows:

$$\boldsymbol{Q} = \boldsymbol{U}\boldsymbol{\Gamma}\boldsymbol{U}^{-1} = \boldsymbol{U}\boldsymbol{\Gamma}\boldsymbol{V} = \sum_{i \in \mathcal{S}} \gamma_i \boldsymbol{E}_i, \tag{16.44}$$

where $\boldsymbol{\Gamma} = \text{diag}[\gamma_0, \gamma_1, \gamma_2, \ldots]$ and the γ_i; $i \in \mathcal{S}$ are the eigenvalues of $\boldsymbol{Q}$; i.e.,

$$\det|\boldsymbol{Q} - \gamma_i \boldsymbol{I}| = 0, \quad i \in \mathcal{S}.$$

The ith column vector $\boldsymbol{u}_i$ of the similarity matrix $\boldsymbol{U} = [\boldsymbol{u}_i;\ i \in \mathcal{S}]$ is the right-eigenvector associated with γ_i. Similarly, the ith row vector $\boldsymbol{v}_i$ of $\boldsymbol{V} = \boldsymbol{U}^{-1}$ is the left-eigenvector associated with γ_i. The matrices

[2] We have seen different types of balance equations, called **detailed balance equations**, in Section 14.2, which will be discussed further in Section 16.3.

$$\boldsymbol{E}_i = \boldsymbol{u}_i \boldsymbol{v}_i^\top, \quad i \in \mathcal{S},$$

are the projection matrices as defined in (15.45). The similarity of $\boldsymbol{U}$, i.e., $\boldsymbol{V}\boldsymbol{U} = \boldsymbol{I}$, implies that $\boldsymbol{v}_i$ and $\boldsymbol{u}_j$ are *biorthonormal*; i.e.,

$$\boldsymbol{v}_i^\top \boldsymbol{u}_j = \delta_{ij}, \quad i, j \in \mathcal{S}.$$

Then the TPM $\boldsymbol{P}(t) = e^{\boldsymbol{Q}t}$ has the spectral expansion given by

$$\boldsymbol{P}(t) = \boldsymbol{U}\boldsymbol{\Lambda}(t)\boldsymbol{V} = \sum_{i \in \mathcal{S}} \lambda_i(t) \boldsymbol{E}_i,$$

where $\lambda_i(t)$ is the ith entry of the diagonal matrix $\boldsymbol{\Lambda}(t)$, and is simply related to the eigenvalue γ_i by

$$\lambda_i(t) = e^{\gamma_i t}. \tag{16.45}$$

The state distribution $\boldsymbol{p}(t)$ of a CTMC $X(t)$ at time t can thus be represented as

$$\boldsymbol{p}(t)^\top = \boldsymbol{p}(0)^\top \boldsymbol{P}(t) = \sum_{i\mathcal{S}} e^{\gamma_i t} \boldsymbol{u}_i \boldsymbol{v}_i^\top = \sum_{i \in \mathcal{S}} a_i e^{\gamma_i t} \boldsymbol{v}_i^\top, \tag{16.46}$$

where a_i is the inner product of vectors $\boldsymbol{p}(0)$ and $\boldsymbol{u}_i$,

$$a_i = \boldsymbol{p}(0)^\top \boldsymbol{u}_i, \quad i \in \mathcal{S}. \tag{16.47}$$

The right-eigenvector associated with the eigenvalue $\gamma_0 = 0$ is a column vector whose entries are all ones; i.e., $\boldsymbol{u}_0^\top = (1, 1, 1, \ldots, 1)$. Then the coefficient a_0 of (16.47) is simply given by $a_0 = 1$. Thus, we have the following expression for the state distribution at time t:

$$\boldsymbol{p}(t) = \boldsymbol{v}_0 + \sum_{i \in \mathcal{S}\setminus\{0\}} a_i e^{\gamma_i t} \boldsymbol{v}_i. \tag{16.48}$$

We see that $\boldsymbol{p}(t)$ converges to $\boldsymbol{v}_0$ as $t \to \infty$. The maximum negative (smallest in magnitude) eigenvalue, which we denote by γ_1 (i.e., $\gamma_n < \gamma_1 < \gamma_0 = 0$ for all $j \geq 2$), determines the rate of convergence. That is, for sufficiently large t, we have

$$\boldsymbol{p}(t) \approx \boldsymbol{v}_0 + a_1 e^{\gamma_1} \boldsymbol{v}_1, \quad \text{for large } t. \tag{16.49}$$

The quantity $|\gamma_1|^{-1}$ is called the *relaxation time* of the system. As time tends to infinity, we find

$$\lim_{t \to \infty} \boldsymbol{p}(t) = \boldsymbol{v}_0; \tag{16.50}$$

therefore, it must hold that

$$\boldsymbol{v}_0 = \boldsymbol{\pi}, \tag{16.51}$$

which is the equilibrium-state distribution of the ergodic Markov chain $X(t)$.

Note that (16.41) implies that the stationary (or invariant) distribution $\boldsymbol{\pi}^\top$ is the left-eigenvector of $\boldsymbol{Q}$ associated with eigenvalue $\gamma_0 = 0$; thus, it confirms (16.51). Note also that the stationary distribution can be expressed as the result of the projection $\boldsymbol{E}_0$ acting on the initial distribution:

$$\boldsymbol{\pi}^\top = \boldsymbol{p}(0)^\top \boldsymbol{E}_0. \tag{16.52}$$

Example 16.2: Ion channel model – continued [62]. Consider the ion channel model of Example 16.1. The eigenvalues of the infinitesimal generator $\boldsymbol{Q}$ of (16.26) can be found from

$$\det|\boldsymbol{Q} - \gamma \boldsymbol{I}| = \det\begin{vmatrix} -\mu_0 - \gamma & \mu_0 \\ \eta\mu_1 & -\eta\mu_1 - \gamma \end{vmatrix} = 0,$$

which gives $\gamma(\gamma + \eta\mu_1 + \mu_0) = 0$. Thus, $\gamma_0 = 0$ and $\gamma_1 = -(\mu_0 + \eta\mu_1)$. The corresponding left- and right-eigenvectors are found as

$$\boldsymbol{v}_0 = \left(\frac{r}{1+r}, \frac{1}{1+r}\right)^\top = \boldsymbol{\pi}, \quad \boldsymbol{u}_0 = (1,\ 1)^\top,$$

$$\boldsymbol{v}_1 = (1,\ -1)^\top, \quad \boldsymbol{u}_1 = \frac{1}{1+r}(1, -r)^\top,$$

where

$$r = \frac{\eta\mu_1}{\mu_0}.$$

Thus, the projection matrices are

$$\boldsymbol{E}_0 = \boldsymbol{u}_0\boldsymbol{v}_0^\top = \frac{1}{1+r}\begin{bmatrix} r & 1 \\ r & 1 \end{bmatrix}, \quad \boldsymbol{E}_1 = \boldsymbol{u}_1\boldsymbol{v}_1^\top = \frac{1}{1+r}\begin{bmatrix} 1 & -1 \\ -r & r \end{bmatrix}. \tag{16.53}$$

Then, the TPM $\boldsymbol{P}(t)$ is given by

$$\boldsymbol{P}(t) = \boldsymbol{E}_0 + e^{-\mu_0(1+r)t}\boldsymbol{E}_1, \quad \text{for all } t \geq 0. \tag{16.54}$$

For small $t > 0$, we have

$$\begin{aligned} \boldsymbol{P}(t) &\approx \boldsymbol{E}_0 + [1 - \mu_0(1+r)t]\boldsymbol{E}_1 \\ &= \frac{1}{1+r}\begin{bmatrix} r + [1 - \mu_0(1+r)t] & 1 - [1 - \mu_0(1+r)t] \\ r - r[1 - \mu_0(1+r)t] & 1 + r - r[\mu_0(1+r)t] \end{bmatrix} = \boldsymbol{I} + \boldsymbol{Q}t, \end{aligned} \tag{16.55}$$

which confirms (16.28).

The time-dependent state distribution $\boldsymbol{p}(t)$ is given from (16.46) as

$$\boldsymbol{p}(t)^\top = \boldsymbol{\pi}^\top + \boldsymbol{p}(0)^\top e^{-\mu_0(1+r)t}\boldsymbol{E}_1, \tag{16.56}$$

where $\boldsymbol{p}(0)$ is the state distribution before the drug is applied, since we can write the eigenvalue $\gamma_1 = -\mu_0(r+1)$. Hence, the probability that the Markov chain is in state 0 (i.e., the channel being open) at time t is

$$p_0(t) = \frac{r}{1+r} + \frac{p_0(0)(1+r) - r}{1+r}e^{-\mu_0(r+1)t}, \quad t \geq 0,$$

where $p_0(0)$ is the probability that the channel is initially open at time $t = 0$. The probability that the channel is closed at time t is

$$p_1(t) = 1 - p_0(t) = \frac{1}{1+r} - \frac{p_0(0)(1+r) - r}{1+r} e^{-\mu_0 (r+1)t}, \quad t \geq 0.$$

□

16.3 Reversible Markov chains

The concept of a **reversible process** often arises in applications of Markov processes. A continuous-time random process $X(t)$ is said to be *reversible* if it has the same finite-dimensional distributions as the process $X(\tau - t)$; i.e., $X(t)$ and $X(\tau - t)$ are statistically identical, for any real value τ. Note that since τ is an arbitrary time-shift, a reversible process is necessarily *stationary*. The process $X(-t)$ is also called the *reversed process* of $X(t)$. Similarly, a stationary discrete-time random process X_n is *reversible* if it is statistically indistinguishable from its reversed process $\{X_{-n}\}$.

16.3.1 Reversible discrete-time Markov chain

Consider an ergodic DTMC X_n defined on a state space $\mathcal{S}$ with TPM $\boldsymbol{P} = [P_{ij}]$, $i, j \in \mathcal{S}$, and stationary distribution $\boldsymbol{\pi} = (\pi_i; i \in \mathcal{S})^\top$. The following theorem characterizes the relationship between X_n and its reversed process $\tilde{X}_n = \{X_{-n}\}$.

THEOREM 16.4 (Reversed balance equations for a DTMC). *The reversed process $\tilde{X}_n$ is an ergodic DTMC with the same stationary distribution $\boldsymbol{\pi}$ and its TPM $\tilde{\boldsymbol{P}} = [\tilde{P}_{ij}]$ satisfies the following* ***reversed balance equations****:*

$$\boxed{\pi_i \tilde{P}_{ij} = \pi_j P_{ji}, \quad i, j \in \mathcal{S}.} \tag{16.57}$$

Proof. To show that the reversed process $\tilde{X}_n$ is a Markov chain, we need to establish that

$$P[\tilde{X}_n = x_0 \mid \tilde{X}_{n-1} = x_1, \ldots, \tilde{X}_{n-m} = x_m] = P[\tilde{X}_n = x_0 \mid \tilde{X}_{n-1} = x_1], \tag{16.58}$$

for any integer $m \geq 1$ and $x_0, x_1, \ldots, x_m \in \mathcal{S}$. Applying the definition of conditional probability, the left-hand side (LHS) of (16.58) can be expressed as

$$\begin{aligned}
\text{LHS of (16.58)} &= \frac{P[\tilde{X}_n = x_0, \tilde{X}_{n-1} = x_1, \ldots, \tilde{X}_{n-m} = x_m]}{P[\tilde{X}_{n-1} = x_1, \ldots, \tilde{X}_{n-m} = x_m]} \\
&= \frac{P[X_{-n} = x_0, X_{-n+1} = x_1, \ldots, X_{-n+m} = x_m]}{P[X_{-n+1} = x_1, \ldots, X_{-n+m} = x_m]} \\
&\overset{(a)}{=} \frac{\pi_{x_0} P_{x_0 x_1} P_{x_1 x_2} \cdots P_{x_{m-1} x_m}}{\pi_{x_1} P_{x_1 x_2} \cdots P_{x_{m-1} x_m}} = \frac{\pi_{x_0} P_{x_0 x_1}}{\pi_{x_1}},
\end{aligned} \tag{16.59}$$

where step (a) follows because X_n is a homogeneous DTMC. Similarly, the right-hand side (RHS) of (16.58) can be expressed as

$$\text{RHS of (16.58)} = \frac{P[\tilde{X}_n = x_0, \tilde{X}_{n-1} = x_1]}{P[\tilde{X}_{n-1} = x_1]} = \frac{\pi_{x_0} P_{x_0 x_1}}{\pi_{x_1}}. \tag{16.60}$$

Since (16.59) and (16.60) imply (16.58), $\tilde{X}_n$ is a DTMC.

Letting $x_0 = i$ and $x_1 = j$, (16.60) leads to the reversed balance equations (16.57). From (16.57) we see that the transition probabilities, $\tilde{P}_{x_1 x_0}$, of $\{\tilde{X}_n\}$, are proportional to the transition probabilities, $P_{x_0 x_1}$, of $\{X_n\}$. Since X_n is an ergodic DTMC, so too must be the reversed chain $\tilde{X}_n$. Next, consider the jth component of the vector $\boldsymbol{\pi}^\top \tilde{\boldsymbol{P}}$, $j \in \mathcal{S}$:

$$[\boldsymbol{\pi}^\top \tilde{\boldsymbol{P}}]_j = \sum_{i \in \mathcal{S}} \pi_i \tilde{P}_{ij} \overset{(a)}{=} \sum_{i \in \mathcal{S}} \pi_j P_{ji} \overset{(b)}{=} \pi_j, \tag{16.61}$$

where (a) follows from the reversed balance equations (16.57) and (b) follows because $\boldsymbol{P}$ is a stochastic matrix. Hence, $\boldsymbol{\pi}^\top \tilde{\boldsymbol{P}} = \boldsymbol{\pi}^\top$, so $\boldsymbol{\pi}$ is the stationary distribution of the reversed chain $\tilde{X}_n$. □

The converse of Theorem 16.4 also holds (Problem 16.12).

THEOREM 16.5 (Converse of reversed balance equations for DTMC). *Let X_n be an ergodic DTMC with TPM $\boldsymbol{P}$. If we can find a TPM $\tilde{\boldsymbol{P}}$ and a probability distribution $\boldsymbol{\pi} = [\pi_i]$, $i \in \mathcal{S}$, such that reversed balance equations* (16.57) *hold, then $\tilde{\boldsymbol{P}}$ is the TPM of the reversed process $\tilde{X}_n = \{X_{-n}\}$ and $\boldsymbol{\pi}$ is the stationary distribution of both X_n and $\tilde{X}_n$.* □

Note that the reversed balance equations (16.57) hold for a DTMC X_n and its reversed chain $\tilde{X}_n$ whether or not X_n is reversible. Reversibility of X_n further requires the transition probabilities of X_n and $\tilde{X}_n$ to be equal; i.e., X_n is reversible if and only if

$$P_{ij} = \tilde{P}_{ij}, \quad i, j \in \mathcal{S}. \tag{16.62}$$

Substituting (16.62) into (16.57), we obtain the following result.

THEOREM 16.6 (Detailed balance equations for DTMC). *An ergodic DTMC X_n is reversible if and only if the following* ***detailed balance equations*** *are satisified:*

$$\boxed{\pi_i P_{ij} = \pi_j P_{ji}, \quad i, j \in \mathcal{S}.} \tag{16.63}$$

Theorems 16.5 and 16.6 imply the following result.

COROLLARY 16.1 (Converse of detailed balance equations for DTMC). *Let X_n be an ergodic DTMC with TPM $\boldsymbol{P}$. If we can find a probability distribution $\boldsymbol{\pi} = [\pi_i]$, $i \in \mathcal{S}$, such that detailed balance equations* (16.63) *hold, then X_n is reversible and $\boldsymbol{\pi}$ is its stationary distribution.*

Besides Theorem 16.6, a convenient test for the reversibility of a homogeneous Markov chain is given by Kolmogorov's criterion (Problem 16.13).

THEOREM 16.7 (Kolmogorov's criterion for reversibility of a DTMC). *An ergodic DTMC X_n is reversible if and only if its transition probabilities satisfy*

$$P_{x_1x_2}P_{x_2x_3}\cdots P_{x_{n-1}x_n}P_{x_nx_1} = P_{x_1x_n}P_{x_nx_{n-1}}\cdots P_{x_3x_2}P_{x_2x_1}$$

for any sequence of states $x_1, x_2, \ldots, x_n$ in the state space $\mathcal{S}$.

Kolmogorov's criterion essentially states that a DTMC is reversible if and only if the product of the transition probabilities along any sequence of states forming a cycle must be the same when the cycle is traversed in either of the two directions.

16.3.2 Reversible continuous-time Markov chain

Analogous to Theorem 16.4, a CTMC and its reversed process satisfy a set of reversed balance equations (Problem 16.14). Consider an ergodic CTMC $X(t)$ defined on a state space $\mathcal{S}$ with infinitesimal generator matrix $\boldsymbol{Q} = [Q_{ij}]$, $i, j \in \mathcal{S}$, and stationary distribution $\boldsymbol{\pi} = (\pi_i; i \in \mathcal{S})$.

THEOREM 16.8 (Reversed balance equations for CTMC). *The reversed process $\tilde{X}(t) = X(-t)$ is an ergodic CTMC with the same stationary distribution $\boldsymbol{\pi}$ and its infinitesimal generator $\tilde{\boldsymbol{Q}} = [\tilde{Q}_{ij}]$ satisfies the following* **reversed balance equations***:*

$$\boxed{\pi_i \tilde{Q}_{ij} = \pi_j Q_{ji}, \quad i, j \in \mathcal{S}.} \tag{16.64}$$

The converse of Theorem 16.64 also holds (Problem 16.15).

THEOREM 16.9 (Converse of reversed balance equations for CTMC). *Suppose we can find an infinitesimal generator matrix $\tilde{\boldsymbol{Q}}$ and a probability distribution $\boldsymbol{\pi} = (\pi_i; i \in \mathcal{S})^\top$ such that the reversed balance equations (16.57) hold. Then $\tilde{\boldsymbol{Q}}$ is the infinitesimal generator of the reversed process $\tilde{X}(t) = X(-t)$ and $\boldsymbol{\pi}$ is the stationary probability distribution of both $X(t)$ and $\tilde{X}(t)$.*

Conditions for reversibility of a CTMC can be stated in a similar manner to Theorems 16.6 and 16.7 for a DTMC (Problem 16.16 and 16.17).

THEOREM 16.10 (Detailed balance equations for a reversible CTMC). *An ergodic CTMC $X(t)$ is reversible if and only if its infinitesimal generator $\boldsymbol{Q} = [Q_{ij}]$, $i, j \in \mathcal{S}$, and stationary probability distribution $\boldsymbol{\pi} = (\pi_i; i \in \mathcal{S})$ satisfy the following detailed balance equations:*

$$\boxed{\pi_i Q_{ij} = \pi_j Q_{ji}, \quad i, j \in \mathcal{S}.} \tag{16.65}$$

Theorems 16.9 and 16.10 imply the following result.

COROLLARY 16.2 (Converse of detailed balance equations for CTMC). *Let X_n be an ergodic CTMC with infinitesimal generator $\boldsymbol{Q}$. If we can find a probability distribution $\boldsymbol{\pi} = [\pi_i]$, $i \in \mathcal{S}$, such that detailed balance equations* (16.65) *hold, then $X(t)$ is reversible and $\boldsymbol{\pi}$ is its stationary distribution.*

Analogous to the discrete-time case, Kolmorogov's criterion provides an alternative test for reversibility of a CTMC.

THEOREM 16.11 (Kolmogorov's criterion for a reversible CTMC). *An ergodic CTMC $X(t)$ is reversible if and only if its infinitesimal generator $\boldsymbol{Q} = [Q_{ij}]$, $i, j \in \mathcal{S}$, and stationary probability distribution $\boldsymbol{\pi} = (\pi_i; i \in \mathcal{S})$ satisfy*

$$Q_{x_1x_2}Q_{x_2x_3}\cdots Q_{x_{n-1}x_n}Q_{x_nx_1} \tag{16.66}$$

for any sequence of states $x_1, x_2, \ldots, x_n \in \mathcal{S}$.

16.4 An application: phylogenetic tree and its Markov chain representation

In the area of bioinformatics, a **phylogenetic tree** (also called an **evolutionary tree**) provides a representation of evolutionary relationships between species. As an example, the *Tree of Life Web Project* is a collaborative effort that aims at classifying all living organisms in terms of a phylogenetic tree called the *tree of life (TOL)* (see [326]). The structure of one instantiation of this tree near the root of the tree is shown in Figure 16.2. We note that viruses are not included in this tree. The tree in Figure 16.2 consists of one root node and three leaf nodes or leaves.[3] The three leaves are labeled as follows: Eubacteria, Eukaryotes, and Archaea. Each of these leaves is in turn the root node of a subtree of the TOL.

Alternative instantiations of the part of the TOL shown in Figure 16.2 are given in Figure 16.3 (a) and (b). In Figure 16.3 (a), the leaf labeled Archaea in the tree of Figure 16.2 has been split into two separate leaves labelled *Euryarchaeota* and

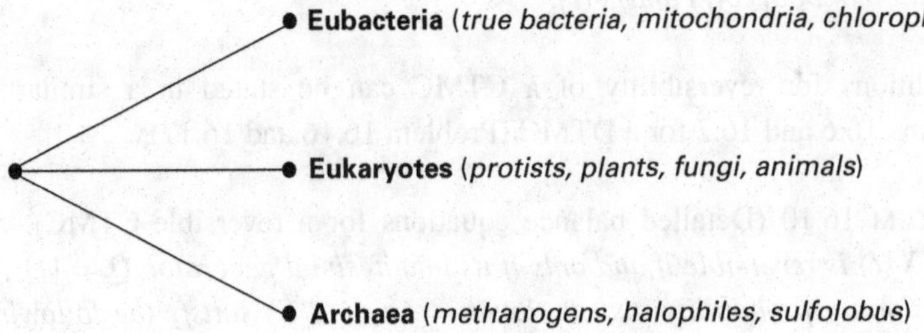

Figure 16.2 Instantiation of subtree near the root of the tree of life.

[3] Often trees are also drawn with the root at the top and leaves at the bottom.

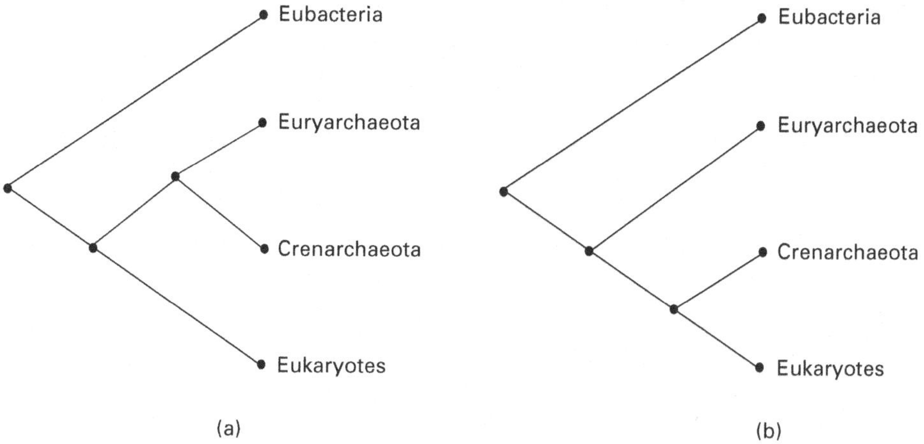

Figure 16.3 Alternative instantiations at the root of the tree of life.

Crenarchaeota (or eocytes). These leaves share a common parent node, which in turn shares a common parent with the leaf labeled Eukaryotes. Figure 16.3 (b) shows a second alternative tree structure in which the leaf labeled *Crenarchaeota* is a sibling to the leaf labelled *Eukyarotes*.

16.4.1 Trees and phylogenetic trees

We now discuss how phylogenetic trees can be modeled mathematically. The abstract structure of a **tree** can be represented as a connected graph $\mathcal{T} = (\mathcal{V}, \mathcal{E})$ with node set $\mathcal{V}$ and edge set $\mathcal{E}$ that contains no cycles; i.e., each pair of nodes is connected by a single, unique path. An edge $e \in \mathcal{E}$ consists of a pair of nodes; e.g., $e = \{i, j\}$, for some nodes $i, j, \in \mathcal{V}$. In a *directed graph*, each edge is an *ordered* pair of nodes; e.g., $e = (i, j)$. Here, the edge is said to be directed from node i to node j, which are also referred to as the *head* and *tail* of edge e.

A **rooted tree** is a tree that has one distinguished node called the **root**, which we shall label as node 0. In this case, the tree is often denoted by $\mathcal{T}^0$ and the edges are viewed as being directed away from the root. Given an edge $e = (u, v)$ in a rooted tree, node u is said to be the *parent* of node v and, conversely, node v is a *child* of node u. We denote the set of child nodes of a given node u in a rooted tree by

$$\text{ch}(u) = \{v \in \mathcal{V} : (u, v) \in \mathcal{E}\}. \tag{16.67}$$

A rooted tree imposes a natural partial ordering of the nodes in $\mathcal{T}$ as follows: for $u, v \in \mathcal{V}$, $u \leq v$ if the (directed) path from the root to node v contains node u. In this case, node u is said to be an **ancestor** of node v and, conversely, node v is said to be a **descendant** of node u. In particular, $u \leq v$ whenever (i, j) is an edge of the tree.

Any node of $\mathcal{T}$ incident on a single edge (i.e., of degree one) is called a **leaf node** or simply a **leaf**, while every other node of $\mathcal{T}$ is called an **interior node**. For a rooted tree $\mathcal{T} = (\mathcal{V}, \mathcal{E})$, we denote the set of leaf nodes of $\mathcal{T}$ by $\tilde{\mathcal{V}}$. Given a rooted tree $\mathcal{T}$

and a node $u \in \mathcal{V}$, the subtree of $\mathcal{T}$ rooted at node u, denoted by $\mathcal{T}^u$, is the tree rooted at node u consisting of all descendants of u and the associated edges. The set of nodes of the subtree $\mathcal{T}^u$ is denoted by $\mathcal{V}^u$. The set of leaf nodes of the subtree $\mathcal{T}^u$ is denoted by $\tilde{\mathcal{V}}^u$.

Formally, a **phylogenetic tree** consists of a tree $\mathcal{T}$ together with a mapping ϕ that sets up a one-to-one correspondence between a set of labels, $\mathcal{L}$, and the set of leaf nodes $\tilde{\mathcal{V}}$; i.e., $\phi : \mathcal{L} \longrightarrow \tilde{\mathcal{V}}$ is a bijective mapping. For example, the label set for the tree in Figure 16.2 is given by $\mathcal{L} = \{$Eubacteria, Eukaryotes, Archaea$\}$ and the label set of the tree in Figure 16.4 is $\{$a, b, c$\}$. The labels may represent different groups of organisms, e.g., species, in accordance with some taxonomy or classification scheme.

Phylogenetic trees are used to characterize evolutionary relationships among a set of such groups with respect to some property or characteristic. For example, a phylogenetic tree could be used to explain the evolutionary relationships among a set of species $\{$a, b, c$\}$ in terms of portions of a DNA sequence common to each species. Figure 16.4 depicts a phylogenetic tree and three DNA strands corresponding to the labels a, b, and c. In this example, each DNA strand consists of ten *sites*, where a site indicates the location of a nucleotide base within a given strand. The nucleotide bases are drawn from a set of four elements, $\mathcal{S} = \{$A, G, C, T$\}$, where the symbols A, C, G, and T represent the four DNA nucleotide bases: adenine, cytosine, guanine, and thymine respectively. If we are interested in protein sequences rather than DNA sequences, the set $\mathcal{S}$ would correspond to a set of amino acids. At this point, it is useful to distinguish between the set $\mathcal{V}$ of nodes in the tree, the set $\mathcal{L}$ of labels assigned to the leaf nodes of the tree, and the set $\mathcal{S}$ of possible characteristics that may be associated with the labels in the tree.

Referring again to Figure 16.4, we see that the three DNA strands are in alignment with respect to sites labeled from 1 to 10. At site 1, species a is associated with the nucleotide base A, species b is associated with C, and species c is associated with T. This mapping between the labels of the tree to the elements of $\mathcal{S}$ is called a *character*, which is defined formally as a one-to-one mapping $\chi : \mathcal{L} \longrightarrow \mathcal{S}$. For example, the character at site 1 is specified by

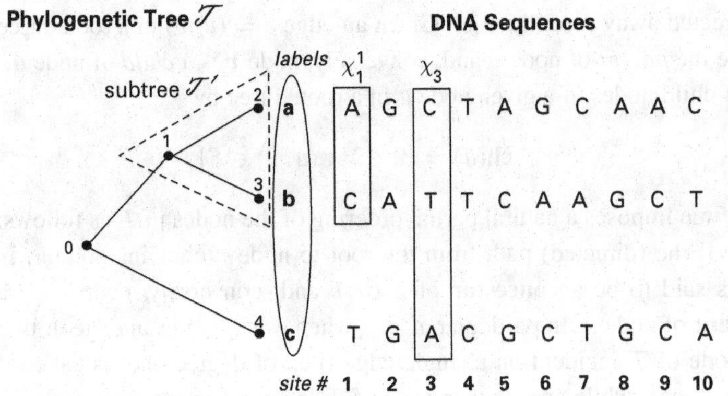

Figure 16.4 Example of a rooted phylogenetic tree.

$$\chi_1(a) = A, \ \chi_1(b) = C, \ \chi_1(c) = T,$$

whereas the character associated with site 3 is specified by

$$\chi_3(a) = C, \ \chi_3(b) = T, \ \chi_3(c) = A.$$

It is often of interest to consider the *restriction*, χ^u, of a character χ to a particular subtree of a tree. Figure 16.4 illustrates the restriction of the character χ_1 to the subtree $\mathcal{T}^1$. This restriction, denoted by χ_1^1, is given by $\chi_1^1 : \mathcal{L}^1 \to \mathcal{S}$, where $\mathcal{L}^1 = \{a, b\}$ and

$$\chi_1^1(a) = A, \ \chi_1^1(b) = C.$$

16.4.2 Markov process on a tree

A Markov process can be used to model the stochastic evolution of characters on a rooted phylogenetic tree $\mathcal{T}$.

Consider an abstract rooted tree $\mathcal{T}$. Assume a total ordering of the nodes of $\mathcal{T}$; i.e., whenever (u, v) is a (directed) edge of $\mathcal{T}$, $u \leq v$. For example, the nodes in the tree of Figure 16.4 are totally ordered by the labeling of the nodes from 0 to 4. A **Markov process on** $\mathcal{T}$ is a set, $\{X_v, \ v \in \mathcal{V}\}$, of RVs associated with the nodes $\mathcal{V}$, taking values in a state space $\mathcal{S}$ such that for any edge $e = (u, v)$ in $\mathcal{T}$

$$P[X_v = j \mid X_w; \ w \in \mathcal{V}, w \leq v] = P[X_v = j \mid X_u], \quad j \in \mathcal{S}. \tag{16.68}$$

Let $P_{ij}(e) = P[X_v = j \mid X_u = i]$. The matrix $\boldsymbol{P}(e) = [P_{ij}(e)]$, $i, j \in \mathcal{S}$ is the TPM corresponding to edge e.

In a Markov process defined on a tree, each node v has an associated (marginal) probability distribution $\pi_i(v) = P[X_v = i]$, $i \in \mathcal{S}$, which can be expressed as a probability vector $\boldsymbol{\pi}(v) = [\pi_i(v)]$, $i \in \mathcal{S}$. Note that a DTMC can be viewed as a Markov process on a rooted tree $\mathcal{T} = (\mathcal{V}, \mathcal{E})$ with root node 0, node set $\mathcal{V} = \{0, 1, 2, \ldots, v, \ldots\}$, and edge set $\mathcal{E} = \{(v, v+1) : \ v \geq 0\}$; i.e., $\mathcal{T}$ is a *path* starting from node 0.

For a Markov process defined on a *phylogenetic tree*, it is of interest to determine the probability that the variables X_v, $v \in \tilde{\mathcal{V}}$, associated with the leaf nodes of the tree assume values specified by a given character χ. This probability can be expressed as follows [296]:

$$P_\chi = P[X_{\phi(\ell)} = \chi(\ell); \ \ell \in \mathcal{L}]. \tag{16.69}$$

The probability P_χ can be interpreted as the probability that the character χ is realized by the phylogenetic tree. This probability can be computed efficiently using the sum-product algorithm (see Problem 21.4 of Chapter 21). Another quantity of interest, the *substitution probability* of an edge $e = (u, v)$, defined by

$$s(e) = P[X_u \neq X_v],$$

indicates the probability that the state changes as the tree is traversed from node u to node v.

In this section, we have been considering a Markov chain model for the stochastic evolution of a fixed site associated with an alignment of sequences, e.g., DNA sequences, with respect to a phylogenetic tree. The model can be summarized by a parameter θ, which consists of the tree $\mathcal{T}$, the stationary probability distribution of the root node, i.e., $\pi(0)$, and the TPMs $\boldsymbol{P}(e)$ assigned to each edge e of $\mathcal{T}$. To explicitly show the dependence on the model parameter, the probability that a given character χ is realized by the phylogenetic tree will be denoted by $P_{\chi;\theta}$.

If we are considering DNA sequences of length N, we can define a **character alignment** as a vector of characters across the sites: $\boldsymbol{\chi} = (\chi_s,\ 1 \le s \le N)$. For reasons of model tractability, it is commonly assumed that site evolution is independent across different sites. Hence, the probability that a character alignment $\boldsymbol{\chi}$ is realized by a Markov phylogenetic tree model parameterized by θ can be expressed as

$$P_{\boldsymbol{\chi};\theta} = \prod_{s=1}^{N} P_{\chi_s,\theta}. \tag{16.70}$$

In the language of statistics, $P_{\boldsymbol{\chi};\theta}$, is called the *likelihood* of the character χ with respect to the model parameterized by θ. In terms of likelihoods, (16.70) can be rewritten as

$$L_{\boldsymbol{\chi}}(\theta) = \prod_{s=1}^{N} L_{\chi_s}(\theta), \tag{16.71}$$

where $L_{\chi_s}(\theta)$ denotes the likelihood of the character χ_s for site s and $L_{\boldsymbol{\chi}}(\theta)$ represents the likelihood of the character alignment $\boldsymbol{\chi}$ across all sites.

16.4.3 Continuous-time Markov chain model

The Markov process model of a tree defined in the previous section consists of a set of node variables, $\{X_v;\ v \in \mathcal{V}\}$, whose stochastic properties are determined by TPMs $\boldsymbol{P}(e) = [P_{uv}(e)]$, $e = (u, v) \in \mathcal{E}$. This model is similar to a nonhomogeneous DTMC model, except that the RVs are indexed by the nodes of $\mathcal{V}$ rather than by time. To capture the element of time in a phylogenetic tree, the evolution of the node variables can be characterized by a CTMC.

For each edge $e = (u, v) \in \mathcal{E}$, we assume that node variable X_u evolves to the node variable X_v in accordance with a CTMC with infinitestimal generator matrix $\boldsymbol{Q}(e)$ over a time period of length $\tau(e)$ with *intensity* $\rho(e)$, which is a dimensionless quantity. The TPM $\boldsymbol{P}(e)$ and the generator matrix $\boldsymbol{Q}(e)$ are related by

$$\boldsymbol{P}(e) = e^{\boldsymbol{Q}(e)\rho(e)\tau(e)}. \tag{16.72}$$

Figure 16.5 illustrates the process of evolution from the state $X_u = \text{A}$ to the state $X_v = \text{T}$ over the time period $\tau(e)$, where $e = (u, v)$. We see that three so-called **mutation** or **substitution** events occur over the time $\tau(e)$: A → C, C → A, A → T. The first two

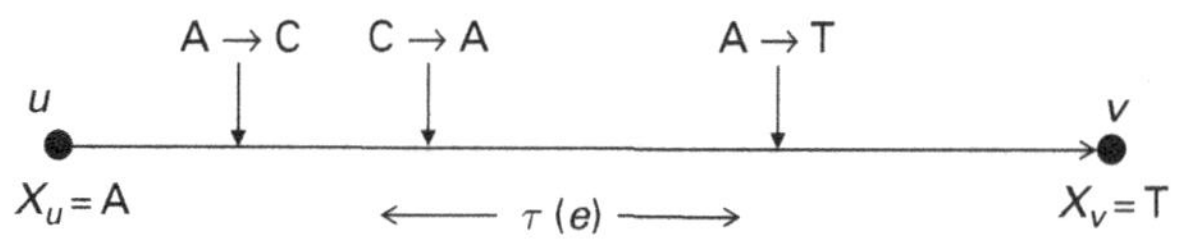

Figure 16.5 Substitution events along an edge $e = (u, v)$ of a phylogenetic tree.

substitutions are said to be *silent*, since we only observe the start and end states; i.e., $X_u = \text{A}$ and $X_v = \text{T}$ respectively.

For reasons of tractability, it is often assumed that the CTMC is homogeneous, such that the Markov process on the tree $\mathcal{T}$ is stationary. More precisely, a Markov process on a rooted phylogenetic tree $\mathcal{T}$ is said to be **stationary** with probability distribution vector $\boldsymbol{\pi}$ if there exists an infinitesimal generator matrix $\boldsymbol{Q}$ such that (i) $\boldsymbol{\pi} = \boldsymbol{\pi}(0)$, where 0 denotes the root node, (ii) $\boldsymbol{\pi}^\top \boldsymbol{Q} = \mathbf{0}$, and (iii) (16.72) holds for all edges e.

A stationary Markov process on a phylogenetic tree $\mathcal{T}$ is said to be *reversible* if the following detailed balance equations hold (cf. (16.65)):

$$\pi_i Q_{ij} = \pi_j Q_{ji}, \quad i, j \in \mathcal{S}. \tag{16.73}$$

Define the diagonal matrix $\boldsymbol{\Pi} = \text{diag}\{\pi_i, \ i \in \mathcal{C}\}$. Then (16.73) is equivalent to requiring that the matrix $\boldsymbol{\Pi Q}$ be symmetric (Problem 16.21). The assumption of reversibility is convenient because, in this case, changing the root node of $\mathcal{T}$ does not affect the resulting TPMs $\boldsymbol{P}(e)$, $e \in \mathcal{E}$. To see this, note that, for an edge $e = (u, v)$,

$$P_{ij}(e) = P[X_u = i \mid X_v = j] \overset{(a)}{=} \frac{P[X_v = j \mid X_u = i]P[X_u = i]}{P[X_v = j]} = \frac{P_{ij}((u, v))\pi_i}{\pi_j}$$
$$\overset{(b)}{=} \frac{P_{ji}((v, u))\pi_j}{\pi_j} = P_{ji}((v, u)), \tag{16.74}$$

where Bayes' theorem is applied in step (a) and step (b) follows from reversibility of the CTMC model (Problem 16.21).

For a reversible stationary process on $\mathcal{T}$, the mean number of substitutions that occur on an edge $e \in \mathcal{E}$ can be expressed as

$$\kappa(e) = -\sum_{i \in \mathcal{S}} \pi_i Q_{ii} \rho(e) \tau(e) = \text{Tr}\{\boldsymbol{\Pi Q}\} \rho(e) \tau(e). \tag{16.75}$$

Consider an arbitrary pair of nodes u, v in the tree. Since a tree does not contain cycles, the nodes are connected by a unique path $u = u_1, u_2, \ldots, u_m = v$. The **evolutionary distance** between u and v is then defined by

$$\kappa((u, v)) = \sum_{n=1}^{m-1} \kappa((u_n, u_{n+1})), \tag{16.76}$$

which gives the mean number of substitutions that occur on the path between nodes u and v. If we further assume that the intensity $\rho(e)$ is constant, then the rate of substitutions per unit time is also constant over all edges.

Example 16.3: Felsensteins's F81 model [41, 101]. Assume that the state set is $\mathcal{C} = \{\mathrm{A, C, G, T}\}$. Felsenstein's F81 model consists of a CTMC with infinitesimal generator matrix given by

$$\boldsymbol{Q} = \begin{bmatrix} -(\pi_\mathrm{C}+\pi_\mathrm{G}+\pi_\mathrm{T}) & \pi_\mathrm{C} & \pi_\mathrm{G} & \pi_\mathrm{T} \\ \pi_\mathrm{A} & -(\pi_\mathrm{A}+\pi_\mathrm{G}+\pi_\mathrm{T}) & \pi_\mathrm{G} & \pi_\mathrm{T} \\ \pi_\mathrm{A} & \pi_\mathrm{C} & -(\pi_\mathrm{A}+\pi_\mathrm{C}+\pi_\mathrm{T}) & \pi_\mathrm{T} \\ \pi_\mathrm{A} & \pi_\mathrm{C} & \pi_\mathrm{G} & -(\pi_\mathrm{A}+\pi_\mathrm{C}+\pi_\mathrm{G}) \end{bmatrix}, \tag{16.77}$$

where the states are ordered alphabetically and π_A, π_C, π_G, and π_T are probabilities that sum to one. One can show that the stationary probability distribution is given by (Problem 16.19)

$$\boldsymbol{\pi} = (\pi_\mathrm{A}, \pi_\mathrm{C}, \pi_\mathrm{G}, \pi_\mathrm{T})^\top. \tag{16.78}$$

Furthermore, the Markov process model specified by $\boldsymbol{Q}$ is reversible (Problem 16.22). □

16.5 Summary of Chapter 16

Semi-Markov process:	$\boldsymbol{P} = [P_{ij}],\ F_{ij}(s),\ i, j \in \mathcal{S},\ s \geq 0$	Def.16.1
Counting process:	$N(t) = \max\{n : t_n \leq t\}$	(16.4)
SMP $X(t)$:	$X(t) = X_{N(t)},\ t \geq 0$	(16.5)
Markov renewal process:	$\{(X_n, t_n)\}$	
Distributions of SMP and EMC:	$\pi_i = \frac{\tilde{\pi}_i E[S_i]}{\sum_{j\in\mathcal{S}} \tilde{\pi}_j E[S_j]},\ i \in \mathcal{S}$	(16.10)
TPMF of a CTMC:	$\boldsymbol{P}(t) = [P_{ij}(t)]$	
where:	$P_{ij}(t) = P[X(s+t) = j \mid X(s) = i]$	(16.18)
Chapman–Kolmogorov eq. for a CTMC:	$\boldsymbol{P}(s+t) = \boldsymbol{P}(s)\boldsymbol{P}(t)$	(16.21)
Infinitesimal generator:	$\boldsymbol{Q} = \left.\frac{d\boldsymbol{P}(t)}{dt}\right\rvert_{t=0}$	(16.22)
Kolmogorov's forward eq.:	$\frac{d\boldsymbol{P}(t)}{dt} = \boldsymbol{P}(t)\boldsymbol{Q}$	(16.34)
The solution:	$\boldsymbol{P}(t) = e^{\boldsymbol{Q}t}$	(16.35)
Kolmogorov's backward eq.:	$\frac{d\boldsymbol{P}(t)}{dt} = \boldsymbol{Q}\boldsymbol{P}(t)$	(16.38)
Balance eq.:	$\boldsymbol{\pi}^\top \boldsymbol{Q} = \boldsymbol{0}^\top$	(16.41)
Reversed balance eqs (DTMC):	$\pi_i \tilde{P}_{ij} = \pi_j P_{ji},\ i, j \in \mathcal{S}$	(16.57)
Detailed balance eqs (DTMC):	$\pi_i P_{ij} = \pi_j P_{ji},\ i, j \in \mathcal{S}$	(16.63)
Reversed balance eqs (CTMC):	$\pi_i \tilde{Q}_{ij} = \pi_j Q_{ji},\ i, j \in \mathcal{S}$	(16.64)
Detailed balance eqs (CTMC):	$\pi_i Q_{ij} = \pi_j Q_{ji},\ i, j \in \mathcal{S}$	(16.65)

16.6 Discussion and further reading

A CTMC is a Markov process, in which the time parameter is continuous and the state space is discrete or finite. In effect, the CTMC is the continuous-time counterpart of the DTMC treated in the previous chapter. An SMP generalizes the DTMC by introducing random, continuous-valued sojourn times in each state. As such, an SMP has sample paths that are piecewise-constant and right-continuous. An SMP $X(t)$ jumps at time t_n according to a DTMC $X_n = \{X(t_n)\}$ specified by a transition probability matrix $\boldsymbol{P}$. The sojourn time, $\tau_n = t_{n+1} - t_n$, in the state of $X(t)$ attained at the nth jump point depends only on the values of the current state $X_n = j$ and the next state $X_{n+1} = i$ and has distribution function $F_{ij}(t)$. While the process $X(t)$ does not satisfy the Markov property in general, the pair process $\{(X_n, \tau_n)\}$ is a Markov process and $\{(X_n, t_n)\}$ is commonly referred to as a **Markov renewal process**. If the sojourn time distribution function $F_{ij}(t)$ is exponentially distributed and does not depend on j, then $X(t)$ satisfies the Markov property and, hence, is a CTMC.

Many physical processes are inherently continuous time and are often most accurately modeled by a CTMC or SMP, as opposed to a DTMC. In some applications, a process is time-sampled and then approximated by a DTMC with some loss of accuracy. In this case, a continuous-time model is often preferred provided the model is tractable, since time sampling is avoided. For reasons of tractability, a CTMC is preferred over an SMP model, but accuracy may be sacrificed in a CTMC-based model by assuming exponential sojourn times between states.

The requirement of reversibility imposes a certain structure on a Markov process, which often leads to a more tractable model in applications. A related concept in queueing theory, called **quasi-reversibility**, was introduced by Kelly [177] to characterize a class of queues having a product-form stationary distribution. Quasi-reversibility is also discussed in the books by Wolff [357] and Kobayashi and Mark [203].

In this chapter we discussed an example application of a CTMC to modeling ion current channels in biology. For further study of ion channel modeling, the reader is referred to papers by Colquhoun and Hawkes [62–64] and Ball *et al.* [12, 13]. We also devoted a subsection to Markovian modeling of phylogenetic or evolutionary trees. Bryant *et al.* [41] provide a survey of likelihood calculation methods in phylogenetics based on a Markov model. We remark that a phylogenetic tree is a special case of a more general structure known as a **phylogenetic network**, which is represented by a general directed graph. Such models are closely related to Bayesian networks pioneered by Pearl [264] and probability trees used to study causality, e.g., Shafer [298]. Bayesian networks and Pearl's belief propagation algorithm are discussed in Section 21.5. More comprehensive treatments of phylogenetics can be found in books by Felsenstein [102] and Semple and Steel [296].

In-depth treatments of Markov renewal theory and SMP can be found in papers by Çinlar [56, 58] and in his textbook [57]. For a further mathematical study of Markov chains, the reader is referred to, for example, Chung [52], Kijima [180], and Rogers and Williams [282] in addition to the books on stochastic processes cited above. It goes without saying that the theory of Markov processes and applications originated in Markov's work; thus, it is appropriate to record his papers and a review article [14, 239, 240].

16.7 Problems

Section 16.1: Semi-Markov process

16.1 Alternating renewal process. Consider a machine which is either "up" or running, or is "down" and gets repaired. The random process U_n represents a sequence of intervals that the machine is up, and D_n represents a sequence of down periods. Assume that the machine starts at $t = 0$, so this machine's behavior can be characterized by $U_1, D_1, U_2, D_2, \ldots$. Let the U_n be i.i.d. RVs with the distribution function $F_U(u)$ and D_n be i.i.d. RVs with the distribution function $F_D(d)$.

(a) Show that we can define an SMP that characterizes this machine's state as a function of time.

(b) The state process for the machine can be defined by $I(t)$, where

$$I(t) = \begin{cases} 1, & \text{if the machine is up at time } t, \\ 0, & \text{otherwise.} \end{cases} \tag{16.79}$$

What is the relation between the process $I(t)$ and the counting process $N(t)$ defined in (16.4)?

(c) What is the probability that the machine is up and running?

16.2* Conditional independence of sojourn times. Show that for any integer $n \geq 1$ and numbers $u_1, \ldots, u_n \geq 0$

$$\begin{aligned} P[\tau_1 \leq u_1, \tau_2 \leq u_2, \ldots, \tau_j \leq u_n \mid X_0, X_1, \ldots] \\ = F_{X_0,X_1}(u_1) F_{X_1,X_2}(u_2) \cdots F_{X_{n-1},X_n}(u_n); \end{aligned}$$

i.e., the sojourn times $\tau_1, \tau_2, \ldots$ are conditionally independent given the Markov chain $\{X_n\}$.

16.3* Semi-Markovian kernel [58]. An SMP $X(t)$ or the associated Markov renewal process (X_n, t_n) can be specified in terms of the family of probabilities

$$Q_{ij}(t) = P[X_{n+1} = j, t_{n+1} - t_n \leq t \mid X_n = i], \quad i, j \in \mathcal{S}, t \geq 0,$$

which is called a *semi-Markovian transition kernel.* Determine the relationship between the semi-Markovian kernel $\boldsymbol{Q}(t) = [Q_{ij}(t)]$, $i, j \in \mathcal{S}$, and the parameters of an SMP given in Definition 16.1: $\boldsymbol{P} = [P_{ij}]$ and $\boldsymbol{F}(t) = [F_{ij}(t)]$, $i, j \in \mathcal{S}$.

16.4 Markov renewal process and renewal process [58]. Consider a Markov renewal process (X_n, t_n).

(a) Let $i \in \mathcal{S}$ be a fixed state and let $S_0(i), S_1(i), \ldots$ be the successive t_n for which $X_n = i$. Show that $\{S_n(i)\}$ forms a (possibly delayed) renewal process (see

Section 14.3). Hence, argue that a renewal process is a special case of a Markov renewal process.

(b) Define

$$Q_{ij}^n(t) = P[X_n = j, t_n \le t \mid X_0 = i].$$

Show that

$$Q_{ij}^0(t) = \delta_{ij}, \tag{16.80}$$

$$Q_{ik}^{n+1}(t) = \sum_{j \in \mathcal{S}} \int_0^t Q_{ij}(u) Q_{jk}^n(t-u)du. \tag{16.81}$$

(c) Let $M_{ij}(t)$ denote the mean number of renewal points t_n for which $X_n = j$ in $(0, t]$, given that $X_0 = i$. The functions $M_{ij}(t)$, $i, j \in \mathcal{S}$, are called **Markov renewal functions** and are analogous to the renewal function of renewal processes (see Section 14.3.1). Show that

$$M_{ij}(t) = \sum_{n=0}^{\infty} Q_{ij}^n(t).$$

Section 16.2: Continuous-time Markov chain (CTMC)

16.5* Markovian property of an SMP. Consider an SMP $X(t)$ with sojourn time distribution functions satisfying (16.17). Show that $X(t)$ is a Markov process.

16.6* CTMC as an SMP. Consider a CTMC $X(t)$ defined on a state space $\mathcal{S}$ with infinitesimal generator $\boldsymbol{Q} = [Q_{ij}]$, $i, j \in \mathcal{S}$. Characterize $X(t)$ as an SMP according to Definition 16.1.

16.7 Semi-Markovian kernel of a CTMC. Consider a CTMC $X(t)$ defined on a state space $\mathcal{S}$ with infinitesimal generator $\boldsymbol{Q} = [Q_{ij}]$, $i, j \in \mathcal{S}$. Determine the semi-Markovian kernel $\boldsymbol{Q}(t) = [Q_{ij}(t)]$ of $X(t)$, as defined in Problem 16.3.

16.8 Nonuniform Markov chain [345]. Show that the series in (16.39) may not converge if $\boldsymbol{Q}$ is not uniform.
Hint: Let $\mathcal{S} = \{0, 1, 2, \ldots\}$ and $Q_{0,j} = j^{-2}$, $Q_{i,0} = i$, for all $i, j \ge 1$. Show that $\boldsymbol{Q}^2$ is not finite.

16.9 Invariant (or stationary) distribution of a CTMC. Show that the invariant distribution of a CTMC with generator matrix $\boldsymbol{Q}$ is the solution of (16.41).

16.10* Balance equations. Derive the global balance equation (16.43).

16.11 Markov-modulated Poisson process. A Markov-modulated Poisson process (MMPP) is a nonhomogeneous Poisson process $N(t)$ whose rate is itself a random process *modulated* by a finite-state CTMC $X(t)$ with state space $\mathcal{S} = \{1, \ldots, r\}$ such that the rate of the Poisson process is λ_i when $X(t) = i$, $i = 1, \ldots, n$. The λ_i are assumed to be distinct and nonnegative. The infinitesimal generator matrix is denoted by $\boldsymbol{Q} = [Q_{ij}], i, j \in \mathcal{S}$.

(a) Let t_n denote the time of the nth arrival event of the Markov-modulated Poisson process and let $X_n = X(t_n)$. Show that (X_n, t_n) is a Markov renewal process.

(b) Let $\boldsymbol{\Lambda} = \text{diag}(\lambda_1, \ldots, \lambda_r)$. Let $\boldsymbol{Q}(t) = [Q_{ij}(t)],\ i, j \in \mathcal{S}$ denote the semi-Markovian kernel for (X_n, t_n). Define the **semi-Markovian kernel density** $\boldsymbol{D}(t) = [D_{ij}(t)], i, j \in \mathcal{S}$, by

$$\boldsymbol{D}(t) = \frac{d\boldsymbol{Q}(t)}{dt}; \tag{16.82}$$

i.e., $D_{ij}(t)$ is the *probability density* that a transition of X_n occurs from state i to state j in time t. Show that $D_{ij}(t)$ satisfies the following equation:

$$D_{ij}(t) = \lambda_i e^{-\lambda_i t} e^{Q_{ii} t} + \int_0^t e^{-\lambda_i s} \sum_{k \neq i} Q_{ik} D_{kj}(t-s) ds, \quad i, j \in \mathcal{S}. \tag{16.83}$$

(c) By differentiating (16.83) with respect to t, show that the semi-Markovian kernel density is given by

$$\boldsymbol{D}(t) = \exp[(\boldsymbol{Q} - \boldsymbol{\Lambda})t]\boldsymbol{\Lambda}. \tag{16.84}$$

(d) A Markov-modulated Poisson process for which $r = 2$ is sometimes called a **switched Poisson process**. If, in addition, $\lambda_1 = 0$, the Markov-modulated Poisson process is called an **interrupted Poisson process (IPP)** (e.g., see [203]). Show that the interrupted Poisson process is a renewal process.

Section 16.3: Reversible Markov chains

16.12* Converse of reversed balance equations for DTMC. Prove Theorem 16.5.

16.13 Kolmogorov's criterion for DTMC. Show that Theorem 16.7 follows from Theorem 16.6.

16.14* Reversed balance equations for CTMC. Prove Theorem 16.8 using the following steps.

(a) For an arbitrary integer $m \geq 1$, let $t_0 < t_1 < \cdots < t_m$ be arbitrary time points and let $x_0, x_1, \ldots, x_0 \in \mathcal{S}$ be arbitrary points in the state space. Show that

$$\begin{aligned} &P[\tilde{X}(t_m) = x_0 \mid \tilde{X}(t_{m-1}) = x_1, \tilde{X}(t_{m-2}) = x_2, \ldots, \tilde{X}(t_0) = x_m] \\ &= P[\tilde{X}(t_m) = x_0 \mid \tilde{X}(t_{m-1}) = x_1] = \frac{\pi_{x_0} P_{x_0 x_1}(t_m - t_{m-1})}{\pi_{x_1}}, \end{aligned} \tag{16.85}$$

where $P_{x_0 x_1}(t)$ is the transition probability function for $X(t)$. Hence, argue that $\tilde{X}(t)$ is an ergodic CTMC with transition probability functions given by

$$\tilde{P}_{ij}(t) = \frac{\pi_j P_{ji}(t)}{\pi_i}, \quad i, j \in \mathcal{S}. \tag{16.86}$$

(b) Using the formula $\tilde{Q}_{ij} = \frac{d\tilde{P}_{ij}}{dt}|_{t=0}$, show that the reversed balance equations (16.64) hold.

16.15 Converse of reversed Balance equations for CTMC. Prove Theorem 16.9.

16.16* Detailed balance equations for CTMC. Prove Theorem 16.10.

16.17 Kolmogorov's criterion for CTMC. Show that Theorem 16.7 follows from Theorem 16.6.

16.18 Reversibility of a BD process. Consider the BD process $N(t)$ discussed in Section 14.2, with generator matrix $\boldsymbol{Q}$ defined by (14.47). Show that $\boldsymbol{Q}$ and the stationary distribution $\boldsymbol{\pi}$ satisfy the detailed balance equations (16.65). Hence, $N(t)$ is a reversible process.

Section 16.4: An application: phylogenetic tree and its Markov chain representation

16.19 Stationary probability distribution of F81 model. Show that the stationary probability vector of the F81 model in Example 16.3 is given by (16.78).

16.20 Transition probabilities of F81 model. Show that the transition probabilities of the F81 model with respect to an edge $e \in \mathcal{E}$ satisfy

$$P_{cd}(e) = \begin{cases} \pi_d + (1 - \pi_d)e^{-\mu(e)}, & \text{if } c = d, \\ \pi_d(1 - e^{-\mu(e)}), & \text{if } c \neq d\,, \end{cases} \tag{16.87}$$

where $c, d \in \mathcal{S}$.
Hint: Apply (16.72) and diagonalize $\boldsymbol{Q}$.

16.21* Reversibility and detailed balance equations.

(a) Show that the detailed balance equations (16.73) hold if and only if the matrix $\boldsymbol{\Pi}\,\boldsymbol{Q}$ is symmetric.
(b) Show that a result analogous to that of part (a) holds for a DTMC with TPM $\boldsymbol{P}$ and stationary probability vector $\boldsymbol{\pi}$.
(c) Show that if a CTMC defined on a state space $\mathcal{C}$ with infinitesimal generator matrix $\boldsymbol{Q}$ and stationary distribution vector $\boldsymbol{\pi}$ is reversible, then

$$\pi_i P_{ij}(\tau) = \pi_j P_{ji}(\tau), \quad i, j \in \mathcal{S},$$

for any $\tau > 0$.

16.22 Reversibility of F81 model. Show that the F81 model in Example 16.3 is reversible.

16.23* Numerical example of F81 model. Consider the F81 model with the following parameter values:

$$\pi_A = 0.1, \; \pi_C = 0.2, \; \pi_G = 0.2, \; \pi_T = 0.5. \tag{16.88}$$

With respect to the phylogenetic tree illustrated in Figure 16.4, assume that the intensities $\lambda(e) = 1$ for each $e \in \mathcal{E}$. Further assume that the time duration associated with each edge is unity; i.e., $\tau(e) = 1$, $e \in \mathcal{E}$.

(a) Calculate the TPM $\boldsymbol{P}(e)$ associated with each edge $e \in \mathcal{E}$.

(b) Calculate the mean number of substitutions $\kappa(e)$ associated with each edge of the tree.

(c) Calculate the likelihood of the character χ_3.

17 Random walk, Brownian motion, diffusion, and Itô processes

In this chapter we discuss three related topics that find important applications in many science and engineering fields. They are random walks, Brownian motion, and diffusion processes.

17.1 Random walk

A **random walk** model appears in the context of many real-world problems, such as a gambling problem, the motion of a particle, the price change in the Dow Jones index, and the dynamic change in network traffic.

Let us imagine that we make a one-dimensional random walk on the real line: we start at some initial position X_0 on the x-axis at time $t = 0$. At $t = 1$, we jump to position X_1, where the step size $S_1 = X_1 - X_0$ is a random variable with some distribution[1] $F(s)$. At time $t = 2$, we jump by another amount S_2, where S_2 is independent of S_1, but has the same distribution $F(s)$. The process continues and our position after n jumps, or at time $t = n$, is thus given by[2]

$$X_n = X_0 + S_1 + S_2 + \cdots + S_n, \tag{17.1}$$

where $\{S_i\}$ is a set of i.i.d. RVs with the common distribution $F(s)$. This discrete time sequence $\{X_n\}$ is called a one-dimensional random walk. A random walk process is a **martingale** (see Definition 10.5).

17.1.1 Simple random walk

A **simple random walk** is defined as a special case of the random walk model, in which only two values are possible for each step S_i; i.e., either $+1$ or -1. The position at time $t = n$ is thus given by

$$X_n = X_0 + \sum_{i=1}^{n} S_i, \quad n = 1, 2, 3, \ldots. \tag{17.2}$$

[1] In this chapter we drop the subscript S of $F_S(s)$, etc. when the omission does not cause any confusion.

[2] In the literature, the symbols X_n and S_n are often switched, where S_n signifies the "summed variable" instead of the "step-size variable" as we define here.

The simple random walk $\{X_n\}$ has the following properties [131]:

1. **Spatial homogeneity**:
$$P[X_n = k|X_0 = a] = P[X_n = k + b|X_0 = a + b]; \tag{17.3}$$
that is, the distribution of $X_n - X_0$ does not depend on the initial value X_0.
2. **Temporal homogeneity**:
$$P[X_n = k|X_0 = a] = P[X_{n+m} = k|X_m = a]; \tag{17.4}$$
that is, $X_{n+m} - X_m$ has the same distribution as $X_n - X_0$ for all $m, n \geq 0$.
3. **Independent increments**: for a set of disjoint intervals $(m_i, n_i]$, $i = 1, 2, \ldots$, the increments $(X_{n_i} - X_{m_i})$ are independent.
4. **Markov property**: the sequence $\{X_n\}$ is a simple Markov chain:
$$P[X_{n+m} = k|X_0, X_1, \ldots, X_n] = P[X_{n+m} = k|X_n], \text{ for any } m \geq 0. \tag{17.5}$$

The proof for each of the above properties is rather straightforward; hence, this is left to the reader as an exercise (see Problem 17.2).

Because the simple random walk is spatially homogeneous, let us assume without loss of generality that
$$X_0 = a = 0.$$

Suppose that, out of n random steps, n_1 steps are taken to the right $(+1)$ and n_2 steps are to the left (-1). These steps are independent, and let us assume the following probabilities:
$$S_i = \begin{cases} +1 & \text{with probability } p, \\ -1 & \text{with probability } q = 1 - p. \end{cases} \tag{17.6}$$

Let the position after the n steps be $X_n = n_1 - n_2 \triangleq k$. Since $n_1 + n_2 = n$, we readily find $n_1 = (n+k)/2$ and $n_2 = (n-k)/2$. Then,
$$\boxed{P[X_n = k] = \binom{n}{\frac{n+k}{2}} p^{(n+k)/2} q^{(n-k)/2}, \quad k = -n, -n+2, \ldots, n-2, n.} \tag{17.7}$$

Note that both $(n+k)$ and $(n-k)$ are even. Thus, for any state k, $P[X_n = k] = 0$ for all n such that $(n+k)$ is odd. Therefore, $\{X_n\}$ is a Markov chain with period $d = 2$.

An alternative way to derive the probability distribution (17.7) is to solve the following difference equations that $p_n(k) \triangleq P[X_n = k]$ must satisfy (Problem 17.1):
$$p_{n+1}(k) = pp_n(k-1) + qp_n(k+1), \tag{17.8}$$
$$p_0(k) = \delta_{k,0}. \tag{17.9}$$

The mean, second moment, and variance of X_n are

$$E[X_n] = \sum_{i=1}^{n} E[S_i] = nE[S_i] = n(p-q), \tag{17.10}$$

$$\begin{aligned} E[X_n^2] &= \sum_{i=1}^{n} E[S_i^2] + \sum_{i \neq j} \sum_{j} E[S_i]E[S_j] = n + (n^2 - n)(p-q)^2 \\ &= 4pqn + n^2(p-q)^2, \end{aligned} \tag{17.11}$$

$$\text{Var}[X_n] = E[X_n^2] - (E[X_n])^2 = 4pqn. \tag{17.12}$$

As stated in property 3, the simple random walk is a process with independent increments. The mean of the increment $X_m - X_n$ is

$$E[X_m - X_n] = (m-n)(p-q). \tag{17.13}$$

The autocorrelation function $R_X(m,n) = E[X_m X_n]$ for $m \geq n$ is obtained as

$$\begin{aligned} R_X(m,n) &= E[(X_m - X_n + X_n)X_n] = E[X_m - X_n]E[X_n] + E[X_n^2] \\ &= (m-n)n(p-q)^2 + n^2(p-q)^2 + 4pqn \\ &= mn(p-q)^2 + 4pqn, \quad m \geq n. \end{aligned} \tag{17.14}$$

Since the autocorrelation function is symmetric, we obtain the general expression whether $m \geq n$ or not:

$$R_X(m,n) = mn(p-q)^2 + 4pq \min\{m,n\}. \tag{17.15}$$

Thus, the covariance between X_m and X_n is

$$\text{Cov}[X_m, X_n] = R_X(m,n) - E[X_m]E[X_n] = 4pq \min\{m,n\} \tag{17.16}$$

and the variance of the increment is

$$\begin{aligned} \text{Var}[X_m - X_n] &= \text{Var}[X_m] + \text{Var}[X_n] - 2\text{Cov}[X_m, X_n] \\ &= 4pq(m+n-2n) = 4pq(m-n), \quad \text{for } m \geq n. \end{aligned} \tag{17.17}$$

Example 17.1: Simple random walk with $p = q = 1/2$. Consider the case in which the probability of our making one step to the right (i.e., $S_i = +1$) is the same as that of jumping to the left ($S_i = -1$); i.e., $p = q = \frac{1}{2}$. Then our expected position after n steps is still our starting position; i.e.,

$$E[X_n] = 0, \; n = 0, 1, 2, \ldots, \tag{17.18}$$

but the variance of our position is

$$\text{Var}[X_n] = n, \; n = 0, 1, 2, \ldots. \tag{17.19}$$

Thus, the standard deviation of X_n is $\sqrt{n}$. This property of the random walk is often referred to as the **square-root law**. The autocorrelation function, which is also the autocovariance function, is found by setting $p = q = \frac{1}{2}$ in (17.14) or (17.16), as

$$\boxed{R_X(m, n) = \min\{m, n\}.} \tag{17.20}$$

The mean and variance of the increment $X_m - X_n$, $m \geq n$, are given, from (17.13) and (17.17), as

$$E[X_m - X_n] = 0 \text{ and } \mathrm{Var}[X_m - X_n] = m - n. \tag{17.21}$$

□

17.1.2 Gambler's ruin

Consider a game in which A plays against B. Suppose that in each trial A wins with probability p, and gets one [dollar] from B, and A loses with probability $q = 1 - p$ and gives one [dollar] to B. The game ends when either A or B has lost all their fortune; i.e., gets ruined.

17.1.2.1 Opponent with infinite capital

Let us assume that A has the initial capital of a dollars and B is infinitely rich. This is a good model when A is a gambler and B is the house at a casino. Let X_n represent A's fortune after n trials. Then the process $\{X_n\}$ can be represented by a one-dimensional random walk similar to the simple random walk model of (17.2); the only difference is that it is restricted to $X_n \geq 0$. Once A's fortune becomes zero, this game (i.e., series of trials) ends. This random walk model and its variants have been classically known as "gambler ruin problems" (e.g., see Feller [99], Chapter 14). Figure 17.1 shows four sample paths of the sequence of A's fortune, assuming their initial capital is $a = 5$ [dollars] and $p = 0.48$ in all four cases.

The random walk can be viewed as a Markov chain with the state space $\mathcal{S} = \{0, 1, 2, \ldots\}$ with state 0 being an absorbing state. Suppose $X_n = i$ and let

$$r_i = P[X_{n'} = 0 \text{ for some } n' > n | X_n = i];$$

i.e., the probability that A, with their fortune i dollars at some point in the game, will **ultimately get ruined**.[3] After one more trial, A will possess $i + 1$ or $i - 1$ [dollars], with probability p and q respectively. Therefore, we have the following difference equation for r_i:

$$\boxed{r_i = p r_{i+1} + q r_{i-1}, \quad i = 1, 2, \ldots,} \tag{17.22}$$

[3] To be consistent with the notation we adopt in (15.59) of Section 15.3, r_i should be written as $f_{i,0} = \sum_{n=0}^{\infty} f_{i,0}^{(n)}$.

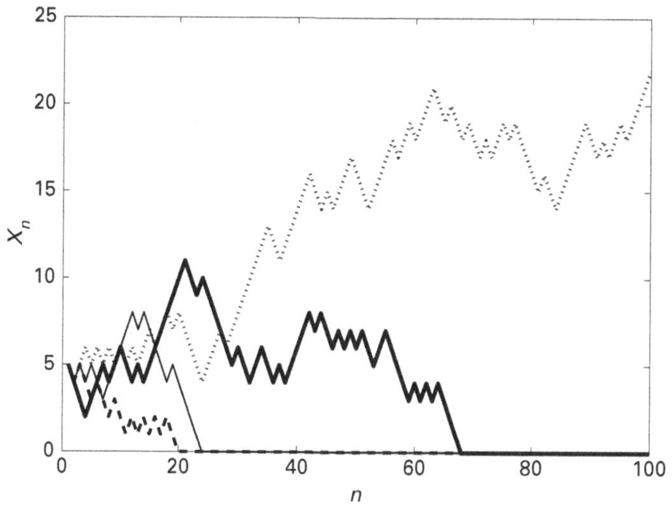

Figure 17.1 Player A's fortune X_n after n trials of the game, with $X_0 = a = 5$ and $p = 0.48$.

with the boundary condition

$$r_0 = 1. \tag{17.23}$$

In order to find r_i for arbitrary $i (> 0)$, define the generating function

$$R(z) = \sum_{i=0}^{\infty} r_i z^i. \tag{17.24}$$

By multiplying both sides of (17.22) by z^i and summing over $i = 1$ to infinity, we have

$$R(z) - 1 = pz^{-1} (R(z) - r_1 z - 1) + qzR(z), \tag{17.25}$$

from which we have

$$R(z)(z-1)(p - qz) = (1 - pr_1)z - p. \tag{17.26}$$

By setting $z = 1$ in the above equation, we find

$$r_1 = \frac{q}{p} \triangleq \gamma. \tag{17.27}$$

Then, we find that the right-hand side of (17.26) becomes $(z - 1)p$, obtaining

$$R(z) = \frac{1}{1 - \gamma z} = \sum_{i=0}^{\infty} \gamma^i z^i. \tag{17.28}$$

Thus,

$$\boxed{r_i = \gamma^i, \quad i = 0, 1, 2, \ldots,} \tag{17.29}$$

where $\gamma = q/p$ is defined in (17.27).

So, player A with initial capital a [dollars] will ultimately be ruined with probability $r_a = \gamma^a$, while with probability $1 - \gamma^a$ A's fortune will increase, in the long run, without bound.

If $p < q$, A's opponent B (i.e., the house) is decidedly in an advantageous position, and will ruin the gambler with probability one. Even if the individual trial is fair, i.e., $p = q = 1/2$, the gambler will be ultimately ruined.

17.1.2.2 Opponent with finite capital

Let us now consider the case where the opponent B also starts with a finite capital b [dollars]. Let

$$c = a + b = \text{sum of the capitals of } A \text{ and } B.$$

Then we may model this game again as a random walk X_n, representing A's fortune after n games. So $(c - X_n)$ is equal to B's fortune after n trials, and X_n is now restricted to $\{0, 1, 2, \ldots, c\}$.

Define r_i again as the probability that A gets ultimately ruined after having i [dollars] at some point in the game. Then this game can be characterized by the same difference equation (17.22), but the boundary condition (17.23) should be replaced by

$$r_0 = 1 \text{ and } r_c = 0. \tag{17.30}$$

Thus, the process $\{X_n\}$ is a finite-state Markov chain with the state space $\mathcal{S} = \{0, 1, 2, \ldots, c\}$, where states 0 and c are absorbing states. The solution is now given (Problem 17.3) as follows:

for $\gamma \neq 1$ (i.e., $p \neq q$),

$$\boxed{r_i = \frac{\gamma^c - \gamma^i}{\gamma^c - 1}, \quad 0 \leq i \leq c;} \tag{17.31}$$

for $\gamma = 1$ (i.e., $p = q$),

$$r_i = 1 - \frac{i}{c} = \frac{c - i}{c}, \quad 0 \leq i \leq c. \tag{17.32}$$

The probability w_i that A ultimately **wins** the game equals the probability that B gets ruined. The latter is obtained from the above formulas by replacing p, q, and i by q, p, and $c - i$ respectively. Thus:

for $\gamma \neq 1$ (i.e., $p \neq q$),

$$w_i = \frac{1 - \gamma^i}{1 - \gamma^c}, \quad 0 \leq i \leq c; \tag{17.33}$$

for $\gamma = 1$ (i.e., $p = q = 1/2$),

$$w_i = 1 - \frac{c - i}{c} = \frac{i}{c}, \quad 0 \leq i \leq c. \tag{17.34}$$

Then it is apparent that

$$r_i + w_i = 1,$$

so an unending game does not exist.

From the last result we find that when the game is fair, i.e., $p = q = 1/2$, the probability of getting ruined is proportional to the initial capital of the opponent; i.e.,

$$P[A \text{ gets ruined}] = r_a = \frac{b}{a+b}, \tag{17.35}$$

$$P[B \text{ gets ruined}] = w_a = \frac{a}{a+b}. \tag{17.36}$$

In other words, the probability of ultimately winning the game is proportional to their own initial capital.

By letting $c \to \infty$ in (17.31) and (17.32), we find

$$\lim_{c\to\infty} r_i = \begin{cases} 1 & \text{if } \gamma \geq 1 \text{ (i.e., } q \geq p), \\ \gamma^i & \text{if } \gamma < 1 \text{ (i.e., } q < p). \end{cases} \tag{17.37}$$

These limits are probabilities of the gambler's ruin in a game against an infinitely rich adversary as discussed in the first case: $b = \infty$ implies $c = a + b = \infty$.

17.2 Brownian motion or Wiener process

Now let us derive a "continuous" random walk model as a limit case of the simple random walk discussed in the previous section.

17.2.1 Wiener process as a limit process

Assume that jumps of the random walk take place at time epochs $h, 2h, 3h, \ldots$, where the interval h is taken to be very small. The step size of each jump is δ; i.e., the binary RV S_i of (17.2) is now modified to

$$S_i = \begin{cases} +\delta, & \text{with probability } \frac{1}{2}, \\ -\delta, & \text{with probability } \frac{1}{2}. \end{cases}$$

For given time $t > 0$, there will be n jumps, where

$$n = \lfloor t/h \rfloor. \tag{17.38}$$

From (17.18) and (17.19) we see that the new RV X_n has mean zero and variance $n\delta^2$. Furthermore, the central limit theorem (CLT) shows that X_n is asymptotically normally distributed; i.e., $X_n \sim N(0, n\delta^2)$. Thus, as we will see below, the continuous analog of simple random walk becomes a Gaussian process in addition to possessing properties 1–4 of $\{S_n\}$ listed in (17.3) through (17.5).

As stated above, taking the limit $h \to 0$ and $\delta \to 0$ will transform the random walk process $\{X_n\}$ into a process that is continuous in both time and displacement. But if we let h and δ approach zero in an arbitrary manner, we cannot expect to have a sensible limit process. Unless we keep h and δ in an appropriate ratio, the process $\{X_n\}$ will degenerate to the limit with its variance tending to zero or infinity. For instance, if we let $\delta/h \to 0$, no motion would result, because the maximum possible displacement in time

t is given by $(t/h)\delta \to 0$. To find the proper ratio, we note that the total displacement in time t is the sum of $n = \lfloor t/h \rfloor$ independent RVs, each having the mean $(p-q)\delta$ and variance $[1-(p-q)^2]\delta^2 = 4pq\delta^2$. Thus, the mean and variance of the total displacement during $(0, t]$ are $t(p-q)\delta/h$ and $4pqt\delta^2/h$ respectively. The finiteness of the variance requires that δ^2/h should remain bounded. Thus, we let $h \to 0$ and $\delta \to 0$, as $n \to \infty$, in such a way that

$$\frac{\delta^2}{h} \to \alpha, \tag{17.39}$$

for some constant $\alpha > 0$.

In Figure 17.2 we plot $X_{\lfloor t/h \rfloor}$ versus t for two cases: (a) $h = 1, \delta = 1$ (i.e., the original simple random walk) and (b) $h = 0.09, \delta = 0.3$. The variance rate parameter is $\alpha = 1$ in both cases.

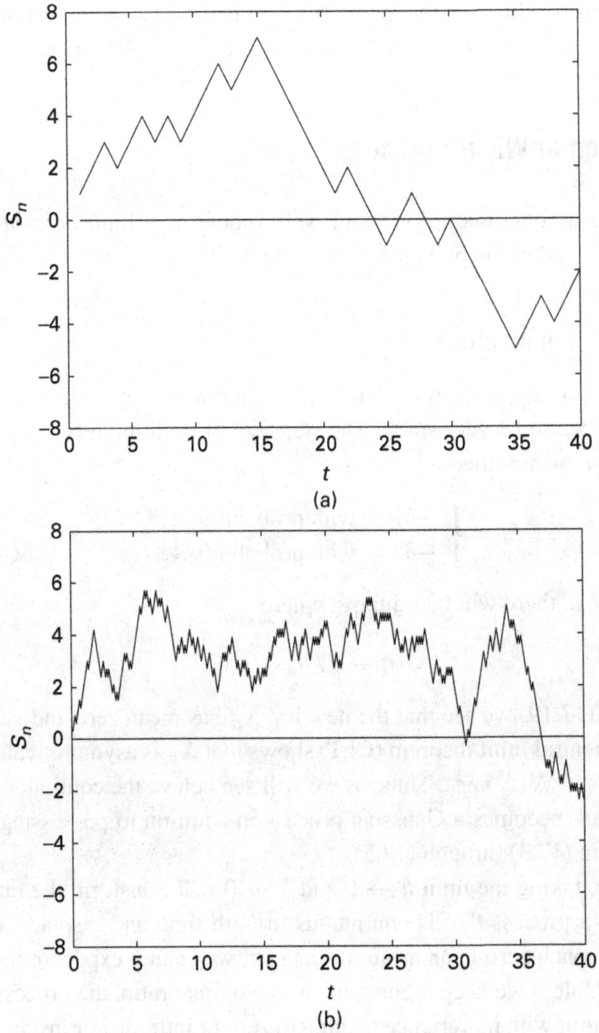

Figure 17.2 Two sample paths of the simple random walk X_n versus t, where $n = \lfloor t/h \rfloor$: (a) $h = 1, \delta = 1$; (b) $h = 0.09, \delta = 0.3$. ($\alpha = 1$ in both cases).

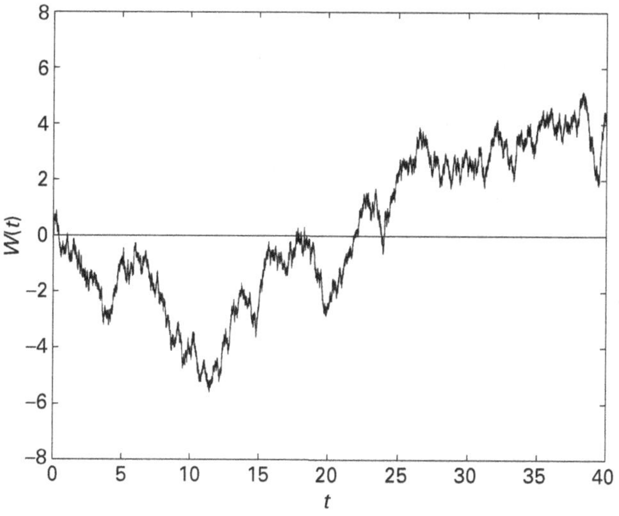

Figure 17.3 Standard Brownian motion or the Wiener process with $\alpha = 1$.

We represent the above limit of the random walk X_n as a function of the continuous time t using the scaling (17.38) and write it as $W(t)$; i.e.,

$$\boxed{\lim_{n\to\infty,h,\delta\to 0} X_n = W(t), \quad \text{while } t = nh \text{ and } \frac{\delta^2}{h} = \alpha.} \tag{17.40}$$

The distribution of the limit process $W(t)$ approaches $N(0, n\delta^2) \to N(0, \alpha nh) \to N(0, \alpha t)$. In Figure 17.3 we show a typical sample path of the limit process $W(t)$, which was obtained by setting $h = 0.009$ and $\delta = 0.03$ in the simple random walk model with α kept to one, as in Figure 17.2.

This limit process is called **Brownian motion** (originally called Brownian movement), in deference to an early nineteenth century botanist, Robert Brown, who studied in 1827 the irregular motion of pollen particles suspended in water. Brownian motion was mathematically formulated independently by a French mathematician Louis Bachelier (1900) and Albert Einstein (1905), and was further investigated mathematically by Norbert Wiener (1923), A. N. Kolmogorov (1931), William Feller (1936), Itô (1950) and others. Brownian motion is often synonymously called a **Wiener process** and, thus, denoted as $B(t)$ or $W(t)$; we adopt the latter notation in this chapter. Because of Bachelier's pioneering work, Feller calls it the Bachelier–Wiener process. When $\alpha = 1$, the process is called the **standard Brownian motion**, which we denote by $W_s(t)$.

17.2.2 Properties of the Wiener process

The process $W(t)$ thus obtained as the limit process of a simple random walk satisfies the four properties of simple random walk plus the Gaussian property:

1. **Spatial homogeneity**:

$$P[W(t) \le w|W(0) = a] = P[W(t) \le w + b|W(0) = a + b]. \tag{17.41}$$

2. **Temporal homogeneity**:

$$P[W(t) \le w|W(0) = a] = P[W(t + s) \le w|W(s) = a]; \tag{17.42}$$

namely, $W(t+s) - W(s)$ has the same distribution as $W(t) - W(0)$ for all $s, t \ge 0$.

3. **Independent increments**: for a set of disjoint intervals $(s_i, t_i]$, $i = 1, 2, \ldots,$ the increments $W(t_i) - W(s_i)$, $i = 1, 2, \ldots,$ are independent.
4. **Markov property**: the process $\{W(t)\}$ is a Markov process:

$$P[W(t + s) \le w|W(u),\ u \le t] = P[W(t + s) \le w|W(t)], \quad \text{for any } s \ge 0. \tag{17.43}$$

5. **Gaussian property**: any increment $W(t) - W(t_0)$ $(t > t_0)$ is normally distributed:

$$\boxed{P[W(t) \le x|W(t_0) = x_0] = \frac{1}{\sqrt{2\pi\alpha(t - t_0)}} \int_{-\infty}^{x - x_0} \exp\left\{-\frac{y^2}{2\alpha(t - t_0)}\right\} dy.} \tag{17.44}$$

From properties 2 and 3 we can show that Var$[W(t)]$ must be of the form Var$[W(t)] = \alpha t$ with some constant α (Problem 17.5 (a)). Furthermore, we can show, for any $t, s \ge 0$, that

$$\text{Var}[W(t) - W(s)] = \alpha|t - s| \tag{17.45}$$

(Problem 17.5 (b)). Since $W(t)$ has zero mean, the autocovariance function is the same as the autocorrelation function. Letting $s \le t$, we have

$$\begin{aligned} R_W(t, s) &= E[W(t)W(s)] = E\left[W(s)\{W(t) - W(s) + W(s)\}\right] \\ &= E[W(s)\{W(t) - W(s)\}] + E[W(s)W(s)] \\ &= E[W^2(s)] = \alpha s \ \text{ for } \ s \le t, \end{aligned} \tag{17.46}$$

where we used the property that $W(s)$ and $W(t) - W(s)$ are independent and have zero mean. Therefore, we obtain

$$\boxed{R_W(t, s) = \alpha \min\{s, t\}, \quad s, t > 0.} \tag{17.47}$$

By equating $t = s + h$ in (17.45) and letting $h \to 0$, we have

$$\lim_{h \to 0} E\left[\{W(s + h) - W(s)\}^2\right] = \lim_{h \to 0} \alpha|h| = 0. \tag{17.48}$$

Therefore, the Wiener process is continuous in the mean-square sense. Furthermore, it can be shown (e.g., [82]) that the sample function of the Wiener process is, with probability one, continuous; that is, sample functions of the Wiener process are almost surely continuous.

17.2.3 White noise

It is often convenient to consider a random process $Z(t)$ with a constant spectral density[4] $N_0/2$:

$$P_Z(f) = \frac{N_0}{2}, \quad -\infty < f < \infty. \tag{17.49}$$

Such a process is called **white noise**. In Figure 17.4 (a) we show a sample path of the "standard" white noise with $\alpha = N_0/2 = 1$.

White noise is not physically realizable, since its mean-square value (power) would be infinite:

$$R_Z(0) = E[Z(t)^2] = \frac{N_0}{2}\int_{-\infty}^{\infty} df = \infty. \tag{17.50}$$

The notion of white noise is found useful, however, in systems analysis, when the noise spectral density is practically flat within the range that includes the signal bandwidth. The autocorrelation function of white noise is given by the impulse function:

$$R_Z(\tau) = \int_{-\infty}^{\infty} P_Z(f)e^{i2\pi f\tau}\, df = \frac{N_0}{2}\delta(\tau), \quad -\infty < \tau < \infty. \tag{17.51}$$

It is clear that the Fourier transform of (17.51) gives the flat spectrum (17.49).

The Wiener process or Brownian motion can be formally represented as a **stochastic integral**[5] of white noise process $Z(t)$. Figure 17.4 (b) shows a sample path of the standard Brownian motion as an integration of white noise; i.e.,

$$\boxed{W(t) = \int_0^t Z(u)\, du,} \tag{17.52}$$

where

$$E[Z(t)Z(s)] = R_Z(t-s) = \frac{N_0}{2}\delta(t-s). \tag{17.53}$$

Hence, the autocorelation function of the Wiener process can be calculated as

[4] Use of $N_0/2$ instead of N_0 is because of the double-sided frequency domain in the Fourier transform. Noise power in the "physical" frequency band $[0, B]$ [hertz] is given by $P_N = N_0 B$ [watts] (or [joule·hertz]). The corresponding "mathematical" frequency band in the Fourier analysis is $[-B, B]$; thus, $\int_{-B}^{B} \frac{N_0}{2}\, df = N_0 B$.

[5] A random process $X(t)$ is said to be stochastically integrable if the limit

$$\int_a^b X(t)\, dt = \lim_{\Delta t_i \to 0} \sum_i X(t_i)\Delta t_i$$

converges in the mean-square sense.

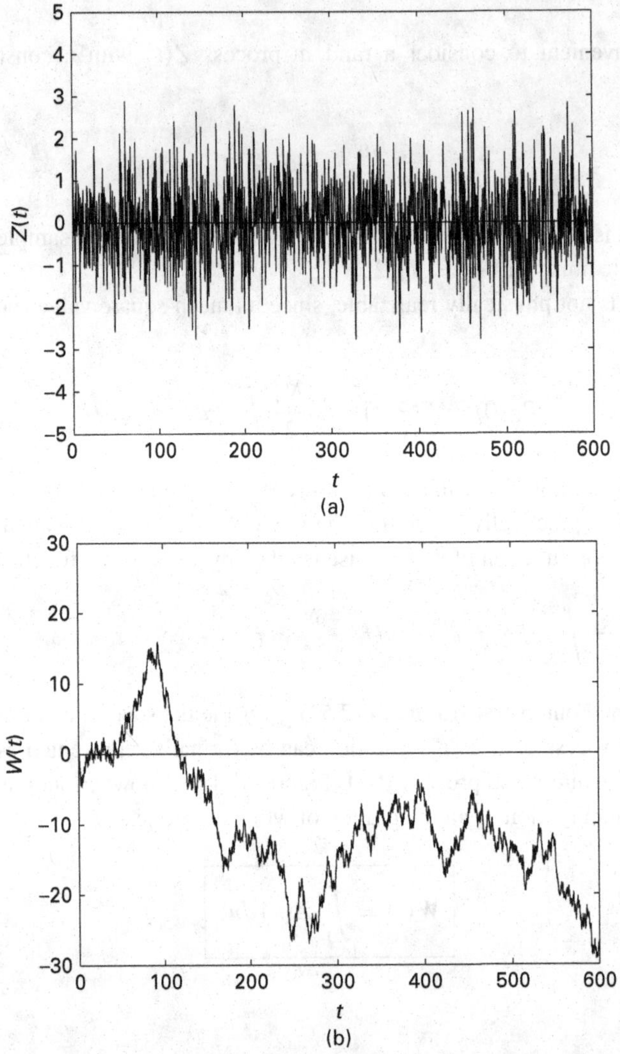

Figure 17.4 (a) Standard white noise $Z_s(t)$ with $\alpha = N_0/2 = 1$; (b) standard Brownian motion $W(t)$.

$$R_W(t, s) = E[W(t)W(s)] = E\left[\int_0^t Z(u)\,du \int_0^s Z(v)\,dv\right]$$
$$= \int_0^t du \int_0^s dv E[Z(u)Z(v)] = \int_0^t du \int_0^s dv \frac{N_0}{2}\delta(u-v)$$
$$= \frac{N_0}{2}\min\{t, s\}, \quad s, t > 0, \tag{17.54}$$

which agrees with (17.47) if we set

$$\frac{N_0}{2} \triangleq \alpha. \tag{17.55}$$

17.2.3.1 Thermal noise in a band-limited system

In communication systems, the dominating noise in a received signal is often thermal noise that is introduced at the receiver front. Thermal noise is well approximated by white noise with spectral density $\alpha = N_0/2$, where $N_0 = k_B T$; k_B is Boltzmann's constant[6] and T is temperature of the thermal noise source in kelvin. If the bandwidth of a signal is B [hertz], then the filter at the receiver should block the noise spectrum outside $[-B, B]$.

The filtered noise output $Z_{out}(t)$ is stationary but nonwhite. The output noise power spectrum is

$$P_{Z_{out}}(f) = \begin{cases} \frac{N_0}{2}, & |f| \le B, \\ 0, & |f| > B, \end{cases} \tag{17.56}$$

and its autocorrelation function is

$$R_{Z_{out}}(\tau) = \frac{N_0}{2} \int_{-W}^{W} e^{i2\pi f \tau}\, df = N_0 W \frac{\sin 2\pi W\tau}{2\pi W\tau}, \tag{17.57}$$

for $-\infty < \tau < \infty$.

Let the receiver input be

$$X(t) = s(t) + Z(t), \tag{17.58}$$

where $s(t)$ is a known deterministic signal and $Z(t)$ is white noise. Then the *matched filter* that yields the maximum SNR at time $t_0 = 0$ is (see Section 13.2.2) given by

$$h_{opt}(t) = s^*(-t) \text{ and } H_{opt}(f) = S^*(f), \tag{17.59}$$

where $S(f) = \mathcal{F}\{s(t)\}$. Then the output noise

$$Z_{out}(t) = \int_{\infty}^{\infty} h_{opt}(u) Z(t-u)\, du \tag{17.60}$$

is stationary, but is nonwhite. The output noise power is

$$\begin{aligned} \text{Var}[Z_{out}] &= \frac{N_0}{2} \int_{-\infty}^{\infty} |H_{opt}(f)|^2\, df = \frac{N_0}{2} \int_{-\infty}^{\infty} |S(f)|^2\, df \\ &= \frac{N_0}{2} \int_0^T |s(t)|^2\, dt = \frac{N_0 E_s}{2}, \end{aligned} \tag{17.61}$$

where $[0, T]$ is the signal duration interval and $E_s = \|s(t)\|^2$ is the signal energy:

$$E_s \triangleq \int_0^T |s(t)|^2\, dt = \int_{-\infty}^{\infty} |S(f)|^2\, df. \tag{17.62}$$

[6] $k_B = 1.3807 \times 10^{-23}$ [joules/kelvin]. Ludwig Boltzmann (1844–1906) derived the relation $S = k_B \ln g$, where S is the entropy of a system and g is the number of ways in which the system's microstates can be arranged. It also defines the relation between the kinetic energy contained in each molecule of an ideal gas. The Boltzmann constant is equal to the ratio of the gas constant to the Avogadro constant.

If we adopt an equivalent *correlation receiver* technique (cf. Section 13.2.2) the output noise power is

$$\begin{aligned}\mathrm{Var}[\tilde{Z}_{\mathrm{out}}] &= E\left[\int_0^T Z(t)s(t)\,dt \int_0^T Z^*(u)s^*(u)\,du\right] \\ &= \frac{N_0}{2}\int_0^T\int_0^T \delta(t-u)s(t)s^*(u)\,dt\,du \\ &= \frac{N_0}{2}\int_0^T |s(t)|^2\,dt = \frac{N_0 E_s}{2},\end{aligned} \tag{17.63}$$

which is, as expected, the same as (17.61). This noise analysis confirms the equivalence of the two techniques.

If, in particular, the signal waveform is a rectangular pulse

$$s(t) = 1 \ \text{ for } 0 \le t \le T,$$

then the output noise of the correlation receiver is the Wiener process evaluated at $t = T$:

$$\boxed{\tilde{Z}_{\mathrm{out}} = \int_0^T Z(t)\,dt = W(T).}$$

17.3 Diffusion processes and diffusion equations

A **diffusion process** is defined as a **continuous-time Markov process** with a **continuous sample path**, and is given as a solution to a stochastic differential equation, called a diffusion equation. Brownian motion discussed in the previous section is a diffusion process. In this section we will introduce a general class of diffusion processes.

17.3.1 Fokker–Planck equation for Brownian motion with drift

Let us return to Brownian motion or the Wiener process $W(t)$ defined as the limit case of the simple random walk $\{X_n\}$, but this time without imposing the symmetric condition $p = q = 1/2$. Recall that the expectation and variance of X_n are

$$E[X_n] = (p-q)\delta n \ \text{ and } \ \mathrm{Var}[X_n] = 4pq\delta^2 n. \tag{17.64}$$

By denoting by h the time interval between two random steps, we consider the limit of the random walk W_n, analogous to (17.40):

$$\boxed{\lim_{n\to\infty,h,\delta\to 0} X_n = X(t), \quad \text{while } t = nh \text{ and } \frac{4pq\delta^2}{h} = \alpha,} \tag{17.65}$$

while we keep $nh = t$ as we did in deriving the Wiener process (see (17.38)). Unlike the symmetric random walk, however, $E[X_n]$ is not zero, as implied by the first expression in (17.64). There is a **drift** proportional to $nh = t$. Thus, from (17.64) we define the drift rate and variance rate, and denote them by β and α:

$$\boxed{\begin{aligned}\beta &\triangleq \frac{E[X_n]}{nh} = \frac{(p-q)\delta}{h},\\ \alpha &\triangleq \frac{\mathrm{Var}[X_n]}{nh} = \frac{4pq\delta^2}{h}.\end{aligned}} \tag{17.66}$$

If the random walk is symmetric, i.e., $p = q = 1/2$, then $\beta = 0$ and $\alpha = \delta^2/h$, as defined for a Wiener process. In general, the sign of β determines the direction of the drift; hence, β is called the **drift coefficient**, whereas α is called the **diffusion coefficient**. From (17.66) and $p + q = 1$, we find

$$p = \frac{1}{2}\left(1 + \frac{\beta h}{\delta}\right), \quad q = \frac{1}{2}\left(1 - \frac{\beta h}{\delta}\right). \tag{17.67}$$

From this and (17.66) we have

$$\alpha = \left(1 - \frac{\beta^2 h^2}{\delta^2}\right)\frac{\delta^2}{h} \to \frac{\delta^2}{h}, \tag{17.68}$$

which is consistent with the limit condition (17.39).

From (17.66) we readily find for the limit process $X(t)$ that

$$E[X(t)] = \lim_{n\to\infty, h,\delta\to 0} E[X_n] = \beta t, \tag{17.69}$$

$$\mathrm{Var}[X(t)] = \lim_{n\to\infty, h,\delta\to 0} \mathrm{Var}[X_n] = \alpha t. \tag{17.70}$$

Because $\{X_n\}$ is asymptotically normally distributed, the limit process $X(t)$ is also normally distributed:

$$F(x, t|x_0, t_0) = P[X(t) \le x|X(0) = 0] = \frac{1}{\sqrt{2\pi}}\int_{-\infty}^{y} e^{-u^2/2}\,du, \tag{17.71}$$

where

$$y = \frac{x - x_0 - \beta(t - t_0)}{\sqrt{\alpha(t - t_0)}}. \tag{17.72}$$

The limit process $X(t)$ thus defined is called **Brownian motion with drift**.

Now we write the limit form of the difference equation (17.8) as

$$f(x, t + h) = pf(x - \delta, t) + qf(x + \delta, t). \tag{17.73}$$

Applying the Taylor series expansions

$$f(x, t + h) = f(x, t) + \frac{\partial f(x, t)}{\partial t}h + o(h), \tag{17.74}$$

$$f(x - \delta, t) = f(x, t) - \delta\frac{\partial f(x, t)}{\partial x} + \frac{\delta^2}{2}\frac{\partial^2 f(x, t)}{\partial x^2} + o(\delta^2), \text{ etc.,} \tag{17.75}$$

we obtain

$$h\frac{\partial f(x,t)}{\partial t} = (q-p)\delta\frac{\partial f(x,t)}{\partial x} + \frac{\delta^2}{2}\frac{\partial^2 f(x,t)}{\partial x^2} + \cdots . \tag{17.76}$$

Using the definitions (17.66) and (17.68), we have in the limit $h, \delta \to 0$

$$\boxed{\frac{\partial f(x,t)}{\partial t} = -\beta\frac{\partial f(x,t)}{\partial x} + \frac{\alpha}{2}\frac{\partial^2 f(x,t)}{\partial x^2}.} \tag{17.77}$$

This partial differential equation is called the **Fokker–Planck** equation or **Kolmogorov's forward equation.**

No equilibrium-state distribution exists (Problem 17.9 (a)), but we can show (Problem 17.9 (b)) the following time-dependent PDF of (17.71) satisfies (17.77):

$$f(x,t) = \frac{1}{\sqrt{2\pi\alpha t}}e^{-(x-\beta t)^2/(2\alpha t)}. \tag{17.78}$$

We define the conditional distribution function and its density function:

$$F(x,t|x_0,t_0) = P[X(t) \le x|X(t_0) = x_0], \tag{17.79}$$

$$f(t,t|x_0,t_0) = \frac{\partial}{\partial x}F(x,t|x_0,t_0). \tag{17.80}$$

Then, the **Chapman–Kolmogorov equation** takes the following expression for the conditional PDF:

$$f(x,t|x_0,t_0) = \int_{-\infty}^{\infty} f(x,t|x_1,t_1)f(x_1,t_1|x_0,t_0)\,dx_1. \tag{17.81}$$

Analogous to (16.40), we also have

$$f(x,t) = \int_{-\infty}^{\infty} f(x,t|x_0,t_0)f(x_0,t_0)\,dx_0. \tag{17.82}$$

It is not difficult to see that the forward diffusion equation (17.77) applies to the conditional PDF (and the conditional distribution function as well):

$$\boxed{\frac{\partial f(x,t|x_0,t_0)}{\partial t} = -\beta\frac{\partial f(x,t|x_0,t_0)}{\partial x} + \frac{\alpha}{2}\frac{\partial^2 f(x,t|x_0,t_0)}{\partial x^2}.} \tag{17.83}$$

The solution is apparent from (17.78):

$$f(x,t|x_0,t_0) = \frac{1}{\sqrt{2\pi\alpha(t-t_0)}}\exp\left\{-\frac{[x-x_0-\beta(t-t_0)]^2}{2\alpha(t-t_0)}\right\}. \tag{17.84}$$

This Brownian motion with drift can be formally expressed by the following **stochastic differential equation** that involves the **standard Brownian motion** $W_s(t)$ that has the distribution $N(0,t)$:

$$\boxed{dX(t) = \beta\,dt + \sqrt{\alpha}\,dW_s(t),} \tag{17.85}$$

with $E[(dW_s(t))^2] = dt$. By assuming the initial values $X(0) = 0$ and $W_s(0) = 0$, the above equation leads to

$$X(t) = \beta t + \sqrt{\alpha} W_s(t). \tag{17.86}$$

17.3.2 Einstein's diffusion equation for Brownian motion

For Brownian motion in which $\beta = 0$, the above diffusion equation equation becomes

$$\frac{\partial f(x,t)}{\partial t} = \frac{\alpha}{2} \frac{\partial^2 f(x,t)}{\partial x^2}. \tag{17.87}$$

Einstein [86, 88] showed that macroscopically the position of the particle $\boldsymbol{X}(t) = (X_1(t), X_2(t), X_3(t))$ should satisfy the following differential equation:

$$\xi \frac{d\boldsymbol{X}(t)}{dt} = \boldsymbol{F}(t), \tag{17.88}$$

where $\boldsymbol{F}(t)$ is the collision force and ξ is the coefficient of friction. The collision force vector process $\boldsymbol{F}(t)$ can be represented as a (three-dimensional) white Gaussian noise process with spectral density $2k_B T \xi$, where $k_B = 1.37 \times 10^{-23}$ [joules/degree] is the Boltzmann constant and T is the temperature in kelvin. Thus, by denoting three-dimensional white Gaussian noise with unit power spectral density as $\boldsymbol{Z}_s(t)$, we have the following **stochastic differential equation**:

$$d\boldsymbol{X}(t) = \frac{\sqrt{2k_B T \xi}}{\xi} \boldsymbol{Z}_s(t)\,dt = \sqrt{D}\boldsymbol{Z}_s(t)\,dt, \tag{17.89}$$

where

$$D = \frac{2k_B T}{\xi} \tag{17.90}$$

is equivalent to the diffusion coefficient α (except for a factor of two) and is sometimes called the "diffusivity." Einstein furthermore showed the concentration density $n(\boldsymbol{x}, t)$ of a large number of Brownian particles should satisfy the following partial differential equation:

$$\frac{\partial n(\boldsymbol{x},t)}{\partial t} = D \sum_{j=1}^{3} \frac{\partial^2 n(\boldsymbol{x},t)}{\partial x_j^2}. \tag{17.91}$$

In the one-dimensional case, Einstein's diffusion equation is equivalent to (17.87), since the density function $n(x_1, t)$ is proportional to the PDF $f_{X_1}(x_1, t)$, and by equating $D = \alpha/2$.

By proper scaling, we may set $D = \frac{1}{2}$, obtaining the partial differential equation

$$\frac{\partial f(\boldsymbol{x},t)}{\partial t} = \frac{1}{2} \sum_{i=1}^{3} \frac{\partial^2 f(\boldsymbol{x},t)}{\partial x_i^2}, \tag{17.92}$$

which is known as the **heat equation** and its solution is given by

$$f(\boldsymbol{x}, t) = \frac{1}{(2\pi t)^{3/2}} \exp\left(-\frac{\|\boldsymbol{x}\|^2}{2t}\right), \quad \text{where } \|\boldsymbol{x}\|^2 = \sum_{i=1}^{3} x_i^2, \tag{17.93}$$

which corresponds to setting $\beta = 0$ and $\alpha = 1$ in the three-dimensional version of (17.78).

17.3.3 Forward and backward equations for general diffusion processes

We now investigate diffusion processes in general. A diffusion process, as stated earlier, is a **continuous-time, continuous-state Markov process**. Consider the **conditional mean** of $X(t)$ given $X(t_0) = x_0$:

$$\mu(t|x_0, t_0) = E[X(t)|X(t_0) = x_0] = \int_{-\infty}^{\infty} x f(x, t|x_0, t_0)\, dx \tag{17.94}$$

and the **conditional variance**

$$\sigma^2(t|x_0, t_0) = \mathrm{Var}[X(t)|X(t_0) = x_0] = \int_{-\infty}^{\infty} [x - \mu(t|x_0, t_0)]^2 f(x, t|x_0, t_0)\, dx. \tag{17.95}$$

Clearly,

$$\lim_{t \to t_0} \mu(t|x_0, t_0) = x_0 \text{ and } \lim_{t \to t_0} \sigma^2(t|x_0, t_0) = 0 \text{ for any } x_0. \tag{17.96}$$

Assuming that these functions are differentiable from the right, we define the slopes of these functions at $t = t_0$:

$$\boxed{\begin{aligned} \beta(x_0, t_0) &\triangleq \left.\frac{\partial \mu(t|x_0, t_0)}{\partial t}\right|_{t=t_0}, \\ \alpha(x_0, t_0) &\triangleq \left.\frac{\partial \sigma^2(t|x_0, t_0)}{\partial t}\right|_{t=t_0}, \end{aligned}}$$

which are called the **drift rate** and **variance rate** respectively.

By writing

$$X(t + dt) - X(t) = dX(t),$$

alternative definitions of $\beta(x, t)$ and $\alpha(x, t)$ are given as

$$\beta(x, t)\, dt = E[dX(t)|X(t) = x], \tag{17.97}$$

$$\begin{aligned} \alpha(x, t)\, dt &= \mathrm{Var}[dX(t)|X(t) = x] \\ &= E[(dX(t) - \beta(x, t)dt)^2|X(t) = x]. \end{aligned} \tag{17.98}$$

Then, by generalizing the result of Fokker–Planck equation (17.77) and (17.87), we state the following theorem due to Kolmogorov.

THEOREM 17.1 (Diffusion equations). *The conditional PDF $f(x,t|x_0,t_0)$ satisfies the* ***forward diffusion equation***

$$\boxed{\frac{\partial f}{\partial t} = -\frac{\partial(\beta(x,t)f)}{\partial x} + \frac{1}{2}\frac{\partial^2(\alpha(x,t)f)}{\partial x^2}} \tag{17.99}$$

and the ***backward diffusion equation***

$$\boxed{\frac{\partial f}{\partial t_0} = -\beta(x_0,t_0)\frac{\partial f}{\partial x_0} - \frac{\alpha(x_0,t_0)}{2}\frac{\partial^2 f}{\partial x_0^2},} \tag{17.100}$$

where $f \triangleq f(x,t|x_0,t_0)$.

Proof. We derive the backward equation (17.100). We write $t_1 = t_0 + h$ with $t_0 < t_1 < t$ in (17.81):

$$f(x,t|x_0,t_0) = \int_{-\infty}^{\infty} f(x,t|x_1,t_0+h)f(x_1,t_0+h|x_0,t_0)\,dx_1, \tag{17.101}$$

where $f(x_1,t_0+h|x_0,t_0)$ is the conditional PDF of RV $X_1 = X(t_0+h)$ with conditional mean, given $x_0 = X(t_0)$,

$$\mu(t_0+h|x_0,t_0) = E[X_1|x_0,t_0] = x_0 + \beta_0 h + o(h) \tag{17.102}$$

and the conditional variance

$$\sigma^2(t_0+h|x_0,t_0) = \text{Var}[X_1|x_0,t_0] = \alpha_0 h + o(h), \tag{17.103}$$

where $\alpha_0 = \alpha(x_0,t_0)$ and $\beta_0 = \beta(x_0,t_0)$. Note that the PDF $f(y,t_0+h|x_0,t_0)$ can be approximated by the following expression as $h \to 0$ (Problem 17.10):

$$f(y,t_0+h|x_0,t_0) = \delta(y-x_0-\beta_0 h) + \delta^{(2)}(y-x_0-\beta_0 h)\frac{\alpha_0 h}{2} + o(h). \tag{17.104}$$

By substituting this into (17.101) and using the property of the rth derivative of the delta function (see the Supplementary Material for the derivation)

$$\boxed{\int_{-\infty}^{\infty} f(x)\delta^{(r)}(x-a)\,dx = (-1)^r f^{(r)}(a),} \tag{17.105}$$

we find

$$\begin{aligned} f(x,t|x_0,t_0) &= f(x,t|x_0+\beta_0 h, t_0+h) + \frac{\partial^2}{\partial x_0^2} f(x,t|x_0+\beta_0 h, t_0)\frac{\alpha_0 h}{2} \\ &= f(x,t|x_0,t_0) + \frac{\partial f}{\partial x_0}\beta_0 h + \frac{\partial f}{\partial t_0}h + \frac{\partial^2}{\partial x_0^2}\left(f + \frac{\partial f}{\partial x_0}\beta_0 h\right)\frac{\alpha_0 h}{2}, \end{aligned} \tag{17.106}$$

where $f = f(x,t|x_0,t_0)$. Then, taking the limit $h \to 0$, we obtain the backward equation (17.100). The derivation of the forward equation (17.99) is left as an exercise (Problem 17.11) □

17.3.4 Ornstein–Uhlenbeck process: Gauss–Markov process

A major deficiency of Brownian motion (with or without drift) as a model of a physical phenomenon such as movement of a particle is that almost all sample paths are nowhere differentiable functions of time t, although they are almost surely continuous functions of t. This is consistent with the mathematical argument that $dW(t)/dt$ is equivalent to a white noise process that can be infinite. We know from Newton's laws that only particles with zero mass can move with infinite speed. Thus, the Brownian motion is not an adequate model in characterizing the movement of a particle over a very short time interval, although it may accurately capture physical movement of the particle over a longer period.

A model that can improve the local behavior of Brownian motion is provided by Uhlenbeck and Ornstein [334]. The basic idea is that the velocity

$$V(t) = \frac{d\boldsymbol{X}(t)}{dt} \tag{17.107}$$

of the particle is modeled by the limit process of some random walk; and its integral represents a sample path $\boldsymbol{X}(t)$ of the particle itself.

When the particle's mass m is taken into account, the differential equation (17.88) should be modified to

$$\boxed{m\frac{d^2\boldsymbol{X}(t)}{dt^2} + \xi\frac{d\boldsymbol{X}(t)}{dt} = \boldsymbol{F}(t),} \tag{17.108}$$

which can be rewritten as the following differential equation:

$$\frac{d\boldsymbol{V}(t)}{dt} + \beta_1\boldsymbol{V}(t) = \sqrt{\alpha_0}\boldsymbol{Z}(t), \tag{17.109}$$

where

$$\alpha_0 \triangleq \frac{2k_B T\xi}{m^2} \text{ and } \beta_1 \triangleq \frac{\xi}{m}. \tag{17.110}$$

The differential equation (17.109) is called the **Langevin equation** in statistical physics. Equations (17.107) and (17.109) are often written as the following stochastic differential equations:

$$d\boldsymbol{X}(t) = \boldsymbol{V}(t)\,dt, \tag{17.111}$$

$$d\boldsymbol{V}(t) = \sqrt{\alpha_0}\,\boldsymbol{Z}(t)\,dt - \beta_1\boldsymbol{V}(t)\,dt, \tag{17.112}$$

where $\boldsymbol{Z}(t) = (Z_1(t), Z_2(t), Z_3(t))$ is the three-dimensional white Gaussian process with zero mean and unit variance. The velocity process $\boldsymbol{V}(t)$ is known as the **Ornstein–Uhlenbeck process**, or the **mean-reverting process**. In the process $\boldsymbol{V}(t)$, a force proportional to the displacement (i.e., $\beta_1\boldsymbol{V}(t)\,dt$) works as a central-restoring force; thus, $V(t)$ is a stable process and the steady-state exists.

17.3.4.1 Analysis of one-dimensional Ornstein–Uhlenbeck process

Since the white noise components $Z_1(t)$, $Z_2(t)$, and $Z_3(t)$ are independent, we analyze the one-dimensional case below. By writing the Fourier transforms of $V(t)$ and $Z(t)$ as $\tilde{V}(f)$ and $\tilde{Z}(f)$, the differential equation (17.109) is transformed to

$$(i2\pi f + \beta_1)\tilde{V}(f) = \sqrt{\alpha_0}\tilde{Z}(f), \tag{17.113}$$

where $i = \sqrt{-1}$. Thus,

$$\tilde{V}(f) = \frac{\sqrt{\alpha_0}}{\beta_1 + i2\pi f}\tilde{Z}(f). \tag{17.114}$$

Hence, the process $V(t)$ can be viewed as the output process when white noise $Z(t)$ is passed into a linear filter with transfer function

$$H(f) = \frac{\sqrt{\alpha_0}}{\beta_1 + i2\pi f}. \tag{17.115}$$

Thus, $V(t)$ is nonwhite (or colored) Gaussian noise with power spectral density

$$\begin{aligned} P_V(f) = |H(f)|^2 &= \frac{\alpha_0}{|\beta_1 + i2\pi f|^2} \\ &= \frac{\alpha_2}{2\beta_1}\left(\frac{1}{\beta_1 + i2\pi f} + \frac{1}{\beta_1 - i2\pi f}\right). \end{aligned} \tag{17.116}$$

The autocorrelation function can be found by taking the inverse Fourier transform of the power spectrum $P_V(f)$, which takes the form of the Cauchy distribution in the f domain, as shown in (17.116). The method of contour integration in the complex plane discussed in Section 8.2 can be applied, and we find

$$\boxed{R_V(\tau) = \frac{\alpha_0}{2\beta_1}e^{-\beta_1|\tau|}.} \tag{17.117}$$

This bilateral (or double-sided) exponential function is characteristic of the Markov property of the Ornstein–Uhlenbeck process. It can be shown that the Ornstein–Uhlenbeck process is the **only** stationary Gaussian Markov process with a continuous autocovariance function.

Since $V(t)$ is the linear filter output of white Gaussian noise, it is obviously a Gaussian process. So in order to find its first-order probability distribution, it is sufficient to find its mean, which is clearly zero, and its variance

$$\sigma_V^2 \triangleq \int_{-\infty}^{\infty} |H(f)|^2\, df = R_V(0) = \frac{\alpha_0}{2\beta_1};$$

hence, the PDF of $V(t)$ for any t in steady state is

$$f_V(v) = \frac{1}{\sqrt{2\pi}\sigma_V}\exp\left(-\frac{v^2}{2\sigma_V^2}\right). \tag{17.118}$$

Now we wish to find the conditional PDF $f(v, t|v_0, t_0)$ of the Ornstein–Uhlenbeck process. The RVs $V(t)$ and $V(t_0)$ are jointly normally distributed with mean zero and common variance $\sigma_V^2 = R_V(0) = \alpha_0/\beta_1$, and the correlation coefficient between them is

$$\rho = \frac{R_V(t - t_0)}{\sigma_V^2} = e^{-\beta_1(t-t_0)}, \quad t \geq t_0.$$

Then using (4.111) concerning the **bivariate normal distribution** in Section 4.3.1, we find that the conditional distribution of $V(t)$ given $V(0)$ is also normal with mean

$$E[V(t)|V(t_0) = v_0] = \rho v_0 = e^{-\beta_1(t-t_0)} v_0, \quad t \geq t_0, \tag{17.119}$$

and variance

$$\text{Var}[V(t)|V(t_0) = v_0] = \sigma_V^2(1 - \rho^2) = \sigma_V^2 \left[1 - e^{-2\beta_1(t-t_0)}\right], \quad t \geq t_0. \tag{17.120}$$

Thus, we can write explicitly the conditional PDF $f(v, t|v_0, t_0)$ as

$$f_V(v, t|v_0, t_0) = \frac{1}{\sigma_V\sqrt{2\pi\left[1 - e^{-2\beta_1(t-t_0)}\right]}} \exp\left\{-\frac{[v - v_0 e^{-\beta_1(t-t_0)}]^2}{2\sigma_V^2\left[1 - e^{-2\beta_1(t-t_0)}\right]}\right\}. \tag{17.121}$$

It will be worth noting (Problem 17.12) that the conditional expectation (17.119) is equivalent to the optimal pure prediction of $V(t)$ given $V(u)$, $u \leq t_0$, of the Gaussian process $V(t)$ (see Chapter 22). Since $V(t)$ is a Markov process, the predicted value depends on the past history $V(u)$ only through the most recent value $V(t_0) = v_0$. The conditional variance is equal to the minimum variance or MSE of such predictor.

From the one-dimensional case of the stochastic differential equation (17.109), we have

$$dV(t) = -\beta_1 V(t)\, dt + \sqrt{\alpha_0} dW_s(t), \tag{17.122}$$

where $W_s(t)$ is the standard Brownian motion. Thus, the drift rate and variance rate of process $V(t)$ are found

$$\beta_V(v, t)\, dt = E[dV(t)|V(t) = v_0] = -\beta_1 v\, dt, \tag{17.123}$$

$$\alpha_V(v, t)\, dt = \text{Var}[dV(t)|V(t) = v] = \alpha_0\, dt, \tag{17.124}$$

where we used the following property of the standard Brownian motion:

$$E[(dW_s(t))^2] = dt. \tag{17.125}$$

Thus, the process $V(t)$ is subject to a drift towards the origin with magnitude proportional to its displacement from the origin, and this creates the **central restoring tendency** and leads to a stable distribution, as given by (17.118).

The forward diffusion equation (17.99) becomes

$$\frac{\partial f_V}{\partial t} = \beta_1 \frac{\partial(v f_V)}{\partial v} - \frac{\alpha_0}{2}\frac{\partial^2 f_V}{\partial v^2}, \tag{17.126}$$

where $f_V = f(v, t|v_0, t_0)$, $\beta_1 = \xi/m$, and $\alpha_0 = 2k_B T/m^2$. Similarly, the backward equation is

$$\frac{\partial f_V}{\partial t_0} = \beta_1 v_0 \frac{\partial f_V}{\partial v_0} - \frac{\alpha_0}{2} \frac{\partial^2 f_V}{\partial v_0^2}. \tag{17.127}$$

The first-order PDF $f_{V(t)}(v)$ in steady state, which we already obtained in (17.118), can be alternatively obtained by setting the left-hand side of the forward equation (17.126) to zero (Problem 17.13).

It is not difficult to solve (17.126) and obtain the time-dependent solution (17.121), which we rewrite as

$$f_V(v, t|v_0, t_0) = \frac{1}{\sqrt{2\pi\sigma_V^2(t|v_0, t_0)}} \exp\left\{ -\frac{[v - \mu_V(t|v_0, t_0)]^2}{2\sigma_V^2(t|v_0, t_0)} \right\}, \tag{17.128}$$

where $\mu_V(t|v_0, t_0)$ and $\sigma_V^2(t|v_0, t_0)$ are the conditional mean and variance of the process $V(t)$ given $V(t_0) = v_0$:

$$\mu_V(t|v_0, t_0) = v_0 e^{-\beta_1(t-t_0)}, \tag{17.129}$$

$$\sigma_V^2(t|v_0, t_0) = \sigma_V^2 \left[1 - e^{-2\beta_1(t-t_0)}\right]. \tag{17.130}$$

As $t \to \infty$, the mean tends to zero and the variance to $\sigma_V^2 = \alpha_0/2\beta_1$

The position $X(t)$ of the particle can be found by integrating the velocity process $V(t)$:

$$X(t) = \int_0^t V(u)\, du, \tag{17.131}$$

where we assume $X(0) = 0$. This implies that $X(t)$ is a Gaussian process with mean zero and variance (Problem 17.15)

$$E[X^2(t)] = \frac{2\sigma_V^2}{\beta_1}\left(t - \frac{1 - e^{-\beta_1 t}}{\beta_1}\right) = \frac{2k_B T}{\xi}\left[t - \frac{m}{\xi}\left(1 - e^{-(\xi/m)t}\right)\right]. \tag{17.132}$$

If time t is sufficiently large, i.e., if $\xi t \gg m$, then, as was shown by Einstein,

$$\boxed{\lim_{t\to\infty} \frac{E[X^2(t)]}{t} = \frac{2k_B T}{\xi} = D = \frac{\alpha}{2}.} \tag{17.133}$$

This result can be obtained directly from the differential equation (17.88):

$$X(t) = \frac{1}{\xi}\int_0^t F(u)\, du, \tag{17.134}$$

where $F(t)$ is white noise with spectral density $2k_B T\xi$. Then

$$E[X^2(t)] = \frac{2k_B T\xi t}{\xi^2} = Dt. \tag{17.135}$$

17.4 Stochastic differential equations and Itô process

By generalizing the stochastic differential equations (17.85), **Itô**[7] introduced the following process, often referred to as the **Itô process** [159]:[8]

$$\boxed{dX(t) = \beta(X(t), t)\,dt + \sqrt{\alpha(X(t), t)}\,dW_s(t).} \tag{17.136}$$

This equation is a shorthand expression of the following integral equation:

$$\boxed{X(t+s) - X(t) = \int_t^{t+s} \beta(X(u), u)du + \int_t^{t+s} \sqrt{\alpha(X(u), u)}\,dW_s(u),} \tag{17.137}$$

where

$$\int_t^{t+s} \sqrt{\alpha(X(u), u)}\,dW_s(u) = \lim_{\max \Delta u_i \to 0} \sum_{i=0}^{n-1} \sqrt{\alpha(X(u_i), u_i)}[W_s(u_{i+1}) - W_s(u_i)] \tag{17.138}$$

and the limit is in the mean square sense.

We often write (17.136) simply as

$$dX = \beta(X, t)\,dt + \sqrt{\alpha(X, t)}\,dW_s. \tag{17.139}$$

The drift rate $\beta(x, t)$ and the variance rate $\alpha(x, t)$ are functions of the underlying variable $X(t) = x$ and time t, and $dW_s \triangleq dW_s(t)$ is the infinitesimal increment of the standard Brownian motion defined by

$$dW_s(t) = W_s(t + dt) - W_s(t) = Z_s(t)\,dt, \tag{17.140}$$

where $Z_s(t)$ is the standard white Gaussian noise defined earlier. The RV $dW_s(t)$ is normally distributed with mean zero and variance

$$\text{Var}[dW_s] = E[dW_s^2] = dt. \tag{17.141}$$

The fourth moment of $dW_s(t)$ is given from the moment formula (4.54) of a normal variable as

$$E[dW_s^4] = 3\left(E[dW_s^2]\right)^2 = 3\,dt^2. \tag{17.142}$$

Thus,

$$\text{Var}[dW_s^2] = E[dW_s^4] - (E[dW_s^2])^2 = 2\,dt^2 = o(dt). \tag{17.143}$$

For any RV X such that $\text{Var}[X] = 0$, we have $X \overset{\text{m.s.}}{=} E[X]$, where "$\overset{\text{m.s.}}{=}$" stands for "equivalent in mean square." We remark that we also have $X \overset{\text{a.s.}}{=} E[X]$ (see

[7] A Japanese mathematician. Kiyoshi Itô (1915–2008). Itô is sometimes spelled as Itō.

[8] In the literature, $\mu(X(t), t)$ and $\sigma(X(t), t)$ are more common than $\beta(X(t), t)$ and $\sqrt{\alpha(X(t), t)}$ that we adopt here. We avoid use of μ and σ, since they were used to define the conditional mean and conditional variance (see (17.94) and (17.95)).

Section 13.1.3). Therefore, dW_s^2 converges to its mean dt as $dt \to 0$. In other words, in the limit dW_s^2 is **no longer stochastic**:

$$\boxed{dW_s^2 \overset{\text{m.s.}}{=} dt.} \tag{17.144}$$

Similarly, we have $E[dW_s(t)\,dt] = E[dW_s]\,dt = 0$ and $E[(dW_s(t)\,dt)^2] = E[dW_s^2]dt^2 = (dt)^3$. Hence, in the limit $dt \to 0$, $dW_s\,dt$ is not stochastic either, and

$$\boxed{dW_s\,dt \overset{\text{m.s.}}{=} 0.} \tag{17.145}$$

Then, for the Itô process $X(t)$, we obtain the following properties:

$$dX\,dt = [\,\beta(X,t)\,dt + \sqrt{\alpha(X,t)}\,dW_s\,]\,dt \overset{\text{m.s.}}{=} o(dt) \tag{17.146}$$

and

$$\boxed{dX^2 = [\,\beta(X,t)\,dt + \sqrt{\alpha(X,t)}\,dW_s\,]^2 \overset{\text{m.s.}}{=} \alpha(X,t)\,dt + o(dt).} \tag{17.147}$$

17.4.1 Itô's formula

Now, for a given Itô process $X(t)$, consider a function of $X(t)$ and t, which we denote by

$$Y = Y(X,t).$$

By applying the Taylor series expansion to this function, we have

$$dY = \frac{\partial Y}{\partial X}dX + \frac{\partial Y}{\partial t}dt + \frac{1}{2}\frac{\partial^2 Y}{\partial X^2}dX^2 + \frac{1}{2}\frac{\partial^2 Y}{\partial t^2}dt^2 + \frac{\partial^2 Y}{\partial X \partial t}dX\,dt + \cdots \tag{17.148}$$

Substituting (17.146) for $dX\,dt$, (17.147) for dX^2, and ignoring terms of order $o(dt)$, we obtain

$$\boxed{dY = \frac{\partial Y}{\partial X}dX + \frac{\partial Y}{\partial t}dt + \frac{1}{2}\frac{\partial^2 Y}{\partial X^2}\alpha(X,t)\,dt,} \tag{17.149}$$

which is often called **Itô's lemma** or **Itô's formula**. Using (17.139) for dX, the above formula can be written as

$$\boxed{dY = \left(\frac{\partial Y}{\partial X}\beta(X,t) + \frac{\partial Y}{\partial t} + \frac{1}{2}\frac{\partial^2 Y}{\partial X^2}\alpha(X,t)\right)dt + \frac{\partial Y}{\partial X}\sqrt{\alpha(X,t)}\,dW_s.} \tag{17.150}$$

This last equation shows that $Y(t)$ is also an Itô process with the drift rate $\frac{\partial Y}{\partial X}\beta(X,t) + \frac{\partial Y}{\partial t} + \frac{1}{2}\frac{\partial^2 Y}{\partial X^2}\alpha(X,t)$ and the variance rate $\left(\frac{\partial Y}{\partial X}\sqrt{\alpha(X,t)}\right)^2$.

Example 17.2: Consider a special case, where

$$\beta(X,t)=0 \text{ and } \alpha(X,t)=1.$$

It follows that

$$dX=dW_s.$$

Then the process $Y(t)$ is simply a function of $W_s(t)$ and t, and (17.150) reduces to

$$\boxed{dY=\left(\frac{\partial Y}{\partial t}+\frac{1}{2}\frac{\partial^2 Y}{\partial X^2}\right)dt+\frac{\partial Y}{\partial X}dW_s.} \tag{17.151}$$

□

17.4.2 Geometric Brownian motion (GBM)

Consider an Itô process $Y(t)$ where

$$\beta(Y(t),t)=\beta_y Y(t) \text{ and } \alpha(Y(t),t)=\alpha Y(t).^9$$

Then the stochastic differential equation (17.136) takes the form

$$\boxed{dY=\beta_y Y\,dt+\sqrt{\alpha}Y\,dW_s, \quad \text{or} \quad \frac{dY}{Y}=\beta_y\,dt+\sqrt{\alpha}\,dW_s,} \tag{17.152}$$

where $W_s(t)$ is the standard Brownian motion. One might be tempted to think that the above equation is equivalent to $dX=\beta\,dt+\sqrt{\alpha}\,dW_s$ with $X=\ln Y$ so that the problem might be reduced to simple Brownian motion. However, Y is a random process and not differentiable in the ordinary sense; thus, we cannot apply the chain rule to conclude $dX(t)=dY(t)/Y(t)$. Instead, we need to apply the above Itô's lemma.

Comparing (17.152) and (17.151) we find the following equations that $Y(t)$ must satisfy:

$$\frac{\partial Y}{\partial t}+\frac{1}{2}\frac{\partial^2 Y}{\partial W_s^2}=\beta_y Y \tag{17.153}$$

and

$$\frac{\partial Y}{\partial W_s}=\sqrt{\alpha}Y. \tag{17.154}$$

Solving this partial differential equation, we have

$$Y(t)=\exp\left[\sqrt{\alpha}W_s(t)+a(t)\right], \tag{17.155}$$

[9] Note that we put the subscript y to the drift rate β but not to the variance rate α. The reason will become clearer in (17.164).

where $a(t)$ is any function of t. By substituting (17.155) into (17.153), we obtain

$$Y(t)\left(a'(t)+\frac{\alpha}{2}-\beta_y\right)=0. \tag{17.156}$$

If $Y(t)=0$ for any t, then $dY(t)=0$, and consequently the process $Y(t)=0$ for all t. Thus, excluding this degenerated case, we have $a'(t)=\beta_y-(\alpha/2)$, leading to

$$a(t)=\left(\beta_y-\frac{\alpha}{2}\right)t+c, \tag{17.157}$$

where c is a constant. Substituting this into (17.155), we have

$$Y(t)=\exp\left[\sqrt{\alpha}W_s(t)+\left(\beta_y-\frac{\alpha}{2}\right)t+c\right]. \tag{17.158}$$

At $t=0$ we have $W_s(0)=0$ and $Y(0)=e^c$; thus,

$$Y(t)=Y(0)\exp\left[\left(\beta_y-\frac{\alpha}{2}\right)t+\sqrt{\alpha}W_s(t)\right], \tag{17.159}$$

or

$$\ln Y(t)=\ln Y(0)+\left(\beta_y-\frac{\alpha}{2}\right)t+\sqrt{\alpha}W_s(t). \tag{17.160}$$

Since $W_s(t)$ is normal, $\ln Y(t)$ is also normal with mean

$$E\left[\ln Y(t)\right]=\ln Y(0)+\left(\beta_y-\frac{\alpha}{2}\right)t \tag{17.161}$$

and variance

$$\mathrm{Var}\left[\ln Y(t)\right]=\alpha t. \tag{17.162}$$

By setting $Y(0)=1$, we can rewrite (17.159) as

$$\boxed{Y(t)=e^{X(t)},} \tag{17.163}$$

where

$$\boxed{X(t)=\beta t+\sqrt{\alpha}W_s(t),\quad \text{with } \beta\triangleq\beta_y-\frac{\alpha}{2}.} \tag{17.164}$$

Since $X(t)$ is Brownian motion with drift rate $\beta=\beta_y-(\alpha/2)$ and diffusion coefficient α, the process $Y(t)$ is called **geometric Brownian motion** (GBM), because of the relation (17.163). In Figure 17.5 we plot a sample path of GBM $Y(t)$ together with the corresponding Brownian motion $X(t)=\ln Y(t)$, when $\beta=0$ and $\alpha=0.25$ (hence $\beta_y=0.125$).

The conditional expectation and variance of $Y(t)$ given $\{Y(u);\ 0\le u\le s\}$, where $s\le t$, can be found (Problem 17.16) as

$$E[Y(t)|Y(u),0\le u\le s]=Y(s)e^{[\beta+(\alpha/2)](t-s)},\quad s\le t, \tag{17.165}$$

and

$$E[Y(t)^2|Y(u),0\le u\le s]=Y(s)^2e^{2(\beta+\alpha)(t-s)},\quad s\le t. \tag{17.166}$$

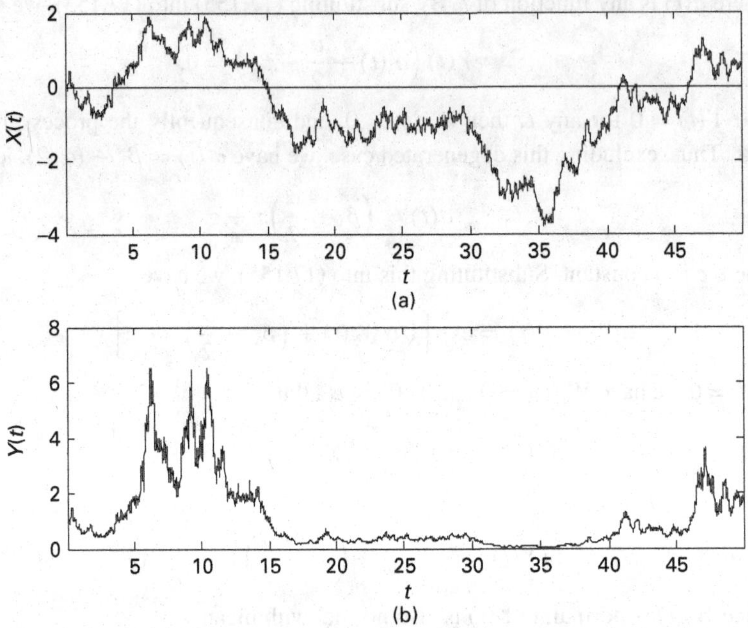

Figure 17.5 (a) Brownian motion $X(t)$ of (17.164); (b) GBM $Y(t) = e^{X(t)}$ of (17.163), $\beta = 0,\ \alpha = 0.25$.

Thus, the conditional variance of $Y(t)$ is

$$\mathrm{Var}[Y(t)|Y(u), 0 \le u \le s] = Y(s)^2 e^{(2\beta+\alpha)(t-s)} \left[e^{\alpha(t-s)} - 1\right], \quad s \le t. \tag{17.167}$$

The conditional PDF for $Y(t)$ given $Y(0) = y_0$ (hence $X(0) = x_0 = \ln y_0$) is given by the following **log-normal distribution**:

$$f_Y(y, t|y_0, 0) = \frac{1}{y\sqrt{2\pi\alpha t}} \exp\left[-\frac{(\ln(y/y_0) - \beta t)^2}{2\alpha t}\right], \quad y > 0. \tag{17.168}$$

The log-normal distribution is discussed in Section 7.4.

The GBM is often used to model the stochastic behavior of stock prices, since stock prices typically exhibit long-term exponential growth. Note also $Y(t)$ is nonnegative, although $X(t)$ may take negative values. Let Y_k denote the price of a stock at the kth day (or any time index). Define the ratio

$$R_k = \frac{Y_k}{Y_{k-1}} = 1 + \epsilon_k, \tag{17.169}$$

where $\epsilon_k \times 100$ represents the percentage change of the stock price from the previous day. Then, we have

$$Y_k = R_k R_{k-1} \cdots R_1 Y_0, \tag{17.170}$$

or

$$\ln\left(\frac{Y_k}{Y_0}\right) = \sum_{i=1}^{k} \ln R_i = \sum_{i=1}^{k} \ln(1+\epsilon_i). \tag{17.171}$$

Assuming that ϵ_i and ϵ_j are uncorrelated for $|i-j| \gg 1$, a generalized CLT implies that $\sum_{i=1}^{k} \ln(1+\epsilon_i)$ is approximately normally distributed; hence, Y_k is log-normally distributed, justifying the use of GBM for stock price movement.

17.4.3 The Black–Scholes model: an application of an Itô process

The application example of this section is quite different from the rest of the examples in this volume. The **Black–Scholes theory**[10] [29] to be outlined below is considered one of the most important developments in **mathematical finance**, and the best-known application of the Itô process and Itô calculus. The main purpose of this section, therefore, is to provide the readers with some idea concerning how the random process theory and stochastic models are applied in the field of finance.

Financial derivatives, as their name implies, are contracts that are based on, or derived from, some underlying assets (e.g., stocks, bonds, commodity), reference rates (e.g., interest rates or currency exchange rates), or indexes. Let $Y(t)$ denote the **price of an underlying asset**, say a stock, and Black and Scholes assume that $Y(t)$ follows the GBM given in (17.152):[11]

$$dY = \beta_y Y\,dt + \sqrt{\alpha}Y\,dW_s. \tag{17.172}$$

We showed there that $X(t) = \ln Y(t)$ is Brownian motion with the drift coefficient $\beta = \beta_y - (\alpha/2)$ and diffusion coefficient α.

Let $V(Y(t), t)$ be the **price of a derivative security**[12] contingent on the underlying stock asset of value $Y(t)$. We assume that V depends differentially on the two independent variables Y and t. Then using the assumption that $Y(t)$ is GBM, Itô's lemma shows (Problem 17.17) that V changes over the infinitesimal time interval dt according to

$$\boxed{dV = \left(\frac{\partial V}{\partial t} + \beta Y\frac{\partial V}{\partial Y} + \frac{\alpha}{2}Y^2\frac{\partial^2 V}{\partial Y^2}\right)dt + \sqrt{\alpha}Y\frac{\partial V}{\partial Y}\,dW_s.} \tag{17.173}$$

Assume that we have a portfolio consisting of one option of the derivative security of value V and N shares of the underlying stock, where the quantity N is yet to be determined, with $N > 0$ for shares held **long** and $N < 0$ for shares held **short**. The **value of the portfolio** at time t, denoted $P(t)$, is the price of the derivative security plus the price of the underlying asset:

[10] Fischer Black (1938–1995) and Myron S. Scholes (1941): Scholes, together with Robert C. Merton (1944), received the 1997 Nobel Prize in Economics.

[11] Notation $S(t)$ is often used instead of $Y(t)$ in the literature. So are μ and σ^2 instead of β and α. The parameter σ, which corresponds to our $\sqrt{\alpha}$, is called **volatility** in mathematical finance.

[12] A derivative security is a financial security, such as an option or future, whose characteristics and value depend on the characteristics and value of an underlying security.

$$P(t) = V(Y,t) + NY(t). \tag{17.174}$$

Its gain over the interval dt is

$$dP = dV + NdY, \tag{17.175}$$

which can be written, using (17.173) and (17.172), as

$$dP = \left(\frac{\partial V}{\partial t} + \beta Y \frac{\partial V}{\partial Y} + \frac{\alpha}{2} Y^2 \frac{\partial^2 V}{\partial Y^2}\right) dt + \sqrt{\alpha} Y \frac{\partial V}{\partial Y} dW_s + N\left(\beta Y\, dt + \sqrt{\alpha} Y\, dW_s\right). \tag{17.176}$$

If we set

$$N = -\frac{\partial V}{\partial Y}, \tag{17.177}$$

the stochastic term dW_s will be eliminated from the above differential equation, yielding

$$\boxed{dP = \left(\frac{\partial V}{\partial t} + \frac{\alpha}{2} Y^2 \frac{\partial^2 V}{\partial Y^2}\right) dt.} \tag{17.178}$$

If the gain in the value $P(t)$ is deterministic, then it would be equivalent to investing at some risk-free interest rate r. To exclude **arbitrage opportunities**,[13] the following relation must hold:

$$\boxed{dP = rP\, dt = r\,(dV + N\, dY)\, t = r\left(V - \frac{\partial V}{\partial Y} Y\right) dt.} \tag{17.179}$$

By equating dP in the last two equations, we find

$$\boxed{\frac{\partial V}{\partial t} + rY \frac{\partial V}{\partial Y} + \frac{\alpha}{2} Y^2 \frac{\partial^2 V}{\partial Y^2} = rV,} \tag{17.180}$$

which is known as the **Black–Scholes differential equation** for option pricing. This equation instructs how to buy or sell assets to maintain a portfolio that grows at the riskless rate. Thus, it provides insurance against downturns in the value of assets held long or protects against a rise in the value of assets held short. Therefore, the portfolio is hedged against losses by having options serving as an insurance policy.

The solution of the differential equation depends on the initial and boundary conditions determined by the specific option contract.

Example 17.3: European call option [173]. There are two types of options: call and put options. A **call option** gives the holder the right to purchase the underlying stock by a certain date, called the *maturity*, for a certain price, called the **strike price**. A **put option**

[13] An arbitrage opportunity is the opportunity to buy an asset at a low price then immediately sell it on a different market for a higher price. This is a riskless profit for the investor/trader.

gives the right to sell the underlying stock by a maturity for a strike price. **American options** can be exercised at any time up to the maturity, whereas **European options** can be exercised only at the maturity. Most options traded on exchanges are American. European options are easier to analyze, and some properties of an American option are often deduced from those of its European counterpart.

Consider a *European call option* in which the holder of the option is entitled to purchase a share of a given stock at an exercise price C at a future time T. The boundary condition is given by

$$V(Y(T), T) = \max\{Y(T) - C, 0\}. \tag{17.181}$$

The price of the option at time $t < T$ will be its expected price at time T discounted back to t. Thus, the option price at time t is given by

$$V(Y(t), t) = e^{-r(T-t)} E[V(Y(T), T)]. \tag{17.182}$$

A direct way to evaluate $E[V(Y(T), T)]$ is by solving the Black–Scholes partial differential equation (17.180) using the boundary condition (17.181).

Another approach is to assume that there exist risk-neutral investors such that the price of the stock follows GBM

$$dY = rY\,dt + \sqrt{\alpha} Y\,dW_s; \tag{17.183}$$

that is, with the risk-free interest rate r replacing β_y in (17.172). Writing $\ln Y(t) = X(t)$ as before, we find (Problem 17.18) $X(T)$ is normally distributed with

$$E[X(T)] = \ln y(t) + \left(r - \frac{\alpha}{2}\right)(T - t) \text{ and } \mathrm{Var}[X(T)] = \alpha(T - t). \tag{17.184}$$

Since $Y(T) = e^{X(T)}$ and

$$\begin{aligned} E[V(Y(T), T)] &= E\left[V\left(e^{X(T)}, T\right)\right] = E\left[\max\left\{e^{X(T)} - C, 0\right\}\right] \\ &= \int_{\ln C}^{\infty} \left(e^x - C\right) f_{X(T)}(x)\,dx, \end{aligned} \tag{17.185}$$

where

$$f_{X(T)}(x) = \frac{1}{\sqrt{2\pi\alpha(T-t)}} \exp\left\{-\frac{\left[x - \ln y(t) - (r - \alpha/2)(T - t)\right]^2}{2\alpha(T - t)}\right\}.$$

Thus, by substituting the above into (17.182), we obtain the following expression for the option price at time t:

$$V(Y(t), t) = e^{-r(T-t)} \int_{\ln C}^{\infty} \left(e^x - C\right) f_{X(T)}(x)\,dx = e^{-r(T-t)}(I_1 - I_2), \tag{17.186}$$

where

$$I_1 = \int_{\ln C}^{\infty} e^x f_{X(T)}(x)\,dx$$
$$= e^{t(T-t)} \int_{\ln C}^{\infty} \frac{1}{\sqrt{2\pi\alpha(T-t)}} \exp\left\{-\frac{\left[x - \ln y(t) - \left(r + \frac{\alpha}{2}\right)(T-t)\right]^2}{2\alpha(T-t)}\right\} dx$$
$$= e^{t(T-t)}[1 - \Phi(-u_1)] = e^{t(T-t)}\Phi(u_1),$$

where $\Phi(u)$ is the distribution function of the unit normal variable U defined in (4.46) and

$$u_1 = \frac{\ln \frac{y(t)}{C} + \left(r + \frac{\alpha}{2}\right)(T-t)}{\sqrt{\alpha(T-t)}}.$$

Similarly, I_2 can be found as

$$I_2 = C \int_{\ln C}^{\infty} f_{X(T)}(x)\,dx = C[1 - \Phi(-u_2)] = C\Phi(u_2),$$

where

$$u_2 = \frac{\ln \frac{y(t)}{C} + \left(r - \frac{\alpha}{2}\right)(T-t)}{\sqrt{\alpha(T-t)}}.$$

Substituting these into (17.186), we finally obtain the solution for the option price:

$$V(Y(t), t) = Y(t)\Phi(u_1) - C\,e^{-r(T-t)}\Phi(u_2). \qquad (17.187)$$

As a numerical example we consider a nondividend-paying stock whose current price is \$100. The exercise price of the option is \$90. The risk-free interest is 10% per annum and the volatility ($\sqrt{\alpha}$) is 20% per annum. The option expires in 6 months. Then substituting $r = 0.10$, $\alpha = 0.04$, $Y(t) = 100$, $C = 90$, and $T - t = 0.5$ into (17.187), we find that the value of the call option is \$15.29.

If we purchase the stock now, we pay \$100. If we purchase the call option and acquire the stock on the option's expiration date, we pay \$90 then. Thus, the stock price has to rise by (\$90 + \$15.29) − \$100 = \$5.29 for the purchaser of the call to break even. The reader is suggested to numerically calculate the option price by choosing different values for volatility and risk-free interest rate (Problem 17.19). □

17.5 Summary of Chapter 17

Simple random walk, properties:	(a) Spatial homogeneity	(17.3)
	(b) Temporal homogeneity	(17.4)
	(c) Independent increment	
	(d) Markov property	(17.5)

Autocorrelation func:	$R_X(m,n) = \min\{m,n\}$	(17.20)
Gambler ruin problem:	$r_i = pr_{i+1} + qr_{i-1}$	(17.22)
Opponent with infinite capital:	$r_i = \alpha^i, \quad i = 0, 1, 2, \ldots$	(17.29)
Opponent with finite capital:	$r_i = \frac{\alpha^c - \alpha^i}{\alpha^c - 1}, \quad 0 \le i \le c$	(17.31)
Wiener process:	$\lim_{n\to\infty, h,\delta\to 0} X_n = W(t), \ t = nh, \delta^2/h = \alpha$	(17.40)
Wiener process; 5th property:	Gaussian property	(17.44)
Autocorrelation of $W(t)$:	$R_W(t,s) = \alpha \min\{s,t\}$	(17.47)
White noise spectrum:	$P_Z(f) = N_0/2, \quad -\infty < f < \infty$	(17.49)
Wiener process vs. white noise:	$W(t) = \int_0^t Z(u)\,du$	(17.52)
Brownian motion with drift:	$\lim_{n\to\infty, h,\delta\to 0} X_n = X(t), (p-q)\delta/h = \beta, 4pq\delta^2/h = \alpha$	(17.65)
Drift coefficient:	$\beta = E[X(t)]/t$	(17.69)
Diffusion coefficient:	$\alpha = \mathrm{Var}[X(t)]/t$	(17.70)
Distribution func. of $X(t)$:	$F(x,t\vert x_0,t_0) = \frac{1}{\sqrt{2\pi}} \int_{-\infty}^{y} e^{-u^2/2}\,du$	(17.71)
where	$y = \frac{x - x_0 - \beta(t-t_0)}{\sqrt{\alpha(t-t_0)}}$	(17.72)
Fokker–Planck equation:	$\frac{\partial f(x,t)}{\partial t} = -\beta \frac{\partial f(x,t)}{\partial x} + \frac{\alpha}{2} \frac{\partial^2 f(x,t)}{\partial x^2}$	(17.77)
Relation to $W_s(t)$:	$dX(t) = \beta dt + \sqrt{\alpha}\, dW_s(t)$	(17.85)
Forward diffusion eq.:	$\frac{\partial f}{\partial t} = -\frac{\partial(\beta(x,t)f)}{\partial x} + \frac{1}{2}\frac{\partial^2(\alpha(x,t)f)}{\partial x^2}$	(17.99)
Backward diffusion eq.:	$\frac{\partial f}{\partial t_0} = -\beta(x_0,t_0)\frac{\partial f}{\partial x_0} - \frac{\alpha(x_0,t_0)}{2}\frac{\partial^2 f}{\partial x_0^2}$	(17.100)
where	$f \triangleq f(x,t\vert x_0,t_0)$	
	$\beta(x_0,t_0) \triangleq \left.\frac{\partial \mu(t\vert x_0,t_0)}{\partial t}\right\vert_{t=t_0}$	(17.97)
	$\alpha(x_0,t_0) \triangleq \left.\frac{\partial \sigma^2(t\vert x_0,t_0)}{\partial t}\right\vert_{t=t_0}$	(17.97)
Itô process:	$dX = \beta(X,t)\,dt = \sqrt{\alpha(X,t)}\,dW_s$	(17.136)
Properties of W_s as $t \to 0$:	$dW_s^2 \overset{\text{m.s.}}{=} dt, \ dW_s\,dt \overset{\text{m.s.}}{=} 0$	(17.144)
Itô's lemma:	$dY = \frac{\partial Y}{\partial X}\,dX + \frac{\partial Y}{\partial t}\,dt + \frac{1}{2}\frac{\partial^2 Y}{\partial X^2}\,\alpha(X,t)\,dt$	(17.149)
GBM $Y(t)$:	$dY = \beta_y Y\,dt + \sqrt{\alpha} Y\,dW_s$	(17.152)
	$Y(t) = e^{X(t)}$	(17.163)
where	$X(t) = \beta t + \sqrt{\alpha} W_s(t), \quad \text{with } \beta \triangleq \beta_y - (\alpha/2)$	(17.164)
Log-normal dist. of GBM:	$f_Y(y,t\vert y_0,0) = \frac{1}{y\sqrt{2\pi\alpha t}} \exp\left\{-\frac{[\ln(y/y_0) - \beta t]^2}{2\alpha t}\right\}, \ y > 0$	(17.168)
Itô's lemma for $V(t)$:	$dV = \left(\frac{\partial V}{\partial t} + \beta Y \frac{\partial V}{\partial Y} + \frac{\alpha}{2} Y^2 \frac{\partial^2 V}{\partial Y^2}\right) dt + \sqrt{\alpha} Y \frac{\partial V}{\partial Y} dW_s$	(17.173)
Black–Scholes diff. eq.:	$\frac{\partial V}{\partial t} + rY\frac{\partial V}{\partial Y} + \frac{\alpha}{2} Y^2 \frac{\partial^2 V}{\partial Y^2} = rV$	(17.180)

17.6 Discussion and further reading

Comprehensive discussions on random walks, Brownian motion, and diffusion processes may be found in Breiman [35], Doob [82], Feller [100], Grimmet and Stirzaker [131], Rogers and Williams [282], Ross [288], Wong and Hajek [359], and others. It will be of historical interest to note that Bachelier [10] formulated Brownian motion in 1900 in formulating the stock price movement, five years before Einstein's famous papers [86, 87] on Brownian motion.

A model that improves Brownian motion, as a model of a physical phenomenon such as movement of a particle, is provided by the Uhlenbeck and Ornstein [334] (see also Cox and Miller [72]). The basic idea is that the velocity of the particle is modeled by the limit process of some random walk; and its integral represents a sample path of the particle itself.

A diffusion process is sometimes adopted to approximate a discrete-state process such as a queue-size process, since the partial differential equation for the diffusion process is often mathematically more amenable than a differential-difference equation that characterizes a typical queueing process. Various diffusion approximation techniques to traffic and queueing problems have been discussed by Cox and Miller [72], Gaver [114], and Newell [255]. See also [194, 195, 203].

Besides GBM and the Itô process [159] as a generalization of Browian motion is fractional Brownian motion (FBM) introduced by Mandelbrot [236]. The FBM is **not** a Markov process and possesses an autocorrelation function that is not summable or integrable. It forms a mathematical basis for the LRD process and self-similar processes. Gubner [133] devotes a chapter on long-range dependent (LRD) models. See also [203] and references therein regarding FBM and its related subjects.

As for a rigorous treatment of stochastic differential equations, stochastic integrals and Itô calculus, the reader is referred to Rogers and Williams [283] and the aforementioned Wong and Hajek [359].

17.7 Problems

Section 17.1: Random walk

17.1 Alternative derivation of (17.7).

(a) Derive the equations (17.8) and (17.9).
(b) Verify that the probability distribution (17.7) is indeed a solution to (17.8) and (17.9).

17.2* Properties of the simple random walk. Prove the four properties 1–4 of a simple random walk $\{X_n\}$ stated in (17.3) through (17.5).

17.3 Gambler ruin problem. Derive the expressions (17.31) and (17.32) for the probability that the gambler with the initial capital a gets ultimately ruined by playing against the house with capital b.

17.4 Expected duration of the game. Consider the gambler ruin problem discussed in Section 17.1.2 in which A with the initial capital a dollars plays a game against B with the capital b dollars. Find the mean expected duration of the game; the game ends when either A or B gets ruined.

Section 17.2: Brownian motion or Wiener process

17.5 Properties of the Wiener process.

(a) Let $\mathrm{Var}[W(t)] = g(t)$. Show that, from properties 2 and 3 alone, $g(t)$ must be of the form $g(t) = \alpha t$.
(b) Show that $\mathrm{Var}[W(t) - W(s)] = \alpha|t - s|$.

17.6 Transformation of a Wiener process. Consider the Wiener process with parameter α as discussed in this section.

(a) Define

$$X(t) = W(t^2).$$

Show that $X(t)$ is a Gaussian process with zero mean and

$$R(t, s) = \alpha \min(t^2, s^2).$$

(b) We define

$$Y(t) = W^2(t).$$

Show that its autocorrelation function is given by

$$R_Y(t_1, t_2) = \alpha^2 t_1(2t_1 + t_2), \quad \text{for } t_1 < t_2.$$

Hint: Note that $W(t)$ is a normal RV.

Section 17.3: Diffusion processes and diffusion equations

17.7 Diffusion equations of Brownian motion with drift. Consider Brownian motion with drift in which

$$\beta(x, t) = \beta \text{ and } \alpha(x, t) = \alpha.$$

Write down the forward and backward diffusion equations.

17.8 Conditional PDFs of the standard Brownian motion. If $0 < t_0 < t$, then the conditional PDF of $W_s(t)$ given $W_s(t_0) = x_0$ is the normal distribution with mean x_0 and variance $t - t_0$, as seen from (17.44).

(a) Consider the case $0 < t < t_0$ and show that the conditional PDF of $W_s(t)$ given $W_s(0) = 0$ and $W_s(t_0) = x_0$ is the normal distribution with mean $(x_0/t_0)t$ and variance $[(t_0 - t)/t_0]t$.
(b) Show that the conditional PDF of $W_s(t)$ for $t_1 < t < t_2$, given $W_s(t_1) = x_1$ and $W_s(t_2) = x_2$, is a normal density with mean

$$x_1 + \frac{x_2 - x_1}{t_2 - t_1}(t - t_1)$$

and variance

$$\frac{(t_2 - t)(t - t_1)}{t_2 - t_1}.$$

17.9 Solution of the Fokker–Plank equation (17.77).

(a) Show that the equation (17.77) does not have a steady-state solution.
Hint: Set the left-hand side of (17.77) to zero and try to find the solution in the limit $t \to \infty$.

(b) Show that (17.78) is the time-dependent solution of the Fokker–Planck equation.

17.10* Derivation of (17.104) and (17.106).

(a) Let X be a random variable with PDF $f(x)$ with mean μ and variance σ^2. Assume that the support of $f(x)$ is a very short interval around its mean, $[\mu - \epsilon, \mu + \epsilon]$; i.e.,

$$f(x) = 0 \text{ for } x \notin [\mu - \epsilon, \mu + \epsilon].$$

Then, for a random variable $Y = g(X)$ with a continuous function $g(X)$, its expectation can be expressed as follows:

$$E[Y] \approx g(\mu) + g''(\mu)\frac{\sigma^2}{2}, \tag{17.188}$$

where $g''(x)$ is the second derivative of $g(x)$. In other words,

$$f(x) = \delta(x - \mu) + \frac{\sigma^2}{2}\delta^{(2)}(x - \mu). \tag{17.189}$$

(b) Using the result of part (a), show the approximation (17.104) and (17.106) for small h.

17.11* Derivation of the forward diffusion equation. Derive Kolmogorov's forward diffusion equation (17.99).
Hint: Start with the following Chapman–Kolmogorov equation:

$$f(x, t + h|x_0, t_0) = \int f(x, t + h|x', t) f(x', t|x_0, t_0)\, dx'.$$

17.12 Conditional expectation and pure prediction. Using Theorem 22.4 and results in Example 22.3 in Section 22.2.4, derive the expressions (17.119) and (17.120).

17.13 First-order PDF of the Ornstein–Uhlenbeck process. Derive the first-order PDF $f_{V(t)}(v)$ in steady state given by (17.118), from the forward equation (17.126).

17.14 Time-dependent solution for the Ornstein–Uhlenbeck process [100, 315]. Derive (17.129) for the time-dependent conditional PDF $f(x, t|x_0, 0)$ of the Ornstein–Uhlenbeck process $X(t)$.

17.15 Variance of the integration of the Ornstein–Uhlenbeck process. Show that the variance of $X(t) = \int_0^t V(u)\,du$ is given by (17.132), where $V(t)$ is the Ornstein–Uhlenbeck process.

Section 17.4: Stochastic differential equations and Itô process

17.16* Conditional mean and variance of the GBM. Derive the conditional mean, the second moment, and the variance of $Y(t)$ that are given by (17.165), (17.166), and (17.167) respectively.

17.17 Itô's lemma applied to GBM. Derive the stochastic differential equation (17.173) for $V(t)$, the price of the derivative security, when the price of the underlying asset, $Y(t)$, is a GBM process as in (17.172), with $\beta_y = \beta + (\alpha/2)$.

17.18 Conditional mean and variance of $X(T)$, given $Y(t)$. Show that the conditional mean and variance of $X(T)$, given $Y(t)$, are given by (17.184).

17.19* European call option. In reference to the model parameters assumed in Example 17.3, how will the option price \$15.29 change if we change one of the model parameters as follows. Keep the other parameters intact.

(a) Suppose we change the risk-free interest from 10% to 5%. Which of the following values is the closest to the option price: (i) \$12.00, (ii) \$13.50, (iii) \$15.00, (iv) \$16.50, (v) \$18.00?
(b) Change volatility from 20% to 30% per annum. Which is the closest option price: (i) \$14.00. (ii) \$15.50, (iii) \$17.00 (iv) \$18.50, (v) \$20.00?
(c) Change the expiration date from 6 months to 1 year. Which is the closest option price: (i) \$14.00, (ii) \$15.50, (iii) \$17.00, (iv) \$18.50, (v) \$20.00?

Part IV

Statistical inference

18 Estimation and decision theory

The study of statistics is concerned with effective use of numerical data available to us, or collected by some experiments, and its theory relies on probability theory, decision theory, and other branches of mathematics. In mathematical statistics, we interpret a set of finite observations $\boldsymbol{x} = (x_1, x_2, \ldots, x_n)$ as a *sample point* of its underlying *RV* $\boldsymbol{X} = (X_1, X_2, \ldots, X_n)$ drawn at random from its *sample space* $\mathcal{X}$.

In this chapter we are primarily concerned with fitting a statistical model to real measurement data. A model is usually described by a set of probability distributions that involve some unknown parameters. Thus, model fitting consists of estimating its parameters from experimental data and assigning some measure of confidence to the model. We will study statistical procedures to estimate such parameters and procedures to test the goodness of fit of the model to the experimental data. We will also investigate computational algorithms for these procedures.

18.1 Parameter estimation

We consider RV $\boldsymbol{X}$ with probability distributions $F(\boldsymbol{x}; \boldsymbol{\theta})$ with parameter $\boldsymbol{\theta}$, which we assume has dimension M:

$$\boldsymbol{\theta} = (\theta_1, \theta_2, \ldots, \theta_M).$$

The value of parameter $\boldsymbol{\theta}$ is unknown and we want to estimate it from observations

$$\boldsymbol{x} = (\boldsymbol{x}_1, \boldsymbol{x}_2, \ldots, \boldsymbol{x}_n) \in \mathcal{X}$$

drawn from the distribution $F(\boldsymbol{x}; \boldsymbol{\theta})$. The number n is called the *sample size*, and the parameter to be estimated is referred to as an **estimand** [143]. We want to find a function $T(\cdot)$ such that $\hat{\boldsymbol{\theta}} = \boldsymbol{T}(\boldsymbol{x})$ is as close to $\boldsymbol{\theta}$ as possible. Such an estimate is called a **point estimate**. But a particular estimate $T(\boldsymbol{x})$ is merely an instance (or a sample point) of the transformed RV $T(\boldsymbol{X})$. Therefore, we cannot assess the quality of an estimate just based on one sample only. We ought to analyze the distribution of the RV $\boldsymbol{T}(\boldsymbol{X})$, which is called a **point estimator** of the parameter.

A function of a sample $\boldsymbol{x}$ is called a **statistic**, where the function is independent of the sample's distribution. The objective of point estimation of parameter $\boldsymbol{\theta}$ is to find an estimator statistic $\hat{\boldsymbol{\theta}}(\boldsymbol{x})$ whose probability distribution, called the **sampling distribution**, is as concentrated around $\boldsymbol{\theta}$ as possible.

We introduce three desirable properties of an estimator:

DEFINITION 18.1 (Unbiasedness, efficiency, and consistency of an estimator).

1. *An estimator $\hat{\boldsymbol{\theta}}(\boldsymbol{X})$ is said to be* ***unbiased*** *if $E[\hat{\boldsymbol{\theta}}(\boldsymbol{X})] = \boldsymbol{\theta}$; otherwise, it is called* ***biased***. *The bias is defined as*

$$b(\boldsymbol{\theta}) = E[\hat{\boldsymbol{\theta}}(\boldsymbol{X})] - \boldsymbol{\theta}. \tag{18.1}$$

2. *An unbiased estimator $\hat{\boldsymbol{\theta}}^*(\boldsymbol{X})$ is said to be* ***efficient*** *if it is a* ***minimum-variance estimator***; *that is, $\mathbf{Var}[\hat{\boldsymbol{\theta}}^*(\boldsymbol{X})] \le \mathbf{Var}[\hat{\boldsymbol{\theta}}(\boldsymbol{X})]$ for any other unbiased estimator $\hat{\boldsymbol{\theta}}(\boldsymbol{X})$.*

3. *A sequence of estimators is said to be* ***consistent*** *if the sequence converges in probability to $\boldsymbol{\theta}$.* □

The above matrix inequality $\mathbf{Var}[\hat{\boldsymbol{\theta}}^*(\boldsymbol{X})] \le \mathbf{Var}[\hat{\boldsymbol{\theta}}(\boldsymbol{X})]$ means that $\mathbf{Var}[\hat{\boldsymbol{\theta}}^*(\boldsymbol{X})] - \mathbf{Var}[\hat{\boldsymbol{\theta}}(\boldsymbol{X})]$ is *negative semidefinite*. It is equivalent to

$$\mathrm{Var}[\boldsymbol{a}^\top \hat{\boldsymbol{\theta}}^*(\boldsymbol{X})] \le \mathrm{Var}[\boldsymbol{a}^\top \hat{\boldsymbol{\theta}}(\boldsymbol{X})] \tag{18.2}$$

for any vector $\boldsymbol{a} \neq 0$ (of the same dimension as $\hat{\boldsymbol{\theta}}(\boldsymbol{X})$) (Problem 18.2).

DEFINITION 18.2 (Sufficient statistic). *A statistic $\boldsymbol{T}(\boldsymbol{X})$ is said to be* ***sufficient*** *for parameter $\boldsymbol{\theta}$, if the conditional probability density (or mass) function of $\boldsymbol{X}$, given $\boldsymbol{T}(\boldsymbol{X}) = \boldsymbol{t}$, does not depend on $\boldsymbol{\theta}$.* □

This means that, given $\boldsymbol{T}(\boldsymbol{x}) = \boldsymbol{t}$, full knowledge of the measurement $\boldsymbol{x}$ does not bring any additional information concerning $\boldsymbol{\theta}$.

Example 18.1: Sufficient statistic for Bernoulli distribution parameter. Consider Bernoulli trials of size n: we define variable X_i to be 1 when the ith trial is a success and 0 when the trial is a failure, $i = 1, 2, \ldots, n$. Let θ be the probability of success, which is unknown and we wish to estimate. The probability distribution of the outcome of n trials $\boldsymbol{X} = (X_1, X_2, \ldots, X_n)$ is given by the *Bernoulli distribution* discussed in Section 3.3.1:

$$p_X(\boldsymbol{x}; \theta) = \prod_{i=1}^{n} \theta^{x_i} (1-\theta)^{1-x_i} = \theta^t (1-\theta)^{n-t}, \text{ for all } \boldsymbol{x} \in \{0, 1\}^n, \tag{18.3}$$

where $t = \sum_{i=1}^{n} x_i \triangleq T(\boldsymbol{x})$ is the number of successes in the trials.

Recall that we also discussed the Bernoulli trials in Example 4.4 of Section 4.5, where we treated the parameter as the random variable Θ and updated the estimate by using Bayes' theorem. Here, we assume θ is a fixed constant, so we are discussing the same estimation problem from the frequentist's point of view.

We will show that $T(\boldsymbol{x})$ is a sufficient statistic; i.e., the conditional probability distribution $p_{X|T}(\boldsymbol{x}|t; \theta)$ does not involve θ. The Bernoulli distribution (18.3) represents, by definition, the probability distribution of the vector variable $\boldsymbol{X}$, and can be viewed as the joint probability distribution of $\boldsymbol{X}$ and $T(\boldsymbol{X})$:

$$p_{\boldsymbol{X},T}(\boldsymbol{x}, t; \theta) = \left(\frac{\theta}{1-\theta}\right)^t (1-\theta)^n, \quad \text{for } \boldsymbol{x} \in \{0, 1\}^n \text{ and } t \in [0, n]. \tag{18.4}$$

The marginal distribution of the variable T is computed as

$$p_T(t; \theta) = \sum_{\boldsymbol{x}:T(\boldsymbol{x})=t} p_{\boldsymbol{X},T}(\boldsymbol{x}, t; \theta) = \binom{n}{t}\left(\frac{\theta}{1-\theta}\right)^t (1-\theta)^n, \tag{18.5}$$

which is the *binomial distribution* defined in (3.62) of Section 3.3.1. Then, the conditional probability $p_{\boldsymbol{X}|T}(\boldsymbol{x}|t; \theta)$ is given as

$$p_{\boldsymbol{X}|T}(\boldsymbol{x}|t; \theta) = \frac{p_{\boldsymbol{X},T}(\boldsymbol{x}, t; \theta)}{p_T(t; \theta)} = \frac{1}{\binom{n}{t}}, \tag{18.6}$$

which does not involve θ. Thus, $T(\boldsymbol{x}) = \sum_{i=1}^n x_i$ is a sufficient statistic for estimating θ. □

In the above example, we conjectured that $T(\boldsymbol{x}) = \sum_{i=1}^n x_i$ is a sufficient statistic and proved it by computing the conditional probability. In general it is not so straightforward to find a sufficient statistic. Fortunately, the following factorization theorem makes this task simpler:

THEOREM 18.1 (Fisher–Neyman factorization theorem). *Let $f_{\boldsymbol{X}}(\boldsymbol{x}; \boldsymbol{\theta})$ be the PDF of a continuous RV $\boldsymbol{X}$, parameterized by $\boldsymbol{\theta}$. A statistic $\boldsymbol{T}(\boldsymbol{x})$ is sufficient for $\boldsymbol{\theta}$ if and only if there exist functions $g(\boldsymbol{T}(\boldsymbol{x}); \boldsymbol{\theta})$ and $h(\boldsymbol{x})$ such that*

$$\boxed{f_{\boldsymbol{X}}(\boldsymbol{x}; \boldsymbol{\theta}) = g(\boldsymbol{T}(\boldsymbol{x}); \boldsymbol{\theta})h(\boldsymbol{x}),} \tag{18.7}$$

for all $\boldsymbol{x} \in \mathcal{X}$.

For a discrete RV $\boldsymbol{X}$, the above factorization should hold for the PMF $p_{\boldsymbol{X}}(\boldsymbol{x}; \boldsymbol{\theta})$.

Proof. We consider the case where $\boldsymbol{X}$ is a discrete RV. Let the joint probability distribution of $(\boldsymbol{X}, \boldsymbol{T}(\boldsymbol{X}))$ be denoted as $p_{\boldsymbol{X},\boldsymbol{T}}(\boldsymbol{x}, \boldsymbol{t}; \boldsymbol{\theta})$. By substituting $\boldsymbol{t} = \boldsymbol{T}(\boldsymbol{x})$ into this probability, we find the probability distribution of $\boldsymbol{X}$; i.e.,

$$p_{\boldsymbol{X},\boldsymbol{T}}(\boldsymbol{x}, \boldsymbol{T}(\boldsymbol{x}); \boldsymbol{\theta}) = p_{\boldsymbol{X}}(\boldsymbol{x}; \boldsymbol{\theta}). \tag{18.8}$$

If $\boldsymbol{T}(\boldsymbol{x})$ is a sufficient statistic for $\boldsymbol{\theta}$, then from the definition, we have

$$p_{\boldsymbol{X}|\boldsymbol{T}}(\boldsymbol{x}|\boldsymbol{t}; \boldsymbol{\theta}) = p_{\boldsymbol{X}|\boldsymbol{T}}(\boldsymbol{x}|\boldsymbol{t}). \tag{18.9}$$

Then,

$$p_{\boldsymbol{X},\boldsymbol{T}}(\boldsymbol{x}, \boldsymbol{t}; \boldsymbol{\theta}) = p_{\boldsymbol{T}}(\boldsymbol{t}; \boldsymbol{\theta})p_{\boldsymbol{X}|\boldsymbol{T}}(\boldsymbol{x}|\boldsymbol{t}; \boldsymbol{\theta}) = p_{\boldsymbol{T}}(\boldsymbol{t}; \boldsymbol{\theta})p_{\boldsymbol{X}|\boldsymbol{T}}(\boldsymbol{x}|\boldsymbol{t}). \tag{18.10}$$

Then, by setting $p_T(t;\theta) = g(t;\theta)$ and $p_{X|T}(x|t) = p_{X|T}(x|T(x)) = h(x)$, we have

$$p_X(x;\theta) = p_{X,T}(x, T(x);\theta) = g(t;\theta)h(x). \tag{18.11}$$

Conversely, let us assume that the probability distribution $p_X(x;\theta)$ satisfies the factorization (18.11) for some $g(t;\theta)$ and $h(x)$. Then,

$$\begin{aligned} p_T(t;\theta) &= \sum_{x:T(x)=t} p_{X,T}(x,t;\theta) = \sum_{x:T(x)=t} p_X(x;\theta) \\ &= \sum_{x:T(x)=t} g(t;\theta)h(x) = g(t;\theta)\sum_{x:T(x)=t} h(x). \end{aligned} \tag{18.12}$$

Then the conditional PDF of $X = x$ given $T = t$ should be expressed as

$$\begin{aligned} p_{X|T}(x|t;\theta) &= \frac{p_{X,T}(x,t;\theta)}{p_T(t;\theta)} = \frac{g(t;\theta)h(x)}{g(t;\theta)\sum_{x:T(x)=t} h(x)} \\ &= \frac{h(x)}{\sum_{x:T(x)=t} h(x)}, \end{aligned} \tag{18.13}$$

which does not depend on θ and represents the conditional PDF $p_{X|T}(x|t)$.

For the case where the random vectors X and T are continuous, the proof is much more involved. The reader is directed to advanced books on mathematical statistics, e.g., [25]. □

Example 18.2: Estimating the unknown mean of a normal distribution. Consider a random variable X with the normal distribution $N(\theta, 1)$ whose variance $\sigma^2 = 1$ is known. We want to estimate the unknown mean θ based on n independent samples, $x = (x_1, x_2, \ldots, x_n)$. We can write the PDF of x as

$$\begin{aligned} f_X(x;\theta) &= \prod_{i=1}^{n} \frac{1}{\sqrt{2\pi}} \exp\left[-\frac{(x_i-\theta)^2}{2}\right] \\ &= \frac{1}{(2\pi)^{n/2}} \exp\left(-\frac{1}{2}\sum_{i=1}^{n} x_i^2\right) \exp\left(\theta\sum_{i=1}^{n} x_i - \frac{n\theta^2}{2}\right). \end{aligned}$$

Then, from the factorization theorem, we readily find that $T_n(x) = \sum_{i=1}^n x_i$ is a sufficient statistic for θ.

Then define

$$\hat{\theta}(x) = \frac{T_n(x)}{n} = \frac{1}{n}\sum_{i=1}^{n} x_i,$$

which is the sample mean $\overline{X}_n$. We readily see that $\hat{\theta}(x) = \overline{X}_n$ is an *unbiased estimate* of θ, since $E[\hat{\theta}(X)] = \frac{1}{n}\sum_{i=1}^n E[X_i] = n\theta/n = \theta$. The variance of this estimate is

$$\mathrm{Var}[\overline{X}_n] = \frac{\sum_{i=1}^n \mathrm{Var}[X_i]}{n^2} = \frac{1}{n}.$$

Then from the weak law of large numbers (Theorem 11.27) (or by applying Chebysehv's inequality) we readily see that the sequence $\{\overline{X}_n - \theta\}$ converges in probability to zero. Thus, the estimator $\hat{\theta}(\boldsymbol{X}) = \overline{X}_n$ is a *consistent* estimator. This estimators also turns out to be *efficient*, but we defer its proof until Section 18.1.3.

In Example 4.6 of Section 4.5 we discussed the problem of estimating the mean parameter from the Bayes inference point of view. □

18.1.1 Exponential family of distributions revisited

In Section 4.4 we introduced the notion of the *canonical* or *natural* **exponential family** of distributions:

$$f_X(\boldsymbol{x};\boldsymbol{\eta}) = h(\boldsymbol{x})\exp[\boldsymbol{\eta}^\top \boldsymbol{T}(\boldsymbol{x}) - A(\boldsymbol{\eta})], \tag{18.14}$$

We readily see, from the Fisher–Neyman factorization theorem, that $\boldsymbol{T}(\boldsymbol{x})$ is a sufficient statistic for the parameter $\boldsymbol{\eta}$. Since $f_X(\boldsymbol{x};\boldsymbol{\eta})$ is a PDF,

$$\int f_X(\boldsymbol{x};\boldsymbol{\eta})\,d\boldsymbol{x} = \int h(\boldsymbol{x})\exp[\boldsymbol{\eta}^\top \boldsymbol{T}(\boldsymbol{x}) - A(\boldsymbol{\eta})]\,d\boldsymbol{x} = 1, \tag{18.15}$$

which yields

$$\exp[A(\boldsymbol{\eta})] = \int h(\boldsymbol{x})\exp[\boldsymbol{\eta}^\top \boldsymbol{T}(\boldsymbol{x})]\,d\boldsymbol{x}. \tag{18.16}$$

Thus, $A(\boldsymbol{\eta})$ is uniquely determined by $h(x)$ and $\boldsymbol{T}(\boldsymbol{x})$.

The MGF for the RV $\boldsymbol{T}(\boldsymbol{X})$ is, from (8.40), given by

$$\begin{aligned} M_{\boldsymbol{T}}(\boldsymbol{u}) &\triangleq \int \exp[\boldsymbol{u}^\top \boldsymbol{T}(\boldsymbol{x})] f_X(\boldsymbol{x};\boldsymbol{\eta})\,d\boldsymbol{x} \\ &= \exp[-A(\boldsymbol{\eta})]\int h(\boldsymbol{x})\exp[\boldsymbol{\eta}+\boldsymbol{u})^\top \boldsymbol{T}(\boldsymbol{x})]\,d\boldsymbol{x} \\ &= \exp[-A(\boldsymbol{\eta})]\exp[A(\boldsymbol{\eta}+\boldsymbol{u})], \end{aligned} \tag{18.17}$$

where we used the relation (18.16) to arrive at the last expression. Thus,

$$\boxed{M_{\boldsymbol{T}}(\boldsymbol{u}) = \exp[A(\boldsymbol{\eta}+\boldsymbol{u}) - A(\boldsymbol{\eta})].} \tag{18.18}$$

The *logarithmic MGF* or *cumulant MGF*, defined in (8.6), is therefore given by

$$m_{\boldsymbol{T}}(\boldsymbol{u}) \triangleq \ln M_{\boldsymbol{T}}(\boldsymbol{u}) = A(\boldsymbol{\eta}+\boldsymbol{u}) - A(\boldsymbol{\eta}). \tag{18.19}$$

We can then find the mean (vector) and the variance (matrix) of $T(\boldsymbol{X})$ by differentiating $m_{\boldsymbol{T}}(\boldsymbol{u})$ once and twice respectively, and by setting $\boldsymbol{u} = \boldsymbol{0}$. Because of the relation (18.19), they can be found by differentiating $A(\boldsymbol{\eta})$ with respect to $\boldsymbol{\eta}$:

$$E[\boldsymbol{T}(\boldsymbol{X})] = \nabla_{\boldsymbol{u}} m_{\boldsymbol{T}}(0) = \nabla_{\boldsymbol{\eta}} A(\boldsymbol{\eta}) \tag{18.20}$$

and

$$\text{Var}[\boldsymbol{T}(\boldsymbol{X})] = \nabla_{\boldsymbol{u}} \nabla_{\boldsymbol{u}}^{\top} m_{\boldsymbol{T}}(\mathbf{0}) = \nabla_{\boldsymbol{\eta}} \nabla_{\boldsymbol{\eta}}^{\top} A(\boldsymbol{\eta}), \tag{18.21}$$

where the operator[1] $\nabla_{\boldsymbol{\eta}}$ represents the the *vector differential operator* or *gradient operator*

$$\nabla_{\boldsymbol{\eta}} A(\boldsymbol{\eta}) \triangleq \left(\frac{\partial A(\boldsymbol{\eta})}{\partial \eta_1}, \frac{\partial A(\boldsymbol{\eta})}{\partial \eta_2}, \ldots, \frac{\partial A(\boldsymbol{\eta})}{\partial \eta_M} \right)^{\top} \tag{18.22}$$

and $\nabla_{\boldsymbol{\eta}} \nabla_{\boldsymbol{\eta}}^{\top} A(\boldsymbol{\eta})$ is the **Hessian matrix** of $A(\boldsymbol{\eta})$:

$$\nabla_{\boldsymbol{\eta}} \nabla_{\boldsymbol{\eta}}^{\top} A(\boldsymbol{\eta}) \triangleq \left[\frac{\partial^2 A(\boldsymbol{\eta})}{\partial \eta_i \partial \eta_j} \right]_{M \times M}. \tag{18.23}$$

18.1.2 Maximum-likelihood estimation

We continue to consider the case where we know the functional form of the PDF $f_X(\boldsymbol{x}; \boldsymbol{\theta})$ (or the PMF $p_X(\boldsymbol{x}; \boldsymbol{\theta})$ for the discrete case), but the value of the parameter $\boldsymbol{\theta}$ is unknown. We wish to estimate $\boldsymbol{\theta}$ based on the data $\boldsymbol{x} = (x_1, x_2, \ldots, x_n)$. Then a reasonable procedure is to find the value of $\boldsymbol{\theta}$ that is most likely to produce this $\boldsymbol{x}$. So we define the following function of $\boldsymbol{\theta}$, with $\boldsymbol{x}$ fixed, which is called the **likelihood function**:

$$L_{\boldsymbol{x}}(\boldsymbol{\theta}) \triangleq \begin{cases} f_X(\boldsymbol{x}; \boldsymbol{\theta}), & \text{for continuous RV } \boldsymbol{X}, \\ p_X(\boldsymbol{x}; \boldsymbol{\theta}), & \text{for discrete RV } \boldsymbol{X}. \end{cases} \tag{18.24}$$

The parameter $\boldsymbol{\theta}$ may be either discrete or continuous. Any value of $\boldsymbol{\theta}$ that maximizes the likelihood function is called a **maximum-likelihood estimate** (MLE) and denoted as $\hat{\boldsymbol{\theta}}$:

$$\hat{\boldsymbol{\theta}} = \arg\max_{\boldsymbol{\theta}} L_{\boldsymbol{x}}(\boldsymbol{\theta}). \tag{18.25}$$

This value may not always be unique, or in some cases may not even exist. If we know that it exists and is unique, then it may be appropriate to call it *the* MLE. The procedure to find an MLE is called **maximum-likelihood estimation**.

The concept of the MLE is simple, and it can be shown (see the next section) that an MLE is *asymptotically* unbiased, efficient, and normally distributed as the sample size $n \to \infty$. In some cases we may be able to find a closed-form expression of the MLE in terms of $\boldsymbol{x}$, but it is generally difficult to find it analytically and we have to resort to some numerical technique, as we shall discuss in Sections 19.1 and 19.2.

[1] This operator, which is called **nabla**, from the Greek word for a Hebrew harp that has a similar shape, was introduced by Sir W. R. Hamilton (1805–1865), who was an Irish physicist, astronomer, and mathematician.

If the likelihood function is differentiable with respect to its parameter, a necessary condition for an MLE to satisfy is

$$\nabla_{\boldsymbol{\theta}} L_{\boldsymbol{x}}(\boldsymbol{\theta}) = \mathbf{0}, \ \text{ i.e., } \frac{\partial L_{\boldsymbol{x}}(\boldsymbol{\theta})}{\partial \theta_m} = 0, \quad m = 1, 2, \ldots, M, \tag{18.26}$$

where the $\nabla_{\boldsymbol{\theta}}$ is the differential operator defined in (18.22). Since the logarithmic function is a monotonically increasing and differentiable function, $\boldsymbol{\theta}$ that satisfies the conditions (18.26) can be found from

$$\frac{\partial \log L_{\boldsymbol{x}}(\boldsymbol{\theta})}{\partial \theta_m} = 0, \quad m = 1, 2, \ldots, M. \tag{18.27}$$

These equations have, in general, multiple roots, and in order to find the MLE $\hat{\boldsymbol{\theta}}$, we must select the solution that yields the largest value of $L_{\boldsymbol{x}}(\boldsymbol{\theta})$.

The function $\log L_{\boldsymbol{x}}(\boldsymbol{\theta})$ is called the **log-likelihood function**, and its partial derivative with respect to $\boldsymbol{\theta}$ is called the **score function**, denoted as $\boldsymbol{s}(\boldsymbol{x}; \boldsymbol{\theta})$:

$$\boxed{\boldsymbol{s}(\boldsymbol{x}; \boldsymbol{\theta}) = \nabla_{\boldsymbol{\theta}} \log L_{\boldsymbol{x}}(\boldsymbol{\theta}) \triangleq \left(\frac{\partial \log L_{\boldsymbol{x}}(\boldsymbol{\theta})}{\partial \theta_1}, \frac{\partial \log L_{\boldsymbol{x}}(\boldsymbol{\theta})}{\partial \theta_2}, \ldots, \frac{\partial \log L_{\boldsymbol{x}}(\boldsymbol{\theta})}{\partial \theta_M} \right)^{\top}.} \tag{18.28}$$

The score function represents the rate at which $\log L_{\boldsymbol{x}}(\boldsymbol{\theta})$ changes as $\boldsymbol{\theta}$ varies. If it can be expanded in a Taylor series around the parameter's true (but unknown) value $\boldsymbol{\theta}_0$, then

$$\boldsymbol{s}(\boldsymbol{x}; \boldsymbol{\theta}) = \boldsymbol{s}(\boldsymbol{x}; \boldsymbol{\theta}_0) + \nabla_{\boldsymbol{\theta}} \boldsymbol{s}^{\top}(\boldsymbol{x}; \boldsymbol{\theta}_0)(\boldsymbol{\theta} - \boldsymbol{\theta}_0) + \boldsymbol{r}(\boldsymbol{x}; \boldsymbol{\theta}), \tag{18.29}$$

where the remainder term $\boldsymbol{r}(\boldsymbol{x}; \boldsymbol{\theta})$ is on the order of $O(\|\boldsymbol{\theta} - \boldsymbol{\theta}_0\|^2)$ and is negligibly small for $\boldsymbol{\theta}$ in the vicinity of $\boldsymbol{\theta}_0$. Then by setting the above equation equal to zero, we find that $\hat{\boldsymbol{\theta}}$ that satisfies (18.27) is given by

$$\begin{aligned} \hat{\boldsymbol{\theta}} &\approx \boldsymbol{\theta}_0 - \left[\nabla_{\boldsymbol{\theta}} \boldsymbol{s}^{\top}(\boldsymbol{x}; \boldsymbol{\theta}_0) \right]^{-1} \boldsymbol{s}(\boldsymbol{x}; \theta_0) \\ &= \boldsymbol{\theta}_0 - \boldsymbol{H}^{-1}(\boldsymbol{x}; \boldsymbol{\theta}_0) \boldsymbol{s}(\boldsymbol{x}; \theta_0), \end{aligned} \tag{18.30}$$

where

$$\boldsymbol{H}(\boldsymbol{x}; \boldsymbol{\theta}_0) \triangleq \nabla_{\boldsymbol{\theta}} \boldsymbol{s}^{\top}(\boldsymbol{x}; \boldsymbol{\theta}_0) = \nabla_{\boldsymbol{\theta}} \nabla_{\boldsymbol{\theta}}^{\top} \log L_{\boldsymbol{x}}(\boldsymbol{\theta}_0) = \left[\frac{\partial^2 \log L_{\boldsymbol{x}}(\boldsymbol{\theta}_0)}{\partial \theta_i \partial \theta_j} \right]_{M \times M} \tag{18.31}$$

is the *Hessian matrix* of $\log L_{\boldsymbol{x}}(\boldsymbol{\theta})$ evaluated at $\boldsymbol{\theta} = \boldsymbol{\theta}_0$. The negative of this Hessian matrix

$$\boxed{\mathcal{J}(\boldsymbol{x}; \boldsymbol{\theta}_0) = -\boldsymbol{H}(\boldsymbol{x}; \boldsymbol{\theta}_0)} \tag{18.32}$$

is called the **observed Fisher information matrix**.

Thus, the above $\hat{\boldsymbol{\theta}}$, which is the MLE if it gives a global maximum of $L_{\boldsymbol{x}}(\boldsymbol{\theta})$, can be written as

$$\hat{\boldsymbol{\theta}} \approx \boldsymbol{\theta}_0 + \mathcal{J}(\boldsymbol{x}; \boldsymbol{\theta}_0)^{-1} \boldsymbol{s}(\boldsymbol{x}; \theta_0). \tag{18.33}$$

It is easy to show that the expectation of the score function is zero (see Problem 18.4) for any $\boldsymbol{\theta}$; i.e.,

$$E[\boldsymbol{s}(\boldsymbol{X};\theta_0)] = \boldsymbol{0}. \tag{18.34}$$

The matrix $\mathcal{J}(\boldsymbol{X};\boldsymbol{\theta})$ is also a random variable and its expectation

$$\boxed{\mathcal{I}_{\boldsymbol{x}}(\boldsymbol{\theta}_0) \triangleq E[\mathcal{J}(\boldsymbol{X};\boldsymbol{\theta}_0)]} \tag{18.35}$$

is called the **Fisher information matrix**.

Example 18.3: MLE of $\boldsymbol{\theta} = (\mu, \sigma^2)$ of a normal distribution. Suppose that n independent samples are taken from a common normal distribution $N(\mu, \sigma^2)$, where both mean and variance are unknown, and we wish to find an MLE of these parameters based on $\boldsymbol{x} = (x_1, x_2, \ldots, x_n)$. By setting $\boldsymbol{\theta} = (\mu, \sigma^2)$, we have the likelihood function

$$L_{\boldsymbol{x}}(\boldsymbol{\theta}) = f(\boldsymbol{x};\boldsymbol{\theta}) = \frac{1}{(2\pi\sigma^2)^{n/2}} \exp\left[-\frac{\sum_{i=1}^n (x_i-\mu)^2}{2\sigma^2}\right].$$

Therefore, the log-likelihood function is

$$\log L_{\boldsymbol{x}}(\boldsymbol{\theta}) = -\frac{n\log(2\pi)}{2} - n\log\sigma - \frac{1}{2\sigma^2}\sum_{i=1}^n (x_i-\mu)^2.$$

Then (18.27) yields

$$\frac{1}{2\sigma^2}\sum_{i=1}^n (x_i-\mu) = 0 \text{ and } -n\sigma + \frac{1}{\sigma^3}\sum_{i=1}^n (x_i-\mu)^2 = 0,$$

from which we have

$$\mu^\star = \frac{\sum_{i=1}^n x_i}{n} \text{ and } \sigma^{2\star} = \frac{\sum_{i=1}^n (x_i-\mu^\star)^2}{n} = \frac{1}{n}\sum_{i=1}^n x_i^2 - \mu^{\star 2}. \tag{18.36}$$

It is not difficult to calculate the Hessian matrix (18.31), obtaining

$$\boldsymbol{H}(\boldsymbol{x};\boldsymbol{\theta}_0) = -\frac{1}{\sigma^2{}_0}\begin{bmatrix} \frac{1}{2} & 0 \\ 0 & 2n \end{bmatrix},$$

which is clearly negative definite. In the case of the normal distribution, the log-likelihood function is a *quadratic equation* in $\boldsymbol{\theta}$. Thus, in the Taylor expansion of the score function, the higher order term $\boldsymbol{r}(\boldsymbol{x};\boldsymbol{\theta})$ is nonexistent. Hence, the "$\approx$" used in (18.30) and (18.33) should be replaced by "$=$," and the solution given by $\boldsymbol{\theta}^\star = (\mu^\star, \sigma^{2\star})$ of (18.36) is a global maximum point; hence, it is the MLE $\hat{\boldsymbol{\theta}}$.

It is worthwhile noting that we will arrive at the same result if we define $\boldsymbol{\theta} = (\mu, \sigma)$ instead of $\boldsymbol{\theta} = (\mu, \sigma^2)$ (Problem 18.5). In other words, $\sqrt{\widehat{\sigma^2}} = \hat{\sigma}$. The MLEs in general possess this property: the same value of a parameter is obtained whether we estimate θ itself or some monotone function $g(\theta)$ of this parameter. □

18.1.2.1 Maximum-likelihood estimate for the exponential family distribution

We gave the definition and basic properties of the *canonical exponential family* of distribution in Section 4.4. Let us now focus on parameter estimation for this class of distributions. Assume that we have n independent samples $\boldsymbol{x} = \{x_1, x_2, \ldots, x_n\}$ drawn from a common distribution $f(x; \boldsymbol{\eta})$ that belongs to the canonical exponential family. In other words, we have

$$f_{\boldsymbol{X}}(\boldsymbol{x}; \boldsymbol{\eta}) = \prod_{i=1}^{n} f(x_i; \boldsymbol{\eta}),$$

where

$$f(x_i; \boldsymbol{\eta}) = h(x_i) \exp\left[\boldsymbol{\eta}^\top \boldsymbol{T}(x_i) - A(\boldsymbol{\eta})\right].$$

Note that each sample X_i may itself be vector-valued. The log-likelihood function for the parameter $\boldsymbol{\eta}$ based on the n sampled data is

$$\log f_{\boldsymbol{X}}(\boldsymbol{x}; \boldsymbol{\eta}) = \sum_{i=1}^{n} \boldsymbol{\eta}^\top \boldsymbol{T}(x_i) - nA(\boldsymbol{\eta}) + r(\boldsymbol{x}), \tag{18.37}$$

where

$$r(\boldsymbol{x}) = \sum_{i=1}^{n} \log h(x_i) \tag{18.38}$$

does not depend on $\boldsymbol{\eta}$. By differentiating (18.37) with respect to $\boldsymbol{\eta}$, and setting it to zero, we find that an MLE $\hat{\boldsymbol{\eta}}$ must satisfy the following equation:

$$\boxed{\nabla_{\boldsymbol{\eta}} A(\hat{\boldsymbol{\eta}}) = \frac{1}{n} \sum_{i=1}^{n} \boldsymbol{T}(x_i).} \tag{18.39}$$

Example 18.4: The MLE of a normal distribution – continued. Consider the normal RV X discussed in Examples 4.3 and 18.3. We write the PDF of each sample x_i $(i = 1, 2, \ldots)$ as

$$\begin{aligned} f(x_i; \boldsymbol{\theta}) &= \frac{1}{\sqrt{2\pi}\sigma} \exp\left[-\frac{(x_i - \mu)^2}{2\sigma^2}\right] \\ &= \frac{1}{\sqrt{2\pi}} \exp\left(-\frac{x_i^2}{2\sigma^2} + \frac{x_i \mu}{\sigma^2} - \frac{\mu^2}{2\sigma^2} - \log\sigma\right), \quad i = 1, 2, \ldots, n. \end{aligned}$$

We showed in Example 4.3 that this belongs to the canonical exponential family with

$$\boldsymbol{\eta} = \begin{bmatrix} \eta_1 \\ \eta_2 \end{bmatrix} = \begin{bmatrix} \frac{1}{\sigma^2} \\ \frac{\mu}{\sigma^2} \end{bmatrix}, \quad \boldsymbol{T}(x) = \begin{bmatrix} -\frac{x^2}{2} \\ x \end{bmatrix},$$

$$h(\boldsymbol{X}) = \frac{1}{\sqrt{2\pi}}, \quad A(\boldsymbol{\eta}) = \frac{\mu^2}{2\sigma^2} + \log\sigma.$$

The original parameters $\boldsymbol{\theta} = (\mu, \sigma)$ can be expressed as $\mu = \eta_2/\eta_1$ and $\sigma^2 = 1/\eta_1$. Hence,

$$A(\boldsymbol{\eta}) = \frac{\eta_2^2}{2\eta_1} - \frac{\log \eta_1}{2}.$$

Maximum-likelihood estimates of η_1 and η_2 and the statistic $\boldsymbol{T}(\boldsymbol{x})$ must satisfy (18.39):

$$\nabla_{\eta} A(\hat{\boldsymbol{\eta}}) = \left[-\frac{\hat{\eta}_2^2}{2\hat{\eta}_1^2} - \frac{1}{2\hat{\eta}_1}, \frac{\hat{\eta}_2}{\hat{\eta}_1} \right]^\top = \frac{1}{n}\sum_{i=1}^{n} \boldsymbol{T}(x_i)$$
$$= \frac{1}{n}\left[-\frac{1}{2}\sum_{i=1}^{n} x_i^2, \sum_{i=1}^{n} x_i \right]^\top.$$

Thus, the $\hat{\eta}_1$ and $\hat{\eta}_2$ are uniquely determined, and they in turn uniquely determine the MLE of μ and σ^2:

$$\hat{\mu} = \frac{\hat{\eta}_2}{\hat{\eta}_1} = \frac{1}{n}\sum_{i=1}^{n} x_i \text{ and } \widehat{\sigma^2} = \frac{1}{\hat{\eta}_1} = \frac{1}{n}\sum_{i=1}^{n} x_i^2 - \hat{\mu}^2.$$

These MLEs indeed agree with those we found in Example 18.3. □

18.1.3 Cramér–Rao lower bound

In the introduction to the present section we defined some desirable properties of an estimator. They are unbiasedness, efficiency (or minimum variance), and consistency. We will study in this subsection what is the minimum variance that any unbiased estimator can possibly achieve.

C. R. Cramér [74] and C. R. Rao [277] showed that for probability distributions that satisfy certain *regularity conditions* (such as differentiability, validity of interchange of differentiation with respect to the parameter θ, and integration with respect to $\boldsymbol{x}$, etc.; see Problem 18.4 (c)), the MSE of a single parameter (i.e., one-dimensional parameter) θ is subject to the following bound, now widely known as the **Cramér–Rao lower bound** (**CRLB**) or Cramér–Rao inequality:

$$\boxed{E[(\hat{\theta} - \theta)^2] \geq \frac{|1 + b'(\theta)|^2}{\mathcal{I}_{\boldsymbol{x}}(\theta)},} \tag{18.40}$$

where $b(\theta) = E[\hat{\theta}(\boldsymbol{X})] - \theta$ is the bias of the estimator defined in (18.1) and $\mathcal{I}_{\boldsymbol{x}}(\theta)$ is the Fisher information (a scalar quantity for a single parameter), defined by (18.35) (see Problem 18.4 (c)):

$$\mathcal{I}_{\boldsymbol{x}}(\theta) = -E\left[\frac{\partial^2 \log L_{\boldsymbol{x}}(\theta)}{\partial \theta^2}\right], \tag{18.41}$$

where θ is the *true value* of the parameter. The CRLB for a one-dimensional parameter θ given by (18.40) can be derived from the Cauchy–Schwarz inequality discussed in Section 10.1.1 (Problem 18.6).

For an unbiased estimator $\hat{\theta}$, the numerator of the right-hand side of (18.40) is unity, and the left-hand side is the variance of the estimate of θ. A generalization to the estimation of a vector parameter $\boldsymbol{\theta}$ is given by Cramér [74].

THEOREM 18.2 (Cramér–Rao lower bound (CRLB)). *Let $\hat{\boldsymbol{\theta}}(\boldsymbol{x})$ be any unbiased estimator of $\boldsymbol{\theta}$. Then the following properties hold.*

1. *The variance matrix of $\hat{\boldsymbol{\theta}}(\boldsymbol{X})$ is bounded from below by the inverse of the Fisher information matrix:*

$$\boxed{\mathbf{Var}[\hat{\boldsymbol{\theta}}(\boldsymbol{X})] = E\left[(\hat{\boldsymbol{\theta}}(\boldsymbol{X}) - \boldsymbol{\theta})(\hat{\boldsymbol{\theta}}(\boldsymbol{X}) - \boldsymbol{\theta})^\top\right] \geq \mathcal{I}_x^{-1}(\boldsymbol{\theta}).} \tag{18.42}$$

2. *The lower bound is attained if and only if $\hat{\boldsymbol{\theta}}(\boldsymbol{X})$ satisfies the following equation*

$$\boldsymbol{s}(\boldsymbol{X};\boldsymbol{\theta}) = \mathcal{I}_x(\boldsymbol{\theta})(\hat{\boldsymbol{\theta}}(\boldsymbol{X}) - \boldsymbol{\theta}), \tag{18.43}$$

where $\boldsymbol{s}(\boldsymbol{X};\boldsymbol{\theta})$ is defined by (18.28).

Proof.

1. First, consider the following simple formula for matrix variances and covariances:

$$\mathbf{Var}[\boldsymbol{A} - \boldsymbol{B}] = \mathbf{Var}[\boldsymbol{A}] + \mathbf{Var}[\boldsymbol{B}] - \mathbf{Cov}[\boldsymbol{A}, \boldsymbol{B}] - \mathbf{Cov}[\boldsymbol{B}, \boldsymbol{A}] \geq \mathbf{0}, \tag{18.44}$$

where the equality in the right-hand side holds if and only if $\boldsymbol{A} - \boldsymbol{B} =$ constant; i.e., if and only if $\boldsymbol{A} - \boldsymbol{B} = E[\boldsymbol{A} - \boldsymbol{B}]$. By setting $\boldsymbol{A} = \hat{\boldsymbol{\theta}}(\boldsymbol{X})$ and $\boldsymbol{B} = \mathcal{I}_x^{-1}\boldsymbol{s}(\boldsymbol{X})$, where for notational brevity we write

$$\mathcal{I}_x \triangleq \mathcal{I}_x(\boldsymbol{\theta}),$$
$$\boldsymbol{s}(\boldsymbol{X}) \triangleq \boldsymbol{s}(\boldsymbol{X};\boldsymbol{\theta}) = \nabla_{\boldsymbol{\theta}} \log f_{\boldsymbol{X}}(\boldsymbol{X};\boldsymbol{\theta}),$$

we have

$$\begin{aligned}\mathbf{Var}[\hat{\boldsymbol{\theta}}(\boldsymbol{X}) - \mathcal{I}_x^{-1}\boldsymbol{s}(\boldsymbol{X})] &= \mathbf{Var}[\hat{\boldsymbol{\theta}}(\boldsymbol{X})] + \mathcal{I}_x^{-1}\mathbf{Var}[\boldsymbol{s}(\boldsymbol{X})]\mathcal{I}_x^{-1} \\ &- \mathbf{Cov}[\hat{\boldsymbol{\theta}}(\boldsymbol{X}), \boldsymbol{s}(\boldsymbol{X})]\mathcal{I}_x^{-1} - \mathcal{I}_x^{-1}\mathbf{Cov}[\boldsymbol{s}(\boldsymbol{X}), \hat{\boldsymbol{\theta}}(\boldsymbol{X})] \geq \mathbf{0}.\end{aligned} \tag{18.45}$$

Using formulas (18.113) and (18.115) proved in Problem 18.4,

$$\boxed{\mathbf{Var}[\boldsymbol{s}(\boldsymbol{X})] = \mathcal{I}_x \text{ and } \mathbf{Cov}[\hat{\boldsymbol{\theta}}(\boldsymbol{X}), \boldsymbol{s}(\boldsymbol{X})] = \mathbf{Cov}[\boldsymbol{s}(\boldsymbol{X}), \hat{\boldsymbol{\theta}}(\boldsymbol{X})] = \boldsymbol{I},} \tag{18.46}$$

we find

$$\mathbf{Var}[\hat{\boldsymbol{\theta}}(\boldsymbol{X}) - \mathcal{I}_x^{-1}\boldsymbol{s}(\boldsymbol{X})] = \mathbf{Var}[\hat{\boldsymbol{\theta}}(\boldsymbol{X})] - \mathcal{I}_x^{-1} \geq \mathbf{0}. \tag{18.47}$$

2. The equality in the above holds if and only if

$$\hat{\boldsymbol{\theta}}(\boldsymbol{X}) - E[\hat{\boldsymbol{\theta}}(\boldsymbol{X})] = \mathcal{I}_{\boldsymbol{x}}^{-1}(\boldsymbol{s}(\boldsymbol{X}) - E[\boldsymbol{s}(\boldsymbol{X})]).$$

Since $\hat{\boldsymbol{\theta}}(\boldsymbol{X})$ is an unbiased estimator and $E[\boldsymbol{s}(\boldsymbol{X})] = 0$ (see (18.113)), we have

$$\hat{\boldsymbol{\theta}}(\boldsymbol{X}) - \boldsymbol{\theta} = \mathcal{I}_{\boldsymbol{x}}^{-1}\boldsymbol{s}(\boldsymbol{X}),$$

which implies the vector equation (18.43).

□

Furthermore, we can show that such $\hat{\boldsymbol{\theta}}(\boldsymbol{X})$ that satisfies the CRLB is a *sufficient statistic* for estimating $\boldsymbol{\theta}$ (Problem 18.7).

Recall that the **minimum variance unbiased estimator** (MVUE) that attains the CRLB is called **efficient**, as defined in Definition 18.1.

Example 18.5: The normal distribution $N(\mu, 1)$. Let us consider again the normal distribution $N(\mu, 1)$ discussed in Example 18.2. We found that the sample mean $\overline{X} = \frac{1}{n}\sum_{i=1}^{n} x_i$ is an unbiased and consistent estimator for the distribution mean μ. In Example 18.3 we showed that this estimator is an MLE. Let us investigate whether it is also an efficient estimator.

The variance of the sum of n i.i.d. RVs is n times the variance of one RV. Thus, the variance of $\frac{1}{n}\sum_{i=1}^{n} X_i$ is

$$\text{Var}[\overline{X}] = \frac{1}{n^2} n \text{Var}[X_i] = \frac{1}{n},$$

which is the left-hand side of the Cramér–Rao inequality.

Since the log-likelihood function is given by

$$\log f_{\boldsymbol{X}}(\boldsymbol{x}; \mu) = -\frac{\sum_{i=1}^{n}(x_i - \mu)^2}{2} - \frac{n}{2}\log(2\pi),$$

we have

$$\frac{\partial \log f_{\boldsymbol{X}}(\boldsymbol{x}; \mu)}{\partial \mu} = \sum_{i=1}^{n}(x_i - \mu) = n(\overline{X} - \mu). \tag{18.48}$$

The Fisher information of n independent samples $\boldsymbol{x} = (x_1, x_2, \ldots, x_n)$ is given by $\mathcal{I}_{\boldsymbol{x}}(\mu) = \sum_{i=1}^{n} \mathcal{I}_{x_i}(\mu) = n$, where

$$\mathcal{I}_{x_i}(\mu) = E[(X - \mu)^2)] = 1$$

is the Fisher information for each sample x_i. Thus, the CRLB for any statistic $T(\boldsymbol{x})$ is given by $\mathcal{I}_{\boldsymbol{x}}^{-1}(\mu) = 1/n$, as expected. Note that (18.48) is equivalent to (18.43). This proves that the sample mean $\overline{X}$ is an efficient estimate.

□

18.1.3.1 Asymptotic unbiasedness, efficiency, and normality of the maximum-likelihood estimate

Now we return to the MLE of Section 18.1.2 and discuss its properties. Recall that the MLE satisfies (18.33). The "$\approx$" in that equation becomes "$=$", as $n \to \infty$, where n is the size of the data

$$\boldsymbol{x} = (x_1, x_2, \ldots, x_n) \triangleq \boldsymbol{x}_n.$$

Since the expected value of the score function is zero, the MLE is **asymptotically unbiased** i.e.,

$$\lim_{n\to\infty} E[\hat{\boldsymbol{\theta}}(\boldsymbol{x}_n)] = \boldsymbol{\theta},$$

where $\boldsymbol{\theta}$ is the true value of the parameter.

From (18.33) we also see that the variance matrix[2] of the MLE of $\boldsymbol{\theta}$ is approximated by

$$\begin{aligned}\mathbf{Var}[\hat{\boldsymbol{\theta}}] &\approx E\left[\boldsymbol{H}^{-1}(\boldsymbol{X}_n;\boldsymbol{\theta})\boldsymbol{s}(\boldsymbol{X}_n;\boldsymbol{\theta})\boldsymbol{s}^\top(\boldsymbol{X}_n;\boldsymbol{\theta})\boldsymbol{H}^{-1}(\boldsymbol{X}_n;\boldsymbol{\theta})\right] \\ &\approx \mathcal{I}_{\boldsymbol{x}}^{-1}(\boldsymbol{\theta})\mathcal{I}_{\boldsymbol{x}_n}(\boldsymbol{\theta})\mathcal{I}_{\boldsymbol{x}_n}^{-1}(\boldsymbol{\theta}) = \mathcal{I}_{\boldsymbol{x}_n}^{-1}(\boldsymbol{\theta}),\end{aligned} \tag{18.49}$$

which is the CRLB, where

$$\boldsymbol{X}_n \triangleq \boldsymbol{X} = (X_1, X_2, \ldots, X_n).$$

Therefore, the MLE $\hat{\boldsymbol{\theta}}(\boldsymbol{x}_n)$ is **asymptotically efficient**. In deriving the above result we used the property $E[\boldsymbol{s}(\boldsymbol{X}_n;\boldsymbol{\theta})\boldsymbol{s}^\top(\boldsymbol{X}_n;\boldsymbol{\theta})] = -E[\boldsymbol{\mathcal{J}}(\boldsymbol{X}_n;\boldsymbol{\theta})] = \mathcal{I}_{\boldsymbol{x}_n}(\boldsymbol{\theta})$ (see (18.113) of Problem 18.4), and the "law of large number" argument that, as n becomes large, $\boldsymbol{H}(\boldsymbol{X}_n;\boldsymbol{\theta}) = -\boldsymbol{\mathcal{J}}(\boldsymbol{X}_n;\boldsymbol{\theta})$ can be well approximated by its expectation $-\mathcal{I}_{\boldsymbol{x}_n}(\boldsymbol{\theta})$.

If the data $x_1, x_2, \ldots, x_n$ are i.i.d. samples taken from a common distribution $f(x;\boldsymbol{\theta})$, i.e.,

$$f_{\boldsymbol{X}}(\boldsymbol{x};\boldsymbol{\theta}) = \prod_{i=1}^{n} f(x_i;\boldsymbol{\theta}),$$

the asymptotic variance of the MLE (18.49) becomes

$$\mathbf{Var}[\hat{\boldsymbol{\theta}}] \approx \frac{1}{n}\mathcal{I}_x^{-1}(\boldsymbol{\theta}), \tag{18.50}$$

where $\mathcal{I}_x(\boldsymbol{\theta}) = \frac{1}{n}\mathcal{I}_{\boldsymbol{x}_n}(\boldsymbol{\theta})$ is the Fisher information matrix based on a single sample x:

$$\mathcal{I}_x(\boldsymbol{\theta}) = -E\left[\frac{\partial^2 \log f(x;\boldsymbol{\theta})}{\partial\theta_i\partial\theta_j}\right] = E\left[\frac{\partial \log f(x;\boldsymbol{\theta})}{\partial\theta_i}\frac{\partial \log f(x;\boldsymbol{\theta})}{\partial\theta_j}\right], \tag{18.51}$$

as given in (18.41) and proved in Problem 18.4 (c).

[2] This matrix is sometimes called the "covariance matrix," but we reserve this term for $\mathbf{Cov}[\boldsymbol{X}, \boldsymbol{Y}] = E[(\boldsymbol{X} - E[\boldsymbol{X}])(\boldsymbol{Y} - E[\boldsymbol{Y}])^\top]$ for two random vectors. Some authors call $\mathbf{Cov}[\boldsymbol{X}, \boldsymbol{Y}]$ the "cross-covariance matrix."

The score function $s(x; \theta)$ in (18.33) is the gradient of the log-likelihood function. For i.i.d. samples,

$$s(X; \theta) = \nabla_\theta \left(\sum_{i=1}^{n} \log f(X_i; \theta) \right). \tag{18.52}$$

Thus, from the central limit theorem, the summed term converges in distribution to a normal RV. Since the ∇_θ is a linear operator, the score function $s(X; \theta)$ converges in distribution to a normal random vector with mean zero. Thus, the MLE $\hat{\theta}$ (of (18.33)) is an **asymptotically normal** random vector with mean θ.

For a more rigorous discussion on the above asymptotic properties of the MLE, the interested reader is referred to, for example, Cramér [74] and Bickel and Doksum [25].

18.1.4 Interval or region estimation

In contrast with point estimation discussed in the preceding sections, *region estimation* uses sampled data $x = (x_1, x_2, \ldots, x_n)$ to estimate a region that covers an unknown population parameter θ. A **confidence region** $R_\gamma(x)$ is a range in the parameter space which is expected to include the estimand θ, and the probability that θ indeed falls in this region $P[\theta \in R_\gamma(x)] = \gamma$ is called the **confidence level** or **confidence coefficient**.

In many cases, the component parameters θ_i of the vector parameter θ may be treated separately. Then, a confidence region will become a Cartesian product of M separate **confidence intervals**. In the rest of this section, therefore, we will focus on the one-dimensional parameter space. An interval $I_\gamma = (c_1, c_2) \in \mathbb{R}$ such that

$$P[c_1 < \theta < c_2] = \gamma \tag{18.53}$$

is called a confidence interval associated with the confidence level γ. In order to have an accurate estimate, we want to make the confidence interval as narrow as possible. Thus, an efficient (i.e., unbiased minimum-variance) estimator will help us design a good interval estimator. It should be clear, however, from our earlier discussion on asymptotically efficient estimators that, in order to have a sufficiently narrow confidence interval, while keeping the confidence level sufficiently high, we need to increase the size n of data $x = (x_1, x_2, \ldots, x_n)$ that we can use for estimation.

Let $x = (x_1, x_2, \ldots, x_n)$ represent data of size n sampled from a common PDF $f(x)$. Let $\hat{\theta}(x)$ be an estimator of θ based on the sample x. Then $\hat{\theta}(X)$ is a random variable and we denote its PDF by $f_{\hat{\theta}(X)}(y; \theta)$. In order to determine the confidence interval I_γ defined by (18.53), consider an interval (d_1, d_2) around the estimate $\hat{\theta}$ such that

$$P[d_1 < \hat{\theta}(X) < d_2] = \int_{d_1}^{d_2} f_{\hat{\theta}(X)}(y; \theta)\, dy = \gamma. \tag{18.54}$$

The above equation, however, does not uniquely determine the interval (d_1, d_2). In fact, there are infinitely many such intervals. But let us specify two percentile parameters $0 < \alpha_1 < 1 - \alpha_2 < 1$ such that $\alpha_1 + \alpha_2 = 1 - \gamma$. Then consider $d_1(\theta) < d_2(\theta)$ such that

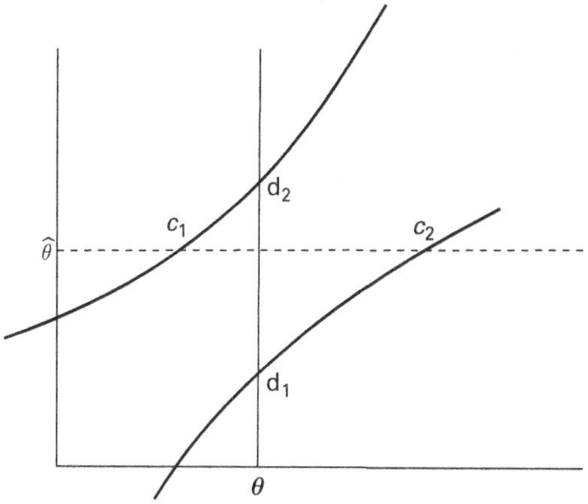

Figure 18.1 Confidence intervals for the unknown parameter θ.

$$\int_{-\infty}^{d_1} f_{\hat{\theta}(\boldsymbol{X})}(y;\theta)\,dy = \alpha_1 \text{ and } \int_{d_2}^{\infty} f_{\hat{\theta}(\boldsymbol{X})}(y;\theta)\,dy = \alpha_2. \tag{18.55}$$

The solutions to these two equations for given θ, denoted $d_1(\theta)$ and $d_2(\theta)$, can be uniquely determined from the α_1 and $1-\alpha_2$ percentiles, denoted as z_{α_1} and $z_{1-\alpha_2}$, of the distribution $f_{\hat{\theta}(\boldsymbol{X})}(y;\theta)$; i.e.,

$$d_1(\theta) = z_{\alpha_1}(\theta) \text{ and } d_2(\theta) = z_{1-\alpha_2}(\theta).$$

The equations in (18.55) can be viewed to represent two curves $\hat{\theta} = d_1(\theta)$ and $\hat{\theta} = d_2(\theta)$ (see Figure 18.1). The interval $(d_1(\theta), d_2(\theta))$ is a *vertical interval* that includes the point estimate $\hat{\theta}(\boldsymbol{X})$.

Suppose that the functions $\hat{\theta} = d_1(\theta)$ and $\hat{\theta} = d_2(\theta)$ are both continuous and monotone increasing functions. Then we can uniquely determine their inverse functions $\theta = d_1^{-1}(\hat{\theta})$ and $\theta = d_2^{-1}(\hat{\theta})$. Then, by writing

$$c_1(\hat{\theta}) = d_1^{-1}(\hat{\theta}) \text{ and } c_2(\hat{\theta}) = d_2^{-1}(\hat{\theta}), \tag{18.56}$$

it is not difficult to show (Problem 18.9) that $c_1(\hat{\theta}(\boldsymbol{X})) < \theta < c_2(\hat{\theta}(\boldsymbol{X}))$ if and only if $d_1(\theta) < \hat{\theta}(\boldsymbol{X}) < d_2(\theta)$. Thus, we have

$$\boxed{P[c_1(\hat{\theta}(\boldsymbol{X})) < \theta < c_2(\hat{\theta}(\boldsymbol{X}))] = P[d_1(\theta) < \hat{\theta}(\boldsymbol{X}) < d_2(\theta)] = \gamma.} \tag{18.57}$$

The interval $(c_1(\hat{\theta}(\boldsymbol{X})), c_2(\hat{\theta}(\boldsymbol{X})))$ on the dashed line in Figure 18.1 represents the confidence interval defined by (18.57).

Example 18.6: Confidence interval of an estimator $\hat{\mu}$ of $N(\mu, 1)$. Let the assumed distribution for the n samples be the normal distribution $N(\mu, 1)$, where the mean μ is unknown. We wish to find the confidence interval of an estimator at confidence level γ.

We know that the MLE for μ is the sample mean: $\hat{\mu}(\boldsymbol{x}) = \frac{1}{n}\sum_i x_i$. Since all the X_i are normal with mean μ and unit variance, the statistic $\hat{\mu}(\boldsymbol{X})$ is also normal with mean μ, with its variance being $1/n$:

$$f_{\hat{\mu}(\boldsymbol{X})}(y, \mu) = \frac{\sqrt{n}}{\sqrt{2\pi}} \exp\left[-\frac{n(y-\mu)^2}{2}\right].$$

Since this distribution is symmetrical with respect to μ, the narrowest confidence interval is obtained if we select $\alpha_1 = \alpha_2$ in the equations of (18.55), which in this case take the form

$$\begin{aligned} \frac{\sqrt{n}}{\sqrt{2\pi}} \int_{-\infty}^{d_1} \exp\left[-\frac{n(y-\mu)^2}{2}\right] dy = \alpha, \\ \frac{\sqrt{n}}{\sqrt{2\pi}} \int_{d_2}^{\infty} \exp\left[-\frac{n(y-\mu)^2}{2}\right] dy = \alpha, \end{aligned} \tag{18.58}$$

where $\alpha = (1-\gamma)/2$. Let u_α be the α percentile of the unit normal distribution $N(0, 1)$. Then the end points of the vertical interval can be written as

$$d_1(\mu) = \mu - \frac{u_\alpha}{\sqrt{n}}, \quad d_2(\mu) = \mu + \frac{u_\alpha}{\sqrt{n}}.$$

Thus, the two curves in Figure 18.1 are two parallel **straight lines** in this example. The end points of the confidence interval I_γ are obtained by solving these equations for μ. Therefore, the confidence interval $I_\gamma = (c_1(\hat{\mu}(\boldsymbol{X})), c_2(\hat{\mu}(\boldsymbol{X})))$ with level γ is given by

$$c_1(\hat{\mu}(\boldsymbol{X})) = \hat{\mu}(\boldsymbol{X}) - \frac{u_\alpha}{\sqrt{n}}, \quad c_2(\hat{\mu}(\boldsymbol{X})) = \hat{\mu}(\boldsymbol{X}) + \frac{u_\alpha}{\sqrt{n}}, \text{ where } \alpha = \frac{1-\gamma}{2}.$$

As we can see, the length of the confidence interval tends to zero as $n \to \infty$. Thus, we can decide how large the sample size n should be to obtain a sufficiently narrow interval. For instance, if we choose the confidence coefficient to be $\gamma = 0.95$, then $\alpha = 0.025$ and from the table of the standard normal distribution (or using the MATLAB function `norminv([0.025 0.975],0,1)`) we find $u_\alpha = 1.96$. The length of the confidence interval is $2u_\alpha/\sqrt{n}$. Thus, if we want the confidence interval to be less than 0.01, then we need a sample size $n \geq 4 \times 196^2 = 153\,664$. □

18.2 Hypothesis testing and statistical decision theory

In Chapter 6 we discussed a statistical procedure of accepting or rejecting a hypothesized probability distribution when we are given experimental data. The techniques we explored in that chapter are primarily based on graphical presentations of data by plotting them, for instance, on the log-normal paper.

18.2.1 Hypothesis testing

In the present section we introduce the theory of statistical hypothesis testing pioneered by J. Neyman and E. Pearson [256] in 1933. We begin with a binary hypothesis test, in which we have two possible hypotheses, denoted H_0 and H_1, one and only one of which is assumed to be true. The hypothesis H_0 is called the **null hypothesis**, whereas the hypothesis H_1 is called the **alternative hypothesis**. In radar or sonar, for instance, H_0 means the absence of a "target," whereas H_1 means its presence. In medical diagnosis, H_0 means that a patient is well, whereas H_1 means that the patient has a specific illness.

Let the hypotheses be concerned with parameter θ of the distribution function $F(\boldsymbol{x};\theta)$, of the observation $\boldsymbol{x}$, and let $\mathcal{S}$ be the parameter space, i.e., $\theta \in \mathcal{S}$. We wish to test the null hypothesis $H_0: \ \theta \in \mathcal{S}_0$ against the alternative hypothesis $H_1: \ \theta \in \mathcal{S}_1$, where $\mathcal{S}_0 \cup \mathcal{S}_1 = \mathcal{S}$ and $\mathcal{S}_0 \cap \mathcal{S}_1 = \emptyset$. If $\mathcal{S}_i \ (i = 0, 1)$ consists of a single point θ_i, the hypothesis is called **simple**; otherwise, it is called **composite**. The null hypothesis H_0 is simple in most cases. In a radar signal detection problem, for example, θ may represent the amplitude of a signal reflected from a target. The assumption $\mathcal{S}_0 = \{\theta = 0\}$ defines H_0, which is simple, whereas $\mathcal{S}_1 = \{\theta > 0\}$ defines H_1, which is composite.

Since we must accept one and only one of the two hypotheses and reject the other, we make two types of possible error in our decision:

1. **Type 1 error** (i.e., *false alarm* or **false positive**); we accept H_1 when H_0 is true.
2. **Type 2 error** (i.e., *miss* or **false negative**); we accept H_0 when H_1 is true.

Since we make a decision solely based on measurement data $\boldsymbol{x}$, we can formally state this statistical decision problem as follows. Let $\mathcal{X}$ be the sample space of the RV $\boldsymbol{X}$. Then a **decision rule** is equivalent to partitioning $\mathcal{X}$ in two regions $\mathcal{R}_c$ and $\overline{\mathcal{R}_c} \triangleq \mathcal{X} \setminus \mathcal{R}_c$, where $\mathcal{R}_c$ is called the **critical region** (or *rejection region*) of H_0 of the decision rule, and its complement $\overline{\mathcal{R}_c}$ is called the **acceptance region** of H_0.

We can then represent a decision rule (also called a *test function* or simply a **test**) $d(\boldsymbol{x})$ as a mapping from $\mathcal{X}$ to $\{0, 1\}$:

$$d(\boldsymbol{x}) = \begin{cases} 1 & (\text{Reject } H_0), \ \text{if } \boldsymbol{x} \in \mathcal{R}_c, \\ 0 & (\text{Accept } H_0), \ \text{if } \boldsymbol{x} \in \overline{\mathcal{R}_c}. \end{cases} \tag{18.59}$$

The probability of the type 1 error,

$$P[\boldsymbol{X} \in \mathcal{R}_c | H_0] = E[d(\boldsymbol{X})|H_0] \triangleq \alpha, \tag{18.60}$$

is called the **level of the test** (also called the **size of the test**). The probability of type 2 error is given as

$$P[\boldsymbol{X} \in \overline{\mathcal{R}_c} | H_1] = 1 - E[d(\boldsymbol{X})|H_1] \triangleq \beta. \tag{18.61}$$

The probability of rejecting H_0 when it is false, i.e., $1 - \beta = E[d(\boldsymbol{X})|H_1]$, is called the **power of the test**, or the *true positive probability*. In the context of signal detection theory it is called the *detection probability*, denoted P_d.

When there are no prior probabilities given, we cannot use the overall probability of error as in the MAP (maximum *a posteriori* probability) estimator, which will be discussed in Section 18.3 (see (18.100)). An alternative criterion that we may adopt in such a circumstance is to keep the probability of type 1 error less than or equal to a prescribed value α and minimize β, the probability of the type 2 error. So we introduce the following notion.

DEFINITION 18.3 (Most powerful test). *Consider a hypothesis testing of H_0 against H_1. Among all tests (or decision rules) with the level at or below $\alpha > 0$, a test with the largest power (i.e., the smallest β) is called the **most powerful** (MP) test at the level α.*

18.2.2 Neyman–Pearson criterion and likelihood ratio test

Now let us consider the case where we test a simple null hypothesis against a simple alternative; i.e.,

$$H_0 : \theta = \theta_0 \text{ versus } H_1 : \theta = \theta_1.$$

In Example 18.7 of Section 18.3, we show that an optimal decision rule is to compare the likelihood ratio function

$$\boxed{\Lambda(\boldsymbol{x}) \triangleq \frac{f(\boldsymbol{x}|\theta_1)}{f(\boldsymbol{x}|\theta_0)}} \tag{18.62}$$

against some threshold. In the MAP decision rule, this threshold is given by the ratio of prior probabilities $p(\theta_0)$ and $p(\theta_1)$ (see (18.104)). Neyman and Pearson [256] showed that the MP test among tests with level α can also be reduced to comparing $\Lambda(\boldsymbol{x})$ with some threshold. Note that $\boldsymbol{X}$ may be a vector RV, whereas $\Lambda \triangleq \Lambda(\boldsymbol{X})$ is always a scalar-valued positive RV.

THEOREM 18.3 (Neyman–Pearson lemma). *If the distribution function of the likelihood ratio variable $\Lambda = \Lambda(\boldsymbol{x})$ is continuous under both **simple hypotheses** H_0 and H_1, then the MP test should take the form of a **likelihood ratio test**:*

$$\boxed{d(\Lambda(\boldsymbol{x})) = u(\Lambda(\boldsymbol{x}) - \lambda_\alpha) = \begin{cases} 1 \text{ (Reject } H_0), & \text{if } \Lambda(\boldsymbol{x}) \geq \lambda_\alpha, \\ 0 \text{ (Accept } H_0), & \text{if } \Lambda(\boldsymbol{x}) < \lambda_\alpha, \end{cases}} \tag{18.63}$$

where $u(\cdot)$ is the unit step function and λ_α is the solution to the equation

$$\alpha = P[\Lambda(\boldsymbol{X}) \geq \lambda_\alpha | H_0] = E\left[u(\Lambda(\boldsymbol{X}) - \lambda_\alpha)|H_0\right]. \tag{18.64}$$

Proof. Let $\mathcal{R}_c$ be the critical region of this MP test; i.e.,

$$\mathcal{R}_c = \{\boldsymbol{x} : \Lambda(\boldsymbol{x}) \geq \lambda_\alpha\}. \tag{18.65}$$

Since we are dealing with two *simple* hypotheses, $H_i: \ \theta = \theta_i; \ \ i = 0, 1$, we denote the condition $|H_i$ simply by $|\theta_i$. Then (18.64) can be rewritten as

$$\alpha = P[\boldsymbol{x} \in \mathcal{R}_c|\theta_0]. \tag{18.66}$$

Let $\mathcal{R}'_c$ be any region in the sample space $\mathcal{X}$ for which

$$P[\boldsymbol{x} \in \mathcal{R}'_c|\theta_0] \leq \alpha. \tag{18.67}$$

Then what we need to show is that the decision rule defined by this $\mathcal{R}'_c$ is less powerful than the Neyman–Pearson test (18.63).

Let $\mathcal{S}$ be the intersection of $\mathcal{R}_c$ and $\mathcal{R}'_c$; i.e.,

$$\mathcal{S} = \mathcal{R}_c \cap \mathcal{R}'_c.$$

Then,

$$f(\boldsymbol{x}|\theta_1) \geq \lambda_\alpha f(\boldsymbol{x}|\theta_0), \quad \text{for all } \boldsymbol{x} \in \mathcal{R}_c \setminus \mathcal{S},$$

which implies

$$P[\boldsymbol{x} \in \mathcal{R}_c \setminus \mathcal{S}|\theta_1] \geq \lambda_\alpha P[\boldsymbol{x} \in \mathcal{R}_c \setminus \mathcal{S}|\theta_0].$$

Thus,

$$\begin{aligned}
P[\boldsymbol{x} \in \mathcal{R}_c|\theta_1] &= P[\boldsymbol{x} \in \mathcal{R}_c \setminus \mathcal{S}|\theta_1] + P[\boldsymbol{x} \in \mathcal{S}|\theta_1] && (18.68)\\
&\geq \lambda_\alpha P[\boldsymbol{x} \in \mathcal{R}_c \setminus \mathcal{S}|\theta_0] + P[\boldsymbol{x} \in \mathcal{S}|\theta_1] \\
&= \lambda_\alpha P[\boldsymbol{x} \in \mathcal{R}_c|\theta_0] - \lambda_\alpha P[\boldsymbol{x} \in \mathcal{S}|\theta_0] + P[\boldsymbol{x} \in \mathcal{S}|\theta_1] && (18.69)\\
&\geq \lambda_\alpha P[\boldsymbol{x} \in \mathcal{R}'_c|\theta_0] - \lambda_\alpha P[\boldsymbol{x} \in \mathcal{S}|\theta_0] + P[\boldsymbol{x} \in \mathcal{S}|\theta_1] && (18.70)\\
&= \lambda_\alpha P[\boldsymbol{x} \in \mathcal{R}'_c \setminus \mathcal{S}|\theta_0] + P[\boldsymbol{x} \in \mathcal{S}|\theta_1] && (18.71)\\
&\geq P[\boldsymbol{x} \in \mathcal{R}'_c \setminus \mathcal{S}|\theta_1] + P[\boldsymbol{x} \in \mathcal{S}|\theta_1] = P[\boldsymbol{x} \in \mathcal{R}'_c|\theta_1] && (18.72)
\end{aligned}$$

In the step from (18.69) to (18.70) we used the following inequality given by (18.64) and (18.67):

$$P[\boldsymbol{x} \in \mathcal{R}'_c|\theta_0] \leq \alpha = P[\boldsymbol{x} \in \mathcal{R}_c|\theta_0].$$

In going from (18.71) to (18.72) we used the fact that points in $\mathcal{R}'_c \setminus \mathcal{S}$ do not belong to $\mathcal{R}_c$; hence,

$$f(\boldsymbol{x}|\theta_1) \leq \lambda_\alpha f(\boldsymbol{x}|\theta_0), \quad \text{for all } \boldsymbol{x} \in \mathcal{R}'_c \setminus \mathcal{S},$$

which implies

$$P[\boldsymbol{x} \in \mathcal{R}_c \setminus \mathcal{S}|\theta_1] \leq \lambda_\alpha P[\boldsymbol{x} \in \mathcal{R}_c \setminus \mathcal{S}|\theta_0], \quad \text{for all } \boldsymbol{x} \in \mathcal{R}'_c \setminus \mathcal{S}.$$

Note that the left-hand side of (18.68) is the power of the MP test (18.63) which has $\mathcal{R}_c$ as its critical region at level α, whereas the right-hand side of (18.72) is the power of the level-α test having $\mathcal{R}'_c$ as its critical region. Thus, we have established that the likelihood ratio test (18.63) is the MP test. □

18.2.3 Receiver operating characteristic (ROC)

The concept of **receiver operating characteristic**, abbreviated as ROC, or the *operating characteristic* [150], was originally introduced in *signal detection theory* [269] to seek an optimal operating point of the radar receiver by trading the *miss probability* (or false negative probability) β for the false alarm probability (or false positive probability or the level) α by setting an appropriate threshold λ_α. In the past few decades, however, the ROC curve has been used in other fields, including psychophysics [127] and clinical medicine [268]. Most recently the ROC has been applied to statistical classification problems in *machine learning* and *data mining*.

The ROC curve is defined by plotting the *detection probability* (true positive probability or the power of the test) $P_d(= 1 - \beta)$ versus the *false alarm probability* $P_{fa} = \alpha$ (e.g., see Figure 18.2 of the example in the next subsection).

18.2.3.1 Properties of the receiver operating characterstic of a Neyman–Pearson test

The ROC curve can be obtained for any decision rule, but we shall derive here some important properties of the Neyman–Pearson decision rule. Similar to (18.64), P_d can be written as a function of λ_α:

$$\begin{aligned} P_d &= \int_{x:\ \Lambda(x)\geq\lambda_\alpha} f(x|H_1)\,dx = \int_{x:\ \Lambda(x)\geq\lambda_\alpha} \Lambda(x) f(x|H_0)\,dx \\ &= E\left[\Lambda(X)u(\Lambda(X) - \lambda_\alpha)|H_0\right]. \end{aligned} \tag{18.73}$$

Differentiating (18.64) and (18.73) with respect to λ_α and using the property that the derivative of the unit step function is Dirac's delta function, we have

$$\frac{d\alpha}{d\lambda_\alpha} = -E\left[\delta(\Lambda(X) - \lambda_\alpha)|H_0\right], \tag{18.74}$$

$$\frac{dP_d}{d\lambda_\alpha} = -E\left[\Lambda(X)\delta(\Lambda(X) - \lambda_\alpha)|H_0\right] = \lambda_\alpha \frac{d\alpha}{d\lambda_\alpha}. \tag{18.75}$$

Taking the ratio of the last two equations, we find the following interesting property:

$$\boxed{\frac{dP_d}{d\alpha} = \frac{dP_d/d\lambda_\alpha}{d\alpha/d\lambda_\alpha} = \lambda_\alpha,} \tag{18.76}$$

which shows that the slope of the ROC curve at any point (α, P_d) is equal to λ_α that specifies this point.

If we set the threshold $\lambda_\alpha = \infty$ (or higher than the maximum possible value of the likelihood function $\Lambda(x)$), then H_1 will never be accepted; thus, $\alpha = 0$ and $P_d = 0$. So the ROC curve starts from the origin point $(0, 0)$. As the threshold λ_α decreases, the operating point moves right and up, as seen in Figure 18.2. The slope of the tangent at the operating point is the largest at the starting point $(0, 0)$ and monotonically decreases as λ_α decreases and the corresponding operating point moves to the right and up on the ROC curve. Hence, the ROC curve is a *concave* (or *convex* $\cap$) function. As the threshold value λ_α approaches zero (or below the smallest possible value of $\Lambda(x)$), both α and P_d approach unity; thus, the ROC curve ends at $(1, 1)$, at which point the slope is the smallest. Thus, the ROC curve of a Neyman–Pearson test always lies above the straight line, $P_{\rm d} = \alpha$, which connects the two end points (0, 0) and (1, 1).

The last property is also evident if we consider a decision rule that ignores the observation data $\boldsymbol{x}$ and randomly selects H_1 with probability p. Then, for this rule, $P_{\rm d} = \alpha = p$, regardless of $\boldsymbol{x}$. Since a Neyman–Pearson decision rule has the largest $P_{\rm d}$ among all rules of its level, it follows that $P_{\rm d} \geq \alpha$.

18.2.4 Receiver operating characteristic application example: radar signal detection

In a radar system, the received discrete-time signal $\boldsymbol{x} = (x_0, x_2, \ldots, x_n)$ is an instance of the RV $\boldsymbol{X}$:

$$H_0 : \boldsymbol{X} = \boldsymbol{Z} \quad \text{and} \quad H_1 : \boldsymbol{X} = A\boldsymbol{s} + \boldsymbol{Z}, \tag{18.77}$$

where $\boldsymbol{Z} = (Z_0, Z_1, Z_2, \ldots, Z_n)$ is random noise, $\boldsymbol{s} = (s_0, s_1, s_2, \ldots, s_n)$ is a signal waveform, and the amplitude A is a positive real number, and is assumed to be known. Otherwise, H_1 becomes a composite hypothesis, which would make the analysis a little more involved. Without loss of generality, we assume $\|\boldsymbol{s}\|^2 \triangleq \sum_{k=0}^{n} s_k^2 = 1$.

Assuming that the Z_k are white Gaussian noise with zero mean and variance σ^2, we can write the PDFs of the RV $\boldsymbol{X}$ under the hypotheses H_0 and H_1 as follows:

$$H_0 : \; f_{\boldsymbol{X}}(\boldsymbol{x}|H_0) = \frac{1}{\left(\sqrt{2\pi}\sigma\right)^n} \exp\left(-\frac{\|\boldsymbol{x}\|^2}{2\sigma^2}\right), \tag{18.78}$$

$$H_1 : \; f_{\boldsymbol{X}}(\boldsymbol{x}|H_1) = \frac{1}{\left(\sqrt{2\pi}\sigma\right)^n} \exp\left(-\frac{\|\boldsymbol{x} - A\boldsymbol{s}\|^2}{2\sigma^2}\right). \tag{18.79}$$

Then the level and power of a test having $\mathcal{R}_{\rm c}$ as its critical region can be written as

$$\alpha = \int_{\boldsymbol{x} \in \mathcal{R}_{\rm c}} f_{\boldsymbol{X}}(\boldsymbol{x}|H_0)\, d\boldsymbol{x}, \tag{18.80}$$

$$P_{\rm d} = \int_{\boldsymbol{x} \in \mathcal{R}_{\rm c}} f_{\boldsymbol{X}}(\boldsymbol{x}|H_1)\, d\boldsymbol{x}. \tag{18.81}$$

Taking the ratio of (18.79) to (18.78), we write the Neyman–Pearson test as

$$\Lambda(\boldsymbol{x}) = \exp\left(-\frac{\|\boldsymbol{x} - A\boldsymbol{s}\|^2 - \|\boldsymbol{x}\|^2}{2\sigma^2}\right) = \exp\left[\frac{A}{\sigma^2}\left(\boldsymbol{s}^\top \boldsymbol{x} - \frac{A}{2}\right)\right] \gtrless \lambda_\alpha, \tag{18.82}$$

where the notation "$\gtrless \lambda_\alpha$" means "Accept H_1 if greater than λ_α and accept H_0 otherwise." On taking the logarithm and after some manipulation, the likelihood ratio test reduces to

$$T(\boldsymbol{x}) \triangleq \boldsymbol{s}^\top \boldsymbol{x} \gtrless t_\alpha, \quad \text{where } t_\alpha \triangleq \frac{A}{2} + \frac{\sigma^2}{A} \log \lambda_\alpha. \tag{18.83}$$

The statistic $T(\boldsymbol{x}) = \boldsymbol{s}^\top \boldsymbol{x}$ is called a *test statistic* and can be interpreted as a **matched filter** output, or a **correlation receiver** output (see Problems 18.11 and 18.12).

Thus, the Neyman–Pearson decision rule (18.63) reduces to

$$d(T(\boldsymbol{x})) = u(T(\boldsymbol{x}) - t_\alpha). \tag{18.84}$$

Since $\boldsymbol{X}$ is a normal RV, its linear functional $T(\boldsymbol{X})$ is also a normal RV. Under H_0, its mean is zero and the variance is

$$\begin{aligned}\mathrm{Var}[T(\boldsymbol{X})|H_0] &= E[T^2(\boldsymbol{X})|H_0] = \boldsymbol{s}^\top E\left[\boldsymbol{X}\boldsymbol{X}^\top|H_0\right]\boldsymbol{s} \\ &= \boldsymbol{s}^\top\sigma^2\boldsymbol{I}\boldsymbol{s} = \sigma^2.\end{aligned} \tag{18.85}$$

Thus, under the hypothesis H_0, $T \sim N(0, \sigma^2)$:

$$f_T(t|H_0) = \frac{1}{\sqrt{2\pi}\sigma}\exp\left(-\frac{t^2}{2\sigma^2}\right). \tag{18.86}$$

For a given level of false alarm probability α, the critical value t_α should be determined by

$$\alpha = \int_{t_\alpha}^{\infty} f_T(t|H_0)\,dt = 1 - \Phi\left(\frac{t_\alpha}{\sigma}\right) = 1 - \Phi(u_\alpha), \tag{18.87}$$

where $\Phi(u)$ is the CDF of the unit (or standard) normal variable U defined in Section 4.2.4, and u_α is the upper $\alpha \times 100$ percentile point of the unit normal distribution. The threshold parameter t_α is given by

$$t_\alpha = \sigma u_\alpha. \tag{18.88}$$

Since $E[T(\boldsymbol{X})|H_1] = \boldsymbol{s}^\top A\boldsymbol{s} = A$, the PDF of $T(\boldsymbol{X})$ under H_1 is $f_T(t - A|H_0)$. Hence,

$$P_{\mathrm{d}} = \int_{t_\alpha}^{\infty} f_T(t - A|H_0)\,dt = 1 - \Phi(u_\alpha - r), \quad \text{where } r \triangleq \frac{A}{\sigma}. \tag{18.89}$$

Figure 18.2 shows the ROC curves of this Neyman–Pearson test for various values of the signal-to-noise ratio (SNR) parameter $r = A/\sigma$. As the threshold t_α moves from $+\infty$ to $-\infty$, the operating point moves from (0, 0) to (1, 1). As SNR r increases, the ROC curve moves further away from the straight line $P_{\mathrm{d}} = \alpha$, which corresponds to the case $r = 0$. In fact, the area between the ROC curve and the straight line is a monotone increasing function of the SNR parameter r (Problem 18.14). The reader is also suggested to verify that this ROC curve satisfies the property (18.76); i.e., the slope of the ROC at any operating point is equal to the critical value of λ_α for the likelihood ratio test (Problem 18.13).

18.3 Bayesian estimation and decision theory

Thus far we have discussed the parameter estimation problem from the frequentists' point of view; namely, we assumed that the parameter $\boldsymbol{\theta}$ is unknown but *fixed*. As we discussed in Sections 1.2.4 and 4.5, however, it is more appropriate in some situations to treat the parameter as a random variable. For example, in a communication system, the

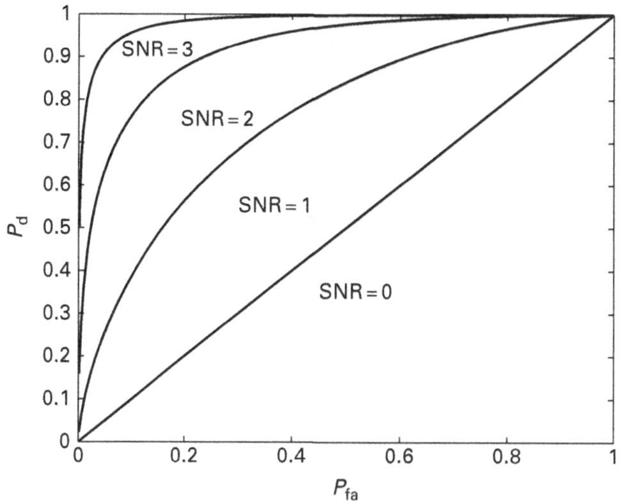

Figure 18.2 The ROC curves for different values of SNR = A/σ.

receiver estimates a transmitted symbol, which we denote for instance as $\boldsymbol{\theta} \in \mathcal{A}$, on the basis of a received signal $\boldsymbol{x}$. Here $\mathcal{A}$ signifies the *signal alphabet*; i.e., the set of all possible symbols. All symbols in the alphabet $\mathcal{A}$ may not be equally likely and we may have some prior probabilities of these symbols. In this case it may be more appropriate to treat the parameter as a random variable $\boldsymbol{\Theta}$ with the **prior probability** distribution $\pi(\boldsymbol{\theta})$.[3]

18.3.1 The Bayes risk

In some applications it may be possible and appropriate to specify the *cost* of an error in estimation. The cost associated with assigning an estimate $\hat{\boldsymbol{\theta}}$ to the parameter when its true value is $\boldsymbol{\theta}$ is denoted as $C(\hat{\boldsymbol{\theta}}, \boldsymbol{\theta})$. The function C may take, for example, the form $C(\hat{\boldsymbol{\theta}}, \boldsymbol{\theta}) = f(\|\hat{\boldsymbol{\theta}} - \boldsymbol{\theta}\|)$, where $f(d)$ is a nonnegative and monotone increasing function of the difference $d = \|\hat{\boldsymbol{\theta}} - \boldsymbol{\theta}\| > 0$.

If we take a sample $\boldsymbol{x}$ when the parameter $\boldsymbol{\Theta}$ takes $\boldsymbol{\theta}$, we write its conditional PDF, or the *likelihood function*, as $f(\boldsymbol{x}|\boldsymbol{\theta})$. Note that in the preceding sections, where we treated θ as an unknown but fixed constant, we wrote the PDF of $\boldsymbol{X}$ as $f(\boldsymbol{x}; \boldsymbol{\theta})$. The cost or risk we incur, by using the estimator $\hat{\boldsymbol{\theta}}(\boldsymbol{x})$ when the true value of the parameter is $\boldsymbol{\theta}$, is given by

$$\boxed{R(\hat{\boldsymbol{\theta}}|\boldsymbol{\theta}) = \int_{\boldsymbol{x}} C(\hat{\boldsymbol{\theta}}(\boldsymbol{x}), \boldsymbol{\theta}) f(\boldsymbol{x}|\boldsymbol{\theta})\, d\boldsymbol{x},} \tag{18.90}$$

which we term **Bayes' risk**. The overall expected cost or risk per estimation is obtained by averaging the above with respect to the prior distribution $\pi(\boldsymbol{\theta})$ of the parameter:

[3] In this section we adopt this **simplified notation**, instead of $p_{\boldsymbol{\Theta}}(\boldsymbol{\theta})$ or $f_{\boldsymbol{\Theta}}(\boldsymbol{\theta})$, in both discrete and continuous cases, as was done in Section 4.5.

$$\overline{R}(\hat{\theta}) = \sum_{\theta} \pi(\theta) R(\hat{\theta}|\theta)$$
$$= \sum_{\theta} \int_{x} \pi(\theta) C(\hat{\theta}(x), \theta) f(x|\theta)\, dx. \tag{18.91}$$

If the parameter takes on continuous values, we replace the summation by integration. We now proceed to find an optimum estimator that should minimize the average risk $\overline{R}$.

18.3.2 The conditional risk

Given the prior probability $\pi(\theta)$ and the measured value x of the RV X, the overall probability and the **posterior probability** $\pi(\theta|x)$[4] of the observation x are given, respectively, by

$$f(x) = \sum_{\theta} \pi(\theta) f(x|\theta), \tag{18.92}$$

$$\pi(\theta|x) = \frac{\pi(\theta) f(x|\theta)}{f(x)}. \tag{18.93}$$

The *conditional risk* $R(\hat{\theta}|x)$ is the average risk for the estimator $\hat{\theta}$ to make when x is obtained; i.e.,

$$R(\hat{\theta}|x) = \sum_{\theta} C(\hat{\theta}, \theta) \pi(\theta|x)$$
$$= \sum_{\theta} \frac{\pi(\theta) C(\hat{\theta}, \theta) f(x|\theta)}{f(x)}. \tag{18.94}$$

The overall expected risk (18.91) can be represented in terms of the conditional risk (18.94) as

$$\overline{R}(\hat{\theta}) = \int_{x} R(\hat{\theta}|x) f(x)\, dx. \tag{18.95}$$

Thus, the overall expected risk can be minimized by making the conditional risk $R(\hat{\theta}|x)$ as small as possible for each value of x. Such an estimator is referred to as a **Bayes estimator**. If the conditional risk is a discrete function of the estimate $\hat{\theta}$, the Bayes estimator $\hat{\theta}^{\star}(x)$ can be found from

$$\boxed{\hat{\theta}^{\star}(x) = \arg\max_{\hat{\theta}} R(\hat{\theta}|x).} \tag{18.96}$$

If the parameter RV $\boldsymbol{\Theta}$ is a continuous RV, the prior distribution $\pi(\theta)$ and the posterior distribution $\pi(\theta|x)$ should be interpreted as the PDF and the conditional PDF of the RV $\boldsymbol{\Theta}$. The conditional risk then also becomes a continuous function of the estimate $\hat{\theta}$. The Bayes estimator $\hat{\theta}^{\star}$ can be found by solving the equation

[4] We adopt here this simpler notation instead of $p_{\Theta|X}(\theta|x)$.

$$\nabla_{\theta} R(\hat{\theta}|x) = \mathbf{0}. \tag{18.97}$$

18.3.3 Maximum *a posteriori* probability (MAP) estimation

When no cost function is available, a natural estimation strategy is to maximize the posterior (or *a posteriori*) PDF $\pi(\theta|x)$. The estimate $\hat{\theta}$ is then the parameter value that is most likely in view of the observation data x. This strategy, however, is equivalent to setting the cost function as follows:

$$C(\hat{\theta}, \theta) = 1 - \delta_{\hat{\theta},\theta} = \begin{cases} 0, & \text{if } \hat{\theta} = \theta, \\ 1, & \text{if } \hat{\theta} \neq \theta, \end{cases} \tag{18.98}$$

where $\delta_{x,y}$ is Kronecker's delta. Then the conditional risk becomes

$$\begin{aligned} R(\hat{\theta}|x) &= \sum_{\theta} \frac{\pi(\theta)[1 - \delta_{\hat{\theta},\theta}] f(x|\theta)}{f(x)} \\ &= \sum_{\theta} \frac{\pi(\theta) f(x|\theta)}{f(x)} - \frac{\pi(\hat{\theta}) f(x|\hat{\theta})}{f(x)} \\ &= 1 - \pi(\hat{\theta}|x), \end{aligned} \tag{18.99}$$

where $\pi(\hat{\theta}|x)$ is the posterior probability of having $\Theta = \hat{\theta}$ given $X = x$, as defined in (18.93).

Then the Bayes estimator (18.96) reduces to

$$\boxed{\hat{\theta}^{\star}(x) = \arg\max_{\hat{\theta}} \pi(\hat{\theta}|x),} \tag{18.100}$$

which is often referred to as the **maximum *a posteriori* probability (MAP) estimator**.[5] Under the cost assignment (18.98), the expected risk is the same as the **expected probability of error**.

Note that due to the Bayes' formula

$$\pi(\theta|x) = \frac{L_x(\theta)\pi(\theta)}{p_X(x)}, \tag{18.101}$$

where $L_x(\theta) = p(x|\theta)$ is the likelihood function, the MAP estimate can be obtained by maximizing $L_x(\theta)\pi(\theta)$ or, equivalently,

$$\hat{\theta}^{\star}(x) = \arg\max_{\theta} \; [\log L_x(\theta) + \log \pi(\theta)]. \tag{18.102}$$

It follows from this equation that the MLE is a special case of the MAP estimate when the parameter is not random or the prior distribution of the parameter is uniform.

[5] In the literature, "maximum *a posteriori*," without "probability", is often assigned to stand for "MAP". However, the word "probability" should be explicitly included in order to make the term technically meaningful.

Example 18.7: Estimation of binary signal. Consider a communication system in which the transmitter sends either symbol $+A$ or symbol $-A$ with probability p and $q = 1 - p$ respectively. The received signal X is corrupted by additive Gaussian noise $Z \sim N(0, \sigma^2)$. Thus, we write

$$X = \Theta + Z,$$

where Θ is the RV that represents the transmitted symbol; i.e., $P[\Theta = +A] = p$ and $P[\Theta = -A] = q$.

Our objective is to design an estimator whose probability of error is minimum in estimating the transmitted symbol. To simplify the argument, we assume that the receiver performs "symbol-by-symbol" estimation. We are concerned about estimation of one symbol, not a sequence of symbols. Minimizing the probability of error is equivalent to minimizing the average risk under the cost function (18.98). Thus, the optimum estimator should be the MAP estimator; i.e.,

$$\hat{\theta}^{\star} = \begin{cases} +A, & \text{if } p_{\Theta|X}(+A|x) \geq p_{\Theta|X}(-A|x), \\ -A, & \text{otherwise.} \end{cases} \tag{18.103}$$

By expressing the posterior probability by using Bayes' formula, the above MAP estimation rule can be expressed as

$$\hat{\theta}^{\star} = \begin{cases} +A, & \text{if } \Lambda(x) \geq \frac{p_{\Theta}(-A)}{p_{\Theta}(+A)} = \frac{q}{p}, \\ -A, & \text{otherwise,} \end{cases} \tag{18.104}$$

where

$$\boxed{\Lambda(x) = \frac{f_{X|\Theta}(x|+A)}{f_{X|\Theta}(x|-A)}} \tag{18.105}$$

is the *likelihood ratio function*. Since the PDF of the additive noise is given by

$$f_Z(z) = \frac{1}{\sqrt{2\pi}\sigma} \exp\left(-\frac{z^2}{2\sigma^2}\right), \tag{18.106}$$

we can readily compute the likelihood ratio function:

$$\Lambda(x) = \frac{f_Z(x - A)}{f_Z(x + A)} = \exp\left(\frac{2Ax}{\sigma^2}\right). \tag{18.107}$$

By substituting this into (18.105) and taking the logarithm of both sides in (18.104), we finally obtain the following simple result:

$$\hat{\theta}^{\star} = \begin{cases} +A, & \text{if } x > \frac{\sigma^2}{2A} \log \frac{q}{p}, \\ -A, & \text{otherwise.} \end{cases} \tag{18.108}$$

□

18.4 Summary of Chapter 18

Likelihood function (continuous):	$L_{\boldsymbol{x}}(\boldsymbol{\theta}) \triangleq f_{\boldsymbol{X}}(\boldsymbol{x};\boldsymbol{\theta})$	(18.24)
MLE:	$\hat{\boldsymbol{\theta}} = \arg\max_{\boldsymbol{\theta}} L_{\boldsymbol{x}}(\boldsymbol{\theta})$	(18.25)
Score function:	$\boldsymbol{s}(\boldsymbol{x};\boldsymbol{\theta}) = \frac{\partial \log L_{\boldsymbol{x}}(\boldsymbol{\theta})}{\partial \boldsymbol{\theta}}$	(18.28)
Observed Fisher information:	$\mathcal{J}(\boldsymbol{x};\boldsymbol{\theta}_0) = -\left[\frac{\partial^2 \log L_{\boldsymbol{x}}(\boldsymbol{\theta}_0)}{\partial\theta_i \partial\theta_j}\right]$	(18.32)
Fisher information matrix:	$\mathcal{I}_{\boldsymbol{x}}(\boldsymbol{\theta}_0) \triangleq E[\mathcal{J}(\boldsymbol{X};\boldsymbol{\theta}_0)]$	
	$= -E\left[\frac{\partial^2 \log L_{\boldsymbol{x}}(\boldsymbol{\theta}_0)}{\partial\theta_i \partial\theta_j}\right]$	(18.35)
MLE $\hat{\boldsymbol{\eta}}$ for exponential family:	$\nabla_{\boldsymbol{\eta}} A(\hat{\boldsymbol{\eta}}) = \frac{1}{n}\sum_{i=1}^{n} \boldsymbol{T}(x_i)$	(18.39)
CRLB for a single parameter:	$E[(\hat{\theta}-\theta)^2] \geq \frac{\left(\frac{\partial E[\hat{\theta}(\boldsymbol{X})]}{\partial\theta}\right)^2}{E\left[\left(\frac{\partial \log f_{\boldsymbol{X}}(\boldsymbol{x};\theta)}{\partial\theta}\right)^2\right]}$	(18.40)
CRLB for multiple parameters:	$\mathbf{Var}[\hat{\boldsymbol{\theta}}(\boldsymbol{X})] \geq \mathcal{I}_{\boldsymbol{x}}^{-1}(\boldsymbol{\theta})$	(18.42)
Likelihood ratio (bipolar signal):	$\Lambda(x) = \frac{f_{X\mid\Theta}(x\mid +A)}{f_{X\mid\Theta}(x\mid -A)}$	(18.105)
Decision rule:	$d(\boldsymbol{x}) = \begin{cases} 1 & \text{(Reject } H_0\text{)}, \text{ if } \boldsymbol{x}\in\mathcal{R}_c \\ 0 & \text{(Accept } H_0\text{)}, \text{ if } \boldsymbol{x}\in\overline{\mathcal{R}_c}\end{cases}$	(18.59)
Likelihood ratio for hypothesis test:	$\Lambda(\boldsymbol{x}) \triangleq \frac{f(\boldsymbol{x}\mid\theta_1)}{f(\boldsymbol{x}\mid\theta_0)}$	(18.62)
Neyman–Pearson test:	$d(\Lambda) = u(\Lambda(bx) - \lambda_\alpha)$	(18.63)
Slope of ROC curve:	$\frac{dP_\mathrm{d}}{d\alpha} = \lambda_\alpha$	(18.76)
Bayes' risk:	$R(\hat{\theta}(\boldsymbol{x})\mid\boldsymbol{\theta}) = \int_{\boldsymbol{x}} C(\hat{\boldsymbol{\theta}},\boldsymbol{\theta}) f(\boldsymbol{x}\mid\boldsymbol{\theta})\, d\boldsymbol{x}$	(18.90)
Bayes' estimate:	$\hat{\boldsymbol{\theta}}^{\star}(\boldsymbol{x}) = \arg\max_{\hat{\boldsymbol{\theta}}} R(\hat{\boldsymbol{\theta}}\mid\boldsymbol{x})$	(18.96)
MAP estimate:	$\hat{\boldsymbol{\theta}}^{\star}(\boldsymbol{x}) = \arg\max_{\hat{\boldsymbol{\theta}}} \pi(\hat{\boldsymbol{\theta}}\mid\boldsymbol{x})$	(18.100)

18.5 Discussion and further reading

Statistical estimation and decision and engineering applications have been discussed in a number of books: Cramér [74], Hald [139], Middleton [248], Helstrom [150], Kendall and Stuart [179], Rao [278], Lehmann [221], Poor [271], and Bickel and Doksum [25]. Among numerous textbooks on probability and random processes, Papoulis and Pillai [262], Leon-Garcia [222], and Fine [105] allocate a chapter on statistical estimation and cover parameter estimation and hypothesis testing.

18.6 Problems

Section 18.1: Parameter estimation

18.1 Sampling distribution. Consider a PDF $f_X(\boldsymbol{x};\theta)$ with a single parameter. Let $T(\boldsymbol{x})$ be an estimator statistic for θ. The distribution of the RV $T(\boldsymbol{X})$ is called the *sampling distribution.*

(a) Show that the sampling distribution function is given by

$$F_T(t) = \int_{-\infty}^{\infty}\int_{-\infty}^{\infty}\cdots\int_{-\infty}^{\infty} I(T(\boldsymbol{x}) \le t)\, f_{\boldsymbol{X}}(\boldsymbol{x};\theta)\, dx_1\, dx_2 \cdots dx_n, \quad (18.109)$$

where $I(A)$ is the indicator function; i.e., $I(A) = 1$ if A is true and $I(A) = 0$ otherwise.

(b) Find an expression for the bias b_T of the estimator $T(\boldsymbol{x})$.

(c) Find an expression for the *sampling variance* σ_T^2.

18.2 Efficient estimator. Show that the unbiased estimator $\hat{\boldsymbol{\theta}}^*(\boldsymbol{X})$ is an efficient estimator of the M dimensional parameter $\boldsymbol{\theta}$ if and only if (18.2) holds for any nonzero vector $\boldsymbol{a}$ of dimension M for any other unbiased estimator $\hat{\boldsymbol{\theta}}(\boldsymbol{X})$.

18.3 More on exponential family of distributions. This is a continuation of Problem 4.26.

(a) Show that the Rayleigh distribution

$$f(x;\sigma) = \frac{x}{\sigma^2}\exp\left(-\frac{x^2}{2\sigma^2}\right), \quad x \ge 0,$$

belongs to the exponential family. Find its canonical form, if possible, and the mean and variance of a sufficient statistic for estimating σ.

(b) Refer to the canonical exponential family defined by (18.14). Show that the MLE of $\boldsymbol{\theta}$ is the solution of the following equation:

$$\nabla_{\boldsymbol{\theta}}\eta^\top(\boldsymbol{\theta})\boldsymbol{T}(\boldsymbol{x}) = \nabla_{\boldsymbol{\theta}} A(\boldsymbol{\theta}). \quad (18.110)$$

18.4* Properties of the score function and the observed Fisher information matrix. The *score function* $\boldsymbol{s}(\boldsymbol{x};\boldsymbol{\theta})$ is defined in (18.28). The *observed Fisher information matrix* $\mathcal{J}(\boldsymbol{x};\boldsymbol{\theta})$ is defined in (18.32). Both are functions of $\boldsymbol{X}$, and hence they are RVs. Derive the following properties:

(a) For *any statistic* $\boldsymbol{T}(\boldsymbol{x};\boldsymbol{\theta})$,

$$E\left[\boldsymbol{s}(\boldsymbol{X};\boldsymbol{\theta})\boldsymbol{T}^\top(\boldsymbol{X};\boldsymbol{\theta})\right] = \nabla_{\boldsymbol{\theta}} E\left[\boldsymbol{T}^\top(\boldsymbol{X};\boldsymbol{\theta})\right] - E\left[\nabla_{\boldsymbol{\theta}}\boldsymbol{T}^\top(\boldsymbol{X};\boldsymbol{\theta})\right]. \quad (18.111)$$

(b) The score function has zero mean:

$$E[\boldsymbol{s}(\boldsymbol{X};\theta)] = \boldsymbol{0}. \quad (18.112)$$

(c) The variance matrix of the score function is equal to the Fisher information matrix $\mathcal{I}(\boldsymbol{\theta}) = E[\mathcal{J}(\boldsymbol{X};\boldsymbol{\theta})]$:

$$E\left[\boldsymbol{s}(\boldsymbol{X};\boldsymbol{\theta})\boldsymbol{s}^\top(\boldsymbol{X};\boldsymbol{\theta})\right] = \mathcal{I}(\boldsymbol{\theta}). \quad (18.113)$$

Show that when $\boldsymbol{\theta}$ is a one-dimensional parameter, the following equality holds:

$$E\left[\left(\frac{\partial \log L_{\boldsymbol{x}}(\theta)}{\partial\theta}\right)^2\right] = -E\left[\frac{\partial^2 \log L_{\boldsymbol{x}}(\theta)}{\partial\theta^2}\right]. \quad (18.114)$$

(d) The covariance of the score function $s(X, \theta)$ with any unbiased estimator $\hat{\theta}$ of θ is equal to the identity matrix:

$$\mathrm{Cov}[s(X, \theta), \hat{\theta}] = \mathrm{Cov}[\hat{\theta}, s(X, \theta)] = I. \tag{18.115}$$

(e) For the canonical exponential family defined by (18.14), show that

$$s(x; \eta) = T(x) - \nabla_\eta A(\eta) \text{ and } \mathcal{I}(\theta) = \nabla_\eta \nabla_\eta^\top A(\eta). \tag{18.116}$$

18.5 MLE of $g(\theta)$. Suppose $g(\cdot)$ is a continuous and monotone function. Show that an MLE of a transformed parameter $\eta = g(\theta)$ is the transformation of an MLE of θ; i.e.,

$$\hat{\eta} = g(\hat{\theta}); \text{ hence, } \hat{\theta} = g^{-1}(\hat{\eta}). \tag{18.117}$$

18.6 The CRLB and the Cauchy–Schwarz inequality. Show that the CRLB (18.40) for the case of one-dimensional parameter θ (i.e., a single-parameter case) can be derived from the Cauchy–Schwarz inequality discussed in Section 10.1.1.

18.7* The CRLB and a sufficient statistic. Show that the unbiased minimum variance estimator $T(X)$ that achieves the CRLB (18.42) is a sufficient statistic for estimating the parameter θ.

18.8 Minimum-variance unbiased linear estimator. Suppose that it is known that the mean of RV X_i is $\mu_i \theta$ $(i = 1, 2, \ldots, n)$, where μ_i are known constants, whereas θ is unknown. Let Σ be the variance matrix of the random vector $X = (X_1, X_2, \ldots, X_n)$.

(a) Show that the minimum-variance unbiased linear estimator of θ is given by

$$\hat{\theta}(x) = \frac{\mu^\top \Sigma^{-1} x}{\mu^\top \Sigma^{-1} \mu}, \tag{18.118}$$

where $\mu^\top = [\mu_1, \mu_2, \ldots, \mu_n]$.

(b) Show that the sampling variance (i.e., the variance of the estimator) is given by

$$\sigma_{\hat{\theta}}^2 = \mathrm{Var}[\hat{\theta}(X)] = \frac{1}{\mu^\top \Sigma^{-1} \mu}. \tag{18.119}$$

18.9 Confidence interval. Derive (18.57), the expression for a confidence interval, given a confidence level.

18.10 Confidence interval of an estimator of the binomial distribution parameter. In Example 18.1 the maximum-likelihood estimator for the parameter p of the binomial distribution is $\hat{p} = k/n$, where k is the number of successes in n independent Bernoulli trials. Assuming that n is sufficiently large, find the confidence interval for this estimator.

(a) Find expressions for $\hat{p} = d_1(p)$ and $\hat{p} = d_2(p)$. Show that they represent lower and upper parts of an ellipse.

(b) Find expressions for the inverse functions $p = d_i^{-1}(\hat{p}) = c_i(\hat{p}), i = 1, 2.$

Section 18.2: Hypothesis testing and statistical decision theory

18.11 Matched filter. Consider the RV $\boldsymbol{X}$ under the hypothesis H_1:

$$\boldsymbol{X} = A\boldsymbol{s} + \boldsymbol{Z}.$$

Apply the received sequence $\boldsymbol{X} = (X_0, X_2, \ldots, X_n)$ to a linear filter with impulse response $\boldsymbol{h} = (h_0, h_1, \ldots, h_n)$. The filter output at time t is given by

$$Y_t = A\tilde{s}_t + \tilde{Z}_t, \quad t = 0, 1, \ldots, n,$$

where

$$\tilde{s}_t = \sum_{k=0}^{t} s_{t-k}h_k \text{ and } \tilde{Z}_t = \sum_{k=0}^{t} Z_{t-k}h_k.$$

We define the signal-to-noise ratio at time $t = n$ by $\text{SNR} = A^2\tilde{s}_n^2/E\left[\tilde{Z}_n^2\right]$.

(a) Show that $E\left[\tilde{Z}_n^2\right] = \|\boldsymbol{h}\|^2\sigma^2$, where $\|\boldsymbol{h}\|^2 = \sum_{k=0}^{n} h_k^2$, and σ^2 is the variance of the white noise Z_k.

(b) Show that SNR is maximized when $h_k = cs_{n-k},\ k = 0, 1, 2, \ldots, n$, where c is any constant. Such a filter is called the **matched filter**; i.e., an optimal filter matched to the signal $\boldsymbol{s} = (s_0, s_1, s_2, \ldots, s_n)$. What is this maximum SNR?
Hint: Use the **Cauchy–Schwarz inequality** (see Section 10.1.1):

$$|\boldsymbol{x}^\top \boldsymbol{y}| \le \|\boldsymbol{x}\| \cdot \|\boldsymbol{y}\|, \quad \text{where } \|\boldsymbol{x}\|^2 = \boldsymbol{x}^\top \boldsymbol{x}.$$

(c) Show that the matched filter output y_n at time $t = n$ is given by $c\boldsymbol{s}^\top\boldsymbol{x} = cT(\boldsymbol{x})$, where $T(\boldsymbol{x})$ is defined in (18.83) and c is the constant chosen in part (b).

18.12 Correlation receiver. In referring to the previous problem, let us consider the following **correlation receiver**:

$$R = \sum_{k=0}^{n} g_k X_k = \boldsymbol{g}^\top \boldsymbol{x};$$

i.e., we multiply the received noisy sequence X_k by some prescribed waveform g_k and sum the product over the signal duration period.

(a) Find an expression for SNR in the correlation receiver output R.

(b) Show that SNR is maximized when $g_k = cs_k;\ k = 0, 1, 2, \ldots, n$; i.e., $\boldsymbol{g} = c\boldsymbol{s}$, where c is an arbitrary scalar constant.

18.13 The slope of the tangent of the ROC curve. Verify that the ROC curves plotted in Figure 18.2 satisfy the property (18.76).

18.14 The area under the ROC curve and SNR parameter r. Show that, in Figure 18.2, the area surrounded by the ROC curve and the straight line $P_d = \alpha$ is a monotone increasing function of the SNR parameter $r = A/\sigma$.

18.15 Exponential distribution with two different parameters. Consider a random variable X, which is exponentially distributed with two different parameters:

$$H_0: F_X(x; \theta_0) = 1 - e^{-\theta_0 x} \text{ and } H_1: F_X(x; \theta_1), \quad x \geq 0, \; \theta_1 > \theta_0.$$

(a) Design a Neyman–Pearson test.
(b) Find the ROC curve.

19 Estimation algorithms

In this chapter we will study statistical methods to estimate parameters and procedures to test the goodness of fit of a model to the experimental data. We are primarily concerned with computational algorithms for these methods and procedures. The expectation-maximization (EM) algorithm for maximum-likelihood estimation is discussed in detail.

19.1 Classical numerical methods for estimation

As we stated earlier, it is often the case that a maximum-likelihood estimate (MLE) cannot be found analytically. Thus, numerical methods for computing the MLE are important. Finding the maximum of a likelihood function is an optimization problem. There are a number of optimization algorithms and software packages. In this and the next sections we will discuss several important methods that are pertinent to maximization of a likelihood function: the *method of moments*, the *minimum χ^2 method*, the *minimum Kullback–Leibler divergence method*, and the *Newton–Raphson algorithm*. In Section 19.2 we give a full account of the **EM algorithm**, because of its rather recent development and its increasing applications in signal processing and other science and engineering fields.

19.1.1 Method of moments

This method is typically used to estimate unknown parameters of a distribution function by equating the sample mean, sample variance, and other higher moments calculated from data to the corresponding moments expressed in the parameters of interest. In Example 18.6 we found that for the normal distribution $N(\mu, \sigma^2)$ the MLE $\hat{\mu}$ of the mean μ is equal to the sample mean $\bar{X}$ and the MLE $\hat{\sigma}^2$ of the variance σ^2 is asymptotically equal to the *sampling variance* σ^2/n.

Although the method of moments in general does not provide an *efficient* (i.e., minimum-variance unbiased) estimate, this method is often used in practice. Sometimes the estimates obtained by the method of moments can be used as a good initial approximation for iterative methods such as the Newton–Raphson method and the EM algorithm to be discussed in the following sections.

Example 19.1: Estimating the mean of a Rayleigh distribution. Let us compare an MLE and an estimate based on the method of moments for the parameter σ of the Rayleigh PDF

$$f(x;\sigma) = \frac{x}{\sigma^2}\exp\left(-\frac{x^2}{2\sigma^2}\right), \quad x \geq 0.$$

The mean $E[X]$ is $\sigma\sqrt{\pi/2}$ (see Problem 7.10 (c)). Thus, the method of moments based on n independent samples $\boldsymbol{x} = (x_1, x_2, \ldots, x_n)$ gives

$$\hat{\sigma} = \frac{\sqrt{2}}{n\sqrt{\pi}}\sum_{i=1}^{n} x_i. \tag{19.1}$$

The log-likelihood function of n independent RVs $\boldsymbol{X} = (X_1, X_2, \ldots, X_n)$ is

$$\log f_{\boldsymbol{X}}(\boldsymbol{x};\sigma) = \sum_{i=1}^{n}\log x_i - 2n\log\sigma - \frac{1}{2\sigma^2}\sum_{i=1}^{n} x_i^2,$$

so that the MLE is found by differentiating the above with respect to σ and setting it to zero, which gives

$$\hat{\sigma} = \sqrt{\frac{1}{2n}\sum_{i=1}^{n} x_i^2}. \tag{19.2}$$

We see that the MLE and the method of moment estimate are quite different. □

19.1.2 Minimum chi-square estimation method

Let the sample space $\mathcal{X}$ of an RV X be partitioned into r disjoint regions:

$$\mathcal{X} = \bigcup_{i=1}^{r}\mathcal{X}_i \text{ and } \mathcal{X}_i \bigcap \mathcal{X}_j = \emptyset, \ i \neq j,$$

and let

$$p_i(\boldsymbol{\theta}) \triangleq P[X = x \in \mathcal{X}_i;\, \boldsymbol{\theta}]$$

be the probability that the RV falls into the i th region $\mathcal{X}_i$. We assume that this probability distribution depends on the unknown parameter $\boldsymbol{\theta}$.

Suppose that we take n independent samples of X, and let n_i be the number of samples that fall in the ith region. Clearly, $\sum_{i=1}^{r} n_i = n$. The **minimum χ^2 estimate** of $\boldsymbol{\theta}$ based on the **grouped data** $\boldsymbol{n} = (n_1, n_2, \ldots, n_r)$ is defined by

$$\hat{\boldsymbol{\theta}} = \arg\min_{\boldsymbol{\theta}} \chi^2(\boldsymbol{n};\boldsymbol{\theta}),$$

where $\chi^2(\boldsymbol{n};\boldsymbol{\theta})$ is a chi-square statistic for the grouped data $\boldsymbol{n}$ defined by

$$\chi^2(\boldsymbol{n};\boldsymbol{\theta}) \triangleq \sum_{i=1}^{r} \frac{(n_i - np_i(\boldsymbol{\theta}))^2}{np_i(\boldsymbol{\theta})} = \sum_{i=1}^{r} \frac{n_i^2}{np_i(\boldsymbol{\theta})} - n. \tag{19.3}$$

This method is similar to the well-known **minimum mean square error** (MMSE) method of fitting parametric distributions. Indeed, an iterative MMSE algorithm can be used to obtain a series of estimates according to

$$\hat{\boldsymbol{\theta}}_{k+1} = \arg\min_{\boldsymbol{\theta}} \sum_{i=1}^{r} \frac{(n_i - np_i(\boldsymbol{\theta}))^2}{np(v_i;\hat{\boldsymbol{\theta}}_k)}. \tag{19.4}$$

The main advantage of the minimum χ^2 method is that we can use not only the estimated parameter $\hat{\boldsymbol{\theta}}$, but also the value of $\chi^2(\hat{\boldsymbol{\theta}})$ to obtain the confidence region by using the χ^2 distribution with $r-p-1$ degrees of freedom, where p is the dimension of the vector parameter $\boldsymbol{\theta}$. In other words, the number of degrees of freedom is reduced by the number of independent parameters that we wish to estimate. This is a very important feature of this method.

A typical situation is as follows: we have measured data and we want to construct a model of a probability distribution (or a model of a random process if the data is a time series) that may explain the data. If we select a parametric model, we must decide how many parameters we need to explain the observed data. For instance, if we choose to use a mixture of distributions, how many component distributions should we use? If we want to fit a Markov process to an observed time series, how many states should we use? If we use an MLE for estimating the model parameter, then the more parameters we choose, the greater will be the corresponding likelihood. If we continue to increase the number of parameters, the model will approximate noise in the data. This is called **overfitting**. Therefore, we need to test a hypothesis that the number of parameters is r versus an alternative hypothesis that it is $r+1$. If we use the $\chi^2(n,\boldsymbol{\theta})$ statistic for the goodness-of-fit test given by (19.4), we can use a level α confidence interval of the χ^2 distribution as an acceptance region. As the number of model parameters p grows, the statistic $\chi^2(n,\boldsymbol{\theta})$ decreases, but both the number of degrees of freedom $r-p-1$ and the confidence interval also decrease.

19.1.3 Minimum Kullback–Leibler divergence method

Let $\boldsymbol{f} = f(\boldsymbol{x})$ and $\boldsymbol{g} = g(\boldsymbol{x})$ be two arbitrary PDFs. The **Kullback–Leibler divergence** (KLD) between the two probability distributions is defined by

$$D(\boldsymbol{f}\|\boldsymbol{g}) \triangleq E_f\left[\log\frac{f(\boldsymbol{X})}{g(\boldsymbol{X})}\right] = \int f(\boldsymbol{x})\log\frac{f(\boldsymbol{x})}{g(\boldsymbol{x})}\,d\boldsymbol{x}. \tag{19.5}$$

It is easy to show (Problem 19.1) that the KLD is nonnegative:

$$D(\boldsymbol{f}\|\boldsymbol{g}) \geq 0, \tag{19.6}$$

where equality holds when $f(\boldsymbol{x}) = g(\boldsymbol{x})$ almost everywhere. Thus, the KLD can serve as a measure of closeness between two distributions.

Similar to the $\chi^2(\boldsymbol{\theta})$ statistic, the KLD can be used for fitting a distribution to the grouped data. The KLD between the probabilities $\boldsymbol{p}(\boldsymbol{\theta}) = \{p_i(\boldsymbol{\theta})\}$ and the corresponding relative frequencies $\boldsymbol{f} = \{f_i\}$ from measured data, where

$$f_i = \frac{n_i}{n}, \quad i = 1, 2, \ldots, r,$$

has the form

$$D(\boldsymbol{f} \| \boldsymbol{p}(\boldsymbol{\theta})) = \sum_{i=1}^{r} f_i \log \frac{f_i}{p_i(\boldsymbol{\theta})}. \tag{19.7}$$

The minimum KLD estimate is found by minimizing the KLD with respect to $\boldsymbol{\theta}$ or, equivalently, the **information criterion** introduced by Kullback [215]:

$$\boxed{I(\boldsymbol{n}; \boldsymbol{\theta}) \triangleq nD(\boldsymbol{f} \| \boldsymbol{p}(\boldsymbol{\theta})) = \sum_{i=1}^{r} n_i \log \left(\frac{n_i}{np_i(\boldsymbol{\theta})} \right),} \tag{19.8}$$

where $\boldsymbol{n} = (n_1, n_2, \ldots, n_r)$. This equation can be rewritten as

$$I(\boldsymbol{n}; \boldsymbol{\theta}) = \sum_{i=1}^{r} n_i \log(n_i/n) - \ell(\boldsymbol{n}; \boldsymbol{\theta}), \tag{19.9}$$

where

$$\ell(\boldsymbol{n}; \boldsymbol{\theta}) = \sum_{i=1}^{r} n_i \log p_i(\boldsymbol{\theta})$$

is the log-likelihood for the *grouped data*. Thus, we can see that estimation based on the minimum Kullback information criterion is equivalent to maximum-likelihood estimation of $\boldsymbol{\theta}$ for the grouped data.

By using the approximation (Problem 19.3 (a))

$$\ln x \approx \frac{1}{2}(x - x^{-1}), \quad \text{for } x \approx 1, \tag{19.10}$$

we can show (see Problem 19.3 (b)) that

$$2I(\boldsymbol{n}; \boldsymbol{\theta})) \approx \chi^2(\boldsymbol{n}; \boldsymbol{\theta}), \quad n \gg 1. \tag{19.11}$$

The statistic $2I(\boldsymbol{n}; \boldsymbol{\theta})$ is *asymptotically* chi-square distributed with $r - 1$ degrees of freedom. If we use minimization of Kullback's information criterion (19.8) to estimate the distribution parameters, then the asymptotic distribution of $2I(\boldsymbol{n}; \hat{\theta})$ has $r - p - 1$ degrees of freedom. These relationships among Kullback's information criterion, the χ^2 statistic, and the log-likelihood function allow us to test the goodness of fit of the MLEs.

19.1.4 Newton–Raphson algorithm for maximum-likelihood estimation

In Section 18.1.2 we defined the score function $\boldsymbol{s}(\boldsymbol{x};\boldsymbol{\theta}) = \nabla_{\boldsymbol{\theta}} \log L_{\boldsymbol{x}}(\boldsymbol{\theta})$ as the gradient of the log-likelihood function, and the MLE $\hat{\boldsymbol{\theta}}$ is a solution of $\boldsymbol{s}(\boldsymbol{x};\hat{\boldsymbol{\theta}}) = \boldsymbol{0}$, which represents a set of M equations (18.27), reproduced here:

$$s_m(\boldsymbol{x};\hat{\boldsymbol{\theta}}) = \frac{\partial \log L_{\boldsymbol{x}}(\boldsymbol{\theta})}{\partial \theta_m} = 0, \quad \text{for } m = 1, 2, \ldots, M. \tag{19.12}$$

The Newton–Raphson algorithm is a generalization of Newton's tangent method for finding roots of general equations. We now apply this technique to find roots of (19.12). The algorithm is based on the Taylor expansion of the score function (18.29), which we can rewrite by using the observed Fisher information matrix (18.32) as

$$\boldsymbol{s}(\boldsymbol{x};\boldsymbol{\theta}) \approx \boldsymbol{s}(\boldsymbol{x};\boldsymbol{\theta}^{(p)}) - \mathcal{J}(\boldsymbol{x};\boldsymbol{\theta}^{(p)})(\boldsymbol{\theta} - \boldsymbol{\theta}^{(p)}), \tag{19.13}$$

where $\boldsymbol{\theta}^{(p)}$ is an approximate estimate of a *likelihood stationary point*, which is defined as a solution of (19.12), obtained after p iterations. We assume that the remainder $\boldsymbol{r}(\boldsymbol{x};\boldsymbol{\theta})$ in (18.29) is negligible when $\boldsymbol{\theta}^{(p)}$ comes close enough to $\boldsymbol{\theta}$. Then the MLE, which is the root of (19.12), can be approximated by the root of the right-hand side of (19.13) being set to zero:

$$\boldsymbol{s}(\boldsymbol{x};\boldsymbol{\theta}^{(p)}) - \mathcal{J}(\boldsymbol{x};\boldsymbol{\theta}^{(p)})(\boldsymbol{\theta} - \boldsymbol{\theta}^{(p)}) = \boldsymbol{0}. \tag{19.14}$$

Denoting the root of this equation as $\boldsymbol{\theta}^{(p+1)}$, we find the following iterative algorithm:

$$\boxed{\boldsymbol{\theta}^{(p+1)} = \boldsymbol{\theta}^{(p)} + \mathcal{J}^{-1}(\boldsymbol{x};\boldsymbol{\theta}^{(p)})\boldsymbol{s}(\boldsymbol{x};\boldsymbol{\theta}^{(p)}), \quad p = 0, 1, 2, \ldots} \tag{19.15}$$

The initial estimate $\boldsymbol{\theta}^{(0)}$ may be found, for instance, by the method of moments. Algorithm 19.1 summarizes the Newton–Raphson algorithm to find an MLE by the above iteration method. In practice, an iterative algorithm must have a *stopping rule* because convergence may be achieved as $p \to \infty$. A stopping rule is applied in Step 4 of the Newton–Raphson algorithm, and the rule is application dependent. For instance, the iteration should stop if $\|\boldsymbol{\theta}^{(p+1)} - \boldsymbol{\theta}^{p}\| < \epsilon$, where ϵ is a sufficiently small number, or if the number of iterations p exceeds a prescribed large number N. Since we are solving the equation $\boldsymbol{s}(\boldsymbol{x};\boldsymbol{\theta}) = \boldsymbol{0}$, we can set the stopping rule: Stop when $\|\boldsymbol{s}(\boldsymbol{x};\boldsymbol{\theta}^{(p)})\| < \epsilon$.

If the log-likelihood is quadratic, the MLE $\hat{\boldsymbol{\theta}}$ is obtained in one step, starting from any initial point $\boldsymbol{\theta}^{(0)}$. If the log-likelihood function is concave, the algorithm converges to the MLE. In general, the algorithm can be very sensitive to the initial value $\boldsymbol{\theta}^{(0)}$. However, if $\boldsymbol{\theta}^{(0)}$ is close enough to $\hat{\boldsymbol{\theta}}$, the iterations will exhibit a *quadratic convergence* i.e.,

$$\| \boldsymbol{\theta}^{(p+1)} - \hat{\boldsymbol{\theta}} \| < k \| \boldsymbol{\theta}^{(p)} - \hat{\boldsymbol{\theta}} \|^2 .$$

In practice, however, we cannot compute with perfect precision, and thus the iterations will exhibit a *sub-quadratic* convergence. The Newton–Raphson algorithm can be computationally expensive: at each step we need to calculate the observed Fisher information matrix $\mathcal{J}$ and invert it. Instead of inverting matrices, we can find the solution of (19.14) with computational steps on the order of $O(M^3)$, where M is the dimension of

Algorithm 19.1 Newton–Raphson algorithm

1: Select the initial value $\boldsymbol{\theta}^{(0)}$ and set the iteration number $p = 0$.
2: Assume the pth estimate $\boldsymbol{\theta}^{(p)}$.
3: Find

$$\boldsymbol{\theta}^{(p+1)} = \boldsymbol{\theta}^{(p)} + \mathcal{J}^{-1}(\boldsymbol{x}; \boldsymbol{\theta}^{(p)})\boldsymbol{s}(\boldsymbol{x}; \boldsymbol{\theta}^{(p)}).$$

4: If any of the stopping conditions is met, go to Step 5. Otherwise, replace $\boldsymbol{\theta}^{(p)}$ by $\boldsymbol{\theta}^{(p+1)}$, set $p \leftarrow p + 1$, and repeat Steps 2 through 4.
5: Output $\boldsymbol{\theta}^{(p)}$ as an approximate MLE, $\mathcal{J}^{-1}(\boldsymbol{x}; \boldsymbol{\theta}^{(p)})$ as an approximate Fisher information matrix, and $\|\boldsymbol{s}(\boldsymbol{x}; \boldsymbol{\theta}^{(p)})\|$ as a measure of convergence.

the parameter vector $\boldsymbol{\theta}$. However, calculating the inverse of the matrix $\mathcal{J}$ would provide an estimate of variance of the asymptotic normal distribution of the MLE.

An alternative to the Newton–Raphson algorithm is the **method of scoring**, in which the observed Fisher information matrix is replaced by the expected Fisher information matrix:

$$\boxed{\boldsymbol{\theta}^{(p+1)} = \boldsymbol{\theta}^{(p)} + \mathcal{I}^{-1}(\boldsymbol{\theta}^{(p)})\boldsymbol{s}(\boldsymbol{x}; \boldsymbol{\theta}^{(p)}), \quad p = 0, 1, 2, \ldots.} \tag{19.16}$$

If we can find an approximate MLE by some slower but more robust algorithm, the fast convergence of the Newton–Raphson algorithm can be exploited. One such algorithm is the EM algorithm to be presented in the next section.

19.2 Expectation-maximization algorithm for maximum-likelihood estimation

The expectation-maximization (EM) algorithm was discussed and given its name by Dempster *et al.* in their 1977 seminal paper [80]. Although the method had been proposed previously by others in special circumstances (e.g., the Baum–Welch algorithm that predates the EM formulation), the 1977 paper generalized the method and provided a solid theory behind this powerful iterative estimation method.

19.2.1 Expectation-maximization algorithm for transformed data

We consider the case where the **observable variable** $\boldsymbol{Y}$ is a **transformed variable** of some variable $\boldsymbol{X}$, which may or may not be completely **unobservable** [80]:

$$\boldsymbol{Y} = T(\boldsymbol{X}), \quad \boldsymbol{X} \in \mathcal{X}, \quad \boldsymbol{Y} \in \mathcal{Y}, \tag{19.17}$$

where $\mathcal{X}$ and $\mathcal{Y}$ are the sample spaces of the variables $\boldsymbol{X}$ and $\boldsymbol{Y}$ respectively. The mapping $T(\boldsymbol{x})$ is generally a many-to-one mapping, so the information included in data $\boldsymbol{x}$ is reduced. Thus, the observed data $\boldsymbol{y}$ contains less information than $\boldsymbol{x}$ concerning the parameter $\boldsymbol{\theta}$. The inverse transformation $T^{-1}(\boldsymbol{y})$ is one-to-many mapping, and hence is not uniquely determinable. Thus, the observable variable Y is called **incomplete**, while X is a **complete** variable.

Ideally, we should seek $\boldsymbol{\theta}$ that would maximize the likelihood function[1] $L_{\boldsymbol{x}}(\boldsymbol{\theta}) = p_X(\boldsymbol{x}; \boldsymbol{\theta})$, but we cannot have the complete data $\boldsymbol{x}$ to evaluate $L_{\boldsymbol{x}}(\boldsymbol{\theta})$. Therefore, we should be content with finding $\boldsymbol{\theta}$ that maximizes the likelihood function $L_{\boldsymbol{y}}(\boldsymbol{\theta}) = p_Y(\boldsymbol{y}; \boldsymbol{\theta})$ based on the observed but incomplete data $\boldsymbol{y}$.

Let us start with the joint variable $(\boldsymbol{X}, \boldsymbol{Y})$ and denote its probability by $p_{X,Y}(\boldsymbol{x}, \boldsymbol{y}; \boldsymbol{\theta})$. We assume that the parameter $\boldsymbol{\theta}$ is fixed. An extension to the case of Bayesian estimation, where the prior distribution $p_{\Theta}(\boldsymbol{\theta})$ is taken into account, as in MAP estimation, is straightforward, and it is left to the reader as an exercise problem (Problem 19.9).

If we substitute $\boldsymbol{y} = T(\boldsymbol{x})$, the above joint probability distribution reduces to the distribution of $\boldsymbol{x}$. Hence,

$$p_X(\boldsymbol{x}; \boldsymbol{\theta}) = p_{X,Y}(\boldsymbol{x}, T(\boldsymbol{x}); \boldsymbol{\theta}). \tag{19.18}$$

Then, the conditional probability of the variable $\boldsymbol{X}$ given the observable $\boldsymbol{Y}$ is

$$p_{X|Y}(\boldsymbol{x}|\boldsymbol{y}; \boldsymbol{\theta}) = \frac{p_{X,Y}(\boldsymbol{x}, \boldsymbol{y}; \boldsymbol{\theta})}{p_Y(\boldsymbol{y}; \boldsymbol{\theta})} = \frac{p_X(\boldsymbol{x}; \boldsymbol{\theta})}{p_Y(\boldsymbol{y}; \boldsymbol{\theta})}, \tag{19.19}$$

from which we obtain the log-likelihood function $\log L_{\boldsymbol{y}}(\boldsymbol{\theta}) = \log p_Y(\boldsymbol{y}; \boldsymbol{\theta})$ as

$$\log L_{\boldsymbol{y}}(\boldsymbol{\theta}) = \log p_X(\boldsymbol{x}; \boldsymbol{\theta}) - \log p_{X|Y}(\boldsymbol{x}|\boldsymbol{y}; \boldsymbol{\theta}). \tag{19.20}$$

Since the complete data $\boldsymbol{x}$ is not available to us, the log-likelihood function $\log L_{\boldsymbol{y}}(\boldsymbol{\theta})$ is a function of the complete variable $\boldsymbol{X}$ given by

$$\log L_{\boldsymbol{y}}(\boldsymbol{X}; \boldsymbol{\theta}) = \log p_X(\boldsymbol{X}; \boldsymbol{\theta}) - \log p_{X|Y}(\boldsymbol{X}|\boldsymbol{y}; \boldsymbol{\theta}). \tag{19.21}$$

Now we take expectations on both sides of (19.21) with respect to the conditional probability $p_{X|Y}(\boldsymbol{x}|\boldsymbol{y}; \boldsymbol{\theta}^{(p)})$ and obtain

$$\log L_{\boldsymbol{y}}(\boldsymbol{\theta}) = E\left[\log p_X(\boldsymbol{X}; \boldsymbol{\theta})|\boldsymbol{y}; \boldsymbol{\theta}^{(p)}\right] - E\left[\log p_{X|Y}(\boldsymbol{X}|\boldsymbol{y}; \boldsymbol{\theta})|\boldsymbol{y}; \boldsymbol{\theta}^{(p)}\right], \tag{19.22}$$

because $E\left[\log L_{\boldsymbol{y}}(\boldsymbol{X}; \boldsymbol{\theta})|\boldsymbol{y}; \boldsymbol{\theta}^{(p)}\right] = \log L_{\boldsymbol{y}}(\boldsymbol{\theta})$. The last equation can be written as

$$\boxed{\log L_{\boldsymbol{y}}(\boldsymbol{\theta}) = Q(\boldsymbol{\theta}|\boldsymbol{\theta}^{(p)}) + H(\boldsymbol{\theta}|\boldsymbol{\theta}^{(p)}),} \tag{19.23}$$

where $Q(\boldsymbol{\theta}|\boldsymbol{\theta}^{(p)})$ is an *auxiliary function*, called the ***Q*-function** [80], defined by

$$Q(\boldsymbol{\theta}|\boldsymbol{\theta}^{(p)}) \triangleq E\left[\log p_X(\boldsymbol{X}; \boldsymbol{\theta})|\boldsymbol{y}; \boldsymbol{\theta}^{(p)}\right], \tag{19.24}$$

and

$$H(\boldsymbol{\theta}|\boldsymbol{\theta}^{(p)}) \triangleq -E\left[\log p_{X|Y}(\boldsymbol{X}|\boldsymbol{y}; \boldsymbol{\theta})|\boldsymbol{y}; \boldsymbol{\theta}^{(p)}\right]. \tag{19.25}$$

[1] For notational simplicity, we consider discrete RVs $\boldsymbol{X}, \boldsymbol{Y}$. If the variables are continuous, all the derivations are valid if we replace $p_{X,Y}(\boldsymbol{x}, \boldsymbol{y}; \boldsymbol{\theta})$, $p_X(\boldsymbol{x}; \boldsymbol{\theta})$, etc. by the corresponding PDFs.

The function H of (19.25) satisfies the following inequality (see Problem 19.1 (a)):

$$\boxed{H(\boldsymbol{\theta}|\boldsymbol{\theta}^{(p)}) \geq H(\boldsymbol{\theta}^{(p)}|\boldsymbol{\theta}^{(p)}) = \mathcal{H}(\boldsymbol{X}|\boldsymbol{y};\boldsymbol{\theta}^{(p)}),} \tag{19.26}$$

where $\mathcal{H}(\boldsymbol{X}|\boldsymbol{y};\boldsymbol{\theta}^{(p)})$ is the **conditional entropy** [300] of the variable $\boldsymbol{X}$ given the observation $\boldsymbol{y}$ and the estimate of $\boldsymbol{\theta}$ after the pth iteration (see also (19.53) of Problem 19.7).

19.2.1.1 Relationship between the log-likelihood function and Q-function

The auxiliary function defined in (19.24) should read as the expectation of $\log p_{\boldsymbol{X}}(\boldsymbol{X};\boldsymbol{\theta})$ under the probability measure $p_{\boldsymbol{X}|\boldsymbol{Y}}(\boldsymbol{x}|\boldsymbol{y};\boldsymbol{\theta}^{(p)})$ for the complete variable $\boldsymbol{X}$, with the observation instance $\boldsymbol{y}$ fixed. Since the Q-function is not symmetric with respect to its two arguments (i.e., $Q(\boldsymbol{\theta}|\boldsymbol{\theta}^{(p)}) \neq Q(\boldsymbol{\theta}^{(p)}|\boldsymbol{\theta})$), the order of the arguments matters. Our final goal is to find an MLE of $\boldsymbol{\theta}$, given an observation instance $\boldsymbol{y}$. Since it is often computationally difficult to find the MLE directly, we will pursue an iterative algorithm to arrive at the MLE. The auxiliary function Q will help us find such a solution, because the following important relation holds between $\log L_{\boldsymbol{y}}(\boldsymbol{\theta})$ and $Q(\boldsymbol{\theta}|\boldsymbol{\theta}^{(p)})$ (Problem 19.4):

$$\boxed{Q(\boldsymbol{\theta}|\boldsymbol{\theta}^{(p)}) - Q(\boldsymbol{\theta}^{(p)}|\boldsymbol{\theta}^{(p)}) \leq \log L_{\boldsymbol{y}}(\boldsymbol{\theta}) - \log L_{\boldsymbol{y}}(\boldsymbol{\theta}^{(p)}),} \tag{19.27}$$

where the equality holds when $\boldsymbol{\theta} = \boldsymbol{\theta}^{(p)}$. Equation (19.26) shows that $H(\boldsymbol{\theta}|\boldsymbol{\theta}^{(p)})$ attains its minimum at $\boldsymbol{\theta} = \boldsymbol{\theta}^{(p)}$, and thus $\nabla_{\boldsymbol{\theta}} H(\boldsymbol{\theta}|\boldsymbol{\theta}^{(p)})|_{\boldsymbol{\theta}=\boldsymbol{\theta}^{(p)}} = \mathbf{0}$. Then, differentiating both sides of (19.23), we obtain

$$\nabla_{\boldsymbol{\theta}} \log L_{\boldsymbol{y}}(\boldsymbol{\theta})|_{\boldsymbol{\theta}=\boldsymbol{\theta}^{(p)}} = \nabla_{\boldsymbol{\theta}} Q(\boldsymbol{\theta})|\boldsymbol{\theta}^{(p)})|_{\boldsymbol{\theta}=\boldsymbol{\theta}^{(p)}}. \tag{19.28}$$

19.2.1.2 Derivation of the E-step and the M-step

Suppose that we have obtained an estimate of $\boldsymbol{\theta}$ after p iterations, which we denote as $\boldsymbol{\theta}^{(p)}$. We want to improve upon this estimate using $Q(\boldsymbol{\theta}|\boldsymbol{\theta}^{(p)})$ as a guide. This Q-function is a function of $\boldsymbol{\theta}$ and the observation $\boldsymbol{y}$, given the current model parameter $\boldsymbol{\theta}^{(p)}$. The variable $\boldsymbol{X}$ is averaged out by taking the expectation:

$$\boxed{Q(\boldsymbol{\theta}|\boldsymbol{\theta}^{(p)}) = E\left[\log p_{\boldsymbol{X}}(\boldsymbol{X};\boldsymbol{\theta})|\, \boldsymbol{y}, \boldsymbol{\theta}^{(p)}\right] \quad \text{(E-step).}} \tag{19.29}$$

The inequality (19.27) implies that if $Q(\boldsymbol{\theta}|\boldsymbol{\theta}^{(p)}) > Q(\boldsymbol{\theta}^{(p)}|\boldsymbol{\theta}^{(p)})$, then $L_{\boldsymbol{y}}(\boldsymbol{\theta}) > L_{\boldsymbol{y}}(\boldsymbol{\theta}^{(p)})$. Therefore, the best choice for the next estimate $\boldsymbol{\theta}^{(p+1)}$, given the current estimate $\boldsymbol{\theta}^{(p)}$, will be found by *maximizing* $Q(\boldsymbol{\theta}|\boldsymbol{\theta}^{(p)})$ with respect to $\boldsymbol{\theta}$:

$$\boxed{\boldsymbol{\theta}^{(p+1)} = \arg\max_{\boldsymbol{\theta}} Q(\boldsymbol{\theta}|\boldsymbol{\theta}^{(p)}) \qquad \text{(M-step).}} \tag{19.30}$$

If $Q(\boldsymbol{\theta}|\boldsymbol{\theta}^{(p)})$ is differentiable, the next estimate $\boldsymbol{\theta}^{(p+1)}$ satisfies $\nabla_{\boldsymbol{\theta}} Q(\boldsymbol{\theta}|\boldsymbol{\theta}^{(p)})|_{\boldsymbol{\theta}=\boldsymbol{\theta}^{(p+1)}}=\mathbf{0}$. Thus, if the algorithm converges to $\boldsymbol{\theta}^*$, $\nabla_{\boldsymbol{\theta}} Q(\boldsymbol{\theta}|\boldsymbol{\theta}^*)|_{\boldsymbol{\theta}=\boldsymbol{\theta}^*} = \mathbf{0}$, which implies, according to (19.28), that

$$\nabla_{\boldsymbol{\theta}} L(\boldsymbol{\theta})|_{\boldsymbol{\theta}=\boldsymbol{\theta}^*} = \mathbf{0}. \tag{19.31}$$

Thus, the algorithm converges to a *stationary point* of the likelihood function. Therefore, an MLE $\hat{\boldsymbol{\theta}}$ can be found by the above iterative procedure of alternating the *expectation step* (E-step) given by (19.29) and the *maximization step* (M-step) given by (19.30). The EM algorithm is summarized as Algorithm 19.2.

Algorithm 19.2 EM algorithm for an MLE

1: Denote the initial estimate of the model parameters as $\boldsymbol{\theta}^{(0)}$. Set the iteration number $p = 0$.
2: Assume the pth estimate $\boldsymbol{\theta}^{(p)}$.
3: Evaluate

$$E\left[\log p_{\boldsymbol{X}}(\boldsymbol{X};\boldsymbol{\theta})|\boldsymbol{y},\boldsymbol{\theta}^{(p)}\right] \triangleq Q(\boldsymbol{\theta}|\boldsymbol{\theta}^{(p)}) \text{ (E-step).}$$

4: Find

$$\boldsymbol{\theta}^{(p+1)} = \arg\max_{\boldsymbol{\theta}} Q(\boldsymbol{\theta}|\boldsymbol{\theta}^{(p)}) \text{ (M-step).}$$

5: If any of the stopping conditions are met, go to Step 6. Otherwise, replace $\boldsymbol{\theta}^{(p)}$ by $\boldsymbol{\theta}^{(p+1)}$, set $p \leftarrow p+1$, and repeat Steps 2 through 5.
6: Output $\boldsymbol{\theta}^{(p+1)}$ as an MLE.

19.2.1.3 Geometrical interpretation of the EM algorithm

If we apply inequality (19.26) to (19.23), we obtain

$$\log L_{\boldsymbol{y}}(\boldsymbol{\theta}) \geq Q(\boldsymbol{\theta}|\boldsymbol{\theta}^{(p)}) + H(\boldsymbol{\theta}^{(p)}|\boldsymbol{\theta}^{(p)}) \triangleq B(\boldsymbol{\theta}|\boldsymbol{\theta}^{(p)}). \tag{19.32}$$

As we can see, the new auxiliary function $B(\boldsymbol{\theta}|\boldsymbol{\theta}^{(p)})$ differs from the $Q(\boldsymbol{\theta}|\boldsymbol{\theta}^{(p)})$ by the term $H(\boldsymbol{\theta}^{(p)}|\boldsymbol{\theta}^{(p)})$, which does not depend on $\boldsymbol{\theta}$. Thus, these functions achieve their extrema at the same point, so they both can be used in the M-step of the algorithm. However, in contrast with the $Q(\boldsymbol{\theta}|\boldsymbol{\theta}^{(p)})$, $B(\boldsymbol{\theta}|\boldsymbol{\theta}^{(p)})$ coincides with the log-likelihood at the point $\boldsymbol{\theta} = \boldsymbol{\theta}^{(p)}$, since

$$\boxed{B(\boldsymbol{\theta}^{(p)}|\boldsymbol{\theta}^{(p)}) = Q(\boldsymbol{\theta}^{(p)}|\boldsymbol{\theta}^{(p)}) + H(\boldsymbol{\theta}^{(p)}|\boldsymbol{\theta}^{(p)}) = \log L_{\boldsymbol{y}}(\boldsymbol{\theta}^{(p)}),} \tag{19.33}$$

as shown in Figure 19.1. According to (19.28), we have

$$\nabla_{\boldsymbol{\theta}} \log L_{\boldsymbol{y}}(\boldsymbol{\theta})|_{\boldsymbol{\theta}=\boldsymbol{\theta}^{(p)}} = \nabla_{\boldsymbol{\theta}} B(\boldsymbol{\theta}|\boldsymbol{\theta}^{(p)})|_{\boldsymbol{\theta}=\boldsymbol{\theta}^{(p)}}, \tag{19.34}$$

which means that $\log L_{\boldsymbol{y}}(\boldsymbol{\theta})$ and its *lower bound* $B(\boldsymbol{\theta}|\boldsymbol{\theta}^{(p)})$ have a common tangential plane, as shown in Figure 19.1, which illustrates the M-step of the EM algorithm.

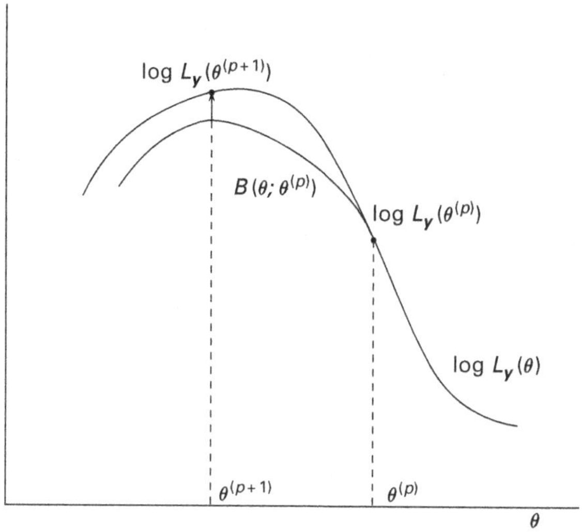

Figure 19.1 The M-step of the EM algorithm.

At $\boldsymbol{\theta} = \boldsymbol{\theta}^{(p)}$, the log-likelihood is $\log L_{\boldsymbol{y}}(\boldsymbol{\theta}^{(p)})$, and the M-step finds the next point $\boldsymbol{\theta}^{(p+1)}$ which maximizes the auxiliary function $B(\boldsymbol{\theta}|\boldsymbol{\theta}^{(p)})$. The log-likelihood at this point $\log L_{\boldsymbol{y}}(\boldsymbol{\theta}^{(p+1)})$ is marked by an arrow. As is illustrated in the figure, the M-step increases the log-likelihood as well:

$$\log L_{\boldsymbol{y}}(\boldsymbol{\theta}^{(p)}) = B(\boldsymbol{\theta}^{(p)}|\boldsymbol{\theta}^{(p)}) \leq B(\boldsymbol{\theta}^{(p+1)}|\boldsymbol{\theta}^{(p)}) \leq \log L_{\boldsymbol{y}}(\boldsymbol{\theta}^{(p+1)}). \tag{19.35}$$

Thus, this equation can be considered as an alternative derivation of the EM algorithm.

Like any iterative algorithm such as the Newton–Raphson algorithm, there is no guarantee that the point of convergence is the true MLE value: it can be a local maximum or a saddle point. But the EM algorithm guarantees that the likelihood-function $L_{\boldsymbol{y}}(\boldsymbol{\theta}^{(p)})$ increases in every step because of (19.35) or (19.27), whereas the Newton–Raphson and other algorithms do not, and can be very unstable until $\boldsymbol{\theta}^{(p)}$ reaches the vicinity of the extremum point. Referring to Figure 19.1, the Newton–Raphson algorithm approximates the log-likelihood function with the quadratic polynomial, which, similar to $B(\boldsymbol{\theta}|\boldsymbol{\theta}^{(p)})$, has the same tangential plane as the $\log L_{\boldsymbol{y}}(\boldsymbol{\theta})$; but, in contrast to the EM algorithm, this polynomial does not necessarily represent the lower bound on the $\log L_{\boldsymbol{y}}(\boldsymbol{\theta})$.

In order to reach a global maximum, we must carefully choose, as in any other iterative algorithm, the initial estimate $\boldsymbol{\theta}^{(0)}$ close to the MLE; if this cannot be arranged, we must try a number of choices as an initial estimate. Wu [362] discusses several convergence properties of the EM algorithm. See also McLachlan and Krishnan [246].

The rate of convergence of the EM algorithm can be faster than conventional first-order iterative algorithms, but is usually slower than the quadratic convergence typically available with Newton-type methods. Dempster *et al.* [80] show that the rate of convergence of the EM algorithm is linear and the rate depends on the proportion of information in the observed data. Thus, if a large portion of data is missing in comparison with the formulated complete-data problem, convergence can be quite slow. Several methods have been proposed to speed up the EM algorithm [246]. The essential idea behind the EM algorithm is that instead of attacking the maximization of $L_y(\boldsymbol{\theta})$ directly, which is often computationally difficult, we take an iterative procedure, and work on maximization of the auxiliary function $Q(\boldsymbol{\theta}|\boldsymbol{\theta}^{(p)})$ in the pth step. When the iterative procedure converges to a stable point, it may converge to a local maximum or a saddle point, instead of the global maximum. The logarithm of $p_X(\boldsymbol{x}; \boldsymbol{\theta})$ may take a simpler form when the probability distribution belongs to the *exponential family* or the *exponential class* (see Section 4.4 and Problem 19.10).

When it is difficult to find $\boldsymbol{\theta}$ that maximizes $Q(\boldsymbol{\theta}|\boldsymbol{\theta}^{(p)})$, we can adopt the Newton–Raphson algorithm or hill-climbing method and choose an appropriate $\boldsymbol{\theta}$ as long as it increases this Q function; i.e.,

$$Q(\boldsymbol{\theta}|\boldsymbol{\theta}^{(p)}) \geq Q(\boldsymbol{\theta}^{(p)}, \boldsymbol{\theta}^{(p)}).$$

It will give a greater likelihood function, i.e., $L_y(\boldsymbol{\theta}) \geq L_y(\boldsymbol{\theta}^{(p)})$. Such a method is often referred to as the **generalized EM** (GEM) method.

19.2.2 Expectation-maximization algorithm for missing data

Let us consider a special case of the model formulated in the previous section; namely, the complete variable $\boldsymbol{X}$ can be written as

$$\boldsymbol{X} = (\boldsymbol{Y}, \boldsymbol{Z}), \tag{19.36}$$

where $\boldsymbol{Y}$ is the **observable variable** and $\boldsymbol{Z}$ is a **latent variable**. The transformation T truncates the second component of $\boldsymbol{x} = (\boldsymbol{y}, \boldsymbol{z})$; i.e.,

$$T(\boldsymbol{x}) = T(\boldsymbol{y}, \boldsymbol{z}) = \boldsymbol{y}. \tag{19.37}$$

An instance $\boldsymbol{x}$ is called **complete data**, $\boldsymbol{y}$ is *observed data* or *incomplete data*, and $\boldsymbol{z}$ is **missing data**. Many practical applications of the EM algorithm, such as parameter estimation in a HMM and parameter estimation for pattern classification, can be formulated under this special case defined by (19.36) and (19.37).[2]

The $\boldsymbol{Q}$-function is defined, similar to (19.24), by:[2]

$$Q(\boldsymbol{\theta}|\boldsymbol{\theta}^{(p)}) \triangleq E\left[\log p_{\boldsymbol{YZ}}(\boldsymbol{y}, \boldsymbol{Z}; \boldsymbol{\theta})|\, \boldsymbol{y}; \boldsymbol{\theta}^{(p)}\right] = \sum_{\boldsymbol{z}} p_{\boldsymbol{Z}|\boldsymbol{Y}}(\boldsymbol{z}|\boldsymbol{y}; \boldsymbol{\theta}^{(p)}) \log p_{\boldsymbol{YZ}}(\boldsymbol{y}, \boldsymbol{z}; \boldsymbol{\theta}), \tag{19.38}$$

which reads as the expectation of $\log p_{\boldsymbol{ZY}}(\boldsymbol{y}, \boldsymbol{Z}; \boldsymbol{\theta})$ under the probability measure $p_{\boldsymbol{Z}|\boldsymbol{Y}}(\boldsymbol{z}|\boldsymbol{y}; \boldsymbol{\theta}^{(p)})$ for the unobserved variable $\boldsymbol{Z}$, with the observation instance $\boldsymbol{y}$ fixed.

This $\boldsymbol{Q}$-function also satisfies the important property (19.27) (Problem 19.5), and an MLE $\hat{\boldsymbol{\theta}}$ can be found by an iterative procedure of alternating the E-step and the

[2] The Q-function defined by Baum *et al.* [16] is slightly different: instead of the conditional probability $p(\boldsymbol{z}|\boldsymbol{y}; \boldsymbol{\theta})$, the joint probability $p(\boldsymbol{z}, \boldsymbol{y}; \boldsymbol{\theta})$ is used as the probability measure.

M-step discussed in the previous section. A slightly different argument to derive the EM algorithm in the presence of the latent variable $\boldsymbol{Z}$ is to consider a tight lower bound function $B^*(\boldsymbol{\theta}|\boldsymbol{\theta}^{(p)})$ defined by (19.53) in Problem 19.7. We argue that maximization of this B^*-function is equivalent to maximization of the Q-function (see Problem 19.8).

19.2.2.1 Bayesian EM algorithm for MAP estimation

Thus far, we assumed that the parameter $\boldsymbol{\theta}$ is unknown but fixed. If we have a *prior probability* distribution $p_{\boldsymbol{\Theta}}(\boldsymbol{\theta})$ of the parameter variable $\boldsymbol{\Theta}$, then we should take the Bayesian approach; namely, we should find the maximum *a posteriori* probability (MAP) estimate of $\boldsymbol{\theta}$ instead of the MLE, i.e., $\hat{\boldsymbol{\theta}}_{\text{MLE}} = \arg\max_{\boldsymbol{\theta}} p_{\boldsymbol{Y}}(\boldsymbol{y};\boldsymbol{\theta})$:

$$\hat{\boldsymbol{\theta}}_{\text{MAP}} = \arg\max_{\boldsymbol{\theta}} p_{\boldsymbol{\Theta}|\boldsymbol{Y}}(\boldsymbol{\theta}|\boldsymbol{y}). \tag{19.39}$$

According to (18.102) the MAP estimate can be obtained by maximizing the sum of the log-likelihood and the prior:

$$\hat{\boldsymbol{\theta}}_{\text{MAP}} = \arg\max_{\boldsymbol{\theta}} \ [\log L_{\boldsymbol{y}}(\boldsymbol{\theta}) + \log \pi(\boldsymbol{\theta})]. \tag{19.40}$$

Therefore, the M-step of (19.30) should be modified (see Problem 19.9 (a)) to

$$\boxed{\boldsymbol{\theta}^{(p+1)} = \arg\max_{\boldsymbol{\theta}} \left[Q(\boldsymbol{\theta}|\boldsymbol{\theta}^{(p)}) + \log p_{\boldsymbol{\Theta}}(\boldsymbol{\theta}) \right].} \tag{19.41}$$

The reader is suggested to write a program similar to Algorithm 19.2 (see Problem 19.9 (b)).

19.2.2.2 Monte Carlo EM algorithm

The EM algorithm is most effective when both the E-step and the M-step can be performed analytically. However, it is not always possible to do so in practical applications. In such a case we need to resort to some kind of simulation method, such as Markov chain Monte Carlo (MCMC), which is discussed in Section 21.7.

The E-step to compute the Q-function of (19.38),

$$Q(\boldsymbol{\theta}|\boldsymbol{\theta}^{(p)}) = E\left[\log p(\boldsymbol{y}, \boldsymbol{Z}; \boldsymbol{\theta})|\boldsymbol{y}, \boldsymbol{\theta}^{(p)}\right], \tag{19.42}$$

can be approximated by generating $N^{(p)}$ samples, $z_1, z_2, \ldots, z_t, \ldots, z_{N^{(p)}}$ (after discarding the initial $B^{(p)}$ burn-in samples) from the conditional distribution $p(z|\boldsymbol{y}, \boldsymbol{\theta}^{(p)})$ and then by substituting the empirical average for the conditional expectation:

$$Q(\boldsymbol{\theta}|\boldsymbol{\theta}^{(p)}) \approx \frac{1}{N^{(p)}} \sum_{t=1}^{N^{(p)}} \log p(\boldsymbol{y}, z_t; \boldsymbol{\theta}). \tag{19.43}$$

Maximization of this function with respect to $\boldsymbol{\theta}$ can be performed using the *simulated annealing* method discussed in Section 21.7.5.

19.3 Summary of Chapter 19

Newton–Raphson algorithm:	$\boldsymbol{\theta}^{(p+1)} = \boldsymbol{\theta}^{(p)} + \mathcal{J}^{-1}(\boldsymbol{x};\boldsymbol{\theta}^{(p)})\boldsymbol{s}(\boldsymbol{x};\boldsymbol{\theta}^{(p)})$	(19.15)
Method of scoring:	$\boldsymbol{\theta}^{(p+1)} = \boldsymbol{\theta}^{(p)} + \mathcal{I}^{-1}(\boldsymbol{\theta}^{(p)})\boldsymbol{s}(\boldsymbol{x};\boldsymbol{\theta}^{(p)})$	(19.16)
Kullback information criterion:	$I(\boldsymbol{n};\boldsymbol{\theta}) \triangleq nD(\boldsymbol{f}\\|\boldsymbol{p}(\boldsymbol{\theta})) = \sum_{i=1}^{r} n_i \log\left(\frac{n_i}{np_i(\boldsymbol{\theta})}\right)$	(19.8)
Q-function for EM:	$Q(\boldsymbol{\theta}\|\boldsymbol{\theta}^{(p)}) \triangleq E\left[\log p_{\boldsymbol{X}}(\boldsymbol{X};\boldsymbol{\theta})\|\boldsymbol{y};\boldsymbol{\theta}^{(p)}\right]$	(19.24)
Q-func. and $\log L_{\boldsymbol{y}}(\boldsymbol{\theta})$:	$\log L_{\boldsymbol{y}}(\boldsymbol{\theta}) = Q(\boldsymbol{\theta}\|\boldsymbol{\theta}^{(p)}) + H(\boldsymbol{\theta}\|\boldsymbol{\theta}^{(p)})$	(19.23)
H-func. inequality:	$H(\boldsymbol{\theta}\|\boldsymbol{\theta}^{(p)}) \geq H(\boldsymbol{\theta}^{(p)}\|\boldsymbol{\theta}^{(p)})$	(19.26)
Key property of Q-function:	$Q(\boldsymbol{\theta}\|\boldsymbol{\theta}^{(p)}) - Q(\boldsymbol{\theta}^{(p)}\|\boldsymbol{\theta}^{(p)})$ $\leq \log L_{\boldsymbol{y}}(\boldsymbol{\theta}) - \log L_{\boldsymbol{y}}(\boldsymbol{\theta}^{(p)})$	(19.27)
E-step:	$Q(\boldsymbol{\theta}\|\boldsymbol{\theta}^{(p)}) = E\left[\log p_{\boldsymbol{X}}(\boldsymbol{X};\boldsymbol{\theta})\|\,\boldsymbol{y},\boldsymbol{\theta}^{(p)}\right]$	(19.29)
M-step:	$\boldsymbol{\theta}^{(p+1)} = \arg\max_{\boldsymbol{\theta}} Q(\boldsymbol{\theta}\|\boldsymbol{\theta}^{(p)})$	(19.30)
M-step for MAP estimation:	$\boldsymbol{\theta}^{(p+1)} = \arg\max_{\boldsymbol{\theta}} \left[Q(\boldsymbol{\theta}\|\boldsymbol{\theta}^{(p)}) + \log p_{\boldsymbol{\Theta}}(\boldsymbol{\theta})\right]$	(19.41)

19.4 Discussion and further reading

Most textbooks on probability and random processes written for engineering students do not cover the topics of this chapter, with the exception of Stark and Woods [310] which discusses the EM algorithm.

In communication system applications the EM algorithm is philosophically similar to maximum-likelihood detection/estimation in the presence of unknown delay, phase, or other unknown parameters [96, 193]. The difference is that the EM algorithm does not use the gradient of the likelihood function, as in the Newton–Rhapson method.

There are two classes of basic applications of the EM algorithm: (i) parameter estimation for an HMM and (ii) parameter estimation of mixture distributions. These application examples are discussed at length in Section 20.6. The literature on applications of the EM algorithm abounds, including such fields as digital communications (e.g., see [233, 253, 331, 369]), econometrics (e.g., see Hamilton [141]), machine translation (e.g., see Koehn [205]), and image processing (e.g., see [176, 303, 333]).

In the field of network performance, the EM algorithm and its variants have been applied to the problem of fitting Markovian arrival processes, Markov-modulated Poisson process, and other phase-type distributions (e.g., see Asmussen *et al.* [7], Rydén [291], Breuer [38], Roberts *et al.* [281]). Turin [332] develops the EM algorithm for the maximization of deterministic functions and also discusses the problem of fitting phase-type distributions.

McLachlan and Krishnan [246] devote an entire volume to the EM algorithm and its extensions, including the **GEM** algorithm and the **variational Bayesian EM** algorithm.

19.5 Problems

Section 19.1: Classical numerical methods for estimation

19.1* Nonnegativity of KLD. Let $\boldsymbol{f} = f(\boldsymbol{x})$ and $\boldsymbol{g} = g(\boldsymbol{x})$ be two arbitrary PDFs. Show that the KLD defined by (19.5) is nonnegative, i.e., $D(\boldsymbol{f}\|\boldsymbol{g}) \geq 0$, where the equality holds if and only if $f(\boldsymbol{x}) = g(\boldsymbol{x})$ for all $\boldsymbol{x}$. Prove the nonnegativity using three different approaches.

(a) The inequality $\log x \leq x - 1, \quad x > 0$;
(b) Jensen's inequality (see Section 10.1.2);
(c) The Lagrangian multiplier method.

19.2 KLD between $\boldsymbol{\theta}$ and $\boldsymbol{\theta}'$. Let

$$f(z) \triangleq p(z|\boldsymbol{y}; \boldsymbol{\theta}) = \frac{p(\boldsymbol{y}, z; \boldsymbol{\theta})}{L_{\boldsymbol{y}}(\boldsymbol{\theta})}, \tag{19.44}$$

$$g(z) \triangleq p(z|\boldsymbol{y}; \boldsymbol{\theta}') = \frac{p(\boldsymbol{y}, z; \boldsymbol{\theta}')}{L_{\boldsymbol{y}}(\boldsymbol{\theta}')}, \tag{19.45}$$

where $L_{\boldsymbol{y}}(\boldsymbol{\theta}) = p(\boldsymbol{y}; \boldsymbol{\theta})$ is the likelihood function. We write $D(\boldsymbol{\theta}\|\boldsymbol{\theta}')$ for (19.5), with $\boldsymbol{f}$ and $\boldsymbol{g}$ given by (19.44) and (19.45) respectively.

Show that

$$D(\boldsymbol{\theta}\|\boldsymbol{\theta}') = \log\frac{L_{\boldsymbol{y}}(\boldsymbol{\theta}')}{L_{\boldsymbol{y}}(\boldsymbol{\theta})} + \sum_{z} p(z|\boldsymbol{y}; \boldsymbol{\theta})\left[\log p(\boldsymbol{y}, z; \boldsymbol{\theta}) - \log p(\boldsymbol{y}, z; \boldsymbol{\theta}')\right]. \tag{19.46}$$

19.3 Approximation of $\ln x$ and Kullback's information criterion $I(\boldsymbol{n}; \boldsymbol{\theta})$, and $\chi^2(\boldsymbol{n}; \boldsymbol{\theta})$ statistic.

(a) Derive the approximation formula (19.10):

$$\ln x \approx \frac{1}{2}(x - x^{-1}), \quad \text{for } x \approx 1.$$

Hint: Use the approximation $\sinh t = \frac{e^t - e^{-t}}{2} \approx t$, near $t = 0$.

(b) Show that $I(\boldsymbol{n}; \boldsymbol{\theta})$ defined in (19.8) is related to the chi-square statistic $\chi^2(\boldsymbol{n}; \boldsymbol{\theta})$ of (19.4) by

$$2I(\boldsymbol{n}; \boldsymbol{\theta}) \approx \chi^2(\boldsymbol{n}; \boldsymbol{\theta}).$$

Section 19.2: Expectation-maximization algorithm for maximum-likelihood estimation

19.4 Relation between Q-function and $\log L$-function. Derive the inequality (19.27).
Hint: Use the equation (19.23) and the inequality (19.26).

19.5 KLD and Q-function.

(a) Show the following relation between the KLD and the Q-function (19.38):

$$D(\boldsymbol{\theta}\|\boldsymbol{\theta}') = \log \frac{L_{\boldsymbol{y}}(\boldsymbol{\theta}')}{L_{\boldsymbol{y}}(\boldsymbol{\theta})} + Q(\boldsymbol{\theta}|\boldsymbol{\theta}) - Q(\boldsymbol{\theta}'|\boldsymbol{\theta}). \tag{19.47}$$

(b) Show that the inequality (19.27) holds also for the Q-function of (19.38).

19.6 Lower bound B to the log-likelihood function. Consider the EM algorithm that involves the latent variable $\boldsymbol{Z}$, as discussed in Section 19.2.2. Define a function $B(\boldsymbol{\theta}; \alpha(\boldsymbol{z}))$ by

$$B(\boldsymbol{\theta}; \alpha(\boldsymbol{z})) \triangleq \sum_{\boldsymbol{z}\in\mathcal{Z}^n} \alpha(\boldsymbol{z}) \log \frac{p(\boldsymbol{y}, \boldsymbol{z}; \boldsymbol{\theta})}{\alpha(\boldsymbol{z})}, \tag{19.48}$$

where $\alpha(\boldsymbol{z})$ is an arbitrary probability distribution of the latent variable $\boldsymbol{Z}$.

(a) *Use of Jensen's inequality.* Show that this B-function provides a lower bound to the log-likelihood function:

$$\log L_{\boldsymbol{y}}(\boldsymbol{\theta}) \geq B(\boldsymbol{\theta}; \alpha(\boldsymbol{z})), \tag{19.49}$$

where the equality can be achieved when

$$\alpha(\boldsymbol{z}) = p(\boldsymbol{z}|\boldsymbol{y}, \boldsymbol{\theta}). \tag{19.50}$$

(b) *Use of the KLD.* Show that the above defined B-function is related to the KLD as follows:

$$B(\boldsymbol{\theta}; \alpha(\boldsymbol{z})) = -D\left(\alpha(\boldsymbol{z})||p(\boldsymbol{z}|\boldsymbol{y}; \boldsymbol{\theta})\right) + \log L_{\boldsymbol{y}}(\boldsymbol{\theta}) \leq L_{\boldsymbol{y}}(\boldsymbol{\theta}). \tag{19.51}$$

(c) *Use of the Lagrangian multiplier method.* Obtain the best distribution (19.50) for $\alpha(\boldsymbol{z})$ by applying the Lagrangian method; i.e., maximize

$$J(\alpha(\boldsymbol{z}), \lambda) = B(\boldsymbol{\theta}; \alpha(\boldsymbol{z})) + \lambda \left(\sum_{\boldsymbol{z}\in\mathcal{Z}} \alpha(\boldsymbol{z}) - 1 \right). \tag{19.52}$$

19.7 The reachable lower bound $B^*(\boldsymbol{\theta}|\boldsymbol{\theta}^{(p)})$ of the likelihood function. Rewrite $B(\boldsymbol{\theta}; \alpha(\boldsymbol{z}))$ of (19.48) as $B^*(\boldsymbol{\theta}|\boldsymbol{\theta}^{(p)})$, when

$$\alpha(\boldsymbol{z}) = p_{\boldsymbol{Z}|\boldsymbol{Y}}(\boldsymbol{z}|\boldsymbol{y}; \boldsymbol{\theta}^{(p)}).$$

(a) Show that

$$B^*(\boldsymbol{\theta}|\boldsymbol{\theta}^{(p)}) = Q(\boldsymbol{\theta}|\boldsymbol{\theta}^{(p)}) + H(\boldsymbol{Z}|\boldsymbol{y}; \boldsymbol{\theta}^{(p)}), \tag{19.53}$$

where $H(\boldsymbol{Z}|\boldsymbol{y}; \boldsymbol{\theta}^{(p)})$ is the **conditional entropy** [300] of $\boldsymbol{Z}$ given $\boldsymbol{y}$ and $\boldsymbol{\theta}^{(p)}$.

(b) Show that the likelihood function and its lower bound B^*-function touch at $\theta = \theta^{(p)}$; i.e.,

$$B^*(\boldsymbol{\theta}^{(p)}|\boldsymbol{\theta}^{(p)}) = \log L_{\boldsymbol{y}}(\boldsymbol{\theta}^{(p)}). \tag{19.54}$$

19.8 An alternative derivation of the EM algorithm. In Problem 19.7 we established an important property that the function $B^*(\boldsymbol{\theta}|\boldsymbol{\theta}^{(p)})$ is bounded from above by the log-likelihood function $\log L_{\boldsymbol{y}}(\boldsymbol{\theta})$ and the two functions are equal at $\boldsymbol{\theta} = \boldsymbol{\theta}^{(p)}$.

We also established the relationship (19.53) between $B^*(\boldsymbol{\theta}|\boldsymbol{\theta}^{(p)})$ and $Q(\boldsymbol{\theta}|\boldsymbol{\theta}^{(p)})$. Derive the EM algorithm using these results.

19.9 EM algorithm for a MAP estimate.

(a) Write a program similar to Algorithm 19.2 to find the MAP estimate.
(b) Show that the MAP estimate of $\boldsymbol{\theta}$ may be obtained by using an EM algorithm whose M-step maximizes the APP (*a posteriori* probability) of the complete variable

$$\boldsymbol{\theta}^{(p+1)} = E[\log p(\boldsymbol{\theta}|\boldsymbol{X})|\boldsymbol{y}, \boldsymbol{\theta}^{(p)}). \tag{19.55}$$

19.10* EM algorithm when the complete variables come from the exponential family of distributions. Let the complete variable $\boldsymbol{X}$ come from the exponential family (or PDF) (see Section 4.4, (4.126)):

$$p_{\boldsymbol{X}}(\boldsymbol{x}; \boldsymbol{\theta}) = h(\boldsymbol{x}) \exp\{\boldsymbol{\eta}^\top(\boldsymbol{\theta})\boldsymbol{T}(\boldsymbol{x}) - A(\boldsymbol{\theta})\}, \tag{19.56}$$

and let the observable data $\boldsymbol{y} = T(\boldsymbol{x})$ come from probability distribution (or PDF) $p_{\boldsymbol{Y}}(\boldsymbol{y}; \boldsymbol{\theta})$, which is not necessarily in the exponential family. We want to find an MLE that maximizes $p_{\boldsymbol{Y}}(\boldsymbol{y}; \boldsymbol{\theta})$.

Show that the E-step and M-step are reduced to

$$\begin{aligned}
\boldsymbol{T}^{(p)} &= E[\boldsymbol{T}(\boldsymbol{x})|\boldsymbol{y}, \boldsymbol{\theta}^{(p)}], \\
\boldsymbol{\theta}^{(p+1)} &= \arg\max_{\boldsymbol{\theta}}\{\boldsymbol{\eta}^\top(\boldsymbol{\theta})\boldsymbol{T}^{(p)} - A(\boldsymbol{\theta})\}.
\end{aligned}$$

Part V

Applications and advanced topics

20 Hidden Markov models and applications

20.1 Introduction

In this chapter we shall discuss **hidden Markov models (HMMs)**, which have been widely applied to a broad range of science and engineering problems, including speech recognition, decoding and channel modeling in digital communications, computational biology (e.g., DNA and protein sequencing), and modeling of communication networks.

In an ordinary Markov model, transitions between the states characterize the dynamics of a system in question, and we implicitly assume that a sequence of states can be directly observed, and the observer may even know the structure and parameters of the Markov model. In some fields, such as speech recognition and network traffic modeling, it is useful to remove these restrictive assumptions and construct a model in which the observable output is a probabilistic function of the underlying Markov state. Such a model is referred to as an HMM.

We shall address the important problems of state and parameter estimation associated with an HMM: What is the likelihood that an observed data is generated from this model? How can we infer the most likely state or sequence of states, given a particular observed output? Given observed data, how can we estimate the most likely value of the model parameters, i.e., their MLEs? We will present in a cohesive manner a series of computational algorithms for state and parameter estimation, including the forward and backward recursion algorithms, the Viterbi algorithm, the BCJR algorithm, and the Baum–Welch algorithm, which is a special case of the EM algorithm discussed in Section 19.2.

We provide some illustrative examples from coding and communication systems, and a parameter estimation problem involving mixture distributions.

20.2 Formulation of a hidden Markov model

20.2.1 Discrete-time Markov chain and hidden Markov model

20.2.1.1 Discrete-time Markov chain (DTMC)

Recall the definition and properties of DTMC defined in Section 15.1. Consider a homogeneous (i.e., time-invariant) DTMC $\{S_t\}$, where the number of states M is finite. We label the states, without loss of generality, by integers $\{0, 1, \ldots, M-1\}$ and let $\mathcal{S}$

denote the *state space*:

$$\mathcal{S} \triangleq \{0, 1, \ldots, M-1\}. \quad (20.1)$$

The subscript t of S_t is the discrete-time index, and we label a set of discrete time epochs by integers $0, 1, 2, \ldots, T$, where T can be infinite, but in practice the observation period is finite:

$$t \in \{0, 1, 2, \ldots, T\} \triangleq \mathcal{T}. \quad (20.2)$$

Let the *initial state distribution* be denoted by

$$\boldsymbol{\pi}_0 = (\pi_0(0), \pi_0(1), \ldots, \pi_0(M-1)) \triangleq (\pi_0(i); i \in \mathcal{S}), \quad (20.3)$$

$$\text{where } \pi_0(i) = P[S_0 = i], i \in \mathcal{S}. \quad (20.4)$$

It is apparent that $\boldsymbol{\pi}_0^\top \mathbf{1} = \sum_{i \in \mathcal{S}} \pi_0(i) = 1$,[1] where $\mathbf{1}$ is a column vector of all ones. In a Markov chain,[2] the current state S_t depends on its past only through its most recent value; i.e.,

$$P[S_t = s_t | S_{t-1} = s_{t-1}, S_{t-2} = s_{t-2}, \ldots] = P[S_t = s_t | S_{t-1} = s_{t-1}], \quad (20.5)$$

or, more compactly, using the conditional probability distribution:

$$p(s_t | s_{t-1}, s_{t-2}, \ldots) = p(s_t | s_{t-1}). \quad (20.6)$$

Thus, the (one-step) **state transition probabilities**[3] $\{a(i, j)\}$ govern the evolution of this homogeneous Markov chain:

$$\mathbf{A} \triangleq [a(i, j); i, j \in \mathcal{S}], \quad (20.7)$$

$$\text{where } a(i, j) \triangleq P[S_t = j | S_{t-1} = i], i, j \in \mathcal{S}. \quad (20.8)$$

Clearly, $\sum_{j \in \mathcal{S}} a(i, j) = 1$; hence, $\mathbf{A1} = \mathbf{1}$.

A graphical representation of the Markov chain is helpful. A homogeneous Markov chain can be represented by a **state transition diagram** of the type given in Figure 15.1, which we reproduce here in Figure 20.1 (a). If we depict the states vertically and the discrete times horizontally, we obtain the **trellis diagram** of Figure 20.1 (b). The notion of a trellis diagram was introduced by Forney [107, 108] and explicitly shows the time progression of the state sequences.

Assuming that we are interested in the state sequence for the period $\mathcal{T}$ defined in (20.2), we introduce a vector variable $\mathbf{S}$ and its instance $\boldsymbol{s}$:[4]

$$\mathbf{S} \triangleq (S_t : t \in \mathcal{T}) \in \mathcal{S}^{|\mathcal{T}|} \text{ and } \boldsymbol{s} \triangleq (s_t : t \in \mathcal{T}),$$

[1] We use $\sum_{i \in \mathcal{S}}$ instead of $\sum_{i=0}^{M-1}$ for notational simplicity.

[2] Throughout this chapter we mean by a Markov chain a *simple* Markov chain or *first-order* Markov chain to differentiate from an nth-order Markov chain. See Definition 15.2.

[3] In Chapter 15 we used the notation P_{ij} or $P(i, j)$ for the state transition probability. The symbol $a(i, j)$ is often used in the HMM-related literature, so we follow this convention.

[4] We may sometimes write S_0^T to explicitly show the starting and ending time indices.

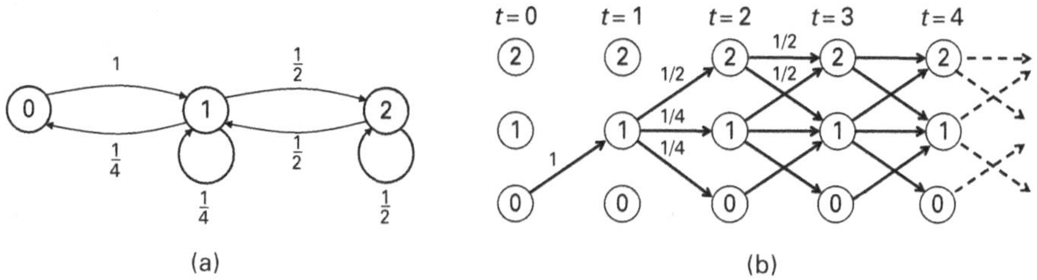

Figure 20.1 (a) The state transition diagram of a three-state Markov chain $\mathcal{S} = \{0, 1, 2\}$ and (b) its trellis diagram.

where $|\mathcal{T}| = T + 1$, the size of set $\mathcal{T}$. For every instance of the state sequence $\boldsymbol{s}$, there is a *unique path* in the trellis diagram, and vice versa. The example of Figure 20.1 (b) shows a case where the initial state is state 0, which is equivalent to setting $\pi_0(0) = 1, \pi_0(1) = \pi_0(2) = 0$.

20.2.1.2 Hidden Markov model

Let us assume that the Markov chain S_t is not observable, but there is another discrete-time random process Y_t, which is a probabilistic function of the chain and is observable. Let us assume that Y_t takes on a *finite set of values*. Then we can assign, without loss of generality, a set of integers $0, 1, 2, \ldots, K - 1$ to represent these output values.

$$\mathcal{Y} \triangleq \{0, 1, 2, \ldots, K - 1\}. \tag{20.9}$$

In the HMM literature, however, the alphabet of output symbols is often written, for instance, as $\mathcal{V} \triangleq \{v_k; k = 1, 2, \ldots, K\}$. But the elements of $\mathcal{V}$, or any finite alphabet of size K, have one-to-one correspondence with the elements of $\mathcal{Y}$; in other words, $\mathcal{V}$ is isomorphic with $\mathcal{Y}$. Thus, writing v_k instead of k is superfluous, making the notation unnecessarily complicated. As far as the HMM-related algorithms we discuss in this chapter, there are no computations of $Y_t(= k)$ such as any addition, multiplication, expectation, maximization. As will be seen below, the value k serves only as the argument of the conditional probabilities such as $c(i; j, k) = P[S_t = j, Y_t = k|S_{t-1} = i]$ and $b(j; k) = P[Y_t = k|S_t = j]$. Of course, when the output symbol Y_t is a continuous RV, the finite alphabet model defined does not apply. (See the remark at the end of this section. See also Problems 20.5 and 20.6).)

Similar to $\boldsymbol{S}$ and $\boldsymbol{s}$ defined above, we denote the **observed sequence variable** over the period $\mathcal{T}$ and its instance by $\boldsymbol{Y}$ and $\boldsymbol{y}$ respectively:

$$\boldsymbol{Y} \triangleq (Y_t; t \in \mathcal{T}) \text{ and } \boldsymbol{y} \triangleq (y_t; t \in \mathcal{T}). \tag{20.10}$$

If the observable process Y_t is a probabilistic function of only S_{t-1} and S_t, then the *pair process*

$$X_t = (S_t, Y_t), t \in \mathcal{T} \tag{20.11}$$

is also a Markov process, since it depends on the past only through $X_{t-1} = (S_{t-1}, Y_{t-1})$. The process Y_t by itself is generally not a Markov process (Problem 20.1).

Now we are ready to present a formal definition of an HMM:

DEFINITION 20.1 (Hidden Markov model). *A Markov process $X_t = (S_t, Y_t)$ is called a* partially observable Markov process *or HMM if its* state transition probability *does not depend on Y_{t-1}; i.e.,*

$$p(x_t|x_{t-1}) = p(x_t|s_{t-1}); \tag{20.12}$$

i.e.,

$$p(s_t, y_t|s_{t-1}, y_{t-1}) = p(s_t, y_t|s_{t-1}). \tag{20.13}$$

Any state $S_t \in \mathcal{S}$ is called a hidden state *and the process $\mathbf{S} = (S_t; t \in \mathcal{T})$ is a hidden process; and $\mathbf{y} = (y_t; t \in \mathcal{T})$ is called an* observation. □

Now we introduce the notion of *model parameter*, which is different from the conventional notion of a *distribution parameter*, such as λ in the Poisson distribution. Later in this chapter we will discuss how to estimate and re-estimate the *model parameters* of a given HMM so that they will best explain observed data.

DEFINITION 20.2 (Model parameters of an HMM). *For a homogeneous HMM, we denote*

$$P[S_0 = i, Y_0 = k] \triangleq \alpha_0(i, k), i \in \mathcal{S}, k \in \mathcal{Y}, \tag{20.14}$$

$$P[S_t = j, Y_t = k|S_{t-1} = i] \triangleq c(i; j, k), i, j \in \mathcal{S}, k \in \mathcal{Y}, \text{for} t = 1, 2, \ldots, T. \tag{20.15}$$

Denote the sets *of initial distribution vectors $\boldsymbol{\alpha}_0(k)$ and the* state transition probability matrices *$\boldsymbol{C}(k)$ by $\boldsymbol{\alpha}_0$ and $\mathbf{C}$ respectively:*

$$\boldsymbol{\alpha}_0 \triangleq (\boldsymbol{\alpha}_0(k); k \in \mathcal{Y}), \tag{20.16}$$

$$\textit{where } \boldsymbol{\alpha}_0(k) \triangleq (\alpha_0(i, k); i \in \mathcal{S}), k \in \mathcal{Y}, \tag{20.17}$$

$$\mathbf{C} \triangleq (\boldsymbol{C}(k); k \in \mathcal{Y}), \tag{20.18}$$

$$\textit{where } \boldsymbol{C}(k) \triangleq [c(i; j, k); i, j \in \mathcal{S}], k \in \mathcal{Y}. \tag{20.19}$$

We then define the **model parameters** *$\boldsymbol{\theta}$ associated with the HMM by the set*

$$\boldsymbol{\theta} = (\boldsymbol{\alpha}_0, \mathbf{C}). \tag{20.20}$$

□

The initial distributions (20.17) and the initial state probability (20.3) are all M-dimensional vectors and are related by

$$\sum_{k \in \mathcal{Y}} \boldsymbol{\alpha}_0(k) = \boldsymbol{\pi}_0. \tag{20.21}$$

Similarly, the $M \times M$ matrices $\boldsymbol{C}(k)$ of (20.19) and $\boldsymbol{A}$ of (20.7) are related by

$$\sum_{k \in \mathcal{Y}} \boldsymbol{C}(k) = \boldsymbol{A}. \tag{20.22}$$

Now we can rewrite the joint conditional probability $p(s_t, y_t|s_{t-1})$ of (20.13) as

$$p(s_t, y_t|s_{t-1}) = p(s_t|s_{t-1})p(y_t|s_{t-1}, s_t). \tag{20.23}$$

Then we define a special type of HMM as follows.

DEFINITION 20.3 (State-based HMM versus transition-based HMM). *A homogeneous HMM is said to be* ***state-based*** *if a special condition*

$$p(y_t|s_{t-1}, s_t) = p(y_t|s_t) \tag{20.24}$$

holds. A state-based HMM is defined by

$$\boldsymbol{\theta} = (\boldsymbol{\pi}_0, \boldsymbol{A}, \boldsymbol{B}), \tag{20.25}$$

where $\boldsymbol{\pi}_0$ *and* $\boldsymbol{A}$ *are defined in* (20.3) *and* (20.7) *respectively and* $\boldsymbol{B}$ *is an* $M \times K$ *matrix given by*

$$\boldsymbol{B} \triangleq [b(j; k)];\ j \in \mathcal{S}, k \in \mathcal{Y}, \tag{20.26}$$

where

$$b(j; k) = P[Y_t = k|S_t = j]. \tag{20.27}$$

If the special condition (20.24) does not hold, the HMM is said to be ***transition based****.* □

Note that state-based HMMs form a subclass of transition-based HMMs, to which all HMMs belong. The joint conditional probability $c(i; j, k)$ can be factored as

$$c(i; j, k) = a(i, j)\tilde{b}(i; j, k), \tag{20.28}$$

where $a(i, j)$ is defined in (20.8) and

$$\tilde{b}(i, j; k) \triangleq P[y_t = k|S_{t-1} = i, S_t = j]. \tag{20.29}$$

The state-based HMM is a special case where the $\tilde{b}(i, j; k)$ are independent of i; i.e.,

$$\tilde{b}(i, j; k) = b(j; k) \text{ for all } i \in \mathcal{S}, \tag{20.30}$$

which is nothing but a restatement of the condition (20.24).

Most existing literature on HMMs, however, deals with only state-based HMMs. This in theory is not a serious defect, because any transition-based HMM can be converted into an equivalent state-based HMM, by defining $\tilde{S}_t \triangleq (S_{t-1}, S_t) \in \mathcal{S} \times \mathcal{S}$ as a new state. However, the corresponding state transition matrix $\tilde{\boldsymbol{A}}$ would become $M^2 \times M^2$, which would make the transition diagram and trellis diagram unbearably too large and complex even for a moderate M, the size of the hidden state space. Thus, clearly there is an advantage in working with a transition-based HMM representation. Thus, we assume a general transition-based HMM, unless stated otherwise.

20.2.1.3 The case where the observable Y_t is a continuous random variable

Thus far, we have assumed that the observed variable Y_t takes on a discrete value from a finite alphabet. If Y_t is a continuous RV, we should replace the conditional joint probability $c(i; j, k)$ of the transition-based HMM by

$$p_{S_t,Y_t|S_{t-1}}(j, y|i)\,dy = P[S_t = j, y < Y_t \le y + dy|S_{t-1} = i]. \tag{20.31}$$

Similarly, in a state-based HMM, the conditional probability $b(j; k)$ should be replaced by a conditional PDF $f_{Y_t|S_t}(y|j)dy$:

$$f_{Y_t|S_t}(y|j)dy = P[y < Y_t \le y + dy|S_t = j]. \tag{20.32}$$

The restrictive condition (20.24) for the state-based model is equivalent to assuming the following factorization:

$$p_{S_t,Y_t|S_{t-1}}(j, y|i) = a(i, j)f_{Y_t|S_t}(y|j). \tag{20.33}$$

20.2.2 Examples of hidden Markov models

Now we will discuss two examples of HMM representations.

Example 20.1: Convolutional encoder and binary symmetric channel. Figure 20.2 shows a schematic diagram of a *convolutional encoder*, which consists of a three-stage shift register[5] and two modulo-2 adders (or binary counters). The encoder takes a binary input sequence sequence $\boldsymbol{I} = (I_1, I_2, \ldots, I_t, \ldots)$ and sends out a binary output sequence $\boldsymbol{O} = (O_0, O_1, \ldots, O_t, \ldots)$, where $O_t = O_t^{(1)}O_t^{(2)}$, where $O_t^{(1)}$ is the output from the upper adder and $O_t^{(2)}$ is the output from the lower adder at discrete time $t, t = 0, 1, 2, \ldots$. We assume that initially the shift register is set to all zeros; i.e., "000." Thus, the upper and lower adders both generate 0. Therefore, 00 will be the output sequence at $t = 0$.

The relation between the input and output of the encoder is given by

$$O_t = O_t^{(1)}O_t^{(2)}, t \ge 0, \tag{20.34}$$

$$\text{where } O_t^{(1)} = I_t \oplus I_{t-1} \oplus I_{t-2} \text{ and } O_t^{(2)} = I_t \oplus I_{t-2}, \tag{20.35}$$

$$I_{-2} = I_{-1} = I_0 = 0, \tag{20.36}$$

[5] In actual implementation, a shift register may not be adopted, but the schematic representation of Figure 20.2 will best illustrate the operation of the encoder.

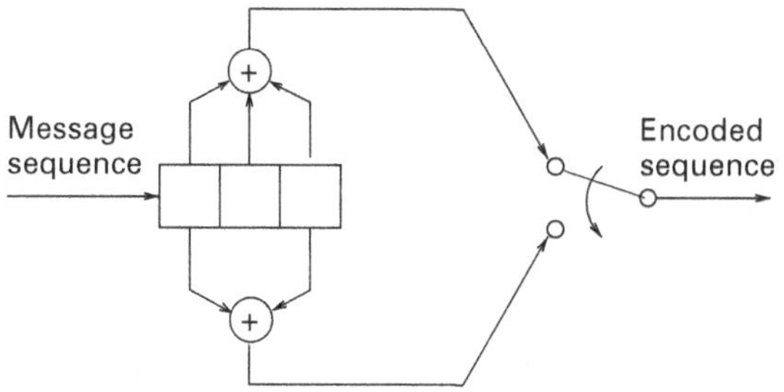

Figure 20.2 A rate 1/2 convolutional encoder with input I_t and output O_t, and a noisy channel with input O_t and output Y_t, which is observable.

where $\oplus$ represents *modulo-2 addition* (which is equivalent to "Exclusive OR" or "XOR" in logic); i.e., $0 \oplus 0 = 0, 0 \oplus 1 = 1, 1 \oplus 0 = 1$ and $1 \oplus 1 = 0$.

Thus, an input information sequence $\boldsymbol{I} = (1, 1, 0, 1, 0, 0, 1, \ldots)$, for instance, will result in an encoded output sequence $\boldsymbol{O} = (11, 01, 01, 00, 10, 11, 11, \ldots)$. The encoded sequence is twice as long as the information sequence, so the amount of information *per binary symbol* in $\boldsymbol{O}$ should be half of that in $\boldsymbol{I}$. Hence, this encoder is called a *rate 1/2* convolutional encoder.

Note that if we set $00 = 0, 01 = 1, 10 = 2$, and $11 = 3$, then the output symbol alphabet becomes simply the $\mathcal{Y}$ of (20.9) with $K = 4$; i.e.,

$$\mathcal{O} = \mathcal{Y} = \{0, 1, 2, 3\}, \text{ with } K = 4.$$

Then, the output sequence can be compactly written as $\boldsymbol{O} = (3, 1, 1, 0, 2, 3, 3, \ldots)$.[6]

A natural choice of the state for this encoder would be to define the three-bit register content as the state; i.e.,

$$S_t = I_t I_{t-1} I_{t-2}, \quad t = 0, 1, 2, \ldots.$$

The reader is suggested to draw the state transition diagram and the trellis diagram (Problem 20.2). We define, instead, state S_t as the two latest message bits; i.e., the two leftmost bits in the three-stage shift register of the encoder:

$$S_t = I_t I_{t-1}, \quad t = 0, 1, 2, \ldots. \tag{20.37}$$

Then, the state space is given by

$$\mathcal{S} = \{00, 01, 10, 11\} = \{0, 1, 2, 3\}.$$

If we assume that $\{I_t\}$ is an i.i.d. sequence, then the state sequence $\{S_t\}$ is a simple Markov chain (Problem 20.3).

Figure 20.3 (a) and (b) shows the state transition diagram and the trellis diagram respectively. We assume the initial condition $I_0 = I_{-1} = 0$; hence, $S_0 = 0$ with probability one. If the encoder is in state $S_{t-1} = 10 = 2$ at time $t - 1$ and if $I_t = 0$, then the next

[6] This is another reason why we define the output alphabet in the form of $\{0, 1, \ldots, K-1\}$ instead of $\{v_0, v_1, \ldots, v_{k-1}\}$ or $\{1, 2, \ldots, K\}$.

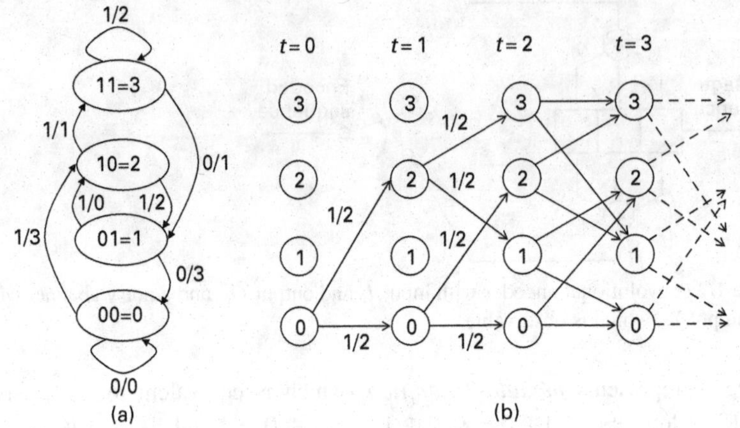

Figure 20.3 (a) The state transition diagram of the rate 1/2 convolutional encoder. (b) The trellis diagram (all transitions have probability 1/2).

state is $S_t = 01 = 1$ and the corresponding output is $O_t = 10 = 2$. We write this relation as "0/2" in the state transition and trellis diagrams. Similarly, state $S_{t-1} = 01 = 1$ with input $I_t = 0$ leads to $S_t = 00 = 0$, generating $O_t = 11 = 3$. Thus, "0/3" is attached to the transition $S_{t-1} = 1 \to S_t = 0$.

If 0 and 1 appear in I_t with equal probability, i.e., $P[I_t = 0] = P[I_t = 1] = 1/2$ for all $t = 1, 2, \ldots$, the state transition matrix associated with the convolutional encoder, denoted as $\boldsymbol{A}$, becomes

$$\boldsymbol{A} \triangleq \begin{bmatrix} 1/2 & 0 & 1/2 & 0 \\ 1/2 & 0 & 1/2 & 0 \\ 0 & 1/2 & 0 & 1/2 \\ 0 & 1/2 & 0 & 1/2 \end{bmatrix}. \tag{20.38}$$

20.2.2.1 Discrete memoryless channel

Suppose that the encoder output sequence O_t is sent over a channel. Because of possible noise in the channel, the channel output Y_t may not necessarily be the same as O_t. We assume that the channel output Y_t has the same alphabet as the channel input; i.e., $\mathcal{Y} = \mathcal{O} = \{00, 01, 10, 11\}$. We can write the relation between the channel input and output as

$$Y_t = O_t \oplus E_t, \tag{20.39}$$

where E_t is called an error pattern, meaning a "1" in E_t will result in an error in the corresponding bit position of Y_t. For instance, if

$$\boldsymbol{O} = (11, 01, 01, 00, 10, 11, 11, \ldots),$$
$$\boldsymbol{E} = (00, 00, 10, 00, 01, 00, 11, \ldots),$$

then

$$Y = (11, 01, 11, 00, 11, 11, 00, \ldots).$$

We assume that errors at different times occur independently of each other. More formally, we assume the following product form for the conditional probabilities of the observable sequence (i.e., the channel output) $\boldsymbol{y}_0^T = (y_0, y_1, \ldots, y_T)$ given the encoder output sequence (i.e., the channel input) $\boldsymbol{o}_0^T = (o_0, o_1, \ldots, o_T)$:

$$p(\boldsymbol{y}_0^T | \boldsymbol{o}_0^T) = \prod_{t=0}^{T} p(y_t | o_t). \tag{20.40}$$

Such a channel is referred to as a *discrete memoryless channel* [300]. Since the channel input (or the encoder output) O_t is a function of the encoder states S_{t-1} and S_t only, the memoryless property implies that the observable channel output Y_t depends probabilistically only on S_{t-1} and S_t. Thus, the pair process $X_t = (S_t, Y_t)$ depends only on S_{t-1}; hence, it is a **hidden Markov process**, as defined in Definition 20.1.

If we further assume that the probability of making an error at a given bit position is ϵ irrespective of the channel content o_t, such a memoryless channel is called a *binary symmetric channel* (BSC).

For instance, a transition from state $s_{t-1} = 0(= 00)$ to state $s_t = 2(= 10)$ will produce the encoder output $o_t = 3(= 11)$. Because of the noise, it can be changed into any of the symbols in $\mathcal{Y} = \mathcal{O} = \{0, 1, 2, 3\}$ at the channel output; i.e.,

$$y_t = \begin{cases} 0(= 00), & \text{with probability } \epsilon^2, \\ 1(= 01), & \text{with probability } \epsilon(1-\epsilon), \\ 2(= 10), & \text{with probability } \epsilon(1-\epsilon), \\ 3(= 11), & \text{with probability } (1-\epsilon)^2. \end{cases}$$

We have the following joint conditional probabilities

$$c(0; 2, k) = \begin{cases} a_{02}\epsilon^2, & \text{for } k = 0, \\ a_{02}\epsilon(1-\epsilon), & \text{for } k = 1, \\ a_{02}\epsilon(1-\epsilon), & \text{for } k = 2, \\ a_{02}(1-\epsilon)^2, & \text{for } k = 3. \end{cases} \tag{20.41}$$

Since there are four possible observable outputs y_t for every one of the eight transition patterns, there are 32 model parameters $\mathbf{C} = [c(i; j, k)]$, so we do not give an exhaustive list here.

Thus, the model parameter of this HMM is given by $\boldsymbol{\theta} = \mathbf{C} = (\boldsymbol{A}, \epsilon)$. If the assumption (20.38) for the state transitions in the convolutional encoder is believed to be correct, as is the case if the binary sequence I_t is completely random, then $\boldsymbol{A}$ will not be an issue for investigation, and we simply have $\boldsymbol{\theta} = \epsilon$. In the model of Figure 20.2 that involves both the convolutional encoder and the BSC, the encoder state sequence

S_t (or equivalently the input information sequence I_t) is a hidden process and Y_t is an observable variable. We will later discuss how to estimate the channel error parameter ϵ as well as the state sequence S_t from the observed output y_t. □

Example 20.2: Gilbert–Elliott channel model for burst errors. In the previous example, error occurrences are assumed to be completely random. But in reality, the channel errors are often not random, but bursty, in the sense that once an error occurs at some time t, then the chance of having errors in subsequent bits tends to be high. Consequently, errors are likely to occur in a burst. The Gilbert–Elliott model [89, 120] described below provides a simple model of burst errors. Let a given channel have two possible states: Good state ("0") and Bad state ("1"); i.e.,

$$\tilde{S}_t \in \{0, 1\}.^7 \tag{20.42}$$

When the channel state $\tilde{S}_{t-1} = 0$, then it will stay in 0 at time t with probability $p_{00} \approx 1$. Once the channel enters the Bad state, it will remain bad with probability p_{11}. Thus, the state transition matrix of $\tilde{\boldsymbol{A}}$ of this Gilbert–Elliott channel is given as

$$\tilde{\boldsymbol{A}} = \begin{bmatrix} p_{00} & 1 - p_{00} \\ 1 - p_{11} & p_{11} \end{bmatrix}. \tag{20.43}$$

Clearly, the sequence $\tilde{S}_t$ is a Markov chain with the equilibrium state probabilities (Problem 20.4 (a))

$$\pi_0 = \lim_{t\to\infty} P[\tilde{S}_t = 0] = \frac{1 - p_{11}}{2 - p_{00} - p_{11}} \text{ and } \pi_1 = \lim_{t\to\infty} P[\tilde{S}_t = 1] = \frac{1 - p_{00}}{2 - p_{00} - p_{11}}. \tag{20.44}$$

The mean state sojourn time (the duration of a channel state) is given as (Problem 20.4 (b))

$$T_0 = \frac{1}{1 - p_{00}} \text{ and } T_1 = \frac{1}{1 - p_{11}}. \tag{20.45}$$

When the channel is in the Bad state, it will cause an error with probability $\epsilon_1 \gg 0$, yielding $E_t = 1$. Similarly, when it is in the Good state, $E_t = 1$ with probability $\epsilon_0 \approx 0$. Thus, we have the conditional probability matrix $\tilde{\boldsymbol{B}}$ of this noise source by

$$\begin{aligned} \tilde{\boldsymbol{B}} &= [b(j;k)] = [p(E_t = k|\tilde{S}_t = j); j \in \{0, 1\}, k \in \{0, 1\}] \\ &= \begin{bmatrix} p(E_t = 0|\tilde{S}_t = 0) & p(E_t = 1|\tilde{S}_t = 0) \\ p(E_t = 0|\tilde{S}_t = 1) & p(E_t = 1|\tilde{S}_t = 1) \end{bmatrix} = \begin{bmatrix} 1 - \epsilon_0 & \epsilon_0 \\ 1 - \epsilon_1 & \epsilon_1 \end{bmatrix}. \end{aligned} \tag{20.46}$$

7 We denote the channel state process by $\tilde{S}_t$ in order to distinguish it from the encoder state process S_t defined earlier.

Since E_t depends probabilistically only on the current state $\tilde{S}_t$, the process $(\tilde{S}_t, E_t)$ is a partially observable Markov process, provided the error pattern process E_t is observable. Then, the Gilbert–Elliott channel model can be characterized as a state-based HMM. The model parameters are $\boldsymbol{\theta} = (\pi_0(0), \tilde{\boldsymbol{A}}, \tilde{\boldsymbol{B}})$, where $\pi_0(0)$ is the probability that the channel is in state 0 at time $t = 0$. Since $\tilde{\boldsymbol{A}}$ is determined by p_{00} and p_{11} and $\tilde{\boldsymbol{B}}$ is determined by ϵ_0 and ϵ_1, the model parameters are $\boldsymbol{\theta} = (\pi_0(0), p_{00}, p_{11}, \epsilon_0, \epsilon_1)$.

If we replace the BSC model by the Gilbert–Elliott model in Figure 20.2, the overall process is $X_t = (S_t, \tilde{S}_t, Y_t)$ is a hidden Markov process, where Y_t is observable, whereas both S_t and $\tilde{S}_t$ are unobservable. □

20.3 Evaluation of a hidden Markov model

In the present and the following three sections, we will consider the following *four basic problems*, which will arise when we apply an HMM to characterize a system of our interest:

1. Model likelihood evaluation (Section 20.3).
2. State sequence estimation (Section 20.4).
3. State estimation (Section 20.5).
4. Model parameter estimation (Section 20.6).

Depending on the application or circumstance, one problem may be more important than the others, or perhaps not all the four problems may be addressed. In these next four sections, however, we will address all the above in that order, and discuss important techniques and computational algorithms associated with the utility of an HMM.

20.3.1 Evaluation of a likelihood function

The first problem, the evaluation of the likelihood of a certain HMM, can be stated as follows.

Given an HMM with the model parameter $\boldsymbol{\theta}$ *and an observation*

$$\boldsymbol{y} = \boldsymbol{y}_0^T = (y_0, y_1, \ldots, y_T),$$

what is the likelihood $p(\boldsymbol{y}; \boldsymbol{\theta})$ *that this* $\boldsymbol{y}$ *will occur in this HMM with this parameter* $\boldsymbol{\theta}$*?* □

The probability $p(\boldsymbol{y}; \boldsymbol{\theta})$[8] can serve as an indicator of how good the HMM and its parameter setting $\boldsymbol{\theta}$ are, where

$$\boldsymbol{\theta} = (\boldsymbol{\alpha}_0, \ \mathbf{C}),$$

[8] Some authors write this probability as $p(\boldsymbol{y}|\boldsymbol{\theta})$, which is appropriate in the Bayesian estimation context, where the parameter is treated as an RV, which we denote by $\boldsymbol{\Theta}$. See Section 21.5.

as defined in (20.20). If there are other competing models or model parameters, the answer to the above question will guide us to accept or reject this particular HMM. The quantity $p(\boldsymbol{y}; \boldsymbol{\theta})$ is, by definition, the probability that the RV $\boldsymbol{Y}$ takes on $\boldsymbol{y}$ for a given parameter value $\boldsymbol{\theta}$. But $p(\boldsymbol{y}; \boldsymbol{\theta})$ can also be interpreted as the *likelihood* that this assumed value of the model parameter $\boldsymbol{\theta}$ may yield this observed instance $\boldsymbol{y}$. Recall that we introduced the notion of likelihood function in Sections 4.5 and 18.1.2 in the context of estimating parameters of probability distribution functions. The same concept can be applied to estimation of the *HMM parameter* we are dealing with in this chapter. Thus, we call $p(\boldsymbol{y}; \boldsymbol{\theta})$ the **likelihood function** of $\boldsymbol{\theta}$ for given $\boldsymbol{y}$, and denote it by $L_{\boldsymbol{y}}(\boldsymbol{\theta})$:

$$\boxed{L_{\boldsymbol{y}}(\boldsymbol{\theta}) \triangleq p(\boldsymbol{y}; \boldsymbol{\theta}).} \tag{20.47}$$

In order to evaluate this likelihood function, we first write it as a marginal probability of the joint probability of $(\boldsymbol{S}, \boldsymbol{Y})$ with the assumed model parameter $\boldsymbol{\theta}$:

$$L_{\boldsymbol{y}}(\boldsymbol{\theta}) = \sum_{\boldsymbol{s} \in \mathcal{S}^{|\mathcal{T}|}} p(\boldsymbol{s}, \boldsymbol{y}; \boldsymbol{\theta}), \tag{20.48}$$

where

$$\mathcal{S}^{|\mathcal{T}|} = \mathcal{S} \times \mathcal{S} \times \cdots \times \mathcal{S}$$

is the state space for the vector variable $\boldsymbol{S} = (S_t : t \in \mathcal{T})$ of size $|\mathcal{T}| = T + 1$.

Using the Markov property of $X_t = (S_t, Y_t)$ and the definition of an HMM given by (20.12), we find the following product form for the joint probability (Problem 20.8 (a)):

$$p(\boldsymbol{s}, \boldsymbol{y}; \boldsymbol{\theta}) = \alpha_0(s_0, y_0) \prod_{t=1}^{T} c(s_{t-1}; s_t, y_t), \tag{20.49}$$

where

$$\alpha_0(i, k) = P[S_0 = i, Y_0 = k],$$
$$c(i; j, k) = P[S_t = j, Y_j = k | S_{t-1} = i], t \geq 1,$$

as defined in (20.14) and (20.15) respectively. Thus, the likelihood function for the *transition-based output model* can be expressed as a *sum of product terms*:

$$L_{\boldsymbol{y}}(\boldsymbol{\theta}) = \sum_{s_0 \in \mathcal{S}} \sum_{s_1 \in \mathcal{S}} \cdots \sum_{s_T \in \mathcal{S}} \alpha_0(s_0, y_0) c(s_0; s_1, y_1) \cdots c(s_{T-1}; s_T, y_T). \tag{20.50}$$

Evaluation of $L_{\boldsymbol{y}}(\boldsymbol{\theta})$ using (20.50) would require computations on the order of $O(TM^T)$. Thus, a brute-force evaluation would become computationally infeasible even for modest sizes of M and/or T. A more efficient computation method can be achieved by converting the above sum of products into a *product of sums*, as will be developed in the next section. In Section 21.5 on Bayesian networks, which can be viewed as a generalization of HMMs, we discuss the **sum-product algorithm**. The algorithm discussed below is a simple instance of the sum-product algorithm (see Section 21.6.1).

20.3.2 Forward recursion algorithm

By defining a row vector $\boldsymbol{\alpha}_0(y_0)^\top$ and an $M \times M$ matrix $\boldsymbol{C}(y_t)$,

$$\boldsymbol{\alpha}_0(y_0)^\top \triangleq [\alpha_0(i, y_0); i \in \mathcal{S}], \boldsymbol{C}(y_t) \triangleq \left[c(i; j, y_t); i, j \in \mathcal{S}\right], y_t \in \mathcal{Y}, t = 1, 2, \ldots, T, \tag{20.51}$$

we obtain a concise expression for the likelihood function:

$$\boxed{L_{\boldsymbol{y}}(\boldsymbol{\theta}) = \boldsymbol{\alpha}_0(y_0)^\top \left(\prod_{t=1}^{T} \boldsymbol{C}(y_t)\right) \mathbf{1},} \tag{20.52}$$

where $\mathbf{1}$ is a column vector whose entries are all one. The expression (20.52) can be rewritten as

$$L_{\boldsymbol{y}}(\boldsymbol{\theta}) = \{[(\boldsymbol{\alpha}_0(y_0)^\top \boldsymbol{C}(y_1))\boldsymbol{C}(y_2)] \cdots \boldsymbol{C}(y_T)\}\mathbf{1}, \tag{20.53}$$

which involves T steps of multiplying a row vector with a matrix (M^2 additions and M^2 multiplication per step), ended by summing up the entries of the final row vector (M additions). This rearrangement leads to a computationally efficient procedure, called the **forward recursion algorithm** (or simply **forward algorithm**) for calculating $L_{\boldsymbol{y}}(\boldsymbol{\theta})$. This forward recursion is a direct consequence of (20.49) for the *joint probability of* $\boldsymbol{S}_t$ *and* $\boldsymbol{Y}_t$ obtained in terms of the product form of $c(i; j, k)$.

An alternative derivation of the above result (20.52) can be obtained in terms of the **forward variable**,[9] which we define as the *joint probability of* S_t *and* $\boldsymbol{y}_0^T$:

$$\boxed{\alpha_t(j, \boldsymbol{y}_0^t) \triangleq P[S_t = j, \boldsymbol{Y}_0^t = \boldsymbol{y}_0^t; \boldsymbol{\theta}],} \tag{20.54}$$

which can be written as

$$\begin{aligned}
\alpha_t(j, \boldsymbol{y}_0^t) &= \sum_{i \in \mathcal{S}} P[S_{t-1} = i, S_t = j, \boldsymbol{Y}_0^{t-1} = \boldsymbol{y}_0^{t-1}, Y_t = y_t; \boldsymbol{\theta}] \\
&= \sum_{i \in \mathcal{S}} P[S_{t-1} = i, \boldsymbol{Y}_0^{t-1} = \boldsymbol{y}_0^{t-1}; \boldsymbol{\theta}] P[S_t = j, Y_t = y_t | S_{t-1} = i, Y_{t-1} = y_{t-1}, \boldsymbol{\theta}] \\
&= \sum_{i \in \mathcal{S}} P[S_{t-1} = i, \boldsymbol{Y}_0^{t-1} = \boldsymbol{y}_0^{t-1}; \boldsymbol{\theta}] P[S_t = j, Y_t = y_t | S_{t-1} = i, \boldsymbol{\theta}] \\
&= \sum_{i \in \mathcal{S}} \alpha_{t-1}(i, \boldsymbol{y}_0^{t-1}) c(i; j, y_t), \; j \in \mathcal{S}.
\end{aligned} \tag{20.55}$$

In obtaining the second line in (20.55), we used the fact that $X_t = (Y_t, S_t)$ is a Markov process, and hence depends on its past only through X_{t-1}. But by definition of an HMM, X_t should depend only on S_{t-1}; hence, we obtain the third line. In other words, if S_{t-1} is

[9] In the HMM literature, it is common to write the forward variable as $\alpha_t(i)$ or $\alpha_i(t)$ by dropping $\boldsymbol{y}_0^t$. We include $\boldsymbol{y}_0^t$ as the second argument to emphasize that the forward variable depends on the observations up to time t. In algorithmic statements, however, we drop this argument.

Algorithm 20.1 Forward algorithm for $L_y(\boldsymbol{\theta})$

1: Compute the forward variables recursively:

$$\boldsymbol{\alpha}_0^\top = (\alpha_0(i, y_0); i \in \mathcal{S}),$$
$$\boldsymbol{\alpha}_t^\top = \boldsymbol{\alpha}_{t-1}^\top \boldsymbol{C}(y_t), t = 1, 2, \ldots, T.$$

2: Compute

$$L_y(\boldsymbol{\theta}) = \boldsymbol{\alpha}_T^\top \mathbf{1}.$$

---

known, the past observation $\boldsymbol{y}_0^{t-1}$ does not provide any additional information regarding S_t and y_t. We define the **forward vector variable** as a row vector

$$\boldsymbol{\alpha}_t(\boldsymbol{y}_0^t)^\top = (\alpha_t(i; \boldsymbol{y}_0^t); i \in \mathcal{S}), t = 1, 2, \ldots, T. \tag{20.56}$$

From (20.55) we indeed obtain the *forward recursion algorithm*:

$$\boxed{\boldsymbol{\alpha}_t(\boldsymbol{y}_0^t)^\top = \boldsymbol{\alpha}_{t-1}(\boldsymbol{y}_0^{t-1})^\top \boldsymbol{C}(y_t), t = 1, 2, \ldots, T,} \tag{20.57}$$

with the initial vector $\boldsymbol{\alpha}_0(y_0)$ defined by (20.51). Equation (20.53) readily follows from this result.

Thus, $L_y(\boldsymbol{\theta})$ can be evaluated, after T steps of the forward recursion, as

$$\boxed{L_y(\boldsymbol{\theta}) = \boldsymbol{\alpha}_T(\boldsymbol{y}_0^T)^\top \mathbf{1} = \sum_{i \in \mathcal{S}} \alpha_T(i, \boldsymbol{y}_0^T).} \tag{20.58}$$

The computational complexity of this evaluation step is merely TM^2 in contrast to $O(TM^T)$ of the direct enumeration method based on (20.50).

We summarize the forward algorithm in Algorithm 20.1. Here, we drop $\boldsymbol{y}_0^t$ from the arguments of the forward variable and simply write them as $\alpha_t(j)$, because, from the algorithmic point of view, the variables α can be stored in a two-dimensional array of size $M \times (T+1)$. The values $\alpha_t(j)$ do certainly depend on the specific observation sequence $\boldsymbol{y}_0^t = (y_0, y_1, \ldots, y_t)$.

20.3.3 Backward algorithm and forward-backward algorithm

Now, let us consider the *time-reversed version* of the above procedure. The matrix multiplications in (20.52) now rearranged, instead of (20.53), as

$$L_y(\boldsymbol{\theta}) = \boldsymbol{\alpha}(y_0)^\top \{\boldsymbol{C}(y_1)[\boldsymbol{C}(y_2) \cdots \boldsymbol{C}(y_T)\mathbf{1})]\}. \tag{20.59}$$

We define the **backward variable** $\beta_t(i; \boldsymbol{y}_{t+1}^T)$ as the probability of the partial observation $\boldsymbol{y}_{t+1}^T = (y_{t+1}, y_{t+2}, \ldots, y_T)$, given that the Markov chain is in state i at time t:

$$\boxed{\beta_t(i; \boldsymbol{y}_{t+1}^T) \triangleq p(\boldsymbol{y}_{t+1}^T | S_t = i; \boldsymbol{\theta}), i \in \mathcal{S}, t = T-1, T-2, \ldots, 0.} \tag{20.60}$$

Then, by defining the **backward vector variable** as a *column vector,*

$$\boldsymbol{\beta}(\boldsymbol{y}_{t+1}^T) = (\beta_t(i; \boldsymbol{y}_{t+1}^T); i \in \mathcal{S})^\top, t = T-1, T-2, \ldots, 0, \tag{20.61}$$

we obtain, analogous to the forward recursion formula (20.57), the following **backward recursion** formula (Problem 20.10):

$$\boxed{\boldsymbol{\beta}_t(\boldsymbol{y}_{t+1}^T) = \boldsymbol{C}(y_{t+1})\boldsymbol{\beta}_{t+1}(\boldsymbol{y}_{t+2}^T), t = T-1, T-2, \ldots, 0.} \tag{20.62}$$

The boundary condition (i.e., the initial condition in the backward recursion) is

$$\boldsymbol{\beta}_T(\emptyset) = \mathbf{1} = (1, 1, \ldots, 1)^\top. \tag{20.63}$$

Specification of a **backward algorithm**, similar to Algorithm 20.1, is left to the reader as an exercise (Problem 20.11).

By combining the forward and backward variables, we find (Problem 20.12)

$$\alpha_t(i, \boldsymbol{y}_0^t)\beta_t(i; \boldsymbol{y}_{t+1}^T) = P[S_t = i, \boldsymbol{Y} = \boldsymbol{y}; \boldsymbol{\theta}], i \in \mathcal{S}, t \in \mathcal{T}. \tag{20.64}$$

Then we find an alternative expression to evaluate $L_{\boldsymbol{y}}(\boldsymbol{\theta})$:

$$\boxed{L_{\boldsymbol{y}}(\boldsymbol{\theta}) = \boldsymbol{\alpha}_t(\boldsymbol{y}_0^t)^\top \boldsymbol{\beta}_t(\boldsymbol{y}_{t+1}^T), \text{ for any } t \in \mathcal{T}.} \tag{20.65}$$

The choice of $t = T$ in the last expression reduces to the forward algorithm (20.58), whereas $t = 0$ gives

$$L_{\boldsymbol{y}}(\boldsymbol{\theta}) = \boldsymbol{\alpha}_0(y_0)^\top \boldsymbol{\beta}_0(\boldsymbol{y}_1^T) = \sum_{i \in \mathcal{S}} \alpha_0(i, y_0)\beta_0(i; \boldsymbol{y}_1^T). \tag{20.66}$$

The reader is suggested (Problem 20.13) to state an algorithm to compute the likelihood function based on (20.65). This algorithm is called the **forward–backward algorithm** (FBA).

Simultaneous use of both forward and backward algorithms may make sense when the computational speed for each recursion is slower than the rate at which observation data are collected. Although the backward recursion cannot commence until all the observations $\boldsymbol{y}_0^T$ are in, all of the forward recursions may not have been completed by then. It may then speed up the computation of $L_{\boldsymbol{y}}(\boldsymbol{\theta})$ if the forward and backward recursions are performed concurrently, after the data $\boldsymbol{y}_0^T$ are gathered. Again, the storage requirements are only $2M$ times the space to store one variable. The same amount of space is needed for backward variables; thus, storage space for $4M$ variables is required altogether. If the individual $\alpha_t(i)$ and $\beta_t(i)$ are needed for later use, then storage for $2M(T+1)$ variables would be needed. The MAP state estimation algorithm (Algorithm 20.3) to be discussed later is such an example.

20.4 Estimation algorithms for state sequence

Now consider a case where we wish to estimate the hidden state sequence $\boldsymbol{s}(=\boldsymbol{s}_0^T)$, based on the observation sequence $\boldsymbol{y}(=\boldsymbol{y}_0^T)$. As will be shown below, an optimum estimate is not unique, because there is more than one choice of optimality condition.

20.4.1 Forward algorithm for maximum *a posteriori* probability state sequence estimation

In Section 18.3 we introduced the notion of the **maximum *a posteriori* probability (MAP) estimation** as one of the estimation criteria. The MAP state sequence estimate $\hat{\boldsymbol{s}}^*$ is defined by

$$\hat{\boldsymbol{s}}^* = \arg\max_{\boldsymbol{s}} \pi(\boldsymbol{s}|\boldsymbol{y}), \tag{20.67}$$

where $\pi(\boldsymbol{s}|\boldsymbol{y})$ is the posterior probability of the state sequence $\boldsymbol{s}$, which can be rewritten, using the formula $\pi(\boldsymbol{s}|\boldsymbol{y}) = p(\boldsymbol{s}, \boldsymbol{y})/p(\boldsymbol{y})$, as

$$\hat{\boldsymbol{s}}^* = \arg\max_{\boldsymbol{s}} p(\boldsymbol{s}, \boldsymbol{y}). \tag{20.68}$$

We introduce the following *auxiliary variables*:

$$\boxed{\tilde{\alpha}_t(j, \boldsymbol{y}_0^t) \triangleq \max_{\boldsymbol{s}_0^{t-1}} P[\boldsymbol{S}_0^{t-1} = \boldsymbol{s}_0^{t-1}, S_t = j, \boldsymbol{Y}_0^t = \boldsymbol{y}_0^t],} \tag{20.69}$$

for $j \in \mathcal{S}, 1 \le t \le T$. The variable $\tilde{\alpha}_t(j, \boldsymbol{y}_0^t)$ is similar to the forward variable $\alpha_t(j, \boldsymbol{y}_0^t)$ of (20.54), which we can rewrite as

$$\alpha_t(j, \boldsymbol{y}_0^t) = P[S_t = j, \boldsymbol{Y}_0^t = \boldsymbol{y}_0^t] = \sum_{\boldsymbol{s}_0^{t-1}} P[\boldsymbol{S}_0^{t-1} = \boldsymbol{s}_0^{t-1}, S_t = j, \boldsymbol{Y}_0^t = \boldsymbol{y}_0^t]. \tag{20.70}$$

So the difference is that the summation in (20.70) is replaced by max in (20.69). Then analogous to the recursion formula (20.57) for the variable $\alpha_t(j, \boldsymbol{y}_0^t)$, we obtain (Problem 20.14) the following recursion formula:

$$\boxed{\tilde{\alpha}_t(j, \boldsymbol{y}_0^t) = \max_{i \in \mathcal{S}} \left\{\tilde{\alpha}_{t-1}(i, \boldsymbol{y}_0^{t-1}) c(i; j, y_t)\right\}, j \in \mathcal{S}, 1 \le t \le T,} \tag{20.71}$$

with the initial value

$$\tilde{\alpha}_0(i, y_0) = \alpha_0(i, y_0) = P[S_0 = i, Y_0 = y_0], i \in \mathcal{S}. \tag{20.72}$$

Note that all the variables $\alpha_t(j, \boldsymbol{y}_0^t)$ and parameters $c(i; j, y_t)$ involved in the above recursion are probabilities; thus, they are nonnegative quantities.

We can show by mathematical induction that the recursion formula leads to the MAP estimation sequence $\hat{s}^*$ of (20.68). First, we consider the case where we know the initial state $s_0 \in \mathcal{S}$. Then, given an observation up to time $t-1$ denoted as $\boldsymbol{y}_0^{t-1} = (y_0, y_1, \ldots, y_{t-1})$, suppose that we have found a most likely state sequence; i.e., the one that maximizes $\tilde{\alpha}_{t-1}(i, \boldsymbol{y}_0^{t-1})$ among all possible state sequences entering state i at time t. We denote this most likely path on the trellis diagram as $\hat{\boldsymbol{s}}_0^{t-1}(i)$, and retain it as a **surviving sequence** and discard all other possible state sequences entering state i from further consideration. This is because they will never be a part of the overall optimal sequence, regardless of the observations in the future, $\boldsymbol{y}_t^T$; since all the quantities involved in (20.71) are nonnegative.

Then, as we observe y_t at time t, we proceed to find a most likely state sequence that maximizes the right-hand side of (20.71) among all possible state sequences entering at state j at time t, so there will be $M(= |\mathcal{S}|)$ surviving sequences, one per each state $j \in \mathcal{S}$. Then the argument of mathematical induction implies that there will be M surviving sequences when we reach the end of the observation period $t = T$. The sequence that has the largest auxiliary variable $\tilde{\alpha}_t(j, \boldsymbol{y}_0^T)$ is the MAP sequence estimate $\hat{s}^*$ we have been after.

As observed above, there are exactly $M(= |\mathcal{S}|)$ survivors at any time t, but these M survivors may share a unique state subsequence up to time $t'(< t)$. In other words, the sequence $s_0\hat{s}_1, \ldots, \hat{s}_{t'}$ is common to all the M survivors at time t. Then we know for sure that this sequence must be a part of the MAP sequence, even though we have not yet seen the future observation data $\boldsymbol{y}_{t+1}^T$.

If the initial state is unknown, we must apply the same procedure to each of the M initial states and proceed in parallel. In the initial period there may be as many as M^2 distinct surviving paths, but as time t progresses, there will be again only one subsequence up to a recent past that is common to all the survivors, similar to the case where the initial state is fixed and known.

20.4.1.1 The logarithmic conversion of probabilities

Since the logarithm is a monotonic and continuous function, the operation $\max_x \log f(x) = \log \max_x f(x)$ may be performed. By applying the logarithmic transformation to (20.71) and (20.72),[10] this leads to the following recursion:

$$\boxed{\vec{\alpha}_t(j, \boldsymbol{y}_0^t) = \max_{i \in \mathcal{S}} \left\{ \vec{\alpha}_{t-1}(i, \boldsymbol{y}_0^{t-1}) + d(i; j, y_t) \right\}, \; j \in \mathcal{S}, 1 \le t \le T,} \tag{20.73}$$

where

$$\vec{\alpha}_t(j, \boldsymbol{y}_0^t) \triangleq \log \tilde{\alpha}_t(j, \boldsymbol{y}_0^t) \tag{20.74}$$

and

$$\boxed{d(i; j, y_t) \triangleq \log c(i; j, y_t) = \log P[S_t = j, Y_t = y_t | S_{t-1} = i\,],} \tag{20.75}$$

[10] We define $\log 0 = -\infty$, or a sufficiently large negative number in algorithmic computations.

with the initial value

$$\tilde{\alpha}_0(i, y_0) = \log(\alpha_0(i, y_0)) = d(\emptyset; i, y_0), i \in \mathcal{S}. \tag{20.76}$$

Note that any positive number, not necessarily 10 or e, can serve as the base of the logarithm.

The maximization operation in (20.71) or (20.73) suggests that we need to retain only one sequence, i.e., a surviving sequence, that enters state j at time t. Thus, at any time t, there are only M surviving sequences, one ending at each node $j \in \mathcal{S}$. The forward variable $\tilde{\alpha}_t(j, y_0^t)$ or $\tilde{\alpha}_t(j, \boldsymbol{y}_0^t)$ represents the "score" of this surviving sequence.

20.4.2 The Viterbi algorithm

Now consider the case where all possible state sequences (i.e., all sequences that have legitimate paths in the trellis diagram of the hidden Markov chain) are considered *equally likely* prior to the occurrence of the specific $\boldsymbol{y}$; i.e.,

$$\pi(\boldsymbol{s}) = \text{constant for all legitimate sequences } \boldsymbol{s}. \tag{20.77}$$

In digital communications, for instance, it is usually assumed that the information sequences are i.i.d. binary sequences so that all paths on the trellis diagram associated with a convolutional encoder (see Example 20.1) or a partial-response channel (Problem 20.6) are equally likely.

If the initial state is given (say $S_0 = 0$ with probability one), some state sequences (i.e., sequences with $s_0 \neq 0$) will never occur. However, all feasible state sequences are still equally likely; i.e., the condition (20.77) holds.

Then, it is apparent that the MAP state sequence estimate obtained in the preceding section should also maximize the *likelihood function* defined[11] by

$$L_{\boldsymbol{y}}(\boldsymbol{s}) \triangleq p(\boldsymbol{y}|\boldsymbol{s}), \tag{20.78}$$

where $\boldsymbol{s} = (s_0, s_1, \ldots, s_T)$. The solution to this problem is called the **maximum-likelihood sequence estimate** (MLSE). See Problem 20.16 for a further discussion regarding the MAP sequence estimation versus the MLSE.

A remark is in order concerning maximum-likelihood estimation. In Section 18.1.2 we discussed the MLE of parameter $\boldsymbol{\theta}$ that maximizes the likelihood function $L_{\boldsymbol{x}}(\boldsymbol{\theta}) = p_{\boldsymbol{X}}(\boldsymbol{x}; \boldsymbol{\theta})$, when $\boldsymbol{x}$ is observed. In Section 19.2 on the EM algorithm, we also discussed the MLE of the parameter $\boldsymbol{\theta}$ of $L_{\boldsymbol{x}}(\boldsymbol{\theta}) = p_{\boldsymbol{X}}(\boldsymbol{x}; \boldsymbol{\theta})$, when the observed sequence is not $\boldsymbol{x}$, but $\boldsymbol{y}$, which is an instance of a transformed variable $\boldsymbol{Y} = T(\boldsymbol{X})$. A special case is where T truncates the complete variable so that $\boldsymbol{X} = (\boldsymbol{Y}, \boldsymbol{Z})$, where $\boldsymbol{Z}$ is a latent or missing variable. As remarked earlier, an HMM is exactly such a case, by identifying $\boldsymbol{Z} = \boldsymbol{S}$, the hidden Markov process. We will discuss algorithms to estimate $\boldsymbol{\theta}$ of an HMM based on $\boldsymbol{y}$ in Section 20.6. Our problem at hand, however, is the MLE of the latent variable $\boldsymbol{S}$, while the parameter $\boldsymbol{\theta}$ is known.

[11] Note that this likelihood function is different from $L_{\boldsymbol{y}}(\boldsymbol{\theta})$ defined by (20.47).

Algorithm 20.2 Forward algorithm for MAP state sequence estimation: the Viterbi algorithm

1: Compute the forward variables recursively:

$$\vec{\alpha}_0(i) = \log \alpha_0(i, y_0), i \in \mathcal{S},$$
$$\vec{\alpha}_t(j) = \max_{i \in \mathcal{S}} \{\vec{\alpha}_{t-1}(i) + d(i; j, y_t)\}, j \in \mathcal{S}, t = 1, 2, \ldots, T.$$

While computing the survivor's score $\vec{\alpha}_t(j)$, keep a pointer to the state $\hat{s}^*_{t-1}$ from which the surviving path emanates; i.e.,

$$\hat{s}^*_{t-1} = \arg\max_{i \in \mathcal{S}} \{\vec{\alpha}_{t-1}(i) + d(i; j, y_t)\}.$$

2: Find the surviving state at $t = T$; i.e.,

$$\arg\max_{j \in \mathcal{S}} \vec{\alpha}_T(j) \triangleq \hat{s}^*_T.$$

3: Starting from $\hat{s}^*_T$, backtrack to obtain the state sequence $(\hat{s}^*_{T-1}, \ldots, \hat{s}^*_t, \ldots, \hat{s}^*_1, \hat{s}^*_0)$, as the pointer to each surviving state indicates.

The algorithms based on (20.73), after the logarithmic conversion of the forward variables, for calculating the MAP state sequence (or the MLSE under the assumption (20.77)) is given in Algorithm 20.2, which is commonly known as the **Viterbi algorithm** [108, 339].

The state sequence

$$\hat{\boldsymbol{s}}^* = (\hat{s}^*_0, \hat{s}^*_1, \hat{s}^*_2, \ldots, \hat{s}^*_t, \ldots, \hat{s}^*_T)$$

thus obtained is the MAP state sequence estimate, which is also the MLSE when prior probabilities for all feasible state sequences are equal; i.e.,

$$\boxed{\hat{\boldsymbol{s}}^* = \arg\max_{\boldsymbol{s}} p(\boldsymbol{s}, \boldsymbol{y}).} \tag{20.79}$$

Note that if the initial state s_0 is known to the observer, as is often done in data transmission, the initial forward value in Step 1 of Algorithm 20.2 should be set as

$$\vec{\alpha}_0(i) = \begin{cases} 0, & \text{for } i = s_0, \\ -\infty, & \text{for } i \neq s_0. \end{cases}$$

In digital communications, an information sequence $\boldsymbol{I}$ (which corresponds directly or indirectly to the hidden state sequence $\boldsymbol{s}$) and its transformed version, i.e., the *encoder output* $\boldsymbol{o} = T(\boldsymbol{s})$, are sent over a noisy channel, and the observation $\boldsymbol{y}$ is a noisy version of $\boldsymbol{o}$. If the channel is subject to *additive white Gaussian noise* (AWGN) (e.g., see (20.124)), the problem can be reduced to a **shortest path problem** on the trellis diagram (see Problem 20.124), where the distance metric is the Euclidean distance. If a noisy channel is characterized by a BSC as discussed in Example 20.1, then the distance metric becomes the Hamming distance.[12] It is in fact in this context that the Viterbi algorithm was initially devised [339]. Omura [259] observed that the Viterbi algorithm is equivalent to **Bellman's dynamic programming** [17]. See also Poor [272, 273].

Because of the assumption (20.77) we usually make concerning the prior distribution of state sequences, the Viterbi algorithm is often referred to as a computationally efficient algorithm for *maximum-likelihood sequence estimation* (e.g., see [109, 200]), although it is more appropriate to call it a MAP sequence estimation algorithm [93, 108], because the algorithm does not require the uniform probability assumption (20.77).

The Viterbi algorithm has been successfully applied, in addition to convolutional decoders, to almost all digital recording systems, which now adopt the so-called **PRML (partial-response, maximum-likelihood)** scheme [55, 198, 199] and a channel with intersymbol interference [109, 200, 260] (see also Problem 20.17).

Analogous to the algorithm of Problem 20.11, the backward recursion algorithm for $L_{\boldsymbol{y}}(\boldsymbol{\theta})$, we can derive the *backward algorithm* for the MAP state sequence as well. A mathematical derivation and its algorithmic specification are left to the reader as an exercise (Problem 20.19). The state sequence thus obtained is also a solution to (20.79). We can also find an interesting relation between the forward and backward variables (Problem 20.21), analogous to the relationship (20.64).

20.5 The BCJR algorithm

Suppose that we wish to find a *MAP* estimate, denoted $\hat{s}_t^*$, of a hidden state s_t at time t (rather than the hidden state sequence $\boldsymbol{s}$) on the basis of the entire observed data $\boldsymbol{y} \triangleq (y_0, y_1, \ldots, y_T)$;[13] i.e.,

$$\hat{s}_t^* = \arg\max_{i \in \mathcal{S}} P[s_t = i | \boldsymbol{Y}_0^T = \boldsymbol{y}],\ t \in \mathcal{T} \triangleq [0, 1, 2, \ldots, T]. \tag{20.80}$$

To this end, we consider the following *a posteriori* probability (APP) of state S_t being in $i \in \mathcal{M}$:

$$\gamma_t(i|\boldsymbol{y}) \triangleq P[S_t = i | \boldsymbol{Y}_0^T = \boldsymbol{y}],\ i \in \mathcal{S},\ t \in \mathcal{T}. \tag{20.81}$$

If we sum this APP across the observation interval $\mathcal{T}$, we can interpret $\sum_{t \in \mathcal{T}} \gamma_t(i|\boldsymbol{y})$ as the expected number of times that $S_t; t \in \mathcal{T}$ enters state i, when $\boldsymbol{y}$ is observed. This should also be equal to the expected number of transitions out of state i.

We will find it more convenient to use the following *joint probability*, instead of the above conditional probability:

$$\lambda_t(i, \boldsymbol{y}) \triangleq P[S_t = i, \boldsymbol{Y}_0^T = \boldsymbol{y}],\ i \in \mathcal{S},\ t \in \mathcal{T}. \tag{20.82}$$

[12] For given two binary sequences of length n, $\boldsymbol{x} = (x_1, x_2, \ldots, x_n)$ and $\boldsymbol{y} = (y_1, y_2, \ldots, y_n)$, the Hamming distance between $\boldsymbol{x}$ and $\boldsymbol{y}$ is defined as $d_{\mathrm{H}}(\boldsymbol{x}, \boldsymbol{y}) = \sum_{i=1}^n x_i \oplus y_i$. If $\boldsymbol{x}$ and $\boldsymbol{y}$ are real-valued vectors, the Euclidean distance is defined by $d_{\mathrm{E}}(\boldsymbol{x}, \boldsymbol{y}) = \sqrt{\sum_{i=1}^n (x_i - y_i)^2}$.

[13] In this section we also drop the model parameter, although we should write $p(s_t = i | \boldsymbol{y}; \boldsymbol{\theta})$, if we wish to emphasize the dependency of the APP on the model parameters, as we shall do in the next section.

For an observation sequence $\boldsymbol{y}$, with the parameter $\boldsymbol{\theta}$ given and fixed, we can evaluate $p(\boldsymbol{y}) \triangleq p(\boldsymbol{y};\boldsymbol{\theta}) = L_{\boldsymbol{y}}(\boldsymbol{\theta})$ using (20.58), (20.65), or (20.66). Then

$$\gamma_t(i|\boldsymbol{y}) = \frac{\lambda_t(i,\boldsymbol{y})}{p(\boldsymbol{y})}, i \in \mathcal{S}, t \in \mathcal{T}. \tag{20.83}$$

The joint probability $\lambda_t(i,\boldsymbol{y})$ can be written in terms of the *forward and backward variables*

$$\begin{aligned}\lambda_t(i,\boldsymbol{y}) &= P[S_t = i, \boldsymbol{Y}_0^t = \boldsymbol{y}_0^t]P[\boldsymbol{Y}_{t+1}^T = \boldsymbol{y}_{t+1}^T|S_t = i, \boldsymbol{Y}_0^t = \boldsymbol{y}_0^t] \\ &= \alpha_t(i,\boldsymbol{y}_0^t)P[\boldsymbol{Y}_{t+1}^T = \boldsymbol{y}_{t+1}^T|S_t = i] = \alpha_t(i,\boldsymbol{y}_0^t)\beta_t(i;\boldsymbol{y}_{t+1}^T),\end{aligned} \tag{20.84}$$

which we already derived in (20.64) (see Problem 20.12).

From (20.83) and (20.84), we find the following relation between the APP $\gamma_t(i|\boldsymbol{y})$ and the forward and backward variables:

$$\boxed{\gamma_t(i|\boldsymbol{y}) = \frac{\alpha_t(i,\boldsymbol{y}_0^t)\beta_t(i;\boldsymbol{y}_{t+1}^T)}{p(\boldsymbol{y})}, i \in \mathcal{S}, t \in \mathcal{T},} \tag{20.85}$$

where the normalization factor $p(\boldsymbol{y})(= L_{\boldsymbol{y}}(\boldsymbol{\theta}))$, as given by (20.65), makes $\gamma_t(i|\boldsymbol{y})$ a bona fide conditional probability so that $\sum_{i\in\mathcal{S}} \gamma_t(i|\boldsymbol{y}) = 1$. Once the APPs $\gamma_t(i|\boldsymbol{y})$ are found for all states $S_t = i \in \mathcal{S}$ at given time t, the MAP estimate (20.80) can then be expressed as

$$\boxed{\hat{s}_t^* = \arg\max_{i\in\mathcal{S}} \left\{\alpha_t(i,\boldsymbol{y}_0^t)\beta_t(i;\boldsymbol{y}_{t+1}^T)\right\}, t \in \mathcal{T}.} \tag{20.86}$$

The above steps to obtain the MAP estimate were found by Bahl, Cocke, Jelinek and Raviv [11]. By taking the initials of the four authors, this computation algorithm is known as the **BCJR algorithm**, which we summarize in Algorithm 20.3, in which the initial probability $\alpha_0(i, y_0) = P[S_0 = i, Y_0 = y_0]$, as given in (20.49).

Algorithm 20.3 Forward–backward algorithm for MAP state estimation: BCJR algorithm

---

1: Compute and save the forward vector variables recursively:

$$\begin{aligned}\boldsymbol{\alpha}_0^\top &= (\alpha_0(i, y_0), i \in \mathcal{S}),\\ \boldsymbol{\alpha}_t^\top &= \boldsymbol{\alpha}_{t-1}^\top \boldsymbol{C}(y_t), t = 1, 2, \ldots, T.\end{aligned}$$

2: Compute the backward vector variables recursively and find the MAP state estimate:

$$\begin{aligned}\boldsymbol{\beta}_T &= \mathbf{1},\\ \boldsymbol{\beta}_t &= \boldsymbol{C}(y_{t+1})\boldsymbol{\beta}_{t+1},\\ \hat{s}_t^* &= \arg\max_{i\in\mathcal{S}} \alpha_t(i)\beta_t(i), t = T-1, T-2, \ldots, 1.\end{aligned}$$

The BCJR algorithm is based on applying the FBA, discussed in Section 20.3.3 for evaluating the likelihood $L_{\boldsymbol{y}}(\boldsymbol{\theta})$, to a related but different problem of estimating a hidden state variable S_t at a given $t \in \mathcal{T}$. We will show in Section 20.6 that the FBA can also be applied to estimate and re-estimate HMM parameters ($\boldsymbol{\alpha}_0$, $\mathbf{C}$) (or ($\boldsymbol{\pi}_0, \boldsymbol{A}, \boldsymbol{B}$)). This algorithm, as applied to parameter estimation, is known as the **Baum–Welch algorithm**. In fact, the auxiliary variables $\lambda_t(i, \boldsymbol{y})$ and $\gamma_t(i|\boldsymbol{y})$ used in the BCJR algorithm were originally introduced in the Baum–Welch algorithm.

20.6 Maximum-likelihood estimation of model parameters

In many situations we may not have prior knowledge about exact values of the model parameter $\boldsymbol{\theta} = (\boldsymbol{\alpha}_0, \ \mathbf{C})$. In such a case, we make an initial estimate of the parameter in some manner. Once we collect the observations $\boldsymbol{y}$, we may learn about the model parameter and find a possibly better estimate of the parameter. This *learning* process is referred to as *parameter estimation and re-estimation*, or simply as *parameter estimation.*

As we discussed in Chapter 18, it is generally too difficult or complex to come up with an analytic expression for an MLE or a Bayes' estimate for a certain parameter associated with some general probability distribution. This is also the case for an estimate of the model parameter associated with an HMM. Therefore, we need to resort to some numerical algorithm to seek an optimal estimate under a given criterion. In Section 19.2.2 on the EM algorithm for missing data, we discussed an iterative method for an MLE of the parameter $\boldsymbol{\theta}$ associated with a probability distribution $p_{\boldsymbol{X}}(\boldsymbol{x}; \boldsymbol{\theta})$, where only $\boldsymbol{Y}$ of $\boldsymbol{X} = (\boldsymbol{Y}, \boldsymbol{Z})$ is observable. Our problem at hand exactly fits into that framework. The hidden Markov process $\boldsymbol{S}$ corresponds to the latent variable $\boldsymbol{Z}$ assumed in that formulation. Thus, it is apparent that the expectation-maximization (EM) algorithm is applicable to maximum-likelihood estimation of the HMM parameters.

20.6.1 Forward–backward algorithm for a transition-based hidden Markov model

In this section we focus on the EM algorithm for estimation and re-estimation of

$$\boldsymbol{\theta} = (\alpha_0(j, k), c(i; j, k), i, j \in \mathcal{S}, k \in \mathcal{Y}).$$

Recall the general auxiliary function derived in (19.38) of Section 19.2.2:

$$\boxed{Q(\boldsymbol{\theta}|\boldsymbol{\theta}^{(p)}) = E\left[\log p(\boldsymbol{S}, \boldsymbol{y}; \boldsymbol{\theta})|\, \boldsymbol{y}; \boldsymbol{\theta}^{(p)}\right] = \sum_{\boldsymbol{s}} p(\boldsymbol{s}|\boldsymbol{y}; \boldsymbol{\theta}^{(p)}) \log p(\boldsymbol{s}, \boldsymbol{y}; \boldsymbol{\theta}),} \tag{20.87}$$

where $\boldsymbol{\theta}^{(p)}$ is the pth estimate of the model parameters, $p = 0, 1, 2, \ldots$.

From (20.49) we have

$$p(\boldsymbol{s}, \boldsymbol{y}; \boldsymbol{\theta}) = \alpha_0(s_0, y_0; \boldsymbol{\theta}) \prod_{t=1}^{T} c(s_{t-1}; s_t, y_t), \tag{20.88}$$

where

$$\boldsymbol{y} \triangleq \boldsymbol{y}_0^T$$

as defined before. By taking the logarithm of the above expression, and replacing $\boldsymbol{s}$ by $\boldsymbol{S}$, we have

$$\log p(\boldsymbol{S}, \boldsymbol{y}; \boldsymbol{\theta}) = \log \alpha_0(S_0, y_0; \boldsymbol{\theta}) + \sum_{t=1}^{T} \log c(S_{t-1}; S_t, y_t). \tag{20.89}$$

Since $\boldsymbol{S}$ is an RV, both sides of the above equation are RVs. Now, taking the expectation with respect to $p(\boldsymbol{s}|\boldsymbol{y}; \boldsymbol{\theta}^{(p)})$, we have

$$\begin{aligned} Q(\boldsymbol{\theta}|\boldsymbol{\theta}^{(p)}) &= E[\log \alpha_0(S_0, y_0; \boldsymbol{\theta})|\boldsymbol{y}, \boldsymbol{\theta}^{(p)}] + \sum_{t=1}^{T} E[\log p(S_t, y_t|S_{t-1}; \boldsymbol{\theta})|\boldsymbol{y}, \boldsymbol{\theta}^{(p)}] \\ &= Q_0(\boldsymbol{\theta}|\boldsymbol{\theta}^{(p)}) + Q_1(\boldsymbol{\theta}|\boldsymbol{\theta}^{(p)}), \end{aligned} \tag{20.90}$$

where we used the definition $c(s_{t-1}; s_t, y_t) = p(s_t, y_t|s_{t-1}; \boldsymbol{\theta})$. The first term can be written as

$$Q_0(\boldsymbol{\theta}|\boldsymbol{\theta}^{(p)}) = \sum_{i \in \mathcal{S}} \log \alpha_0(i, y_0; \boldsymbol{\theta}) \gamma_0(i|\boldsymbol{y}; \boldsymbol{\theta}^{(p)}), \tag{20.91}$$

where $\gamma_0(i|\boldsymbol{y}; \boldsymbol{\theta}^{(p)}) = P[S_0 = i|\boldsymbol{y}; \boldsymbol{\theta}^{(p)}]$ is the APP of the initial hidden state S_0 defined by (20.85). Similarly, we can write the second term of (20.90) as

$$Q_1(\boldsymbol{\theta}|\boldsymbol{\theta}^{(p)}) = \sum_{t=1}^{T} \sum_{i \in \mathcal{S}, j \in \mathcal{S}} \log p(j, y_t|i; \boldsymbol{\theta}) \xi_{t-1}(i, j|\boldsymbol{y}; \boldsymbol{\theta}^{(p)}), \tag{20.92}$$

where $\xi_{t-1}(i, j|\boldsymbol{y}; \boldsymbol{\theta}^{(p)}) = P[S_{t-1} = i, S_t = j|\boldsymbol{y}; \boldsymbol{\theta}^{(p)}]$ is the conditional joint probability of the pair of states (S_{t-1}, S_t). Similar to the derivation of (20.85), it is not difficult to show (see Problem 20.22 (c)) that

$$\xi_{t-1}^{(p)}(s_{t-1}, s_t|\boldsymbol{y}) = \frac{\alpha_{t-1}^{(p)}(s_{t-1}, \boldsymbol{y}_0^{t-1}) c^{(p)}(s_{t-1}; s_t, y_t) \beta_t^{(p)}(s_t; \boldsymbol{y}_{t+1}^T)}{L_{\boldsymbol{y}}(\boldsymbol{\theta}^{(p)})}, \tag{20.93}$$

where we use the simplified notation

$$\begin{aligned} \xi_{t-1}^{(p)}(s_{t-1}, s_t|\boldsymbol{y}) &\triangleq \xi_{t-1}(s_{t-1}, s_t|\boldsymbol{y}; \boldsymbol{\theta}^{(p)}), \\ \alpha_{t-1}^{(p)}(s_{t-1}, \boldsymbol{y}_0^{t-1}) &\triangleq \alpha_{t-1}(s_{t-1}, \boldsymbol{y}_0^{t-1}; \boldsymbol{\theta}^{(p)}), \\ \beta_t^{(p)}(s_t; \boldsymbol{y}_{t+1}^T) &\triangleq \beta_t(s_t; \boldsymbol{y}_{t+1}^T; \boldsymbol{\theta}^{(p)}). \end{aligned}$$

Since these conditional probabilities are expressed in terms of the forward and backward variables, they can be calculated using the forward-backward algorithm (FBA) as was done in Algorithm 20.3. Note that, because $\gamma_t(i|\boldsymbol{y};\boldsymbol{\theta}^{(p)}) = \sum_{j\in\mathcal{S}} \xi_t(i,j|\boldsymbol{y};\boldsymbol{\theta}^{(p)})$ (see Problem 20.24), it is sufficient to calculate $\xi_t(s_{t-1}, s_t|\boldsymbol{y};\boldsymbol{\theta}^{(p)})$ only.

Thus, the E-step of the EM algorithm can be performed using the FBA discussed in Section 20.3.3. In the forward part we compute and save the forward variables $\boldsymbol{\alpha}(\boldsymbol{y}_0^t;\boldsymbol{\theta}^{(p)})$, and in the backward part we compute $\boldsymbol{\beta}(\boldsymbol{y}_t^T;\boldsymbol{\theta}^{(p)})$ and, using the saved forward variables, compute $\xi_{t-1}(s_{t-1}, s_t|\boldsymbol{y};\boldsymbol{\theta}^{(p)})$ and accumulate the sums to evaluate $Q_0(\boldsymbol{\theta}|\boldsymbol{\theta}^{(p)})$ of (20.91) and $Q_1(\boldsymbol{\theta}|\boldsymbol{\theta}^{(p)})$ of (20.92).

20.6.1.1 The M-step of the EM algorithm for a transition-based hidden Morkov model

Recall that, in a transition-based HMM, the model parameter is $\boldsymbol{\theta} = (\boldsymbol{\alpha}, \boldsymbol{C})$, where

$$\boldsymbol{\alpha} = (\alpha_0(j,k);\, j\in\mathcal{S}, k\in\mathcal{Y}),\, \boldsymbol{C} = [c(i;j,k);\, i,j\in\mathcal{S}, k\in\mathcal{Y}],$$

and

$$\begin{aligned}\alpha_0(j,k) &= P[S_0 = j, Y_0 = k],\ j\in\mathcal{S}, k\in\mathcal{Y},\\ c(i;j,k) &= P[S_t = j, Y_t = k|S_{t-1} = i],\ i,j\in\mathcal{S}, k\in\mathcal{Y}, t = 1,2,\ldots,T.\end{aligned}$$

Since these parameters are the joint probability and conditional joint probability distributions, they must satisfy constraints

$$\sum_{j\in\mathcal{S},k\in\mathcal{Y}} \alpha_0(j,k) = 1, \tag{20.94}$$

$$\sum_{j\in\mathcal{S},k\in\mathcal{Y}} c(i;j,k) = 1,\quad \text{for all } i\in\mathcal{S}. \tag{20.95}$$

We readily see from (20.91) and (20.92) that $Q_0(\boldsymbol{\theta}|\boldsymbol{\theta}^{(p)})$ depends only on $\boldsymbol{\alpha}_0$ and $Q_1(\boldsymbol{\theta}|\boldsymbol{\theta}^{(p)})$ depends on $\boldsymbol{C}$, but not on $\boldsymbol{\alpha}_0$. Hence, we can maximize them separately. Applying the log-sum inequality (10.21), we find that $Q_0(\boldsymbol{\theta}|\boldsymbol{\theta}^{(p)})$ is maximized at the $(p+1)$st step when $\alpha_0(j, y_0;\boldsymbol{\theta})$ is set to

$$\boxed{\alpha_0^{(p+1)}(j,y_0) = \gamma_0(i|\boldsymbol{y};\boldsymbol{\theta}^{(p)}) = \frac{\alpha_0^{(p)}(j,y_0)\beta_0^{(p)}(j;\boldsymbol{y}_1^T)}{L_{\boldsymbol{y}}(\boldsymbol{\theta}^{(p)})}.} \tag{20.96}$$

Similarly, $Q_1(\boldsymbol{\theta}|\boldsymbol{\theta}^{(p)})$ can be maximized at

$$\boxed{c^{(p+1)}(i;j,k) = \frac{\sum_{t=1}^T \xi_{t-1}^{(p)}(i,j|\boldsymbol{y})\delta_{y_t,k}}{\sum_{j\in\mathcal{S}}\sum_{t=1}^T \xi_{t-1}^{(p)}(i,j|\boldsymbol{y})\delta_{y_t,k}}.} \tag{20.97}$$

Because of the multiplier $\delta_{y_t,k}$, summation is done with respect to those t for which $y_t = k$. The last equation can also be presented more explicitly by using (20.140):

$$c^{(p+1)}(i; j, k) = \frac{\sum_{t=1}^{T} \alpha_{t-1}^{(p)}(i, \boldsymbol{y}_0^{t-1}) c^{(p)}(i; j, y_t) \beta_t^{(p)}(j; \boldsymbol{y}_{t+1}^{T}) \delta_{y_t,k}}{\sum_{j \in \mathcal{S}} \sum_{t=1}^{T} \alpha_{t-1}^{(p)}(i, \boldsymbol{y}_0^{t-1}) c^{(p)}(i; j, y_t) \beta_t^{(p)}(j; \boldsymbol{y}_{t+1}^{T})}. \tag{20.98}$$

Algorithm 20.4 implements the EM algorithm discussed above. The forward part of the E-step is the same as in Algorithms 20.1 and 20.3, and we use the vector-matrix notation as before. The backward part is basically the same as in Algorithm 20.3, as far as the computation of the backward vector variables $\boldsymbol{\beta}_t^{(p)}$ is concerned. However, we need to compute

$$S^{(p)}(i, j, k) = \sum_{t=1}^{T} \alpha_{t-1}^{(p)}(i, \boldsymbol{y}_0^{t-1}) c^{(p)}(i; j, y_t) \beta_t^{(p)}(j; \boldsymbol{y}_{t+1}^{T}) \delta_{y_t,k}, \tag{20.99}$$

Algorithm 20.4 EM algorithm for a transition-based HMM

1: Set $p \leftarrow 0$, and denote the initial estimate of the model parameters as $\boldsymbol{\alpha}_0^{(0)} = [\alpha_0^{(0)}(i, y_0), i \in \mathcal{S}]$ and $\boldsymbol{C}^{(0)}(y_0) = [c^{(0)}(i; j, y_0); i, j \in \mathcal{S}, k \in \mathcal{Y}]$.

2: **The forward algorithm in the E-step:** Compute and save the forward vector variables $\boldsymbol{\alpha}_t^{(p)}$ recursively:

$$\boldsymbol{\alpha}^{(p)}{}_t^{\top} = \boldsymbol{\alpha}^{(p)}{}_{t-1}^{\top} \boldsymbol{C}^{(p)}(y_t),\ t = 1, 2, \ldots, T.$$

3: Compute the likelihood function: $L^{(p)} = \mathbf{1}^{\top} \boldsymbol{\alpha}_T^{(p)}$.

4: **The backward algorithm in the E-step:**Compute the backward vector variables $\boldsymbol{\beta}_t^{(p)}$ recursively. Compute and accumulate $\alpha_{t-1}^{(p)}(i) c^{(p)}(i; j, k) \beta_t^{(p)}(j)$.

  1. Set $\boldsymbol{\beta}_T^{(p)} = \mathbf{1}$and $S^{(p)}(i, j, k) = 0, i, j \in \mathcal{S}, k \in \mathcal{Y}$.
  2. For $t = T - 1, T - 2, \ldots, 0$:
     a. Compute $\boldsymbol{\beta}_t^{(p)} = \boldsymbol{C}^{(p)}(y_{t+1}) \boldsymbol{\beta}_{t+1}^{(p)}$.
     b. Compute

$$S^{(p)}(i, j, k) \leftarrow S^{(p)}(i, j, k) + \alpha_{t-1}^{(p)}(i) c^{(p)}(i; j, y_t) \beta_t^{(p)}(j) \delta_{k,y_t}.$$

5: **The M-step:** Update the model parameters:

$$\alpha_0^{(p+1)}(j) \longleftarrow \frac{\alpha_0^{(p)}(j) \beta_0^{(p)}(j)}{L^{(p)}}, \text{ for all } j \in \mathcal{S},$$

$$c^{(p+1)}(i; j, k) \longleftarrow \frac{S^{(p)}(i, j, k)}{\sum_{j \in \mathcal{S}} S^{(p)}(i, j, k)} \text{ for all } i, j \in \mathcal{S}, k \in \mathcal{Y}.$$

6: If any of the stopping conditions are met, stop the iteration and output the estimated $\boldsymbol{\alpha}_0^{(p+1)}$ and $\boldsymbol{C}^{(p+1)}$; else set $p \leftarrow p + 1$ and repeat Steps 2 through 5.

which is required to calculate $c^{(p)}(i; j, k)$ of (20.98). For the parameter variables used in the algorithm, we explicitly show the superscript (p), although we suppress the observed data $\boldsymbol{y}$. If we do not need to keep all the computation results in the iterative procedure, we can overwrite the parameter values of the previous iteration and can suppress (p).

20.6.2 The Baum–Welch algorithm

Let us turn our attention to the *state-based output model*, which is commonly assumed in the HMM literature; namely, a special case where both the initial distribution $\alpha_0(j, k) = P[S_0 = j, Y_0 = k]$ and the conditional probability $c(i; j, k) = P[S_t = j, Y_t = k|S_{t-1} = i]$ can be factored as follows (see Definition 20.3 of Section 20.2.1):

$$\alpha_0(j, k) = \pi_0(j)b(j; k),\ j \in \mathcal{S}, k \in \mathcal{Y}, \tag{20.100}$$

$$c(i; j, k) = a(i, j)b(j; k), i, j \in \mathcal{S}, k \in \mathcal{Y}. \tag{20.101}$$

Then the model parameter is $\boldsymbol{\theta} = (\boldsymbol{\pi}_0, \boldsymbol{A}, \boldsymbol{B})$, where

$$\boldsymbol{\pi}_0 = (\pi_0(i); i \in \mathcal{S}), \boldsymbol{A} = [a(i, j); i, j \in \mathcal{S}], \boldsymbol{B} = [b(j; k); j \in \mathcal{S}, k \in \mathcal{Y}].$$

The EM algorithm for this type of HMM is known as the Baum–Welch algorithm [346].

By proceeding in a fashion similar to the steps that we took to arrive at the M-step formula (20.97), we can find M-step formulas for the model parameters $\boldsymbol{A}$, $\boldsymbol{B}$, and $\boldsymbol{\pi}_0$. However, rather than starting from the definition of the Q-function given in (20.87), we make use of the expression (20.90) obtained for the transition-based HMM. Using (20.100), we can write Q_0 of (20.91) as

$$Q_0(\boldsymbol{\theta}|\boldsymbol{\theta}^{(p)}) = \sum_{i\mathcal{S}} \log \pi_0(i)\gamma^{(p)}(i|\boldsymbol{y}) + \sum_{i\in\mathcal{S}} \log b(i; y_0)\gamma_0^{(p)}(i|\boldsymbol{y}), \tag{20.102}$$

where

$$\gamma_0^{(p)}(i|\boldsymbol{y}) = \gamma_0(i|\boldsymbol{y}; \boldsymbol{\theta}^{(p)}).$$

Similarly, using (20.101), we write Q_1 of (20.92) as

$$\begin{aligned} Q_1(\boldsymbol{\theta}|\boldsymbol{\theta}^{(p)}) &= \sum_{t=1}^{T} \sum_{i\in\mathcal{S}, j\in\mathcal{S}} \log a(i, j)\xi_{t-1}^{(p)}(i, j|\boldsymbol{y}) + \sum_{t=1}^{T} \sum_{i\in\mathcal{S}, j\in\mathcal{S}} \log b(j; y_t)\xi_{t-1}^{(p)}(i, j|\boldsymbol{y}) \\ &= \sum_{t=1}^{T} \sum_{i\in\mathcal{S}, j\in\mathcal{S}} \log a(i, j)\xi_{t-1}^{(p)}(i, j|\boldsymbol{y}) + \sum_{t=1}^{T} \sum_{j\in\mathcal{S}} \log b(j; y_t)\gamma^{(p)}(j|\boldsymbol{y}), \end{aligned} \tag{20.103}$$

where we used the identity (Problem 20.24 (c)) $\sum_{i\in\mathcal{S}} \xi_{t-1}(i, j|\boldsymbol{y}) = \gamma_t(j|\boldsymbol{y})$. Then, from (20.102) and (20.103), we find

$$Q(\boldsymbol{\theta}|\boldsymbol{\theta}^{(p)}) = \sum_{iS} \log \pi_0(i)\gamma^{(p)}(i|\mathbf{y}) + \sum_{t=1}^{T} \sum_{i\in\mathcal{S}, j\in\mathcal{S}} \log a(i,j)\xi_{t-1}^{(p)}(i,j|\mathbf{y})$$
$$+ \sum_{t=0}^{T} \sum_{j\in\mathcal{S}} \log b(j;y_t)\gamma^{(p)}(j|\mathbf{y}). \tag{20.104}$$

It is clear that we can maximize the three summed terms separately, by applying the equality condition for the log-sum inequality as done in the previous section, and find the following M-step formulas:

$$\boxed{\begin{aligned}
\pi_0^{(p+1)}(i) &= \gamma_0^{(p)}(i|\mathbf{y}),\ i \in \mathcal{S},\\
a^{(p+1)}(i;j) &= \frac{\sum_{t=1}^{T} \xi_{t-1}^{(p)}(i,j\,|\mathbf{y})}{\sum_{t=1}^{T} \gamma_{t-1}^{(p)}(i\,|\mathbf{y})},\ i,j \in \mathcal{S},\\
b^{(p+1)}(j;k) &= \frac{\sum_{t=0}^{T} \gamma_t^{(p)}(j|\mathbf{y})\delta_{k,y_t}}{\sum_{t=0}^{T} \gamma_t^{(p)}(j|\mathbf{y})},\ j \in \mathcal{S}.
\end{aligned}} \tag{20.105}$$

The reader is suggested to derive the above updating formula directly from the definition of the auxiliary function $Q(\boldsymbol{\theta}|\boldsymbol{\theta}^{(p)})$ of (20.87) (Problem 20.26), instead of starting from (20.90) as was done here, which saved a number of derivation steps.

Algorithm 20.5 shows the EM algorithm for a *state-based HMM*, which is widely known as the Baum–Welch algorithm, as stated earlier. In order to simplify the notation, we define an $M \times M$ matrix $\boldsymbol{C}(y_t)$, similar to (20.51), by

$$\boldsymbol{C}(y_t) = [a(i,j)b(j;y_t);\ i,j \in \mathcal{S}],\ y_t \in \mathcal{Y}, \tag{20.106}$$

and the forward vector variables, also as in (20.51):

$$\boldsymbol{\alpha}_t = (\alpha_t(i);\ i \in \mathcal{S})^\top. \tag{20.107}$$

Then the forward part can be written as a simple vector-matrix equation as shown in the algorithm.

We introduce the following arrays to represent the numerators and denominators of $a^{(p+1)}(i;j)$ and $b^{(p+1)}(j;k)$. For notational brevity, we drop the superscript (p).

$$m_t(i;j) \triangleq \alpha_{t-1}(i, \mathbf{y}_0^{t-1})a(i;j)b(j;y_t)\beta_t(j;\mathbf{y}_{t+1}^T),\ i,j \in \mathcal{S},$$
$$M(i;j) = \sum_{t=1}^{T} m_t(i,j),\ i,j \in \mathcal{S},$$
$$n_t(j,k) \triangleq \alpha_t(j,\mathbf{y}_0^t)\beta_t(j;\mathbf{y}_{t+1}^T)\delta_{k,y_t} = \lambda_t(j,\mathbf{y})\delta_{k,y_t},\ j \in \mathcal{S},\ y_t \in \mathcal{Y},$$
$$N(j,k) = \sum_{t=0}^{T} n_t(j,y_t),\ j \in \mathcal{S},\ k,\ y_t \in \mathcal{Y},$$

where $\lambda_t(y,\mathbf{y}) = \alpha_t(j,\mathbf{y}_0^t)\beta_t(j;\mathbf{y}_{t+1}^T)$, as defined in (20.82) (see also (20.84)).

Algorithm 20.5 Baum–Welch algorithm: the EM algorithm for a state-based HMM

1: Set $p \leftarrow 0$, and denote the initial estimate of the model parameters as $\boldsymbol{\theta}^{(0)} = (\boldsymbol{\pi}_0^{(0)}, \boldsymbol{A}^{(0)}, \boldsymbol{B}^{(0)}(y_0))$, and define an $M \times M$ matrix $\boldsymbol{C}(y_0) = [a(i, j)b(j; y_0); i, j \in \mathcal{S}]$.

2: **The forward algorithm in the E-step:** Compute and save the forward vector variables $\boldsymbol{\alpha}_t^{(p)}$ recursively:

$$\boldsymbol{\alpha}_0^{(p)\top} = (\alpha_0^{(p)}(i); i \in \mathcal{S}), \text{ where } \alpha_0^{(p)}(i) = \pi_0^{(p)}(i)b^{(p)}(i; y_0),$$
$$\boldsymbol{\alpha}_t^{(p)\top} = \boldsymbol{\alpha}_{t-1}^{(p)\top}\boldsymbol{C}^{(p)}(y_t), t = 1, 2, \ldots, T.$$

3: Compute the likelihood function: $L^{(p)} = \mathbf{1}^\top \boldsymbol{\alpha}_T^{(p)}$.

4: **The backward algorithm in the E-step:** Compute the backward vector variables $\boldsymbol{\beta}_t^{(p)}$ recursively. Compute and accumulate the $m_t^{(p)}(i, j)$ and $n_t^{(p)}(j, k)$.

1. Set

$$\boldsymbol{\beta}_T^{(p)} = \mathbf{1}, M^{(p)}(i, j) = 0, \text{ and}$$
$$N(j, k) = \alpha_T^{(p)}(j)\delta_{k,y_T}, i, j \in \mathcal{S}, k \in \mathcal{Y}.$$

2. For $t = T - 1, T - 2, \ldots, 1$:
   a. Compute $\boldsymbol{\beta}_t^{(p)} = \boldsymbol{C}^{(p)}(y_{t+1})\boldsymbol{\beta}_{t+1}^{(p)}$.
   b. Compute $m_t^{(p)}(i, j) = \alpha_{t-1}^{(p)}(i)a^{(p)}(i; j)b^{(p)}(j; y_t)\beta_t^{(p)}(j)$ and add to $M^{(p)}(i, j)$:

$$M^{(p)}(i, j) \leftarrow M^{(p)}(i, j) + m_t^{(p)}(i, j), i, j \in \mathcal{S}, k \in \mathcal{Y}.$$

   c. Compute $n_t^{(p)}(j, k) = \alpha_t^{(p)}(j)\beta_t^{(p)}(j)\delta_{k,y_t}$ and add to $N^{(p)}(j, k)$:

$$N^{(p)}(j, k) \leftarrow N^{(p)}(j, k) + n_t^{(p)}(j, y_t).$$

5: **The M-step:** Update the model parameter:

$$\pi_0^{(p+1)}(i) \leftarrow \gamma_0^{(p)}(i), i \in \mathcal{S},$$
$$a^{(p+1)}(i; j) \leftarrow \frac{M^{(p)}(i, j)}{\sum_{j \in \mathcal{S}} M^{(p)}(i, j)}, i, j \in \mathcal{S},$$
$$b^{(p+1)}(j; k) \leftarrow \frac{N^{(p)}(j, k)\delta_{k,y_t}}{N^{(p)}(j, y_t)}, j \in \mathcal{S}, y_t \in \mathcal{Y}.$$

6: If any of the stopping conditions are met, stop the iteration and output the estimated the model parameter $\boldsymbol{\theta}^{(p+1)} = (\boldsymbol{\pi}_0^{(p+1)}, \boldsymbol{A}^{(p+1)}, \boldsymbol{B}^{(p+1)})(\boldsymbol{y})$; else set $p \leftarrow p + 1$, and repeat Steps 2 through 5.

Note that $m_t(i; j)$ is the probability of a transition from state $S_{t-1} = i$ to $S_t = j$. Thus, $M(i; j)$ is the expected number of transitions from state i into state j in $\boldsymbol{S}_0^T$. Similarly, $n_t(j, k)$ is the joint probability that $S_t = i$, $y_t = k$; hence, $N(j, k)$ is the expected number of occurrences of $(S_t = j, y_t = k)$ in $(\boldsymbol{S}_0^T, \boldsymbol{y}_0^T)$. Here, $\gamma_0(i)$ is the probability that the system is in state i at time $t = 0$. Then,

$$a^{(p+1)}(i; j) \leftarrow \frac{M^{(p)}(i, j)}{\sum_{j \in \mathcal{S}} M^{(p)}(i, j)}, i, j \in \mathcal{S},$$

$$b^{(p+1)}(j; k) \leftarrow \frac{N^{(p)}(j, k)\delta_{k, y_t}}{N^{(p)}(j, y_k)}, j \in \mathcal{S}, k, y_k \in \mathcal{Y}.$$

Needless to say, the same storage space can be assigned to all of these variables and arrays; hence, the superscripts become superfluous and can be dropped in the algorithm and its implementation. But if we are interested in observing the convergence behavior of the iterative steps, it will be worthwhile to store all these intermediate results. The same holds when you are in the debugging stage of implementing the algorithm.

20.7 Application: parameter estimation of mixture distributions

As an example of application of the results of the preceding section, consider the problem of estimating parameters of a mixture of M probability distributions whose PDF has the form

$$f_Y(y; \boldsymbol{\theta}) = \sum_{i \in \mathcal{S}} \pi_i f_i(y; \boldsymbol{\phi}_i), \text{ where } \mathcal{S} \triangleq 0, 1, \ldots, M-1. \quad (20.108)$$

In this equation, each $f_i(y; \boldsymbol{\phi}_i)$, $i \in \mathcal{S}$, represents a PDF with an unknown (vector) parameter $\boldsymbol{\phi}_i$. The mixture coefficients π_i are positive numbers whose sum is equal to unity; otherwise, $f_Y(y; \boldsymbol{\theta})$ would not be a bona fide PDF.

Our goal is to estimate the parameter

$$\boldsymbol{\theta} = (\pi_0, \ldots, \pi_{M-1}, \boldsymbol{\phi}_0, \ldots, \boldsymbol{\phi}_{M-1}) \triangleq (\boldsymbol{a}, \boldsymbol{\phi}) \quad (20.109)$$

of this distribution from a sequence of independent samples $\boldsymbol{y} = (y_0, y_1, \ldots, y_t, \ldots, y_T)$ taken from the distribution (20.108).

The likelihood function of $\boldsymbol{\theta}$, given the observation $\boldsymbol{y}$, is

$$L_{\boldsymbol{y}}(\boldsymbol{\theta}) = f_{\boldsymbol{Y}}(\boldsymbol{y}; \boldsymbol{\theta}) = \prod_{t \in \mathcal{T}} f_Y(y_t; \boldsymbol{\theta}). \quad (20.110)$$

It is clear that direct minimization of the likelihood function is generally a difficult task, unless $f_{\boldsymbol{Y}}(\boldsymbol{y}; \boldsymbol{\theta})$ has a nice closed-form expression. Therefore, we may have to resort to a simulation or an iterative algorithm to numerically obtain an MLE of $\boldsymbol{\theta}$. We now show that we can formulate the above problem as a parameter estimation problem for an HMM.

We introduce an i.i.d. random sequence $\{S_t\}$ that takes on $i \in \mathcal{S}$ with probability π_i, where the set $\mathcal{S}$ is defined in (20.108). An i.i.d. sequence can be considered as a zeroth order Markov chain with state transition probabilities of the form

$$a(i, j) = \pi_j, \text{ for all } i \in \mathcal{S}, j \in \mathcal{S}. \tag{20.111}$$

So we can treat S_t as a hidden Markov process such that, if $S_t = i$, the sample y_t is drawn from the distribution associated with the state i. Thus, we have a state-based HMM.

The likelihood function of $\boldsymbol{\theta} = (\boldsymbol{\pi}, \boldsymbol{\phi})$ given the complete data $(\boldsymbol{s}, \boldsymbol{y})$ is given by

$$L_{\boldsymbol{s},\boldsymbol{y}}(\boldsymbol{\theta}) \triangleq \prod_{t=0}^{T} \pi_{s_t} f_{s_t}(y_t; \phi_{s_t}). \tag{20.112}$$

Thus, the log-likelihood function is

$$\log L_{\boldsymbol{s},\boldsymbol{y}}(\boldsymbol{\theta}) = \sum_{t=0}^{T} \log \pi_{s_t} + \sum_{t=0}^{T} \log f_{s_t}(y_t; \phi_{s_t}). \tag{20.113}$$

We denote the estimate of $\boldsymbol{\theta}$ obtained after the pth iteration as $\boldsymbol{\theta}^{(p)} = (\boldsymbol{\pi}^{(p)}, \boldsymbol{\phi}^{(p)})$. Then, by taking the expectation of (20.113) under the conditional probability $p(\boldsymbol{s}|\boldsymbol{y}; \boldsymbol{\theta}^{(p)})$, we find that the conditional probability of $\boldsymbol{S}$, given $\boldsymbol{y}$ and $\boldsymbol{\theta}^{(p)}$, is

$$P[\boldsymbol{S}|\boldsymbol{y}; \boldsymbol{\theta}^{(p)}] = \prod_{t=0}^{T} \gamma_t^{(p)}(s_t|y_t),$$

where

$$\gamma_t^{(p)}(s_t|y_t) \triangleq P[S_t|y_t; \boldsymbol{\theta}^{(p)}].$$

We find the auxiliary function of (19.24) as

$$Q(\boldsymbol{\theta}|\boldsymbol{\theta}^{(p)}) = Q_1(\boldsymbol{\pi}|\boldsymbol{\theta}^{(p)}) + Q_2(\boldsymbol{\phi}|\boldsymbol{\theta}^{(p)}), \tag{20.114}$$

where

$$\begin{aligned} Q_1(\boldsymbol{\pi}|\boldsymbol{\theta}^{(p)}) &= \sum_{t=0}^{T} \sum_{s_t \in \mathcal{S}} \gamma_t^{(p)}(s_t|y_t) \log \pi_{s_t} \\ &= \sum_{i \in \mathcal{S}} \left\{ \sum_{t=0}^{T} \gamma_t^{(p)}(i|y_t) \right\} \log \pi_i \end{aligned} \tag{20.115}$$

and

$$\begin{aligned} Q_2(\boldsymbol{\phi}|\boldsymbol{\theta}^{(p)}) &= \sum_{t=0}^{T} \sum_{s_t \in \mathcal{S}} \gamma_t^{(p)}(s_t|y_t) \log f_{s_t}(y_t; \phi_{z_t}) \\ &= \sum_{i \in \mathcal{S}} \sum_{t=0}^{T} \gamma_t^{(p)}(i|y_t) \log f_i(y_t; \phi_i). \end{aligned} \tag{20.116}$$

We can maximize the Q_1 and Q_2 of the above expression separately. The condition to maximize Q_1 can be found by using once again the *log-sum inequality* of (10.21):

$$\pi_i = \frac{\sum_{t=0}^{T} \gamma_t^{(p)}(i|y_t)}{\sum_{i\in\mathcal{S}}\sum_{t=0}^{T} \gamma_t^{(p)}(i|y_t)}, \; i \in \mathcal{S}. \tag{20.117}$$

Noting that the denominator of the right-hand side is equal to $T+1$, we find that the M-step to update an estimate of probability π_i should be

$$\pi_i^{(p+1)} = \frac{1}{T+1}\sum_{t=0}^{T} \gamma_t^{(p)}(i|y_t), \; i \in \mathcal{S}. \tag{20.118}$$

The conditional probability in the above equation can be computed, using Bayes' formula, as

$$\gamma_t^{(p)}(i|y_t) = \frac{\pi_i^{(p)} f_i(y_t; \boldsymbol{\phi}_i^{(p)})}{f_Y(y_t; \boldsymbol{\theta}^{(p)})}, \tag{20.119}$$

where $f_Y(y_t; \boldsymbol{\theta}^{(p)})$ is given by setting $\boldsymbol{\theta} = \boldsymbol{\theta}^{(p)}$ in (20.108). Thus, we have

$$\boxed{\pi_i^{(p+1)} = \frac{\pi_i^{(p)}}{T+1}\sum_{t=0}^{T} \frac{f_i(y_t; \boldsymbol{\phi}_i^{(p)})}{f_Y(y_t; \boldsymbol{\theta}^{(p)})}.} \tag{20.120}$$

In order to find the parameters ϕ_i that maximize Q_2, we write from (20.116) and (20.119)

$$Q_2(\boldsymbol{\phi}|\boldsymbol{\theta}^{(p)}) = \sum_{i\in\mathcal{S}}\sum_{t=0}^{T} \frac{\pi_i^{(p)} f_i(y_t; \boldsymbol{\phi}_i^{(p)})}{f_Y(y_t; \boldsymbol{\theta}^{(p)})} \log f_i(y_t; \phi_i). \tag{20.121}$$

Since ϕ_i's are independent, we can maximize each summand of the sum over $i \in \mathcal{S}$ independently. Thus, a general solution for the M-step for the parameter $\boldsymbol{\phi}_i$ is given by

$$\boxed{\boldsymbol{\phi}_i^{(p+1)} = \arg_{\boldsymbol{\phi}_i} \max \sum_{t=0}^{T} \frac{\pi_i^{(p)} f_i(y_t; \boldsymbol{\phi}_i^{(p)})}{f_Y(y_t; \boldsymbol{\theta}^{(p)})} \log f_i(y_t; \phi_i).} \tag{20.122}$$

Analytic expressions for $\boldsymbol{\phi}_i^{(p+1)}$ may be obtained, however, if we assume the exponential family distribution for the PDFs $f_i(y; \phi)$, $i \in \mathcal{S}$, such as in an often-assumed mixture of Gaussian distributions (see Problems 20.28 and 20.29; see also Section 21.3.2).

20.8 Summary of Chapter 20

State space of S_t:	$\mathcal{S} = \{0, 1, \ldots, M-1\}$	(20.1)
Alphabet of output y_t	$\mathcal{Y} = \{0, 1, 2, \ldots, K-1\}$	(20.9)
Definition of HMM:	$p(s_t, y_t \mid s_{t-1}, y_{t-1}) = p(s_t, y_t \mid s_{t-1})$	(20.13)
State-based HMM:	$p(y_t \mid s_{t-1}, s_t) = p(y_t \mid s_t)$.	(20.24)
Transition-based HMM:	$p(y_t \mid s_{t-1}, s_t) \neq p(y_t \mid s_t)$	Def. 20.3
HMM parameter:	$c(i; j, k) = P[S_t = j, Y_t = k \mid S_{t-1} = i]$	(20.15)
Likelihood function:	$L_{\boldsymbol{y}}(\boldsymbol{\theta}) = p(\boldsymbol{y}; \boldsymbol{\theta})$	(20.47)
Matrix rep. of $L_{\boldsymbol{y}}(\boldsymbol{C})$:	$L_{\boldsymbol{y}}(\boldsymbol{C}) = \boldsymbol{\alpha}_0(y_0)^\top \left(\prod_{t=1}^T \boldsymbol{C}(y_t)\right) \mathbf{1}$	(20.52)
Forward variable:	$\alpha_t(j, \boldsymbol{y}_0^t) = P[S_t = j, \boldsymbol{Y}_0^t = \boldsymbol{y}_0^t; \boldsymbol{C}]$	(20.54)
Forward recursion:	$\boldsymbol{\alpha}_t(\boldsymbol{y}_0^t)^\top = \boldsymbol{\alpha}_{t-1}(\boldsymbol{y}_0^{t-1})^\top \boldsymbol{C}(y_t)$	(20.57)
Forward evaluation of $L_{\boldsymbol{y}}(\boldsymbol{C})$:	$L_{\boldsymbol{y}}(\boldsymbol{C}) = \boldsymbol{\alpha}_T(\boldsymbol{y}_0^T)^\top \mathbf{1} = \sum_{i \in \mathcal{S}} \alpha_T(i, \boldsymbol{y}_0^T)$	(20.58)
Backward variable:	$\beta_t(i; \boldsymbol{y}_{t+1}^T) = p(\boldsymbol{y}_{t+1}^T \mid S_t = i, \boldsymbol{C})$	(20.60)
Backward recursion:	$\boldsymbol{\beta}_t(\boldsymbol{y}_{t+1}^T) = \boldsymbol{C}(y_{t+1}) \boldsymbol{\beta}_{t+1}(\boldsymbol{y}_{t+2}^T)$	(20.62)
Forward–backward evaluation:	$L_{\boldsymbol{y}}(\boldsymbol{C}) = \boldsymbol{\alpha}_t(\boldsymbol{y}_0^t)^\top \boldsymbol{\beta}_t(\boldsymbol{y}_{t+1}^T)$	(20.65)
Forward variable for MAP sequence estimation:	$\tilde{\alpha}_t(j, \boldsymbol{y}_0^t) \triangleq \max_{s_0^{t-1}} P[\boldsymbol{S}_0^{t-1} = \boldsymbol{s}_0^{t-1}, S_t = j, \boldsymbol{Y}_0^t = \boldsymbol{y}_0^t]$	(20.69)
Viterbi algorithm:	$\tilde{\alpha}_t(j, \boldsymbol{y}_0^t) = \max_{i \in \mathcal{S}} \left\{\tilde{\alpha}_{t-1}(i, \boldsymbol{y}_0^{t-1}) c(i; j, y_t)\right\}$	(20.71)
APP of $S_t = i$, given $\boldsymbol{y}$:	$\gamma_t(i \mid \boldsymbol{y}) = \frac{\alpha_t(i, \boldsymbol{y}_0^t)\beta_t(i; \boldsymbol{y}_{t+1}^T)}{p(\boldsymbol{y})}$	(20.85)
MAP estimate (BCJR):	$\hat{s}_t = \arg\max_{i \in \mathcal{S}} \{\alpha_t(i, \boldsymbol{y}_0^t)\beta_t(i; \boldsymbol{y}_{t+1}^T)\}$	(20.86)
Q-function for HMM:	$Q(\boldsymbol{\theta} \mid \boldsymbol{\theta}^{(p)}) = E\left[\log p(\boldsymbol{S}, \boldsymbol{y}; \boldsymbol{\theta}) \mid \boldsymbol{y}; \boldsymbol{\theta}^{(p)}\right] = \sum_{\boldsymbol{s}} p(\boldsymbol{s} \mid \boldsymbol{y}; \boldsymbol{\theta}^{(p)}) \log p(\boldsymbol{s}, \boldsymbol{y}; \boldsymbol{\theta})]$	(20.87)
EM algorithm for HMM:	$c^{(p+1)}(i; j, k) = \frac{\sum_{t=1: y_t = k}^T \xi_{t-1}^{(p)}(i, j \mid \boldsymbol{y})}{\sum_{t=1}^T \gamma_{t-1}^{(p)}(i \mid \boldsymbol{y})}$	(20.97)
where	$\xi_t^{(p)}(i, j \mid \boldsymbol{y}) = \frac{\alpha_t^{(p)}(i, \boldsymbol{y}_0^t) c^{(p)}(i; j, y_{t+1}) \beta_{t+1}^{(p)}(j; \boldsymbol{y}_{t+2}^T)}{L_{\boldsymbol{y}}(\boldsymbol{C}^{(p)})}$	(20.137)
	$\gamma_t^{(p)}(i \mid \boldsymbol{y}) = \frac{\alpha_t^{(p)}(i, \boldsymbol{y}_0^t) \beta_t^{(p)}(i; \boldsymbol{y}_{t+1}^T)}{L_{\boldsymbol{y}}(\boldsymbol{C}^{(p)})}$	(20.85)
Baum–Welch algorithm:	$\pi_0^{(p+1)}(i) = \gamma_0^{(p)}(i \mid \boldsymbol{y})$	(20.105)
	$a^{(p+1)}(i; j) = \frac{\sum_{t=1}^T \xi_{t-1}^{(p)}(i, j \mid \boldsymbol{y})}{\sum_{t=1}^T \gamma_{t-1}^{(p)}(i \mid \boldsymbol{y})}$	(20.105)
	$b^{(p+1)}(j; k) = \frac{\sum_{t \in \mathcal{T}} \gamma_t^{(p)}(j \mid \boldsymbol{y}) b^{(p)}(j; y_t) \delta_{y_t, k}}{\sum_{t \in \mathcal{T}} \gamma_t^{(p)}(j \mid \boldsymbol{y}) b^{(p)}(j; k)}$	(20.105)

20.9 Discussion and further reading

It is difficult to trace back the history of HMMs, because similar concepts have been developed in diverse fields; hence, the different nomenclatures: e.g., a *stochastic sequential machine*, a *Markov function*, a *probabilistic automata*, a *state-space system*, and a *partially observable Markov process*. Markov considers, in his 1912 paper [239], *unobserved events* A, B, and C representing a (Markov) chain, and events E and F occurring associated with them with different probabilities. Romanovsky [284, 285] generalizes Markov's results to a chain with n states. Shannon introduced in his seminal paper [300] a finite-state channel, which is an HMM. Burke and Rosenblatt [43] studied the problem of HMM state reduction. Baum and coworkers [15, 16] used the term "probabilistic functions of Markov chains."

As we shall discuss in Chapter 21, the forward recursion algorithm discussed in Section 20.3.2 is a simplest example of the sum-product algorithm or Pearl's belief propagation algorithm discussed in Section 21.5 for the Bayesian network, which can be viewed as a generalization of the HMM.

The Viterbi algorithm [339, 340], originally devised by A. Viterbi in 1967 as an optimal decoding scheme for convolutional codes, is perhaps the most widely practiced algorithm in HMM applications. See Kobayashi [200] and Forney [109] for its early applications to digital communication channels with intersymbol interference, and Kobayashi [198, 199] for its application to high-density digital recording. See also Forney [108], and Hayes *et al.* [148], and Poor [272, 273] on the Viterbi and related algorithms.

The BCJR algorithm was proposed [11] as an optimal scheme for the *minimum symbol error-rate decoding* of convolutional and linear codes. Chang and Hancock [49] derived an algorithm equivalent to the BCJR algorithm in the context of optimal reception of a signal over a channel with intersymbol interference (ISI). The BCJR algorithm, however, requires substantially more computations than the Viterbi algorithm, because the matrix multiplications in (20.57) and (20.62) require significantly more computations than the max operation of (20.73), yet the performance gain of the BCJR algorithm over the Viterbi algorithm may be insignificant. Thus, it had not been used in practice until its utility was found in the mid 1990s when it was found it could constitute an important part of an iterative decoding scheme known as the *Turbo decoding* algorithm [22]. See also Hagenauer *et al.* [138].

Application of HMM to speech recognition has been explored by Jelinek and coworkers at IBM Research since the early 1970s [163–165]. Rabiner and coworkers [274–276] at Bell Telephone Laboratories, who also worked on speech recognition, seem to have popularized the term "hidden Markov model." Ephraim [92] discusses the HMM in the context of statistical-model-based speech-enhancement systems. The HMM formulation and the Viterbi algorithm have also been practiced in computational biology; e.g., DNA sequence analysis (e.g., see Durbin *et al.* [85]) and protein sequencing (e.g., see Kumar and Cowen [216] and reference therein). See also Brown *et al.* [39].

The Viterbi algorithm and the BCJR algorithm differ in a fundamental manner. The former computes an MLE of a state sequence, which is a *point estimate* in the state space S^T, whereas the latter computes the *posterior probability* for each message bit. The real value of the BCJR algorithm is that the posterior probability can be used as a part of Turbo decoding in an iterative fashion. Turbo decoding can be viewed as a *Bayesian learning* procedure; i.e., an instance of Pearl's *belief propagation algorithm* or the *sum-product algorithm* discussed in Section 21.6.1 (see also [245]).

The BCJR algorithm can also be viewed as an extension of *smoothing*, which will be discussed in the context of Wiener filtering in Section 22.2, in the sense that it provides an optimal estimate of s_t on the basis of the observations y_0^T received prior to and after time t. In Wiener filtering, Y_t and S_t are restricted to the case where $Y_t = S_t + N_t$, and both S_t and N_t are WSS processes. As we shall discuss in Section 22.5, a continuous-time algorithm based on forward-filtering and backward smoothing, similar to the BCJR algorithm, was devised by Rauch–Tung–Striebel (RTS) [279]. See also Cappé *et al.* [46] on this and other inference problems in HMMs.

As for model parameter estimation for the HMM, the Baum–Welch algorithm [16, 346] is perhaps the best known and most frequently used algorithm. Turin [332] provides a comprehensive treatment of HMMs as applied to digital communication systems. Ephraim and Merhav [93] give an extensive review of HMM theory and related literature. Kschischang *et al.* [213] show that the forward–backward algorithm (FBA) is a sum-product algorithm and they relate the algorithm to the Tanner graph and other factor graphs. See Section 21.6.1 for more on these subjects.

The EM algorithm and its variants have also been applied to the problem of fitting Markovian arrival processes, Markov modulated Poisson process (MMPP) (see Problems 16.11 and 20.30) and other phase-type distributions (e.g., see Asmussen *et al.* [7], Rydén [291], Breuer [38], and Roberts *et al.* [281]). See also Problems 20.30–20.32 of this chapter. The aforementioned book by Turin [332] discusses the problem of fitting phase-type distributions using the EM algorithm.

In the discrete-time HMM discussed in this chapter, we assumed that the *duration* of any state is either constant (i.e., the unit time in a discrete-time model) or geometrically distributed. If we allow the state duration to have a general distribution, such a model is referred to as an explicit-duration HMM or a hidden semi-Markov model (HSMM) [224, 250]. Ferguson [103] seems to be the first to investigate an estimation algorithm for the HSMM. Yu and Kobayashi [366, 367] discuss an HSMM with missing data and multiple observation sequences, and an efficient FBA. See also Yu [365] for an excellent survey on the HSMM and its applications.

20.10 Problems

Section 20.2: Formulation of a hidden Markov model

20.1* Observable process $Y(t)$. Show that the observable process defined by (20.12) is not a simple Markov chain.

20.2 An HMM representation of the convolutional encoder with a state-based output model. Apply the state-based output model to the convolutional encoder discussed in Example 20.1. How do you define an HMM in this case? What is the state space? Draw the state-transition diagram and the trellis diagram. What is the model parameter $\boldsymbol{\theta}$ in this model?

20.3 S_t of the convolutional encoder is a Markov chain. Show that the state sequence S_t defined by (20.37) forms a Markov chain when the input variables I_t are independent for different t.

20.4 Equilibrium state distribution and mean state sojourn time. Derive the equilibrium state distribution (20.44) and the mean state sojourn times (20.45) in the Gilbert–Elliott channel model.

20.5 A convolutional encoder and an AWGN channel. Assume that the encoded sequence is sent as a bipolar signal over a channel where noise N_t is added. When N_t are i.i.d. RVs drawn from $N(0, \sigma^2)$, such noise is called *white Gaussian noise* and such a channel is referred to as an *additive white Gaussian noise (AWGN) channel.*

The signal amplitude is $+A$ or $-A$ depending on a binary bit in the encoded sequence is "**1**" or "**0**." Find the model parameter $\boldsymbol{\theta}$ of this HMM.

20.6* Partial-response channel. A partial-response (PR) channel [204, 214] representation plays an important role in digital recording as well as in digital communications [198]. An example of a PR channel is one that transforms a binary sequence into a ternary sequence $\{X_t\}$, where

$$X_t = A(I_t - I_{t-1}), \text{ with } I_t, I_{t-1} \in \{0, 1\}, t = 0, 1, 2, \ldots, \tag{20.123}$$

where A is the amplitude of the signal X_t. We assume the initial condition $I_{-1} = 0$. It is then apparent that $\{X_t; t \geq 1\}$ is a sequence of $\{+A, 0, -A\}$.

Because of noise, an observable signal is a *continuous RV* Y_t:

$$Y_t = X_t + N_t, \tag{20.124}$$

where we assume that N_t is *white Gaussian noise* with zero mean and variance σ^2.

(a) Represent this PR channel by a transition-based output HMM.
(b) Do the same using a state-based HMM.

Section 20.3: Evaluation of a hidden Markov model

20.7* Likelihood function as a sum of products. Show that, for the state-based output model, the likelihood function is also expressed as a sum of products, similar to (20.50).

20.8 Derivation of the product forms.

(a) Derive (20.49) for the HMM (S_t, Y_t).
(b) Show that the process $\boldsymbol{X}_0^t = (\boldsymbol{S}_0^t, \boldsymbol{Y}_0^t)$ satisfies

$$p(\boldsymbol{s}_0^t, \boldsymbol{y}_0^t; \boldsymbol{\theta}) = p(\boldsymbol{s}_0^{t-1}, \boldsymbol{y}_0^{t-1}; \boldsymbol{\theta}) c(s_{t-1}; s_t, y_t). \tag{20.125}$$

20.9* Forward recursion formula when Y_t is a continuous RV. Suppose Y_t is a continuous RV. Then, how do you define the functions $c(i; j, y_t)$? Will this affect the forward recursion algorithm?

20.10 Backward recursion formula. Derive the backward recursion formula (20.62).

20.11 Backward algorithm. Write the backward recursion formula as a program similar to Algorithm 20.1.

20.12 Forward–backward formula. Derive the formula (20.64).

20.13 Forward–backward algorithm. Write the forward–backward recursion formula as a program similar to Algorithm 20.1 and the answer to Problem 20.11.

Section 20.4: Estimation algorithms for state sequence

20.14* The Viterbi algorithm. Derive the recursion formula (20.71).

20.15 The Viterbi algorithm for a convolutional encoded sequence sent over a BSC. Consider the communication system discussed in Example 20.1 of Section 20.2.2. For a given error rate ϵ of the BSC, discuss how the Viterbi algorithm works to obtain an MLE of the information sequence I_t based on the observation y_0^T at the output of the BSC channel.

20.16 MAP state estimation versus maximum-likelihood sequence estimation. We showed in the text that when prior probabilities $\pi(s)$ of possible state sequences are equal, the MAP state estimation algorithm can be used to find an MLSE. Let us assume that the prior probabilities $\pi(s)$ are not equal. Then the MLSE is clearly different from the MAP sequence estimation algorithm. Show that the auxiliary variable (20.69) should be modified as

$$\tilde{\alpha}_t(j, y_0^t) \triangleq \max_{s_0^{t-1}} P[Y_0^t = y_0^t | S_0^{t-1} = s_0^{t-1} S_t = j] \tag{20.126}$$

and (20.71) should be modified to

$$\tilde{\alpha}_t(j, y_0^t) = \max_{i \in \mathcal{S}} \{\tilde{\alpha}_{t-1}(i, y_0^{t-1}) \tilde{b}(i; j, y_t)\}, j \in \mathcal{S}, 1 \le t \le T, \tag{20.127}$$

with the initial value

$$\tilde{\alpha}_0(i, y_0) = b(i, y_0), i \in \mathcal{S}, \tag{20.128}$$

where $b(i, y_0)$ and $\tilde{b}(i, j; y_t)$ are defined by (20.27) and (20.28), respectively.

20.17 The Viterbi algorithm for an AWGN channel. Consider a communication channel where an information sequence $\{I_t\}$ is transformed into a Markov chain $\{S_t\}$ in some fashion. We assume that this transformation is invertible in the sense that, once we recover the state sequence $\{S_t\}$, we can reconstruct the information sequence $\{I_t\}$.

The transmitter sends a signal X_t which is uniquely determined by the state transition $S_{t-1} = i \to S_t = j$, which we denote as

$$X_t = X_t(i, j).$$

The communication channel is an AWGN channel:

$$Y_t = X_t(i, j) + N_t,$$

where $N_t \sim N(0, \sigma^2)$.

(a) Find $f_{Y_t|S_{t-1}S_t}(y_t|i, j)$ and $c(i; j, y_t)$ asked in Problem 20.9.
(b) Show that the recursion formula (20.71) can be replaced by the following equation for a new forward variable $\breve{\alpha}_t(j, \boldsymbol{y}_0^t)$, $j \in \mathcal{S}$, $t = 1, 2, \ldots, T$:

$$\breve{\alpha}_t(j, \boldsymbol{y}_0^t) = \min_{i \in \mathcal{S}} \left\{ \breve{\alpha}_{t-1}(i, \boldsymbol{y}_0^{t-1}) + [y_t - x_t(i, j)]^2 - 2\sigma^2 \ln a(i, j) \right\}, \quad (20.129)$$

with the initial condition

$$\breve{\alpha}_0(j, y_0) = \begin{cases} 0, & \text{for } j = s_0^*, \\ \infty, & \text{for } j \neq s_0^*. \end{cases}$$

20.18* The Viterbi algorithm for a partial-response channel [199, 200]. Consider the partial-response channel of Problem 20.6. The state transition probability matrix $\boldsymbol{A} = [a(i, j)]$ is given by

$$\boldsymbol{A} = \left[a(i, j)\right] = \begin{bmatrix} 1/2 & 1/2 \\ 1/2 & 1/2 \end{bmatrix}. \quad (20.130)$$

Assume that the information sequence $I_t (= S_t)$ is an i.i.d. binary sequence. Assume that initially the transmitter state is set to $S_0 = 0$.

(a) Show that the recursion equation (20.129) is further simplified to

$$\vec{\alpha}_t(j, \boldsymbol{y}_0^t) = \max_{i \in \mathcal{S}} \left\{ \vec{\alpha}_{t-1}(j, \boldsymbol{y}_0^{t-1}) + y_t x_t(i, j) - \frac{x_t^2(i, j)}{2} \right\}, \quad (20.131)$$

where

$$\vec{\alpha}_0(i, y_0) = \begin{cases} 0, & \text{for } i = 0, \\ -\infty, & \text{for } i = 1. \end{cases}$$

(b) Further simplify the above recursion formula by noting that $x(0, 0) = x(1, 1) = 0$, $x(0, 1) = +A$ and $x(1, 0) = -A$.

20.19 Backward version of the Viterbi algorithm and its program.

(a) Show that a backward version of the auxiliary variable for the Viterbi algorithm is given by

$$\tilde{\beta}_t(i; \boldsymbol{y}_{t+1}^T) = \max_{\boldsymbol{s}_{t+1}^T} P[\boldsymbol{S}_{t+1}^T = \boldsymbol{s}_{t+1}^T, \boldsymbol{Y}_{t+1}^T = \boldsymbol{y}_{t+1}^T | S_t = i]. \quad (20.132)$$

(b) Show that the variable $\tilde{\beta}_t(i; \mathbf{y}_{t+1}^T)$ satisfies the following recursion:

$$\tilde{\beta}_t(i; \mathbf{y}_{t+1}^T) = \max_{j \in \mathcal{S}}\{c(i; j, y_{t+1})\tilde{\beta}_{t+1}(j; \mathbf{y}_{t+2}^T)\}. \tag{20.133}$$

(c) State the backward version of the Viterbi algorithm. Assume that the final state s_T is known; i.e., $S_T = s_T$ with probability one.

(d) Write an algorithmic specification similar to Algorithm 20.2 for the backward version of the Viterbi algorithm.

20.20 Backward Viterbi algorithms for an AWGN channel and the partial-response channel.

(a) Show that, for an AWGN channel, the backward recursion (20.133) is replaced by the following recursion:

$$\check{\beta}_t(i; \mathbf{y}_{t+1}^T) = \min_{j \in \mathcal{S}}\left\{\check{\beta}_{t+1}(j; \mathbf{y}_{t+2}^T) + [y_{t+1} - x_{t+1}(i, j)]^2 - 2\sigma^2 \ln a(i, j)\right\}, \tag{20.134}$$

with the initial condition

$$\check{\beta}_T(j, \emptyset) = 0 \text{ for all } j \in \mathcal{S}.$$

Note that if the terminal state s_T is known to the receiver, the above initial condition should be replaced by

$$\check{\beta}_T(j, \emptyset) = \begin{cases} 0, & \text{for } j = s_T, \\ \infty, & \text{for } j \neq s_T. \end{cases}$$

(b) Consider the partial-response channel discussed in Problems 20.6 and 20.18. Assume that the terminal state is known to be zero; i.e., $S_T = 0$. Show that the above recursion (20.134) can be replaced by

$$\vec{\beta}_t(i; \mathbf{y}_{t+1}^T) = \max_{j \in \mathcal{S}}\left\{\vec{\beta}_{t+1}(j; \mathbf{y}_{t+2}^T) + y_{t+1}x_{t+1}(i, j) - \frac{x_{t+1}^2(i, j)}{2}\right\}, \tag{20.135}$$

where

$$\vec{\beta}_T(j; \emptyset) = \begin{cases} 0, & \text{for } j = 0, \\ -\infty, & \text{for } j = 1. \end{cases}$$

Further simplify the above recursion formula by noting that $x(0, 0) = x(1, 1) = 0, x(0, 1) = +A$ and $x(1, 0) = -A$.

20.21 Forward and backward variables of the Viterbi algorithm. Show that the auxiliary variables $\tilde{\alpha}_t(j, \mathbf{y}_0^t)$ of (20.69) and $\tilde{\beta}_t(j; \mathbf{y}_{t+1}^T)$ of (20.132) satisfy

$$\tilde{\alpha}_t(j, \mathbf{y}_0^T)\tilde{\beta}_t(j; \mathbf{y}_{t+1}^T) = \max_{\mathbf{s}_0^{t-1}, \mathbf{s}_{t+1}^T} P[\mathbf{S}_0^{t-1} = \mathbf{s}_0^{t-1}, S_t = j, \mathbf{S}_{t+1}^T = \mathbf{s}_{t+1}^T, \mathbf{Y}_0^T = \mathbf{y}_0^T]. \tag{20.136}$$

Section 20.5: The BCJR algorithm

20.22 APP $\xi_t(i, j|y)$ and joint probability $\sigma_t(i; j, y)$.

(a) Consider the APP of observing a transition $S_t = i \to S_{t+1} = j$, given the observation $\boldsymbol{y}$:

$$\xi_t(i, j \,|\boldsymbol{y}) \triangleq P[S_t = i, S_{t+1} = j \,|\boldsymbol{Y}_0^T = \boldsymbol{y}]. \tag{20.137}$$

What does $\sum_{t=0}^{T} \xi_t(i, j \,|\boldsymbol{y})$ represent?

(b) Consider the joint probability

$$\sigma_t(i; j, \boldsymbol{y}) \triangleq P[S_t = i, S_{t+1} = j, \boldsymbol{Y}_0^T = \boldsymbol{y}]. \tag{20.138}$$

Show that

$$\sigma_t(i; j, \boldsymbol{y}) = \alpha_t(i, \boldsymbol{y}_0^t)c(i; j, y_{t+1})\beta_{t+1}(j; \boldsymbol{y}_{t+2}^T). \tag{20.139}$$

(c) Show that

$$\xi_t(i, j \,|\boldsymbol{y}) = \frac{\alpha_t(i, \boldsymbol{y}_0^t)c(i; j, y_{t+1})\beta_{t+1}(j; \boldsymbol{y}_{t+2}^T)}{p(\boldsymbol{y})}. \tag{20.140}$$

20.23 Alternative derivation of the MAP estimate. Obtain an alternative method to find the MAP estimate by focusing on state transitions as follows.

(a) Obtain the following backward recursion formula for the variable $\gamma_t(i|\boldsymbol{y})$ defined in (20.85):

$$\gamma_t(i|\boldsymbol{y}) = \gamma_{t+1}(i|\boldsymbol{y}) + \sum_{j\in\mathcal{S}\setminus i} [\xi_t(i, j \,|\boldsymbol{y}) - \xi_t(j, i \,|\boldsymbol{y})], \tag{20.141}$$

with the initial condition

$$\gamma_T(i|\boldsymbol{y}) = \frac{\alpha_T(i, \boldsymbol{y}_0^T)}{p(\boldsymbol{y})} = \frac{\alpha_T(i, \boldsymbol{y}_0^T)}{\sum_{i\in\mathcal{S}} \alpha_T(i, \boldsymbol{y}_0^T)}, \tag{20.142}$$

where the symbol $\setminus$ denotes the "set difference" operator; i.e., $\mathcal{S} \setminus i$ denotes the set of all elements in $\mathcal{S}$ excluding i.
Hint: Use the following identity:

$$\begin{aligned} P[S_t = i, S_{t+1} = i] &= P[S_t = i] - \sum_{j\in\mathcal{S}\setminus i} P[S_t = i, S_{t+1} = j] \\ &= P[S_{t+1} = i] - \sum_{j\in\mathcal{S}\setminus i} P[S_t = j, S_{t+1} = i]. \end{aligned} \tag{20.143}$$

(b) Show that the MAP estimate can be expressed as

$$\hat{S}_t^* = \arg\max_{i\in\mathcal{S}} \gamma_t(i|\boldsymbol{y}), \text{ for } t = T, T-1, \ldots, 1, 0. \tag{20.144}$$

Section 20.6: Maximum-likelihood estimation of model parameters

20.24 Derivation of (20.85). Define $\xi_t(i, j|\boldsymbol{y})$ by

$$\xi_t(i, j|\boldsymbol{y}) = \frac{\alpha_t(i, \boldsymbol{y}_0^t)c(i; j, y_{t+1})\beta_{t+1}(j; \boldsymbol{y}_{t+2}^T)}{L_{\boldsymbol{y}}(\boldsymbol{\theta})}, i, j \in \mathcal{S}, t \in \mathcal{T}. \quad (20.145)$$

(a) Show that it is the APP of observing a transition $S_t = i \to S_{t+1} = j$, given the observations $\boldsymbol{y}$.

(b) Show that another APP $\gamma_t(i|\boldsymbol{y})$ defined by (20.85) is related to $\xi_t(i, j|\boldsymbol{y})$ by

$$\gamma_t(i|\boldsymbol{y}) = \sum_{j \in \mathcal{S}} \xi_t(i, j|\boldsymbol{y}) = \frac{\alpha_t(i, \boldsymbol{y}_0^t)\beta_t(i; \boldsymbol{y}_{t+1}^T)}{L_{\boldsymbol{y}}(\boldsymbol{\theta})}. \quad (20.146)$$

(c) Show the following identity:

$$\gamma_t(j|\boldsymbol{y}) = \sum_{i \in \mathcal{S}} \xi_{t-1}(i, j|\boldsymbol{y}). \quad (20.147)$$

20.25* Alternative derivation of the FBA for the transition-based HMM. The FBA for the transition-based HMM discussed in this section can be derived in an alternative way as follows.

Define $M(i; j, k)$ as the number of times that $(S_t; S_{t+1}, y_{t+1}) = (i; j, k)$ is found in the sequence $(\boldsymbol{S}_0^T, \boldsymbol{y}_0^T)$. Since $M(i; j, k)$ is a function of $\boldsymbol{S}_0^T$, it is an RV.

(a) Show that

$$Q_0(\boldsymbol{\theta}|\boldsymbol{\theta}^{(p)}) = E\left[\log \alpha_0(S_0, y_0)|\boldsymbol{y}; \boldsymbol{\theta}^{(p)}\right], \quad (20.148)$$

$$Q_1(\boldsymbol{\theta}|\boldsymbol{\theta}^{(p)}) = \sum_{i,j \in \mathcal{S}, k \in \mathcal{Y}} E\left[M(i; j, k)|\boldsymbol{y}, \boldsymbol{\theta}^{(p)}\right] \log c(i; j, k). \quad (20.149)$$

where $Q_i(\boldsymbol{\theta}|\boldsymbol{\theta}^{(p)})$; $i = 0, 1$, correspond to the two terms in (20.90).

(b) Show

$$\begin{aligned}\overline{M}^{(p)}(i; j, k|\boldsymbol{y}) &\triangleq E\left[M(i; j, k)|\boldsymbol{y}, \boldsymbol{\theta}^{(p)}\right] \\ &= \frac{\sum_{t=1}^T \alpha_{t-1}^{(p)}(i, \boldsymbol{y}_0^{t-1})c^{(p)}(i; j, k)\beta_t^{(p)}(j; \boldsymbol{y}_{t+1}^T)\delta_{y_t,k}}{L_{\boldsymbol{y}}(\boldsymbol{\theta}^{(p)})}. \end{aligned} \quad (20.150)$$

Hint: Use equation (20.139) of Problem 20.22.

(c) Show how the maximization step is performed, and verify the solutions obtained in (20.96) and (20.97).

20.26* Alternative derivation of the Baum–Welch algorithm. Derive the Baum–Welch algorithm by using the following expression for $p(\boldsymbol{S}, \boldsymbol{y}; \boldsymbol{\theta})$ in the definition of the auxiliary function $Q(\boldsymbol{\theta}|\boldsymbol{\theta}^{(p)})$ of (20.87):

$$p(\boldsymbol{S}, \boldsymbol{y}; \boldsymbol{\theta}) = p(\boldsymbol{S}|\boldsymbol{\theta})p(\boldsymbol{y}|\boldsymbol{S}; \boldsymbol{\theta}) \quad (20.151)$$

and

$$p(\boldsymbol{S}|\boldsymbol{\theta}) = \pi_0(S_0)\prod_{t=1}^{T} a(S_{t-1}, S_t) = \pi_0(S_0)\prod_{i,j\in\mathcal{S}} a(i,j)^{M(i,j)}, \tag{20.152}$$

$$p(\boldsymbol{y}|\boldsymbol{S};\boldsymbol{\theta}) = \prod_{t=0}^{T} b(S_t; y_t) = \prod_{j\in\mathcal{S},k\in\mathcal{S}} b(j;k)^{N(j,k)}, \tag{20.153}$$

where the RV $M(i, j)$ represents the number of times that $(S_{t-1}, S_t) = (i, j)$ occurs in the state sequence $\boldsymbol{S}_0^T$. Similarly, the RV $N(j, k)$ is the number of times that $(S_t, Y_t) = (j, k)$ appears in the Markov process $(\boldsymbol{S}_0^T, \boldsymbol{Y}_0^T)$.

Section 20.7: Application: parameter estimation in mixture distributions

20.27 Derivation of (20.118). Find the condition (20.118) by applying the Lagrangian method to Q_1 of (20.115).

20.28 Mixture of Gaussian distribution. In the example of Section 20.7, assume that the M distributions are all Gaussian distributions $N(\mu_i, \sigma_i^2)$, $i = 0, 1, \ldots, M-1$. Find the EM algorithm in a form to obtain an MLE of the model parameters μ_i, σ_i as well as π_i, the probability that the ith distribution is chosen, $i = 0, 1, \ldots, M-1$. Also state the algorithm similar to Algorithm 20.5.

20.29 Mixture of exponential family distribution. Show that the updating formula for the parameter $\boldsymbol{\phi}_i$ can take a simpler form if the PDFs $f_i(\boldsymbol{y}; \boldsymbol{\phi}_i)$ are all from the canonical exponential family distributions. Verify this result in the case of Problem 20.28.

20.30 Markov modulated Poisson process and Markov modulated Poisson sequence. Consider a CTMC $S(t)$; $0 \le t < \infty$ with the state space $\mathcal{S} = \{0, 1, \ldots, M-1\}$, and its infinitesimal generator $\boldsymbol{Q}$ (16.22) given by

$$\boldsymbol{Q} = \left[Q_{ij}\right], i, j \in \mathcal{S}. \tag{20.154}$$

Associated with the process $S(t)$ is a Poisson counting process $N(t)$, whose rate depends on the state $\mu_{S(t)}$. In other words,

$$P[S(t+h) = j, N(t+h) = n \mid S(t) = i, N(t) = m]$$
$$= \begin{cases} Q_{ij}h(1 - \mu_i h + o(h)), & \text{if } n = m, \\ Q_{ij}h(\mu_i h + o(h)), & \text{if } n = m+1, \\ o(h), & \text{if } n \ge m+2. \end{cases} \tag{20.155}$$

The process $(S(t), N(t))$ is called **Markov modulated Poisson process (MMPP)** (see also Problem 16.11). If the CTMC is hidden and only $N(t)$ is observable, then the process $(S(t), N(t))$ is an HMM.

Let us divide the time axis into disjoint and contiguous subintervals of length Δ and transform the process $S(t)$ into a discrete-time state sequence S_k, $k = 0, 1, 2, \ldots,$ such that

$$S_k = S(k\Delta), k = 0, 1, 2, \ldots \tag{20.156}$$

We define the observation sequence Y_k as the increment of the counting process $N(t)$ in the interval $(k\Delta, (k+1)\Delta]$; i.e.,

$$Y_k = N((k+1)\Delta) - N(k\Delta), k = 0, 1, 2, \ldots \tag{20.157}$$

(a) Find an expression for the transition probability matrix $\boldsymbol{A}$ of its discrete-time counterpart, which we term a **Markov modulated Poisson sequence (MMPS)**.
(b) Find the MMPS parameter $\boldsymbol{\lambda} = [\lambda_i; i \in \mathcal{S}]$ in terms of the time-continuous model parameters.

20.31 Traffic modeling based on MMPS. Consider the problem of modeling packet traffic in terms of an MMPS defined in Problem 20.30, in which we denote a hidden Markov chain as S_k; $k = 0, 1, 2, \ldots, S_k \in \mathcal{S} \triangleq \{0, 1, 2, \ldots, M-1\}$. The packet arrival process Y_k is observable and is known to be Poisson distributed with mean λ_i, when the underlying hidden Markov process is in state i; i.e.,

$$P[Y_k = n | S_k = i] = \frac{\lambda_i^n}{n!} e^{-\lambda_i}, n = 0, 1, 2, \ldots, i \in \mathcal{S}. \tag{20.158}$$

The model parameter we want to estimate are

$$\boldsymbol{\theta} = (\boldsymbol{\pi}_0, \boldsymbol{A}, \boldsymbol{\lambda}), \tag{20.159}$$

where $\boldsymbol{\pi}_0 = (\pi_0(i); i \in \mathcal{S})$ is the initial state probability vector; $\boldsymbol{A} = [a(i, j)], i, j \in \mathcal{S}$, is the state transition matrix of the hidden Markov chain S_k and

$$\boldsymbol{\lambda} = \{\lambda_i; i \in \mathcal{S}\} \triangleq (\lambda_0, \lambda_1, \ldots, \lambda_{M-1}). \tag{20.160}$$

Derive a Baum–Welch-type algorithm to obtain an MLE $\hat{\boldsymbol{\theta}}$ of the parameter $\boldsymbol{\theta}$ of (20.159).

20.32 Markov modulated Bernoulli sequence (MMBS). Consider the following variant of the MMPS model discussed in Problem 20.30. At a given discrete-time, the number of packets that arrive in the tth interval is at most one; i.e.,

$$P[Y_t = n | S_t = i] = \begin{cases} 1 - b_i, & \text{for } n = 0, \\ b_i, & \text{for } n = 1, \\ 0, & \text{for } n \geq 2, \end{cases} \tag{20.161}$$

where $0 \leq b_i \leq 1$ for all $i \in \mathcal{S} = \{0, 1, 2 \ldots, M-1\}$. The model parameter we need to specify and estimate is

$$\boldsymbol{\theta} = (\boldsymbol{\pi}_0, \boldsymbol{A}, \boldsymbol{b}),$$

where $\boldsymbol{b} = [b_0, b_1, \ldots, b_{M-1}]$.

Find the re-estimation formula for the parameter $\boldsymbol{\theta} = (\boldsymbol{\pi}_0, \boldsymbol{A}, \boldsymbol{b})$.

21 Probabilistic models in machine learning

21.1 Introduction

Machine learning refers to the design of computer algorithms for gaining new knowledge, improving existing knowledge, and making predictions or decisions based on empirical data. Applications of machine learning include speech recognition [164, 275], image recognition [60, 110], medical diagnosis [309], language understanding [50], biological sequence analysis [85], and many other fields. The most important requirement for an algorithm in machine learning is its ability to make accurate predictions or correct decisions when presented with instances or data not seen before.

Classification of data is a common task in machine learning. It consists of finding a function $z = G(\boldsymbol{y})$ that assigns to each data sample $\boldsymbol{y}$ its *class label* z. If the range of the function is discrete, it is called a *classifier*, otherwise it is called a *regression* function. For each class label z, we can define the acceptance region A_z such that $\boldsymbol{y} \in A_z$ if and only if $z = G(\boldsymbol{y})$. An error occurs if the classifier assigns a wrong class to $\boldsymbol{y}$. The probability of classification error

$$\mathcal{E}(G) = P[Z \neq G(\boldsymbol{Y})] \tag{21.1}$$

is called the *generalization error* in machine learning, where Z denotes the actual class to which the observation variable $\boldsymbol{Y}$ belongs. The classifier that minimizes the generalization error is called the *Bayes classifier* and the minimized $\mathcal{E}(G)$ is called the *Bayes error.* In practical applications, we generally do not know the probability distribution of $(\boldsymbol{Y}, Z)$. Hence, we cannot find an analytic expression for the Bayes classifier. We can, however, construct an empirical solution for the classifier using a statistical model for the probability distribution. One successful approach to machine learning is what is termed *statistical learning*, which consists of three steps: (i) observe instances $(\boldsymbol{y}_1, z_1), (\boldsymbol{y}_2, z_2), \ldots, (\boldsymbol{y}_n, z_n)$ of training data with correct labels;[1] (ii) construct a probabilistic (or statistical) model from a set of observed data, a step that is called *training*; and (iii) make predictions or decisions, using the model.

One important requirement for a model is its agreement with observed data. But, given data, it is always possible to construct a model that fits the data exactly, but such a model may not serve a useful purpose in predicting or classifying future instances.

[1] Learning using labeled training data is called *supervised learning* or *learning with a teacher.*

Generality of the model is more important. The model should be a particular case of a broad family of models. The menu of models should be rich enough to enable us to choose one that agrees with the data (i.e., extract sufficient knowledge from the data) so that the model application will give credible results.

In this chapter we will present some important concepts and techniques useful in probabilistic machine learning. They are MAP (maximum *a posteriori* probability) recognition algorithms, clustering algorithms, sequence recognition algorithms, Bayesian networks, factor graphs, and Markov chain Monte Carlo (MCMC) methods. We will illustrate the application of HMMs to several tasks of machine learning (speech recognition, handwriting recognition, and bioinformatics). Our goal is to show similarities among problems in these diverse fields and to discuss statistical methods for their solutions. A Bayesian network (BN) is a generalization of the HMM. A factor graph representation and the sum-product algorithm are useful to find solutions for the BN by augmenting the methods for the HMMs discussed in Chapter 20. The MCMC method is a very powerful tool for Bayesian models in general, including machine learning formulated in the Bayesian framework.

There are many other successful approaches and techniques developed for machine learning that are not discussed in this chapter. *Support vector machines* (SVMs) [338] perform regression and classification by constructing the boundaries of the acceptance region A_z in a multidimensional space. The most popular functions to determine the boundaries include linear, polynomial, and *radial basis functions* (RBFs).[2] They are also useful in regression analysis and time series prediction or forecasting. Other topics in machine learning not covered in this chapter include. (i) *Boosting*, which refers to a class of heuristic algorithms for supervised learning, in which a set of *weak learners* are combined, typically in an iterative manner to form a *strong learner* [242, 293]. (ii) The *decision tree algorithm*, which predicts the value of a target variable based on input variables using a decision tree. Each interior node corresponds to a value of the input variable, and each leaf node represents a value of the target variable given the input variables' instance represented by the path from the root to the leaf. (iii) The *artificial neural network* (ANN) [149] finds a classifier $G(\boldsymbol{y})$ as a composition of interconnecting artificial neurons (programming constructs) that simulate properties of biological neurons [154]. ANNs are trained by minimizing the difference between the observed class z and that of the the network output. ANNs have been applied successfully to speech recognition, image analysis, robotics, and in other fields. Mathematical theory and algorithms used in ANNs include statistical estimation theory, optimization, and control theory. (iv) The *genetic algorithm* (GA) [241], which refers to a class of heuristic algorithms for search, optimization, and machine learning, using techniques inspired by natural evolution, such as inheritance, mutation, selection, and crossover.

Closely related to machine learning is the rapidly developing field of **data mining** [356]. Data mining refers to the process of extracting relevant patterns from current and past data and then analyzing and predicting future trends. It is routinely practiced in a

2 An RBF $f(\boldsymbol{x})$ is a real-valued function whose value depends only on the distance from the origin; i.e., $f(\boldsymbol{x}) = f(\|\boldsymbol{x}\|)$.

wide range of applications, including marketing, surveillance, medicine, genetics, and bioinformatics. Many of the tools and techniques used in data mining are common to those developed for machine learning.

21.2 Maximum *a posteriori* probability (MAP) recognition

Suppose that we want a machine to recognize *handwritten symbols* drawn from an alphabet $\mathcal{A} = \{a_1, a_2, ..., a_m\}$ based on their images. Let $\boldsymbol{Y}$ represent the image of a symbol to be recognized. Then the MAP decision rule (or estimation procedure) (see Section 18.3) corresponds to the following symbol recognition procedure:

$$a^* = \arg\max_{a \in \mathcal{A}} p(a|\boldsymbol{y}), \tag{21.2}$$

where $p(a|\boldsymbol{y})$ is the *posterior* probability of a, given an instance $\boldsymbol{y}$ (i.e., an observed image data) of the RV $\boldsymbol{Y}$. Using Bayes' theorem we can present the previous equation in the equivalent form

$$a^* = \arg\max_{a \in \mathcal{A}} p(a) f(\boldsymbol{y}|a), \tag{21.3}$$

where $p(a)$ is the *prior probability* of the symbol a and $f(\boldsymbol{y}|a)$ is the conditional PDF of the observation variable $\boldsymbol{Y}$ when it comes from symbol $a \in \mathcal{A}$. Needless to say, $f(\boldsymbol{y}|a)$ should be replaced by the conditional PMF $p(\boldsymbol{y}|a)$ when the observation takes on discrete values.

For a given image $\boldsymbol{y}$, the function $L_{\boldsymbol{y}}(a_i) \triangleq f(\boldsymbol{y}|a_i)$ is called the *likelihood function* of a_i (see Section 18.1.2). In order to adopt the MAP recognition rule, which is designed to minimize the overall expected probability of error (see (18.100)), we need to have *probabilistic models*, such as HMMs or Gaussian models, for the observation variable $\boldsymbol{Y}$ so that we can compute the right-hand side of (21.3). The parameters of such models are estimated using training data $\boldsymbol{y}_{i,k} (k = 1, 2, ..., n_i)$, where n_i is the size of the training data for the model of symbol a_i. If the $m(= |\mathcal{A}|)$ symbols appear with equal probability (i.e., $p(a_i) = 1/m$ for all $a_i \in \mathcal{A}$), the above MAP decision rule (21.3) will reduce to the *maximum-likelihood decision* rule.

If we have a sufficient number of training samples for each symbol, we can create the symbol models by fitting the PDFs $f(\boldsymbol{y}|a_i)$ to the training data. For example, we can choose a *Gaussian model*

$$f(\boldsymbol{y}|\boldsymbol{\theta}_i) = \frac{1}{(2\pi)^{k/2}|\boldsymbol{\Sigma}_i|^{1/2}} \exp\left[-\frac{1}{2}(\boldsymbol{y} - \boldsymbol{\mu}_i)^\top \boldsymbol{\Sigma}_i^{-1}(\boldsymbol{y} - \boldsymbol{\mu}_i)\right], i \in \mathcal{M}, \tag{21.4}$$

where $\mathcal{M} = \{1, 2, \ldots, m\}$ is the index set for the alphabet $\mathcal{A} = \{a_i; i \in \mathcal{M}\}$. The model parameters $\boldsymbol{\theta}_i = (\boldsymbol{\mu}_i, \boldsymbol{\Sigma}_i)$ are the mean vector and covariance matrix of the image variable $\boldsymbol{Y}$, when it is generated by symbol a_i. In actual situations, we may not know true parameter values, so we need to obtain their estimates $\hat{\boldsymbol{\theta}}_i$ from the training data.

Once the model parameters are estimated, the MAP recognizer (21.3) should be designed as follows. We partition the space $\mathcal{Y}$ of the observation variable $\boldsymbol{Y}$ into m acceptance regions $\mathcal{Y}_i$ as follows:

$$\mathcal{Y}_i = \{\boldsymbol{y} : p(a_i)f(\boldsymbol{y}|\hat{\boldsymbol{\theta}}_i) > p(a_j)f(\boldsymbol{y}|\hat{\boldsymbol{\theta}}_j) \text{ for all } j \neq i\}, \text{ for } i \in \mathcal{M}. \tag{21.5}$$

If an image $\boldsymbol{y}$ falls into $\mathcal{Y}_i$, we recognize it as a_i. If $\boldsymbol{y}$ falls on the boundary of $n(\leq m)$ regions, we can resolve the tie by a random decision; i.e., we choose one of the n regions with equal probability $1/n$, or, more properly, according to their prior probabilities, i.e., $\propto p(a_i)$. If the observation $\boldsymbol{Y}$ is a continuous RV, such a tie situation occurs with probability zero.

In particular, if we use Gaussian models whose covariance matrices are the identity matrices ($\boldsymbol{\Sigma}_i = \boldsymbol{I}$ for all i) and if the prior probabilities are all equal (i.e., $p(a_i) = 1/m$ for all $a_i \in \mathcal{A}$), then the decision rule (21.5) reduces to the following **minimum distance decision rule**:

$$\mathcal{Y}_i = \{\boldsymbol{y} : \|\boldsymbol{y} - \hat{\boldsymbol{\mu}}_i\| < \|\boldsymbol{y} - \hat{\boldsymbol{\mu}}_j\| \text{ for all } j \neq i\}, i, j \in \mathcal{M}. \tag{21.6}$$

where $\|\boldsymbol{y} - \boldsymbol{\mu}\|$ denotes the Euclidean distance between $\boldsymbol{y}$ and $\boldsymbol{\mu}$. These regions are called *Voronoi regions* and the collection of these regions is called a **Voronoi diagram**. The image $\boldsymbol{y}$ is recognized as symbol a_i if $\boldsymbol{y} \in \mathcal{Y}_i$ (i.e., $\hat{\boldsymbol{\mu}}_i$ is the closest of all the $\hat{\boldsymbol{\mu}}_i$ to $\boldsymbol{y}$).

Now we are ready to evaluate the recognition performance, using test data. The recognizer's performance measure can be the average cost of misclassification:

$$\bar{C} = \sum_{i,j} \frac{c_{i,j} n_{i,j}}{n}, \tag{21.7}$$

where $c_{i,j}$ is the cost or risk associated with misclassifying a_i as a_j (with $c_{i,i} = 0$ for correct classification), n_{ij} is the number of tests in which a_i is classified as a_j, and $n = \sum_{i,j \in \mathcal{M}} n_{i,j}$ is the total number of tests. Note that the average cost $\bar{C}$ of (21.7) is an empirical estimate of the overall **Bayes' risk** (18.91) defined in Section 18.3, because $n_{i,j}/n$ is an empirical estimate of $P[\boldsymbol{y} \in \mathcal{Y}_j | a_i]$. As pointed out in Section 18.3, if we set $c_{i,j} = \delta_{i,j}$ (Kronecker's delta), the average cost $\bar{C}$ reduces to the overall probability of misclassification per test.

The above example illustrates the steps involved in a **typical machine learning procedure**. However, there are many issues related to its practical implementation:

- The scanned images must be preprocessed to remove noise and distortions. The images should be represented by their *attributes* or *feature vectors* to reduce the dimensionality of $\boldsymbol{Y}$. Such data reduction may be achieved, for instance, by using the principal component analysis (PCA) or the singular value decomposition (SVD) techniques discussed in Section 13.3. The selection of the representative attributes is not an easy task. The attributes are often selected by using some background information in combination with trial and error.
- The shapes of the acceptance regions $\mathcal{Y}_i$ should be selected to optimize the performance. A **discriminative learning** approach in which the points close to the boundaries play a more important role than the other points is often more efficient

than other methods. The SVMs use this approach to create special types of separating boundaries between the regions by mapping them into higher dimensional spaces in which they become hyperplanes.

- The dimension of the symbol space can be very large (e.g., if we treat whole words or sentences as symbols), necessitating the use of some sophisticated **parametric models**, such as HMMs.

The *MAP recognition* of handwritten symbol recognition formulated in terms of (21.2) and (21.3) should remain valid for a general MAP recognition problem. The alphabet $\mathcal{A}$ should be interpreted as a set of "symbols," where a symbol may represent a word, sequence, or DNA molecule, depending on applications, and $\boldsymbol{y}$ is an observation, which is called an **evidence** in the machine learning literature.

21.3 Clustering

The type of learning considered in the previous section is called **supervised learning** (or learning with a teacher) because we know the symbol a_i of image data we use for training the machine. However, often we do not know the symbol a_i and we need to obtain this knowledge from the observable data. This type of learning is called **unsupervised learning**. A typical example of unsupervised learning is **data clustering**, which consists of partitioning the training data into nonintersecting subsets (clusters) based on some similarity measure (or distance measure) so that the members of the same cluster are more similar (or closer) than the elements of different clusters.

Clustering has applications in various diverse fields, including machine learning, data mining, image analysis, and bioinformatics. For example, it is used for data classification and compression. This is a very important task, because in many applications the dimensionality of the observation vectors is enormous. In bioinformatics, clustering is used, for instance, to group homologous[3] sequences into gene families and to investigate possible subtypes.[4] Classification of healthy and diseased samples is critical in diagnostics and forensic biology. In signal processing, clustering is used for lossy compression of a signal. For example, analog-to-digital (A/D) conversion can be viewed as a form of clustering.

There are many different algorithms for clustering. We consider here three such algorithms: the K-means algorithm, the EM algorithm, and the k-nearest neighbor rule.

21.3.1 *K*-means algorithm

The K**-means algorithm** (also known as Lloyd's algorithm) is an *iterative* algorithm for creating K *clusters* of observed data $\boldsymbol{y}_1, \boldsymbol{y}_2, \ldots, \boldsymbol{y}_n$. The ith cluster $\mathcal{C}_i, i = 1, 2, ..., K$, is a subset of the data points. Denote as $\boldsymbol{c}_i, i = 1, 2, ..., K$, the **centroid** of the cluster $\mathcal{C}_i$,

[3] In biology, similarity due to shared ancestry is called *homology*.
[4] Subtypes are descendants of a biological type (e.g., a disease) in a *phylogenetic tree* (see Section 16.4).

where the centroid is defined as the average point or the center of gravity for the cluster.

The iterative clustering algorithm finds centroids $\boldsymbol{c}_i^{(p)}$ of the clusters $\mathcal{C}_i^{(p)}$ obtained after p iterations. Then it forms new clusters $\mathcal{C}_i^{(p+1)}$ based on the centroids and a distance measure $d(\boldsymbol{y}_k, \boldsymbol{c}_i^{(p)})$ between the data points and centroids; $\boldsymbol{y}_k \in \mathcal{C}_i^{(p+1)}$ if and only if $\boldsymbol{c}_i^{(p)}$ is the closest centroid to $\boldsymbol{y}_k$:

$$d(\boldsymbol{y}_k, \boldsymbol{c}_i^{(p)}) = \min_j d(\boldsymbol{y}_k, \boldsymbol{c}_j^{(p)}). \tag{21.8}$$

In other words,

$$\boxed{\boldsymbol{y}_k \in \mathcal{C}_i^{(p+1)} \quad \text{if and only if} \quad \boldsymbol{c}_i^{(p)} = \arg\min_j d(\boldsymbol{y}_k, \boldsymbol{c}_j^{(p)}).} \tag{21.9}$$

The K-means algorithm uses the *mean values* of the cluster points as its centroids and can be described as follows.

1. **Initialization.** The iteration count p is set to zero and the initial centroids $\boldsymbol{c}_i^{(0)}$ are selected based on some background information or at random.
2. **Assignment step.** Data points are assigned to K clusters $\mathcal{C}_i^{(p+1)}$ according to (21.9).
3. **Update step.** New centroids $\boldsymbol{c}_i^{(p+1)}$ are found as the means of the data points in the clusters $\mathcal{C}_i^{(p+1)}$:

$$\boldsymbol{c}_i^{(p+1)} = \frac{1}{n_i^{(p+1)}} \sum_{\boldsymbol{y}_k \in \mathcal{C}_i^{(p+1)}} \boldsymbol{y}_k,$$

where $n_i^{(p+1)}$ is the number of data points that belong to the ith cluster at the $(p+1)$st step. (Clearly, $\sum_{i=1}^K n_i^{(p)} = n$ for all $p = 0, 1, 2 \ldots$.)

4. **Repeat** the **Assignment step** and the **Update step** until the assignments do not change.

21.3.2 EM algorithm for clustering

Clustering can be performed by fitting a mixture of distributions to experimental data. For example, if we apply a **Gaussian mixture model** (GMM) (see Section 20.7, Problem 20.28), then

$$f(\boldsymbol{y}; \boldsymbol{\theta}) = \sum_{i=1}^K \frac{\pi_i}{(2\pi)^{r/2}|\boldsymbol{\Sigma}_i|^{1/2}} \exp\left[-\frac{1}{2}(\boldsymbol{y} - \boldsymbol{\mu}_i)^\top \boldsymbol{\Sigma}_i^{-1}(\boldsymbol{y} - \boldsymbol{\mu}_i)\right], \tag{21.10}$$

where π_i is the probability or weight assigned to the ith distribution; hence $\sum_{i=1}^K \pi_i = 1$ and r is the dimension of the observation vector $\boldsymbol{y}$. Then we can use the $\boldsymbol{\mu}_i$ as the centroids of the clusters while the other parameters allow us to define the boundaries of the clusters. Note that clustering based on the GMM is similar to the *MAP recognition problem* with the Gaussian model (21.4) in which π_i is the prior probability assigned to the ith symbol. As we discussed in Section 20.7, the parameters of a GMM can be estimated efficiently using the *EM algorithm*. This method of clustering is called **GMM/EM clustering** in the machine learning literature.

21.3.3 The k-nearest neighbor rule

For a sample $\boldsymbol{y}$ that needs to be classified, the *k-nearest neighbor (k-NN)* classifier first finds its k nearest neighbors among the samples $\boldsymbol{y}_1, \boldsymbol{y}_2, \ldots, \boldsymbol{y}_n$ and then takes a majority vote over the class labels of these nearest neighbors and assigns it to the sample $\boldsymbol{y}$. The distance is usually the Euclidean distance in the feature vector space.

The following asymptotic result is known [68]: as the sample size $n \to \infty$, the probability of error of the k-NN rule is bounded above by twice the Bayes error. Thus, the generalization error (21.1) of any classifier based on an infinite data set cannot be smaller than one half of that of the k-NN rule.

21.4 Sequence recognition

In principle, the methods for symbol recognition can be applied to recognize sequences of symbols by treating the sequences as new symbols. However, this approach is not practical in general because the dimensionality of the problem grows exponentially as the lengths of the sequences grow. Therefore, we need to use a **dynamical model** for sequence recognition. The ARMA time series (13.236) driven by white Gaussian noise $\boldsymbol{e}_n$ is a Gauss–Markov dynamical model. Similarly, the state space model (22.166) is a Markov process dynamical model. The finite state machine representation of a convolutional encoder (see Example 20.1 in Section 20.2.2) is another example of a dynamical model, and an optimal decoding of the convolutional code is done by the Viterbi decoding algorithm, which is also known as dynamic programming (Section 20.4.2). However, in the case of a convolutional code, the encoded sequence model is known, whereas in a machine learning problem the learning algorithm needs to *extract the sequence model* from training data.

In this section, we consider two major examples in which HMMs are applied to **sequence modeling**. One is speech recognition and the other is biological sequence alignment. However, the methods to be described here will also be relevant to other applications, such as handwriting recognition and signature verification.

21.4.1 Speech recognition

In speech recognition, a continuous speech signal is converted into a sequence of (typically 26-dimensional) *feature vectors* $\boldsymbol{a}_1, \boldsymbol{a}_2, \ldots$, which are used by the learning algorithm. For each phoneme in the selected *phonetic alphabet*, an HMM, called a **phoneme HMM**, is constructed and its model parameter is estimated by using, for example, the Baum–Welch algorithm. For each word in the vocabulary $\mathcal{W}$, the word model is constructed by concatenating the corresponding phoneme HMMs. The use of **null transitions**, which are state transitions that do not emit any output, can add flexibility in aligning the sequences. The word models are then concatenated with pause models to form a sentence model.

21.4.1.1 Isolated-word recognition

In order to perform isolated-word recognition, we define the vocabulary $\mathcal{W}$ of legitimate words for a particular application. For each word $\boldsymbol{w} \in \mathcal{W}$ we build a distinct HMM by combining its component phoneme HMMs stated above, and by obtaining its model parameter, denoted $\hat{\boldsymbol{\theta}}(\boldsymbol{w})$. The word recognition procedure consists of obtaining its observation sequence $\boldsymbol{y}$ and computing the likelihood function $L_{\boldsymbol{y}}(\boldsymbol{w}) = p(\boldsymbol{y}; \hat{\boldsymbol{\theta}}(\boldsymbol{w}))$ for each word $\boldsymbol{w} \in \mathcal{W}$. The word having the maximum likelihood is selected as the recognizer's output:

$$\boldsymbol{w}^* = \arg\max_{\boldsymbol{w}} L_{\boldsymbol{y}}(\boldsymbol{w}). \tag{21.11}$$

An optimal solution $\boldsymbol{w}^*$ is usually found using the Viterbi algorithm (see Section 20.4.2). The computational burden of the Viterbi algorithm is determined by $\sum_{j \in \mathcal{S}} n_j$, where n_j is the number of edges that are incident on node $j \in \mathcal{S}$ in the **trellis diagram**. In Figure 20.3 of Section 20.2.2, for instance, $n_j = 2$ for all $j \in \mathcal{S} = \{0, 1, 2, 3\}$ (see (20.38)). The Viterbi algorithm has to perform n_j comparisons to determine a surviving path that ends at state j at time t. This n_j is equal to the number of nonzero $a_{(i,j)}$ in the state transition matrix $\boldsymbol{A} = a_{(i,j)}$ for given j. Therefore, we often assume a **left–right HMM**, where each transition takes place from a state to itself or only to neighboring states, say up to r states on the right; i.e., $a_{(i,j)} = 0$ for all $i > j$ and $i < j - r$, which guarantees that $n_j < \min\{j, r\}$ (see Figure 21.1). In the left–right HMM, the number of states is often taken to be the length of the word.

21.4.1.2 Connected-word recognition

Isolated-word recognition capability may be sufficient in applications that accept spoken commands, one at a time. In order to process **continuous speech**, we need to create a sentence model by concatenating the word HMMs and inserting transition states between the words and pause models.

The MAP speech recognition procedure is based on (21.3), in which $\boldsymbol{a}$ is a sentence to be recognized based on the observation sequence $\boldsymbol{y}$. The prior probabilities $p(\boldsymbol{a})$ represent the statistical properties of the language that impose certain constraints (i.e., the grammar and syntax) on word sequences. We can also take advantage of the fact

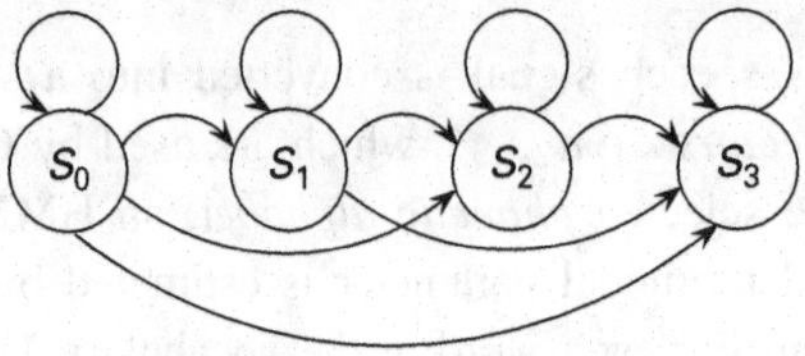

Figure 21.1 A left–right HMM.

that the utterance of a phoneme depends on its adjacent phonemes. This phenomenon is known as **coarticulation**, which is often modeled by **biphones** or **triphones**; i.e., sets of two or three adjacent phonemes selected as the basic units for the HMM at the phonetic level.

Statistical dependency in a sequence of words $\boldsymbol{a}_1, \boldsymbol{a}_2, \ldots, \boldsymbol{a}_t, \ldots$ can be expressed by a Markov chain. Typically, a second-order Markov chain is used, and such a language model at the word level is called a **trigram**. In this case, we have

$$p(\boldsymbol{a}_t|\boldsymbol{a}_1, \boldsymbol{a}_2, ..., \boldsymbol{a}_{t-1}) = p(\boldsymbol{a}_t|\boldsymbol{a}_{t-2}, \boldsymbol{a}_{t-1}). \tag{21.12}$$

Suppose that a word is broken into N_p phonemes (or biphones or triphones) on the average, and each phoneme HMM has N_s hidden phoneme states on the average. Then a word model involves on average $N_p M_p^{N_s}$ hidden states, where M_p is the number of phonemes (or biphones or triphones). In connected-word recognition with the trigram language model, the number of states is N_w^2, where N_w is the number of words in a given application. Then the total number of hidden states at the acoustic level is $N_w^2 N_p M_p N_s$, which becomes an astronomical number even for modest values of N_w and M_p. Thus, a brute-force application of the Viterbi algorithm to find a MAP state sequence at the phoneme level is out of the question. Therefore, we need to resort to a suboptimal algorithm, which searches an optimal sequence in a hierarchical manner; for instance, by using the Viterbi algorithm at the word sequence level, on top of the individual word recognition achieved by the Viterbi algorithm at the phoneme (or triphone) sequence level. In order to reduce the number of state sequence searches, a heuristic algorithm called the **beam search Viterbi algorithm** is often used, which considers only highly plausible sequences, by pruning out less promising sequences.

21.4.2 Biological sequence analysis

It is well known that similarity between genes and proteins strongly suggests that they have a common ancestor. (In biology, similarity due to shared ancestry is called *homology*). A **DNA molecule** is represented by a sequence of **nucleotides** from the four-letter alphabet {A, G, C, T} (nucleotide bases: A for adenine, G for guanine, C for cytosine, and T for thymine), whereas a **protein** is represented by a sequence of 20 *amino acids* that are coded by **codons**, which are three nucleotides, called *trinucleotides*. There are $64(= 4^3)$ possible codons, but only 20 different amino acids exist; thus, some distinct codons must represent the same amino acid (e.g., see Durbin *et al.* [85]).

Random mutations in the sequences that have a common ancestor accumulate over time. Therefore, the analysis of sequence similarity plays an important role in discovery of any evolutionary relationship between protein and gene sequences and in the creation of phylogenetic trees (see Section 16.4). Their similarity can be discovered by aligning the sequences and scoring similarities. Multiple sequence alignments are used to identify conserved sequence regions across a group of sequences that are assumed to be evolutionarily related. The sequences are represented as rows (see Figure 16.4 in Section 16.4) and gaps are inserted between the characters so that identical or similar characters are aligned in successive columns. This method can be used for short sequences.

Alternatively, an HMM is trained on sequences that are members of the same family. A typical HMM for sequences consists of three hidden states ("m" for *match*, "d" for *delete*, and "i" for *insert*). The model is defined by the state transition probability matrix $\boldsymbol{A}$ and the matrix $\boldsymbol{B}$ of state-conditional observation probabilities (i.e., probabilities of observing the letters A, C, G, or T given the hidden state "m" or "i") with no observations in a "d" state. It is always assumed that there is one initial state and one terminal state. This HMM allows us to decide whether a test sequence belongs to a particular family by evaluating the corresponding likelihood (e.g., by using the forward algorithm (20.53)). The Viterbi algorithm allows us to align sequences that belong to the same family by means of match and insert states. This alignment identifies regions of conserved structures that are characteristic of the family. The model parameter is estimated using the Baum–Welch algorithm.

21.5 Bayesian networks

A **Bayesian network** (**BN**) [166] is a probabilistic graphical model introduced in machine learning to represent a broad class of complex stochastic systems with many dependent random variables. A BN may be viewed as a generalization of an HMM. It has been successfully applied to such diverse fields as medical diagnosis [309], image recognition [60], speech recognition [371], language understanding [50], and turbo decoding [245].

We begin with some basic definitions of graph theory required for defining a BN. Recall that we already used some of these concepts in Section 16.4.

DEFINITION 21.1 (Directed graph). *A* directed graph *is an ordered pair* $\mathcal{G} = (\mathcal{V}, \mathcal{E})$, *where* $\mathcal{V}$ *is a set of nodes (or vertices) and* $\mathcal{E}$ *is a set of ordered pairs* (u, v) *of nodes called directed edges. For given nodes* $u, v \in \mathcal{V}$, *we denote the directed edge from* u *to* v *as* (u, v). □

An example of a directed graph is depicted in Figure 21.2. The set of *parent nodes* of a node $v \in \mathcal{V}$ is defined by (see Section 16.4)

$$\mathrm{pa}(v) \triangleq \{u \in \mathcal{V} : (u, v) \in \mathcal{E}\}. \tag{21.13}$$

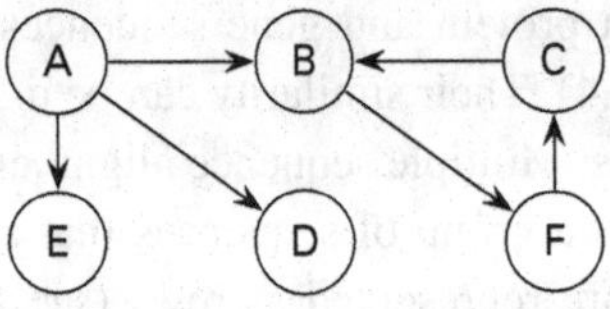

Figure 21.2 Directed graph.

In Figure 21.2, node A does not have a parent, while the rest of the nodes have one parent. The set of *child nodes* of a node $u \in \mathcal{V}$ is defined by (see also Section 16.4)

$$\text{ch}(u) \triangleq \{v \in \mathcal{V} : (u, v) \in \mathcal{E}\}. \tag{21.14}$$

In Figure 21.2, node A has three children, nodes B, C, and F have one child, while nodes D and E do not have a child.

In a *cycle graph* the sequence of all nodes with edges $(v_1, v_2), (v_2, v_3), \ldots, (v_n, v_1)$ represents a cycle. In Figure 21.2, the subgraph[5] consisting of nodes B, F, and C is a cycle graph. If a cycle graph occurs as a subgraph of another graph, it defines a cycle of that graph. A directed graph with no cycles is called a **directed acyclic graph** (DAG).The graph in Figure 21.2 is not acyclic because it has a cycle B, F, C. However, the subgraph consisting of nodes A, B, D, E, F is acyclic.

DEFINITION 21.2 (Bayesian network). *A Bayesian network (BN) is a DAG $\mathcal{G} = (\mathcal{V}, \mathcal{E})$ whose nodes have their associated RVs X_v with the joint probability distribution of the form*

$$p(\boldsymbol{x}) = \prod_{v \in \mathcal{V}} p(x_v | \boldsymbol{x}_{\text{pa}(v)}), \tag{21.15}$$

where $\boldsymbol{x} = (x_v;\, v \in \mathcal{V})$ is an instance of the random vector $\boldsymbol{X} = (X_v;\, v \in \mathcal{V})$ and is called a state of $\boldsymbol{X}$ and $\boldsymbol{x}_{\text{pa}(v)}$ denotes an instance of the RVs $\boldsymbol{X}_{\text{pa}(v)}$ associated with the parent nodes of v. □

Edges represent conditional dependencies between the RVs associated with the nodes; the absence of an edge between two given nodes means that the corresponding RVs are conditionally independent of each other. It follows from this definition that it is sufficient to know the conditional probabilities $p(x_v|\boldsymbol{x}_{\text{pa}(v)})$ to describe a BN. Thus, a BN can be defined as a triplet $(\mathcal{V}, \mathcal{E}, \boldsymbol{P})$, where $\boldsymbol{P}$ defines a set of conditional probability distributions (CPDs) $p(x_v|\boldsymbol{x}_{\text{pa}(v)})$, which are also called **local probability distributions**. These distributions are estimated from experimental data or provided by experts of a given application field. The latter possibility represents a major advantage of the BN: the entire BN or a part of the BN can be constructed without using any experimental data. The CPD is usually represented by the conditional probability table (CPT), which lists the probabilities for values at a node conditioned on the values of its parents. If a node does not have parents (i.e., $\text{pa}(v) = \emptyset$), the CPD reduces to $p(x_v|\boldsymbol{x}_{\text{pa}(v)}) = p(x_v)$, the prior distribution. If the RV X_v is continuous, $p(x_v|\boldsymbol{x}_{\text{pa}(v)})$ of (21.15) should be replaced by the PDF $f(x_v|\boldsymbol{x}_{\text{pa}(v)})$.

Example 21.1: Consider the subgraph consisting of the nodes A, B, D, E, and F in Figure 21.2. This subgraph is a DAG and, according to its structure, we can write

[5] A *subgraph* of a graph G is a graph whose vertex set is a subset of G, and whose adjacency relation is a subset of that of G restricted to this subset.

$$P(A, B, D, E, F) = P(A)P(B|A)P(D|A)P(E|A)P(F|B), \quad (21.16)$$

where we denote the RVs by the corresponding node symbols (e.g., $P(A)$ and $P(B|A)$ instead of $p(x_A)$ and $p(x_B|x_A)$ respectively) for notational brevity. This BN is characterized by the probabilities in the right-hand side of (21.16). □

A set of nodes ∂v is called a **Markov blanket** of node v if x_v is conditionally independent[6] of the value $\boldsymbol{x}_{\mathcal{U}}$ of any subset $\mathcal{U}$ of other nodes in the network (i.e., $\mathcal{U} \subset \mathcal{V}, \mathcal{U} \cap \partial v = \emptyset$), given its blanket:

$$p(x_v|\boldsymbol{x}_{\partial v}, \boldsymbol{x}_{\mathcal{U}}) = p(x_v|\boldsymbol{x}_{\partial v}), \quad (21.17)$$

where $\boldsymbol{x}_{\partial v}$ denotes the variables that belong to the Markov blanket of v. This equation represents the Markovian property of the BNs. Similar to Markov chains, the conditional probability $p(x_v|\boldsymbol{x}_{-v})$ of the node variable x_v given the values of the rest of the nodes in the BN (denoted as $\boldsymbol{x}_{-v}$) satisfies

$$p(x_v|\boldsymbol{x}_{-v}) = p(x_v|\boldsymbol{x}_{\partial v}). \quad (21.18)$$

Using (21.15) we can show (see Problem 21.3) that

$$p(x_v|\boldsymbol{x}_{-v}) = C_v p(x_v|\boldsymbol{x}_{\text{pa}(v)}) \prod_{k \in \text{ch}(v)} p(x_k|\boldsymbol{x}_{\text{pa}(k)}), \quad (21.19)$$

where

$$C_v = \left[\sum_{x_v} p(x_v|\boldsymbol{x}_{\text{pa}(v)}) \prod_{k \in \text{ch}(v)} p(x_k|\boldsymbol{x}_{\text{pa}(k)}) \right]^{-1} \quad (21.20)$$

is the normalization factor. It follows from this equation that the Markov blanket of a node consists of its parents, its children, and the other parents (co-parents) of its children.

A **dynamic Bayesian network** (DBN) is a BN of a random process X_t. A state-based HMM $X_t = (S_t, Y_t)$ is an example of DBN, and can be presented by the directed graph shown in Figure 21.3. The first row of nodes in this graph represents hidden state variables S_t, whereas the observation Y_t is a child of the state variable S_t, and so is

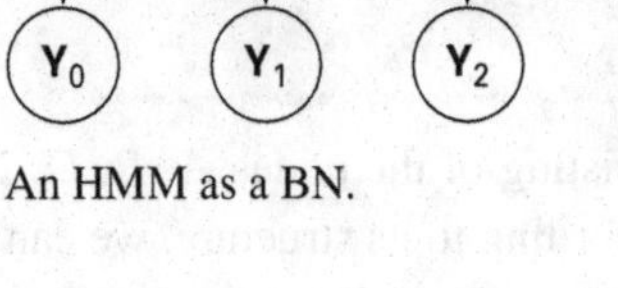

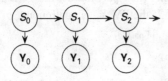

Figure 21.3 An HMM as a BN.

[6] Recall that we say RVs X and Y are conditionally independent, given Z, if $P[X = x, Y = y|Z = z] = P[X = x|Z = z]P[Y = y|Z = z]$, or, equivalently, if $P[X = x|Y = y, Z = z] = P[X = x|Z = z]$ for all x, y, and z.

S_{t+1}. Note that the case where the S_t are continuous RVs can also be represented by this DBN.

21.6 Factor graphs

In Section 20.3, we presented several efficient algorithms (e.g., the forward–backward algorithm (FBA) and the Viterbi algorithm) for solving various estimation problems associated with HMMs. In order to generalize these algorithms for BNs, we need to introduce the notion of a **factor graph**.

Problems that we encounter in computation algorithms often involve a function $g(\boldsymbol{x})$ of multivariates $\boldsymbol{x} = (x_1, x_2, \ldots, x_n)$, which can be factored into several functions $f_i(\boldsymbol{x}_i)$, where $\boldsymbol{x}_i$ denotes an arbitrary subset of the variables $\boldsymbol{x}$:

$$g(x_1, x_2, \ldots, x_n) = \prod_{i=1}^{K} f_i(\boldsymbol{x}_i). \tag{21.21}$$

The joint probability distribution given by (21.15) is an example of such $g(\boldsymbol{x})$.

DEFINITION 21.3 (Factor graph). *A* factor graph $G = (\mathcal{X}, \mathcal{F}, \mathcal{E})$ *is an undirected* ***bipartite graph***[7] *representing the* factorization *(21.21). It consists of* ***variable nodes*** $x_i \in \mathcal{X}$, ***factor nodes*** $f_i \in \mathcal{F}$, *and* ***edges*** $(i, j) \in \mathcal{E}$ *between factor node* f_i *and variable node* x_j, *where* x_j *is an argument of* f_i. *A factor node is also called an* operational node. □

In other words, a factor graph represents the relation between a set of functions and their arguments in terms of a bipartite graph. In order to apply the sum-product algorithm to be discussed below, we consider a class of factor graphs that can be represented as a tree. In such a factor graph, for any given pair of nodes, a path from a node to the other node is unique. It is a convention to show the variable nodes in circles and the factor nodes by squares.

Example 21.2: A factor graph. Let

$$g(x_1, x_2, x_3, x_4, x_5) = f_1(x_1, x_2) f_2(x_2) f_3(x_2, x_3, x_4) f_4(x_4, x_5).$$

The corresponding factor graph is shown in Figure 21.4, and this factor graph has a tree structure. If the function f_1 contains the variable x_3, i.e., $f_1(x_1, x_2, x_3)$, the corresponding factor graph does not allow a tree representation, because between the factor f_1 and the variable x_3 there are two paths: $f_1 \to x_3$ and $f_1 \to x_2 \to f_3 \to x_3$. □

[7] A bipartite graph is a graph whose nodes $\mathcal{V}$ can be partitioned into two disjoint sets $\mathcal{V}_1$ and $\mathcal{V}_2$ such that every edge connects a node in $\mathcal{V}_1$ to one in $\mathcal{V}_2$.

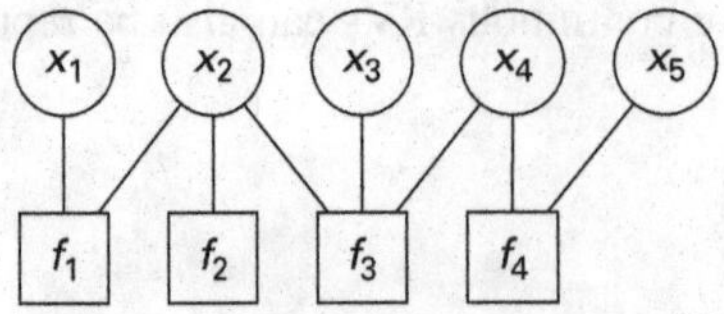

Figure 21.4 Factor graph for the product $f_1(x_1, x_2)f_2(x_2)f_3(x_2, x_3, x_4)f_4(x_4, x_5)$.

21.6.1 Sum-product algorithm

Consider now the problem of finding an efficient algorithm to compute a sum of product terms such as (21.21) over a certain subset of the state space $\mathbb{R}^n$ of the n-dimensional variables $\boldsymbol{x} = (x_1, x_2, \ldots, x_n)$. Computation of a marginal distribution of the joint probability given by the product form (21.15) is such an instance.

Suppose that we want to find the following sum-product:

$$g_s(x_s) = \sum_{\mathcal{X}_{-s}} g(x_1, x_2, ..., x_n) = \sum_{\mathcal{X}_{-s}} \prod_{i=1}^{K} f_i(\boldsymbol{x}_i), \tag{21.22}$$

where $\mathcal{X}_{-s} \triangleq \{\boldsymbol{x} : -\infty < x_i < \infty \text{ for all } i = 1, 2, \ldots, s-1, s+1, \ldots, n\}$; i.e., the space of an $(n-1)$-dimensional variable $\boldsymbol{x}_{-s}$. To perform this summation efficiently, we take advantage of the factorization form of $g(\boldsymbol{x})$. As noted earlier, a factor graph allows a tree representation where any node can serve as its root. So we consider a tree having the node associated with the variable x_s as its root. The variable nodes that belong to different subtrees represent mutually exclusive subsets of $\mathcal{V}$. Thus, we can perform the summations over the variables in different subtrees independently. First, the factor node f_i adjacent to a leaf node v (where x_v is one of the arguments of the function f_i) "eliminates" or "marginalizes out" x_v by computing $\sum_{x_v} f_i$, and passes this marginal sum as a "message" to its next factor node by way of the variable node in the path towards the root. A factor node that receives messages from multiple variable nodes computes the sum of the product of the messages over these variables, and passes the result as a message to its next factor node towards the root. This process of *node elimination* propagates down to the factor node connected to the root node s.

Before we develop the general algorithm, let us consider a simple example.

Example 21.3: Sum-product algorithm. Consider the problem of finding

$$g_3(x_3) = \sum_{\mathcal{X}_{-3}} g(x_1, x_2, x_3, x_4, x_5) = \sum_{x_1, x_2, x_4, x_5} g(x_1, x_2, x_3, x_4, x_5) \tag{21.23}$$

for the function of Example 21.2

$$g(x_1, x_2, x_3, x_4, x_5) = f_1(x_1, x_2)f_2(x_2)f_3(x_2, x_3, x_4)f_4(x_4, x_5).$$

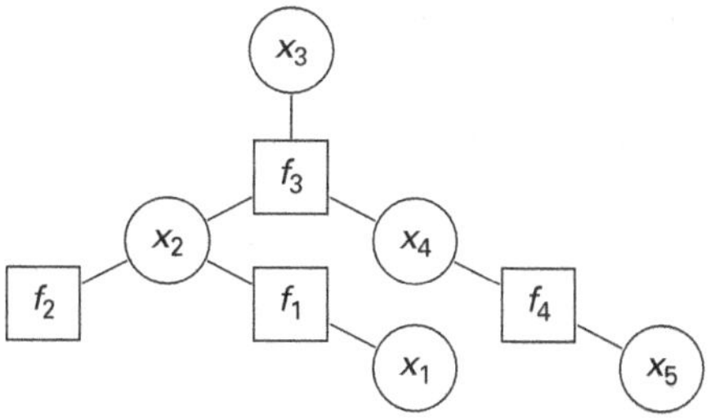

Figure 21.5 Factor graph of Figure 21.4 presented as a tree.

To obtain the marginal $g_3(x_3)$, we translate the bipartite graph of Figure 21.4 into a tree graph whose root node is x_3, as shown in Figure 21.5.[8] In order to compute the sum-product (21.23) we apply the following simple sum-product formula, which can be viewed as the reverse of the *distributive law*:

$$\sum_{i=1}^{n} a \cdot b_i = a \cdot \sum_{i=1}^{n} b_i. \tag{21.24}$$

It is clear that the right-hand side of this equation requires fewer computations (one multiplication and n additions) than the left-hand side (n multiplications and n additions). This reduction in computations is the essence of the sum-product algorithm. Thus, we can rearrange (21.23) as

$$g_3(x_3) = \sum_{x_2,x_4} f_3(x_2, x_3, x_4) f_2(x_2) \sum_{x_1} f_1(x_1, x_2) \sum_{x_5} f_4(x_4, x_5). \tag{21.25}$$

As the first step, the factor node f_4, adjacent to the leaf node x_5, computes $\sum_{x_5} f_4(x_4, x_5) \triangleq \mu_{f_4 \to x_4}(x_4)$ and passes this message to the variable node x_4, which simply rewrites the message as $\mu_{x_4 \to f_3}(x_4)$ and passes to f_3:

$$\mu_{x_4 \to f_3}(x_4) \triangleq \mu_{f_4 \to x_4}(x_4) = \sum_{x_5} f_4(x_4, x_5). \tag{21.26}$$

Similarly, f_1 computes $\sum_{x_1} f_1(x_1, x_2) = \mu_{f_1 \to x_2}(x_2)$ and passes this message to the node x_2. The node x_2 also receives the message $f_2(x_2) \triangleq \mu_{f_2 \to x_2}(x_2)$ from f_2 and forwards the product of these two messages to f_3:

$$\mu_{x_2 \to f_3}(x_2) \triangleq \mu_{f_2 \to x_2}(x_2)\mu_{f_1 \to x_2}(x_2) = f_2(x_2) \sum_{x_1} f_1(x_1, x_2). \tag{21.27}$$

The factor node f_3 combines the messages from x_2 and x_4 and its locally generated message $f_3(x_2, x_3, x_4)$ and sums the product term over $(x_2, x_4) \in \mathbb{R}^2$ and passes it to the root node x_3 as message $\mu_{f_3 \to x_3}(x_3)$:

[8] In Figure 21.5 we use x_v to denote the node v associated with the variable x_v. The reader should not be confused by this slight abuse of notation.

$$\begin{aligned}\mu_{f_3\to x_3}(x_3) &\triangleq \sum_{x_2,x_4} f_3(x_2,x_3,x_4)\mu_{x_2\to f_3}(x_2)\mu_{x_4\to f_3}(x_4)\\ &= \sum_{x_2,x_4} f_3(x_2,x_3,x_4)\mu_{f_2\to x_2}(x_2)\mu_{f_1\to x_2}(x_2)\mu_{f_4\to x_4}(x_4)\\ &= \sum_{x_2,x_4} f_3(x_2,x_3,x_4)f_2(x_2)\sum_{x_1} f_1(x_1,x_2)\sum_{x_5} f_4(x_4,x_5),\end{aligned} \tag{21.28}$$

which is indeed the desired function $g_3(x_3)$ of (21.25).

The algorithm described above is called the **sum-product algorithm**. We have explained it in terms of message passing: the algorithm starts from leaf nodes (i.e., f_2, x_1, x_5 in the above example); a factor node sends its message to its adjacent variable node towards the root; a variable node processes and forwards the message to its adjacent factor node towards the root; this process is repeated until all the messages arrive at the root node associated with the argument of the desired marginal (i.e., x_3 of $g_3(x_3)$ in this example). □

Consider now a general factor graph. As in Example 21.3, we start the algorithm at leaf nodes. To formally state the start of the algorithm, we define the initial messages from leaf nodes as follows. If a leaf node is a factor node f_i, it passes the message $\mu_{f_i\to x_i}(x_i) = f_i(x_i)$ to the variable node x_i; if the leaf node is a variable node x_j, its message to its adjacent factor node f_m (i.e., the function f_m has x_j as one of its arguments) is defined to be unity; in other words, $\mu_{x_j\to f_m}(x_j) = 1$.

Since the summation of the products in (21.22) is obtained by repeated use of the formula (21.24), we derive the following recursion (similar to the recursion (21.28)) to express the message to be sent from a factor node $f(x, z_1, z_2, ..., z_k)$ to the variable node x:

$$\sum_{z_1,\ldots,z_k} f(x, z_1, z_2, ..., z_k)\mu_{z_1\to f}(z_1)\cdots\mu_{z_k\to f}(z_k) \triangleq \mu_{f\to x}(x), \tag{21.29}$$

where $\mu_{z_i\to f}(z_i)$ is the message that the factor node f has received from the variable node z_i. If the variable z_i, an argument of the function f, is also an argument of m other functions $h_1, h_2, ..., h_m$, the message $\mu_{z_i\to f}(z_i)$ is the product of the m messages that z_i received from the factor nodes $h_1, h_2, ..., h_m$:

$$\mu_{z_i\to f}(z_i) \triangleq \mu_{h_1\to z_i}(z_i)\cdots\mu_{h_m\to z_i}(z_i). \tag{21.30}$$

This equation generalizes (21.27) of Example 21.3. Repeat these recursions until all messages reach the root node x_s, which delivers the marginal $g_s(x_s)$.

If we need to compute more than one marginal, we can reuse some of the messages computed to obtain the previous marginals. For instance, suppose we want to compute $g_2(x_2)$ after $g_3(x_3)$ has been computed. We now form a tree with x_2 as its root. Then we just need to compute

$$\mu_{f_3\to x_2}(x_2) = \sum_{x_3,x_4} f_3(x_2,x_3,x_4)\mu_{x_4\to f_3}(x_4)\mu_{x_3\to f_3}(x_3),$$

where $\mu_{x_3 \to f_3}(x_3) = 1$ (since x_3 is a leaf node). Then, by using $\mu_{f_2 \to x_2}(x_2)\mu_{f_1 \to x_2}(x_2) = \mu_{x_2 \to f_3}(x_2)$ obtained in (21.27), we find

$$g_2(x_2) = \mu_{f_2 \to x_2}(x_2)\mu_{f_1 \to x_2}(x_2)\mu_{f_3 \to x_2}(x_2) = \mu_{x_2 \to f_3}(x_2)\mu_{f_3 \to x_2}(x_2).$$

Repeating this procedure, we can find a complete set of marginals $g_i(x_i)$ for all i in any order.

Some important remarks are in order:

- If the variables are continuous variables, the sum-product algorithm discussed above becomes the **integral-product algorithm**, by replacing the summation by integration.
- It is clear that the sum-product algorithm can be applied to any operations that follow the distributive law [244]. In particular, the Viterbi algorithm can be viewed as a special case of such an algorithm, if we replace the summation by maximization (see (20.71)). The Viterbi algorithm may therefore be called the **max-product algorithm**, where the message defined by (21.29) is replaced by

$$\mu_{f \to x}(x) \triangleq \max_{z_1, \ldots, z_k} f(x, z_1, z_2, \ldots, z_k)\mu_{z_1 \to f}(z_1) \cdots \mu_{z_k \to f}(z_k). \tag{21.31}$$

21.6.2 Applications of factor graphs to Bayesian networks

We now apply factor graphs to BNs. Since (21.15) is a special case of (21.21), we can apply the sum-product algorithm to find marginal distributions of any BN, if it is a tree network. If it is not a tree, we can construct an equivalent tree BN by replacing joint variables by new variables, which define new nodes. Such a tree network is called a **junction tree**. For example, a transition-based HMM (see Definition 20.3) can be represented by the BN depicted in Figure 21.6, which is not a tree. However, as we remarked concerning Definition 20.3, any transition-based HMM can be converted into an equivalent state-based HMM, by defining $\tilde{S}_t \triangleq (S_{t-1}, S_t)$ as a new state, whose state space has dimension M^2, where M is the dimension of the original state space. This straightforward translation of a transition-based HMM to an equivalent state-based HMM might make the corresponding BN too large, and required computations too intensive, even for a moderate M. In some cases, however, the factor graph representing a BN may form a tree, even if the BN is not a tree. For example, the aforementioned transition-based HMM can be represented by the factor graph of Figure 21.7, in which factor nodes f_t correspond to the state-transition probabilities $p_t(s_t, \mathbf{y}_t|s_{t-1})$. Using the sum-product algorithm, we can find all the marginal distributions or a joint distribution

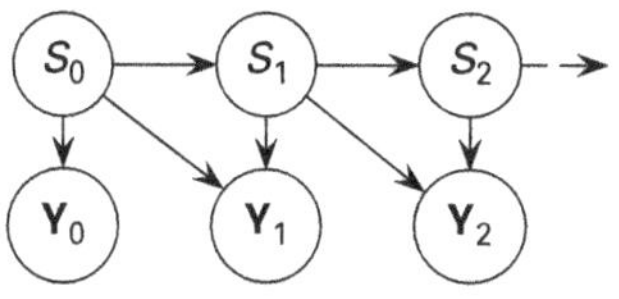

Figure 21.6 State-transition based HMM.

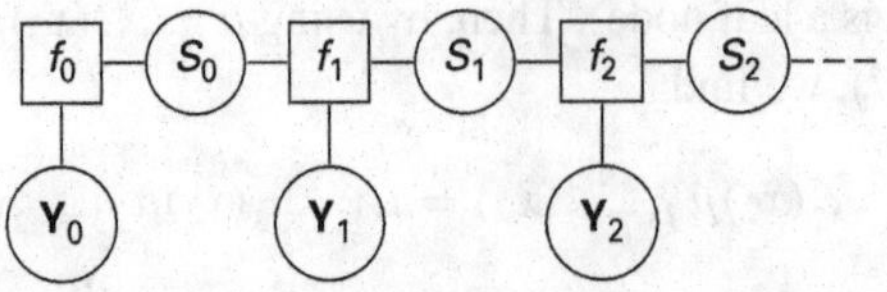

Figure 21.7 Factor graph for the state-transition based HMM.

of a subset of variables. For example, if some states of a BN are unobservable, we might want to evaluate the probability distribution of the observable states only. To find this distribution we marginalize out the unobserved states using the sum-product algorithm. Thus, this algorithm generalizes the FBA for computing the distribution of an HMM observation sequence described in Sections 20.3.2 and 20.3.3.

A final remark is in order. In contrast with the FBA for an HMM, where the computation of the forward or backward variables must be performed in a specific order, there is no explicit ordering of nodes in the sum-product algorithm for a BN. Thus, any operation to be performed by a factor node can be done at any time after all necessary messages have arrived at this node. Therefore, message-passing scheduling is an important part of the sum-product algorithm.

21.7 Markov chain Monte Carlo (MCMC) methods

Rarely can the analysis of a complex system be performed explicitly. Typically, many unrealistic assumptions have to be made in order to obtain analytical solutions. Often, a better approach to solving such a problem is to use a complex but realistic model that can represent accurately and then try to find an approximate solution to the problem at hand. The Monte Carlo simulation method, therefore, has become an important technique in the analysis of complex systems. In the context of analyzing a dynamical system, Monte Carlo simulation is also referred to as self-driven simulation [203].

As we discussed in Section 5.4, Monte Carlo simulation often refers to a numerical technique for evaluating a certain integration expression (e.g., a marginal distribution) by introducing a random variable whose distribution is related to the solution of the original problem. In many applications, including Bayesian statistics, computational physics, computational biology, and computational linguistics, we often need to evaluate such quantities as $E[g(X)]$, where X is a random variable with PDF $f_X(x)$. The random variate generation methods in classical Monte Carlo simulation discussed in Section 5.4 can provide independent samples from the **target distribution** $f_X(x)$. However, even when $g(X)$ is a simple function, such as $g(X) = X$ or $g(X) = X^2$, it may not necessarily be easy to evaluate the integral $\int g(x) f_X(x)\, dx$ with sufficient accuracy using the classical Monte Carlo technique if the target distribution $f_X(x)$ is a complicated distribution or a multidimensional distribution. You may recall that in the *acceptance–rejection* algorithm (Algorithm 5.1) and its variants, we need to generate, on the average, N random numbers to successfully generate one usable random variate, where N depends on the shape of the target PDF $f_X(x)$.

In a modern Monte Carlo simulation method, called **Markov chain Monte Carlo (MCMC)**, we design a Markov chain in such a way that its stationary distribution $\pi(x)$ is equivalent to the target distribution $f_X(x)$. We resort to simulating such a Markov chain, because we do not know how to draw samples from the target distribution $f_X(x)(= \pi(x))$.

21.7.1 Metropolis–Hastings algorithm

Consider a discrete-time Markov chain (DTMC) $\{X_0, X_1, X_2, \ldots, X_t, \ldots\}$, which takes on values from a finite or countably infinite set of states $\{s_1, s_2, s_3, \ldots\}$, which we represent as $\{s_i; i \in \mathcal{S}\}$, with $\mathcal{S}$ representing an index set

$$\mathcal{S} = \{1, 2, 3, \ldots\}.$$

This notation simplifies the DTMC representation and its analysis. Assume that the DTMC $\{X_t\}$ is a homogeneous (or stationary) Markov chain with transition probability matrix (TPM) (see Definition 15.3)

$$\boldsymbol{P} = [P_{ij}; i, j \in \mathcal{S}], \quad \text{where } P_{ij} = P[X_t = s_j | X_{t-1} = s_i].$$

We showed in Section 15.3.4 that if the chain is ergodic (i.e., irreducible and aperiodic with all states being positive recurrent), there exists a unique stationary distribution $\boldsymbol{\pi} = (\pi_j; j \in \mathcal{S})^\top$ such that

$$\boldsymbol{\pi}^\top = \boldsymbol{\pi}^\top \boldsymbol{P} \tag{21.32}$$

and

$$\lim_{t\to\infty} P_{i,j}^{(t)} = \pi_j, \quad \text{for all } i, j \in \mathcal{S}. \tag{21.33}$$

Now suppose that we are interested in generating random variates $x_1, x_2, \ldots, x_t, \ldots$ with the distribution $\boldsymbol{p}_X \triangleq (p_X(x_i); i \in \mathcal{S})$. The MCMC method designs a DTMC with TPM $\boldsymbol{P}$ such that its stationary distribution $\boldsymbol{\pi}$ is equal to the target distribution $\boldsymbol{p}_X$. Suppose that the number of states is finite; i.e., $|\mathcal{S}| = M$. Then (21.32) represents less than M independent linear constraints on the P_{ij} (since some of the equations are linearly dependent) and the stochastic property of $\boldsymbol{P}$ (see (15.14)) gives another M constraints; hence, the total is less than $2M$ constraints. Thus, there are infinitely many choices for choosing the M^2 entries of the TPM $\boldsymbol{P}$ of an ergodic Markov chain (see Problem 21.5).

In solving (21.32) for the unknown TPM $\boldsymbol{P}$ for a given target stationary distribution $\boldsymbol{\pi}$, let us limit ourselves to a class of **time-reversible** (or simply **reversible**) Markov chains (Section 16.3.1). Theorem 16.4 implies that the necessary and sufficient condition for reversibility of an ergodic DTMC is for the **detailed balance equations** to hold:

$$\pi_i P_{ij} = \pi_j P_{ji}, i, j \in \mathcal{S}. \tag{21.34}$$

Reversible Markov chains are common in the MCMC method, because the detailed balance equations (21.34) necessarily imply that, in the Markov chain thus constructed,

$\boldsymbol{\pi}$ is a stationary distribution and it is the steady-state distribution if the chain is ergodic. Note that we work on the class of reversible Markov chains to simplify the design of the TPM. The simplicity of (21.34) is a key element in developing an MCMC algorithm, which finds the solution for P_{ij} easily. The restriction imposed by (21.34), however, might lead to the design of a Markov chain with poor mixing properties. Some non-reversible Markov chain may exhibit better convergence behavior. Combining several reversible Markov chains may deliver a Markov chain which is generally nonreversible, but has a higher rate of convergence (see Problem 21.6).

The **Metropolis–Hastings (MH) algorithm**, to be described below, provides a general approach to constructing a time-reversible Markov chain. Furthermore, this algorithm can be used even when we know only the shape of the target distribution $\{g_i; i \in \mathcal{S}\}$. The algorithm does not require us to evaluate the normalization constant $G = \sum_{j \in \mathcal{S}} g_j$ to find the target distribution $\pi_i = g_i/G, i \in \mathcal{S}$. Thus, the MH algorithm is especially useful, for instance, when g_i is a product of several functions and the state space $\mathcal{S}$ is too large for a brute-force enumeration, as found in the discussion that led to the sum-product algorithm of the previous section. A similar problem occurs when we need to compute the normalization constant G in a product-form queueing network or loss network model (e.g., see [203, Chapter 8]). In applications to Bayesian inference (see Section 4.5) the RVs of interest are the parameter $\boldsymbol{\theta}$ and the target distribution, which is the posterior distribution $\pi(\boldsymbol{\theta}|\boldsymbol{y})$, where $\boldsymbol{y}$ represents data. The posterior distribution is updated according to (e.g., see (4.144)) $\pi(\boldsymbol{\theta}|\boldsymbol{y}) \propto p(\boldsymbol{y}|\boldsymbol{\theta})\pi(\boldsymbol{\theta})$, where $p(\boldsymbol{y}|\boldsymbol{\theta})$ is the likelihood function and $\pi(\boldsymbol{\theta})$ is the prior distribution. Again, the evaluation of the normalization constant to obtain the posterior distribution $\pi(\boldsymbol{\theta}|\boldsymbol{y})$ is often computationally too expensive.

In the first step for obtaining a desired TPM $\boldsymbol{P}$, we begin with an (arbitrary) Markov chain with TPM $\boldsymbol{Q}$ defined[9] over the same set of states $\{s_i; i \in \mathcal{S}\}$:

$$\boldsymbol{Q} = \left[Q_{ij}; i, j \in \mathcal{S}\right]. \tag{21.35}$$

The ith row of this matrix represents the probability distribution Q_{ij}; $j \in \mathcal{S}$ of choosing the next state s_j when the current state is s_i, and is referred to as the **proposal distribution**. The Markov chain with TPM $\boldsymbol{Q}$ is, in general, not a reversible chain, nor is its stationary distribution equal to the target distribution $\boldsymbol{\pi}$. Therefore, we consider a chain with state transition probabilities of the form

$$P_{ij} = \begin{cases} Q_{ij}\alpha_{ij}, & \text{for } j \neq i, \\ Q_{ii} + \sum_{k(\neq i) \in \mathcal{S}} Q_{ik}(1 - \alpha_{ik}), & \text{for } j = i. \end{cases} \tag{21.36}$$

The Markov chain with TPM $\boldsymbol{P}$ thus defined satisfies the detailed balance equations (21.34) if and only if

$$\pi_i Q_{ij}\alpha_{ij} = \pi_j Q_{ji}\alpha_{ji}, \quad i, j \in \mathcal{S}. \tag{21.37}$$

[9] This $\boldsymbol{Q}$ should not be confused with the transition rate matrix or infinitesimal generator $\boldsymbol{Q}$ of a CTMC defined in Chapter 16.

These balance equations are satisfied if we choose

$$\alpha_{ij} = \min\left\{\frac{\pi_j Q_{ji}}{\pi_i Q_{ij}}, 1\right\}, \tag{21.38}$$

because if $\alpha_{ij} = \pi_j Q_{ji}/\pi_i Q_{ij}$, then $\alpha_{ji} = 1$ and vice versa.

Equation (21.36) is interpreted as follows: if the present state of the Markov process X_t is s_i, generate the next state s_j according to the proposal probability $Q_{ij} = P[X_{t+1} = s_j | X_t = s_i]$ and *accept* a proposed sample $s_j (\neq s_i)$ with probability α_{ij}. If this is rejected (which happens with probability $1 - \alpha_{ij}$), the chain remains at s_i; i.e., $X_{t+1} = s_i$. The parameter α_{ij} of (21.38) is called the **acceptance ratio**.

Equation (21.38) can be written as

$$\alpha_{ij} = \min\left\{\frac{g_j Q_{ji}}{g_i Q_{ij}}, 1\right\}, \tag{21.39}$$

where $\{g_i \propto \pi_i; i \in \mathcal{S}\}$ is an **unnormalized** target distribution; the normalization constant $G = \sum_{j\in\mathcal{S}} g_j$ is immaterial in determining α_{ij}, and this confirms our earlier statement that the MH algorithm greatly simplifies the Monte Carlo simulation. Since the target distribution is the steady-state distribution, we need to discard the initial states generated by the algorithm until we believe that the Markov chain has converged to its equilibrium state.

In Monte Carlo simulation, the initial period in which samples are discarded is referred to as the **burn-in period**. This is closely related to the notion of **mixing time** in the theory of a Markov chain (e.g., see Levin *et al.* [223]), which is defined as the time until the state distribution of the Markov chain becomes "close" to its steady-state distribution. It should be recognized, however, that the steady state is a limiting condition that may be approached but, in general, never attained exactly in finite time. There is no single point in time beyond which the Markov chain is in equilibrium. But we can choose some point beyond which we are willing to neglect the *error* that is made by considering the system to be in equilibrium. It has been shown that some families of Markov chains exhibit a sharp transition from "unmixed phase" to a nearly completely mixed phase after a specific amount of time. Such a phase transition is called the "cut-off" phenomenon (see [223], Chapter 18).

A practical, but laborious, way to estimate the mixing time or burn-in time is to make a number of preliminary pilot runs from the *same starting point* and compare the observed distribution of the state of the system at various ages (e.g., see [203], pp. 689–690). If the burn-in period is estimated to be $t \in [0, B]$, then we discard the first $B + 1$ samples $x_0, x_1, \ldots, x_B$ and use the remaining samples $x_{B+1}, x_{B+2}, \ldots, x_{B+N}$ as usable N samples drawn from the target distribution $\boldsymbol{\pi}$. The MH algorithm, given an unnormalized target distribution $g_i \propto \pi_i; i \in \mathcal{S}$, and the proposal distributions Q_{ij}, is summarized in Algorithm 21.1.

In order to accept the proposed variate $X_t = s_j$ with the probability α_{ij} (21.39) so that the detailed balance equations (21.37) can be met, we use a uniform random variable $U(0, 1)$. Note that the uniform variate u is less than one, so the condition $u \leq \alpha_{ij} = \min\{g_j Q_{ji}/g_i Q_{ij}, 1\}$ reduces to $u \leq g_j Q_{ji}/g_i Q_{ij}$. Step 3 in Algorithm 21.1 shows

Algorithm 21.1 MH algorithm

1: Choose the initial state $x_0 \in \mathcal{S}$, set $t \leftarrow 1$, and $s_i \leftarrow x_0$.
2: Generate a variate s_j according to $P[X = s_j] = Q_{ij}$ and a uniform variate $u \sim U(0, 1)$.
3: If $u \leq \frac{g_j Q_{ji}}{g_i Q_{ij}}$, then set the next state $x_t \leftarrow s_j$; else $x_t \leftarrow s_i$.
4: Set $t \leftarrow t + 1$ and $s_i \leftarrow x_t$.
5: If $t < N + B$ repeat Steps 2–4; else return $x_{B+1}, x_{B+2}, \ldots, x_{B+N}$.

the proposed next state $x_{t+1} = s_j$ will be rejected with probability $1 - \alpha_{ij}$; thus, sometimes this algorithm is interpreted as a variant of the acceptance–rejection algorithm discussed in Section 5.4.2. However, in contrast to the acceptance–rejection algorithm, which actually discards some samples, the MH algorithm *does not* discard any samples (beyond the burn-in period): it selects either $x_t = s_j$ or $x_t = s_i$. So no random number generated will be wasted.

21.7.1.1 Choices of proposal distributions

The art of designing a good MH algorithm lies primarily in finding an appropriate proposal distribution Q_{ij}.

- If the proposal distribution is symmetric: $Q_{ij} = Q_{ji}$, then (21.42) becomes

$$\alpha_{ij} = \min \left\{ \frac{\pi_j}{\pi_i}, 1 \right\}. \tag{21.40}$$

  The algorithm with this α_{ij} is the original **Metropolis algorithm** [247]. Later, Hastings generalized this algorithm to accommodate nonsymmetric distributions.
- If $Q_{ij} = q_j$, for all $i \in \mathcal{S}$, then

$$\alpha_{ij} = \min \left\{ \frac{\eta_j}{\eta_i}, 1 \right\}, \quad \text{for all } i \in \mathcal{S}, \tag{21.41}$$

  where $\eta_i = \pi_i / q_i$.
- If the proposal distribution has the form $Q_{ij} = q_{j-i}$, then we can write $s_j = s_i + w_k$, where $w_k \sim q_k$. This case is most popular in applications and is called the **random walk MH algorithm**. If, in addition, the distribution is symmetric, then we also have a simple expression for α_{ij} given by (21.40).

The length of the burn-in period B and the way in which the chain **mixes** (meaning that the chain moves to different parts of the state space) depend critically on the proposal distribution. Convergence will be slow and mixing properties will be poor if the proposed transitions are mostly between nearby states in the state space. However, if we choose a proposal distribution with a wide support aiming at distant transitions, it may result in a lower acceptance ratio, which leads to slow convergence and poor mixing. Thus, the proposal distribution should be chosen in such a way as to allow both distant transitions and a high acceptance ratio. One way to achieve this is to alternate different proposal distributions in light of sampled elements.

Suppose that we have several proposal distributions $Q_{ij}^{(1)}, Q_{ij}^{(2)}, \ldots, Q_{ij}^{(m)}$ for the same target distribution π_j. Then we can design the corresponding MH algorithms with different acceptance ratios $\alpha_{ij}^{(1)}, \alpha_{ij}^{(2)}, \ldots, \alpha_{ij}^{(m)}$:

$$\alpha_{ij}^{(k)} = \min\left\{\frac{\pi_j Q_{ji}^{(k)}}{\pi_i Q_{ij}^{(k)}}, 1\right\}, k = 1, 2, \ldots, m. \tag{21.42}$$

We assume that the conditions for convergence are satisfied in all algorithms. Then we can use these algorithms in any order. By alternating the proposals, we may be able to improve the algorithm's mixing properties (see Problem 21.6). As for the convergence properties of the MH algorithm, see Tierney [320, 321] for example.

21.7.2 Metropolis–Hastings algorithm for continuous variables

If the target distribution is a continuous (and multidimensional) distribution, the MH algorithm presented in the previous section still applies with a rather straightforward modification in the notation. Suppose we wish to generate samples from a multivariate PDF $\pi(\boldsymbol{x})$, where $\boldsymbol{X} = (X_1, X_2, \ldots, X_m)$ is an m-dimensional random vector:

$$P[\boldsymbol{X} \in \mathcal{A}] = \int_{\boldsymbol{x}\in\mathcal{A}} \pi(\boldsymbol{x})\, d\boldsymbol{x}, \tag{21.43}$$

where $\mathcal{A}$ is any region in $\mathbb{R}^m$. Then we want to design a **discrete-time continuous-space** Markov process, which we specify in terms of the transitional (or conditional) PDF $f(\boldsymbol{x}; \boldsymbol{y})$ (also referred to as the **transition kernel**) such that

$$P[\boldsymbol{X}_{t+1} \in \mathcal{B} | \boldsymbol{X}_t = \boldsymbol{x}] = \int_{\boldsymbol{y}\in\mathcal{B}} f(\boldsymbol{x}; \boldsymbol{y})\, d\boldsymbol{y}, \tag{21.44}$$

where $\mathcal{B} \subseteq \mathbb{R}^m$. The requirement that this Markov process is time-reversible is given by the following detailed balance equation:

$$\pi(\boldsymbol{x}) f(\boldsymbol{x}; \boldsymbol{y}) = \pi(\boldsymbol{y}) f(\boldsymbol{y}; \boldsymbol{x}), \quad \boldsymbol{x}, \boldsymbol{y} \in \mathbb{R}^m. \tag{21.45}$$

Then,

$$\begin{aligned} P[\boldsymbol{X}_{t+1} \in \mathcal{B}] &= \int_{\boldsymbol{y}\in\mathcal{B}} \int_{\boldsymbol{x}\in\mathbb{R}^m} \pi(\boldsymbol{x}) f(\boldsymbol{x}; \boldsymbol{y})\, d\boldsymbol{x}\, d\boldsymbol{y} \\ &= \int_{\boldsymbol{y}\in\mathcal{B}} \int_{\boldsymbol{x}\in\mathbb{R}^m} \pi(\boldsymbol{y}) f(\boldsymbol{y}; \boldsymbol{x})\, d\boldsymbol{x}\, d\boldsymbol{y} = \int_{\boldsymbol{y}\in\mathcal{B}} \pi(\boldsymbol{y})\, d\boldsymbol{y}, \end{aligned} \tag{21.46}$$

which shows that $\pi(\boldsymbol{x})$ is the stationary distribution of this reversible Markov chain with the transitional PDF $f(\boldsymbol{x}; \boldsymbol{y})$. Then, by defining an ergodic Markov chain with the transitional PDF $q(\boldsymbol{x}; \boldsymbol{y})$ (the proposal PDF) and the acceptance ratio

$$\alpha(\boldsymbol{x}, \boldsymbol{y}) = \begin{cases} \min\left\{\frac{\pi(\boldsymbol{y})q(\boldsymbol{y};\boldsymbol{x})}{\pi(\boldsymbol{x})q(\boldsymbol{x};\boldsymbol{y})}, 1\right\}, & \text{if } \pi(\boldsymbol{x})q(\boldsymbol{x}; \boldsymbol{y}) \neq 0, \\ 1, & \text{otherwise}, \end{cases} \tag{21.47}$$

Algorithm 21.2 MH algorithm for continuous variables

1: Choose the initial value $\boldsymbol{x}_0 \in \mathbb{R}^m$, set $t \leftarrow 1$ and $\boldsymbol{x} \leftarrow \boldsymbol{x}_0$.
2: Generate a proposed variate $\boldsymbol{y}$ according to the transitional PDF $q(\boldsymbol{x}; \boldsymbol{y})$ and a uniform variate $u \sim U(0, 1)$.
3: If $u \leq \frac{\pi(\boldsymbol{y})q(\boldsymbol{y};\boldsymbol{x})}{\pi(\boldsymbol{x})q(\boldsymbol{x};\boldsymbol{y})}$, then set the next value $\boldsymbol{x}_t \leftarrow \boldsymbol{y}$; else $\boldsymbol{x}_t \leftarrow \boldsymbol{x}$.
4: Set $t \leftarrow t + 1$ and $\boldsymbol{x} \leftarrow \boldsymbol{x}_t$.
5: If $t < N + B$ repeat Steps 2–4; else return $x_{B+1}, x_{B+2}, \ldots, x_{B+N}$.

we can design a reversible discrete-time continuous-space Markov process with the transitional PDF

$$f(\boldsymbol{x}; \boldsymbol{y}) = q(\boldsymbol{x}; \boldsymbol{y})\alpha(\boldsymbol{x}, \boldsymbol{y}) + r(\boldsymbol{x})\delta(\boldsymbol{y} - \boldsymbol{x}), \tag{21.48}$$

where

$$r(\boldsymbol{x}) = 1 - \int_{\boldsymbol{y}\in\mathbb{R}^m} q(\boldsymbol{x}; \boldsymbol{y})\alpha(\boldsymbol{x}, \boldsymbol{y})d\boldsymbol{y}. \tag{21.49}$$

As we can see, the transition kernel is a mixture of continuous and discrete components. Given the current state $\boldsymbol{X}_t = \boldsymbol{x}$, the Markov process $\boldsymbol{X}_{t+1}$ moves to the next state $\boldsymbol{y} \neq \boldsymbol{x}$ with PDF $q(\boldsymbol{x}; \boldsymbol{y})\alpha(\boldsymbol{x}, \boldsymbol{y})$ and does not move from $\boldsymbol{x}$; i.e., $\boldsymbol{X}_{t+1} = \boldsymbol{x}$ with probability $r(\boldsymbol{x})$. The condition $\pi(\boldsymbol{x})q(\boldsymbol{x}; \boldsymbol{y}) \neq 0$ of (21.47) is satisfied if the range of the function $q(\boldsymbol{x}; \boldsymbol{y})$ is within the support of the target distribution $\pi(\boldsymbol{x})$. Note that, in computing $\alpha(\boldsymbol{x}, \boldsymbol{y})$, an unnormalized target density $g(\boldsymbol{x}) \propto \pi(\boldsymbol{x})$ is sufficient as in the discrete-state case, because $\pi(\boldsymbol{y})/\pi(\boldsymbol{x}) = g(\boldsymbol{y})/g(\boldsymbol{x})$.

In Algorithm 21.2 we present the MH algorithm for generating $B + N$ samples of continuous variates $\boldsymbol{x}_0, \boldsymbol{x}_1, \ldots, \boldsymbol{x}_t, \ldots, \boldsymbol{x}_{B+N}$, for a given unnormalized target density $g(\boldsymbol{x}) \propto \pi(\boldsymbol{x})$, and the proposal PDF $q(\boldsymbol{x}; \boldsymbol{y})$. The first $B + 1$ samples $\boldsymbol{x}_0, \ldots, \boldsymbol{x}_B$ should be discarded, since they are the "burn-in" samples.

21.7.3 Block Metropolis–Hastings algorithm

Suppose that we want to sample from a joint distribution $\pi(\boldsymbol{x}_1, \boldsymbol{x}_2)$, where the variables appear in two blocks, both of which may be vectors. Furthermore, suppose that $\pi(\boldsymbol{x}_1|\boldsymbol{x}_2)$ and $\pi(\boldsymbol{x}_2|\boldsymbol{x}_1)$ are the conditional distributions, for which sampling algorithms are known. We want to show that, by applying the MH algorithm to samples from the conditional PDFs $\pi(\boldsymbol{x}_1|\boldsymbol{x}_2)$ and $\pi(\boldsymbol{x}_2|\boldsymbol{x}_1)$, we can obtain samples from the joint PDF $\pi(\boldsymbol{x}_1, \boldsymbol{x}_2)$.

We need to design two Markov chains with transitional PDFs $f_1(\boldsymbol{x}_1; \boldsymbol{y}_1|\boldsymbol{x}_2)$ and $f_2(\boldsymbol{x}_2; \boldsymbol{y}_2|\boldsymbol{x}_1)$ such that their stationary distributions are equal to $\pi_1(\boldsymbol{x}_1|\boldsymbol{x}_2)$ and $\pi_2(\boldsymbol{x}_2|\boldsymbol{x}_1)$ respectively. That is,

$$\pi_1(\boldsymbol{y}_1|\boldsymbol{x}_2) = \int_{\boldsymbol{x}_1} f_1(\boldsymbol{x}_1; \boldsymbol{y}_1|\boldsymbol{x}_2)\pi_1(\boldsymbol{x}_1|\boldsymbol{x}_2)\, d\boldsymbol{x}_1, \tag{21.50}$$

$$\pi_2(\boldsymbol{y}_2|\boldsymbol{x}_1) = \int_{\boldsymbol{x}_2} f_2(\boldsymbol{x}_2; \boldsymbol{y}_2|\boldsymbol{x}_1)\pi_2(\boldsymbol{x}_2|\boldsymbol{x}_1)\, d\boldsymbol{x}_2. \tag{21.51}$$

Algorithm 21.3 MH algorithm with two blocks

1: Choose the initial value $(\boldsymbol{x}_1^{(0)}, \boldsymbol{x}_2^{(0)})$, set $t \leftarrow 1$ and $(\boldsymbol{x}_1, \boldsymbol{x}_2) \leftarrow (\boldsymbol{x}_1^{(0)}, \boldsymbol{x}_2^{(0)})$.
2: Generate $\boldsymbol{y}_1$ from $q_1(\boldsymbol{x}_1; \boldsymbol{y}_1|\boldsymbol{x}_2)$ and $u_1 \sim U(0, 1)$.
3: If $u_1 \leq \frac{\pi(\boldsymbol{y}_1|\boldsymbol{x}_2)q_1(\boldsymbol{y}_1;\boldsymbol{x}_1|\boldsymbol{x}_2)}{\pi(\boldsymbol{x}_1|\boldsymbol{x}_2)q_1(\boldsymbol{x}_1;\boldsymbol{y}_1|\boldsymbol{x}_2)}$, then set $\boldsymbol{x}_1^{(t)} \leftarrow \boldsymbol{y}_1$; else $\boldsymbol{x}_1^{(t)} \leftarrow \boldsymbol{x}_1$.
4: Generate $\boldsymbol{y}_2$ from $q_2(\boldsymbol{x}_2; \boldsymbol{y}_2|\boldsymbol{x}_1^{(t)})$ and $u_2 \sim U(0, 1)$.
5: If $u_2 \leq \frac{\pi(\boldsymbol{y}_2|\boldsymbol{x}_1^{(t)})q_2(\boldsymbol{y}_2;\boldsymbol{x}_2|\boldsymbol{x}_1^{(t)})}{\pi(\boldsymbol{x}_2|\boldsymbol{x}_1^{(t)})q_2(\boldsymbol{x}_2;\boldsymbol{y}_2|\boldsymbol{x}_1^{(t)})}$, set $\boldsymbol{x}_2^{(t)} \leftarrow \boldsymbol{y}_2$; else $\boldsymbol{x}_2^{(t)} \leftarrow \boldsymbol{x}_2$.
6: Set $t \leftarrow t + 1$ and $(\boldsymbol{x}_1, \boldsymbol{x}_2) \leftarrow (\boldsymbol{x}_1^{(t)}, \boldsymbol{y}_2^{(t)})$.
7: If $t < N + B$ repeat Steps 2–6; else return $x_{B+1}, x_{B+2}, \ldots, x_{B+N}$.

Then it can be shown (Problem 21.7) that a Markov chain with the transitional PDF $f_1(\boldsymbol{x}_1, \boldsymbol{y}_1|\boldsymbol{x}_2)f_2(\boldsymbol{x}_2, \boldsymbol{y}_2|\boldsymbol{y}_1)$ has $\pi(\boldsymbol{x}_1, \boldsymbol{x}_2) = \pi_1(\boldsymbol{x}_1)\pi_2(\boldsymbol{x}_2|\boldsymbol{x}_1)$ as its stationary distribution:

$$\int_{\boldsymbol{x}_1}\int_{\boldsymbol{x}_2} f_1(\boldsymbol{x}_1; \boldsymbol{y}_1|\boldsymbol{x}_2)f(\boldsymbol{x}_2; \boldsymbol{y}_2|\boldsymbol{y}_1)\pi(\boldsymbol{x}_1, \boldsymbol{x}_2)\, d\boldsymbol{x}_1\, d\boldsymbol{x}_2 = \pi_1(\boldsymbol{y}_1, \boldsymbol{y}_2). \tag{21.52}$$

We now denote the Markov sequence that the MH algorithm with two blocks generates as $(\boldsymbol{x}_1^{(0)}, \boldsymbol{x}_2^{(0)}), (\boldsymbol{x}_1^{(1)}, \boldsymbol{x}_2^{(1)}), \ldots, (\boldsymbol{x}_1^{(t)}, \boldsymbol{x}_2^{(t)}), \ldots$, and the algorithm is shown in Algorithm 21.3. Note that Step 2 generates a proposed value $\boldsymbol{y}_1$ conditioned on the current value $\boldsymbol{x}_1$ in the same block and the current value $\boldsymbol{x}_2$ in the other block. Step 4 generates a proposed value $\boldsymbol{y}_2$ conditioned on the current value $\boldsymbol{x}_2$ in the same block and the updated value $\boldsymbol{x}_1^{(t)}$ in the other block. The block MH algorithm consists of alternating the MH algorithms for each conditional distribution.

Extension of the MH algorithm to m blocks of variables

$$\boldsymbol{x} = (\boldsymbol{x}_1, \boldsymbol{x}_2, \ldots, \boldsymbol{x}_m)$$

(Algorithm 21.4) is possible when we know how to generate random variates from proposal densities $q_i(\boldsymbol{x}_i; \boldsymbol{y}_i|\boldsymbol{x}_{-i}^{(t)})$, where $\boldsymbol{x}_i$ is the previous ith block (i.e., $\boldsymbol{X}_i^{(t-1)} = \boldsymbol{x}_i$) and $\boldsymbol{y}_i$ is a proposed value for $\boldsymbol{X}_i^{(t)}$. The variate $\boldsymbol{x}_{-i}$ contains all the blocks of $\boldsymbol{x}$ except for $\boldsymbol{x}_i$; i.e.,

$$\boldsymbol{x}_{-i}^{(t)} = (\boldsymbol{x}_1^{(t)}, \boldsymbol{x}_2^{(t)}, \ldots, \boldsymbol{x}_{i-1}^{(t)}, \boldsymbol{x}_{i+1}^{(t)}, \ldots, \boldsymbol{x}_m^{(t-1)}),\ i = 1, 2, \ldots, m. \tag{21.53}$$

21.7.4 The Gibbs sampler

The **Gibbs algorithm**, or **Gibbs sampler** as it is often called [118], is a special case of the MH algorithm, and is perhaps the most widely used MCMC algorithm. Let us consider the Gibbs algorithm with two blocks (Algorithm 21.5).

We want to sample from a joint distribution $\pi(\boldsymbol{x}_1, \boldsymbol{x}_2)$, where $\boldsymbol{x}_1$ and $\boldsymbol{x}_2$ may be vector variables. We assume that we know simulation algorithms for the two conditional distributions $\pi_1(\boldsymbol{x}_1|\boldsymbol{x}_2)$ and $\pi_2(\boldsymbol{x}_2|\boldsymbol{x}_1)$. Let $\boldsymbol{x} = (\boldsymbol{x}_1, \boldsymbol{x}_2)$ be the state at time t, i.e.,

Algorithm 21.4 MH algorithm with m blocks

1: Choose the initial values $\boldsymbol{x}^{(0)} = (\boldsymbol{x}_1^{(0)}, \boldsymbol{x}_2^{(0)}, \ldots, \boldsymbol{x}_m^{(0)})$. Set $t \leftarrow 1$ and $\boldsymbol{x} \leftarrow \boldsymbol{x}^{(0)}$.
2: For $i = 1, 2, \ldots, m$:

1. Generate a proposed sample $\boldsymbol{y}_i$ from $q_i(\boldsymbol{x}_i; \boldsymbol{y}_i | \boldsymbol{x}_{-i})$ and $u_i \sim U(0, 1)$.
2. If $u_i \leq \frac{\pi(\boldsymbol{y}_i|\boldsymbol{x}_{-i}^{(t)}) q_i(\boldsymbol{y}_i; \boldsymbol{x}_i|\boldsymbol{x}_{-i}^{(t)})}{\pi(\boldsymbol{x}_i|\boldsymbol{x}_{-i}^{(t)}) q_i(\boldsymbol{x}_i; \boldsymbol{y}_i|\boldsymbol{x}_{-i}^{(t)})}$, then set $\boldsymbol{x}_i^{(t)} \leftarrow \boldsymbol{y}_i$; else $\boldsymbol{x}_i^{(t)} \leftarrow \boldsymbol{x}_i$.

3: Set $t \leftarrow t + 1$ and $\boldsymbol{x} \leftarrow \boldsymbol{x}^{(t)}$, where $\boldsymbol{x}^{(t)} = (\boldsymbol{x}_1^{(t)}, \boldsymbol{x}_2^{(t)}, \ldots, \boldsymbol{x}_m^{(t)})$.
4: If $t < N + B$ repeat Steps 2 and 3; else return $x_{B+1}, x_{B+2}, \ldots, x_{B+N}$.

Algorithm 21.5 The Gibbs algorithm with two blocks

1: Choose the initial value $(\boldsymbol{x}_1^{(0)}, \boldsymbol{x}_2^{(0)})$. Set $t \leftarrow 1$ and $(\boldsymbol{x}_1, \boldsymbol{x}_2) \leftarrow (\boldsymbol{x}_1^{(0)}, \boldsymbol{x}_2^{(0)})$.
2: Generate $\boldsymbol{x}_1^{(t)}$ from $\pi_1(\boldsymbol{x}_1|\boldsymbol{x}_2)$.
3: Generate $\boldsymbol{x}_2^{(t)}$ from $\pi_2(\boldsymbol{x}_2|\boldsymbol{x}_1)$.
4: Set $t \leftarrow t + 1$ and $(\boldsymbol{x}_1, \boldsymbol{x}_2) \leftarrow (\boldsymbol{x}_1^{(t)}, \boldsymbol{x}_2^{(t)})$.
5: If $t < N + B$ repeat Steps 2–4; else return $x_{B+1}, x_{B+2}, \ldots, x_{B+N}$.

$\boldsymbol{X}_t = \boldsymbol{x}$, and let $\boldsymbol{y} = (\boldsymbol{y}_1, \boldsymbol{y}_2)$ be the next value, i.e., $\boldsymbol{X}_{t+1} = \boldsymbol{y}$. The Gibbs algorithm sets the transitional kernel as

$$f(\boldsymbol{x}; \boldsymbol{y}) = \pi_1(\boldsymbol{y}_1|\boldsymbol{x}_2)\pi_2(\boldsymbol{y}_2|\boldsymbol{y}_1), \tag{21.54}$$

from which we can readily show (Problem 21.8)

$$\int \pi(\boldsymbol{x}) f(\boldsymbol{x}; \boldsymbol{y})\, d\boldsymbol{x} = \pi(\boldsymbol{y}). \tag{21.55}$$

Thus, $\pi(\boldsymbol{x})$ is the stationary distribution of a Markov chain with the transitional PDF $f(\boldsymbol{x}; \boldsymbol{y})$ defined by (21.54).

In the algorithm, there are no steps that correspond to Steps 3 and 5 of Algorithm 21.3, where the proposed samples are compared against the acceptance ratio, because in the Gibbs algorithm the transitional kernel takes the following form:

$$q_1(\boldsymbol{x}_1; \boldsymbol{y}_1|\boldsymbol{x}_2) = \pi_1(\boldsymbol{y}_1|\boldsymbol{x}_2) \text{ for all } \boldsymbol{x}_1,$$
$$q_2(\boldsymbol{x}_2; \boldsymbol{y}_2|\boldsymbol{x}_1) = \pi_2(\boldsymbol{y}_2|\boldsymbol{x}_1) \text{ for all } \boldsymbol{x}_2,$$

which gives

$$\alpha_1(\boldsymbol{x}_1, \boldsymbol{y}_1|\boldsymbol{x}_2) = \frac{\pi_1(\boldsymbol{y}_1|\boldsymbol{x}_2)\pi_2(\boldsymbol{x}_2|\boldsymbol{x}_2)}{\pi_2(\boldsymbol{x}_2|\boldsymbol{x}_2)\pi_1(\boldsymbol{y}_1|\boldsymbol{x}_2)} = 1,$$
$$\alpha_2(\boldsymbol{x}_2, \boldsymbol{y}_2|\boldsymbol{x}_1) = 1.$$

Extension of the Gibbs algorithm to m blocks of variables $\boldsymbol{x} = (\boldsymbol{x}_1, \boldsymbol{x}_2, \ldots, \boldsymbol{x}_m)$ (Algorithm 21.6) can be done in a similar fashion, when we know how to generate

Algorithm 21.6 Gibbs algorithm with m blocks

1: Choose the initial values $\boldsymbol{x}^{(0)} = (\boldsymbol{x}_1^{(0)}, \boldsymbol{x}_2^{(0)}, \ldots, \boldsymbol{x}_m^{(0)})$. Set $t \leftarrow 1$ and $\boldsymbol{x} \leftarrow \boldsymbol{x}^{(0)}$.
2: For $i = 1, 2, \ldots, m$:
3: Generate $\boldsymbol{x}_i^{(t)}$ from $\pi_i(\boldsymbol{x}_i|\boldsymbol{x}_{-i})$.
4: Set $\boldsymbol{x} \leftarrow \boldsymbol{x}^{(t)}$, where $\boldsymbol{x}^{(t)} = (x_1^{(t)}, x_2^{(t)}, \ldots, x_m^{(t)})$ and set $t \leftarrow t + 1$.
5: If $t < N + B$ repeat Steps 2 and 3; else return $x_{B+1}, x_{B+2}, \ldots, x_{B+N}$.

random variates from the conditional densities $\pi(\boldsymbol{x}_i|\boldsymbol{x}_{-i})$, where $\boldsymbol{x}_{-i}$ contains all the blocks of $\boldsymbol{x}$ except for $\boldsymbol{x}_i$, as defined in (21.53).

21.7.4.1 Gibbs sampling in Bayesian networks

Gibbs sampling is the most frequently used MCMC method for sampling from the joint probability distribution of a BN $(\mathcal{V}, \mathcal{E}, \boldsymbol{P})$, where each full conditional distribution depends only on its Markov blanket, as shown in (21.18):

$$p(x_v|\boldsymbol{x}_{-v}) = p(x_v|\boldsymbol{x}_{\partial v}) \propto p(x_v|\boldsymbol{x}_{\mathrm{pa}(v)}) \prod_{k \in \mathrm{ch}(v)} p(x_k|\boldsymbol{x}_{\mathrm{pa}(k)}). \tag{21.56}$$

This Markovian property simplifies sampling from the full conditionals. Then, by making the number of blocks m be equal to $V = |\mathcal{V}|$, the number of node variables (hence, each block variable is a scalar variable), the above Gibbs sampling with V blocks can generate samples from the joint distribution $p(\boldsymbol{x})$ of the BN, where $\boldsymbol{x} = (x_1, x_2, \ldots, x_V)$. This approach is used in the Bayesian updating with Gibbs sampling (BUGS) software package [121].

21.7.5 Simulated annealing

In many applications, we need to find the value $\boldsymbol{x}$ that maximizes a given PDF $\pi(\boldsymbol{x})$ (or PMF $p_{\boldsymbol{x}}$). Such a need arises when we want to find an MLE or MAP estimate. In principle, we can achieve this by sampling from the given distribution and finding a sample $\boldsymbol{x}$ that maximizes $\pi(\boldsymbol{x})$:

$$\boldsymbol{x}_{\max} = \arg \max_{\boldsymbol{x} \in \mathbf{R}^m} \pi(\boldsymbol{x}). \tag{21.57}$$

However, this approach may not always work well, when the distribution is broad or nearly flat near the point of maximum or when the number of samples is not large enough. To force the Markov process $\boldsymbol{X}^{(t)}$ to move to the vicinity of the global maximum, we apply the MH algorithm to the distribution $g(\boldsymbol{x})^{T_t^{-1}}$, where $g(\boldsymbol{x})$ is a distribution that has the same shape as the target distribution; i.e., $g(\boldsymbol{x}) \propto \pi(\boldsymbol{x})$. T_t is called the system's temperature at time t, and is a monotonically decreasing sequence, converging to zero: $\lim_{t\to\infty} T_t^{-1} = \infty$. The sequence $\{T_t\}$ is called the *cooling schedule*.

As T_t deceases, the distribution $g(\boldsymbol{x})^{T_t^{-1}}$ becomes concentrated around $\boldsymbol{x}_{\max}$. Thus, if we apply an MCMC simulation method to this skewed distribution, we will be sampling

Algorithm 21.7 Simulated annealing algorithm

1: Choose the initial value $\boldsymbol{x}^{(0)}$ and T_0. Set $\boldsymbol{x} \leftarrow \boldsymbol{x}^{(0)}$, $T \leftarrow T^{(0)}$, and $t \leftarrow 1$.
2: Generate the proposed next state $\boldsymbol{y}$ from $q(\boldsymbol{x}; \boldsymbol{y})$ and $u \sim U(0, 1)$
3: If $u \le \frac{g(\boldsymbol{y})^{T^{-1}} q(\boldsymbol{y};\boldsymbol{x})}{g(\boldsymbol{x})^{T^{-1}} q(\boldsymbol{x};\boldsymbol{y})}$, set $\boldsymbol{x}^{(t)} = \boldsymbol{y}$; else $\boldsymbol{x}^{(t)} = \boldsymbol{x}$.
4: Set $\boldsymbol{x} \leftarrow \boldsymbol{x}^{(t)}$, $T \leftarrow T_t$ (according to the cooling schedule), and set $t \leftarrow t + 1$.
5: If $t < N + B$ repeat Steps 2–4; else return $x_{B+1}, x_{B+2}, \ldots, x_{B+N}$.

from the vicinities of $\boldsymbol{x}_{\max}$. This method is called **simulated annealing** because, as the temperature goes to zero, the Markov process will "freeze" at states around $\boldsymbol{x}_{\max}$.

The acceptance ratio is modified according to

$$\alpha^{(t)}(\boldsymbol{x}, \boldsymbol{y}) = \min \left\{ \frac{g(\boldsymbol{y})^{T_t^{-1}} q(\boldsymbol{y}; \boldsymbol{x})}{g(\boldsymbol{x})^{T_t^{-1}} q(\boldsymbol{x}; \boldsymbol{y})}, 1 \right\}, \tag{21.58}$$

where $g(\boldsymbol{x}) \propto \pi(\boldsymbol{x})$. The maximization algorithm is summarized in Algorithm 21.7. Discard the first $B + 1$ burn-in samples and use $\boldsymbol{x}^{(B+1)}, \ldots, \boldsymbol{x}^{(B+N)}$ as usable N samples.

The **M-step** in the EM algorithm, for example, can be done by using this maximization algorithm, when an analytic method is not available, as discussed in Section 19.2.2.

21.8 Summary of Chapter 21

Parent nodes of v:	$\text{pa}(v) \triangleq \{u \in \mathcal{V} : u \to v\}$	(21.14)
Children nodes of u:	$\text{ch}(u) \triangleq \{v \in \mathcal{V} : u \to v\}$	(21.14)
Bayesian network:	$p(\boldsymbol{x}) = \prod_{v \in \mathcal{V}} p(x_v \vert \boldsymbol{x}_{\text{pa}(v)})$	(21.15)
Markov blanket:	$p(x_v \vert \boldsymbol{x}_{\partial v}, \boldsymbol{x}_{\mathcal{U}}) = p(x_v \vert \boldsymbol{x}_{\partial v})$	(21.17)
Markov property:	$p(x_v \vert x_{-v}) = p(x_v \vert \boldsymbol{x}_{\text{pa}(v)}) \prod_{k \in \text{ch}(v)} p(x_k \vert \boldsymbol{x}_{\text{pa}(k)})$	(21.19)
Message $f \to x$:	$\mu_{f \to x}(x) \triangleq \sum_{z_1, \ldots, z_k} f(x, z_1, \ldots, z_k)$ $\cdot \mu_{z_1 \to f}(z_1) \cdots \mu_{z_k \to f}(z_k)$	(21.29)
Message $z_i \to f$:	$\mu_{z_i \to f}(z_i) \triangleq \mu_{h_1 \to z_i}(z_i) \cdots \mu_{h_m \to z_i}(z_i)$	(21.30)
MH acceptance rate:	$\alpha(\boldsymbol{x}, \boldsymbol{y}) = \min \left\{ \frac{\pi(\boldsymbol{y}) q(\boldsymbol{y};\boldsymbol{x})}{\pi(\boldsymbol{x}) q(\boldsymbol{x};\boldsymbol{y})}, 1 \right\}$	(21.47)
Gibbs sampler:	$q_i(\boldsymbol{x}_i; \boldsymbol{y}_i \vert \boldsymbol{x}_{-i}) = \pi_i(\boldsymbol{y}_i \vert \boldsymbol{x}_{-i}), i = 1, 2, \ldots, m$	Alg. 21.6

21.9 Discussion and further reading

Bayesian networks are becoming increasingly popular in various applications because of their generality and ease of use. Applications of BNs include image modeling, Markov random fields, fast Fourier transforms, and turbo codes.

Several problems are important in applications of BNs: evaluation of the posterior joint distribution, finding the most probable path in the network, and estimating the BN parameters. They are solved by algorithms that generalize those used for HMMs, such as the belief propagation algorithm and the EM algorithm. As we mentioned before, Pearl's belief propagation algorithm [265] was the first algorithm to introduce the notion of message passing between the nodes. This algorithm was designed to solve the problem of finding the conditional joint distribution of the BN node values given the values of some other nodes. However, it was later shown [213] that the sum-product algorithm and Pearl's algorithm are equivalent. We presented the sum-product algorithm because it is easier to derive and understand.

Other powerful models that are popular in machine learning include SVMs [337, 338] and ANNs. Support vector machines perform data separation with some surfaces using discriminative learning. The ANNs use sophisticated multivariable and multilayered mapping to model unknown functions using the values of their input (arguments) and the corresponding outputs. In many cases it is possible to use any of these models to solve the same problems. Bayesian networks, however, are more suitable for modeling purposes if we have some prior knowledge about the data.

Computationally efficient methods based on MCMC have made the Bayesian approach practical, and are increasingly used not only in machine learning (Koller and Friedman [206] and Kononenko [211]), but also in information theory (MacKay [234]), econometrics (Greenberg [128]), and bioinformatics (Wilkinson [352]).

21.10 Problems

Section 21.5: Bayesian networks

21.1 Markov chain as a DBN. Present a Markov chain as a DBN and find a Markov blanket of its state.

21.2 An HMM as a DBN.

(a) Find Markov blankets for the observed and hidden states.
(b) Derive the FBA as a special case of the sum-product algorithm.
(c) Derive the Viterbi algorithm as a special case of the sum-product algorithm.

Section 21.6: Factor graphs

21.3 Conditional probability of a Markov blanket. Prove (21.19).

21.4* Sum-product algorithm for a phylogenetic tree. Develop a sum-product algorithm to compute the probability given in (16.69) for a Markov process defined on a phylogenetic tree as discussed in Section 16.4.2.

Section 21.7: Markov chain Monte Carlo (MCMC) methods

21.5* Second-order Markov chains. Show that there are infinitely many second-order ergodic Markov chains whose stationary distribution is $\pi = (\pi_1, \pi_2)$.

21.6 Nonreversible Monte Carlo.

(a) Show that Markov chains with TPMs

$$P_1 = \begin{bmatrix} 0.1 & 0.2 & 0.7 \\ 0.2 & 0.6 & 0.2 \\ 0.7 & 0.2 & 0.1 \end{bmatrix} \quad \text{and} \quad P_2 \begin{bmatrix} 0.2 & 0.3 & 0.5 \\ 0.3 & 0.5 & 0.2 \\ 0.5 & 0.2 & 0.3 \end{bmatrix}$$

are both reversible and ergodic chains, but their product $P_1 P_2$ represents a nonreversible ergodic Monte Carlo.

(b) Prove that a Markov chain with TPM $P_1 P_2$ converges faster than a chain with TPM P_1 or P_2.

21.7* Stationary distribution in the block MH algorithm. Derive (21.52).

21.8* Stationary distribution in the Gibbs sampler. Show that $\pi(x)$ is the stationary distribution of a Markov chain with the transitional PDF $f(x; y)$ defined by (21.54).

21.9 Gibbs sampler for multidimensional normal distribution. Design a Gibbs sampler to sample from the multidimensional normal distribution.

22 Filtering and prediction of random processes

The estimation of a random variable or process by observing other random variables or processes is an important problem in communications, signal processing and other science and engineering applications. In Chapter 18 we considered a *partially observable RV* $\boldsymbol{X} = (\boldsymbol{Y}, \boldsymbol{Z})$, where the unobservable part $\boldsymbol{Z}$ is called a *latent variable*. In this chapter we study the problem of estimating the unobserved part using samples of the observed part. In Chapter 18 we also considered the problem of estimating RVs, called *random parameters* using the maximum *a posteriori* probability (MAP) estimation procedure. When the prior distribution of the random parameter is unknown, we normally assume a uniform distribution, and then the MAP estimate reduces to the maximum-likelihood estimate (MLE) (see Section 18.1.2): if the prior density is not uniform, the MLE is not optimal and does not possess the nice properties described in Section 18.1.2. If an estimator $\hat{\boldsymbol{\theta}} = \boldsymbol{T}(\boldsymbol{X})$ has a Gaussian distribution $N(\boldsymbol{\mu}, \boldsymbol{\Sigma})$, its log-likelihood function is a quadratic form $(\boldsymbol{t} - \boldsymbol{\mu})^\top \boldsymbol{\Sigma}^{-1}(\boldsymbol{t} - \boldsymbol{\mu})$,[1] and the MLE is obtained by minimizing this quadratic form. If the variance matrix is diagonal, the MLE becomes what is called a *minimum weighted square error* (MWSE) estimate. If all the diagonal terms of $\boldsymbol{\Sigma}$ are equal, the MWSE becomes the *minimum mean square error* (MMSE) estimate. Thus, the MMSE estimator is optimal only under these assumptions cited above.

In this chapter, we will first discuss an important relation between MMSE estimation and the *conditional expectation*, and then introduce *regression analysis*, since its criterion, the *least squares*, is the statistical counterpart of the MSE. We then discuss the theory of filtering and prediction of wide-sense stationary (WSS) processes, which was pioneered by Norbert Wiener [350]. In the final section we will discuss a recursive algorithm for predicting a discrete-time Gauss Markov processes, widely known as the Kalman filtering algorithm [171, 172]. We formulate this problem as a hidden Markov model (HMM).

22.1 Conditional expectation, minimum mean square error estimation and regression analysis

A fundamental problem in statistics and statistical signal processing is the estimation of a random variable given a set of observed RVs. First, we define the concept of mean

[1] The square root of this quadratic form $d_{\mathrm{M}}(\boldsymbol{t}, \boldsymbol{\mu}) = \sqrt{(\boldsymbol{t} - \boldsymbol{\mu})^\top \boldsymbol{\Sigma}^{-1}(\boldsymbol{t} - \boldsymbol{\mu})}$ is sometimes called the *Mahalanobis distance*.

square error (MSE) and a minimum mean square error (MMSE) estimator. Then we look at linear estimators that minimize the MSE. Finally, we discuss the relationship between MMSE estimation and conditional expectation.

22.1.1 Minimum mean square error estimation

In many applications, we wish to estimate the value of an unobservable random variable S in terms of the observed values of a set of RVs

$$\boldsymbol{X} = (X_1, X_2, \ldots, X_m)^\top.$$

For example, we may want to estimate S, a transmitted signal, based on the received signal $\boldsymbol{X}$, which is corrupted by noise and multipath effects of the wireless channel. The estimate $\hat{S}$ is a function of the observed random vector $\boldsymbol{X}$. The vector variable $\boldsymbol{X}$ is often called an *independent variable*[2] or a *predictor variable* and S is called a *dependent variable* or *response variable*, especially in the context of **regression analysis**,[3] which will be discussed in Section 22.1.4.

As we defined in Chapter 18, an *estimate* is a function $T(\boldsymbol{x})$ of a sample $\boldsymbol{x}$, whereas an *estimator* $\hat{S}$ is a function of the RV $\boldsymbol{X}$:

$$\hat{S} = T(\boldsymbol{X}), \tag{22.1}$$

where $T(\cdot)$ has the dimension m. Clearly, $\hat{S}$ is also a random vector, since $\boldsymbol{X}$ is a random vector. The mean square error (MSE) (also called mean squared error) of the estimate $\hat{S}$ is defined as

$$\boxed{\mathcal{E} = E[|S - \hat{S}|^2].} \tag{22.2}$$

It is of interest to determine T such that the MSE is minimized. Such a T is called a minimum mean square error (MMSE) estimator, or a *least squares estimator*.

In order to properly define an *error* of estimating a vector-valued complex RV, we must recall the *inner product* of RVs and the *norm* (see Definition 10.3 of Section 10.1.1). The *inner product* of RVs X and Y was defined by (10.7):

$$\boxed{\langle X, Y\rangle \triangleq E[XY^*].} \tag{22.3}$$

The scalar value $\|X\| \triangleq \sqrt{\langle X, X\rangle} = \sqrt{E[|X|^2]}$ is called the *norm* of the RV X. Recall that in Section 11.2.4 we also defined the notion of *mean square equivalence*. Given two RVs X and Y, if $\|X - Y\|^2 = E[\|X - Y\|^2] = 0$, we say that they are equivalent in mean square and denote this relationship by

$$X \overset{\text{m.s.}}{=} Y.$$

[2] Although $\boldsymbol{X}$ is a vector, it is conventional to call it an independent *variable* rather than an independent *vector*. We may use the terms random vectors and RV interchangeably.

[3] As for the meaning of the concept "regression" in statistical analysis, see the historical remark given in Section 1.2.4.

Thus, if $\mathcal{E} = 0$, we say that S and $\hat{S}$ are *equivalent in mean square* or S equals $\hat{S}$ in mean square. Similarly, $\|X\| = 0$ if and only if $X \overset{\text{m.s.}}{=} 0$.

22.1.2 Linear minimum mean square error estimation

Suppose we restrict ourselves to linear estimators; i.e.,

$$\hat{S} = T(\boldsymbol{X}) = \sum_{i=1}^{m} \beta_i X_i, \tag{22.4}$$

where the β_i are scalar coefficients.[4] If we define a row vector $\boldsymbol{\beta}^\top = (\beta_1, \beta_2, \ldots, \beta_n)$, then (22.4) can be expressed as

$$\hat{S} = \boldsymbol{\beta}^\top \boldsymbol{X}. \tag{22.5}$$

The MSE is a function of the coefficients $\boldsymbol{\beta}$:

$$\begin{aligned}\mathcal{E}(\boldsymbol{\beta}) &= E[|S - \hat{S}|^2] = E[|S|^2] - \boldsymbol{\beta}^{\mathrm{H}} E[S^* \boldsymbol{X}] - E[\boldsymbol{X}^{\mathrm{H}} S]\boldsymbol{\beta}^* + \boldsymbol{\beta}^\top E[\boldsymbol{X}\boldsymbol{X}^{\mathrm{H}}]\boldsymbol{\beta}^* \\ &= r_{ss} - \boldsymbol{\beta}^\top \boldsymbol{r}_{sx} - \boldsymbol{r}_{sx}^{\mathrm{H}} \boldsymbol{\beta}^* + \boldsymbol{\beta}^\top \boldsymbol{R}_{xx} \boldsymbol{\beta}^*,\end{aligned} \tag{22.6}$$

where

$$r_{ss} = E[|S|^2],\ \boldsymbol{r}_{sx} = E[S^* \boldsymbol{X}], \text{ and } \boldsymbol{r}_{xx} = E[\boldsymbol{X}\boldsymbol{X}^{\mathrm{H}}], \tag{22.7}$$

r_{ss} is a real scalar, $\boldsymbol{r}_{sx}$ is a complex-valued column vector of size m, and $\boldsymbol{R}_{xx}$ is an $m \times m$ Hermitian matrix. It is easy to show (Problem 22.1) that the quadratic polynomial (22.6) is minimized when

$$\boxed{\boldsymbol{R}_{xx} \boldsymbol{\beta}^* = \boldsymbol{r}_{sx}.} \tag{22.8}$$

Thus, if $\boldsymbol{R}_{xx}$ is positive definite, the optimal coefficient vector is given as

$$\boldsymbol{\beta}_{\text{opt}} = \boldsymbol{R}_{xx}^{-1} \boldsymbol{r}_{sx}^*. \tag{22.9}$$

Thus, the *linear MMSE estimate* or *linear least squares estimate* is

$$\hat{S}_{\text{opt}} = \boldsymbol{\beta}_{\text{opt}}^\top \boldsymbol{X} = \boldsymbol{r}_{sx}^{\mathrm{H}} \boldsymbol{R}_{xx}^{-1} \boldsymbol{X}, \tag{22.10}$$

where we used the self-adjoint (or Hermitian) property of the correlation matrix; i.e., $\boldsymbol{R}_{xx}^{\mathrm{H}} = \boldsymbol{R}_{xx}$. Then, the MMSE can be found by substituting (22.9) into (22.6):

$$\mathcal{E}_{\min} = \mathcal{E}(\boldsymbol{\beta}_{\text{opt}}) = r_{ss} - \boldsymbol{r}_{sx}^{\mathrm{H}} \boldsymbol{R}_{xx}^{-1} \boldsymbol{r}_{sx}. \tag{22.11}$$

The condition (22.8) can be rewritten as (see Problem 22.2)

$$\boxed{E[(\hat{S} - S)\boldsymbol{X}^*] = \boldsymbol{0},} \tag{22.12}$$

whose geometric interpretation is given below (see (22.23)).

[4] Alternatively, we can denote the linear coefficients by β_i^* instead of β_i, which might somewhat simplify some of the subsequent equations, by denoting $*\top$ by a single symbol H standing for Hermitian.

22.1.2.1 Unbiased linear minimum mean square error estimate

The linear MMSE estimate of the form (22.4) has the *bias*

$$b = E[\hat{S}] - E[S] = \boldsymbol{\beta}^\top E[\boldsymbol{X}] - \mu_s. \tag{22.13}$$

In order to obtain an unbiased linear estimate, we replace (22.4) by

$$\hat{S} = \beta_0 + \boldsymbol{\beta}^\top \boldsymbol{X}. \tag{22.14}$$

An estimator of the form (22.14) can actually be seen as a special case of the linear form (22.5), obtained by introducing a random variable X_0 that is identically equal to the constant one. In this case $\hat{S} = \sum_{i=0}^{m} \beta_i X_i$. One can then show that, to minimize the MSE, β_0 must satisfy the equation

$$\mu_s = \beta_0 + \boldsymbol{\beta}^\top E[\boldsymbol{X}]; \tag{22.15}$$

that is,

$$\beta_0 = \mu_s - \boldsymbol{\beta}^\top E[\boldsymbol{X}] = -b, \tag{22.16}$$

which means that the constant β_0 cancels the bias (22.13).

Subtracting (22.15) from (22.14), we obtain

$$\hat{S} - \mu_s = \boldsymbol{\beta}^\top (\boldsymbol{X} - E[\boldsymbol{X}]), \tag{22.17}$$

which is essentially (22.5), with $\hat{S}$ replaced by $\hat{S} - \mu_s$ and $\boldsymbol{X}$ replaced by $\boldsymbol{X} - E[\boldsymbol{X}]$. Thus, (22.14) gives a linear MMSE estimator of S as a function of $\boldsymbol{X}$ if and only if (22.16) holds, and $\boldsymbol{\beta}$ satisfies (Problem 22.4)

$$\boldsymbol{C_{xx}} \cdot \boldsymbol{\beta} = \boldsymbol{c_{sx}}, \tag{22.18}$$

where $\boldsymbol{c_{sx}}$ is a *vector* whose tth element is $\text{Cov}(S, X_i)$ and $\boldsymbol{C_{xx}}$ is the *covariance matrix* of the vector variable $\boldsymbol{X}$:

$$\boldsymbol{c_{sx}} = \boldsymbol{r_{sx}} - \mu_s E[\boldsymbol{X}], \tag{22.19}$$

$$\boldsymbol{C_{xx}} = \boldsymbol{R_{xx}} - E[\boldsymbol{X}]E[\boldsymbol{X}^\top]. \tag{22.20}$$

The condition (22.18) implies (see Problem 22.2)

$$\text{Cov}[\hat{S} - S, \boldsymbol{X}] = \boldsymbol{0}. \tag{22.21}$$

Recall the notion of *orthogonality* between two RVs defined in Definition 10.3 of Section 10.1.1. The RVs X and Y are said to be *orthogonal* if $\langle X, Y \rangle = E[XY^*] = 0$. Then the condition (22.12) can be rewritten as

$$\langle \hat{S} - S, X_i \rangle = 0, \quad \text{for all } i = 1, 2, \ldots, n, \tag{22.22}$$

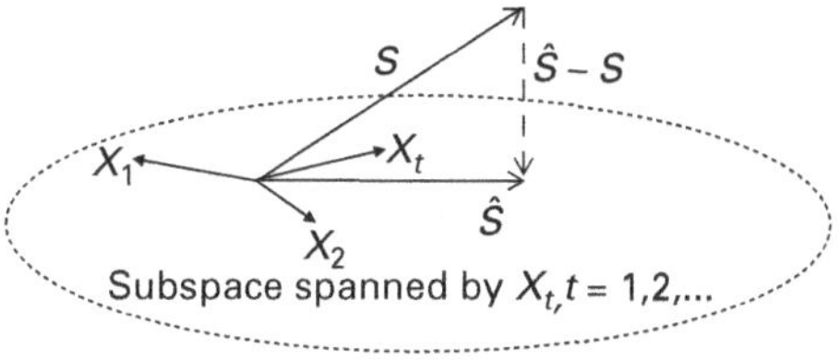

Figure 22.1 $\hat{S}$ as the projection of S onto the subspace spanned by $\boldsymbol{X}$.

which we express in a compact form

$$\boxed{\langle \hat{S} - S, \boldsymbol{X} \rangle = \boldsymbol{0}.} \tag{22.23}$$

The orthogonality expression (22.22), or equivalently (22.23), means that $\hat{S} - S$ is orthogonal to all X_i, $i = 1, 2, \ldots, n$. Therefore $\hat{S} - S$ is orthogonal to the linear subspace spanned by the X_i. Then it is evident that $\hat{S}$ is the **projection** of S onto this linear subspace, as illustrated in Figure 22.1. Recall that, in a vector space with inner product, a *projection* $\boldsymbol{T}$ is a linear transformation from the vector space to itself such that $\boldsymbol{TT} = \boldsymbol{T}$. This definition of projection formalizes the "graphical projection" illustrated in Figure 22.1. The transformation $\boldsymbol{T}$ such that $\boldsymbol{T}S = \hat{S}$ surely satisfies $\boldsymbol{TT} = \boldsymbol{T}$ because $\boldsymbol{T}\hat{S} = \hat{S}$. Recall that we have already discussed the projection or projection matrix a few times in this book: (i) in the eigenvector expansion of the correlation matrix $\boldsymbol{R} = \sum_i \lambda_i \boldsymbol{v}\boldsymbol{v}^{\mathrm{H}} = \sum_i \lambda_i \boldsymbol{E}_i$ (see (13.83) of Section 13.2.1), where $\boldsymbol{E}_i$ are projection matrices; (ii) the transition probability matrix (TPM) $\boldsymbol{P}$ of a discrete-time Markov chain (DTMC) $\boldsymbol{P} = \sum_i \lambda_i \boldsymbol{u}\boldsymbol{v}^{\top} = \sum_i \lambda E_i$ has E_i as projection matrices, and similarly (iii) in a continuous-time Markov chain (CTMC), the projection matrices associated with an eigenvalue of the *infinitesimal generator* (or *transition rate matrix*) $\boldsymbol{Q}$ have essentially the same expansion and their orthogonal projection defined in the same manner (see Example 16.2).

The MMSE $\mathcal{E}_{\min}$ (22.11) is the norm square of the estimation error:

$$\mathcal{E}_{\min} = \|S - \hat{S}\|^2 = \|S\|^2 - \|\hat{S}\|^2, \tag{22.24}$$

which is equivalent to (22.11).

22.1.3 Conditional expectation and minimum mean square error estimation

If all of the available information on S is summarized in the marginal distribution of S, then the MMSE is simply the *mean*, or the *prior expectation*, $E[S]$ (Problem 22.6). If S is stochastically related to another RV $\boldsymbol{X}$ whose value can be observed, then it should not be surprising to assert that the MMSE estimate, given $\boldsymbol{X}$, is the **conditional expectation** or *posterior expectation*, $E[S|\boldsymbol{X}]$. In order to formally prove this, we use the following property of the conditional expectation.

LEMMA 22.1 (Property of the conditional expectation). *Let $h(\boldsymbol{X})$ be any scalar function of the RV $\boldsymbol{X}$. Then the RVs $S - E[S|\boldsymbol{X}]$ and $h(\boldsymbol{X})$ are* orthogonal, *i.e.,*

$$\boxed{\langle S - E[S|\boldsymbol{X}], h(\boldsymbol{X})\rangle = 0.} \tag{22.25}$$

Proof.

$$\begin{aligned}\langle S - E[S|\boldsymbol{X}], h(\boldsymbol{X})\rangle &= E\left[(S - E[S|\boldsymbol{X}])h^*(\boldsymbol{X})\right] \\ &= \int\int (s - E[S|\boldsymbol{x}])h^*(\boldsymbol{x}) f_{S\boldsymbol{X}}(s, \boldsymbol{x})\, ds\, d\boldsymbol{x} \\ &= \int \left(\int (s - E[S|\boldsymbol{x}]) f_{S|\boldsymbol{X}}(s|\boldsymbol{x})\, ds\right) h^*(\boldsymbol{x}) f_{\boldsymbol{X}}(\boldsymbol{x})\, d\boldsymbol{x} = 0,\end{aligned} \tag{22.26}$$

because the term inside the parentheses is zero for all $\boldsymbol{x}$:

$$\begin{aligned}\int (s - E[S|\boldsymbol{x}]) f_{S|\boldsymbol{X}}(s|\boldsymbol{x})\, ds &= \int s f_{S|\boldsymbol{X}}(s|\boldsymbol{x})\, ds - E[S|\boldsymbol{x}] \int f_{S|\boldsymbol{X}}(s|\boldsymbol{x})\, ds \\ &= E[S|\boldsymbol{x}] - E[S|\boldsymbol{x}] = 0.\end{aligned} \tag{22.27}$$

Another, but essentially equivalent, proof can be derived by using the *law of iterated expectations* (see Problem 22.3). □

THEOREM 22.1 (Equivalence between the MMSE estimator and the conditional expectation). *Let $T(\boldsymbol{X})$ be any estimator of S; i.e., any scalar function of $\boldsymbol{X}$. Then the following inequality holds:*

$$\boxed{E\left[(S - T(\boldsymbol{X}))^2\right] \geq E\left[(S - E[S|\boldsymbol{X}])^2\right],} \tag{22.28}$$

where the equality holds if and only if $T(\boldsymbol{X}) \overset{\text{m.s.}}{=} E[S|\boldsymbol{X}]$. Thus, the MMSE estimator and the conditional expectation are equivalent in mean square.

Proof.

$$\begin{aligned}\mathcal{E} = E\left[(S - T(\boldsymbol{X}))^2\right] &= E\left[\{(S - E[S|\boldsymbol{X}]) + (E[S|\boldsymbol{X}] - T(\boldsymbol{X}))\}^2\right] \\ &= E\left[(S - E[S|\boldsymbol{X}])^2\right] + 2E\left[(S - E[S|\boldsymbol{X}])(E[S|\boldsymbol{X}] - T(\boldsymbol{X}))\right] \\ &\quad + E\left[(T(\boldsymbol{X}) - E[S|\boldsymbol{X}])^2\right] \\ &= E\left[(S - E[S|\boldsymbol{X}])^2\right] + E\left[(T(\boldsymbol{X}) - E[S|\boldsymbol{X}])^2\right],\end{aligned} \tag{22.29}$$

where the second term in (22.29) can be seen to be zero by defining $h(\boldsymbol{X}) = T(\boldsymbol{X}) - E[S|\boldsymbol{X}]$ in the above lemma. Therefore, in order to minimize the MSE $\mathcal{E}$, we must have

$$E\left[(T(\boldsymbol{X}) - E[S|\boldsymbol{X}])^2\right] = 0.$$

In other words, $E[S|\boldsymbol{X}]$ is mean-square equivalent to $T(\boldsymbol{X})$. □

Example 22.1: Additive noise model. Consider an additive noise model

$$X = S + N,$$

where $S \sim N(\mu_s, \sigma_s^2)$, $N \sim N(0, \sigma_n^2)$. The input or transmitted signal S and the noise N are independent and X is the observed or received signal. Then X and S are jointly normally distributed with

$$\mu_x = E[X] = \mu_s, \sigma_x^2 = \text{Var}[X] = \sigma_s^2 + \sigma_n^2, \text{ and } \text{Cov}[S, X] = \sigma_s^2.$$

The conditional distribution of S given X is also normal with the mean (Problem 22.7)

$$E[S|X] = \mu_s + \rho_{sx}\frac{\sigma_s}{\sigma_x}(X - \mu_x), \tag{22.30}$$

where ρ_{sx} is the correlation coefficient between S and X, and can be obtained as

$$\rho_{sx} = \sqrt{\frac{\sigma_s^2}{\sigma_s^2 + \sigma_n^2}}. \tag{22.31}$$

Thus, the MMSE estimator of S given X can then be written from (22.30) as

$$\hat{S}(X) = E[S|X] = \mu_s + \frac{\text{SNR}}{1 + \text{SNR}}(X - \mu_s), \tag{22.32}$$

where

$$\text{SNR} = \frac{\sigma_s^2}{\sigma_n^2}.$$

Thus, if $\text{SNR} \gg 1$, $\hat{S}(X) \approx X$, whereas if $\text{SNR} \ll 1$, $\hat{S}(X) \approx \mu_s$. In other words, if σ_s^2 is much greater than the σ_n^2, then the MMSE estimator of S given X is approximately X itself. For the opposite extreme, the MMSE estimator of S given X is approximately μ_s, which is the *prior expectation*; i.e., in this case, the observed signal is so corrupted by noise that the *posterior expectation* $E[S|X]$ differs very little from μ_s. □

22.1.4 Regression analysis

The theory of regression [278] is concerned with prediction of a variable Y on the basis of information provided by observable variables $\boldsymbol{X} = (X_1, X_2, \ldots, X_m)$.[5] It is customary to call $\boldsymbol{X}$ an *independent* or *predictor* variable and Y a *dependent* or *response variable*. Sometimes the term "regressor" is used for $\boldsymbol{X}$.

Let us define a statistic $M(\boldsymbol{x})$ by

$$M(\boldsymbol{x}) \triangleq E[Y|\boldsymbol{x}], \tag{22.33}$$

[5] The following theory can be generalized to the case where Y is a vector variable, but here we assume a scalar variable for simplicity of presentation. It seems that simple matrix expressions do not hold when Y, as well as $\boldsymbol{X}$, is a multivariate and when we have n samples of $\boldsymbol{X}$.

which is is called the **regression function**, or simply **regression**, of Y on $\boldsymbol{X} = (X_1, X_2, \ldots, X_m)$. Theorem (22.1) shows that, for any statistic $T(\boldsymbol{x})$, $E\left[Y - T(\boldsymbol{X})|^2\right] \geq E\left[|Y - M(\boldsymbol{X})|^2\right]$; that is, the best predictor that minimizes the MSE is the conditional expectation $M(\boldsymbol{X})$. The error of the MMSE predictor is thus the conditional variance of Y given $\boldsymbol{X}$, denoted as

$$\sigma^2_{Y|\boldsymbol{X}} \triangleq E\left[|Y - M(\boldsymbol{X})|^2\right]. \tag{22.34}$$

It can be shown (Problem 22.9) that the correlation coefficient between any predictor $T(\boldsymbol{X})$ and Y, denoted $\rho_{T,Y}$,

$$\rho_{T,Y} = \frac{\operatorname{Cov}[T, Y]}{\sigma_T \sigma_Y}, \tag{22.35}$$

satisfies the inequality

$$\rho^2_{T,Y} \leq \rho^2_{M,Y}, \tag{22.36}$$

where the equality holds if and only if T is a linear function of M.

The variance of Y can be decomposed as (Problem 22.10)

$$E\left[|Y - E[Y]|^2\right] = E\left[|Y - M|^2\right] + E\left[|M - E[M]|^2\right], \tag{22.37}$$

which can be compactly expressed as

$$\sigma^2_Y = \sigma^2_{Y|\boldsymbol{X}} + \sigma^2_M, \tag{22.38}$$

which is the fundamental equation for **analysis of variance** (ANOVA), as will be explored later. The ratio

$$\eta^2_{Y\boldsymbol{X}} \triangleq \frac{\sigma^2_M}{\sigma^2_Y} = 1 - \frac{\sigma^2_{Y|\boldsymbol{X}}}{\sigma^2_Y} \tag{22.39}$$

is called the **correlation ratio** and approaches unity as the error of the MSE predictor $\sigma^2_{Y|\boldsymbol{X}}$ approaches zero. Thus, $\eta^2_{Y\boldsymbol{X}}$ provides a measure of association between Y and $\boldsymbol{X}$.

22.1.4.1 Linear regression

Now let us investigate a special case, where the regression function $M(\boldsymbol{x})$ is linear in $\boldsymbol{x}$. We consider an arbitrary linear predictor

$$T(\boldsymbol{x}) = \beta_0 + \beta_1 x_1 + \cdots + \beta_m x_m = \beta_0 + \boldsymbol{\beta}^\top \boldsymbol{x}$$

and determine the coefficients that minimize the prediction error

$$\begin{aligned}\mathcal{E} &= E\left[\left|Y - \beta_0 - \boldsymbol{\beta}^\top \boldsymbol{X}\right|^2\right] \\ &= \sigma^2_Y + |b|^2 + \boldsymbol{\beta}^\top \boldsymbol{C}_{xx} \boldsymbol{\beta}^* - \boldsymbol{\beta}^\top \boldsymbol{c}_{xy} - \boldsymbol{c}_{xy} \boldsymbol{\beta}^*,\end{aligned} \tag{22.40}$$

where

$$b = \beta_0 - E[Y] + \boldsymbol{\beta}^\top E[\boldsymbol{X}], \tag{22.41}$$

$$\boldsymbol{c}_{\boldsymbol{x}y} = E\left[(\boldsymbol{X} - E[\boldsymbol{X}])(Y - E[Y])^*\right], \tag{22.42}$$

and $\boldsymbol{C}_{\boldsymbol{xx}}$ is the covariance matrix defined in (22.20). The optimal choices of b, $\boldsymbol{\beta}$, and β_0 are (Problem 22.11) given, respectively, by

$$b_{\text{opt}} = 0, \quad \boldsymbol{C}_{\boldsymbol{xx}}\boldsymbol{\beta}^*_{\text{opt}} = \boldsymbol{c}_{\boldsymbol{x}y}, \quad \text{and} \quad \beta_{0,\text{opt}} = E[Y] - \boldsymbol{\beta}^\top_{\text{opt}} E[\boldsymbol{X}]. \tag{22.43}$$

Then the minimum value of (22.40) is

$$\mathcal{E}_{\min} = \sigma_Y^2 - \boldsymbol{c}^{\text{H}}_{\boldsymbol{x}y}\boldsymbol{C}^{-1}_{\boldsymbol{xx}}\boldsymbol{c}_{\boldsymbol{x}y}, \tag{22.44}$$

which is essentially the same as (22.11). From (22.43) we see that the linear regression is completely specified by mean values, variances, and covariances of the variables.

The correlation ratio of (22.39) can be written for the linear regression as

$$\eta^2_{y\boldsymbol{x}} = 1 - \frac{\sigma^2_{y|\boldsymbol{x}}}{\sigma^2_y} = \frac{\boldsymbol{c}^{\text{H}}_{\boldsymbol{x}y}\boldsymbol{C}^{-1}_{\boldsymbol{xx}}\boldsymbol{c}_{\boldsymbol{x}y}}{\sigma^2_y}. \tag{22.45}$$

22.1.4.2 Best linear predictors

We have seen above that if the regression $M(\boldsymbol{x}) = E[Y|\boldsymbol{x}]$ is a linear function of $\boldsymbol{x}$, the best predictor can be determined in terms of means, variances and covariances of the concerned variables. Then if the linear assumption of regression function does not hold, what is the best linear predictor? As a matter of fact, we already investigated this question in Section 22.1.2 and found that the best linear MMSE estimate should satisfy the orthogonality equation given by (22.23). The reader is suggested to show the equivalence of that solution and the linear regression determined by (22.43) (Problem 22.12). We state this important fact as a theorem.

THEOREM 22.2. *The linear predictor $T_{\text{opt}}(\boldsymbol{X}) = \beta_{\text{opt}} + \boldsymbol{\beta}^\top_{\text{opt}}\boldsymbol{X}$ given by (22.43) is the MMSE linear predictor of Y. It is also the linear predictor that has the maximum correlation with Y.*

22.1.4.3 Statistical analysis of regression[6]

With the theoretical background given above, we are now ready to present regression analysis often applied to empirical data in a variety of fields, including biostatistics, econometrics, and machine learning, in which empirical models are important.

We assume that the response variable Y and the predictive variables $\boldsymbol{X} = (X_1, X_2, \ldots, X_m)$ are related by

$$Y = f(\boldsymbol{X}; \boldsymbol{\theta}) + \epsilon, \tag{22.46}$$

[6] The material of this section is based on Kobayashi [197] pp. 377–393.

where the functional form $f(\boldsymbol{X};\boldsymbol{\theta})$ is known or assumed except for the parameter

$$\boldsymbol{\theta} = (\theta_1, \theta_2, \ldots, \theta_p).$$

The quantity ϵ represents the effect of uncontrolled factors or "noise" inherent in the measurement. Our prior knowledge of the structured model describes $f(\boldsymbol{X})$ except for certain parameters that are to be estimated from the data. When the theory is insufficient to provide this information, we may resort to the empirical expedient of using some arbitrarily selected function that seems to have about the right shape; this procedure may be useful for *interpolation*, but it cannot be safely used for *extrapolation*.

Suppose that there are n observations y_i, where

$$y_j = f(\boldsymbol{x}_j;\boldsymbol{\theta}) + \epsilon,\ j = 1, 2, \ldots, n, \tag{22.47}$$

and $\boldsymbol{x}_j = (x_{1j}, x_{2j}, \ldots, x_{mj})$ is the jth sample of the predictor multivariable $\boldsymbol{X}$. Thus, the data of the vector RV $\boldsymbol{X}$ may be presented in a two-dimensional array, or matrix, denoted as $\mathbf{X}$, which is called *panel data* in econometrics:

$$\mathbf{X} = \begin{bmatrix} x_{11} & x_{12} & \cdots & x_{1n} \\ x_{21} & x_{22} & \cdots & x_{2n} \\ \vdots & \vdots & \ddots & \vdots \\ x_{m1} & x_{m2} & \cdots & x_{mn} \end{bmatrix}. \tag{22.48}$$

Recall that we used the same representation of panel data in principal component analysis (PCA) and singular value decomposition (SVD) discussed in Section 13.3. The estimation of the parameters $\theta_1, \theta_2, \ldots, \theta_p$ is usually done by the **method of least squares** (see Section 1.2.4 for a historical remark on this method), which can be viewed as the statistical analysis counterpart of the MMSE discussed in the preceding section. We let

$$Q = \sum_{j=1}^{n} \left[y_j - f(\boldsymbol{x}_j;\boldsymbol{\theta})\right]^2 \tag{22.49}$$

represent, as a function of the unknown parameters $\boldsymbol{\theta}$, the sum of the squares of deviation of the observed points from the functional curve. The least-square estimates of the parameters are those that minimize Q. Estimating an equation of the form (22.46) is equivalent to fitting a curve through a *scatter diagram* of plotting y_i versus $\boldsymbol{x}_i$, and this is called the *regression* of y on $\boldsymbol{x}$. This is consistent with our earlier definition of regression function or regression given in (22.33), where the regression of the response variable Y on $\boldsymbol{X}$ is defined as the conditional expectation $E[Y|\boldsymbol{X}] \triangleq M(\boldsymbol{X})$.

We denote the sample average of y_j by

$$\overline{y} = \frac{\sum_{j=1}^{n} y_i}{n}, \tag{22.50}$$

and the *predicted value*, that is, the value on the *regression curve* corresponding to $\boldsymbol{x}_j$, by

$$\hat{y}_j = f(\boldsymbol{x}_j;\boldsymbol{\theta}),\ j = 1, 2, \ldots, n. \tag{22.51}$$

Then the **total sum of the squares** of deviation from the predicted value is given by

$$Q \triangleq \sum_{j=1}^{n}(y_j - \hat{y})^2. \tag{22.52}$$

In view of the ANOVA suggested by (22.38), the above Q can be decomposed as

$$Q = \sum_{j=1}^{n}(y_j - \hat{y}_j)^2 + \sum_{j=1}^{n}(\hat{y}_j - \overline{y})^2 \tag{22.53}$$

$$= Q_{\mathrm{aR}} + Q_{\mathrm{bR}}, \tag{22.54}$$

where the term Q_{aR} represents the variation *about the regression curve* and Q_{bR} is the sum of the squares of the deviations *between the values on the regression curve*; that is, the variation of Y explained by regression.

Example 22.2: Linear regression. Consider, for instance, the following simple regression equation:

$$f(\boldsymbol{x}) = \beta_0 + \sum_{i=1}^{m} \beta_i x_i = \beta_0 + \boldsymbol{\beta}^\top \boldsymbol{x}, \tag{22.55}$$

where $\boldsymbol{\beta} = (\beta_1, \beta_2, \ldots, \beta_m)$. Then the sum of the squares of the deviations is given by

$$Q = \sum_{j=1}^{n}(y_j - \beta_0 - \boldsymbol{\beta}^\top \boldsymbol{x}_j)^2, \tag{22.56}$$

and we obtain the least-square estimates of *regression coefficients* as

$$\hat{\beta} = \frac{\sum_{j=1}^{n}(x_j - \overline{x})y_j}{\sum_{j=1}^{n}(x_j - \overline{x})^2} \tag{22.57}$$

and

$$\hat{\beta}_0 = \overline{y} - \hat{\boldsymbol{\beta}}^\top \overline{\boldsymbol{x}}, \tag{22.58}$$

where

$$\overline{\boldsymbol{x}} = \frac{\sum_{j=1}^{n} \boldsymbol{x}_j}{n} \tag{22.59}$$

and $\overline{y}$ is as given in (22.50). Therefore, we obtain the following regression equation:

$$f(\boldsymbol{x}) = \overline{y} + \hat{\boldsymbol{\beta}}^\top (\boldsymbol{x} - \overline{\boldsymbol{x}}). \tag{22.60}$$

It can be shown (Problem 22.13) that the estimates $\hat{\beta}_0$ and $\hat{\boldsymbol{\beta}}$ have the following properties:

$$E[\hat{\beta}_0] = \beta_0, \mathrm{Var}[\hat{\beta}_0] = \frac{\sigma_\epsilon^2}{m}\boldsymbol{x}^{\mathrm{H}} \left[\sum_{j=1}^{n}(\boldsymbol{x}_i - \overline{\boldsymbol{x}})(\boldsymbol{x}_i - \overline{\boldsymbol{x}})^{\mathrm{H}}\right]^{-1} \boldsymbol{x}, \tag{22.61}$$

$$E[\hat{\boldsymbol{\beta}}] = \boldsymbol{\beta}, \operatorname{Var}[\hat{\boldsymbol{\beta}}] = \sigma_\epsilon^2 \left[\sum_{j=1}^{n} (\boldsymbol{x}_i - \overline{\boldsymbol{x}})(\boldsymbol{x}_i - \overline{\boldsymbol{x}})^{\mathrm{H}} \right]^{-1}, \tag{22.62}$$

where the expectation is taken with respect to Y for given $\boldsymbol{X}$; σ_ϵ^2 is the variance of the error ϵ.

The quantity Q_{bR} is the variance of Y explained by the regression and Q is the total variation of Y, and their ratio

$$r_{yx}^2 = \frac{Q_{\mathrm{bR}}}{Q} \tag{22.63}$$

is equal to the square of the sample correlation coefficient between Y and $\boldsymbol{X}$, and is also called the *coefficient of determination*, since it represents the proportion of the total variance of Y explained by fitting the linear regression.

From the relation established by (22.53) and (22.63), the following null hypothesis test on $\boldsymbol{\beta}$ may be constructed: the question is whether the ratio of the explained variance Q_{bR} to the unexplained variance Q_{aR} is sufficiently large to reflect the hypothesis that Y is unrelated to $\boldsymbol{X}$. Specifically, a test of the hypothesis

$$H_0 : \boldsymbol{\beta} = \boldsymbol{0} \tag{22.64}$$

involves forming the ratio

$$F = \frac{Q_{\mathrm{bR}}}{Q_{\mathrm{aR}}/(n-2)} = \frac{\hat{\boldsymbol{\beta}}^{\mathrm{H}} \sum_{j=1}^{n} (\boldsymbol{x}_i - \overline{\boldsymbol{x}})(\boldsymbol{x} - \overline{\boldsymbol{x}})^{\mathrm{H}} \hat{\boldsymbol{\beta}}}{s_{\mathrm{aR}}^2}, \tag{22.65}$$

which has an F-distribution with degrees of freedom 1 and $n-2$ (see Section 7.3 for Fisher's F distribution). The *mean squared dispersion* s_{aR}^2 defined in the last expression is an estimate of σ_ϵ^2. □

For a further discussion on regression analysis, including *multiple regression analysis*, *analysis of covariance*, and case study examples, the reader is directed to, for example, Kobayashi [197] (pp. 377–413) and references therein.

22.2 Linear smoothing and prediction: Wiener filter theory

Classical filters that we study in linear circuit theory are designed in the frequency domain. Norbert Wiener [350] took a different approach: he assumed that a signal process and additive noise are both WSS random processes with known spectral characteristics or autocorrelation and cross-correlation functions. Then he sought a filter that is *physically realizable* (or *causal*) under the criterion of MMSE. Wiener presented his theory for continuous-time processes, but the theory is equally applicable to the *discrete-time case*, as we shall discuss below. The discrete-time case was independently investigated by Kolmogorov and was published in 1941 [209]. Thus, this filter theory is

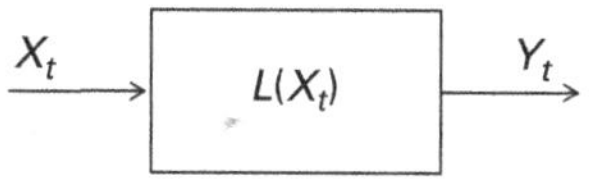

Figure 22.2 A linear system with random process input X_t and output process Y_t.

often referred to as the Wiener–Kolmogorov filter theory. We begin with a brief review of linear time-invariant systems with random inputs.

22.2.1 Linear time-invariant system with random input

Consider the system of Figure 22.2 with a random process X_t as its input and $Y_t = L(X_t)$ as its output process, where t is a *discrete time index*; i.e., $t = 0, 1, 2, \ldots$. The *operator* $L(\cdot)$ that acts on the process X_t is a *linear operator.*

We say that the linear system $L(\cdot)$ is *time invariant* if, for any sample function x_t of the random process X_t and for any finite time index k,

$$y_t = L(x_t) \Longrightarrow L(x_{t+k}) = y_{t+k}. \tag{22.66}$$

In other words, in a time-invariant linear system the only effect of a delay k in the input is a corresponding delay in the output.

A linear time-invariant system can be characterized by its *impulse response* $h[t]$;[7] $t = \ldots, -2, -1, 0, 1, 2, \ldots$, which is the output sequence when the input is the unit pulse at $t = 0$; i.e., $\delta_{t,0}$:

$$h[t] = L(\delta_{t,0}), t = \ldots, -2, -1, 0, 1, 2, \ldots. \tag{22.67}$$

Its Z-transform

$$H(z) = \sum_{t=-\infty}^{\infty} h[t]z^{-t} \tag{22.68}$$

is called the *system function* or *transfer function.*

In a linear time-invariant system, the output random process Y_t is given as the convolution sum between the input random process X_t and the impulse response $h[k]$:

$$Y_t = \sum_{k=-\infty}^{\infty} h[k]X_{t-k} = \sum_{k=-\infty}^{\infty} h[t-k]X_k,^{8} \tag{22.69}$$

where the process X_t, Y_t, and the $h[k]$ may be *complex-valued.*

[7] We use the notation $h[t]$ instead of h_t, because subscripts such as opt cannot be added to h_t. Thus, we adopt mixed notation: X_t, Y_t, etc. for the discrete-time random processes, and $h[t]$, $h[k]$, etc. for discrete-time impulse response sequences.

[8] This expression should be interpreted as the limit in mean square or l.i.m.; that is, $Y_t = \text{l.i.m.}_{m,n\to\infty} \sum_{k=-m}^{n} h[k]X_{t-k}$, meaning $\lim_{m,n\to\infty} E\left[\left|Y_t - \sum_{k=-m}^{n} h[k]X_{t-k}\right|^2\right] = 0$. See Section 11.2.4 on convergence in mean square.

A system is called **causal**, or *physically realizable*, if its output Y_t depends only on its past and present inputs $X_t, X_{t-1}\ldots$; i.e., if and only if

$$h[k] = 0, \text{ for } k < 0, \tag{22.70}$$

which is referred to as the *causality condition*. A system whose output depends on future inputs is called *noncausal* or *physically unrealizable*. If the output depends only on future inputs, the system is called *anti-causal*.

If the input process is stationary, we have

$$E[Y_t] = \sum_{k=-\infty}^{\infty} h[k]E[X_{t-k}] = \mu_x \sum_{k=-\infty}^{\infty} h[k]. \tag{22.71}$$

The autocorrelation function of the output process Y_t is[9]

$$\begin{aligned} R_{yy}[t_1, t_2] &= E[Y_{t_1} Y_{t_2}^*] \\ &= E\left[\sum_{k=-\infty}^{\infty} h[k]X_{t_1-k} \sum_{j=-\infty}^{\infty} h^*[j]X_{t_2-j}^*\right] \\ &= \sum_{k=-\infty}^{\infty}\sum_{j=-\infty}^{\infty} h[k]h^*[j]R_{xx}[t_1-k, t_2-j], \end{aligned} \tag{22.72}$$

with Y_t^* and $h^*[j]$ denoting the complex conjugates of Y_t and $h[j]$ respectively. If X_t is WSS,[10]

$$R_{xx}[t_1-k, t_2-j] \triangleq R_{xx}[d+j-k], \quad \text{where } d = t_1 - t_2.$$

Thus,

$$R_{yy}[t_1, t_2] = \sum_{k=-\infty}^{\infty}\sum_{j=-\infty}^{\infty} h[k]h^*[j]R_{xx}[d+j-k]. \tag{22.73}$$

Hence, $R_{yy}[t_1, t_2]$ is also a function of $d = t_1 - t_2$ only; thus, Y_t is also WSS, and we can write $R_{yy}[t_1, t_2] = R_{yy}[t_1 - t_2]$.

If the linear system can be characterized by an FIR, for instance,

$$\boldsymbol{h}^\top = (h[0], h[1], \ldots, h[n]),$$

(22.73) can have the following vector-matrix representation:

$$R_{yy}[d] = \boldsymbol{h}^\top \boldsymbol{R}_{xx}[d]\boldsymbol{h}^*, \tag{22.74}$$

[9] We use square brackets for the discrete-time argument of the autocorrelation R_{yy}. When we deal with the continuous-time process, we use parentheses; i.e., $R_{yy}(t_1, t_2)$, $Y(t)$, etc.

[10] Recall that we say a continuous-time process $X(t)$ is WSS if its autocorrelation function $R_{xx}(t_1, t_2)$ is a function of the time difference $t_1 - t_2$ only, and we write it as $R_{xx}(t_1 - t_2)$.

where

$$\boldsymbol{R}_{xx}[d] = \begin{bmatrix} R_{xx}[d] & R_{xx}[d+1] & \cdots & R_{xx}[d+n] \\ R_{xx}[d-1] & R_{xx}[d] & \cdots & R_{xx}[d+n-1] \\ \vdots & \vdots & \ddots & \vdots \\ R_{xx}[d-n] & R_{xx}[d-n+1] & \cdots & R_{xx}[d] \end{bmatrix}. \tag{22.75}$$

The reader who wonders why the right-hand side of (22.74) is not given in the form $\boldsymbol{h}^{\mathrm{H}}\boldsymbol{R}_{xx}[d]\boldsymbol{h}$ (where H signifies the complex conjugate and transpose) is suggested to work on Problem 22.14.

We denote the Z-transform of the autocorrelation function (22.73) by $P_{yy}(z)$:

$$P_{yy}(z) = \sum_{k=-\infty}^{\infty} R_{yy}[k]z^{-k} = \sum_{k=-\infty}^{\infty} h[k] \sum_{j=-\infty}^{\infty} h^*[j] \sum_{d=-\infty}^{\infty} R_{xx}[d+j-k]z^{-d}. \tag{22.76}$$

This function is simply related to the *power spectrum* or *(power) spectral density* $P_y(\omega)$ defined as the Fourier transform of the autocorrelation function (see (13.49) of Chapter 13):

$$P_y(\omega) = \frac{1}{2\pi}P_{yy}(e^{i\omega}) = \frac{1}{2\pi}\sum_{k=-\infty}^{\infty} R[k]e^{-ik\omega}, -\pi \le \omega \le \pi, \tag{22.77}$$

where $i = \sqrt{-1}$. Therefore, $P_{yy}(z)$ of (22.76) is also often called the **power spectrum** of the WSS process $Y(t)$.

Letting $p = d + j - k$ or $d = k - j + p$, we can write

$$\begin{aligned} P_{yy}(z) &= \sum_{k=-\infty}^{\infty} h[k]z^{-k} \sum_{j=-\infty}^{\infty} h^*[j]z^{j} \sum_{p=-\infty}^{\infty} R_{xx}[p]z^{-p} \\ &= H(z)H^*(z^{-1})P_{xx}(z). \end{aligned} \tag{22.78}$$

Therefore, we find

$$\boxed{P_y(\omega) = |H(e^{i\omega})|^2 P_x(\omega),} \tag{22.79}$$

where $P_x(\omega)$ is the spectral density of the input process $X(t)$.

22.2.2 Optimal smoothing and prediction of a stationary process

Let the input X_t be a superposition of a stochastic signal S_t and noise process N_t:

$$X_t = S_t + N_t. \tag{22.80}$$

We assume that S_t and N_t are both WSS processes and $E[N_t] = 0, -\infty < t < \infty$.

We want to design a linear filter with an impulse response $h[k]$ that acts on the input in such a way that the output Y_t is the best estimate of S_{t+p} in the mean square sense (Figure 22.3); i.e.,

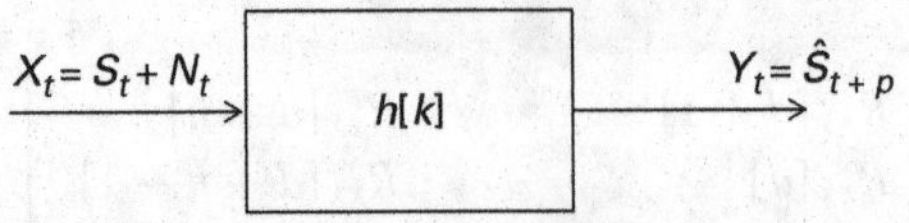

Figure 22.3 A linear predictor $h[k]$ for the signal S_{t+p} based on its noisy past input X_{t-k}, $k \geq 0$.

$$\min_{h[k]} E\left[|Y_t - S_{t+p}|^2\right], \tag{22.81}$$

where

$$Y_t = \sum_{k=-\infty}^{\infty} h[k]X_{t-k}. \tag{22.82}$$

When $p > 0$ in (22.81), the Y_t is a **predicted value** of the signal at p [time units] in the future. When $p \leq 0$, Y_t is called either a **smoothed** estimate or **filtered** estimate of S_{t+p}, depending on the range of k in (22.82). There seems to be no universally accepted precise definition of smoothing versus filtering. But we give here the definition adopted in a majority of the literature. The aim of *filtering* is to find a good estimate of S_{t+p} (for $p \leq 0$) based on the information available at time t; hence, the range of the running variable k in (22.82) is $[0, \infty)$. Thus, when we say filtering, it usually means physically realizable in real-time. The aim of *smoothing* is to estimate S_{t+p} (for $p \leq 0$) by taking account of the information available after time t as well as before t. In this definition, a smoother is equivalent to a physically unrealizable filter, for the case $p \leq 0$. The term "filtering" usually assumes the presence of noise or error in the observed data, whereas smoothing may be applied even to noise-free data. The moving-average procedure, often used in econometrics, may be called smoothing (e.g., "exponentially weighted smoothing"), but such estimation is seldom called filtering, whether it is causal or not. A precise distinction of the two terms, however, is not of our concern here, so we may use the two terms interchangeably for the purpose of this chapter.

In order for a filter to be actually implementable, we must impose the *causality condition* (22.70). With this condition we denote the MSE $\mathcal{E}$ as

$$\mathcal{E} = E\left\{\left|\sum_{k=0}^{\infty} h[k]X_{t-k} - S_{t+p}\right|^2\right\}, \tag{22.83}$$

which can be expanded as

$$\mathcal{E} = \sum_{k=0}^{\infty}\sum_{j=0}^{\infty} h[k]R_{xx}[j-k]h^*[j] - 2\Re\left\{\sum_{j=0}^{\infty} R_{sx}[p+j]h^*[j]\right\} + R_{ss}[0], \tag{22.84}$$

where $R_{xx}[k]$ is the autocorrelation function of $X_t = S_t + N_t$ and $R_{sx}[k]$ is the cross-correlation function between S_t and X_t:

$$R_{xx}[k] = E[X_{t+k}X_t^*] \text{ and } R_{sx}[k] = E[S_{t+k}X_t^*], \tag{22.85}$$

and $R_{ss}[k]$ is the autocorrelation of the stochastic signal S_t; hence, $R_{ss}[0] = E[|S_t|^2]$.

Let $g[k]$ be an arbitrary causal filter (i.e., $g[k]=0, k<0$). Then $h[k]+\delta g[k]$ is also causal. If we replace $h[k]$ by $h[k]+\delta g[k]$ in the above expression for $\mathcal{E}$,

$$\mathcal{E}(\delta)=\mathcal{E}+\delta^2\sum_{k=0}^{\infty}\sum_{j=0}^{\infty}g[k]R_{xx}[j-k]\overline{g}[j] + 2\delta\Re\left\{\sum_{k=0}^{\infty}\sum_{j=0}^{\infty}h[k]R_{xx}[j-k]\overline{g}[j]\right\}-2\delta\Re\left\{\sum_{j=0}^{\infty}R_{sx}[p+j]\overline{g}[j]\right\}. \tag{22.86}$$

In order for $\{h[k]\}$ to be an optimal linear predictor, it is necessary and sufficient that $\mathcal{E}\leq\mathcal{E}(\delta)$ holds for any $\{g[k]\}$ and any real number δ; i.e.,

$$\delta^2\sum_{k=0}^{\infty}\sum_{j=0}^{\infty}g[k]R_{xx}[j-k]\overline{g}[j] + 2\delta\Re\left[\sum_{k=0}^{\infty}\sum_{j=0}^{\infty}h[k]R_{xx}[j-k]\overline{g}[j]-\sum_{j=0}^{\infty}R_{sx}[p+j]\overline{g}[j]\right]\geq 0. \tag{22.87}$$

Then a necessary and sufficient condition for the above inequality to hold is that the expression in the square brackets is zero; that is,

$$\sum_{j=0}^{\infty}\left[\sum_{k=0}^{\infty}h[k]R_{xx}[j-k]-R_{sx}[p+j]\right]\overline{g}[j]=0. \tag{22.88}$$

Since we assume that g_j is an arbitrary function, the above equation holds if and only if

$$\boxed{\sum_{k=0}^{\infty}h[k]R_{xx}[j-k]=R_{sx}[j+p],\quad \text{for all } j\geq 0.} \tag{22.89}$$

Thus, the optimal linear predictor's impulse response $\{h[k]\}$ is given as a solution to the above equation. The continuous-time version of this equation is the following integral equation:

$$\boxed{\int_0^{\infty}h(u)R_{xx}(t-u)\,du=R_{sx}(t+\lambda),\quad \text{for all } t\geq 0,} \tag{22.90}$$

which is known as an integral equation of **Wiener–Hopf type**. The solution $h(t)$ of this integral equation is the impulse response function of an optimum filter that predicts the signal value at time $t+\lambda$ (where $\lambda>0$) based on the observation $\{X(t'); t'\leq t\}$. When $\lambda\leq 0$, the solution is an optimum smoother or filter. An alternative derivation of the above equations (22.89) and (22.90) can be done by using the *orthogonality principle* defined in Definition 10.3 of Section 10.1.1 and also refreshed in arriving at Lemmma 22.1 in the preceding section (see also Problem 22.15).

Our objective is to find a causal filter $\{h[k]\}$ that satisfies the equation (22.89) subject to the condition for the filter to be physically realizable; i.e., $h[k] = 0, k < 0$. This causality condition makes the task of solving (22.89) nontrivial. Thus, we first consider noncausal filters.

22.2.3 Noncausal (or physically unrealizable) filter

Suppose we drop the restriction $h[t] = 0, t < 0$. Then the above equation is replaced by

$$R_{sx}[t+p] = \sum_{k=-\infty}^{\infty} h[k]R_{xx}[t-k], -\infty < t < \infty. \tag{22.91}$$

Taking the Z-transform of both sides, we have

$$\begin{aligned} \text{LHS} &= \sum_{t=-\infty}^{\infty} R_{sx}[t+p]z^{-t} = \sum_{t'=-\infty}^{\infty} R_{sx}[t']z^{-t'+p} \\ &= z^p P_{sx}(z), \end{aligned} \tag{22.92}$$

and similarly,

$$\text{RHS} = \sum_{t=-\infty}^{\infty}\left(\sum_{k=-\infty}^{\infty} h[k]R_{xx}[t-k]\right) z^{-t} = H(z)P_{xx}(z). \tag{22.93}$$

Hence, we can obtain the optimal (but noncausal) filter's transfer function as

$$\boxed{H_{\text{opt}}(z) = \frac{P_{sx}(z)z^p}{P_{xx}(z)}.} \tag{22.94}$$

If the signal process S_t and noise N_t are uncorrelated, we can simplify the cross-correlation function

$$\begin{aligned} R_{sx}[k] &= E[S_{t+k}X_t] = E\left[S_{t+k}\left(S_t + N_t\right)\right] \\ &= R_{ss}[k] + E[S_{t+k}]E[N_t] = R_{ss}[k], \end{aligned} \tag{22.95}$$

where we used the property $E[N_t] = 0$. Hence, by taking the Z-transform of the above equation we find

$$P_{sx}(z) = P_{ss}(z) \text{ for all } z. \tag{22.96}$$

Similarly, we find

$$\begin{aligned} R_{xx}[k] &= E[X_{t+k}X_t] = E\left[\left(S_{t+k} + N_{t+k}\right)\left(S_t + N_t\right)\right] \\ &= R_{ss}[k] + R_{nn}[k], \end{aligned} \tag{22.97}$$

from which we obtain

$$P_{xx}(z) = P_{ss}(z) + P_{nn}(z). \tag{22.98}$$

By substituting the last result into (22.94), we have

$$\boxed{H_{\text{opt}}(z) = \frac{P_{ss}(z)z^p}{P_{ss}(z) + P_{nn}(z)} = \frac{z^p}{1 + \frac{P_{nn}(z)}{P_{ss}(z)}}.} \tag{22.99}$$

The multiplication factor z^p corresponds to advancing the input signal by p time units.

22.2.4 Prediction of a random signal in the absence of noise

Now we consider the special case where the noise is absent, i.e., $N_t = 0, -\infty < t < \infty$, and our task is to design a physically realizable filter $h[k]$ that gives the MMSE in predicting S_{t+p} based on the past input $\{X_{t-k} = S_{t-k}, k \geq 0\}$ (Figure 22.4).

All the information we can use is that the signal process is WSS with its autocorrelation function $R_{ss}[k]$. Thus, if the stochastic signal S_t is replaced by another WSS process Z_t with the same autocorrelation function as $R_{ss}[k]$, the MSE $\mathcal{E}$ will remain unchanged. Hence, the optimal predictor for predicting Z_{t+p} should be the same as the optimal predictor for S_{t+p}.

Now consider a discrete-time white noise sequence[11] W_t of unit spectral density; i.e.,

$$P_{ww}(z) = 1 \text{ for all } z; \text{ equivalently, } R_{ww}[k] = \delta_{k,0}.$$

Let W_t be applied as input to a causal filter $g_s[k]$ to produce the random process Z_t whose second-order statistics (i.e., autocorrelation function or, equivalently, power spectrum) are the same as that of S_t. This is schematically shown in Figure 22.5:

$$Z_t = \sum_{k=-\infty}^{t} g_s[t-k]W_k. \tag{22.100}$$

We choose the Z-transform $G_s(z)$ of the filter impulse response $\{g_s[k]\}$ such that

$$\boxed{G_s(z)G_s^*(z^{-1}) = P_{ss}(z).} \tag{22.101}$$

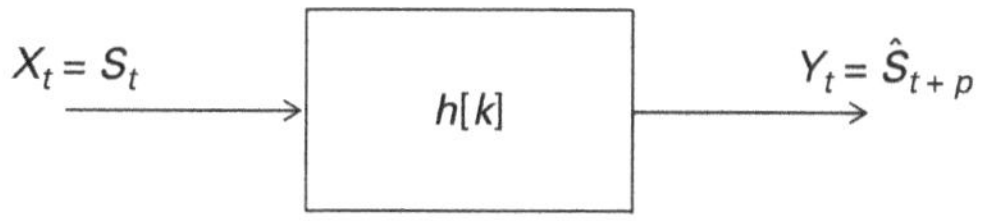

Figure 22.4 A pure predictor $h[k]$ for the signal S_{t+p} based on its noiseless input $X_t = S_t$.

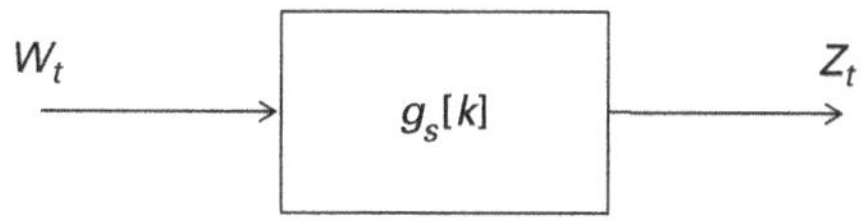

Figure 22.5 White noise W_t passes through a linear filter $\{g_s[k]\}$ to produce Z_t that is statistically equivalent to S_t, the signal of our concern.

[11] In Chapter 17 we used the symbol $W(t)$ for the Wiener process and $Z(t)$ for white noise.

or, equivalently,

$$\boxed{|G_s(e^{i\omega})|^2 = 2\pi P_s(\omega).} \tag{22.102}$$

Then, the power spectrum of the output noise Z_t is given by (Problem 22.16)

$$P_{zz}(z) = G_s(z)P_{ww}(z)G_s^*(z^{-1}) = P_{ss}(z) \text{ for all } z. \tag{22.103}$$

Thus, the stochastic signal S_t and the filtered noise Z_t possess the same autocorrelation function; hence, they are statistically equivalent as far as their predictability is concerned.

Therefore, the signal process S_t is representable as

$$S_t = \sum_{k=-\infty}^{t} g_s[t-k]W_k, \tag{22.104}$$

where $G_s(z) = \mathcal{Z}\{g_s[k]\}$ should satisfy (22.101). From (22.104), we can write the future value of S_t as

$$\begin{aligned} S_{t+p} &= \sum_{k=-\infty}^{t+p} g_s[t+p-k]W_k \\ &= \sum_{k=-\infty}^{t} g_s[t+p-k]W_k + \sum_{k=t+1}^{t+p} g_s[t+p-k]W_k. \end{aligned} \tag{22.105}$$

Note that $\{W_k, -\infty < k < t\}$ and $\{W_k, t < k < t+p\}$ are independent, and a causal filter must predict the value of S_{t+p} solely based on the past input, i.e., $\{S_k, -\infty < k < t\}$, which in turn is a function of $\{W_j, -\infty < j < k < t\}$.

Thus, it is not difficult to conjecture that the optimally predicted value should be the first term of the above equation.

THEOREM 22.3 (Optimum linear pure predictor). *For a WSS signal S_t with autocorrelation function $R_{ss}[k]$, an optimal linear pure prediction of S_{t+p} based on the input $\{S_{t'}, -\infty < t' < t\}$ is given by*

$$\hat{S}_{t+p} = \sum_{k=-\infty}^{t} g_s[t+p-k]W_k, \tag{22.106}$$

where $g_s[k] = \mathcal{Z}^{-1}\{G_s(z)\}$ such that $G_s(z)G_s^*(z^{-1}) = P_{ss}(z) = \mathcal{Z}\{R_{ss}[k]\}$, and W_t is a white noise sequence with unit spectral density.

The minimized MSE is

$$\mathcal{E}_{\min} = E\left[|S_{t+p} - \hat{S}_{t+p}|^2\right] = E\left[\left|\sum_{k=t}^{t+p} g_s[t+p-k]W_k\right|^2\right]. \tag{22.107}$$

Proof. Let

$$\tilde{S}_{t+p} = \sum_{k=-\infty}^{t} f[t-k]W_k \tag{22.108}$$

be an arbitrary linear prediction of S_{t+p}. Then we can show that

$$E\left[|S_{t+p} - \tilde{S}_{t+p}|^2\right] = E\left[|S_{t+p} - \hat{S}_{t+p}|^2\right] + E\left[\left|\sum_{k=-\infty}^{t} (f[t-k] - g_s[t+p-k])W_k\right|^2\right]. \tag{22.109}$$

Thus, an optimum prediction is achieved when

$$\boxed{f[k] = f_s[k] \triangleq g_s[k+p], k \geq 0,} \tag{22.110}$$

or, equivalently, when

$$\tilde{S}_{t+p} = \hat{S}_{t+p} = \sum_{k=-\infty}^{t} g_s[t+p-k]W_k, \tag{22.111}$$

and the MSE is given by the first term of the RHS of (22.109):

$$\mathcal{E}_{\min} = E\left[|S_{t+p} - \hat{S}_{t+p}|^2\right], \tag{22.112}$$

which is equivalent to (22.107). □

From (22.110), the optimal causal filter $f_s[k]$ that acts on white noise W_t is related to $g_s[k]$ according to

$$f_s[k] = \begin{cases} 0, & k < 0, \\ g_s[k+p], & k \geq 0. \end{cases} \tag{22.113}$$

Then, the optimal predicted value of S_{t+p} is obtained by passing the white noise sequence to the filter $f_s[k]$:

$$\hat{S}_{t+p} = \sum_{k=-\infty}^{t} f_s[t-k]W_k. \tag{22.114}$$

In order to represent the relation between $f_s[k]$ and $g_s[k]$ in the transformed domain, we introduce a new notation $[]_+$ as follows:

$$\boxed{B(z) = \mathcal{Z}\{b_k\} \Longrightarrow [B(z)]_+ \triangleq \mathcal{Z}\{b[k]u[k]\},} \tag{22.115}$$

where $u[k]$ is the unit step function defined in the discrete-time domain; i.e.,

$$u[k] = \begin{cases} 1, & k = 0, 1, 2, \ldots, \\ 0, & k = \ldots, -2, -1. \end{cases}$$

Thus, for given $B(z)$, the corresponding $[B(z)]_+$ is found by taking the inverse Z-transform to obtain $\{b[k]\}$, discarding its negative time part, and then taking its Z-transform. If $\{b[k]\}$ is an impulse response function, then $b[k]u[k]$ is its physically realizable part and $[B(z)]_+$ is the corresponding transfer function.

Then, we write (22.113) as

$$\boxed{F_s(z) = \left[G_s(z)z^p\right]_+.} \tag{22.116}$$

Figure 22.6 illustrates these relationships. Since the white noise W_t can be created by passing the signal S_t into a filter with transfer function $1/G_s(z)$, we finally find that the transfer function $H_{\text{opt}}(z)$ of the optimal pure predictor is given by

$$\boxed{H_{\text{opt}}(z) = \frac{1}{G_s(z)}\left[G_s(z)z^p\right]_+,} \tag{22.117}$$

which is schematically shown in Figure 22.7.

Therefore, the impulse response $h_{\text{opt}}[k]$ of the optimal pure predictor is given by the inverse Z-transform of $H_{\text{opt}}(z)$:

$$\boxed{h_{\text{opt}}[k] = \mathcal{Z}^{-1}\left\{\frac{F_s(z)}{G_s(z)}\right\}.} \tag{22.118}$$

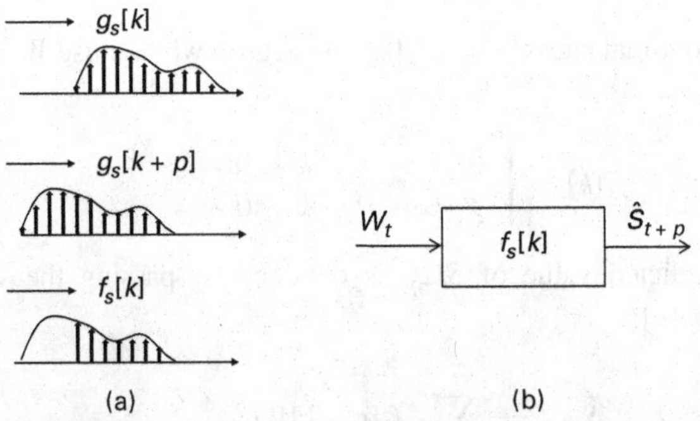

Figure 22.6 (a) The relation between $g_s[k]$ and $f_s[k]$; (b) an optimal predictor $f_s[k] = g_s[k + p]$ that acts on white noise.

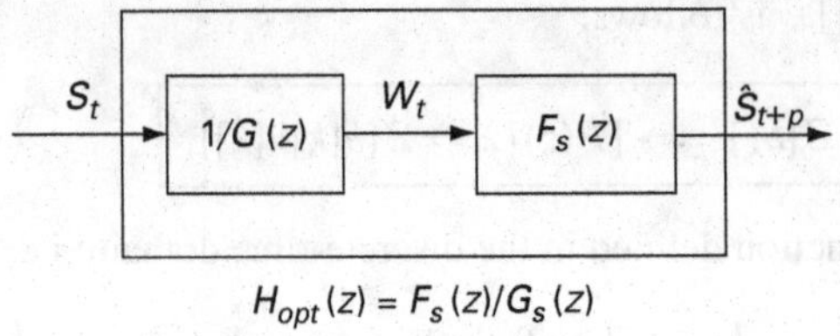

Figure 22.7 An optimal predictor $H_{\text{opt}}(z)$ that acts on the input signal S_t.

In referring to Figure 22.6, we note that

$$\sum_{k=0}^{\infty} |g_s[k]|^2 = E\left[|S_{t+p}|^2\right] = \|S_{t+p}\|^2 \tag{22.119}$$

and

$$\sum_{k=0}^{\infty} |f_s[k]|^2 = \sum_{k=p}^{\infty} |g_s[k]|^2 = E\left[|\hat{S}_{t+p}|^2\right] = \|\hat{S}_{t+p}\|^2 \tag{22.120}$$

represent the norm square of the random variables S_{t+p} and $\hat{S}_{t+p}$ respectively (see Figure 22.17 of Problem 22.15). Thus, the norm square of the prediction error e_{t+p} is given by

$$\|e_{t+p}\|^2 = \|S_{t+p}\|^2 - \|\hat{S}_{t+p}\|^2 = \sum_{k=0}^{p-1} |g_s[k]|^2. \tag{22.121}$$

We provide a more formal derivation of these important results at the end of Section 22.2.5.

Example 22.3: Pure predictor of an autoregressive signal. We consider a pure prediction problem; i.e., $X_t = S_t$. Suppose the signal spectrum takes the form

$$P_{ss}(z) = \frac{|A|^2}{(1-\alpha z^{-1})(1-\overline{\alpha} z)}$$

or, equivalently,

$$P_s(\omega) = \frac{|A|^2}{2\pi\left(1 + |\alpha|^2 - 2\Re\{\alpha\}\cos\omega\right)}. \tag{22.122}$$

Then a physically realizable filter $G_s(z)$ such that $G_s(z)G_s^*(z^{-1}) = P_{ss}(z)$ is found:

$$G_s(z) = \frac{A}{1-\alpha z^{-1}}.$$

Then, by referring to Figure 22.5, the filtered noise $\{Z_t\}$ and the white noise $\{W_t\}$ are related by

$$Z_t = \alpha Z_{t-1} + A W_t, \tag{22.123}$$

and Z_t is an *autoregressive sequence* of first order, often denoted as AR(1) (see Section 13.4.3). If the white noise W_t is Gaussian, then (22.123) implies that Z_t is Gaussian as well as Markovian, and hence it is a Gauss-Markov process (GMP), as defined in Section 13.4.3. Thus, the stochastic signal S_t, which is statistically equivalent to Z_t, is also a GMP. Figure 22.8 shows an example of white Gaussian noise $W_t \sim N(0,1)$, and GMP Z_t (in a solid line), which is statistically equivalent to the signal process S_t. We set $A = 1$ and $\alpha = 2^{-1/4}$.

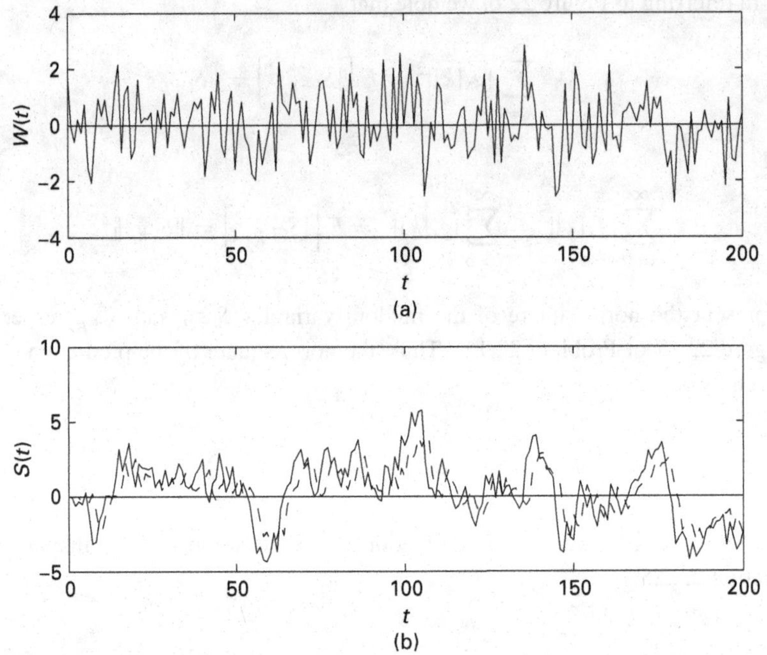

Figure 22.8 (a) White noise W_t that generates (b) a Gauss–Markov signal S_t with a specified power spectral density (22.122). The dashed curve is the output of a pure Wiener predictor, which estimates S_t based on S_{t-2}.

The impulse response function $\{g_s[k]\}$ is readily found as

$$g_s[k] = \begin{cases} 0, & k < 0. \\ A\alpha^k, & k = 0, 1, 2, \ldots. \end{cases} \tag{22.124}$$

Hence, we find

$$f_s[k] = \begin{cases} 0, & k < 0, \\ A\alpha^{k+p}, & k \geq 0. \end{cases}$$

By taking the Z-transform, we obtain the transfer function

$$F_s(z) = \frac{A\alpha^p}{1 - \alpha z^{-1}}.$$

Then, combining this and $1/G_s(z) = A^{-1}(1 - \alpha z^{-1})$, we find that the optimal pure predictor is given by

$$H_{\text{opt}}(z) = \frac{F_s(z)}{G_s(z)} = \alpha^p. \tag{22.125}$$

Thus,

$$\hat{S}_{t+p} = \alpha^p S_t.$$

In the lower figure of Figure 22.8 the dashed curve is the pure prediction $\hat{S}_t = \alpha^p S_{t-p}$, where $p = 2$ time units and $\alpha = 2^{-1/4}$. The predicted waveform is a scaled (by $\alpha^p = 2^{-1/2}$) and shifted (by $p = 2$ time units) version of the actual signal waveform.

It will be instructive to note that Theorem 22.4 (to be presented below) implies that the best linear predictor for this example problem is also the optimal predictor (linear or nonlinear) in the MMSE sense, which is given by the conditional expectation $E[S_{t+p}|S_k, k \leq t]$. The latter reduces to the conditional expectation $E[S_{t+p}|S_t]$, because S_t here is a *Markov process*. Furthermore, S_{t+p} and S_t are bivariate normal RVs, both having zero mean and variance σ_s^2. The correlation coefficient between them is $\rho = R_{ss}[p]/R_{ss}[0] = \alpha^p$. Applying the formula (4.111) of Section 4.3.1 to the bivariate RV (S_t, S_{t+p}), we find that the conditional distribution of S_{t+p} given S_t is also normal with mean

$$E[S_{t+p}|S_t] = \rho S_t = \alpha^p S_t,\ p \geq 0, \tag{22.126}$$

and variance

$$\text{Var}[S_{t+p}|S_t] = \sigma_s^2(1-\rho^2) = \sigma_s^2\left(1-\alpha^{2p}\right),\ p \geq 0. \tag{22.127}$$

In order to find σ_s^2, we take the absolute square of both sides of (22.123):

$$|Z_t|^2 = |\alpha Z_{t-1}|^2 + |AW_t|^2 + 2\Re\{\alpha \overline{A} Z_{t-1} \overline{W}_t\}.$$

Taking the expectations of both sides and using the independence between Z_{t-1} and W_t, we readily find

$$\sigma_s^2 = |\alpha|^2\sigma_s^2 + |A|^2.$$

Thus,

$$\sigma_s^2 = \frac{|A|^2}{1-|\alpha|^2}. \tag{22.128}$$

The last expression can be also found from the spectrum $P_{ss}(z)$ or $P_s(\omega)$ (Problem 22.18).

The continuous-time version of this GMP has a power spectrum of the form

$$P_{ss}(f) = \frac{|A|^2}{1+|a|^2 f^2}. \tag{22.129}$$

The reader is suggested to obtain the filter $g_s(t)$ and optimum predictor $f_s(t)$ (Problem 22.19). □

Thus far, we have limited ourselves to a class of linear predictors. If we expand to include nonlinear predictors, we should be able to do better; i.e., a nonlinear predictor may provide a smaller mean square prediction error. However, if the stochastic signal S_t is Gaussian, we can claim the following.

THEOREM 22.4 (Optimality of the best linear pure predictor for a Gaussian process). *If the signal S_t is a Gaussian process, the best linear pure predictor $\hat{S}_{t+p}$ is as good as any predictor, linear or nonlinear, in the MMSE sense.*

Proof. Recall that uncorrelated Gaussian RVs are statistically independent. If S_t is Gaussian, the whitened noise W_t is also *Gaussian*. Let Y_t be *any* prediction of S_{t+p} based on the input prior to time t; i.e., $\{S_{t'}, t' \le t\}$. Then Y_t is independent of the future value of the white noise $\{W_k, k > t\}$. Since

$$Z_t = Y_t - \hat{S}_{t+p} = Y_t - \sum_{k=-\infty}^{t} g_s[t+p-k]W_k, \tag{22.130}$$

Z_t is also independent of $\{W_k, k > t\}$. Furthermore, by writing $S_{t+p} - \hat{S}_{t+p}$ as

$$S_{t+p} - \hat{S}_{t+p} = \sum_{k=t+1}^{t+p} g_s[t+p-k]W_k,$$

we find that Z_t and $S_{t+p} - \hat{S}_{t+p}$ are independent and, hence, orthogonal. Thus, the MSE of Y_t as a predictor can be written as

$$\begin{aligned} E\left[|S_{t+p} - Y_t|^2\right] &= E\left[\left|S_{t+p} - \hat{S}_{t+p} - Z_t\right|^2\right] \\ &= E\left[\left|S_{t+p} - \hat{S}_{t+p}\right|^2\right] + E\left[|Z_t|^2\right] \\ &\ge E\left[\left|S_{t+p} - \hat{S}_{t+p}\right|^2\right]. \end{aligned} \tag{22.131}$$

Hence, we have shown that the MSE of any prediction Y_t cannot be made smaller than the error of the best linear prediction; i.e., $\hat{S}_{t+p} = \sum_{k=0}^{\infty} h[k]S_{t-k}$. □

In the context of our pure prediction problem, the optimum MSE predictor of S_{t+p} given S_t is the *conditional expectation* $E[S_{t+p}|S_k, u \le t]$, which will be in general a nonlinear function of $S_k, u \le t$. The above theorem states that if S_t is a Gaussian process, then $E[S_{t+p}|S_k, u \le t] = \sum_{k=0}^{\infty} h[k]S_{t-k}$ with some linear predictor $h[k], k \ge 0$.

22.2.5 Prediction of a random signal in the presence of noise

Now let us return to the original model, i.e., the input X_t is signal plus noise, as given in (22.80). The impulse response of the MMSE predictor, $\{h[k]\}$, must satisfy (22.89), the discrete-time version of the Wiener–Hopf integral equation. The main difficulty in solving the above equation, as we pointed out earlier, is to take into account the causality condition; i.e., $h[k] = 0$ for $k < 0$. Were it not for this restriction, the equation would be easily solved by the Z-transform method (or the Fourier transform method for the continuous-time case), as discussed in Section 22.2.3.

The first step in solving the general case is similar to the one adopted in the pure prediction problem discussed in the preceding section; that is, to factor out $P_{xx}(z)$:

$$\boxed{P_{xx}(z) = G_x(z)G_x^*(z^{-1}),} \tag{22.132}$$

where $G_x(z)$ is a causal filter; i.e., $g_x[k] = \mathcal{Z}^{-1}\{G_x(z)\} = 0, k < 0$. Recall that in the pure prediction problem $P_{xx}(z)$ reduces to $P_{ss}(z)$, so (22.132) becomes (22.101).

The factorization of the general power spectrum function $P_{xx}(z)$ may be generally a formidable task, but it becomes somewhat easier if $P_{xx}(z)$ is a *rational function* of z.

Example 22.4: Factorization of $P_{xx}(z)$. Let us assume that both signal and noise are AR(1); i.e., autoregressive sequences of the first-order, as defined in Example 22.3:

$$S_t = \alpha S_{t-1} + AW_t, N_t = \beta N_{t-1} + BW'_t, \tag{22.133}$$

where W_t and W'_t are independent white noise with zero mean and unit variance (not necessarily Gaussian). In Figure 22.9 we plot a sample of S_t, N_t, and their superposition $X_t = S_t + N_t$, by assuming W_t and W'_t are white Gaussian noise $\sim N(0, 1)$ and

$$|A|^2 = |B|^2 = 1, \alpha = 2^{-1/4} \approx 0.8409, \beta = 2^{-1} = 0.5. \tag{22.134}$$

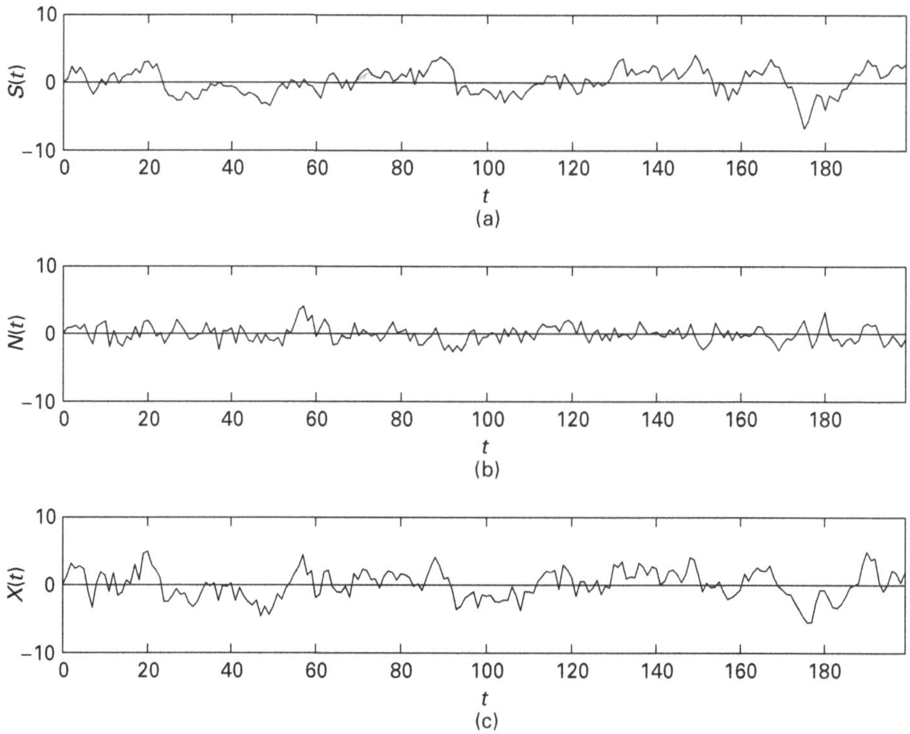

Figure 22.9 (a) Gauss–Markov signal S_t, (b) noise N_t, (c) their superposition $X_t = S_t + N_t$, where SNR is 2.56 (or 4.08 dB).

In this case both signal and noise are GMPs. Using the result of σ_s^2 of (22.128), the signal-to-noise ratio (SNR) is given by

$$\text{SNR} = \frac{\sigma_s^2}{\sigma_n^2} = \frac{|A|^2(1-|\beta|^2)}{|B|^2(1-|\alpha|^2)} = \frac{3}{4(1-2^{-1/2})} = 2.56 \approx 4.08 \text{ dB}.$$

The power spectrums take the following form:

$$P_{ss}(z) = \frac{|A|^2}{(1-\alpha z^{-1})(1-\overline{\alpha} z)}, \; P_{nn}(z) = \frac{|B|^2}{(1-\beta z^{-1})(1-\overline{\beta} z)}. \tag{22.135}$$

The power spectral density of the signal process is

$$P_s(\omega) = \frac{1}{2\pi} P_{ss}(e^{i\omega}) = \frac{|A|^2}{2\pi|1-\alpha e^{-i\omega}|^2} = \frac{1}{2\pi(1-2\Re\{\alpha\}\cos\omega + |\alpha|^2)},$$

where $-\pi \le \omega \le \pi$, and there is a similar expression for the noise process. In Figure 22.10 we plot $P_s(\omega) = P_{ss}(e^{i\omega})/2\pi$ and $P_n(\omega) = P_{nn}(e^{i\omega})/2\pi$ for the parameters set in (22.134).

Since $\log_2 \alpha / \log_2 \beta = 1/4$, we expect the signal bandwidth is to be one-fourth of the noise bandwidth, and clearly Figure 22.10 confirms this observation. The power spectrum of the superposed process $X_t = S_t + N_t$ is

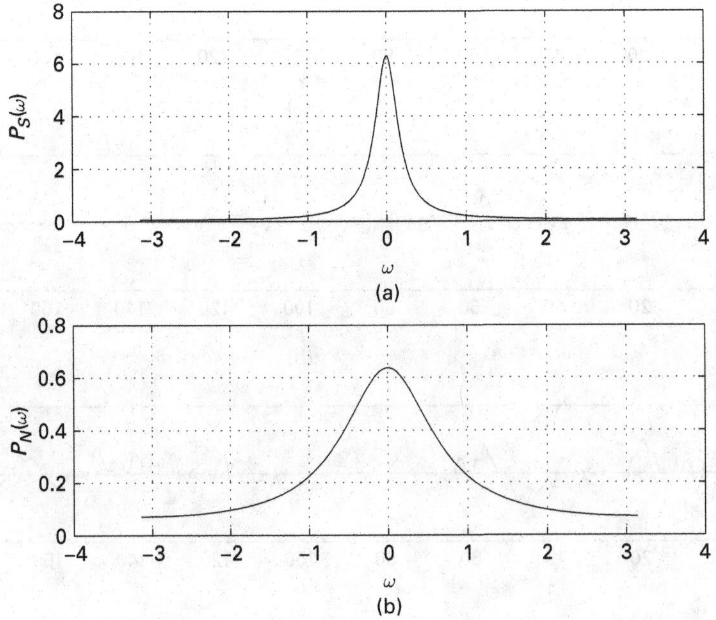

Figure 22.10 (a) Signal and (b) noise power spectrums $P_s(\omega)$ and $P_n(\omega)$, where SNR is 2.56 (or 4.08 dB) and the noise bandwidth is four times as broad as the signal bandwidth.

$$
\begin{aligned}
P_{xx}(z) = P_{ss}(z) + P_{nn}(z) &= \frac{|B|^2(1-\alpha z^{-1})(1-\overline{\alpha} z) + |A|^2(1-\beta z^{-1})(1-\overline{\beta} z)}{(1-\alpha z^{-1})(1-\overline{\alpha} z)(1-\beta z^{-1})(1-\overline{\beta} z)} \\
&= \frac{|C|^2(1-\gamma z^{-1})(1-\overline{\gamma} z)}{(1-\alpha z^{-1})(1-\overline{\alpha} z)(1-\beta z^{-1})(1-\overline{\beta} z)},
\end{aligned}
$$

where $|C|$ and γ are determined by

$$
\begin{aligned}
&|C|^2(1+|\gamma|^2) = |B|^2(1+|\alpha|^2) + |A|^2(1+|\beta|^2), \\
&|B|^2\alpha + |A|^2\beta = |C|^2\gamma.
\end{aligned}
$$

For the parameters chosen above, we find

$$\gamma = 0.5175 \text{ and } |C|^2 = 2.3326.$$

We can factor $P_{xx}(z)$:

$$P_{xx}(z) = G_x(z)G_x^*(z^{-1}), \tag{22.136}$$

where

$$G_x(z) = \frac{C(1-\gamma z^{-1})}{(1-\alpha z^{-1})(1-\beta z^{-1})}.$$

The inverse Z-transform gives

$$g_x[k] = \frac{C(\alpha-\gamma)}{\alpha-\beta}\alpha^k + \frac{C(\gamma-\beta)}{\alpha-\beta}\beta^k,\ k = 0, 1, 2, \ldots.$$

If we choose C to be real and positive among infinitely many possibilities, we find $C = 1.5273$, and we find

$$g_x[k] = 1.4489 \times 2^{-k/4} + 0.0784 \times 2^{-k},\ k = 0, 1, 2, \ldots,$$

which is the sum of two geometrically decaying series. Clearly, the second term dies down four times as fast as the first term as time k progresses. □

Note that all zeros ($z = 0, \gamma$) and poles ($z = \alpha, \beta$) of $G_x(z)$ are within the unit circle ($|z| = 1$), so that the inverse filter $1/G_x(z)$ should be also physically realizable. Such a linear system is called a *minimum-phase system*.

Let us denote the inverse Z-transform of $G_x^*(z^{-1})$ as $g_x^{(-)}[k]$:

$$
\begin{aligned}
g_x^{(-)}[k] &= \mathcal{Z}^{-1}\left\{G_x^*(z^{-1})\right\} = \overline{\mathcal{Z}^{-1}\left\{G_x(z^{-1})\right\}} \\
&= \overline{g_x[-k]} = \overline{g}_x[-k].
\end{aligned}
$$

Therefore, we have established that

$$g_x^{(-)}[k] = 0,\ k > 0. \tag{22.137}$$

This filter is quite opposite to the causal filter $g_x[k]$, in the sense that its output depends only on the future and present inputs. Such a filter is referred to as an *anti-causal filter*, as defined earlier.

Since $P_{xx}(z) = G_x(z)G_x^*(z)$, the inverse Z-transform yields

$$R_{xx}[k] = g_x[k] \circledast g_x^{(-)}[k] = \sum_{j=-\infty}^{0} g_x[k-j]g_x^{(-)}[j]. \tag{22.138}$$

Define $A(z)$ such that

$$\boxed{P_{sx}(z) = A(z)G_x^*(z^{-1}).} \tag{22.139}$$

By taking the inverse Z-transform, we have

$$R_{sx}[k] = a[k] \circledast g_x^{(-)}[k] = \sum_{j=-\infty}^{0} a[k-j]g_x^{(-)}[j], \tag{22.140}$$

where

$$\boxed{a[k] = \mathcal{Z}^{-1}\{A(z)\}.} \tag{22.141}$$

Note that, in the pure prediction problem, $A(z)$ reduces to $G_x(z)$.

By substituting (22.138) and (22.140) into (22.90), we find

$$\sum_{j=-\infty}^{0} a[t+p-j]g_x^{(-)}[j] = \sum_{k=0}^{\infty} h[k] \sum_{j=-\infty}^{0} g_x[t-k-j]g_x^{(-)}[j], t \geq 0, \tag{22.142}$$

or

$$\sum_{j=-\infty}^{0} g_x^{(-)}[j] \left[a[t+p-j] - \sum_{k=0}^{\infty} h[k]g_x[t-k-j] \right] = 0, t \geq 0. \tag{22.143}$$

The last equation is satisfied if the expression in [ ] vanishes for all $j < 0$ and $t \geq 0$; i.e., if

$$a[t+p-j] = \sum_{k=0}^{\infty} h[k]g_x[t-k-j], t \geq 0, j < 0. \tag{22.144}$$

By writing $t - j = t' > 0$, we have

$$\boxed{a[t'+p] = \sum_{k=0}^{\infty} h[k]g_x[t'-k], t' > 0.} \tag{22.145}$$

This equation looks similar to the Wiener–Hopf equation (22.89), but there is an important difference; namely, $g_x[k] = 0$ for $k < 0$ (whereas $R_{xx}[k] \neq 0$ for $k < 0$). Thus,

(22.145) can be solved by using the steps taken for the physically unrealizable case of Section 22.2.3.

We find the Z-transform of the left-hand side of (22.145) to be

$$\begin{aligned} \text{LHS} &= \mathcal{Z}\{a[t'+p]u[t']\} = \left[A(z)z^p\right]_+ \\ &= \left[\frac{P_{sx}(z)z^p}{G_x^*(z^{-1})}\right]_+ \triangleq F_{sx}(z). \end{aligned} \tag{22.146}$$

Since the right-hand side of (22.145) is the convolution of the two causal functions $h[k]$ and $g_x[k]$, its Z-transform is readily found as

$$\text{RHS} = \mathcal{Z}\left\{\sum_{k=0}^{\infty} h[k]g_x[t-k]\right\} = H(z)G_x(z). \tag{22.147}$$

From the last two equations we find the transfer function of the optimal linear predictor:

$$\boxed{H_{\text{opt}}(z) = \frac{F_{sx}(z)}{G_x(z)} = \frac{1}{G_x(z)}\left[\frac{P_{sx}(z)z^p}{G_x^*(z^{-1})}\right]_+.} \tag{22.148}$$

Let us examine the above solution for some special cases.

Pure prediction: If $N_t = 0$ for all t as in the pure prediction problem, then

$$P_{sx}(z) = P_{ss}(z) = G_s(z)G_s^*(z^{-1}).$$

Thus, (22.148) becomes

$$H_{\text{opt}}(z) = \frac{1}{G_s(z)}\left[\frac{P_{ss}(z)z^p}{G_s^*(z^{-1})}\right]_+ = \frac{1}{G_s(z)}\left[G_s(z)z^p\right]_+, \tag{22.149}$$

which is (22.117).

Smoothing: If $p = -d \leq 0$, then it is no longer a prediction problem, but instead a "smoothing" problem; i.e., to filter out the noise N_t as much as possible and obtain the best possible estimate of the signal S_{t-d}. Such an optimal smoothing filter with delay d is found by setting $p = -d$ in (22.148).

Uncorrelated signal and noise: If the noise N_t is uncorrelated with the signal process S_t, then

$$P_{sx}(z) = P_{ss}(z).$$

Then, an optimum p-step predictor is given by

$$H_{\text{opt}}(z) = \frac{1}{G_x(z)}\left[\frac{P_{ss}(z)z^p}{G_x^*(z^{-1})}\right]_+. \tag{22.150}$$

Similarly, an optimum smoothing filter is given by setting $p = -d \leq 0$.

Example 22.5: Optimal predicting filter for uncorrelated signal and noise. Let us consider an optimal prediction problem when the AR signal and AR noise are uncorrelated and their spectrums are given in (22.135) of Example 22.4. In that example, we found that the spectrum $P_{xx}(z) = P_{ss}(z) + P_{nn}(z)$ could be factored as $P_{xx}(z) = G_x(z)G_x^*(z^{-1})$, where

$$G_x(z) = \frac{C(1-\gamma z^{-1})}{(1-\alpha z^{-1})(1-\beta z^{-1})}, \quad G_x^*(z^{-1}) = \frac{\overline{C}(1-\overline{\gamma} z)}{(1-\overline{\alpha} z)(1-\overline{\beta} z)}.$$

Thus,

$$\frac{P_{ss}(z)z^p}{G_x^*(z^{-1})} = \frac{|A|^2}{\overline{C}} \frac{(1-\overline{\beta}z)z^p}{(1-\alpha z^{-1})(1-\overline{\gamma}z)} = \frac{|A|^2}{\overline{C}} \frac{z^{p+1}(1-\overline{\beta}z)}{(z-\alpha)(1-\overline{\gamma}z)}$$
$$= \frac{|A|^2}{\overline{C}} \left[\frac{az^p}{1-\alpha z^{-1}} + \frac{bz^{p+1}}{1-\overline{\gamma}z} \right], \tag{22.151}$$

where

$$a = \frac{1-\alpha\overline{\beta}}{1-\alpha\overline{\gamma}} \text{ and } b = \frac{\overline{\gamma}-\overline{\beta}}{1-\alpha\overline{\gamma}}.$$

If $p \geq 1$, the first term in the brackets [ ] of (22.151) contributes to the causal part, whereas the second term is all noncausal (Problem 22.17). For $p \geq 0$, we find

$$\left[\frac{P_{ss}(z)z^p}{G_x^*(z^{-1})}\right]_+ = \frac{a|A|^2}{\overline{C}} \left[\frac{z^p}{1-\alpha z^{-1}}\right]_+ = \frac{a|A|^2}{\overline{C}} \sum_{k=0}^{\infty} \alpha^k z^{-(k-p)} u_{k-p}$$
$$= \frac{a|A|^2\alpha^p}{\overline{C}} \sum_{t=0}^{\infty} \alpha^t z^{-t} = \frac{a|A|^2\alpha^p}{\overline{C}} \frac{1}{1-\alpha z^{-1}}. \tag{22.152}$$

Thus, the optimal predicting filter is from (22.150) given as

$$\boxed{H_{\text{opt}}(z) = c\alpha^p \frac{1-\beta z^{-1}}{1-\gamma z^{-1}},} \tag{22.153}$$

where

$$c = \frac{a|A|^2}{|C|^2} = \frac{(1-\alpha\overline{\beta})|A|^2}{(1-\alpha\overline{\gamma})|C|^2}.$$

By taking the Z-transform, we obtain the impulse response of the optimal predictor:

$$h_{\text{opt}}[k] = \begin{cases} c\alpha^p, & k=0, \\ c(\gamma-\beta)\alpha^p\gamma^{k-1}, & k=1,2,\ldots, \end{cases} \tag{22.154}$$

which is an almost geometrically decaying function; i.e., $h_{\text{opt}}[k+1]/h_{\text{opt}}[k] = \gamma < 1$ for all $k \geq 1$, but $h_{\text{opt}}[1]/h_{\text{opt}}[0] = \gamma - \beta < \gamma < 1$. The multiplication factor α^p is the same as in the pure predictor (22.125) for the AR signal obtained earlier.

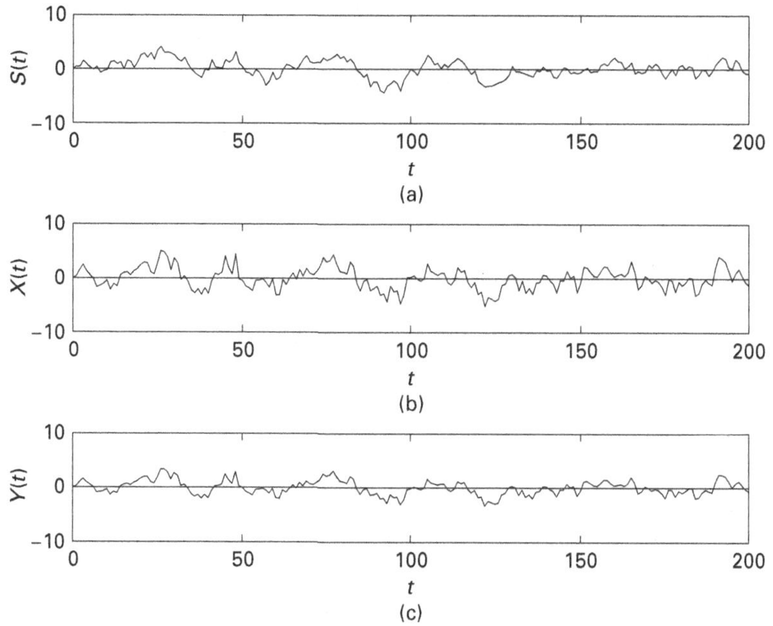

Figure 22.11 (a) The GMP signal S_t, (b) its noisy version X_t, with the same power spectrums as in Figure 22.9, and (c) the filter output $Y_t = \hat{S}_t$ with $p = 0$.

In Figure 22.11 we plot the GMP signal S_t and noisy version $X_t = S_n + N_t$ and the filtered output Y_t, where we set $p = 0$. It is certainly clear that the causal optimal filter (22.154) filters out much of the noise process. Let us examine the above solution for two special cases.

White noise: If we let $\beta = 0$, then the noise spectrum becomes $P_{nn}(z) = B$ for all z; hence, the noise is white. The predicting filter is found from the last equation as

$$H_{\text{opt}}(z) = \frac{c\alpha^p}{1 - \gamma z^{-1}}.$$

Thus, we find the impulse response function of the optimum predicting filter:

$$h_{\text{opt}}[k] = \begin{cases} c\alpha^p \gamma^k, & k = 0, 1, 2, \ldots, \\ 0, & k < 0, \end{cases} \tag{22.155}$$

which is a geometrically decaying function for all $k \geq 0$.

Pure prediction: Now let us consider the case where there is no noise; i.e., $X_t = S_t$. We want to find $h[k]$ that gives the best prediction of S_{t+p}. By setting $B = 0$, we have

$$P_{xx}(z) = P_{ss}(z) = \frac{|A|^2}{(1 - \alpha z^{-1})(1 - \overline{\alpha} z)} \text{ and } G_s(z) = \frac{A}{1 - \alpha z^{-1}}, \; G_s^*(z^{-1}) = \frac{\overline{A}}{1 - \overline{\alpha} z}.$$

Thus, the optimum pure predictor is found from (22.149) as

$$H_{\text{opt}}(z) = (1 - \alpha z^{-1}) \left[\frac{z^p}{1 - \alpha z^{-1}} \right]_+ .$$

Following the derivation step in (22.152), we have

$$\left[\frac{z^p}{1 - \alpha z^{-1}} \right]_+ = \frac{\alpha^p}{1 - \alpha z^{-1}}.$$

Thus, we find

$$H_{\text{opt}}(z) = (1 - \alpha z^{-1}) \frac{\alpha^p}{1 - \alpha z^{-1}} = \alpha^p,$$

which is essentially an "attenuator" by the factor α^p, as obtained in (22.125) of Example 22.3.

When $p = -d < 0$, the above problem becomes an optimal smoothing problem. This is left to the reader as an exercise (Problem 22.20). □

Evaluation of MMSEs: Now that we have found the optimum predictor or smoothing filter, we wish to obtain their performance in terms of the MMSE:

$$\mathcal{E}_{\text{min}} = E\left[\left| S_{t+p} - \hat{S}_{t+p} \right|^2 \right]. \tag{22.156}$$

By substituting the optimally predicted value

$$\hat{S}_t = \sum_{k=0}^{\infty} h_{\text{opt}}[k] X_{t-k}, \tag{22.157}$$

we have

$$\mathcal{E}_{\text{min}} = R_{ss}[0] - 2\Re \left\{ \sum_{j=0}^{\infty} R_{sx}[j+p] h^*_{\text{opt}}[j] \right\} + \sum_{k=0}^{\infty} \sum_{j=0}^{\infty} h_{\text{opt}}[k] R_{xx}[j-k] h^*_{\text{opt}}[j]. \tag{22.158}$$

By substituting

$$R_{sx}[j+p] = \sum_{k=0}^{\infty} h_{\text{opt}}[k] R_{xx}[j-k] \tag{22.159}$$

into the second term of (22.158) we have

$$\mathcal{E}_{\text{min}} = R_{ss}[0] - \sum_{k=0}^{\infty} \sum_{j=0}^{\infty} h_{\text{opt}}[k] R_{xx}[j-k] h^*_{\text{opt}}[j]. \tag{22.160}$$

Using Parseval's formula, we can write the above (Problem 22.21) as

$$\mathcal{E}_{\text{min}} = \frac{1}{2\pi i} \oint P_{ss}(z) \frac{dz}{z} - \frac{1}{2\pi i} \oint H_{\text{opt}}(z) P_{xx}(z) H^*_{\text{opt}}(z^{-1}) \frac{dz}{z}. \tag{22.161}$$

By substituting (22.148) into the above and using $G_x(z)G_x^*(z^{-1}) = P_{xx}(z)$, we finally obtain

$$\boxed{\mathcal{E}_{\min} = \frac{1}{2\pi i}\oint P_{ss}(z)\frac{dz}{z} - \frac{1}{2\pi i}\oint F_{sx}(z)\overline{F}_{sx}(z^{-1})\frac{dz}{z},} \tag{22.162}$$

where $F_{sx}(z)$ was defined in (22.146):

$$\boxed{F_{sx}(z) = \left[\frac{P_{sx}(z)z^p}{G_x^*(z^{-1})}\right]_+ .} \tag{22.163}$$

Pure prediction case: In the *pure prediction* problem, i.e., in the absence of noise, we have

$$P_{sx}(z) = P_{xx}(z) = P_{ss}(z) = G_s(z)G_s^*(z^{-1}),$$

such that

$$g_s[t] = 0, t < 0.$$

Then the minimum prediction error is

$$\mathcal{E}_{\min} = \oint G_s(z)G_s^*(z^{-1})\frac{dz}{z} - \oint F_s(z)\overline{F}_s(z^{-1})\frac{dz}{z}, \tag{22.164}$$

where $F_s(z)$ is an optimum predictor for the whitened signal as shown in Figure 22.6:

$$F_s(z) = \left[G_s(z)z^p\right]_+ .$$

Using the inverse formula $\mathcal{Z}^{-1}\{G_s(f)z^p\} = g_s[k+p]$, we find

$$f_s[k] = \mathcal{Z}^{-1}\{F_s(z)\} = g_s[k+p]u[k].$$

Applying Parseval's formula again, we have an expression for the minimum error in the time-domain:

$$\boxed{\mathcal{E}_{\min} = \sum_{k=0}^{\infty}|g_s[k]|^2 - \sum_{k=0}^{\infty}|f_s[k]|^2 = \sum_{k=0}^{p-1}|g_s[k]|^2.} \tag{22.165}$$

In referring to Figure 22.6 (a), the MMSE of the pure prediction is given by the norm square of the noncausal part of the second impulse response $g_s[k+p]$.

22.3 Kalman filter

The Wiener filter approach discussed in the preceding section has two disadvantages: (i) WSS of the signal and noise is assumed; (ii) solving the Wiener–Hopf equation, or obtaining the corresponding factorization in the Z- or f-domain, is generally difficult, and can be computationally expensive, if not impossible, as the size of the problem grows.

In this section, we will consider the **Kalman filter**, which is a recursive filtering and prediction approach, by assuming a certain structure between successive system states and the observation on which the estimation is based. Our exposition of the Kalman filter is different from a majority of textbooks, in that we provide a useful insight to the Kalman filter model by relating it to the hidden Markov model and its estimation algorithm, the topics we explored in Chapters 18 and 20.

In this section we assume a discrete-time model, although the theory has been developed for the continuous-time case as well. We begin with a discussion of a discrete time state-space model.

22.3.1 State space model

A discrete-time **state-space model** is defined by the following two equations:

$$\boxed{\begin{aligned} \boldsymbol{S}_t &= G_t(\boldsymbol{S}_{t-1}, \boldsymbol{W}_t), \\ \boldsymbol{Y}_t &= H_t(\boldsymbol{S}_t, \boldsymbol{N}_t), \end{aligned}} \tag{22.166}$$

where the sequence $\boldsymbol{S}_t$ is the **system state** process and $\boldsymbol{Y}_t$ represents the **observation** or **output** process. In the context of control theory, $\boldsymbol{W}_t$ and $\boldsymbol{N}_t$ can be *input* or *control* sequences and are often deterministic functions in the absence of disturbance or noise. Thus, in the control theory literature it is more common to represent these processes by lower case symbols; i.e., $\boldsymbol{w}_t$, $\boldsymbol{n}_t$, etc.

In our case, however, we assume that $\boldsymbol{W}_t$ and $\boldsymbol{N}_t$ are independent **white noise sequences**, but they need not be identically distributed for different t. Unlike in the Wiener filter theory, we can assume **nonstationarity** for these sequences, and consequently for the state and observation processes as well. The function G_t in (22.166) determines the structure of how the current state $\boldsymbol{S}_t$ is related to the previous state $\boldsymbol{S}_{t-1}$ and the transition noise. Clearly, $\boldsymbol{S}_t$ is a Markov process. The function H_t determines the structural relation between the observation $\boldsymbol{Y}_t$, the current state $\boldsymbol{S}_t$, and the observation noise $\boldsymbol{N}_t$. The processes $(\boldsymbol{S}_t, \boldsymbol{Y}_t)$ thus defined constitute an HMM, since the state process $\boldsymbol{S}_t$ is a hidden Markov process, and we wish to estimate or predict its value based on the observation process $\boldsymbol{Y}_t$. We assume that the processes $\boldsymbol{S}_t$, $\boldsymbol{Y}_t$, $\boldsymbol{W}_t$, and $\boldsymbol{N}_t$ are vector-valued and their dimensions can be different. Of course, they can be scalars as well. It is important to recognize that the Kalman filter can be discussed in the **framework of an HMM**, which is duly discussed in Chapter 20.

An alternative approach to the above state-space model representation is to characterize stochastic behavior of the processes $\boldsymbol{S}_t$ and $\boldsymbol{Y}_t$ in terms of their *joint and conditional PDFs*. The first equation of (22.166) can be replaced by the state transition PDF $f_{\boldsymbol{S}_t|\boldsymbol{S}_{t-1}}(\boldsymbol{s}_t|\boldsymbol{s}_{t-1})$, while the second equation of (22.166) can be replaced by the conditional PDF $f_{\boldsymbol{Y}_t|\boldsymbol{S}_t}(\boldsymbol{y}_t|\boldsymbol{s}_t)$. If the the state space $\mathcal{S}$ is discrete and finite, the state transition dynamic can be represented by the TPM $\boldsymbol{P}_t = [p_{ij}(t)]$, where $i, j \in \mathcal{S}$ are the state indices.

Our goal is to find an optimal estimate of the state $\boldsymbol{S}_t$ on the basis of an instance of observations $\boldsymbol{y}_0^t = (\boldsymbol{y}_0, \boldsymbol{y}_1, \ldots, \boldsymbol{y}_t)$ (filtering or smoothing) or based on the past output $\boldsymbol{y}_0^{t-1}$ (one-step prediction). Here, the criterion for optimality is the MMSE, which corresponds to the conditional expectation. In the case of *Gaussian processes*, which will be our primary focus in this section, the MAP estimate is indeed equivalent to the MMSE estimate or conditional expectation.

For the discrete-state HMM this problem is solved in Chapter 20 using the *forward algorithm* (Algorithm 20.1). A similar algorithm can be developed for the continuous-state HMM as follows. We denote the **forward variable** by the joint PDF of $(\boldsymbol{S}_t, \boldsymbol{Y}_0^t)$, similar to (20.54) of Chapter 20:

$$\boxed{\alpha_t(\boldsymbol{s}_t, \boldsymbol{y}_0^t) = f_{\boldsymbol{S}_t, \boldsymbol{Y}_0^t}(\boldsymbol{s}_t, \boldsymbol{y}_0^t).} \tag{22.167}$$

By replacing the summation in the forward recursion equation (20.55) by integration, we obtain

$$\alpha_t(\boldsymbol{s}_t, \boldsymbol{y}_0^t) = \int_{\mathcal{S}} f_{\boldsymbol{S}_t, \boldsymbol{Y}_t | \boldsymbol{S}_{t-1}}(\boldsymbol{s}_t, \boldsymbol{y}_t | \boldsymbol{s}_{t-1}) \alpha_{t-1}(\boldsymbol{s}_{t-1}, \boldsymbol{y}_0^{t-1})\, d\boldsymbol{s}_{t-1}, \tag{22.168}$$

where $d\boldsymbol{s}_{t-1}$ should be interpreted as an infinitesimal volume and should be distinguished from a similar notation often used as the tangential infinitesimal vector in the contour integrals (e.g., see $d\boldsymbol{x}$ in (18.16) of Chapter 18).

The MAP estimate of the state at time t based on the observation $\boldsymbol{y}_0^t$, for instance, can be obtained by maximization of the above joint PDF:

$$\hat{\boldsymbol{s}}_t(\boldsymbol{y}_0^t) = \arg\max_{\boldsymbol{s}_t} \alpha_t(\boldsymbol{s}_t, \boldsymbol{y}_0^t). \tag{22.169}$$

The main difficulty associated with this estimation is that we need to compute multidimensional integrals in (22.168).

For the **jointly Gaussian processes** $(\boldsymbol{S}_t, \boldsymbol{Y}_t)$, however, these integrals can be computed analytically, because the Gaussian distribution is completely defined by their means, variances, and covariance. The **Kalman filter algorithm** to be described below can be viewed as an efficient recursive procedure to evaluate the means and variance matrices in the *forward algorithm*.

Before we present the algorithm, it will be useful to transform the forward algorithm (22.168) into equivalent expressions using conditional PDFs instead of the joint PDFs. By dividing both sides of (22.168) by $f_{\boldsymbol{Y}_0^{t-1}}(\boldsymbol{y}_0^{t-1})$ and integrating over the variable $\boldsymbol{y}_t$, we obtain (see Problem 22.22)

$$\boxed{f_{\boldsymbol{S}_t | \boldsymbol{Y}_0^{t-1}}(\boldsymbol{s}_t | \boldsymbol{y}_0^{t-1}) = \int_{\mathcal{S}} f_{\boldsymbol{S}_t | \boldsymbol{S}_{t-1}}(\boldsymbol{s}_t | \boldsymbol{s}_{t-1}) f_{\boldsymbol{S}_{t-1} | \boldsymbol{Y}_0^{t-1}}(\boldsymbol{s}_{t-1} | \boldsymbol{y}_0^{t-1})\, d\boldsymbol{s}_{t-1}.} \tag{22.170}$$

This equation represents what is termed the **propagation step** of the filter algorithm. The maximization of this function gives $\hat{s}_{t|t-1}$, the MAP or MMSE estimate of s_t on the basis of y_0^{t-1}; i.e., *one-step prediction.*

To obtain $\hat{s}_t(y_0^t)$, we use the following equation for the corresponding PDF, which is called the **update step** of the Kalman filter algorithm:

$$f_{S|Y_0^t}(s_t|y_0^t) = \frac{f_{Y_t|S_t}(y_t|s_t) f_{S_t|Y_0^{t-1}}(s_t|y_0^{t-1})}{f_{Y_t|Y_0^{t-1}}(y_t|y_0^{t-1})}. \tag{22.171}$$

This follows from Bayes's theorem (see Problem 22.23). The maximization of this density gives us the filtering estimate $\hat{s}_t(y_0^t)$.

22.3.2 Derivation of the Kalman filter

Consider now a linear discrete-time state-space system that characterizes the state sequence S_t and the observation sequence Y_t:

$$\begin{aligned} S_t &= A_t S_{t-1} + W_t, \\ Y_t &= B_t S_t + N_t, \end{aligned} \tag{22.172}$$

where we assume that W_t and N_t are *independent* and are both *white Gaussian noise*[12] with zero mean and covariance matrices given by

$$E[W_t W_{t'}^\top] = C_{w_t}\delta_{t,t'} \text{ and } E[N_t N_{t'}^\top] = C_{n_t}\delta_{t,t'} \tag{22.173}$$

respectively. We further assume that the initial state variable S_0 is also Gaussian; i.e., $S_0 \sim N(\hat{s}_0, P_0)$, where

$$\hat{s}_0 \triangleq E[S_0],\ P_0 \triangleq \text{Var}[S_0].^{13}$$

The first equation in (22.172) can be viewed as a multivariate analog of the AR signal (22.123) discussed in Example 22.3. Recall that the state-space representation of an ARMA process is given in (13.236) and its relation to the HMM and the Kalman filter

[12] A random process $N(t)$ (a continuous-time process) or N_t (a discrete-time process) is called **white Gaussian noise** if the probability distribution of $N(t)$ or N_t is Gaussian distributed for any given t and its power spectral density is flat; i.e., $P_n(f) = \text{const.}\ \ -\infty < f < \infty$ (for the continuous-time case) or $P_n(\omega) = \frac{1}{2\pi} P_{nn}(e^{i\omega}) = \text{const.}\ \ -\pi < \omega < \pi$. The definition of power spectrum $P_y(\omega)$ of a discrete-time process Y_t and its relation to $P_{yy}(z)$, the Z-transform of the autocorrelation sequence $R_{yy}[k]$, are provided in (22.77).

[13] We define the *variance* of a random vector X by $C_X = E\left[(X - E[X])(X - E[X])^{\text{H}}\right]$, although this matrix is often called the *covariance matrix* in the context of multivariate analysis; i.e., we write $C_X = \left[C_{ij}\right]$, where $C_{ij} = E\left[(X_i - E[X_i])(X_j - E[X_j])^*\right]$ is usually called the covariance between X_i and X_j. We reserve the *covariance* of random vectors to mean $\text{Cov}[X, Y] = E\left[(X - E[X])(Y - E[Y])^{\text{H}}\right]$. Use of P for the variance matrix is unconventional, but P is often used in the Kalman filter literature, because probably the variance corresponds to the "power" of the signal.

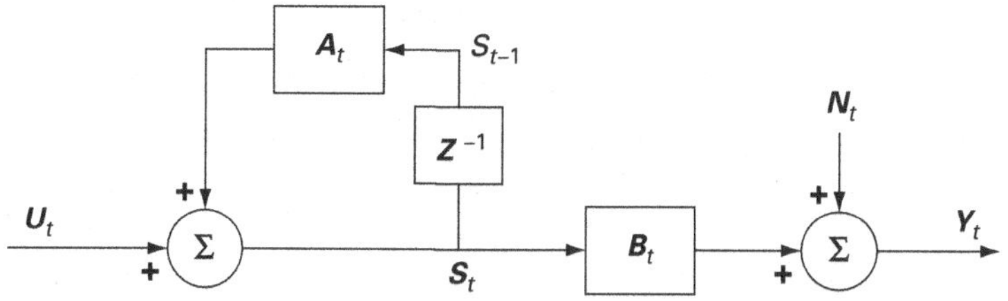

Figure 22.12 A schematic representation of the stochastic signal $\boldsymbol{S}_t$ and the noisy observation $\boldsymbol{Y}_t$ represented by the state-space system (22.172).

is described in Section 13.4.3. In Figure 22.12 we show a schematic structure of the stochastic signal $\boldsymbol{S}_t$ and the noisy observation $\boldsymbol{Y}_t$.

We can show by induction that $(\boldsymbol{Y}_t, \boldsymbol{S}_t)$ are jointly Gaussian. Since $\boldsymbol{S}_t$ is Markovian by definition (see the first equation of (22.172)), $\boldsymbol{S}_t$ is a GMP. As pointed out in the preceding section, the pair process, often called the *complete process*, $\boldsymbol{X}_t = (\boldsymbol{S}_t, \boldsymbol{Y}_t)$ is a hidden Markov process or a partially observable process (see Chapter 20 for details), and in this case it is a Gaussian process, so we may term $\boldsymbol{X}_t$ a **hidden Gauss-Markov process**. The observation sequence $\boldsymbol{Y}_t$ is not Markovian.

According to (22.172) with $t = 0$, $\boldsymbol{S}_0$ and $\boldsymbol{Y}_0 (= \boldsymbol{B}_0\boldsymbol{S}_0 + \boldsymbol{W}_0)$ are jointly Gaussian with mean and covariance matrix given, respectively, by (Problem 22.24)

$$\begin{bmatrix} \hat{\boldsymbol{s}}_0 \\ \boldsymbol{B}_0\hat{\boldsymbol{s}}_0 \end{bmatrix} \text{ and } \begin{bmatrix} \boldsymbol{P}_0 & \boldsymbol{P}_0\boldsymbol{B}_0^\top \\ \boldsymbol{B}_0\boldsymbol{P}_0 & \boldsymbol{B}_0\boldsymbol{P}_0\boldsymbol{B}_0^\top + \boldsymbol{C}_{\boldsymbol{w}_0} \end{bmatrix}. \tag{22.174}$$

22.3.2.1 Kalman filter estimate

Once we have found the bivariate normal distribution, we can find the conditional PDF $f_{\boldsymbol{S}_0|\boldsymbol{Y}_0}(\boldsymbol{s}_0|\boldsymbol{y}_0)$ using the formulae (4.124) and (4.125) of Section 4.3.2. It is also a normal distribution with mean

$$\hat{\boldsymbol{s}}_{0|0} = \hat{\boldsymbol{s}}_0 + \boldsymbol{K}_0\left(\boldsymbol{y}_0 - \hat{\boldsymbol{y}}_0\right) \tag{22.175}$$

and covariance

$$\boldsymbol{P}_{0|0} = \boldsymbol{P}_0 - \boldsymbol{K}_0 \cdot \mathbf{Cov}[\boldsymbol{Y}_0, \boldsymbol{S}_0], \tag{22.176}$$

where

$$\hat{\boldsymbol{y}}_0 \triangleq E[\boldsymbol{Y}_0], \hat{\boldsymbol{s}}_{0|0} \triangleq E[\boldsymbol{S}_0|\boldsymbol{y}_0], \text{ and } \boldsymbol{P}_{0|0} \triangleq \mathbf{Cov}[\boldsymbol{S}_0|\boldsymbol{y}_0],$$

and

$$\begin{aligned} \boldsymbol{K}_0 &= \mathbf{Cov}[\boldsymbol{S}_0, \boldsymbol{Y}_0] \cdot \mathbf{Var}^{-1}[\boldsymbol{Y}_0] \\ &= \boldsymbol{P}_0\boldsymbol{B}_0^\top(\boldsymbol{B}_0\boldsymbol{P}_0\boldsymbol{B}_0^\top + \boldsymbol{C}_{\boldsymbol{n}_0})^{-1} \end{aligned} \tag{22.177}$$

is called the **Kalman gain**.

By substituting these results into (22.174), we *update* the estimate of $\boldsymbol{s}_0$ (an instance of the initial state variable $\boldsymbol{S}_0$) based on the observation $\boldsymbol{y}_0$:

$$\hat{s}_{0|0} = \hat{s}_0 + K_0(y_0 - B_0\hat{s}_0). \tag{22.178}$$

The term $(y_0 - B_0\hat{s}_0)$ is the difference between the observation y_0 and its predicted value based on the prior estimate $\hat{s}_0$ and is called the **innovation**. Thus, the amount of the adjustment in *updating* the estimate of S_0 is the innovation multiplied by the Kalman gain K_0.

The observation y_0 should also help update the covariance matrix of S_0 from the prior value P_0 to the following posterior estimate:

$$P_{0|0} = (I - K_0 B_0)\, P_0, \tag{22.179}$$

where I is the identity matrix. Clearly, the greater the Kalman gain, the greater the reduction in uncertainty regarding the variance estimate.

By setting $t = 0$ in the first equation of (22.172), we find that $S_1 = A_1 S_0 + W_1$ is also Gaussian; hence, (S_1, S_0) are jointly Gaussian distributed. Then, the state transition PDF $f_{S_1|S_0}(s_1|s_0)$ is also normal $\sim N\left(A_1\hat{s}_0, C_{w_1}\right)$. Based on this result and the result for the conditional PDF $f_{S_0|Y_0}(s_0|y_0) \sim N\left(\hat{s}_{0|0}, P_{0|0}\right)$, we find (see Problem 22.26) that $f_{S_1|Y_0}(s_1|y_0) \sim N\left(\hat{s}_{1|0}, P_{1|0}\right)$ with

$$\hat{s}_{1|0} = A_1\hat{s}_{0|0} \text{ and } P_{1|0} = A_1 P_{0|0} A_1^\top + C_{w_1}, \tag{22.180}$$

where $\hat{s}_{1|0}$ is called the **predicted estimate** of s_1 based on the observation y_0.

From these results and (22.172), it follows that the joint variables (S_1, Y_1), conditioned on the observation $Y_0 = y_0$, are also normally distributed with the conditional mean and conditional covariance matrix given, respectively, by

$$\begin{bmatrix} \hat{s}_{1|0} \\ B_1\hat{s}_{1|0} \end{bmatrix} \text{ and } \begin{bmatrix} P_{1|0} & P_{1|0}B_1^\top \\ B_1 P_{1|0} & B_1 P_{1|0} B_1^\top + C_{n_1} \end{bmatrix}. \tag{22.181}$$

Note the similarity between this equation and (22.174). Hence, by repeating the previous derivations, we obtain, similar to (22.178) and (22.179), the following **posterior estimates** (i.e., updated estimates) of S_1 and its covariance:

$$\hat{s}_{1|1} = \hat{s}_{1|0} + K_1\left(y_1 - B_1\hat{s}_{1|0}\right), \tag{22.182}$$

$$P_{1|1} = (I - K_1 B_1)\, P_{1|0}, \tag{22.183}$$

where

$$\hat{s}_{1|1} = E[S_1|y_0^1],\ P_{1|1} = \text{Var}[S_1|y_0^1], \text{ and } y_0^1 = (y_0, y_1).$$

The Kalman gain K_1 is given by

$$K_1 = P_{1|0}B_1^\top\left(B_1 P_{1|0} B_1^\top + C_{n_1}\right)^{-1}. \tag{22.184}$$

Continuing this procedure, we obtain the recursive algorithm for updating and predicting the state estimates and their covariances. By repeating updating and prediction steps in a recursive manner, we obtain

Algorithm 22.1 Kalman filter

1: Initialize:

$$\hat{s}_{0|-1} \triangleq \hat{s}_0 = E[S_0], \quad P_{0|-1} \triangleq P_0 = \mathrm{Var}[S_0].$$

2: For $t = 0, 1, 2, \ldots,$

- Update step:

$$\begin{aligned} K_t &= P_{t|t-1} B_t^\top \left(B_t P_{t|t-1} B_t^\top + C_{n_t} \right)^{-1}, \\ \hat{s}_{t|t} &= \hat{s}_{t|t-1} + K_t \left(y_t - B_t \hat{s}_{t|t-1} \right), \\ P_{t|t} &= \left(I - K_t B_t \right) P_{t|t-1}, \end{aligned}$$

- Prediction step:

$$\begin{aligned} \hat{s}_{t+1|t} &= A_{t+1} \hat{s}_{t|t}, \\ P_{t+1|t} &= A_{t+1} P_{t|t} A_{t+1}^\top + C_{w_{t+1}}. \end{aligned}$$

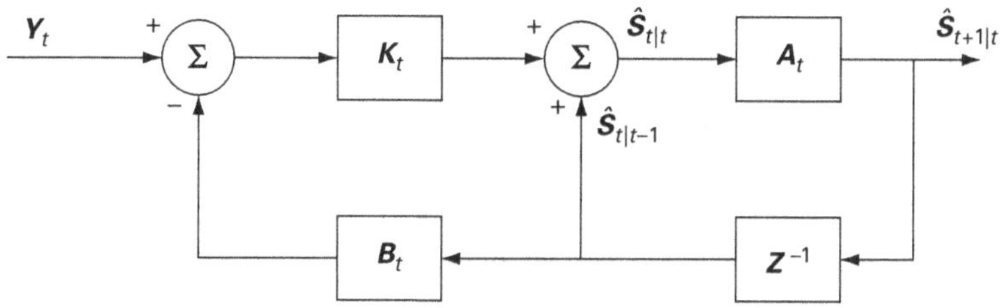

Figure 22.13 The Kalman filtering and prediction system of Algorithm 22.1.

$$\hat{s}_{t|t} = E[S_t | y_0^t], \; P_{t|t} = \mathrm{Var}[S_t | y_0^t], \tag{22.185}$$

$$\hat{s}_{t+1|t} = E[S_{t+1} | y_0^t], \text{ and } P_{t+1|t} = \mathrm{Var}[S_{t+1} | y_0^t], \; t = 0, 1, 2, \ldots, \tag{22.186}$$

where $y_0^t = (y_0, y_1, \ldots, y_t)$. We present this recursive algorithm in Algorithm 22.1.

Note that the state estimate $\hat{s}_{t|t}$ depends on the present observation y_t as well as past observations y_0^{t-1}, but the Kalman gain K_t and the updated covariance $P_{t|t}$ do not depend on the current observation y_t. Similarly, the predicted covariance $P_{t+1|t}$ does not depend on y_t either. Therefore, these matrices can be pre-computed and stored just based on y_0^{t-1}, prior to the arrival of y_t. Thus, this algorithm trades memory for computation time. In Figure 22.13 we show a schematic diagram of the Kalman filtering and prediction algorithm that corresponds to Algorithm 22.1.

Example 22.6: Gauss–Markov process signal and AWGN. Consider the following state-space model

$$S_t = \alpha S_{t-1} + W_t, \; t = 0, 1, 2, \ldots, \tag{22.187}$$

$$Y_t = S_t + N_t, \; t = 0, 1, 2, \ldots, \tag{22.188}$$

where $W_t \sim N(0, \sigma_w^2)$ and $N_t \sim N(0, \sigma_n^2)$ for all t; i.e., stationary white noise. The stochastic signal S_t specified by (22.187) is an AR sequence of first order, denoted AR(1), which was discussed in Examples 22.3, 22.4, and 22.5. The parameters A and B in those examples are now represented by σ_s and σ_n respectively. The parameter β in (22.133) is now equal to zero, since we assume N_t is white. Since we assume W_t is white *Gaussian* noise, the resulting S_t is a GMP as discussed there. We wish to find the Kalman predictor and its MSE $\mathcal{E}(\hat{S}_t)$ as $t \to \infty$.

By referring to (22.172), we find

$$\boldsymbol{A}_t = \alpha, \boldsymbol{B}_t = 1, \quad \text{for all } t = 0, 1, 2, \ldots.$$

Then the Kalman filtering algorithm is summarized as follows:

- Update step:

$$K_t = \frac{P_{t|t-1}}{P_{t|t-1} + \sigma_n^2}, \tag{22.189}$$

$$\hat{s}_{t|t} = \hat{s}_{t|t-1} + K_t \left(y_t - \hat{s}_{t|t-1} \right), \tag{22.190}$$

$$P_{t|t} = (1 - K_t) P_{t|t-1}. \tag{22.191}$$

- Prediction step:

$$\hat{s}_{t+1|t} = \alpha \hat{s}_{t|t}, \tag{22.192}$$

$$P_{t+1|t} = \alpha^2 P_{t|t} + \sigma_w^2. \tag{22.193}$$

Figure 22.14 shows a MATLAB simulation run of the AR(1) signal S_t, the observation process Y_t, and the Kalman filter estimate $\hat{S}_{t|t}$. Since $\sigma_s^2 = \sigma_w^2/(1-\alpha^2) = 1/(1-2^{-1/2})$ and $\sigma_n^2 = 1$, the SNR$= \sqrt{2}/(\sqrt{2}-1) = 3.412 \approx 5.33\,\text{dB}$, which is somewhat larger than the SNR of Examples 22.4 and 22.5, where the noise is not white but is a GMP with the correlation parameter $\beta = 1/2$, which makes $\sigma_n^2 = 4/3$.

Figure 22.15 shows another run, where the bottom curve is the Kalman predictor output $\hat{S}_{t|t-1}$; i.e., an estimate of S_t based on the observation up to time $t-1$. Because of the attenuation factor α as given in (22.192), the Kalman predictor $\hat{S}_{t|t-1}$ is smaller than the Kalman filter $\hat{S}_{t-1|t-1}$ by this factor. This relation between estimation and prediction is similar to what we found in the Wiener filter (see (22.154) with $p = 0$ versus $p = 1$).

From (22.189), (22.191), and (22.193), we readily find that for $\alpha^2 = 2^{-1/2}$, $\sigma_w = \sigma_n = 1$,

$$\lim_{t\to\infty} P_{t|t-1} = \sqrt{2} \approx 1.4142, \text{ and } \lim_{t\to\infty} K_t = \lim_{t\to\infty} P_{t|t} = \frac{\sqrt{2}}{\sqrt{2}+1} \approx 0.5858.$$

Figure 22.16 plots the first 20 time units of the Kalman gain K_t, the estimation error variance $P_t|t$, and the prediction error variance $P_{t|t-1}$. These values converge to their limit values rather quickly; i.e., by $t = 3$.

We find, from (22.190) and (22.192), that the predictor takes the following recursive form:

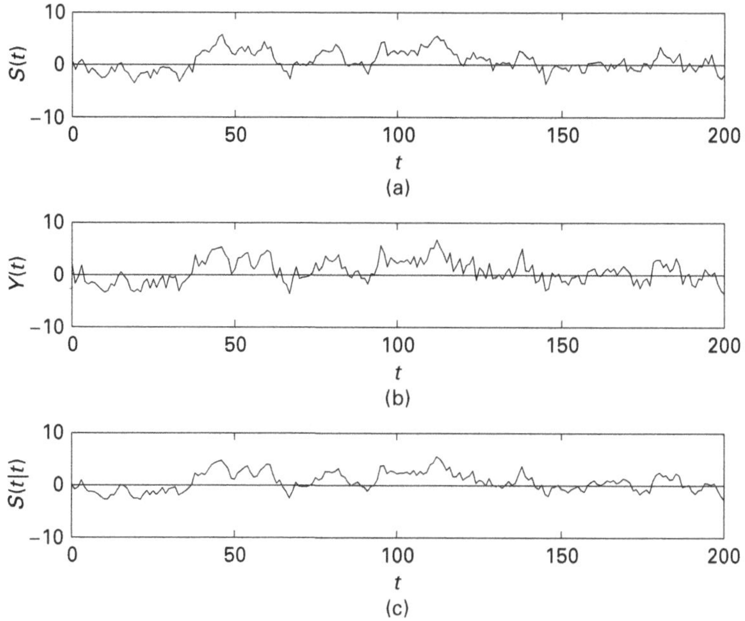

Figure 22.14 (a) GMP signal S_t, (b) the observation sequence Y_t and (c) the Kalman filter output $\hat{S}_{t|t}$.

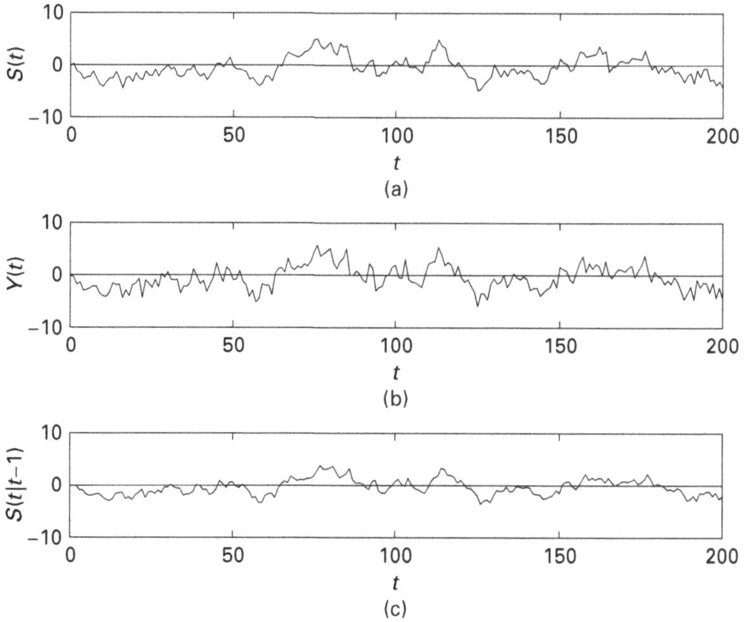

Figure 22.15 (a) GMP signal S_t, (b) the observation sequence Y_t, (c) and the Kalman predictor output $\hat{S}_{t|t-1}$.

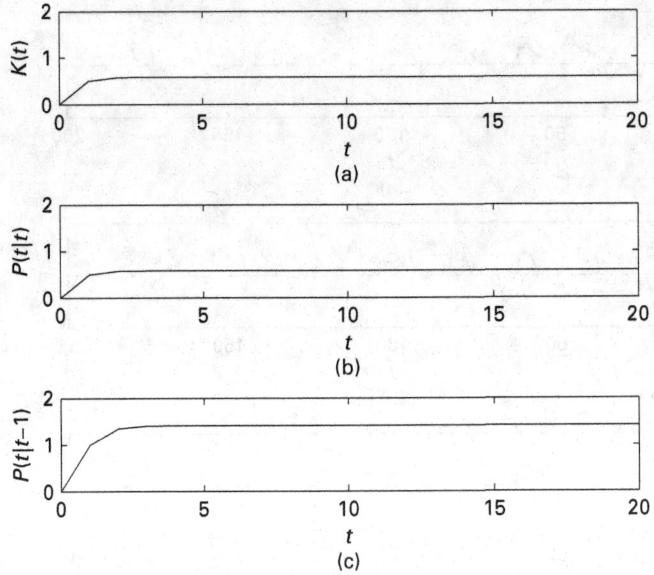

Figure 22.16 (a) Kalman gain K_t, (b) the observation sequence Y_t, (c) and the Kalman predictor output $\hat{S}_{t|t-1}$.

$$\begin{aligned}\hat{s}_{t+1|t} &= \alpha\left[\hat{s}_{t|t-1} + K_t\left(y_t - \hat{s}_{t|t-1}\right)\right] \\ &= \alpha(1-K_t)\hat{s}_{t|t-1} + \alpha K_t y_t.\end{aligned} \tag{22.194}$$

The predictor's MSE is give by

$$\mathcal{E}_{t+1} \triangleq P_{t+1|t}, \tag{22.195}$$

which satisfies the following recursive form:

$$\mathcal{E}_{t+1} = \alpha^2(1-K_t)\mathcal{E}_t + \sigma_w^2 = \frac{\alpha^2\sigma_n^2\mathcal{E}_t}{\mathcal{E}_t + \sigma_n^2} + \sigma_w^2, \tag{22.196}$$

from which we find the MSE in the steady state:

$$\mathcal{E}_\infty = \frac{\sigma_w^2 - (1-\alpha^2)\sigma_n^2 + \sqrt{[\sigma_w^2 - (1-\alpha^2)\sigma_n^2]^2 + 4\sigma_w^2\sigma_n^2}}{2}. \tag{22.197}$$

If $\sigma_n = 0$, i.e., $N_t = 0$, then

$$\mathcal{E}_\infty = \sigma_w^2. \tag{22.198}$$

If $\sigma_w = 0$, i.e., $W_t = 0$, then

$$\mathcal{E}_\infty = \begin{cases} 0, & \text{if } |\alpha| < 1, \\ (\alpha^2-1)\sigma_n^2, & \text{if } |\alpha| \geq 1. \end{cases} \tag{22.199}$$

For sufficiently large t, the Kalman predictor takes the following form:

$$\hat{s}_{t+1|t} = \frac{\alpha[\sigma_n^2 \hat{s}_{t|t-1} + \mathcal{E}_\infty y_t]}{\mathcal{E}_\infty + \sigma_n^2}, \tag{22.200}$$

If $\sigma_n = 0$, then $\hat{s}_{t+1|t} = \alpha y_t$, as expected from (22.187) and (22.188) with $N_t = 0$. □

22.4 Summary of Chapter 22

MSE:	$\mathcal{E} = E[(S - \hat{S})^2]$	(22.2)
$X \overset{\text{m.s.}}{=} Y$:	$E[(X - Y)^2] = 0$	Sec. 22.1.1
Orthogonality principle:	$\langle \hat{S} - S, \boldsymbol{X} \rangle = 0$	(22.23)
Orthogonality of $E[S\|\boldsymbol{X}]$:	$E\,[(S - E[S\|\boldsymbol{X}])\,h(\boldsymbol{X})] = 0$	(22.25)
$E[S\|\boldsymbol{X}]$ is an MMSE estimate:	$E\left[(S - \phi(\boldsymbol{X}))^2\right] \geq E\left[(S - E[S\|\boldsymbol{X}])^2\right]$	(22.28)
Wiener–Hopf eq. (discrete-time):	$\sum_{k=0}^{\infty} h[k] R_{xx}[j-k] = R_{sx}[j+p]$	(22.89)
Wiener–Hopf eq. (cont.-time):	$\int_0^\infty h(u) R_{xx}(t-u)\,du = R_{sx}(t+\lambda)$	(22.90)
Optimal noncausal filter:	$H_{\text{opt}}(z) = \frac{P_{sx}(z) z^p}{P_{xx}(z)}$	(22.94)
Optimal pure predictor:	$H_{\text{opt}}(z) = \frac{1}{G_s(z)}\left[G_s(z) z^p\right]_+$	(22.117)
where	$G_s(z) G_s^*(z^{-1}) = P_{ss}(z)$	(22.101)
Optimal causal filter:	$H^{(\text{opt})}(z) = \frac{1}{G(z)}\left[\frac{P_{sx}(z) z^p}{G_x^*(z^{-1})}\right]_+$	(22.148)
where	$G_x(z) G_x^*(z^{-1}) = P_{xx}(z)$	(22.132)
MMSE in pure prediction:	$\mathcal{E}_{\min} = \sum_{k=0}^{p-1} \|g_s[k]\|^2$	(22.165)
Linear SSM (state transition):	$\boldsymbol{S}_t = \boldsymbol{A}_t \boldsymbol{S}_{t-1} + \boldsymbol{W}_t$	(22.172)
Linear SSM (observable output):	$\boldsymbol{Y}_t = \boldsymbol{B}_t \boldsymbol{S}_t + \boldsymbol{N}_t$	(22.172)
Kalman update:	$\hat{\boldsymbol{s}}_{t\|t} = \hat{\boldsymbol{s}}_{t\|t-1} + \boldsymbol{K}_t\left(\boldsymbol{y}_t - \boldsymbol{B}_t \hat{\boldsymbol{s}}_{t\|t-1}\right)$	Alg. 22.1
Kalman prediction:	$\hat{\boldsymbol{s}}_{t+1\|t} = \boldsymbol{A}_{t+1} \hat{\boldsymbol{s}}_{t\|t}$	Alg. 22.1
Kalman gain:	$\boldsymbol{K}_t = \boldsymbol{P}_{t\|t-1} \boldsymbol{B}_t^\top \left(\boldsymbol{B}_t \boldsymbol{P}_{t\|t-1} \boldsymbol{B}_t^\top + \boldsymbol{C}_{\boldsymbol{n}_t}\right)^{-1}$	Alg. 22.1
Covariance prediction:	$\boldsymbol{P}_{t\|t-1} = \boldsymbol{A}_t \boldsymbol{P}_{t-1\|t-1} \boldsymbol{A}_t^\top + \boldsymbol{C}_{\boldsymbol{w}_t}$	Alg. 22.1
Covariance update:	$\boldsymbol{P}_{t\|t} = (\boldsymbol{I} - \boldsymbol{K}_t \boldsymbol{B}_t)\,\boldsymbol{P}_{t\|t-1}$	Alg. 22.1
Initialization:	$\hat{\boldsymbol{s}}_{0\|-1} = E[\hat{\boldsymbol{S}}_0]$, $\boldsymbol{P}_{0\|-1} = \text{Cov}[\hat{\boldsymbol{S}}_0]$	Alg. 22.1

22.5 Discussion and further reading

Although we have primarily presented the discrete-time version of the Wiener filter theory, the original work by Wiener [350] is concerned with the continuous-time model. Davenport and Root [77] and Thomas [319] treat this topic in more detail than most textbooks on random processes written for electrical engineering students. Kailath *et al.* [170] discuss the subject in great depth. Several books on random processes written by mathematicians also discuss the Wiener filter theory; e.g., Breiman [36], Doob [82], Gikhman and Skorokhod [119], and Karlin and Taylor [175].

The Kalman filter is a recursive MMSE estimator of a signal in additive noise. The Kalman filter can adapt itself to nonstationary environments, whereas the Wiener filter is applicable only to WSS processes. On the other hand, the Kalman filter theory requires that the processes $(\boldsymbol{S}_t, \boldsymbol{Y}(t))$ form a GMP to be solvable, while the the Wiener filter theory does not have these requirements. Even if the signal and/or noise are non-Gaussian, the Kalman filter is still the best linear MMSE estimator. Several textbooks on probability and random processes written for engineering students (e.g., Papoulis and Pillai [262], Stark and Woods [310], and Fine [105]) discuss the Kalman filter to varying degrees, but the aforementioned book by Kailath *et al.* [170] and Hänsler [137] treat this subject most thoroughly.

As we discussed in Section 22.3.1, the Kalman filter can be formulated as a problem of estimating or predicting hidden states in an HMM setting, and the algorithm can be viewed as an efficient recursive procedure to evaluate the mean vectors and variance matrices in the **forward algorithm**. As we discussed in Chapter 20, a **smoothing estimate** of S_t based on the observations y_0^T (where $t \in [0, T]$) can be efficiently computed by a forward–backward algorithm (FBA), called the Bahl–Cooke–Jelinek–Raviv (BCJR) algorithm. A similar FBA, which is called the Rauch–Tung–Striebel (RTS) smoother, has been developed for continuous-state HMMs [279].

22.6 Problems

Section 22.1: Conditional expectation, MMSE estimation and regression analysis

22.1 The condition for MSE. Show that the solution (22.8) minimizes (22.6).

22.2 Linear MMSE condition and the orthogonality. Show that (22.8) is equivalent to (22.12) and that (22.18) is equivalent to (22.21).

22.3* Alternative proof of Lemma 22.1. Apply the law of iterated expectations (cf. (3.38) and (4.106)) to prove Lemma 22.1.

22.4 Derivation of (22.18). Derive (22.18).

22.5 Alternative derivation of (22.18). Define $(n+1)$-dimensional column vectors

$$\tilde{\boldsymbol{X}} = \begin{bmatrix} 1 \\ \boldsymbol{X} \end{bmatrix} \text{ and } \tilde{\boldsymbol{\beta}} = \begin{bmatrix} a_0 \\ \boldsymbol{\beta} \end{bmatrix}.$$

(a) Show that the condition for $\tilde{\boldsymbol{\beta}}$ for the linear MSE estimate is

$$E[\tilde{\boldsymbol{X}}\tilde{\boldsymbol{X}}^{\top}]\tilde{\boldsymbol{\beta}} = E[S\tilde{\boldsymbol{X}}]. \tag{22.201}$$

(b) Show that the above condition is equivalent to (22.15) and (22.18).

22.6 The MMSE estimate and the expectation. Show that if all the available information about S is given in terms of its distribution function $F_S(s)$ or its density function $f_S(s)$, the estimate $\hat{S}$ that minimizes the MSE $\mathcal{E} = E\left[(\hat{S} - S)^2\right]$ is given by the expectation $\hat{S} = \mu_s$.

22.7 Example 22.1: Additive noise model. With respect to Example 22.1, show the following:

(a) S and X are jointly normally distributed.
(b) The correlation coefficient of S and X is given by (22.31).
(c) The MMSE estimator $\hat{S}(X)$ is given by (22.32).

22.8 Independent normal variables and their product. Suppose X and Y are independent RVs, $X \sim N(1, 2)$, $Y \sim N(-1, 1)$, and $Z = XY$.

(a) Evaluate $E[Z]$ and Var$[Z]$.
(b) Evaluate Cov$[X, Z]$.
(c) Evaluate the linear MMSE estimator $\hat{X} = a_0 + a_1 Z$ of X given Z.
(d) Evaluate the MMSE estimator $\hat{Z} = T(X)$ of Z given X.

22.9 Correlation coefficient between the response variable and a predictor. Show that the correlation coefficient $\rho_{T,Y}$ between any predictor $T(X)$ and Y satisfies the inequality (22.36).

22.10 ANOVA equation. Show that variance of Y can be decomposed according to (22.37), which is the fundamental equation for ANOVA.

22.11 Optimal choice of regression coefficients. Show that the optimal choice of b, $\boldsymbol{\beta}$, and β_0 are given by (22.43).

22.12 Equivalence of the regression and the solution from the orthogonality equation. Show that the linear regression determined by (22.43) and the best linear MMSE estimate obtained from the orthogonality equation (22.23) are equivalent.

22.13* Regression coefficient estimates. Verify the properties (22.61) and (22.62) of the regression coefficient estimates.

Section 22.2: Linear smoothing and prediction: Wiener filter theory

22.14* An alternative expression for (22.74). A reader who is familiar with the following expression, instead of (22.74), may wonder where this discrepancy comes from:

$$R_{yy}[d] = \boldsymbol{h}^{\mathrm{H}} \boldsymbol{R}_{xx}[d] \boldsymbol{h}.$$

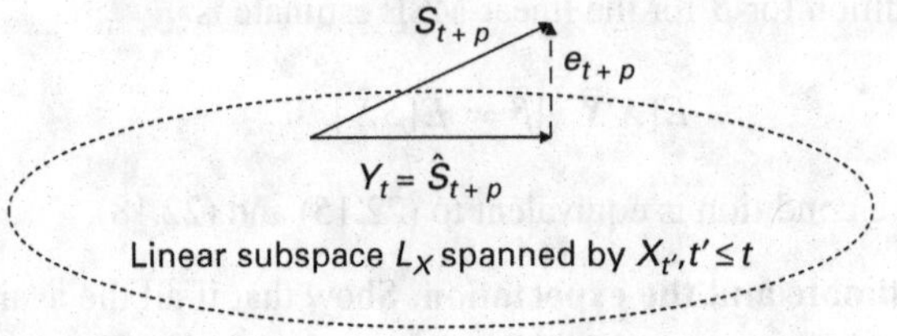

Figure 22.17 The orthogonality principle: the prediction error e_{t+p} should be orthogonal to the linear subspace L_X spanned by the input process $X_{t-k}, k \geq 0$.

Resolve this question by going back to the original definition of the input and output relation of the linear system and the definition of autocorrelation matrix.

22.15 Alternative derivation of the Wiener–Hopf equation (22.89) and (22.90).

(a) In referring to Figure 22.17, the predictor output $Y_t = \hat{S}_{t+p}$ must lie in the linear subspace L_X spanned by the past input $\{X_{t-k}, k \geq 0\}$, since

$$Y_t = \hat{S}_{t+p} = \sum_{k=0}^{\infty} h[k] X_{t-k}$$

and the prediction error e_{t+p} is given by

$$\begin{aligned} e_{t+p} &\triangleq S_{t+p} - \hat{S}_{t+p} \\ &= S_{t+p} - \sum_{k=0}^{\infty} h[k] X_{t-k}. \end{aligned} \tag{22.202}$$

Derive (22.89) by observing that the MSE $\mathcal{E} = E\left[e_{t+p}^2\right] = \|e_{t+p}\|^2$ will be minimum if the error e_{t+p} is orthogonal to the linear subspace L_X.

(b) By making a similar observation, derive the Wiener–Hopf integral equation (22.90).

22.16 Derivation of (22.103). Show that the spectrum of the filtered white noise is given by (22.103).

22.17 Physically unrealizable filter. Explain why the second term in the square brackets in the last expression of (22.151) is physically unrealizable.

22.18 Signal power σ_s^2. Derive the signal power σ_s^2 (22.128) from $P_{ss}(z)$ or $P_s(\omega)$.

22.19 Continuous-time Gauss-Markov process (GMP). Consider the Gaussian process $S(t)$ with the following power spectrum:

$$P_{ss}(f) = \frac{1}{1 + a^{-2} f^2}.$$

(a) Find a causal filter $G_s(f)$ such that $|G_s(f)|^2 = P_{ss}(f)$.

(b) Show that the impulse response function is given by

$$g_s(t) = \begin{cases} 0, & t < 0, \\ 2\pi a\, e^{-2\pi a t}, & t \geq 0. \end{cases}$$

(c) Show that the optimum predictor is given by

$$H_{\text{opt}}(f) = \frac{F_s(f)}{G_s(f)} = e^{-2\pi a \lambda} \leq 1. \tag{22.203}$$

(d) What is the optimal prediction of $S(t+\lambda)$ given $S(t)$?

(e) The conditional distribution of $S(t+\lambda)$ given $S(t)$ is Gaussian. What are the conditional mean $E[S(t+\lambda)|S(t)]$ and conditional variance $\text{Var}[S(t+\lambda)|S(t)]$? Note that this *GMP* $S(t)$ is the **Ornstein–Uhlenbeck process** discussed in Section 17.3.4.

22.20 Optimum smoothing filter for uncorrelated signal and noise. Consider an optimum smoothing problem for the signal and noise model assumed in Example 22.5.

(a) Let $p = -d < 0$. Then find the causal part of (22.151); i.e.,

$$\frac{P_{ss}(z)z^{-d}}{G_x^*(z^{-1})} = \frac{|A|^2}{\overline{C}} \left[\frac{az^{-d}}{1 - \alpha z^{-1}} + \frac{bz^{-d+1}}{1 - \overline{\gamma} z} \right]. \tag{22.204}$$

(b) Assume white noise and find the optimal smoothing filter. Verify that the case $d = 0$ reduces to (22.155) in Example (22.151).

(c) Assume that noise is absent. Find an optimal smoothing filter. What is the error of the estimate?

22.21 Conservation of inner products and expressions for the MMSE.

(a) Show the following identity of the inner product defined in the discrete-time domain and the z-domain:

$$\langle \mathbf{g}^{(1)}, \mathbf{g}^{(2)} \rangle = \langle G^{(1)}, G^{(2)} \rangle;$$

or, equivalently,

$$\sum_{k=-\infty}^{\infty} g^{(1)}[k]\overline{g}^{(2)}[k] = \frac{1}{2\pi i} \oint G^{(1)}(z) G^{*(2)}(z^{-1}) \frac{dz}{z},$$

where

$$G^{(i)}(z) = \sum_{k=-\infty}^{\infty} g^{(i)}[k] z^{-k} \text{ and } g^{(i)}[k] = \frac{1}{2\pi i} \oint G^{(i)}(z) \frac{dz}{z}.$$

(b) Derive the expression (22.161).

Section 22.3: Kalman filter

22.22 Derivation of the propagation step. Derive (22.170), the propagation step of the Kalman filter algorithm for a continuous-state HMM.

22.23 Derivation of the update step. Derive (22.171), the update step of the Kalman filter algorithm for a continuous HMM.

22.24 Derivation of the mean and covariance matrix (22.174). Derive (22.174), the mean and covariance of the joint Gaussian variables $(\boldsymbol{S}_0, \boldsymbol{Y}_0)$.

22.25 Kalman gain for scalar variables $\boldsymbol{S}_0$ and $\boldsymbol{Y}_0$. If both $\boldsymbol{S}_0$ and $\boldsymbol{Y}_0$ are scalar RVs, find an expression for the Kalman gain.

22.26 Derivation of the Kalman's predicted estimate. Derive (22.180), the predicted estimate and variance of $\boldsymbol{s}_1$ based on the observation $\boldsymbol{y}_0$.

23 Queueing and loss models

23.1 Introduction

In this chapter we will provide a brief overview of queueing and loss models as an application of Markov process theory (Chapter 16) and birth–death (BD) processes in particular (Chapter 14). *Queueing theory* was originally developed in telephone traffic engineering. Its origin goes back to the paper published in 1917 by Erlang [94], a Danish mathematician and engineer. Today, queueing theory is well established as a branch of applied probability, pertaining to traffic congestion analysis, queueing, and scheduling of various services and logistic systems. Many queueing theory formulas obtained under Markovian assumptions, such as Poisson processes and exponential distributions, have been found useful, despite some of these assumptions being far from reality. Such surprising results can be explained by the *robustness* or *insensitivity* of these formulas to distributional forms of the RVs involved. We provide intuitive interpretations of such important properties, leaving rigorous mathematical arguments to advanced books and relevant literature. In this sense our treatment may be quite unique by calling the reader's attention to recently developed useful results in the subject field.

Before we discuss specific queueing and loss models, we introduce a simple, yet most important formula, called *Little's formula*, which holds in a very general setting, not just in queueing or loss systems.

23.2 Little's formula: $\overline{L} = \lambda \overline{W}$

Consider a "system" into which "customers" arrive, and stay there for some finite duration, and eventually depart. A system may be any well-defined facility (e.g., a gasoline station, a runway at an airport, a communication link in a network) or a surrounding/environment (e.g., a park, a shopping mall, a country). We assume that the system is *stable* in the sense that the population in the system remains finite at all times. Consider, for instance, some university as a system. The customers may be students, professors, administrators, etc. If we focus on students as the customers, they typically enter the university as freshmen (i.e., as first-year students) and then leave (i.e., graduate from or drop out of) the university after some number of years. Then the mean number of students $\overline{L}$, the number of entering students per year λ, and the mean number of years that a student stays in the university $\overline{W}$ must satisfy the following simple relation:

THEOREM 23.1 (Little's Formula). *Let $\overline{L}$ be the mean number of customers found in the system, λ be the arrival rate, and $\overline{W}$ the mean duration that a customer stays in the system. Then, the following relation holds among these quantities:*

$$\boxed{\overline{L} = \lambda \overline{W}.} \tag{23.1}$$

This is known as **Little's formula**, *Little's theorem*, or *Little's law*, because its formal proof was given by Little [228] in 1961. This formula is perhaps the most frequently used formula in queueing analysis. If the system is a queue, then $\overline{L}$ is the mean queue length and $\overline{W}$ is the mean waiting time.

Proof. We provide the proof assuming that the system in question is a queue to simplify the terminology, but the proof applies to any stable system. Let the RV W_j be the waiting time of the jth customer, $j = 1, 2, \ldots$, and let $A(t)$ and $D(t)$ be the cumulative counts of arrivals and departures in the interval $[0, t]$, respectively. The assumption that the queue is stable means that it becomes empty infinitely often. Let $t = 0$ and $t = T$ be two such epochs; i.e., $A(0) = D(0)$ and $A(T) = D(T)$ (see Figure 23.1).

We further define the following two RVs:

$$n(T) \triangleq A(T) - A(0) = \text{total number of arrivals in } (0, T] \tag{23.2}$$

and

$$\lambda(T) \triangleq \frac{n(T)}{T} = \text{mean arrival rate during } (0, T]. \tag{23.3}$$

Note that $n(T)$ also represents the total count of departures in $(0, T]$, i.e., $n(T) = D(T) - D(0)$, and $\lambda(T)$ is the average departure rate during $(0, T]$.

The shaded area of Figure 23.1 can be decomposed into $n(T)$ horizontal strips of height one and length W_j, $j = 1, 2, \ldots, n(T)$. Thus, the total shaded area can be represented in two different ways:

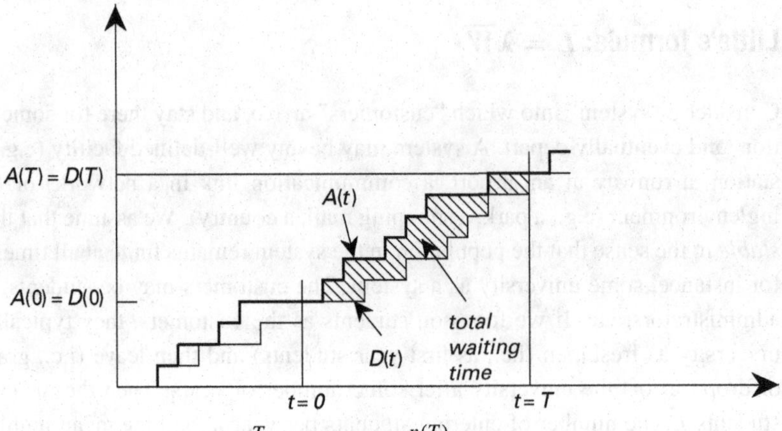

Figure 23.1 The total waiting time as $\int_0^T L(t)\,dt$ and $\sum_{j=1}^{n(T)} W_j$.

$$\text{Shaded area} = \int_0^T [A(t) - D(t)]\, dt = \sum_{j=1}^{n(T)} W_j. \tag{23.4}$$

Since $A(t) - D(t) = L(t)$, the above equation leads to

$$\overline{L}(T) = \frac{\int_0^T L(t)\, dt}{T} = \frac{\sum_{j=1}^{n(T)} W_j}{T} = \frac{n(T)}{T} \frac{\sum_{j=1}^{n(T)} W_j}{n(T)}. \tag{23.5}$$

The strong law of large numbers (see Section 11.3.3) implies that $\lambda(T)$ converges almost surely (i.e., with probability one) to λ, as $T \to \infty$. Similarly, the time average $\overline{L}(T)$ converges almost surely to the ensemble average $\overline{L}$ as $T \to \infty$. Then the average waiting time per customer, $\overline{W}_{n(T)}$ must converge almost surely; i.e.,

$$\overline{W}_{n(T)} = \frac{\sum_{j=1}^{n(T)} W_j}{n(T)} \xrightarrow{\text{a.s.}} \overline{W}. \tag{23.6}$$

It is apparent that $\overline{W}$ is a constant determined by λ and $\overline{L}$ through (23.1). □

23.3 Queueing models

In this section we will derive several queueing and loss models from the results on **BD processes** discussed in Section 14.2.

23.3.1 M/M/1: the simplest queueing model

Consider a queueing model which consists of a *single server* and its *queue* (see Figure 23.2). We make the following assumptions:

1. Customers arrive according to a *Poisson process* with rate λ [s^{-1}].
2. Service times S_j are i.i.d. RVs and *exponentially distributed* with mean $E[S_j] = \mu^{-1}[s]$, $j = 1, 2, \ldots$.
3. Customers waiting for service will form a queue, and it can accommodate an infinite number of customers.

If no customers are found in this *system* (i.e., the server plus the queue), a newly arrived customer receives service immediately. Otherwise, it joins the end of the queue and waits until all the previous customers are served and cleared. Once this customer, say the jth customer, denoted C_j, enters service, C_j receives service for the service interval S_j [s]. Here we use [s] as the unit of service or work, but we can assume any work

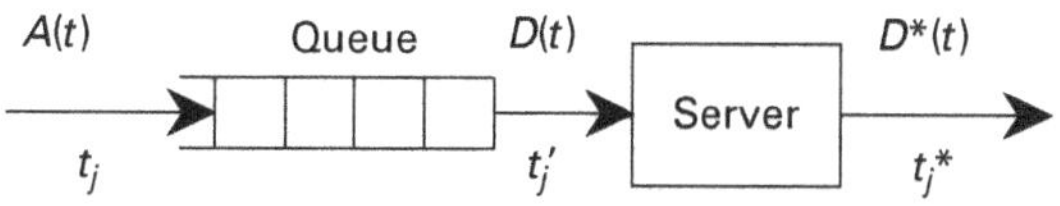

Figure 23.2 A single-server queue and its counting processes $A(t)$, $D(t)$, and $D^*(t)$.

unit, such as [bits], [packets], etc. Then the server's capacity must be described using an appropriate unit for its service rate, i.e., [bits/s], [packets/s], etc.

Because of the memoryless property of the exponential distribution, the probability that this customer's service completes in the next small interval h is given by $\lambda h + o(h)$, regardless of how long the customer has been in service. The constant parameter λ is thus called the *service completion rate*.

This simple single-server model is usually denoted as an "**M/M/1**" queue. The first letter "M" signifies "Markovian" arrivals, which in this case means a Poisson process.[1] The second "M" signifies a "Markovian" service time, which means here an exponentially distributed service time. The last symbol "1" means that there is only one server.

The arrival rate λ is the *birth rate*, and the service completion rate is the *death rate*. Thus, assuming that the system is empty at time $t = 0$, the number of customers $N(t)$ in the system is a *counting process*:

$$N(t) = A(t) - D^*(t), \tag{23.7}$$

where $D^*(t)$ is the departure process at the output of the server, whereas $D(t)$ is the departure process at the queue; i.e., the arrival process to the server. Thus,

$$L(t) = A(t) - D(t) \tag{23.8}$$

is the queue length at time t.

The dynamics of the M/M/1 queue can be studied by analyzing the BD process $N(t)$ with

$$\lambda_n = \lambda, \quad n = 0, 1, 2, \ldots, \tag{23.9}$$

$$\mu_n = \mu, \quad n = 1, 2, 3, \ldots. \tag{23.10}$$

It should be noted that the process $L(t)$, despite its similarity to $N(t)$, is not a BD process (Problem 23.6).

The **transition rate matrix**, or the **infinitesimal generator**, of $N(t)$ defined in (14.42) takes the following form:

$$Q_{\mathrm{M/M/1}} = \begin{bmatrix} -\lambda & \lambda & 0 & 0 & \cdots \\ \mu & -\lambda-\mu & \lambda & 0 & \cdots \\ 0 & \mu & -\lambda-\mu & \lambda & \cdots \\ 0 & 0 & \mu & -\lambda-\mu & \cdots \\ \vdots & \vdots & \vdots & \vdots & \ddots \end{bmatrix}. \tag{23.11}$$

Then the steady-state distribution $\boldsymbol{\pi} = (\pi_n, n = 0, 1, 2, \ldots)^\top$, where $\pi_n = \lim_{t\to\infty} P[N(t) = n]$, is given as the solution of

$$\boldsymbol{\pi}^\top Q_{\mathrm{M/M/1}} = \mathbf{0}^\top. \tag{23.12}$$

[1] In the queueing literature the term "Markovian arrival" generally represents a broader class of arrival processes than a Poisson process.

If we expand the above matrix equation, we obtain

$$-\lambda_0 + \mu\pi_1 = 0, \tag{23.13}$$

$$\lambda\pi_{n-1} - (\lambda + \mu)\pi_n + \mu\pi_{n+1} = 0, \tag{23.14}$$

from which we find

$$-\lambda\pi_n + \mu\pi_{n-1} = -\lambda\pi_{n-1} + \mu\pi_{n-2} = \cdots = -\lambda\pi_0 + \mu\pi_1 = 0. \tag{23.15}$$

Then we can readily find

$$\pi_n = \rho\pi_{n-1} = \rho^n\pi_0, \quad n = 0, 1, 2, \ldots, \tag{23.16}$$

where

$$\rho = \frac{\lambda}{\mu} = \lambda E[S_j] \tag{23.17}$$

is called the *traffic intensity* and represents the expected number of arrivals during the service of a customer. The parameter ρ is also called the *utilization factor* or *server utilization*, since it represents the fraction of time on average that the server is busy.

If $\rho < 1$, then the series (14.57) converges to the constant

$$G = \sum_{n=0}^{\infty} \rho^n = \frac{1}{1-\rho}, \tag{23.18}$$

and thus

$$\pi_0 = 1 - \rho, \tag{23.19}$$

yielding

$$\boxed{\pi_n = (1-\rho)\rho^n, \quad n = 0, 1, 2, \ldots.} \tag{23.20}$$

If $\rho > 1$, the series (14.57) diverges, reflecting the situation in which the customers arrive, on average, faster than the server can handle, and thus the queue grows without bound. If we apply the concept of state classification in a Markov chain discussed in Section 15.3, we find that when $\rho > 1$, all states in the state space, $\mathbb{Z}^+ = \{0, 1, 2, \ldots\}$, are transient, when $\rho = 1$, all states are null-recurrent, and when $\rho < 1$, all states are ergodic (Problem 23.5). Thus, when $\rho < 1$, the mean and variance of $N(t)$ are computed, respectively, as (Problem 23.7)

$$\overline{N} = E[N(t)] = \sum_{n=0}^{\infty} n\pi_n = \frac{\rho}{1-\rho}, \tag{23.21}$$

$$\sigma_N^2 = \sum_{n=0}^{\infty} n^2\pi_n - \overline{N}^2 = \frac{\rho}{(1-\rho)^2}. \tag{23.22}$$

Recall that $L(t)$ of (23.8) represents the length of the queue at time t (excluding the customer in service); thus, its average and variance are given, respectively, by (also Problem 23.7)

$$\overline{L} = \lim_{t\to\infty} E[L(t)] = \sum_{n=1}^{\infty}(n-1)\pi_n = \frac{\rho^2}{1-\rho}, \tag{23.23}$$

$$\sigma_L^2 = \sum_{n=1}^{\infty}(n-1)^2\pi_n - \overline{L}^2 = \frac{\rho^2(1+\rho-\rho^2)}{(1-\rho)^2}. \tag{23.24}$$

Note that

$$\overline{N} = \overline{L} + \rho, \tag{23.25}$$

which is not unexpected: the server utilization ρ (< 1) can be interpreted as the expected number of customers within the server.

23.3.1.1 Waiting and system times

The mean waiting time can be readily obtained from Little's formula as

$$\overline{W} = \frac{\overline{L}}{\lambda} = \frac{\rho^2}{\lambda(1-\rho)} = \frac{\rho}{\mu(1-\rho)}. \tag{23.26}$$

If we apply Little's formula to the system (i.e., the queue plus the server), we have

$$\overline{N} = \lambda\overline{T}. \tag{23.27}$$

Thus, from this and (23.21), we find

$$\boxed{\overline{T} = \frac{\overline{N}}{\lambda} = \frac{1}{\mu(1-\rho)},} \tag{23.28}$$

which could be alternatively derived from $\overline{T} = \overline{W} + \overline{S} = \frac{\rho}{\mu(1-\rho)} + \frac{1}{\mu}$.

Since $\mu^{-1} = E[S_j]$ is the mean service time, the quantities $\mu\overline{W}$ and $\mu\overline{T}$ are the normalized mean waiting and system times and are dimensionless. In Figure 23.3, we plot $\mu\overline{W}$ and $\mu\overline{T}$ versus ρ.

The above results obtained for the M/M/1 in terms of the mean values $\overline{N}, \overline{L}, \overline{T}$, and $\overline{W}$ do not depend on the **queue discipline** (or **scheduling discipline**) as long as it is work-conserving. A queue discipline is said to be *work-conserving* if the work or service demand of each customer is not affected by the queue discipline and if the server is not idle when there are customers waiting for service. Work-conserving queue disciplines include FCFS (first-come, first served), LCFS (last-come, first-served), or random, or even a discipline that interrupts or preempts the customer in service.

The performance metrics $\overline{N}, \overline{L}, \overline{T}$, and $\overline{W}$ of the M/M/1 are *insensitive* to a specific queue discipline, as long as it is work-conserving, because the exponential service distribution has the memoryless property (see (4.26)). Even when a customer in service is interrupted and a new customer enters the server, the behavior of $N(t)$ is unaffected, because the remaining service time of the interrupted customer and the service time of

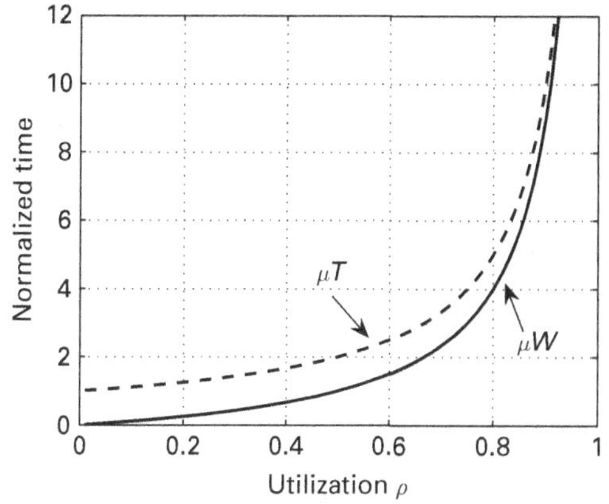

Figure 23.3 The normalized mean waiting time $\mu\overline{W}$ and mean system time $\mu\overline{T}$ versus the server utilization (or traffic intensity) ρ.

the interrupting customer are statistically equivalent, both being exponential RVs with completion rate μ.

Although the mean waiting time $\overline{W}$ and system time $\overline{T}$ can be derived directly from $\overline{L}$ and $\overline{N}$, using Little's formula, the distribution functions of W and T require us to start from scratch.

Now we assume the FCFS discipline. Suppose a customer arrives at the queue to find n customers in the system; that is, one customer in service and $n-1$ customers in queue, $n \geq 1$. Then its waiting time is

$$W = R_1 + S_2 + \cdots + S_n, \tag{23.29}$$

where the RVs R_1 is the remaining service time of the customer in service and $S_2, \ldots, S_n$ are the service times in queue. These n random variables are i.i.d. with a common exponential distribution with mean μ^{-1}. Thus, the conditional distribution of W given that there are n customers are in the system, not including the newly arrived customer in question, is given by that of the n-stage Erlang distribution defined in (4.163):

$$F_W(x|n) = 1 - e^{-\mu x} \sum_{i=0}^{n-1} \frac{(\mu x)^i}{i!} = 1 - Q(n-1; \mu x), \quad x \geq 0, \tag{23.30}$$

where $Q(n; a)$ is the cumulative Poisson distribution; i.e.,

$$Q(n; a) \triangleq \sum_{i=0}^{n} P(i; a), \; n = 0, 1, 2, \ldots, \tag{23.31}$$

$$P(n; a) \triangleq \frac{a^n}{n!} e^{-a}, \; n = 0, 1, 2, \ldots \tag{23.32}$$

Then the PDF of $F_W(x|n)$ is given by

$$f_W(x|n) = \mu P(n-1; \mu x) = \frac{(\mu x)^{n-1}}{(n-1)!}\mu\, e^{-\mu x}, \quad x \geq 0. \tag{23.33}$$

We can show (Problem 23.24) that when the arrival process is Poisson, the probability distribution $\{a_n\}$ of the number of customers $N(t)$ in the system observed by an arriving customer is the same as the long-run time-average distribution $\{\pi_n\}$. This interesting and useful property is called the **PASTA (Poisson arrivals see time averages)** property. This property holds for the case where the service time distribution is general, i.e., for an M/G/1 queue, as well.

Thus, the distribution function of W is given by

$$F_W(x) = \sum_{n=0}^{\infty} a_n F_W(x|n) = \sum_{n=0}^{\infty} \pi_n F_W(x|n). \tag{23.34}$$

The first term, $F_W(x|0)$, represents the waiting time distribution of a customer that finds the system to be empty. Then the waiting time is zero; that is,

$$F_W(x|0) = \begin{cases} 1, & x \geq 0, \\ 0, & x < 0. \end{cases} \tag{23.35}$$

By noting that $\pi_0 = 1 - \rho$, we find

$$\begin{aligned} F_W(x) &= 1 - \rho + (1-\rho)\sum_{n=1}^{\infty} \rho^n \left[1 - e^{-\mu x}\sum_{i=0}^{n-1}\frac{(\mu x)^i}{i!}\right] \\ &= 1 - \rho + (1-\rho)\sum_{n=1}^{\infty} \rho^n [1 - Q(n-1; \mu x)], \quad x \geq 0. \end{aligned} \tag{23.36}$$

By arranging the double summation of the second term, and with some algebraic manipulations (Problem 23.8), we obtain the following surprisingly simple result:

$$\boxed{F_W(x) = 1 - \rho\, e^{-\mu(1-\rho)x}, \quad x \geq 0.} \tag{23.37}$$

Thus, the probability that a customer must wait longer than x is

$$F_W^c(x) = P[W > x] = \rho\, e^{-\mu(1-\rho)x}, \quad x \geq 0. \tag{23.38}$$

Therefore, the mean waiting time is obtained as

$$E[W] = \int_0^{\infty} F_W^c(x)\, dx = \frac{\rho}{(1-\rho)\mu}, \tag{23.39}$$

which, as expected, agrees with $\overline{W}$ of (23.26). Note that while the mean waiting time remains the same for any work-conserving queue discipline, the distribution function obtained above is valid only for FCFS.

The system time of the jth customer is by definition $T_j = W_j + S_j$. For FCFS, the random variables W_j and S_j are independent; thus, the distribution function of T_j can be obtained by convolution:

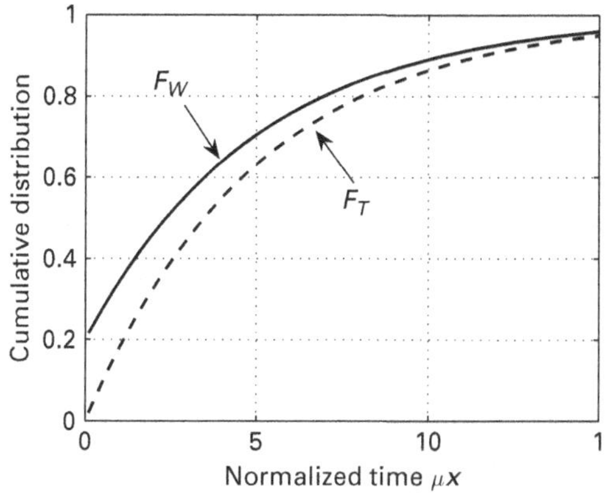

Figure 23.4 The waiting time distribution function and the system time distribution function in M/M/1. The horizontal axis is μx, the time normalized by the mean service time μ^{-1}, and the server utilization is $\rho = \lambda/\mu = 0.8$

$$F_T(x) = \int_0^\infty F_W(x-y) f_S(y)\, dy$$
$$= \int_0^\infty \left[1 - \rho\, e^{-\mu(1-\rho)(x-y)}\right] \mu\, e^{-\mu y}\, dy. \quad (23.40)$$

Then we have

$$\boxed{F_T(x) = 1 - e^{-\mu(1-\rho)x}, \quad x \geq 0,} \quad (23.41)$$

which is also surprisingly simple. It is the exponential distribution with mean $1/(\mu(1-\rho))$. In Figure 23.4 we plot the curves $F_W(x)$ and $F_T(x)$ versus μx when traffic intensity $\rho = 0.8$ [erlangs].

23.3.2 M/M/∞ and M/G/∞: infinite servers

Suppose that there are *ample* servers in parallel and, whenever a new customer arrives, a server is immediately assigned to the customer. We assume as before that customers arrive according to a Poisson process with rate λ. Their service times are exponentially distributed with mean $1/\mu$. When there are n customers in the center, n servers will work; hence, there is no queue to be formed. The *service completion rate* of thiscenter is $n\mu$. Since the number n can become arbitrarily large, the service center must have, in theory, infinitely many servers. Hence, we call such a system an *infinite server* (IS) queue (although actually a queue never forms) and denote it by M/M/∞.

Let $N(t)$ be, as before, the number of customers in this M/M/∞ queue at time t. The process $N(t)$ is again a BD process, now with birth rate

$$\lambda_n = \lambda, \quad n = 0, 1, 2, \ldots, \quad (23.42)$$

and death rate

$$\mu_n = n\mu, \quad n = 1, 2, 3, \ldots. \tag{23.43}$$

Substitution of the above results into (14.57) leads to the following *normalization constant*:

$$G = 1 + a + \frac{a^2}{2!} + \cdots + \frac{a^n}{n!} + \cdots = e^a, \tag{23.44}$$

where the parameter

$$a = \lambda/\mu \tag{23.45}$$

is called the *traffic intensity*, *offered load*, or *traffic load*, similar to ρ defined by (23.17) for a single-server queue, and represents the rate at which work enters the system. In telephone engineering, the unit [erlang] or [erl] is often used for traffic load a in deference to Erlang. The offered load a can also be interpreted as the average number of call arrivals during the service of a call. However, unlike in the M/M/1 queue, this quantity *cannot* be interpreted as server utilization. It is instead the average number of busy servers, as will be shown below.

From (14.55) and (23.44), we obtain the equilibrium-state distribution or stationary distribution:

$$\boxed{\pi_n = \frac{a^n}{n!} e^{-a} = P(n\,;a), \quad n = 0, 1, 2, \ldots,} \tag{23.46}$$

which is the Poisson distribution with mean a.

23.3.2.1 M/ G/∞ model and formula (23.46)

It is important to know that the equilibrium-state distribution (23.46) holds also for a general M/G/∞ model, where "G" stands for a *general* service time distribution. In other words, the distribution (23.46) is **insensitive** to the form of service time distribution $F_S(t)$. One way to prove this property is to use the notion of **quasi-reversibility** introduced by Kelly [177]. Another direct proof is suggested here as an exercise (Problem 23.11), which shows that, given that the system is empty at $t = 0$, the probability $\pi_n(t) = P[N(t) = n]$ is given by

$$\boxed{\pi_n(t) = \frac{\left[\int_0^t (1 - F_S(u))\,du\right]^n}{n!} \exp\left[-\lambda \int_0^t (1 - F_S(u))\,du\right], \quad n = 0, 1, 2, \ldots} \tag{23.47}$$

So the time-dependent solution $\pi_n(t)$ is also Poisson distributed, with a parameter that depends on $F_S(t)$. In the limit $t \to \infty$, however, this converges to the equilibrium-state distribution (23.46) regardless of $F_S(t)$, since $\int_0^\infty (1 - F_S(x))\,dx = E[S] = \mu^{-1}$. With a little extension of the above result, we can show (Problem 23.12) that the departure process is also a Poisson process with rate λ.

We provide here a simple and intuitive interpretation of the formula (23.46). In a stable system the distribution $\{a_n\}$ seen by arriving customers is equal to the distribution $\{d_n\}$ seen by departing customers. Then, using the PASTA property, the probability distribution $\{\pi_n\}$ of M/G/∞ is equivalent to $\{a_n\}$, and hence to $\{d_n\}$ as well. The latter distribution should be, on average, equal to the distribution of the number of customers who arrive during the expected service time $E[S] = \mu^{-1}$ of an arbitrarily chosen customer. From the definition of Poisson process, we know that the number of arrivals during the interval $E[S]$ is Poisson distributed with mean $\lambda E[S] = \lambda/\mu = a$.

We will observe this type of *insensitivity property* in loss models and a *processor shared* queueing model to be discussed in subsequent sections. Interestingly enough, this insensitivity property can always be explained by identifying a set of (equivalent) infinite servers involved in such models. The reader will gain a better understanding of this important feature of many queueing or loss models, as we will repeat similar arguments in the following.

23.3.3 M/M/m: multiple server queue

Now we consider an M/M/m queue; i.e., the case where there are m parallel servers, sharing a common queue. The M/M/1 and M/M/∞ systems are two extreme cases of M/M/m. If $N(t) = n \le m$, then the n customers are simultaneously served by n parallel servers. If $N(t) = n > m$, then $n - m$ customers will be present in the queue. Since both the arrival and service processes are memoryless, $N(t)$ is again a BD process, with the following BD rates:

$$\lambda_n = \lambda, \quad n = 0, 1, 2, \ldots, \tag{23.48}$$

$$\mu_n = \min\{n, m\}\mu, \quad n = 1, 2, 3, \ldots \tag{23.49}$$

If $a = \lambda/\mu < m$, then the normalization constant G of (14.57) is given by

$$G = \sum_{n=0}^{m-1} \frac{a^n}{n!} + \frac{a^m}{m!}\frac{1}{1-\rho}, \tag{23.50}$$

where

$$\rho = \frac{a}{m} = \frac{\lambda}{m\mu} < 1 \tag{23.51}$$

is the utilization per server. Thus, the equilibrium-state distribution of $N(t)$ is

$$\boxed{\begin{aligned} &\pi_0 = G^{-1}, \\ &\pi_n = \begin{cases} \frac{a^n}{n!}\pi_0, & n = 1, 2, \ldots, m, \\ \rho^{n-m}\pi_m, & n = m+1, m+2, \ldots \end{cases} \end{aligned}} \tag{23.52}$$

It will be instructive to observe that the distribution in the region $0 \le n \le m$ takes the form of a Poisson distribution (as in M/M/∞), whereas in the range $n > m$ it is a geometric distribution with parameter ρ (as in M/M/1). These two distributions are connected at $n = m$.

Using the PASTA property of the Poisson arrivals defined in the section on M/M/1, we see that the probability that an arriving customer finds n customers in queue is π_{m+n}. Thus, the probability that a customer cannot be immediately served is

$$F^{\mathrm{c}}_W(0) = P[W > 0] = \sum_{n=0}^{\infty} \pi_{m+n} = \frac{\pi_m}{1-\rho}$$
$$= \frac{a^m}{m!}\frac{m}{m-a}\pi_0 \triangleq C(m, a), \tag{23.53}$$

which is known as **Erlang's delay formula**, *Erlang's second formula*, or the *Erlang C formula.* We then rearrange $G = \pi_0^{-1}$ of (23.50) and find (Problem 23.14)

$$\boxed{C(m, a) = \frac{mB(m, a)}{m - a(1 - B(m, a))},} \tag{23.54}$$

where $B(m, a)$ is

$$\boxed{B(m, a) \triangleq \frac{a^m/m!}{\sum_{i=0}^{m} \frac{a^i}{i!}},} \tag{23.55}$$

which is called *Erlang's loss formula, Erlang's first formula*, or the *Erlang B formula.* This formula will be further discussed in Section 23.4.1.

The distribution of waiting time under the FCFS discipline can be derived in the same way as was done for the M/M/1 queue, and is left to the reader as an exercise (Problem 23.15):

$$\boxed{F_W(x) = 1 - C(m, a)e^{-m\mu(1-\rho)x}.} \tag{23.56}$$

For $m = 1$, the above expression indeed reduces to (23.37), the waiting time distribution for the M/M/1 queue.

From (23.149) we obtain the mean waiting time for M/M/m :

$$\boxed{E[W] = \int_0^\infty F^{\mathrm{c}}_W(x)\,dx = \frac{C(m, a)}{m\mu(1-\rho)}.} \tag{23.57}$$

Needless to say, the mean waiting time could have been obtained more readily by calculating the mean queue length first and then applying Little's formula (see Problem 23.13).

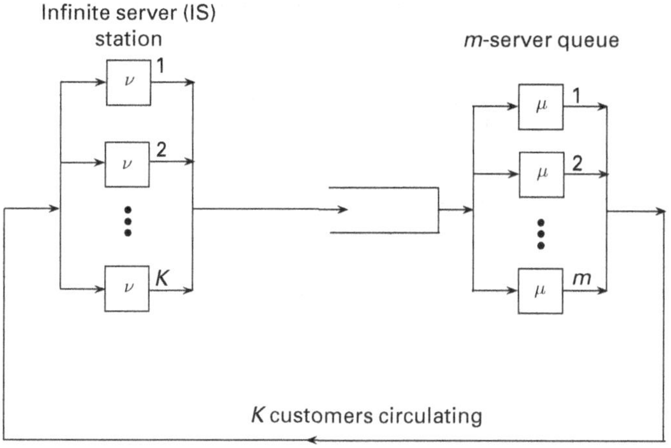

Figure 23.5 Multiple access model: G(K)/M/m.

23.3.4 M(K)/M/m and G(K)/M/m: multiple access models

We now consider a situation where customers are generated from *finite sources*, as shown in Figure 23.5.

This is a proper model to represent, for instance, a *multiple access* communication system, in which m servers represent channels or links of m units; customers are messages or packets to be transmitted by K users.

In the queueing theory literature this model with $m = 1$ has been extensively discussed under the name *machine repairman model*, *machine-servicing model* [99], or *machine interference model* [73]: a repairman (the server) maintains a group of K machines, and each machine is either "up" (running) or "down" (requiring repair service). When a machine breaks down, it joins the queue (i.e., a logical queue: the machine cannot walk!) for repair. Let a random variable U represent the amount of time that the machine is up and S be the amount of time the repairman spends to repair the broken machine. Thus, the case with $m > 1$ may be aptly called the "multiple repairmen model" [203].

First, we assume that both U and S are exponentially distributed:

$$F_U(x) = 1 - e^{-\nu x}, \tag{23.58}$$

$$F_S(x) = 1 - e^{-\mu x}, \tag{23.59}$$

where ν^{-1} and μ^{-1} are the mean "user" time (or "up" time) and the mean service time respectively. This queueing model is denoted as M/M/m/K/K according to the conventional notation (called Kendall's notation) in the queueing theory literature. Instead, we adopt the notation M(K)/M/m [4, 203], which will be easier to remember: "M(K)" means K Markovian sources.

The arrival process from finite sources with the exponential source time (23.58) is called a *quasi-random* arrival process (as opposed to completely random arrivals or

Poisson arrivals). Then the number of customers $N(t)$ found in m servers or their common queue at time t can be represented as a BD process with birth rate

$$\lambda_n = \begin{cases} (K-n)\nu, & 0 \leq n \leq K, \\ 0, & n > K, \end{cases} \tag{23.60}$$

and death rate

$$\mu_n = \min\{n, m\}\mu, \quad 0 \leq n \leq K. \tag{23.61}$$

The normalization constant G of (14.57) becomes (Problem 23.16)

$$G(K) = \pi_0^{-1}(K) = \sum_{n=0}^{m-1} \binom{K}{n} r^n + \sum_{n=m}^{K} \frac{K!}{(K-n)!m!} m^{m-n} r^n, \tag{23.62}$$

where r is defined as

$$r = \frac{\nu}{\mu} = \frac{E[S]}{E[U]}. \tag{23.63}$$

Then, the distribution of the number of customers found in the servers or their common queue in the G(K)/M/m system is (Problem 23.17)

$$\pi_n(K) = \begin{cases} \binom{K}{n} r^n \pi_0(K), & 0 \leq n \leq m, \\ \frac{K!}{(K-n)!m!} m^{m-n} r^n \pi_0(K), & m \leq n \leq K, \\ 0, & n > K. \end{cases} \tag{23.64}$$

which will take a simpler form when $m = 1$, to be shown below. Before we proceed, we will discuss the *robustness* of the above formula.

23.3.4.1 G(K)/ M/ m and formula (23.64)

Although we have assumed for the sake of analysis that the user time (or the machine's uptime) variable U is exponentially distributed, the formula (23.64) should remain unchanged, even if the distribution of U were a *general* distribution, insofar as its mean is $E[U] = \nu^{-1}$. Although its mathematical proof can be found elsewhere (e.g., [203, Chapter 5]), we will provide here an intuitive argument by extending the similar property we found for M/G/∞ in Section 23.3.2.

The model of Figure 23.5 can be viewed as a two-stage cyclic queueing system, which consists of an IS (infinite server) station and a queue with m parallel servers, with K customers circulating. By extending the argument we presented for M/G/∞, the distributional form of the time spent in the IS station is immaterial. All that counts is its mean value, in this case $E[U] = \nu^{-1}$. If we consider the case $m = 1$, this analogy will become even clearer.

23.3.4.2 G(K)/ M/1: The machine repairman model (*m* = 1).

As we remarked earlier, the case $m = 1$ is historically called the machine repairman model and has been applied to modeling of a time-shared computer system (e.g., [197]) and other systems. The formula (23.64) becomes

$$\pi_n(K) = \frac{\frac{r^n}{(K-n)!}}{\sum_{i=0}^{K} \frac{r^i}{(K-i)!}}, \quad n = 0, 1, 2, \ldots, \tag{23.65}$$

which leads, after some manipulation, to the following expression (Problem 23.20):

$$\boxed{\pi_n(K) = \frac{P(K-n; r^{-1})}{Q(K; r^{-1})}, \quad n = 0, 1, 2, \ldots.} \tag{23.66}$$

This shows that $K - n$ (the number of machines up running) has a truncated Poisson distribution. We will revisit this result in Section 23.4.1 on Erlang loss model.

It can be shown that if we let $K \to \infty$ and $\nu \to 0$, while keeping $K\nu = \lambda$ (hence, $Kr = K\nu/\mu \to \rho$), then the system G(K)/M/1 approaches the M/M/1 queue. The distribution (23.66) then should converge to the geometric distribution $(1-\rho)\rho^n$, $n = 0, 1, 2, \ldots$ (Problem 23.21). This limiting argument also implies that the merged stream of a sufficient number K of independent sources converges to a Poisson stream. It is important to note that each sub-stream need not be a Poisson stream. This is analogous to the central limit theorem studied earlier: the sum of K independent RVs converges to a normal variable as $K \to \infty$, even when the component variables are not normal, as long as any of their contributions are negligibly small compared with the summed variable.

23.3.4.3 Waiting time distribution in M(K)/M/m.

Now we proceed to derive the waiting time distribution $F_W(x)$. The results we obtain here will apply only to M(K)/M/m, unlike (23.64), which holds for G(K)/M/m in general.

First, we need to obtain $\{a_n\}$, the probability distribution of the number of customers found by an arriving customer. The PASTA property does not hold for the quasi-random input. However, a simple relation can be found between the distributions $\{\pi_n(K)\}$ and $\{a_n(K)\}$, as we shall see below.

Over a long time interval $(0, T)$ the average number of customers who arrive when the system is in state n (i.e., when $N(t) = n$) is given by $\lambda_n T \pi_n(K)$, $n = 0, 1, 2, \ldots, K-1$. As $T \to \infty$, the proportion of these customer arrivals to all customer arrivals in this period converges almost surely (i.e., with probability one) to the constant

$$a_n(K) = \frac{\lambda_n T \pi_n(K)}{\sum_{i=0}^{K-1} \lambda_i T \pi_i(K)} = \frac{\lambda_n \pi_n(K)}{\sum_{i=0}^{K-1} \lambda_i \pi_i(K)}, \quad 0 \le n \le K-1. \tag{23.67}$$

Note that the last expression holds for *any state-dependent arrivals*, not just for the quasi-random arrival. Substituting (23.60) into (23.67), we find

$$a_n(K) = \frac{(K-n)\pi_n(K)}{\sum_{i=0}^{K-1}(K-i)\pi_i(K)}, \quad 0 \le n \le K-1, \tag{23.68}$$

and with (23.64) we arrive at (Problem 23.18)

$$\boxed{a_n(K) = \pi_n(K-1), \quad 0 \le n \le K-1.} \tag{23.69}$$

This result suggests that the distribution seen by an arriving customer is what would be observed at a randomly chosen instant in steady state, with that particular source removed from the system.

The relation (23.69) is a special case of the **arrival theorem** [220, 297] that holds in any *Markovian closed queueing network*. Since it is known [47] that any closed Markovian network can have an equivalent two-stage cyclic queueing system with state-dependent arrivals and service rates, it suffices to prove (23.69) for a state-arrival rate of the form $\lambda_j = f(K-j)$ with an arbitrary function $f(\cdot)$ and $f(0) = 0$, instead of the specific linear form as in (23.60), and for arbitrary state-dependent service rates μ_j, instead of the specific form as in (23.61) (Problem 23.19).

In order to obtain the waiting time distribution, we proceed in a way similar to what we did for the M/M/m. We can show that the complementary function of the waiting time distribution is given (Problem 23.22 (a)) by

$$F_W^c(x) = \sum_{n=0}^{K-m} \pi_{n+m}(K-1)F_W^c(x|n+m), \tag{23.70}$$

where $F_W^c(x|n+m)$ is given by

$$F_W^c(x|n+m) = e^{-\mu x}\sum_{j=0}^{n}\frac{(\mu x)^j}{j!} = Q(n; m\mu x), \; 0 \le n \le K-m. \tag{23.71}$$

The function $Q(n; a)$ in the last expression is the cumulative Poisson distribution defined in (23.31).

Thus, the waiting time distribution function and PDF of waiting time are given by (Problem 23.22 (b))

$$F_W(x) = 1 - \frac{m^m}{m!}\pi_0(K-1)\frac{Q(K-m-1; m\mu(x+\frac{1}{\nu}))}{P(K-1; mr^{-1})} \tag{23.72}$$

and

$$f_W(x) = F_W(0)\delta(x) + \frac{m^m\mu}{m!}\pi_0(K-1)\frac{P(K-m-1; m\mu(x+\frac{1}{\nu}))}{P(K-1; mr^{-1})}. \tag{23.73}$$

23.3.5 M/G/1 queueing model

If we retain the Poisson arrival assumption but remove the assumption of the exponential service time, the resulting single-server queueing system is denoted as M/G/1.

The M/G/1 queueing model has found many applications. In packet transmission in a computer network, for instance, packet traffic may be adequately characterized as a Poisson process, but its service time, which is proportional to the packet length, may be far from being exponentially distributed. Thus, any result obtained using an M/M/1 queueing model may be suspect.

When the service time is not exponential, the transition probabilities (14.41) are no longer independent of t. It depends on how long the system has been in the current state. Thus, $N(t)$ is no longer a BD process. There are a couple of ways to deal with this non-Markov queueing process. One of them is called the method of *embedded Markov chain* (EMC).

Let N_k be the number of customers found in the system just after the service completion of customer C_k, and let A_k be the number of customers who arrive while C_k is served (Figure 23.6). Clearly, $A_1, A_2, \dots$ are i.i.d. RVs. Furthermore, under the FCFS discipline, A_k is independent of $N_1, N_2, \dots, N_{k-1}$. However, N_k and A_k are *not* independent, but related by the following equation:

$$N_k = \begin{cases} A_k, & \text{if } N_{k-1} = 0, \\ N_{k-1} + A_k - 1, & \text{if } N_{k-1} > 0. \end{cases} \tag{23.74}$$

By defining the left-continuous unit step function[2] $U(t)$,

$$U(x) \triangleq \begin{cases} 1, & x > 0, \\ 0, & x \leq 0, \end{cases} \tag{23.75}$$

we can rewrite (23.74) in a single line:

$$\boxed{N_k = N_{k-1} - U(N_{k-1}) + A_k.} \tag{23.76}$$

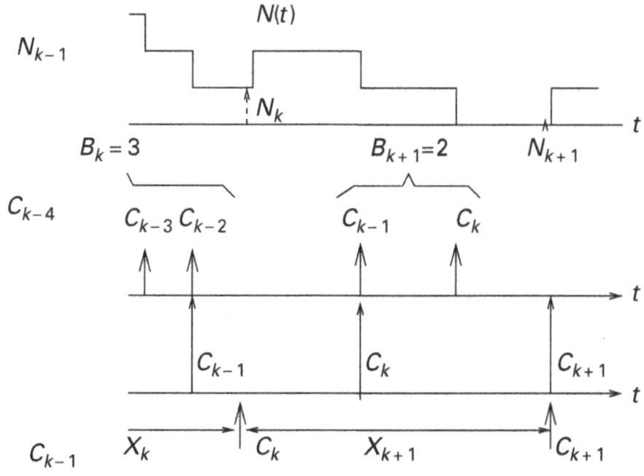

Figure 23.6 The Markov chain $\{N_k\}$ embedded in the process $N(t)$.

[2] Note that $U(t)$ is different from the right-continuous unit step function $u(t)$ used elsewhere.

From this expression it is apparent that the sequence $\{N_k\}$ is a DTMC. Since this chain is *embedded* in the original time-continuous (non-Markov) process $N(t)$, we say that N_k is an EMC associated with $N(t)$.

The *equilibrium-state* distribution

$$d_n = \lim_{k\to\infty} P[N_k = n], \quad n = 0, 1, 2, \ldots, \tag{23.77}$$

is the distribution seen by departing customers. But we are primarily interested in the *equilibrium-state* distribution of the process $N(t)$; i.e.,

$$\pi_n = \lim_{t\to\infty} P[N(t) = n], \quad n = 0, 1, 2, \ldots, \tag{23.78}$$

which is also the long-run time-average distribution. How are $\{d_n\}$ and $\{\pi_n\}$ related?

In any stable system, $\{d_n\}$ should be equal to $\{a_n\}$, the distribution seen by arriving customers. Because of the PASTA property in an M/G/1 system, these distributions should be identical to $\{\pi_n\}$.

23.3.5.1 Queue distribution in M/G/1

Let us denote the PGF of $\{\pi_n\}$ by $P(z)$:

$$\boxed{P(z) = \sum_{n=0}^{\infty} \pi_n z^n = \sum_{n=0}^{\infty} d_n z^n = \lim_{k\to\infty} E[z^{N_k}].} \tag{23.79}$$

Then, substituting (23.76) into the above equation and noting that A_k and N_{k-1} are independent, we find

$$\begin{aligned} P(z) &= \lim_{k\to\infty} E[z^{A_k}]E[z^{N_{k-1}-U(N_{k-1})}] = A(z)\sum_{n=0}^{\infty}\left[\pi_n\left(z^{n-U(n)}\right)\right] \\ &= A(z)\left[\pi_0 + \sum_{n=1}^{\infty}\pi_n z^{n-1}\right] = A(z)\left[\pi_0 + z^{-1}(P(z)-\pi_0)\right], \end{aligned} \tag{23.80}$$

from which we readily obtain

$$\boxed{P(z) = \frac{\pi_0 A(z)(z-1)}{z - A(z)},} \tag{23.81}$$

where $A(z)$ is the PGF of A_k:

$$\begin{aligned} A(z) &= \sum_{n=0}^{\infty} P[A_k = n]z^n = \sum_{n=0}^{\infty}\int_0^{\infty}\frac{e^{-\lambda t}(\lambda t z)^n}{n!} f_S(t)\,dt \\ &= \int_0^{\infty} e^{-\lambda t}e^{\lambda t z} f_S(t)\,dt = f_S^*(\lambda - \lambda z). \end{aligned} \tag{23.82}$$

Here, $f_S^*(\cdot)$ is the Laplace transform of $f_S(t)$.

The constant π_0 is obtained by letting $z \to 1$ in the above and applying l'Hôpital's rule (Problem 23.25):

$$\pi_0 = 1 - \lambda E[S] = 1 - \rho, \tag{23.83}$$

which is as expected. Thus, we finally have

$$\boxed{P(z) = \frac{(1-\rho)(z-1) f_S^*(\lambda - \lambda z)}{z - f_S^*(\lambda - \lambda z)},} \tag{23.84}$$

which is called the *Pollaczek–Khintchine transform equation* or just the **Pollaczek–Khintchine formula**.

Example 23.1: M/D/1 queue. If the service time is constant or deterministic, i.e., $S = \mu^{-1}$ with probability one, then we denote such an M/G/1 queue by M/D/1. Using the right-continuous unit step function $u(x)$, we can write the service time distribution function as

$$F_S(x) = u(x - \mu^{-1}) \tag{23.85}$$

and its density function as

$$f_S(x) = \delta(x - \mu^{-1}), \tag{23.86}$$

where $\delta(x)$ is the *Dirac delta function* . The Laplace transform of $f_S(x)$ is therefore

$$f_S^*(s) = \int_0^\infty e^{-sx} \delta(x - \mu^{-1})\, dx = e^{-s/\mu}. \tag{23.87}$$

By substituting the above result into (23.84), we obtain

$$\boxed{P(z) = \frac{(1-\rho)(1-z)}{1 - z\, e^{\rho(1-z)}}.} \tag{23.88}$$

The Taylor series expansion of (23.88) gives

$$P(z) = (1-\rho)(1-z) \sum_{j=0}^{\infty} z^j e^{j\rho} e^{-j\rho z}. \tag{23.89}$$

By equating the coefficients of z^n on both sides, we obtain

$$\pi_n = (1-\rho) \sum_{j=0}^{n} (-1)^{n-j} e^{j\rho} \left[\frac{(j\rho)^{n-j}}{(n-j)!} + \frac{(j\rho)^{n-j-1}}{(n-j-1)!} \right]. \tag{23.90}$$

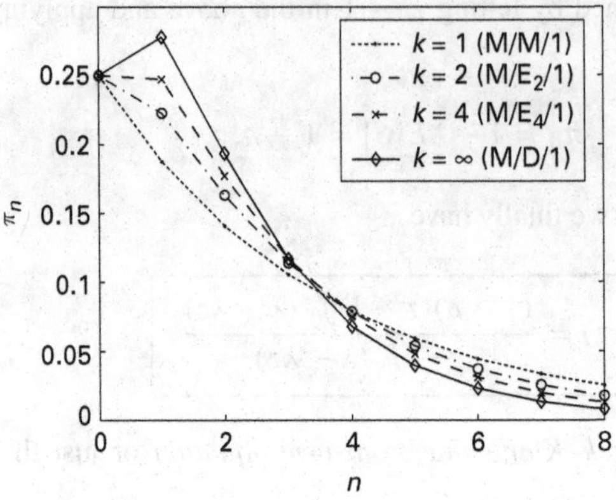

Figure 23.7 The queue distribution of the M/E$_k$/1 system for $k = 1, 2, 4, \infty$.

Figure 23.7 shows the distribution (23.90), to which the solution for the M/E$_k$/1 system converges as $k \to \infty$. □

23.3.5.2 Pollaczek–Khintchine mean value formula

By differentiating (23.84) and letting $z \to 1$, we obtain the average number of customers in the M/G/1 system (Problem 23.28):

$$\boxed{\overline{N} = \rho + \frac{\rho^2(1 + c_S^2)}{2(1-\rho)},} \tag{23.91}$$

where c_S is the coefficient of variation of the service time. Equation (23.91) is called the *Pollaczek–Khintchine mean value formula*. There is another way to derive this important formula (see Problem 23.29).

If the service time distribution is exponential, then $c_S = 1$ and (23.91) reduces to

$$\overline{N} = \frac{\rho}{1-\rho} \quad \text{(M/M/1)}. \tag{23.92}$$

If the service time is constant, then $c_S = 0$ and

$$\overline{N} = \frac{\rho(1 - \frac{\rho}{2})}{1-\rho} \quad \text{(M/D/1)}. \tag{23.93}$$

Thus, under a *heavy traffic* condition ($\rho \approx 1$), the ratio of the mean queue size of M/M/1 to that of M/D/1 is two to one.

Since the utilization factor ρ of a single-server system can be interpreted as the mean number of customers in service, we can write $\overline{N}$ as the sum of ρ and the mean number $\overline{Q}$ of customers in queue. Thus,

$$\overline{N} = \rho + \overline{Q}. \tag{23.94}$$

From (23.91) and (23.94) we have

$$\boxed{\overline{Q} = \frac{\lambda^2 E[S^2]}{2(1-\rho)}.} \tag{23.95}$$

The expected waiting time under the FCFS queue discipline is now readily obtainable using Little's formula:

$$E[W_{\text{FCFS}}] = \frac{\overline{Q}}{\lambda} = \frac{\lambda E[S^2]}{2(1-\rho)} = \frac{\rho E[S](1+c_S^2)}{2(1-\rho)}, \tag{23.96}$$

where $c_S = \sqrt{\text{Var}[S]}/E[S]$ is the coefficient of variation. Similarly, we find the expected system time under the FCFS discipline:

$$E[T_{\text{FCFS}}] = E[S] + \frac{\rho E[S](1+c_S^2)}{2(1-\rho)} = \frac{E[S]}{1-\rho}\left[1 - \frac{\rho(1-c_S^2)}{2}\right]. \tag{23.97}$$

23.3.5.3 Waiting time distribution in M/G/1

Under the FCFS (first come, first served) scheduling discipline, N_k is equivalent to the number of arrivals during the system time of customer C_k (see Figure 23.6). Thus, we find the following parallel relation:

A_k = number of arrivals during the service time S_k,
N_k = number of arrivals during the system time T_k.

Then the relation between $P(z)$ and the system time distribution $F_T(t)$ is parallel to that between $A(z)$ and the service time distribution $F_S(t)$ (see also Problem 23.30). Therefore, the relation (23.82) implies

$$P(z) = f_T^*(\lambda - \lambda z), \tag{23.98}$$

where $f_T^*(\cdot)$ is the Laplace transform of $f_T(t)$, which we wish to obtain. By equating (23.84) and (23.98) and setting $s = \lambda - \lambda z$, we find

$$\boxed{f_T^*(s) = \frac{(1-\rho)s f_S^*(s)}{s - \lambda + \lambda f_S^*(s)} \quad \text{(FCFS).}} \tag{23.99}$$

Under the FCFS scheduling rule, the system time T_k, waiting time W_k, and service time S_k of customer C_k are simply related according to

$$T_k = W_k + S_k. \tag{23.100}$$

Since W_k and S_k are independent RVs, the above equation implies

$$f_T^*(s) = f_W^*(s) f_S^*(s). \tag{23.101}$$

Then, from (23.99) and (23.101), we readily have

$$\boxed{f_W^*(s) = \frac{(1-\rho)s}{s - \lambda + \lambda f_S^*(s)} \quad \text{(FCFS).}} \tag{23.102}$$

Then by applying the inverse Laplace transform to $f_T^*(s)$ and $f_S^*(s)$, we can derive the PDFs $f_T(t)$ and $f_W(t)$ respectively.

The expression for $f_W^*(s)$ becomes somewhat simpler if we use $f_R^*(s)$, the Laplace transform of the residual lifetime (or forward recurrence time) R associated with the service time S (Problem 23.31), yielding

$$\boxed{f_W^*(s) = \frac{(1-\rho)}{1-\rho f_R^*(s)} = (1-\rho)\sum_{n=0}^{\infty} \rho^n \left(f_R^*(s)\right)^n \quad \text{(FCFS)}.} \tag{23.103}$$

This formula is often referred to as the *Pollaczek–Khintchine formula for the waiting time distribution.*

23.3.6 M/G/1 with processor sharing (PS)

The FCFS scheduling is often assumed as a default scheduling in many queueing models, but it is not necessarily a *fair* scheduling strategy, since small customers behind a big customer may suffer unreasonably. The notion of **processor sharing** (PS) was originally introduced by Kleinrock [188] as the limiting case of round robin (RR), in which the time quantum (or time slice) is allowed to approach zero. Thus, under this discipline, the server divides the service among all of the customers presently in the system. So, whenever there are n customers in the system, each customer will receive service at the rate of $1/n$ work unit per unit time.

The PS discipline was used to model the scheduling policy adopted in time-shared computing systems. The recent renewed interest in PS stems from its applicability to (i) modeling of statistically multiplexed traffic over a high-speed link, (ii) modeling of Web servers, and (iii) modeling of links congested with transmission control protocol (TCP) traffic. Job schedulers in Web servers often employ PS-based algorithms to achieve fairness. Roughly speaking, TCP attempts to provide each session (TCP flow) with an equal share of the bandwidth of a given link, and this is exactly what the PS discipline provides.

M/G/1 with PS is similar to M/G/∞ in the sense all the customers in the system simultaneously receive services without being to forced to wait. The difference is that the total service rate at the service station is n work units per time unit in M/G/∞, whereas it is one work unit per unit time in the M/G/1 with PS. Thus, it should not be so surprising to find that the equilibrium state of the number of customers in the system $\{\pi_n\}$ depends only on the mean service time, and not on the distributional form of $F_S(t)$, as in M/G/∞. That is, it has the same distribution as that for M/M/1 with PS. But the probability distribution $\{\pi_n\}$ of M/M/1 is independent of the scheduling discipline as long as it is *work-conserving*, which the PS discipline is. Therefore, the steady-state distribution of $N(t)$ is the same as that of M/M/1; i.e.,

$$\pi_n = (1-\rho)\rho^n, \quad n = 0, 1, 2, \ldots, \quad \text{where } \rho = \lambda E[S]. \tag{23.104}$$

Because of the inherent fairness of the PS discipline, it is easy to see that the system time T should be proportional to its service demand S. If we define the conditional system time variable

$$T_{\mathrm{PS}}(S) = \text{time that a customer with service demand } S \text{ spends in the system}, \tag{23.105}$$

then, as was shown by Sakata *et al.* [292] and Coffman *et al.* [59], its expectation is simply given by

$$E[T_{\mathrm{PS}}(S)|S] = \frac{S}{1-\rho}, \quad x > 0. \tag{23.106}$$

Then the unconditional expectation of the system time is

$$\overline{T}_{\mathrm{PS}} = E_S[E[T_{\mathrm{PS}}(S)|S]] = \frac{E[S]}{1-\rho}, \tag{23.107}$$

which agrees with (23.28) as expected.

If we define the waiting time variable of a customer, whose service time requirement is S, by

$$W_{\mathrm{PS}}(S) = T_{\mathrm{PS}}(S) - S, \tag{23.108}$$

then we find

$$E[W_{\mathrm{PS}}(S)|S] = E[T_{\mathrm{PS}}(S)|S] - S = \frac{\rho S}{1-\rho}, \tag{23.109}$$

and the unconditional mean waiting time is

$$\overline{W}_{\mathrm{PS}} = \frac{\rho E[S]}{1-\rho}, \tag{23.110}$$

which agrees with (23.26).

23.4 Loss models

In the queueing models studied in the preceding sections, all customers are eventually served no matter how large the queue may develop temporarily, provided the long-run average rate of service requests is less than the server capacity. We shall now consider the other extreme, where there is no room for a queue to form and arriving customers are denied service when all the servers are busy. This is typically the case in a *circuit-switched* telephone system, including a cellular phone system. A denied call attempt will be *lost*. So a *loss model* is useful in analyzing CAC (call admission control or connection admission control) in connection-oriented network services. *Congestion* of TCP flows in the Internet can be analyzed by formulating the system as a loss system. Thus, in a loss system model, a customer is usually a "call" and servers are "lines" or "circuits." The service time is often called the *(call) holding time.* A lost call is sometimes referred to as a *blocked* call.

In this section we discuss two classical loss models, the Erlang and Engset loss models, which are again special cases of the BD process model. We also remark on generalizations of these classical loss models.

23.4.1 M/M/m(0) and M/G/m(0): Erlang loss models

Consider a service center with m parallel servers, but no waiting room, as shown in Figure 23.8. Customers or calls arrive according to a Poisson process with rate λ, and the service time S is exponentially distributed with mean μ^{-1}, as assumed in the queueing models studied in the preceding sections. We denote this loss system as M/M/m(0), where (0) means no room for a queue. The conventional notation is $M/M/m/m$, where the last m represents the maximum number of customers that can be accommodated in the system. The system M/M/m(0) or $M/M/m/m$ is called the *Erlang loss model* in deference to A. K. Erlang.

Let $N(t)$ be the number of calls in service at time t. Clearly, $0 \le N(t) \le m$. When $N(t) = m$, a new call is denied service and is lost. The process $N(t)$ is a BD process with

$$\lambda_n = \lambda, \quad n = 0, 1, 2, \ldots, m-1, \quad \text{and} \quad \mu_n = n\mu, \quad n = 1, 2, \ldots, m. \qquad (23.111)$$

The *normalization constant* G of (14.57) becomes

$$G = \pi_0^{-1} = \sum_{n=0}^{m} \frac{a^n}{n!}, \qquad (23.112)$$

where the dimensionless parameter

$$a = \lambda E[S] = \frac{\lambda}{\mu} \text{ [erl]} \qquad (23.113)$$

is the traffic intensity defined in (23.45), but is often referred to as the **offered load**. Hence,

$$\rho = \frac{a}{m} = \frac{\lambda}{m\mu} \qquad (23.114)$$

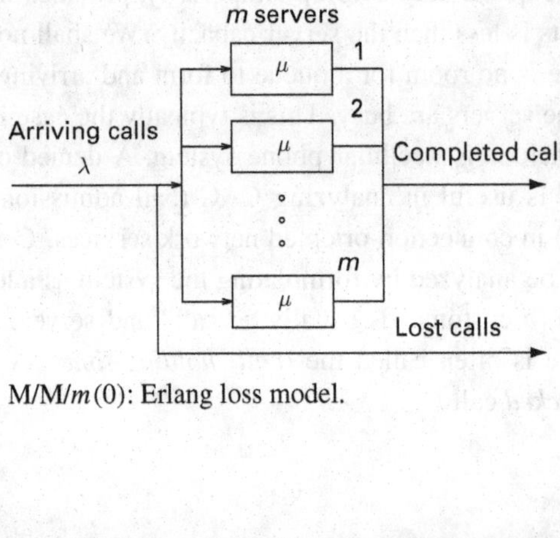

Figure 23.8 M/M/m(0): Erlang loss model.

is the *offered load per server*. As defined earlier in (23.113), the offered load a [erl] represents the total average load placed on the service station per unit time, or, equivalently, the average number of call arrivals during service of a call. The parameter ρ is *not* a measure of server utilization or productivity in a loss system, because some of the offered calls will be lost and never served.

The equilibrium state distribution of $N(t)$

$$\boxed{\pi_n = \frac{a^n/n!}{\sum_{i=0}^{m} \frac{a^i}{i!}}, \quad \text{for } n = 0, 1, 2, \ldots, m,} \tag{23.115}$$

is called the **Erlang distribution.**[3] As $m \to \infty$, this Erlang distribution (23.115) approaches the Poisson distribution (23.46). This is because, in the limit $m \to \infty$, the systems M/M/$m(0)$ and M/M/∞ are equivalent: the waiting room is immaterial when infinitely many servers are available.

We can express the Erlang distribution as a *truncated* Poisson distribution

$$\boxed{\pi_n = \frac{P(n; a)}{Q(m; a)}, \quad \text{for } n = 0, 1, 2, \ldots, m,} \tag{23.116}$$

where $P(n; a)$ and $Q(n; a)$ are the Poisson and cumulative Poisson distributions defined in (23.32) and (23.31), respectively.

23.4.1.1 Generalized Erlang loss model: M/ G/ $m(0)$

Now we make an important observation that the Erlang distribution (23.115) or (23.116) is insensitive to the form of the service time distribution $F_S(t)$, similar to the property we found for the systems M/G/∞ and G(K)/M/m. A formal proof can be provided by applying the notion of *quasi-reversibility* of a *symmetric queue* as discussed in Kelly [177] (see also [203] Section 5.2). See also Ross [287] for a formal proof.

Here, we provide an intuitive argument by relating the the generalized Erlang loss model M/G/$m(0)$ to the machine repairman model G(m)/M/1 (with the parameter K now replaced by m), for which we showed the insensitivity of the queue distribution to the distribution of U, the interval that a machine is up and running. As far as the queue distribution at the m-parallel exponential servers is concerned, the lost calls in the system M/G/$m(0)$ have no effect: a lost call does not change the system state. So if we consider only accepted and served calls, then we can construct a two-stage cyclic *closed* system that is equivalent to the system M/G/$m(0)$, although the latter is an *open* system. This equivalent closed system consists of an infinite server (IS) station (where a customer stays according to $F_S(t)$), and an exponential server (with mean service time λ^{-1}), with m customers or calls circulating.

[3] This should not be confused with the k-stage Erlang or Erlangian distribution E_k discussed in Section 4.2.3.

The probability that n out of m machines are in service by the repairman in the system G(m)/M/1 is given from (23.66) as

$$\pi_n(m) = \frac{P(m-n, r^{-1})}{Q(m, r^{-1})}, \tag{23.117}$$

where r is defined in (23.63). This is equal to the probability that $m - n \triangleq n'$ machines are up and running (i.e., in the IS station side). In the corresponding Erlang loss model, the probability that n' calls are in service is given from (23.116) by

$$\pi_{n'} = \frac{P(n', a)}{Q(m, a)}, \tag{23.118}$$

where $n' = m - n$ and $a = \mu/\nu = r^{-1}$. Clearly, these two probabilities are equivalent. Thus, the insensitivity of the Erlang loss model with respect to service time distribution is reduced to the insensitivity of a G(m)/M/1 system.

23.4.1.2 Blocking probability versus loss probability

The probability π_m in (23.116) represents the proportion of time that all the m servers are occupied. In telephony, this long-run time-average probability is called the *blocking probability* or *time congestion* [316]. A related quantity is the *loss probability*, *call loss probability*, or *call congestion*, denoted by L, which is the probability that an arriving call finds all the m servers (or lines) occupied, and hence will be lost.

Because of the PASTA property, $\{a_n\}$, the probability distribution of the number of calls in progress observed by an arriving call should be equal to $\{\pi_n\}$, the long-run time-average distribution of $N(t)$ (Problem 23.35). Thus, the call congestion L is equal to the time congestion π_m:

$$\boxed{L = \pi_m = \frac{a^m/m!}{\sum_{i=0}^{m} \frac{a^i}{i!}} = \frac{P(m; a)}{Q(m; a)} \triangleq B(m, a),} \tag{23.119}$$

which is called **Erlang's loss formula**, *Erlang's first formula*, or the *Erlang B formula*.

Figure 23.9 shows L as a function of traffic load a with the number of circuits m fixed. Four different values of m are assumed: $m = 10, 20, 40$, and 80. Telephone networks are usually designed and operated so that $L \approx 0.01$ is attained for an assumed value of a.

23.4.1.3 Carried load

We now introduce the notion of **carried load** or *carried traffic*, denoted a_c, which is the average number of servers occupied in the steady state. It is equivalent to *throughput*:

$$\boxed{a_c = \lim_{t\to\infty} E[N(t)].} \tag{23.120}$$

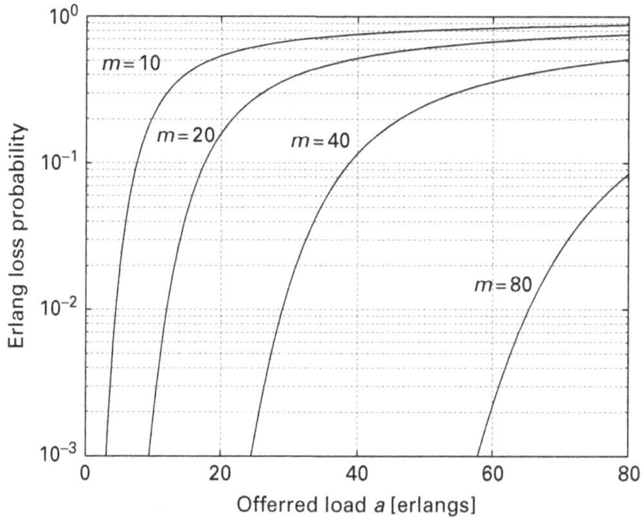

Figure 23.9 Erlang's loss formula: the loss probability L versus offered load a: the number of circuits $m = 10, 20, 40,$ and 80 ([203] Reprinted with permission from Pearson Education, Inc., Upper Saddle River, NJ.).

For the Erlang loss model we have

$$a_c = \sum_{n=1}^{m} n\pi_n = \frac{a \sum_{n=1}^{m} \frac{a^{n-1}}{(n-1)!}}{\sum_{i=0}^{m} \frac{a^i}{i!}} = a[1 - B(m, a)]. \tag{23.121}$$

The loss probability L is, by definition, the proportion of calls that cannot be carried; hence, we can write

$$\boxed{L = 1 - \frac{a_c}{a},} \tag{23.122}$$

and with (23.121) we have

$$L = \frac{a - a[1 - B(m, a)]}{a} = B(m, a), \tag{23.123}$$

which is Erlang's loss formula (23.119), obtained earlier.

23.4.2 M(*K*)/M/*m*(0) and G(*K*)/G/*m*(0): Engset loss models

We now consider a loss system with K sources and m parallel servers, as shown in Figure 23.10. This model can represent a circuit-switched telephone system, in which K input lines (sources) are served by m output lines. It is also a proper model for a cellular radio system in which K mobile users are served by m channels that belong to a base station. Similar to the M(K)/M/m queueing model of Section 23.3.4, each source spends a random time interval U before generating a customer.

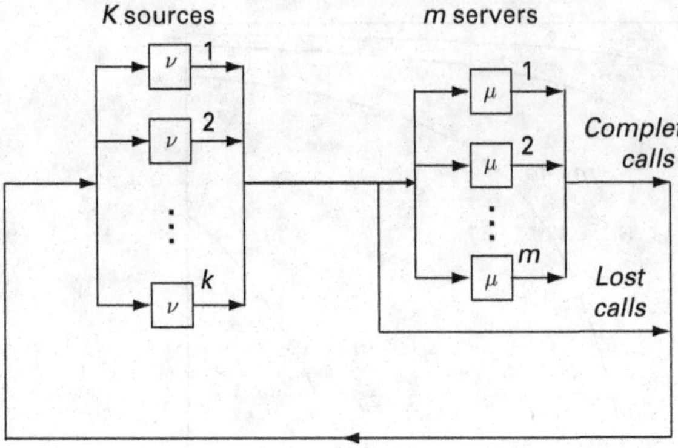

Figure 23.10 M(K)/M/m(0): Engset loss model.

By extending our argument concerning the two-stage cyclic queueing system representation of the Erlang loss model, this finite source loss model can also be viewed as a two-stage cyclic system which is composed of an IS station and a station with m-parallel servers, with K customers (or calls) circulating. Therefore, the queue distribution in the steady state is *insensitive* to the distribution of time spent in the sources (the IS station) $F_U(t)$ as well as the service time distribution $F_S(t)$. A mathematical argument to support this statement can be found in, for example, [203].

We first assume, however, that $F_U(t) = 1 - e^{-\nu t}$. The call arrivals form a *quasi-random* arrival process, as was defined in Section 23.3.4. The service time or call holding time is exponentially distributed; i.e., $F_S(t) = 1 - e^{-\mu t}$. This loss system model, a hybrid between the multiple access model (or multiple repairman model) M(K)/M/m and the Erlang loss model M/M/m(0), is called the **Engset loss model** in deference to Engset [91], denoted as M(K)/M/m(0). The conventional notation in the queueing literature is M/M/$m/m/K$.

Then $N(t)$, the number of calls in service at time t, is a BD process with birth rate (23.60) and death rate (23.111), and the normalization (14.57) becomes

$$G = \sum_{n=0}^{m} \frac{K(K-1)\cdots(K-n+1)}{n!}\frac{\nu^n}{\mu^n} = \sum_{n=0}^{m}\binom{K}{n} r^n, \tag{23.124}$$

where r is as defined in (23.63). Therefore, the stationary distribution of $N(t)$ is

$$\boxed{\pi_n(K) = \frac{\binom{K}{n} r^n}{\sum_{i=0}^{m}\binom{K}{i} r^i} \triangleq E(m, K, r), \quad 0 \le n \le m,} \tag{23.125}$$

which is a *truncated binomial distribution*. This distribution is known as the *Engset distribution*. Then the blocking probability or time congestion is given by

$$\pi_m(K) = \frac{\binom{K}{m} r^m}{\sum\limits_{i=0}^{m} \binom{K}{i} r^i} \triangleq E(m, K, r), \tag{23.126}$$

which we call the **Engset blocking probability**.

There are on average $K - E[N(t)]$ sources that are eligible to generate new customers with intensity ν/μ. Thus, the offered load is expressed as

$$a = (K - E[N(t)])\frac{\nu}{\mu} = r\sum_{n=0}^{m}(K-n)\,\pi_n(K) = rK\frac{\sum\limits_{n=0}^{m}\binom{K-1}{n} r^n}{\sum\limits_{i=0}^{m}\binom{K}{i} r^i}. \tag{23.127}$$

An alternative way to derive the first equality above is to recognize that the average service time for a given customer is $E[S] = \mu^{-1}$. During its service, the expected arrival rate of new customers is given by $(K - E[N(t)])\nu$. The offered load is by definition the product of these two quantities.

The carried load a_c, the expected number of servers that are occupied at given time, can be readily obtained as

$$a_c = \sum_{n=1}^{m} n\pi_n(K) = rK\frac{\sum\limits_{j=0}^{m-1}\binom{K-1}{j} r^j}{\sum\limits_{i=0}^{m}\binom{K}{i} r^i}. \tag{23.128}$$

From the last two equations we can obtain the *Engset loss probability* or *call congestion* as

$$\boxed{L(K) = 1 - \frac{a_c}{a} = \frac{\binom{K-1}{m} r^m}{\sum\limits_{i=0}^{m}\binom{K-1}{i} r^i} = E(m, K-1, r),} \tag{23.129}$$

which is called **Engset's loss formula**.

23.4.2.1 Probability distribution seen by arrivals

Unlike in the Erlang loss model, the PASTA property does not hold for the Engset loss model. However, the "arrival theorem" we derived in Section 23.3.4 holds here as well, since the Engset loss model is a Markovian queueing model. Thus, similar to (23.69), we find

$$\boxed{a_n(K) = \pi_n(K-1) = \frac{\binom{K-1}{n} r^n}{\sum\limits_{i=0}^{m}\binom{K-1}{i} r^i}, \quad \text{for } 0 \le n \le m.} \tag{23.130}$$

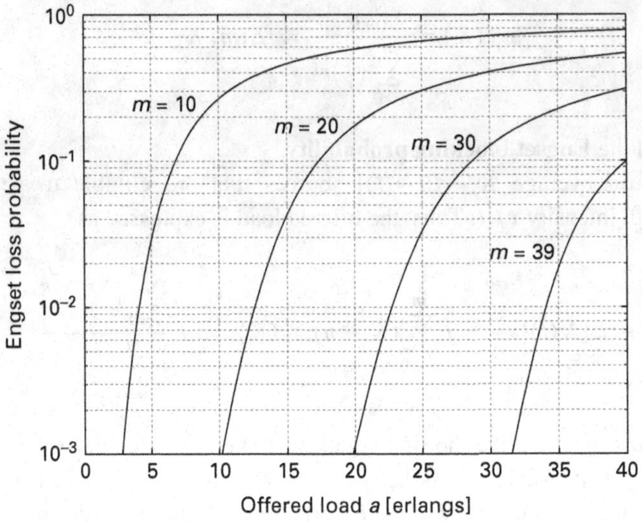

Figure 23.11 Engset loss formula with $K = 40$ sources: the loss probability $L(K)$ versus offered load a for different values of the number of servers $m = 10, 20, 30, 39$ ([203] Reprinted with permission from Pearson Education, Inc., Upper Saddle River, NJ.).

The loss probability $L(K)$ is by definition $a_n(K)$ at $n = m$:

$$L(K) = a_m(K) = \pi_m(K-1) = E(m, K-1, r), \tag{23.131}$$

which is Engset's loss formula (23.129) obtained earlier from a and a_c.

Figure 23.11 shows the loss probability $L(K)$ as a function of the offered load a in a loss system with $K = 40$ sources. Four different values of m are considered: $m = 10, 20, 30,$ and 39. The case $m = 40$ servers does not show up here, since the loss probability should always be zero if $m \geq K$. The parameter $r = \nu/\mu$ is implicit, but is related to the offered load a through (23.127). As r increases, the offered load approaches $(K - m)r$ [erl], as is evident from (23.127).

We define the *trunk efficiency* or *link efficiency* as the utilization factor per server (i.e., output line or trunk):

$$\eta = \frac{a_c}{m}. \tag{23.132}$$

We can then show from Engset's loss formula that the trunk efficiency increases as m increases for any K, when the loss probability $L(K)$ is fixed.

23.4.3 Loss network model and generalized loss station

In this section we will introduce the notion of *loss networks* introduced by Kelly [178], which may be viewed as loss system counterparts of *queueing networks*. A queueing network has been applied to analyze a packet-switched network (e.g., see Kleinrock [190]), whereas a loss network model can represent, for instance, (i) a circuit-switched network such as a telephone network and (ii) the Internet network at the flow level with a resource reservation protocol.

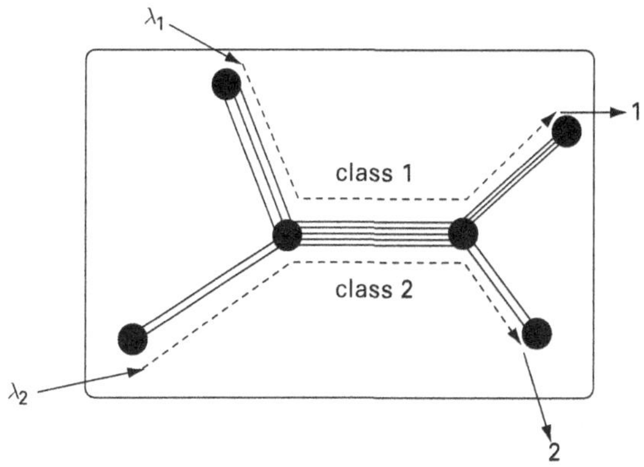

Figure 23.12 Open loss network ([203] Reprinted with permission from Pearson Education, Inc., Upper Saddle River, NJ.).

23.4.3.1 Open loss network

First we consider a loss network, as illustrated in Figure 23.12. We call this type of loss network an *open loss network* (OLN). The Erlang loss model is a simple example of OLN, whereas the Engset loss model is an example of *closed loss network* (CLN).

We define the following notation and properties of OLN.

1. Let $\mathcal{L}$ denote the set of links in the network. There are $L = |\mathcal{L}|$ links in the network; link $\ell \in \mathcal{L}$ contains m_l units for service, where the service unit may be a channel or a time slot.
2. A call class $r \in \mathcal{R}$ is defined as a pair (c, τ); i.e.,

$$r = (c, \tau), \quad c \in \mathcal{C}, \quad \tau \in \mathcal{T}, \tag{23.133}$$

   where $\mathcal{C}$ is the set of *routing chains* or *paths* and $\mathcal{T}$ is the set of *call types*. Thus, the set $\mathcal{R} = \mathcal{C} \times \mathcal{T}$.
3. The arrival pattern of class-r calls to the OLN is a Poisson process with rate λ_r, $r \in \mathcal{R}$.
4. A class $r = (c, \tau)$ call seeks to simultaneously acquire $A_{\ell,r}$ service units for all links ℓ in its route c.
5. The holding time of a class-r call is a general distribution with mean $1/\mu_r$.

This OLN model provides a general model for a *circuit-switched network* that carries *multirate* traffic: different bandwidth requirements correspond to different *types* of call. The bandwidths of different links are generally different. We assume that a class $r = (c, \tau)$ call requires $A_{\ell,r}$ units (i.e., servers) from link ℓ. Although Figure 23.12 might give a false impression that the OLN model can represent only unidirectional traffic, that is not the case. The model can handle bidirectional flows by assigning different class identifications to traffic in the reverse directions. The reverse traffic for a given pair of nodes may have different bandwidth requirements. The path for the reverse traffic need not be the reverse path of the forward path.

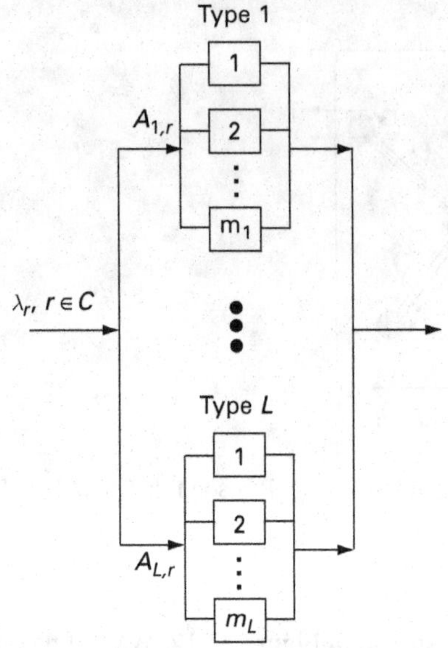

Figure 23.13 GLS which is equivalent to the OLN ([203] Reprinted with permission from Pearson Education, Inc., Upper Saddle River, NJ.).

23.4.3.2 Generalized loss station (GLS)

We now consider what we term a *generalized loss station* (GLS) of Figure 23.13, which consists of multiple *server types* $\ell \in \mathcal{L}$. This loss station is a generalization of the loss stations defined in the previous sections (i.e., the Erlang and Engset loss stations), in the sense that it has multiple types of server. Class-r customers arrive at the GLS as a birth process with rate $\lambda_r\, a_r(n_r)$. Every class-r customer requires $A_{\ell,r}$ servers of type $\ell \in \mathcal{L}$, all simultaneously; thus, a class-r customer holds a total of $\sum_{\ell\in\mathcal{L}} A_{\ell,r}$ servers.

Let $N_r^{(\mathrm{GLS})}(t)$ be the number of class-r customers being served by the GLS at time t. Then the process $N^{(\mathrm{GLS})}(t) = \{N_r^{(\mathrm{GLS})}(t);\ \ r \in \mathcal{C}\}$ is *equivalent* to the process $N^{(\mathrm{OLN})}(t)$ defined over the OLN. We state this relationship as a theorem.

THEOREM 23.2 (Equivalence of an OLN and its associated GLS). *The OLN defined above can be reduced to a GLS with* multiple server types.[4]

Let $N_r(t)$ represent the number of class-r calls in progress at time t in an OLN or the number of class-r customers in service at its equivalent GLS. Then, the process $\boldsymbol{N}(t) = \{N_r(t); r \in \mathcal{C}\}$ is reversible and has a stationary distribution of product form:

$$\boxed{\pi_{\boldsymbol{N}}(\boldsymbol{n}) = \frac{1}{G(\boldsymbol{m})} \prod_{r\in\mathcal{R}} \frac{a_r^{n_r}}{n_r!}, \quad \boldsymbol{n} \in \mathcal{F}_{\boldsymbol{N}}(\boldsymbol{m}),} \tag{23.134}$$

[4] Note that the *link index* $\ell \in \mathcal{L}$ in the OLN corresponds to the *server type* in its equivalent GLS. Thus, the number of servers of type ℓ is m_ℓ in the equivalent GLS, $\ell \in \mathcal{L}$.

where $a_r = \lambda_r / \mu_r$ *and*

$$G(\boldsymbol{m}) \triangleq \sum_{\boldsymbol{n} \in \mathcal{F}_N()} \prod_{r \in \mathcal{R}} \frac{a_r^{n_r}}{n_r!}, \tag{23.135}$$

$$\mathcal{F}_N(\boldsymbol{m}) \triangleq \{\boldsymbol{n} \geq \boldsymbol{0} : \sum_{r \in \mathcal{R}} A_{\ell,r} n_r \leq m_\ell, \ \ \ell \in \mathcal{L}\}, \tag{23.136}$$

where m_ℓ *is the number of type-*ℓ *servers.*

A formal proof of this theorem makes use of the *quasi-reversibility* of the GLS. Interested readers are referred to [201, 203] for further details.

We now introduce a CLN by replacing the set of Poisson streams by a set of finite number K_r of sources, $r \in \mathcal{R}$. The generalized Engset loss model is an example of a CLN. We classify the *chains* of a loss network into the set of open subchains $\mathcal{C}_{\text{op}}$ and the set of closed subchains $\mathcal{C}_{\text{cl}}$. Type-τ customers in an open chain $c \in \mathcal{C}_{\text{op}}$ arrive to an OLN according to a Poisson process with rate λ_r, where $r = (c, \tau)$, whereas type-τ customers in a closed chain $c \in \mathcal{C}_{\text{cl}}$ enter the equivalent GLS from $K_r - n_r$ *ready* sources, each of which sends out calls at rate ν_r.

A *mixed loss network* (MLN), as depicted in Figure 23.14, has both kinds of routes. The MLN generalizes both the Erlang and Engset stations of the previous section. To simplify notation, for a given class membership $r = (c, \tau)$, we write $r \in \mathcal{R}_{\text{op}}$ if $c \in \mathcal{C}_{\text{op}}$. Similarly, $r \in \mathcal{R}_{\text{cl}}$ and $c \in \mathcal{C}_{\text{cl}}$ are equivalent. Denote the population vector process in the loss network proper (excluding those in the sources) by $\boldsymbol{N}(t) = [\boldsymbol{N}_{\text{op}}(t), \boldsymbol{N}_{\text{cl}}(t)]$, with $\boldsymbol{N}_{\text{op}}(t) = (N_r(t) : r \in \mathcal{R}_{\text{op}})$ and $\boldsymbol{N}_{\text{cl}}(t) = (N_r(t) : r \in \mathcal{R}_{\text{cl}})$, where $r = (c, \tau)$. We have the following result for the MLN.

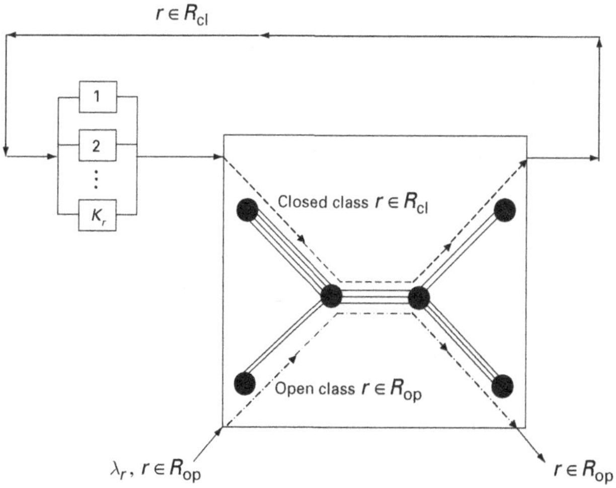

Figure 23.14 Mixed loss network ([203] Reprinted with permission from Pearson Education, Inc., Upper Saddle River, NJ.).

THEOREM 23.3 (Mixed loss network and reversibility). *The population vector process of the MLN is a reversible process with equilibrium distribution given by*

$$\boxed{\pi_N(\boldsymbol{n}|\boldsymbol{m},\boldsymbol{K}) = \frac{1}{G(\boldsymbol{m},\boldsymbol{K})}\pi_{\text{op}}(\boldsymbol{n}_{\text{op}})\pi_{\text{cl}}(\boldsymbol{n}_{\text{cl}}|\boldsymbol{K}), \quad \boldsymbol{n} \in \mathcal{F}_N(\boldsymbol{m},\boldsymbol{K}),} \tag{23.137}$$

where $\boldsymbol{n} = (\boldsymbol{n}_{\text{op}}, \boldsymbol{n}_{\text{cl}})$ *and*

$$\boxed{\pi_{\text{op}}(\boldsymbol{n}_{\text{op}}) = \prod_{r\in\mathcal{R}_{\text{op}}} \frac{a_r^{n_r}}{n_r!}, \quad \pi_{\text{cl}}(\boldsymbol{n}_{\text{cl}}|\boldsymbol{K}) = \prod_{r\in\mathcal{R}_{\text{cl}}} \binom{K_r}{n_r} r_r^{n_r},} \tag{23.138}$$

with $a_r = \lambda_r/\mu_r$ $(r \in \mathcal{R}_{\text{op}})$, $r_r = \nu_r/\mu_r$ $(r \in \mathcal{R}_{\text{cl}})$,

$$\mathcal{F}_N(\boldsymbol{m},\boldsymbol{K}) = \{\boldsymbol{n} \geq \boldsymbol{0} : \sum_{r\in\mathcal{R}} A_{\ell,r} n_r \leq m_\ell,\ \ell \in \mathcal{L};\ n_r \leq K_r,\ r \in \mathcal{R}_{\text{cl}}\}, \tag{23.139}$$

and

$$G(\boldsymbol{m},\boldsymbol{K}) = \sum_{\boldsymbol{n}\in\mathcal{F}_N(\boldsymbol{m},\boldsymbol{K})} \pi_{\text{op}}(\boldsymbol{n}_{\text{op}})\pi_{\text{cl}}(\boldsymbol{n}_{\text{cl}}|\boldsymbol{K}). \tag{23.140}$$

□

A proof of this result can be found in [203].

From the stationary distribution of the MLN obtained above, we can express the *time congestion* and *call congestion* in terms of the normalization constant $G(\boldsymbol{m},\boldsymbol{K})$ as follows:

1. For a class-r customer in an open route $r \in \mathcal{R}_{\text{op}}$:

$$\boxed{\begin{aligned} B_r(\boldsymbol{m},\boldsymbol{K}) &= 1 - \frac{G(\boldsymbol{m}-\boldsymbol{A}_r,\boldsymbol{K})}{G(\boldsymbol{m},\boldsymbol{K})}, \\ L_r(\boldsymbol{m},\boldsymbol{K}) &= B_r(\boldsymbol{m},\boldsymbol{K}), \end{aligned}} \tag{23.141}$$

   where $\boldsymbol{A}_r$ is the rth column of the matrix $\boldsymbol{A} = [A_{\ell,r}]$. The last equation is due to the PASTA property.

2. For a class-r customer in a closed route $r \in \mathcal{R}_{\text{cl}}$:

$$\boxed{\begin{aligned} B_r(\boldsymbol{m},\boldsymbol{K}) &= 1 - \frac{G(\boldsymbol{m}-\boldsymbol{A}_r,\boldsymbol{K})}{G(\boldsymbol{m},\boldsymbol{K})}, \\ L_r(\boldsymbol{m},\boldsymbol{K}) &= B_r(\boldsymbol{m},\boldsymbol{K}-\boldsymbol{e}_r), \end{aligned}} \tag{23.142}$$

   where $\boldsymbol{e}_r$ denotes the unit vector, whose rth component is unity, with all other components being zero.

The above formulae for the time and call congestions are generalizations of the formulas obtained for the Erlang and Engset models. Finally, as $\boldsymbol{m} \to \infty$, $G(\boldsymbol{m},\boldsymbol{K})$ increases monotonically to the normalization constant for a network with all IS stations:

$$G(\infty, \boldsymbol{K}) = \exp\left(\sum_{r\in\mathcal{R}_{\text{op}}} a_r\right) \prod_{r\in\mathcal{R}_{\text{cl}}} (1 + r_r)^{K_r}. \qquad (23.143)$$

23.4.3.3 Computational algorithms for loss networks

Although computation of the Erlang loss formula (23.119) and Engset loss formula (23.129) should be relatively straightforward, their generalizations given by (23.141) and (23.142) may be computationally difficult or infeasible, as some of the dimensions of the set of links $\mathcal{L}$, the numbers of lines m_ℓ, $\ell \in \mathcal{L}$, the number of call classes in $\mathcal{R}_{\text{op}}$ and $\mathcal{R}_{\text{cl}}$, and the number of sources K_r, $r \in \mathcal{R}_{\text{cl}}$, become somewhat large, because the number of terms that appear in the the expressions for the normalization constant (23.135) or (23.140) may be enormous.

There are several computationally efficient algorithms for the normalization constant G: (i) recursive algorithms, (ii) numerical evaluation of an inversion integral for a generating function representation of the normalization constant, (iii) asymptotic approximation of the inversion integral, and (iv) reduced load approximation. In the interest of space, the interested reader is directed to [201, 203] and references therein.

23.5 Summary of Chapter 23

Little's law:	$\overline{L} = \lambda \overline{W}$	(23.1)
M/M/1:		
Prob. distribution:	$\pi_n = (1-\rho)\rho^n, \quad n = 0, 1, 2, \ldots$	(23.20)
Mean system time:	$\overline{T} = \frac{\overline{N}}{\lambda} = \frac{1}{\mu(1-\rho)}$	(23.28)
Waiting time dist. function:	$F_W(x) = 1 - \rho\, e^{-\mu(1-\rho)x}, \quad x \geq 0$	(23.37)
System time dist. function:	$F_T(x) = 1 - e^{-\mu(1-\rho)x}, \quad x \geq 0$	(23.41)
M/G/∞:		
Prob. distribution:	$\pi_n = \frac{a^n}{n!} e^{-a} = P(n\,; a), \quad n = 0, 1, 2, \ldots$	(23.46)
Transient. prob. distr.:	$\pi_n(t) = \frac{\left(\int_0^t (1-F_S(u))\,du\right)^n}{n!} e^{-\lambda\int_0^t (1-F_S(u))\,du}$	(23.47)
M/M/m:		
Prob. distribution:	$\pi_0 = G^{-1}$	
	$\pi_n = \begin{cases} \frac{a^n}{n!}\pi_0, & n = 1, 2, \ldots, m \\ \rho^{n-m}\pi_m, & n = m+1, m+2, \ldots \end{cases}$	(23.52)
Erlang B formula:	$B(m, a) \triangleq \frac{a^m/m!}{\sum_{i=0}^m \frac{a^i}{i!}}$	(23.55)
Erlang C formula:	$C(m, a) = \frac{mB(m,a)}{m - a(1-B(m,a))}$	(23.54)
Waiting time dist. function:	$F_W(x) = 1 - C(m, a)e^{-m\mu(1-\rho)x}$	(23.149)
Mean waiting time:	$E[W] = \frac{C(m,a)}{m\mu(1-\rho)}$	(23.57)

G(K)/M/1: $\pi_n(K) = \frac{P(K-n;r^{-1})}{Q(K;r^{-1})}, \quad n = 0, 1, 2, \ldots$ (23.66)

M(K)/M/m: $a_n(K) = \frac{\lambda_n \pi_n(K)}{\sum_{i=0}^{K-1} \lambda_i \pi_i(K)}, 0 \le n \le K-1$ (23.67)

Arrival theorem: $a_n(K) = \pi_n(K-1), \ 0 \le n \le K-1$ (23.69)

M/G/1:

Recurrence relation: $N_k = N_{k-1} - U(N_{k-1}) + A_k$ (23.76)

PGF of $\{\pi_n\}$: $P(z) = A(z)\left[\pi_0 + z^{-1}(P(z) - \pi_0)\right]$ (23.80)

PGF of $\{A_k\}$: $A(z) = f_S^*(\lambda - \lambda z)$ (23.82)

P–K transform formula: $P(z) = \frac{(1-\rho)(z-1) f_S^*(\lambda-\lambda z)}{z - f_S^*(\lambda-\lambda z)}$ (23.84)

P–K mean value formula: $\overline{N} = \rho + \frac{\rho^2(1+c_S^2)}{2(1-\rho)}$ (23.91)

Laplace transform of system time (FCFS): $f_T^*(s) = \frac{(1-\rho) s f_S^*(s)}{s-\lambda+\lambda f_S^*(s)}$ (23.99)

M/G/1 with PS:

Prob. distribution: $\pi_n = (1-\rho)\rho^n, \ n = 0, 1, \ldots$ (23.104)

Mean waiting time: $\overline{W}_{\mathrm{PS}} = \frac{\rho E[S]}{1-\rho}$ (23.110)

(Generalized) Erlang loss model:

Erlang distribution: $\pi_n = \frac{a^n/n!}{\sum_{i=0}^{m} \frac{a^i}{i!}}, \quad n = 0, 1, \ldots, m$ (23.115)

Erlang B loss formula: $L = \pi_m = \frac{a^m/m!}{\sum_{i=0}^{m} \frac{a^i}{i!}} = \frac{P(m;a)}{Q(m;a)} \triangleq B(m, a)$ (23.119)

(Generalized) Engset loss model:

Engset distribution: $\pi_n(K) = \frac{\binom{K}{n} r^n}{\sum_{i=0}^{m} \binom{K}{i} r^i} \triangleq E(m, K, r), 0 \le n \le m$ (23.125)

Engset loss formula: $L(K) = 1 - \frac{a_c}{a} = \frac{\binom{K-1}{m} r^m}{\sum_{i=0}^{m} \binom{K-1}{i} r^i} = E(m, K-1, r)$ (23.129)

23.6 Discussion and further reading

A general and rigorous proof of Little's formula, a seemingly obvious formula, is rather involved mathematically (e.g., see Wolff [358]). In this section we followed Jewell [168], by restricting ourselves to the situation in which the queue empties itself infinitely often. Little's formula has been generalized to $H = \lambda G$, where H and G are respectively time and customer averages of quantities that bear a certain relationship to each other but are otherwise unspecified. For instance, if G represents the average cost per customer and λ is the customer arrival rate, then H represents the time average cost per unit time. This law was first proved by Brumelle [40] in a stochastic setting and then by Heyman and Stidham [151] as a sample-path law. See Stidham and El-Taha [311] for a comprehensive discussion on this subject.

The BD process-based queueing models, M/M/1, M/M/∞, M/M/m, and M(K)/M/1 and M/G/1 are discussed in most books on queueing theory; e.g., Cohen [61],

Cooper [67], Cox and Smith [73], Gross *et al.* [132], Kleinrock [189], Lipsky [227], Morse [251], Nelson [254], Syski [316], Takács [317], Wolff [358], and others. Loss models, however, are not so extensively covered in these, an exception being Syski's book. Our treatment on the GLS and its equivalence to loss networks draws from our earlier publications [201–203].

In the interest of space, we entirely skipped discussion on G/M/m, G/G/1, and *queueing networks*, which are extensively covered in many books on queueing theory and its applications; e.g., Bolch *et al.* [33], Chandy and Sauer [48], Daigle [76], Gelenbe and Mitrani [116], Haverkort [146], Hayes and Ganesh Babu [147], Kelly [177], King [183], Kobayashi [197], Kobayashi and Mark [203], Lavenberg [219], Tijms [322], and others.

23.7 Problems

Section 23.2: Little's formula: $\overline{L} = \lambda \overline{W}$

23.1 Little's formula [197]. Determine whether the following statements concerning the formula $\overline{L} = \lambda \overline{W}$ are true or false.

(a) The formula $\overline{L} = \lambda \overline{W}$ holds only when the arrival process is a Poisson process with rate λ.
(b) It is not necessary to assume a Poisson process, but interarrival times must be statistically independent variables.
(c) The formula $\overline{L} = \lambda \overline{W}$ is valid under any queueing discipline. Thus, the average queue length $\overline{L}$ is invariant under different queue disciplines.
(d) The formula holds even when the arrival rate at a given time is dependent on the congestion of the queueing system.

23.2 Little's formula for multiple customer types [197]. Suppose that there are R types of customer. The arrival rates are given by $\lambda_r, r = 1, 2, \ldots, R$. How do you generalize the formula $\overline{L} = \lambda \overline{W}$? If the queue discipline is FCFS, what can we say about the average queue sizes of different types?

23.3 Distributions seen by arrivals and departures [197]. Choose a sufficiently long observation interval $(0, T)$ such that $Q(t) = A(t) - D(t) = 0$ at both $t = 0$ and $t = T$. With the aid of the diagram of $Q(t)$, show that the distribution of queue size seen by arriving customers is the same as that seen by departing customers.

Section 23.3: Queueing models

23.4 Choice of work unit in a queueing model. Suppose that a server is a communication link and customers are packets. Assume that the communication link has a bandwidth or speed of B [bits/s] and packet lengths are random variable L_n [bits]. The interarrival times between packets are i.i.d. RVs T_n [s].

(a) Find the arrival rate λ and the service completion rate μ.
(b) What assumptions do we have to make in order to formulate this problem as an M/M/1 queue?

23.5 State classification in the M/M/1 queue. Show that, in the M/M/1 queue, the states are all transient if $\rho = \lambda/\mu > 1$, null-recurrent if $\rho = 1$, and ergodic if $\rho < 1$.

23.6 $L(t)$ in the M/M/1 queue analysis. Show that the queue-length process $L(t)$ defined by (23.8) is not a BD process, despite its similarity to $N(t)$ defined by (23.7). *Hint*: Consider the state $L(t) = 0$. How does a new arrival affect the state?

23.7 Mean and variance in the M/M/1 queue. Show that the mean and variance of $N(t)$ in the M/M/1 queue are given by (23.21) and (23.22). Show also that the mean and variance of the length of the waiting line $L(t)$ are given by (23.23) and (23.24), respectively.

23.8* Derivation of the waiting time distribution (23.37). Derive $F_W(x)$ of (23.37).

23.9 Cumulative Poisson distribution $Q(n; a)$ [203]. Show the following properties of $Q(k\,; a)$:

(a)

$$\sum_{j=0}^{k} P(k-j\,; a_1)Q(j\,; a_2) = Q(k\,; a_1 + a_2).$$

(b)

$$Q(k\,; a) = \int_a^\infty P(k\,; y)\,dy = \frac{\Gamma(k+1, a)}{k!},$$

where $\Gamma(p, z) = \int_z^\infty y^{p-1}e^{-y}\,dy$ is called the *upper incomplete gamma function* (cf. (7.126)).

(c)

$$Q(k\,; a) = \frac{aQ(k-1; a)}{k+a+1} + \frac{(k+1)Q(k+1; a)}{k+a+1}.$$

(d)

$$\sum_{j=0}^{k-1} Q(j\,; a) = kQ(k\,; a) - aQ(k-1; a).$$

(e)

$$\int_a^\infty Q(k-1; y)\,dy = kQ(k\,; a) - aQ(k-1; a).$$

23.10* Time-dependent solution for a certain BD process. Consider a BD process with $\lambda_n = \lambda$ for all $n \geq 0$ and $\mu_n = n\mu$ for all $n \geq 1$. This process represents the

M/M/∞ queue. Find the partial differential equation that $G(z,t)$ must satisfy. Show that the solution to this equation is

$$G(z,t) = \exp\left[\frac{\lambda}{\mu}(1 - e^{-\mu t})(z-1)\right].$$

Show that the solution for $p_n(t)$ is given as

$$p_n(t) = \frac{\left[\frac{\lambda}{\mu}(1 - e^{-\mu t})\right]^j}{j!} \exp\left[-\frac{\lambda}{\mu}(1 - e^{-\mu t})\right], \quad 0 \le n < \infty. \tag{23.144}$$

23.11 Time-dependent solution for M/G/∞ is also Poisson [203, 249]. Show that $N(t)$ in an M/G/∞ is Poisson distributed, following the steps given below. Assume that the system is initially empty; i.e., $N(0) = 0$. Pick any customer who has arrived in the interval $(0, t)$. From the *uniformity property* of the Poisson process (e.g., Section 14.1.2 see [203]), the arrival time U of this customer is uniformly distributed over $(0, t)$.

(a) Show that the probability that this randomly picked customer is still in service at time t is

$$p(t) = \frac{\int_0^t (1 - F_S(t-u))\,du}{t}, \tag{23.145}$$

where $F_S(t)$ is the service time distribution.

(b) Suppose that N customers have arrived in $(0, t)$. What is the probability that $n(\le N)$ customers are still in service at time t?

(c) Show that the probability that there are n customers in service at time t is given by (23.47).

23.12 The departure process of M/G/∞ is Poisson. Continue the above problem to show that the departure process of M/G/∞ is also a Poisson process with rate λ in the equilibrium state.

(a) Find the probability that the number of customers m who have received service and departed by time t.

(b) Show that the number of customers that depart in $(t, t+s)$ is independent of the number of customers that have departed in $(0, t)$.

(c) Show that the probability of departure in the time increment $(t, t+h)$ approaches $\lambda h + o(h)$ in the limit $t \to \infty$.

23.13 Mean queue length and mean waiting time.

(a) Show that the mean queue length for M/M/m is given by

$$\overline{Q} = \frac{\rho C(m,a)}{(1-\rho)}.$$

(b) Derive the mean waiting time $E[W]$ of (23.57), using the above result.

23.14 Relations between the two Erlang formulas. Derive (23.54) for Erlang's delay formula $C(m, a)$ in terms of Erlang's loss formula $B(m, a)$.

23.15* Waiting time distribution in the M/M/*m* queue.

(a) Show

$$F_W^c(x) = P[W > x] = \sum_{n=0}^{\infty} a_{m+n} F_W^c(x|m+n)$$
$$= F_W^c(0)(1-\rho)\sum_{n=0}^{\infty} \rho^n F_W^c(x|m+n), \tag{23.146}$$

where

$$F_W^c(x|m+n) = P[\ W > x \mid (m+n) \text{ customers found by an arrival }]. \tag{23.147}$$

Hint: $a_{m+n} = \pi_{m+n}$ due to PASTA.

(b) Let T_i be the interval between the $(i-1)$st service completion and the ith completion, $i = 2, 3, \ldots, n+1$. Show that $T_1, T_2, \ldots, T_{n+1}$ are i.i.d. RVs with an exponential distribution of mean $1/m\mu$.

(c) Show then that

$$F_W(x|m+n) = 1 - e^{-m\mu x}\sum_{j=0}^{n} \frac{(m\mu x)^j}{j!}, \tag{23.148}$$

which is an $(n+1)$-stage Erlang distribution.

(d) Derive

$$\boxed{F_W(x) = 1 - C(m, a)e^{-m\mu(1-\rho)x}.} \tag{23.149}$$

23.16 Normalization constant (23.62). Derive the normalization constant $G(K)$ of (23.62).

23.17 Steady-state distribution of the number of customers in the system G(*K*)/M/*m*. Show that the steady-state distribution $\pi_n(K)$ is given by (23.64).

23.18 Derivation of (23.69). For $\lambda_n = (K-n)\nu$, derive the formula (23.69):

$$a_n(K) = \pi_n(K-1), \quad 0 \le n \le K-1.$$

23.19 The arrival theorem. Show that the relationship (23.69) between $\{a_n\}$ and $\{\pi_n\}$ still holds for general service rates μ_j and the arrival rate λ_j is an arbitrary function of $K-j$; i.e.,

$$\lambda_j = f(K-j),$$

where $f(i)$ is an arbitrary positive function for $i > 0$, but $f(0) = 0$.

23.20 Derivation of the distribution (23.66). Show that (23.65) leads to (23.66).

23.21 Limit of the machine repairman model. Show that in the limit $K \to \infty$, $\nu \to 0$ while $K\nu = \lambda$, the distributions (23.65) and (23.66) converge to that of the M/M/1.

23.22* The waiting time distribution in M(K)/M/m. Derive the complementary waiting time distribution (23.70).

23.23 The waiting time distribution in M(K)/M/m – continued. Derive (23.72) and (23.73).

23.24 PASTA in an M/G/1 queue. Consider an M/G/1 queue. Let $\{\pi_n\}$ be the probability distribution of the number of customers found in the system at a randomly chosen time.

(a) Let the Poisson arrival rate be λ. What is the expected number of arriving customers during the interval $(0, T]$ that find exactly n customers in the system?
(b) Show then that $\{a_n\}$, the probability distribution seen by arriving customers, is equal to $\{\pi_n\}$.

23.25 Determination of π_0. Derive (23.83).

23.26 M/G/1 with hyperexponential service time distribution [203]. Consider an M/G/1 queue whose service time distribution is a two-stage hyperexponential distribution:

$$F_S(t) = 1 - p_1 e^{-\mu_1 t} - \pi_2 e^{-\mu_2 t}, \tag{23.150}$$

where $p_1 + p_2 = 1$. Find the equilibrium-state distribution $\{\pi_n\}$.

23.27 M/G/1 with Erlangian service time distribution [203]. Consider an M/G/1 queue whose service time distribution is a k-stage Erlang distribution with mean $1/\mu$.

(a) Find the equilibrium-state distribution $\{\pi_n\}$.
(b) Plot the distributions $\{\pi_n\}$ for the cases $k = 1, 2$, and 4 with traffic intensity $\rho = 0.75$.

23.28 Derivation of Pollaczek–Khintchine formula. Show the derivation steps from (23.84) to (23.91).

23.29 Alternative derivation of Pollaczek–Khintchine formula [203]. Derive the Pollaczek–Khintchine formula as instructed below.

(a) Take the expectation of (23.76) and find the probability that the server is busy in the equilibrium state.
(b) Square both sides of (23.76) and then take the expectation. Derive the formula (23.91).

23.30 Derivation of (23.98). Discuss the relation between $P(z)$ and $F_T(t)$ and derive the expression (23.98).

23.31* Derivation of waiting time distribution (23.103). Derive the Pollaczek–Khintchine formula (23.103) for waiting time distribution in M/G/1.

23.32 System time distribution of an M/H$_2$/1 queue. Show that the system time distribution $F_T(t)$ of an M/H$_2$/1 system (see Problem 23.26) is also a two-phase hyperexponential distribution.

23.33 Conditional mean system time under FCFS. Show that the conditional mean response time for a customer with service time S in an M/G/1 system with FCFS is given by

$$E[T_{\text{FCFS}}(S)|S] = S + \frac{\rho E[S](1 + c_S^2)}{2(1-\rho)}. \tag{23.151}$$

23.34 Comparison of $\overline{T}_{\text{PS}}$ and $\overline{T}_{\text{FCFS}}$. What determines $\overline{T}_{\text{PS}} \gtrless \overline{T}_{\text{FCFS}}$?

Section 23.4: Loss models

23.35 PASTA in the Erlang loss model. Consider an M/M/m(0) loss system and let p_n be the probability distribution of the number of calls in service. If we choose a sufficiently long interval $(0, T)$, the portion of time that the system contains n calls is, on average, $\pi_n T$.

(a) Assume a Poisson arrival process with rate λ. What is the expected number of arriving calls during the interval $(0, T)$ that find exactly n calls in the system? Show then that $\{a_n\}$, the probability distribution seen by an arriving call, is equal to $\{\pi_n\}$.
(b) What can you say about the relation between $\{\pi_n\}$ and $\{d_n\}$, the distribution seen by a completing call?
(c) If the arrival process is not Poisson, the above result is no longer true in general. Find an example to illustrate this fact.

23.36 Erlang loss model and the machine repairman model. Study the relationship between the Erlang loss model M/G/m(0) and the machine repairman model G(m)/M/1 by answering the following questions.

(a) By drawing a schematic diagram, show that the machine repairman model G(m)/M/1 can be represented as a two-stage cyclic system with an IS station and a single exponential server (i.e., the repairman), with m machines circulating. Let U, the period that a machine is up and running, have mean ν^{-1} and assume that S, the time for repair, is exponentially distributed with mean μ^{-1}.
(b) Observe that the service completions (or departures) form a Poisson process with rate μ insofar as the repairman is busy. When the repairman becomes idle, there will be no departure of repaired machines. Hence, the departure process from the repairman is an *interrupted Poisson process* (IPP) (see Problem 16.11(d)). How many servers in the IS station are busy (i.e., how many machines are up and running) when this interruption occurs in the departing Poisson process?

(c) Find the relation between the above IPP and lost calls out of Poisson arrivals to the Erlang loss model M/G/$m(0)$.
(d) Explain the equivalence between (23.116) and (23.66).

23.37 Recursive computation of Erlang's loss formula. Show that Erlang's loss formula $B(m, a)$ of (23.119) satisfies the following recursion:

$$B(m, a) = \frac{aB(m-1, a)}{m + aB(m-1, a)}. \qquad (23.152)$$

23.38 Bounds for Erlang's loss formula [155].

(a) Show that Erlang's loss formula $B(m, a)$ has the following upper and lower bounds:

$$\frac{a^m}{m!}e^{-a} \leq B(m, a) \leq \sum_{i=m}^{\infty} \frac{a^i}{i!}e^{-a}.$$

(b) Show that, for $i \geq m$,

$$\frac{a^i}{i!} \leq \left(\frac{a}{m}\right)^{i-m} \frac{a^m}{m!}.$$

(c) Derive the following upper bound for $B(m, a)$:

$$B(m, a) \leq \frac{1}{1 - \frac{a}{m}} \frac{a^m}{m!} e^{-a}. \qquad (23.153)$$

23.39 TDMA-based digital cellular network [203]. Consider the following call-connection problem for a *time division multiple access* (TDMA)-based digital cellular network. For the uplink there are m time slots per frame. Suppose that there are two classes of users: a class-1 user occupies only one time slot per frame, whereas a class-2 user requires two time slots per frame. When the base station cannot accommodate an arriving call, the call will be lost.

Let K_r be the number of class-r users, where $r = 1$ or 2, $1/\nu_r$ be the expected inactive period during which a class-r user does not make a new call after the completion or loss of its previous call, and $1/\mu_r$ be the mean holding time of a class-r call. For simplicity, we assume load-independent service rates; i.e., $\beta_r(n_r) = 1$. Let $\boldsymbol{N}(t) = (N_1(t), N_2(t))$ be the number of class-1 and class-2 calls in service at time t.

(a) Define the feasible set of states $\mathcal{F}_{\boldsymbol{N}}(m, \boldsymbol{K})$ for $\boldsymbol{N}(t)$.
(b) Find the stationary distribution of $\boldsymbol{N}(t)$.
(c) Let us assume the following system parameters:

$$K_1 = 3,\ 1/\mu_1 = 3\,[\text{min}],\ 1/\nu_1 = 6\,[\text{min}],\ \text{hence } r_1 = \frac{1}{2};$$

$$K_2 = 2,\ 1/\mu_2 = 6\,[\text{min}],\ 1/\nu_2 = 9\,[\text{min}],\ \text{hence } r_2 = \frac{2}{3}.$$

Find the $\mathcal{F}_{\boldsymbol{N}}(m, \boldsymbol{K})$ and the corresponding normalization constants $G(m, \boldsymbol{K})$ for $m = 4, 5$, and 6.

23.40 TDMA-based digital cellular network – continued. [203]. We continue Problem 23.39.

(a) Find the blocking probabilities for both classes of calls.
(b) Find the call loss probabilities for both classes of calls.

23.41* Differential-difference equation for the Engset model [203]. Consider the Engset model with K sources and m output lines (or servers), where $m \leq K$. Let ν be the call generation rate from a source that is not engaged in service. Assume that the holding times are exponentially distributed with mean $1/\mu$.

(a) Let $n(t)$ be the number of lines engaged in service at time t. Let

$$p_n(t; K) = P[N(t) = n], \quad n = 0, 1, \ldots, m.$$

Find the differential-difference equation that $p_n(t; K)$ must satisfy.
(b) Find the balance equation that must hold in the equilibrium state. Then find the equilibrium distribution $\{\pi_n(K)\}$, where $\pi_n(K) = \lim_{t\to\infty} p_n(t; K)$.
(c) Let B_n represent the event that n servers are in service and A be the event that a new call is generated in a small interval δt. Then it should be clear that $P[A|B_n] = (K - n)\nu\delta t$. Let $\{a_n(K)\}$ be the probability that an arriving call finds that n servers are busy. Find $a_n(K)$ in terms of the probabilities associated with events A and B_n. Then relate this probability to $\pi_n(K)$ defined in part (b).
(d) Consider the limit case $K \to \infty$ in the result of part (c). Make any additional assumptions required to make the limit meaningful. What does this limit case mean?

23.42 Engset model without restriction $K > m$ [203]. In the text we made the assumption $K > m$. Suppose we drop this restriction.

(a) How should the formula (23.125) be changed?
(b) How about the formula (23.130)?

23.43* Link efficiency [203]. The carried traffic load that each output line carries is called the *efficiency* of the trunk or link as defined in (23.132).

(a) Show that for the Engset loss model, the trunk efficiency is given by

$$\eta = \frac{a(1 - L(K))}{m}. \tag{23.154}$$

(b) Plot the link efficiency η as a function of m, the number of output lines or circuits. Choose the number of input sources to be $K = 10$, 20, 40, 60, 80, and 100. The traffic parameter r should be chosen so that the loss rate is maintained at $L = 0.01$.

23.44* Example of MLN [203]. Consider a cellular network that can be represented as an MLN shown in Figure 23.15 with $m = 6$ channels and two classes of calls: class-1 calls from a small number ($K_1 = 3$) of talkative subscribers forming a closed chain and class-2 calls from a large number of other ordinary subscribers forming an open

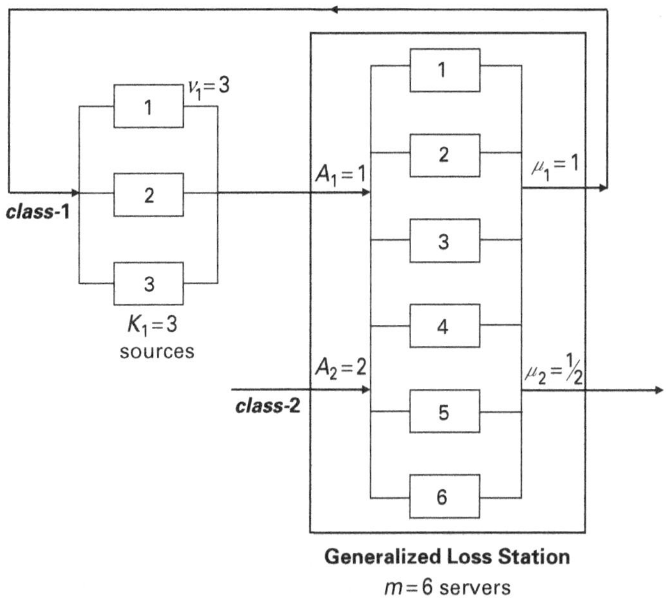

Figure 23.15 An example of an MLN ([203] Reprinted with permission from Pearson Education, Inc., Upper Saddle River, NJ.).

chain. The system provides preferential treatment to class-2 calls by providing more bandwidth per call.

Assume the following parameters:

$$1/\mu_1 = 6 \text{ [min/call]},\ 1/\nu_1 = 12 \text{ [min between calls]},\ A_1 = 1 \text{ [channel/call]};$$
$$1/\mu_2 = 2 \text{ [min/call]},\ \lambda_2 = 0.25 \text{ [calls/min]},\ A_2 = 2 \text{ [channels/call]}.$$

(a) Find r_1 and a_2.
(b) Find the steady-state distribution of $\boldsymbol{N}(t)$.
(c) Fix $K_1 = 3$. Find $\mathcal{F}(m, 3)$ for $m = 0, 1, 2, \ldots, 6$.
(d) Find the blocking probability (or time congestion) and call congestion for class-2 customers in the open route.
(e) Find the call loss probability (or call congestion) for class-1 calls.

References

[1] J. Abate, G. L. Choudhury, and W. Whitt. Numerical inversion of multidimensional Laplace transforms by the Laguerre method. *Performance Evaluation*, **31**: (1998), 3 & 4, 229–243. (Cited on p. 234.)

[2] J. Abate and W. Whitt. Numerical inversion of Laplace transforms of probability distributions. *ORSA Journal on Computing*, **7**:1 (1995), 36–40. (Cited on p. 234.)

[3] M. Abramowitz and I. A. Stegun. *Handbook of Mathematical Functions: With Formulas, Graphs, and Mathematical Tables* (New York: Dover, 1977). (Cited on p. 261.)

[4] H. Akimaru and K. Kawashima. *Teletraffic: Fundamentals and Applications* (Tokyo: Institute of Electrical Communications, 1990). (In Japanese.) (Cited on p. 707.)

[5] R. Arens. Complex process for envelopes of normal noise. *IRE Transactions on Information Theory*, **IT-3** (1957), 204–207. (Cited on p. 340.)

[6] S. Asmussen. *Applied Probability and Queues* (New York: Springer, 2003). (Cited on p. 268.)

[7] S. Asmussen, O. Nerman, and M. Olsson. Fitting phase-type distributions via the EM algorithm. *Scandinavian Journal of Statistics*, **23**:4 (1996), 419–441. (Cited on pp. 566, 606.)

[8] K. Azuma. Weighted sums of certain dependent random variables. *Tohoku Mathematics Journal*, **19**:3 (1967), 357–367. (Cited on p. 272.)

[9] L. Bachelier. Théorie de la spéculation. *Annales scientifiques de l'École Normale Supérieure, 3e série*, **17** (1900), 21–86. (Cited on pp. 5, 11.)

[10] L. Bachelier. *Louis Bachelier's Theory of Speculation: The Origins of Modern Finance* (Translated and with Commentary by Mark Davis and Alison Etheridge) (Princeton, NJ: Princeton University Press, 2007). (Cited on pp. 5, 11, 516.)

[11] L. R. Bahl, J. Cocke, F. Jelinek, and J. Raviv. Optimal decoding of linear codes for minimizing symbol error rate. *IEEE Transactions on Information Theory*, **IT-20** (1974), 284–287. (Cited on pp. 593, 605.)

[12] F. G. Ball, R. K. Milne, and G. F. Yeo. Continuous-time Markov chains in a random environment, with applications to ion channel modelling. *Advances in Applied Probability*, **26**:4 (1994), 919–946. (Cited on p. 477.)

[13] F. G. Ball, R. K. Milne, and G. F. Yeo. Marked continuous-time Markov chain modelling of burst behaviour for single ion channels. *Journal of Applied Mathematics and Decision Sciences*, (2007), 1–14. (Cited on p. 477.)

[14] G. P. Barsharin, A. N. Langville, and V. A. Naumov. The life and work of A. A. Markov. *Linear Algebra and Its Applications*, **386** (2004), 3–26. (Cited on pp. 319, 478.)

[15] L. E. Baum and T. Petrie. Statistical inference for probabilistic functions of finite state Markov chains. *Annals of Mathematical Statististics*, **37** (1966), 1559–1563. (Cited on p. 605.)

[16] L. E. Baum, T. Petrie, G. Soules, and N. Weiss. A maximization technique occurring in the statistical analysis of probabilistic functions of Markov chains. *Annals of Mathematical Statistics*, **41**:1 (1970), 164–171. (Cited on pp. 564, 605, 606.)

[17] R. E. Bellman. *Dynamic Programming* (Princeton, NJ: Princeton University Press, 1957). (Cited on p. 592.)

[18] J. O. Berger. *Statistical Decision Theory and Bayesian Analysis*, 2nd edition (Springer-Verlag, 1995). (Cited on p. 104.)

[19] J. M. Bernardo and A. F. M. Smith. *Bayesian Theory* (Wiley, 1994). (Cited on p. 104.)

[20] P. L. Bernstein. *Against the Gods: The Remarkable Story of Risk* (John Wiley & Sons, 1996). (Cited on p. 12.)

[21] S. N. Bernstein. On certain modifications of Chebyshev's inequality. *Doklady Akademii Nauk USSR*, **17**:6 (1937), 275–277. (Cited on p. 271.)

[22] C. Berrou and A. Glavieux. Near optimum error correcting coding and decoding: Turbo codes. *IEEE Transactions on Communications*, **44** (1996), 1261–1271. (Cited on p. 605.)

[23] M. Berry, S. Dumais, and G. O'Brien. Using linear algebra for intelligent information retrieval. *SIAM Review*, **37**:4 (1995), 573–595. (Cited on pp. 372, 395.)

[24] D. Bertsekas and J. N. Tsitsiklis, *Introduction to Probability*, 2nd edition (Athena Scientific, 2008). (Cited on p. 37.)

[25] P. J. Bickel and K. A. Doksum. *Mathematical Statistics: Basic Ideas and Selected Topics*, Vol. 1 (Prentice Hall, 2001). (Cited on pp. 526, 536, 549.)

[26] P. Billingsley. *Probability and Measure*, 3rd edn (John Wiley & Sons, 1995). (Cited on pp. 293, 305, 308.)

[27] P. Billingsley. *Convergence of Probability Measures* (New York: Wiley, 1999). (Cited on p. 202.)

[28] G. Birkhoff and S. MacLane. *A Survey of Modern Algebra*, 5th edn (New York: Macmillan, 1996). (Cited on p. 241.)

[29] F. Black and M. Scholes. The pricing of options and corporate liabilities. *Journal of Political Economy*, **81** (1973), 637–654. (Cited on pp. 5, 511.)

[30] R. B. Blackman and J. W. Tukey. *The Measurement of Power Spectra, from the Point of View of Communication Engineering* (New York: Dover, 1959). (Cited on p. 357.)

[31] I. F. Blake. *An Introduction to Applied Probability* (New York: John Wiley & Sons, Inc., 1979). (Cited on p. 37.)

[32] R. Blossey. *Statistical Mechanics for Biologists* (Chapman & Hall/CRC, 2006). (Cited on p. viii.) xxviii.)

[33] G. Bolch, S. Greiner, H. de Meer, and K. S. Trivedi. *Queueing Networks and Markov Chains: Modeling and Performance Evaluation with Computer Science Applications* (John Wiley & Sons, 2006). (Cited on p. 731.)

[34] E. Borel. Les probabilités dénombrables et leurs applications arithmétiques. *Rendiconti del Circolo Matematico di Palermo*, **27** (1909), 247–271. (Cited on p. 300.)

[35] L. Breiman. *Probability* (Reading, MA: Addison-Wesley, 1968). (Cited on pp. 308, 516.)

[36] L. Breiman. *Probability and Stochastic Processes with a View Toward Applications* (Boston: Houghton Mifflin, 1969). (Cited on p. 690.)

[37] L. Bresslau, P. Cao, L. Fan, G. Phillips, and S. Shenker. Web caching and Zipf-like distributions: Evidence and implications. In *Proceedings of INFOCOM'99*, pp. 126–134 (1999). (Cited on p. 63.)

[38] L. Breuer. An EM algorithm for batch Markovian arrival processes and its comparison to simpler estimation procedure. *Annals of Operations Research*, **112** (2002), 123–138. (Cited on pp. 566, 606.)

[39] M. P. Brown, R. Hughey, A. Krogh *et al.* Using Dirichlet mixture priors to derive hidden Markov models for protein families. In *Proceedings of First International Conference on Intelligent Systems for Molecular Biology*, L. Hunter, D. Searts and J. Shavlile (eds), pp. 47–55 (1993). (Cited on p. 605.)

[40] S. L. Brumelle. Some inequalities for parallel server queues. *Operations Research*, **19**:2 (1971), 402–413. (Cited on p. 730.)

[41] D. Bryant, N. Galtier, and M. A. Poursat. Likelihood calculation in molecular phylogenetics. In O. Gascuel, ed., *Mathematics of Evolution and Phylogeny* (Oxford University Press, 2004). (Cited on pp. 476, 477.)

[42] J. A. Bucklew. *Large Deviation Techniques in Decision, Simulation and Estimation* (New York: John Wiley & Sons, Inc., 1990). (Cited on p. 268.)

[43] C. J. Burke and M. A. Rosenblatt. Markovian function of a Markov chain. *Annals of Mathematical Statistics*, **29**:4 (1958), 1112–1122. (Cited on p. 605.)

[44] P. I. Butzer, P. J. S. G. Ferreira, J. R. Higgins *et al.* Interpolation and sampling: E.T. Whittaker, K. Ogura and their followers. *Journal of Fourier Analysis and Applications, Online First*™, 2010. (Cited on p. 353.)

[45] F. A. Campbell and R. M. Foster. *Fourier Integrals for Practical Applications* (Princeton, NJ: Van Nostrand Company, Inc., 1948). (Cited on p. 194.)

[46] O. Cappé, E. Moulines, and T. Rydén. *Inference in Hidden Markov Models* (Springer, 2005). (Cited on p. 606.)

[47] K. M. Chandy, U. Herzog, and L. Woo. Parametric analysis of queuing networks. *IBM Journal of Research and Development*, **19**:1 (1975), 43–49. (Cited on p. 710.)

[48] K. M. Chandy and C. H. Sauer. Computational algorithms for product form queueing networks. *Communications of the ACM*, **23**:10 (1980), 573–583. (Cited on p. 731.)

[49] R. W. Chang and J. C. Hancock. On receiver structures for channels having memory. *IEEE Transactions on Information Theory*, **IT-12**:4 (1966), 463–468. (Cited on p. 605.)

[50] E. Charniak and R. P. Goldman. A semantics for probabilistic quantifier-free first-order languages with particular application to story understanding. In *Proceedings of the 1989 International Joint Conference on Artificial Intelligence*, pp. 1074–1079 (1989). (Cited on pp. 615, 624.)

[51] C. Chatfield. *The Analysis of Time Series: Theory and Practice* (London: Chapman & Hall, 1975). (Cited on p. 357.)

[52] K. L. Chung. *Markov Chains: With Stationary Transition Probabilities* (New York: Springer-Verlag, 1967). (Cited on p. 477.)

[53] K. L. Chung. *A Course in Probability Theory* (New York: Bruce & World, 1968). (Cited on p. 202.)

[54] K. L. Chung. *A Course in Probability Theory*, 2nd edn (New York: Academic Press, 1974). (Cited on p. 308.)

[55] R. D. Cideciyan, F. Dolivo, R. Hermann, W. Hirt, and W. Schott. A PRML system for digital magnetic recording. *IEEE Journal of Special Areas in Communications*, **JSAC-10**:1 (1992), 3856. (Cited on p. 592.)

[56] E. Çinlar. Markov renewal theory. *Advances in Applied Probability*, **1** (1969), 123–187. (Cited on pp. 457, 477.)

[57] E. Çinlar. *Introduction to Stochastic Processes* (Englewood Cliffs, NJ: Prentice Hall, 1975). (Cited on pp. 418, 451, 477.)

[58] E. Çinlar. Markov renewal theory: a survey. *Management Science*, **21**:7 (1975), 727–752. (Cited on pp. 457, 477, 478.)

[59] E. G. Coffman, R. R. Muntz, and H. Trotter. Waiting time distributions for processor-sharing systems. *Journal of ACM*, **17**:1 (1970), 123–130. (Cited on p. 717.)

[60] I. Cohen, N. Sebe, F. G. Cozman, M. C. Cirelo, and T. S. Huang. Learning Bayesian network classifiers for facial expression recognition both labeled and unlabeled data. In *Proceedings 2003 IEEE Computer Society Conference on Computer Vision and Pattern Recognition*, vol. 1, pp. I-595–I-601 (2003). (Cited on pp. 615, 624.)

[61] J. W. Cohen. *The Single Server Queue*, 2nd edn (Amsterdam: North-Holland, 1982). (Cited on p. 730.)

[62] D. Colquhoun and A. G. Hawkes. Relaxation and fluctuations of membrane currents that flow through drug-operated channels. *Proceedings of the Royal Society of London, Series B*, **199** (1977), 231–262. (Cited on pp. 461, 466, 477.)

[63] D. Colquhoun and A. G. Hawkes. On the stochastic properties of single ion channels. *Proceedings of the Royal Society of London, Series B*, **211** (1981), 205–235. (Cited on p. 477.)

[64] D. Colquhoun and A. G. Hawkes. On the stochastic properties of bursts of single ion channel openings and of clusters of bursts. *Proceedings of the Royal Society of London, Series B*, **300** (1982), 1–59. (Cited on p. 477.)

[65] J. W. Cooley, P. A. W. Lewis, and P. D. Welch. The fast transform algorithm: programming considerations in the calculation of sine, cosine and Laplace transforms. *Proceedings of Cambridge Philosophical Society*, **12**:3 (1970), 315–337. (Cited on p. 234.)

[66] J. W. Cooley and J. W. Tukey. An algorithm for the machine computation of complex Fourier series. *Mathematics of Computation*, **19** (1965), 297–301. (Cited on p. 357.)

[67] R. B. Cooper. *Introduction to Queueing Theory*, 2nd edn (New York: North-Holland, 1981). (Cited on p. 731.)

[68] T. M. Cover and P. E. Hart. Nearest neighbor pattern classification. *IEEE Transactions on Information Theory*, **IT-13**:1 (1967), 21–27. (Cited on p. 621.)

[69] T. M. Cover and J. A. Thomas. *Elements of Information Theory* (New York: John Wiley & Sons, Inc., 1991). (Cited on p. 247.)

[70] D. R. Cox. *Renewal Theory* (Methuen, 1962). (Cited on p. 418.)

[71] D. R. Cox and P. A. W. Lewis. *The Statistical Analysis of the Series of Events* (London: Methuen, 1966). (Cited on pp. 145, 153, 418.)

[72] D. R. Cox and H. D. Miller. *The Theory of Stochastic Processes* (New York: John Wiley & Sons, Inc., 1965). (Cited on p. 516.)

[73] D. R. Cox and W. L. Smith. *Queues* (London: Methuen, 1961). (Cited on pp. 707, 731.)

[74] H. Cramér. *Mathematical Methods of Statistics* (Princeton, NJ: Princeton University Press, 1946). (Cited on pp. 305, 308, 532, 533, 536, 549.)

[75] K. S. Crump. Numerical inversion of Laplace transforms using a Fourier series approcimation. *Journal of the ACM*, **23**:1 (1976), 89–96. (Cited on p. 234.)

[76] J. N. Daigle. *Queueing Theory for Telecommunications* (Reading, MA: Addison-Wesley, 1992). (Cited on p. 731.)

[77] W. B. Davenport, Jr. and W. L. Root. *An Introduction to the Theory of Random Signals and Noise* (New York: McGraw-Hill, 1958). (Cited on pp. 37, 394, 690.)

[78] F. N. David. *Games, Gods and Gambling* (London: Charles Griffin & Co., 1962). (Cited on pp. 7, 14.)

[79] G. Del Corso, A. Gulli, and F. Romani. Fast PageRank computation via a sparse linear system. *Internet Mathematics*, **2**:3 (2005), 251–273. (Cited on pp. 384, 395.)

[80] A. P. Dempster, N. M.Laird, and D. B. Rubin. Maximum likelihood from incomplete data via the EM algorithm. *Journal of the Royal Statistical Society Series B*, **39**:1 (1977), 1–38. (Cited on pp. 13, 559, 560, 563.)

[81] P. A. M. Dirac. *Principles of Quantum Mechanics* (Oxford, UK: Oxford University Press, 1935). (Cited on p. 46.)

[82] J. L. Doob. *Stochastic Processes* (New York: John Wiley & Sons, Inc., 1953). (Cited on pp. 268, 308, 323, 331, 347, 451, 492, 516, 690.)

[83] H. Dubner and J. Abate. Numerical inversion of Laplace transforms by relating them to the finite Fourier cosine transform. *Journal of the ACM*, **15**:1 (1968), 115–123. (Cited on p. 234.)

[84] J. Dugundji. Envelope and pre-envelope of real waveforms. *IRE Transactions on Information Theory*, **IT-4** (1958), 53–57. (Cited on p. 340.)

[85] R. Durbin, S. R. Eddy, A. Krogh, and G. J. Mitchison. *Biological Sequence Analysis: Probabilistic Models of Proteins and Nucleic Acids* (Cambridge, UK: Cambridge University Press, 1998). (Cited on pp. 605, 615, 623.)

[86] A. Einstein. Über die von der molekularkinetischen Theorie der Wärme geforderte Bewegung von in ruhenden Flüssigkeiten suspendierten Teilchen (On the movement of small particles suspended in a stationary liquid demanded by the molecular-kinetic theory of heat). *Annalen der Physik*, **17** (1905), 549–560. (Cited on pp. 11, 499, 516.)

[87] A. Einstein. On the theory of Brownian motion. *Annalen der Physik*, **19** (1906), 371–381. (Cited on p. 516.)

[88] A. Einstein. *Investigations of the Theory of the Brownian Movement* (translated by A. D. Cowper) (New York: Dover, 1956). (Cited on p. 499.)

[89] E. O. Elliott. Estimates of error rates for codes on burst-noise channels. *Bell System Technical Journal*, **42** (1963), 1977–1997. (Cited on p. 582.)

[90] R. Ellis. *Entropy, Large Deviations, and Statistical Mechanics* (New York: Springer-Verlag, 2006). (Cited on p. 268.)

[91] T. Engset. Die Wahrscheinlichkeitsrechung zur Bestimmung der Wähleranzahl in automatischen Fernsprechämtern. *Electrotechnische Zeitschrift*, **31** (1918), 304–305. (Cited on p. 722.)

[92] Y. Ephraim. Statistical-model-based speech enhancement systems. *Proceedings of the IEEE*, **80**:10 (1992), 1526–1555. (Cited on p. 605.)

[93] Y. Ephraim and N. Merhav. Hidden Markov processes. *IEEE Transactions on Information Theory*, **48**:6 (2002), 1518–1569. (Cited on pp. 592, 606.)

[94] A. K. Erlang. The theory of probabilities and telephone conversations. *Nyt Tidsskrift for Matematik B*, **20** (1909), 33–39. (Cited on p. 695.)

[95] A. K. Erlang. Solution of some problems in the theory of probabilities of significance in automatic telephone exchanges. *The Post Office Electrical Engineer's Journal*, **10** (1917–1918), 189–197. (Cited on p. 6.)

[96] M. Feder and J. Catipovic. Algorithms for joint channel estimation and data recovery – application to equalization in underwater communications. *IEEE Journal of Oceanic Engineering*, **16** (1991), 42–55. (Cited on p. 566.)

[97] W. Feller. Über den zentralen Grenzwertsatz der Wahrscheinlichkeitsrechnung. *Mathematische Zeitschrift*, **40** (1935), 521–559. (Cited on p. 305.)

[98] W. Feller. A direct proof of Stirling's formula. *The American Mathematical Monthly*, **74**:10 (1967), 1223–1225. (Cited on p. 275.)

[99] W. Feller. *Introduction to Probability and Its Applications*, Vol. I, 3rd edn (New York: John Wiley & Sons, Inc., 1968). (Cited on pp. 19, 34, 37, 38, 40, 59, 66, 77, 80, 82, 83, 84, 85, 104, 202, 235, 275, 302, 303, 308, 451, 486, 707.)

[100] W. Feller. *Introduction to Probability and Its Applications*, Vol. II, 2nd edn (New York: John Wiley & Sons, Inc., 1971). (Cited on pp. 202, 235, 251, 303, 304, 305, 308, 418, 451, 516, 518.)

[101] J. Felsenstein. Evolutionary trees from DNA sequences. *Journal of Molecular Evolution*, **17** (1981), 368–376. (Cited on p. 476.)

[102] J. Felsenstein. *Inferring Phylogenies* (Sunderland, MA: Sinauer Associates, 2004). (Cited on p. 477.)

[103] J. D. Ferguson. Variable duration models for speech. In *Symposium on the Application of Hidden Markov Models to Text and Speech*, pp. 143–179, Institute for Defense Analyses, Princeton, NJ, October 1980. (Cited on p. 606.)

[104] A. M. Ferrenberg, D. P. Landau, and Y. J. Wong. Monte Carlo simulations: hidden errors from "good" random number generators. *Physical Review Letters*, **69**:23 (1992), 3382–3384. (Cited on p. 131.)

[105] T. L. Fine. *Probability and Probabilistic Reasoning for Electrical Engineering* (Upper Saddle River, NJ: Pearson Prentice Hall, 2006). (Cited on pp. 14, 37, 131, 549, 690.)

[106] R. A. Fisher. *Design of Experiments*, vol. 1, 3rd edn (Edinburgh: Oliver and Boyd, 1935). (Cited on p. 164.)

[107] G. D. Forney, Jr. Review of random tree codes. In *Final Report on Contract NAS2-3637, NASA CR73176*. NASA Ames Research Center, Ames, CA (December 1967). (Cited on p. 574.)

[108] G. D. Forney, Jr. The Viterbi algorithm (invited paper). *Proceedings of the IEEE*, **IT-9**:61 (1973), 268–278. (Cited on pp. 574, 591, 592, 605.)

[109] G. D. Forney, Jr. Maximum likelihood sequence estimation of digital sequences in the presence of intersymbol interference. *IEEE Transactions on Information Theory*, **IT-18** (1972), 363–378. (Cited on pp. 592, 605.)

[110] D. Fox, W. Burgard, and S. Thrun. Markov localization for mobile robots in dynamic environments. *Journal of Artificial Intelligence Research*, **11** (1999), 391–427. (Cited on p. 615.)

[111] J. Franklin. *The Science of Conjecture: Evidence and Probability before Pascal* (Baltimore: The John Hopkins Press, 2001). (Cited on p. 14.)

[112] H. Freeman. *Discrete-Time Systems* (New York: John Wiley & Sons, Inc., 1965). (Cited on p. 211.)

[113] R. G. Gallager. *Information Theory and Reliable Communications* (New York: John Wiley & Sons, Inc., 1968). (Cited on p. 268.)

[114] D. P. Gaver. Diffusion approximation methods for certain congestion problems. *Journal of Applied Probability*, **5** (1968), 607–623. (Cited on p. 516.)

[115] D. P. Gaver, S. S. Lavenberg, and T. G. Price, Jr. Exploratory analysis of access path length data for a data base management system. *IBM Journal of Research and Development*, **20**:5 (1976), 449–464. (Cited on p. 146.)

[116] E. Gelenbe and I. Mitrani. *Analysis and Synthesis of Computer Systems* (Academic Press, 1980). (Cited on p. 731.)

[117] A. Gelman, J. B. Carlin, H. S. Stern, and D. B. Rubin. *Bayesian Data Analysis*, 2nd edn (Boca Raton, FL: Chapman and Hall/CRC, 2003). (Cited on p. 104.)

[118] S. Geman and D. Geman. Stochastic relaxation, Gibbs distributions and the Bayesian restoration of images. *IEEE Transactions on Pattern Analysis and Machine Intelligence*, **6**:6 (1984), 721–741. (Cited on p. 639.)

[119] I. I. Gikhman and A. V. Skorokhod. *Introduction to The Theory of Random Processes* (W. B. Saunders Company, 1969). (Cited on p. 690.)

[120] E. N. Gilbert. Capacity of a burst-noise channel. *Bell System Technical Journal*, **39** (1960), 1253–1265. (Cited on p. 582.)

[121] W. R. Gilks, A. Thomas, and D. J. Spiegelhalter. A language and program for complex Bayesian modelling. *The Statistician*, **43** (1994), 169–178. (Cited on p. 641.)

[122] B. V. Gnedenko. *Theory of Probability* (New York: Chelsea, 1962). (Cited on p. 308.)

[123] B. V. Gnedenko and N. Kolmogorov. *Limit Distributions for Sums of Independent Random Variables* (Reading, MA: Addison Wesley, 1954). (Cited on p. 202.)

[124] G. H. Golub and C. F. Van Loan. *Matrix Computations* (Baltimore: The Johns Hopkins University Press, 1996). (Cited on pp. 361, 362, 370, 381, 395.)

[125] R. M. Gray. *Toeplitz and Circulant Matrices: A Review* (Norwell, MA: Now Publishers, 2006). (Cited on pp. 361, 362.)

[126] R. M. Gray and L. Davisson. *An Introduction to Statistical Signal Processing* (Cambridge University Press, 2004). (Cited on p. 37.)

[127] D. Green and J. Swets. *Signal Detection Theory and Psychophysics* (New York: John Wiley and Sons Inc., 1966). (Cited on p. 542.)

[128] E. Greenberg. *Introduction to Bayesian Econometrics* (Cambridge University Press, 2008). (Cited on pp. xxviii, 14, 643.)

[129] U. Grenander and G. Szego. *Toeplitz Forms and Their Applications*, 2nd edn (New York: Chelsea, 1984). (Cited on p. 361.)

[130] T. L. Grettenberg. A representation theorem for complex normal process. *IEEE Transactions on Information Theory*, **IT-11** (1965), 395–306. (Cited on pp. 177, 332.)

[131] G. R. Grimmett and D. R. Stirzaker. *Probability and Random Processes*, (Oxford: Oxford University Press, 1992). (Cited on pp. 34, 37, 66, 104, 205, 235, 265, 271, 284, 287, 302, 303, 308, 310, 324, 325, 349, 418, 484, 516.)

[132] D. Gross, J. F. Shortle, J. M. Thompson, and C. M. Harris. *Fundamentals of Queueing Theory*, 4th edn (John Wiley & Sons, Inc., 2008). (Cited on p. 731.)

[133] J. A. Gubner. *Probability and Random Processes for Electrical and Computer Engineers* (Cambridge University Press, 2006). (Cited on pp. 37, 66, 104, 394, 516.)

[134] B. Gueye, A. Ziviani, M. Crovella, and S. Fdida. Constraint-based geolocation of Internet hosts. *IEEE/ACM Transactions on Networking*, **14**:6 (2006), 1219–1232. (Cited on pp. 150, 151.)

[135] I. Hacking. *The Emergence of Probability* (New York: Cambridge University Press, 1975). (Cited on p. 14.)

[136] I. Hacking. *An Introduction to Probability and Inductive Logic* (New York: Cambridge University Press, 2001). (Cited on pp. 14, 37.)

[137] E. Haensler. *Statische Signale: Grundlagen und Anwendungen*. 3. Auflage (Berlin: Springer Verlag, 2001). (Cited on p. 690.)

[138] J. Hagenauer, E. Offer, and L. Parke. Iterative decoding of binary block and convolutional codes. *IEEE Transactions on Information Theory*, **42**:2 (1996), 429–445. (Cited on p. 605.)

[139] A. Hald. *Statistical Theory with Engineering Applications* (New York: John Wiley & Sons, Inc., 1952). (Cited on pp. 142, 153, 549.)

[140] A. Hald. *A History of Probability and Statistics, and Their Applications before 1750* (New York: Wiley, 1990 and 2003). (Cited on p. 14.)

[141] J. D. Hamilton. *Time Series Analysis* (Princeton, NJ: Princeton University Press, 1994). (Cited on pp. 322, 389, 395, 566.)

[142] J. D. Hamilton. Regime-switching models. In S. Durlauf and L. Blume, eds, *New Palgrave Dictionary of Economics* (Palgrave McMillan Ltd., 2008). (Cited on p. 5.)

[143] J. M. Hammersley and D. C. Handscomb. *Monte Carlo Methods* (London: Methuen, 1964). (Cited on p. 523.)

[144] E. J. Hannan. *Time-Series Analysis* (London: Methuen, 1960). (Cited on p. 357.)

[145] A. C. Harvey. *Forecasting, Structural Time Series Models and the Kalman Filter* (Cambridge University Press, 1989). (Cited on pp. 322, 389, 395.)

[146] B. R. Haverkort. *Performance of Computer Communication Systems: A Model-Based Approach* (New York: John Wiley & Sons, Inc., 1998). (Cited on p. 731.)

[147] J. F. Hayes and T. V. J. Ganesh Babu. *Modeling and Analysis of Telecommunications Networks* (Hoboken, NJ: John Wiley & Sons, Inc., 2004). (Cited on p. 731.)

[148] J. F. Hayes, T. M. Cover, and J. B. Riera. Optimal sequence detection and optimal symbol-by-symbol detection: similar algorithms. *IEEE Transactions on Commmunications*, **COM-30**:1 (1982), 152–157. (Cited on p. 605.)

[149] S. Haykin. *Neural Networks: A Comprehensive Foundation* (Upper Saddle River, NJ: Prentice Hall, 1999). (Cited on pp. 3, 616.)

[150] C. W. Helstrom. *Statistical Theory of Signal Detection* (Pergamon Press, 1960). (Cited on pp. 177, 340, 542, 549.)

[151] D. P. Heyman and S. Stidham, Jr. The relation between customers and time averages in queues. *Operations Research*, **28** (1980), 983–994. (Cited on p. 730.)

[152] W. Hoeffding. Probability inequalities for sums of bounded random variables. *Journal of the American Statistical Association*, **58**:1 (1963), 13–30. (Cited on p. 272.)

[153] K. Hoffman and R. Kunze. *Linear Algebra* (Englewood Cliffs, NJ: Prentice Hall, 1961). (Cited on p. 241.)

[154] J. J. Hopfield. Neural networks and physical systems with emergent collective computational abilities. *Proceedings of the National Academy of Sciences, USA*, **79** (1982), 2554–2558. (Cited on p. 616.)

[155] J. Y. Hui. *Switching and Traffic Theory for Integrated Broadband Networks* (Boston, MA: Kluwer, 1990). (Cited on pp. 268, 737.)

[156] IBM. *IBM Subroutine Library – Mathematics, User's Guide, SH12-5300-1*, 2nd edn (IBM Corporation, 1974). (Cited on p. 234.)

[157] K. Itô. Stochastic integral. *Proceedings of the Imperial Academy (Tokyo)*, **20** (1944), 519–524. (Cited on p. 11.)

[158] K. Itô. *Selected Papers (Edited by D. W. Stroock and S. R. Sriniva Varadhan)* (New York: Springer, 1987). (Cited on p. 11.)

[159] K. Itô and J. H. P. McKean. *Diffusion Processes and Their Sample Paths* (Berlin: Springer, 1965). (Cited on pp. 506, 516.)

[160] D. L. Jagerman. An inversion technique for the Laplace transform with applications. *Bell System Technical Journal*, **57** (1978), 669–710. (Cited on p. 234.)

[161] D. L. Jagerman. An inversion technique for the Laplace transform. *Bell System Technical Journal*, **61** (1982), 1995–2002. (Cited on p. 234.)

[162] D. L. Jagerman, B. Melamed, and W. Willinger. Stochastic modeling of traffic processes. In J. H. Dshalalow, ed., *Frontiers in Queueing: Models and Applications in Science and Engineering* pp. 271–320. (Boca Raton, FL: CRC Press, 1997). (Cited on pp. 393, 395.)

[163] F. Jelinek. Continuous speech recogniton by statistical methods. *Proceedings of the IEEE*, **64** (1976), 532–556. (Cited on p. 605.)

[164] F. Jelinek. *Statistical Methods for Speech Recognition* (MIT Press, 1998). (Cited on pp. 425, 605, 615.)

[165] F. Jelinek, L. Bahl, and R. Mercer. Design of a linguistic statistical decoder for the recognition of continuous speech. *IEEE Transactions on Information Theory*, **21** (1975), 250–256. (Cited on p. 605.)

[166] F. V. Jensen and T. D. Nielsen. *Bayesian Networks and Decision Graphs*, Information Science and Statistics Series, 2nd edn (New York: Springer-Verlag, 2007). (Cited on p. 624.)

[167] M. C. Jeruchim, P. Balaban, and K. S. Shanmugan. *Simulation of Communication Systems: Modeling, Methodology, and Techniques* (Kluwer Academic/Plenum Publishers, 2000). (Cited on p. 268.)

[168] W. S. Jewell. A simple proof of $L = \lambda W$. *Operations Research*, **15**:6 (1967), 1109–1116. (Cited on p. 730.)

[169] I. T. Jolliffe. *Principal Component Analysis*, 2nd edn (New York: Springer, 2002). (Cited on p. 395.)

[170] T. Kailath, A. Sayed, and B. Hassibi. *Linear Prediction* (Prentice-Hall, 2000). (Cited on p. 690.)

[171] R. Kalman. A new approach to linear filtering and predicition problems. *Journal of Basic Engineering*, **82** (1960), 35–45. (Cited on pp. 13, 645.)

[172] R. Kalman and R. S. Bucy. New results in linear filtering and predicition theory. *Journal of Basic Engineering*, **83** (1961), 95–107. (Cited on p. 645.)

[173] E. P. C. Kao. *An Introduction to Stochastic Processes* (Belmont, CA: Duxbury Press, 1979). (Cited on pp. 418, 512.)

[174] K. Karhunen. Über linearen Methoden in der Wahrscheinlichkeitsrechnung. *Annales Academiae Scientarum Fennicae, Series A 1, Mathematica–Physica*, **37** (1947), 3–79. (Cited on p. 365.)

[175] S. Karlin and H. M. Taylor. *A First Course in Stochastic Processes*, 2nd edn (Academic Press, 1975). (Cited on pp. 324, 325, 340, 341, 418, 443, 690.)

[176] J. Kay. The EM algorithm in medical imaging. *Statistical Methods in Medical Research*, **6**:1 (1975), 55–75. (Cited on p. 566.)

[177] F. P. Kelly. *Reversibility and Stochastic Networks* (John Wiley & Sons, Inc., 1979). (Cited on pp. 477, 704, 719, 731.)

[178] F. P. Kelly. Loss networks (invited paper). *The Annals of Applied Probability*, **1** (1991), 319–378. (Cited on p. 724.)

[179] M. G. Kendall and A. Stuart. *The Advanced Theory of Statistics, Vol. II: Inference and Relationship* (London: Charles Griffin, 1961). (Cited on p. 549.)

[180] M. Kijima. *Markov Processes for Stochastic Modeling* (Chapman & Hall, 1997). (Cited on p. 477.)

[181] F. W. King. *Hilbert Transforms*, Vol. 1 (Cambridge University Press, 2009). (Cited on p. 340.)

[182] F. W. King. *Hilbert Transforms*, Vol. 2 (Cambridge University Press, 2010). (Cited on p. 340.)

[183] P. J. B. King. *Computer and Communication System Performance Modeling* (Englewood Cliffs, NJ: Prentice Hall, 1990). (Cited on p. 731.)

[184] J. F. C. Kingman. A martingale inequality in the theory of queues. *Proceedings of the Cambridge Philosophical Society*, **59** (1964), 359–361. (Cited on p. 273.)

[185] J. F. C. Kingman. Inequalities in the theory of queues. *Journal of Royal Statistical Society, B* **32** (1970), 102–110. (Cited on p. 273.)

[186] J. F. C. Kingman. *Poisson Processes* (Clarendon Press, 1992). (Cited on p. 10.)

[187] J. Kleinberg. Authoritative sources in a hyperlinked environment. *Journal of the ACM*, **46**:5 (1999), 604–632. (Cited on pp. 384, 395.)

[188] L. Kleinrock. Time-shared systems: a theoretical treatment. *Journal of the ACM*, **14** (1967), 242–261. (Cited on p. 716.)

[189] L. Kleinrock. *Queueing Systems, Vol. I: Theory* (New York: John Wiley & Sons, Inc., 1975). (Cited on pp. 235, 419, 451, 731.)

[190] L. Kleinrock. *Queueing Systems, Vol. II: Computer Applications* (New York: John Wiley & Sons, Inc., 1976). (Cited on pp. 235, 724.)

[191] D. E. Knuth. *The Art of Computer Programming: Vol. 2. Seminumerical Algorithms*, 3rd edn (Upper Saddle River, NJ: Addison Wesley, 1998). (Cited on pp. 126, 130, 131.)

[192] H. Kobayashi. Representation of complex-valued vector processes and their application to estimation and detection. Ph.D. Thesis, Princeton University, August 1967. (Cited on pp. 177, 340.)

[193] H. Kobayashi. A simultaneous adaptive estimation and decision algorithm for carrier modulated data transmission systems. *IEEE Transactions on Communication Technology*, **COM-19**:3 (1971), 268–279. (Cited on pp. 322, 566.)

[194] H. Kobayashi. Application of the diffusion approximation to queueing networks I: equilibrium queue distributions. *Journal of the ACM*, **21**:2 (1974), 316–328. (Cited on p. 516.)

[195] H. Kobayashi. Application of the diffusion approximation to queueing networks II: nonequilibrium distributions and applications to computer modeling. *Journal of the ACM*, **21**:3 (1974), 459–469. (Cited on p. 516.)

[196] H. Kobayashi. Bounds for the waiting time in queueing systems. In E. Gelenbe and R. Mahl, eds, *Computing Architectures and Networks* (Amsterdam: North-Holland Publishing Company, February 1974) pp. 163–274. (Cited on p. 273, 432.)

[197] H. Kobayashi. *Modeling and Analysis: An Introduction to System Performance Evaluation Methodology* (Reading, MA: Addison-Wesley, 1978). (Cited on pp. 153, 235, 419, 653, 656, 708, 731.)

[198] H. Kobayashi. Partial-response coding, maximum-likelihood decoding: capitalizing on the analogy between communication and recording. *IEEE Communications Magazine*, **47**:3 (2009), 14–17. (Cited on pp. 592, 605, 607.)

[199] H. Kobayashi. Application of probabilistic decoding to digital magnetic recording systems. *IBM Journal of Research and Development*, **15**:1 (1971), 69–74. (Cited on pp. 592, 605, 609.)

[200] H. Kobayashi. Correlative level coding and maximum likelihood decoding. *IEEE Transactions on Information Theory*, **IT-17**:5 (1971), 586–594. (Cited on pp. 592, 605, 609.)

[201] H. Kobayashi and B. L. Mark. Product-form loss networks. In J. H. Dshalalow, ed., *Frontiers in Queueing: Models and Applications in Science and Engineering* (New York: CRC Press, 1997) pp. 147–196. (Cited on pp. 727, 729, 731.)

[202] H. Kobayashi and B. L. Mark. Generalized loss models and queueing-loss networks. *International Transactions on Operational Research*, **9**:1 (2002), 97–112. (Cited on p. 731.)

[203] H. Kobayashi and B. L. Mark. *System Modeling and Analysis: Foundations for System Performance Evaluation* (Prentice Hall, 2009). (Cited on pp. xxviii, 123, 124, 125, 126, 130, 235, 268, 350, 419, 451, 477, 480, 516, 632, 634, 635, 707, 708, 719, 721, 722, 724, 725, 726, 727, 728, 729, 731, 732, 733, 735, 737, 738, 739.)

[204] H. Kobayashi and D. T. Tang. Application of partial-response channel coding to magnetic recording systems. *IBM Journal of Research and Development*, **14**:4 (1970), 368–75. (Cited on p. 607.)

[205] P. Koehn. *Statistical Machine Translation* (Cambridge University Press, 2010). (Cited on p. 566.)

[206] D. Koller and N. Friedman. *Probabilistic Graphical Models* (Cambridge, MA: MIT Press, 2009). (Cited on pp. 3, 14, 643.)

[207] A. N. Kolmogorov. Sur la loi forte des grands nombres. *Comptes Rendus des Séances de l'Académie des Sciences*, **191** (1930), 910–912. (Cited on p. 302.)

[208] A. N. Kolmogorov. *Grundbegriffe der Wahrscheinlichkeitsrechnung* (Berlin: Julius Springer, 1933). (Cited on pp. 9, 20.)

[209] A. N. Kolmogorov. Interpolation and extrapolation. *Bulletin de l'Academie des Sciences de U.S.S.R., Series Mathematics*, **5** (1941), 3–14. (Cited on pp. 13, 656.)

[210] A. N. Kolmogorov. *Foundations of the Theory of Probability* (translated by Nathaniel Morrison) (New York: Chelsea, 1950). (Cited on p. 9.)

[211] I. Kononenko and M. Kukar. *Machine Learning and Data Mining: Introduction to Principles and Algorithms* (Chichester: Horwood Publishing, Ltd, 2007). (Cited on pp. xxviii, 14, 643.)

[212] V. Kotelnikov. On the capacity of 'ether' and cables in electrical communications (in Russian). In *Proceedings of the First All-Union Conference on Questions of Communications*, Moscow, 1933. (Cited on p. 352.)

[213] F. R. Kschischang, B. J. Frey, and H. A. Loeliger. Factor graphs and the sum-product algorithm. *IEEE Transactions on Information Theory*, **47**:2 (2001), 498–519. (Cited on pp. 606, 643.)

[214] E. R. Ktretzmer. Generalization of a technique for binary data transmission. *IEEE Transactions on Communications Technology*, **COM-14** (1966), 67–68. (Cited on p. 607.)

[215] S. Kullback. *Information Theory and Statistics* (New York, NY: John Wiley & Sons, Inc., 1959). (Cited on p. 557.)

[216] A. Kumar and L. Cowen. Augmented training of hidden Markov models to recognize remote homologs via simulated evolution. *Bioinformatics*, **25**:13 (2009), 1602–1608. (Cited on p. 605.)

[217] S. Y. Kung, K. S. Arun, and D. V. B. Rao. State space and singular value decomposition based approximation methods for harmonic retrieval. *Journal of the Optical Society of America*, **73**:12 (1983), 1799–1811. (Cited on p. 395.)

[218] A. Langville and C. Meyer. A survey of eigenvector methods for Web information retrieval. *SIAM Review*, **47**:1 (2005), 135–161. (Cited on pp. 384, 395, 451.)

[219] S. S. Lavenberg, ed. *Computer Performance Modeling Handbook* (Orlando, FL: Academic Press, 1983). (Cited on p. 731.)

[220] S. S. Lavenberg and M. Reiser. Stationary state probabilities of arrival instants for closed queueing networks with multiple types of customers. *Journal of Applied Probability*, **17** (1980), 1048–1061. (Cited on p. 710.)

[221] E. L. Lehmann. *Testing Statistical Hypotheses* (New York: Springer, 1986). (Cited on p. 549.)

[222] A. Leon-Garcia. *Probability and Random Processes for Electrical Engineering*, 2nd edn (Reading, MA: Addison-Wesley, 1994). (Cited on pp. 37, 549.)

[223] D. A. Levin, Y. Peres, and E. L. Wilmer. *Markov Chains and Mixing Times* (American Mathematical Society, 2008). (Cited on p. 635.)

[224] S. Levinson. Continuously variable duration hidden Markov models for automatic speech recognition. *Computer Speech and Langauge*, **1**:1 (1986), 29–45. (Cited on p. 606.)

[225] P. A. W. Lewis and G. S. Shedler. Statistical analysis of non-stationary series of events in a data base system. *IBM Journal of Research and Development*, **20**:5 (1976), 429–528. (Cited on p. 146.)

[226] J. W. Lindeberg. Eine neue Herleitung des Exponentialgesetzes in der Wahrscheinlichkeitsrechnung. *Mathematische Zeitschrift*, **15** (1922), 211–225. (Cited on p. 305.)

[227] L. Lipsky. *Queueing Theory: A Linear Algebraic Approach* (New York: MacMillan, 1992). (Cited on p. 731.)

[228] J. D. C. Little. A proof of the queueing formula $L = \lambda W$. *Operations Research*, **9** (1961), 383–387. (Cited on p. 696.)

[229] M. Loève. Sur les fonctions aléatoires du second ordre. *Revue Scientifique*, **83** (1945), 297–303. (Cited on p. 365.)

[230] M. Loève. *Probability Theory* (Princeton, NJ: D. Van Nostrand, 1955). (Cited on p. 308.)

[231] A. Lyapunov. Sur une proposition de la théorie des probabilités. *Bulletin de l'Academie Impériale des Sciences de St. Petersbourg*, **13** (1900), 359–386. (Cited on p. 304.)

[232] A. Lyapunov. Nouvelle forme de la théoreme dur la limite de probabilité. *Mémoires de l'Academie Impériale des Sciences de St. Petersbourg*, **12** (1901), 1–24. (Cited on p. 304.)

[233] X. Ma, H. Kobayashi, and S. C. Schwartz. EM-based channel estimation algorithms for OFDM. *EURASIP Journal on Applied Signal Processing*, **10** (2004), 1460–1477. (Cited on p. 566.)

[234] D. J. C. MacKay. *Information Theory, Inference, and Learning Algorithms* (Cambridge University Press, 2003). (Cited on pp. 14, 643.)

[235] L. E. Maistrov. *Probability Theory: A Historical Sketch* (New York: Academic Press, 1974). (Cited on p. 14.)

[236] B. B. Mandelbrot and J. V. Ness. Fractional Brownian motions, fractional noise and applications. *SIAM Review*, **10** (1968), 422–437. (Cited on p. 516.)

[237] A. A. Markov. Rasprostranenie zakona bol'shih chisel na velichiny, zavisyaschie drug ot druga. *Izvestiya Fiziko-matematicheskogo obschestva pri Kazanskom universitete, 2-ya seriya*, **15** (1906), 135–156. (Cited on pp. 318, 319.)

[238] A. A. Markov. Investigations of an important case of dependent trials (in Russian). *Izvestiya Academii, Nauk, Series 6 (St. Petersburg)*, **1**:3 (1907), 61–80. (Cited on pp. 3, 9, 10.)

[239] A. A. Markov. Ob ispytaniyah, svyazannyh v cep ne nablyudaemymi sobytiyami (on trials associated into a chain by unobserved events). *Izvestiya Akademii Nauk, SPb (News of the Academy of Sciences, St. Petersburg), VI seriya* **6**:98 (1912), 551–572. (Cited on pp. 319, 478, 605.)

[240] A. A. Markov. Extension of the limit theorems of probability theory to a sum of variables connected in a chain (translated by S. Petelin). In R. A. Howard, ed., *Dynamic Probabilities Systems*, Vol. 1 (New York: Wiley, 1971) pp. 552–576. (Cited on pp. 319, 478.)

[241] W. N. Martin and W. M. Spears, eds. *Foundations of Genetic Algorithms* (Morgan and Kaufmann/Academic Press, 2001). (Cited on p. 616.)

[242] L. Mason, J. Baxter, P. Bartlett, and M. Frean. Boosting algorithms as gradient descent. In S. A. Solla, T. K. Leen, and K.-R. Muller, eds, *Advances in Neural Information Processing Systems* (MIT Press, 2000) pp. 512–518. (Cited on p. 616.)

[243] J. H. Matthews and R. W. Howell. *Complex Analysis for Mathematics and Engineering* (Jones & Bartlett Publishers, Inc., 2006). (Cited on p. 195.)

[244] R. J. McEliece and S. M. Aji. The generalized distributive law. *IEEE Transactions on Information Theory*, **46**:2 (2000), 325–343. (Cited on p. 631.)

[245] R. J. McEliece, D. J. C. MacKay, and J. F. Cheng. Turbo decoding as an instance of Pearl's 'belief propagation' algorithm. *IEEE Journal on Selected Areas in Communications*, **16**:2 (1998), 140–152. (Cited on pp. 606, 624.)

[246] G. McLachlan and T. Krishnan. *The EM Algorithm and Exensions* (John Wiley & Sons, 1997). (Cited on pp. 563, 564, 566.)

[247] N. Metropolis, A. W. Rosenbluth, M. N. Rosenbluth, A. H. Teller, and E. Teller. Equations of state calculations by fast computing machines. *Journal of Chemical Physics*, **21**:6 (1953), 1087–1092. (Cited on p. 636.)

[248] D. Middleton. *An Introduction to Statistical Communication Theory* (New York: McGraw-Hill, 1960). (Cited on p. 549.)

[249] N. M. Mirasol. The output of an M/G/∞ queue is Poisson. *Operations Research*, **11** (1963), 282–284. (Cited on p. 733.)

[250] C. Mitchell, M. Harper, and L. Jamieson. On the complexity of explicit duration HMMs. *IEEE Transactions on Speech and Audio Processing*, **3**:2 (1995), 213–217. (Cited on p. 606.)

[251] P. M. Morse. *Queues, Inventries and Maintenance* (New York: Wiley & Sons, 1958). (Cited on p. 731.)

[252] M. E. Munroe. *Introduction to Measure and Integration* (Reading, MA: Addison-Wesley, 1953). (Cited on p. 293.)

[253] L. B. Nelson and H. V. Poor. Iterative multiuser receivers for CDMA channels: an EM-based approach. *IEEE Transactions on Communications*, **44** (1996), 1700–1710. (Cited on p. 566.)

[254] R. Nelson. *Probability, Stochastic Processes, and Queueing Theory* (New York: Springer-Verlag, 1995). (Cited on pp. 37, 66, 104, 275, 418, 419, 451, 731.)

[255] G. F. Newell. *Applications of Queueing Theory* (London: Chapman & Hall, 1971). (Cited on p. 516.)

[256] J. Neyman and E. Pearson. On the problem of the most efficient tests of statistical hypotheses. *Philosophical Transactions of the Royal Society of London. Series A, Containing Papers of a Mathematical or Physical Character*, **231** (1933), 289–337. (Cited on pp. 539, 540.)

[257] H. Nyquist. Certain topics in telegraph transmission theory. *Transactions of the AIEE*, **47** (1928), 363–390. (Cited on p. 353.)

[258] K. Ogura. On a certain transcendental integral function in the theory of interpolation. *Tohoku Mathematical Journal*, **17** (1920), 64–72. (Cited on p. 352.)

[259] J. K. Omura. On the Viterbi decoding algorithm. *IEEE Transactions on Information Theory*, **IT-15** (1969), 77–179. (Cited on p. 592.)

[260] J. K. Omura. Optimal receiver design for convolutional codes and channels with memory via control theoretic concepts. *Information Science*, **3** (1971), 243–266. (Cited on p. 592.)

[261] M. F. M. Osborne. Brownian motion in the stock market. *Operations Research*, **7**:2 (1959), 145–173. (Cited on p. 5.)

[262] A. Papoulis and U. Pillai. *Probability, Random Variables, and Stochastic Processes*, 4th edn (New York: McGraw-Hill, 2002). (Cited on pp. 37, 66, 104, 108, 131, 325, 394, 549, 690.)

[263] E. Parzen. *Stochastic Processes* (San Francisco: Holden-Day, Inc., 1962). (Cited on p. 323.)

[264] J. Pearl. Bayesian networks: A model of self-activated memory for evidential reasoning. In *UCLA Computer Science Department Technical Report 850021; Proceedings, Cognitive Science Society*, pp. 329–334, UC Irvine, August 1985. (Cited on p. 477.)

[265] J. Pearl. *Probabilistic Reasoning in Intelligent Systems: Networks of Plausible Inference* (San Mateo, CA: Morgan Kaufmann Publishers, 1988). (Cited on p. 643.)

[266] K. Pearson. On a criterion that a system of deviations from the probable in the case of a correlated system of variables in such that it can be reasonably supposed to have arisen in random sampling. *Philosophical Magazine*, **50** (1900), 157–175. (Cited on p. 157.)

[267] K. Pearson. On lines and planes of closest fit to systems of points in space. *Philosophical Magazine*, **6**:2 (1901), 559–572. (Cited on p. 372.)

[268] M. Pepe. *The Statistical Evaluation of Medical Tests for Classification and Prediction* (New York: Oxford University Press, 2003). (Cited on p. 542.)

[269] W. W. Peterson, T. G. Birdsall, and W. C. Fox. The theory of signal detectability. *Transactions of I.R.E.*, **PGIT-4** (1954), 171–212. (Cited on p. 542.)

[270] J. Piasecki. Centenary of Marian Smoluchowski. *Acta Physica Polonica B*, **38**:5 (2007), 1623–1629. (Cited on p. 11.)

[271] H. V. Poor. *An Introduction to Signal Detection and Estimation* (Springer, 1994). (Cited on p. 549.)

[272] H. V. Poor. Sequence detection: backward and forward in time. In R. E. Blahut and R. Koetter, eds, *Codes, Graphs and Systems* (Boston, MA: Kluwer, 2002) pp. 93–112. (Cited on pp. 592, 605.)

[273] H. V. Poor. Dynamic programming in digital communications: Viterbi decoding to turbo multiuser detection. *Journal of Optimization Theory and Applications*, **115**:3 (2002), 629–657. (Cited on pp. 592, 605.)

[274] L. R. Rabiner. A tutorial on hidden Markov models and selected application in speech recognition. *Proceedings of the IEEE*, **77**:2 (1989), 257–286. (Cited on p. 605.)

[275] L. R. Rabiner and B. H. Juang. *Fundamentals of Speech Recogntion* (Prentice Hall, 1993). (Cited on pp. 605, 615.)

[276] L. R. Rabiner, S. E. Levinson, and M. M. Sondhi. On the application of vector quantization and hidden Markov models to speaker-independent, isolated word recogition. *Bell System Technical Journal*, **62**:4 (1983), 1075–1105. (Cited on p. 605.)

[277] C. R. Rao. Information and the accuracy attainable in the estimation of statistical parameters. *Bulletin of the Calcutta Mathematical Society*, **37** (1945), 81–91. (Cited on p. 532.)

[278] C. R. Rao. *Linear Statistical Inference and Its Applications* (New York: John Wiley & Sons, Inc., 1965). (Cited on pp. 270, 302, 303, 304, 305, 308, 549, 651.)

[279] H. E. Rauch, F. Tung, and C. T. Striebel. Maximum likelihood estimates of linear dynamic systems. *AIAA Journal*, **3**:8 (1965), 1445–1450. (Cited on pp. 606, 690.)

[280] S. O. Rice. Statistical properties of a sine wave plus random noise. *Bell System Technical Journal*, **27** (1948), 109–157. (Cited on p. 170.)

[281] W. Roberts, Y. Ephraim, and E. Dieguez. On Rydén's EM algorithm for estimating MMPPs. *IEEE Signal Processing Letters*, **13**:6 (2006), 373–376. (Cited on pp. 566, 606.)

[282] L. C. G. Rogers and D. Williams. *Diffusions, Markov Processes and Martingales: Volume 1, Foundations* (Cambridge University Press, 2000). (Cited on pp. 268, 477, 516.)

[283] L. C. G. Rogers and D. Williams. *Diffusions, Markov Processes and Martingales: Volume 2, Itô Calculus* (Cambridge University Press, 2000). (Cited on p. 516.)

[284] V. Romanovsky. *Diskretnye tsepi Markova* (Moscow: Gostekhizdat, 1949). (Cited on p. 605.)

[285] V. Romanovsky. *Discrete Markov Chain* (translated by E. Senata) (Groningen: Wolters-Noordhoff, 1970). (Cited on p. 605.)

[286] M. Rosenblatt. *Random Processes* (New York: Oxford University Press, 1962). (Cited on p. 451.)

[287] S. M. Ross. *Stochastic Processes* (New York: John Wiley & Sons, Inc., 1983). (Cited on p. 719.)

[288] S. M. Ross. *Stochastic Processes*, 2nd edn (New York: John Wiley & Sons, Inc., 1996). (Cited on pp. 272, 273, 418, 451, 516.)

[289] S. M. Ross. *A First Course in Probability*, 6th edn (Prentice Hall, 2002). (Cited on pp. 37, 66, 104, 268, 271.)

[290] H. L. Royden. *Real Analysis* (Prentice Hall, 1988). (Cited on p. 293.)

[291] T. Rydén. An EM algorithm for estimation in Markov-modulated Poisson process. *Communications in Statististical Data Analysis*, **21** (1996), 431–447. (Cited on pp. 566, 606.)

[292] M. Sakata, S. Noguchi, and J. Oizumi. Analysis of a processor-sharing queueing model for time-sharing systems. In *Proceedings of 2nd Hawaii International Conference on System Science*, pp. 625–628 (1969). (Cited on p. 717.)

[293] R. Schapire. Strength of weak learnability. *Machine Learning*, **5**:2 (1990), 197–227. (Cited on p. 616.)

[294] A. Schuster. On lunar and solar periodities of earthquakes. *Proceedings of the Royal Society*, **61** (1897), 455–465. (Cited on p. 357.)

[295] M. Schwartz, W. R. Bennet, and S. Stein. *Communication Systems and Techniques* (New York: Wiley, 1995). (Cited on pp. 336, 340.)

[296] C. Semple and M. Steel. *Phylogenetics*, volume 24 of *Oxford Lecture Series in Mathematics and Its Applications* (Oxford University Press, 2003). (Cited on pp. 473, 477.)

[297] K. Sevcik and I. Mitrani. The distribution of queueing network states at input and output instants. *Journal of the ACM*, **28**:2 (1981), 358–371. (Cited on p. 710.)

[298] G. Shafer. *The Art of Causal Conjecture* (MIT Press, 1996). (Cited on p. 477.)

[299] G. Shafer and V. Vovk. *Probability and Finance: It's Only a Game!* (John Wiley & Sons, 2001). (Cited on pp. xxviii, 14, 268, 304, 308.)

[300] C. E. Shannon. A mathematical theory of communications. *Bell System Technical Journal*, **27** (1948), 379–423, 623–656. (Cited on pp. 3, 246, 256, 257, 352, 425, 427, 451, 561, 568, 581, 605.)

[301] C. E. Shannon. Communication in the presence of noise. *Proceedings of the Institute of Radio Engineers*, **37**:1 (1949), 10–21. (Also in *Proceedings of the IEEE*, **86**:2 (1998), 447–457.) (Cited on p. 352, 353.)

[302] W. T. Shaw. *Complex Analysis with Mathematica* (Cambridge University Press, 2006). (Cited on p. 195.)

[303] L. A. Shepp and Y. Vardi. Maximum likelihood reconstruction for emission tomography. *IEEE Medical Imaging*, **MI-1**:2 (1983), 113–122. (Cited on p. 566.)

[304] A. Shwartz and A. Weiss. *Large Deviations for Performance Analysis* (Chapman & Hall, 1995). (Cited on p. 268.)

[305] C. A. Sims. Macroeconomics and reality. *Econometrica*, **48**:1 (1980), 1–48. (Cited on p. 5.)

[306] D. Skillicorn. *Understanding Complex Datasets: Data Mining with Matrix Decomposition* (Chapman & Hall/CRC, 2007). (Cited on p. 395.)

[307] M. Smoluchowski. Essai d'une théorie cinétique du mouvement Brownien et des milieux troubles (Outline of the kinetic theory of Brownian motion of suspensions). *Bulletin International de l'Académie des Sciences de Cracovie*, (1906), 577–602. (Cited on p. 11.)

[308] I. Someya. *Waveform Transmission* (in Japanese) (Tokyo: Shukyosha, 1949). (Cited on p. 352.)

[309] D. J. Spiegelhalter, R. Franklin, and K. Bull. Assessment, criticism, and improvement of imprecise probabilities for a medical expert system. In *Proceedings of the Fifth Conference on Uncertainty in Artificial Intelligence*, pp. 285–294 (1989). (Cited on pp. 615, 624.)

[310] H. Stark and J. W. Woods. *Probability and Random Processes with Applications to Signal Processing*, 3rd edn (Upper Saddle River, NJ: Prentice Hall, 2002). (Cited on pp. 37, 566, 690.)

[311] S. Stidham, Jr. and M. El-Taha. Sample-path techniques in queueing theory. In J. H. Dshalalow, ed., *Advances in Queueing: Theory, Methods, and Open Problems* (CRC Press, 1995), pp. 119–166. (Cited on p. 730.)

[312] S. M. Stigler. *The History of Statistics: The Measurement of Uncertainty before 1900* (Cambridge, MA: Harvard University Press, 1986). (Cited on p. 14.)

[313] R. L. Stratonovich. Application of the Markov process theory to optimal filtering. *Radio Engineering and Electronic Physics*, **5**:11 (1960), 1–19. (Cited on p. 12.)

[314] Student (W. S. Gosset). The probable error of a mean. *Biometrika*, **6**:1 (1908), 1–25. (Cited on p. 162.)

[315] A. L. Sweet and J. C. Hardin. Solutions for some diffusion processes with two barriers. *Journal of Applied Probability*, **7** (1970), 423–431. (Cited on p. 518.)

[316] R. Syski. *Introduction to Congestion Theory in Telephone Systems*, 2nd edn (Amsterdam: North-Holland, 1986). (Cited on pp. 720, 731.)

[317] L. Takács. *Introduction to the Theory of Queues* (New York: Oxford University Press, 1962). (Cited on p. 731.)

[318] S. Tezuka. *Uniform Random Numbers: Theory and Practice* (Norwell, MA: Kluwer Academic Publishers, 1995). (Cited on p. 131.)

[319] J. B. Thomas. *An Introduction to Applied Probability and Random Processes* (New York: John Wiley & Sons, Inc., 1971). (Cited on pp. 37, 131, 278, 302, 303, 304, 308, 309, 311, 323, 394, 690.)

[320] L. Tierney. Markov chains for exploring posterior distributions. *Annals of Statistics*, **22**:4 (1994), 1701–1728. (Cited on p. 637.)

[321] L. Tierney. A note on Metropolis–Hastings kernels for general state spaces. *Annals of Applied Probability*, **8**:1 (1998), 1–9. (Cited on p. 637.)

[322] H. C. Tijms. *Stochastic Modeling and Analysis* (John Wiley & Sons, Inc., 1986). (Cited on p. 731.)

[323] E. C. Titchmarsh. *Theory of Functions* (London: Oxford University Press, 1939). (Cited on p. 194.)

[324] E. C. Titchmarsh. *Introduction to the Theory of Fourier Integrals* (London: Oxford University Press, 1948). (Cited on p. 345.)

[325] I. Todhunter. *A History of the Mathematical Theory of Probability from the Time of Pascal to that of Laplace* (New York: Chelsea, 1949, 1965). (Originally published by Macmillan in 1865.) (Cited on p. 14.)

[326] Tree of Life Web Project. WWW page, August 2010. http://tolweb.org. (Cited on p. 470.)

[327] K. S. Trivedi. *Probability & Statistics with Reliability, Queueing and Computer Science Applications*, 2nd edn (New York: John Wiley & Sons, Inc., 2002). (Cited on pp. 37, 418, 419.)

[328] J. W. Tukey. *Exploratory Data Analysis* (Reading, MA: Addison-Wesley, 1977). (Cited on p. 153.)

[329] G. L. Turin. On optimal diversity reception. *IRE Transactions on Information Theory*, **IT-7** (1961), 154–167. (Cited on pp. 177, 322.)

[330] W. Turin. *Digital Transmission Systems: Performance Analysis and Modeling* (McGraw Hill, 1999). (Cited on p. 432.)

[331] W. Turin. MAP decoding in channels with memory. *IEEE Transactions on Communications*, **48**:5 (2000), 757–763. (Cited on p. 566.)

[332] W. Turin. *Performance Analysis and Modeling of Digital Transmission Systems* (Kluwer Academic/Plenum Pulishers, 2004). (Cited on pp. 456, 566, 606.)

[333] W. Turin and R. Boie. Bar code recovery via the EM algorithm. *IEEE Transactions on Signal Processing*, **46**:2 (1998), 354–363. (Cited on p. 566.)

[334] G. E. Uhlenbeck and L. S. Ornstein. On the theory of Brownian motion. *Physical Review*, **36** (1930), 823–841. (Cited on pp. 502, 516.)

[335] P. Valkó. Numerical inversion of Laplace transform: a challenge for developers of numerical methods. `http://pumpjack.tamu.edu/~valko/Nil/` (2003). (Cited on pp. 233, 234.)

[336] A. J. van der Veen. Algebraic method for deterministic blind beamforming. *Proceedings of IEEE*, **86**:10 (1998), 1987–2008. (Cited on p. 395.)

[337] V. Vapnik. *The Nature of Statistical Learning Theory* (New York: Springer-Verlag, 1995). (Cited on p. 643.)

[338] V. Vapnik. *Statistical Learning Theory* (New York: John Wiley, 1998). (Cited on pp. 3, 616, 643.)

[339] A. J. Viterbi. Error bounds for convolutional codes and asymptotically optimum decoding algorithm. *IEEE Transactions on Information Theory*, **IT-13** (1967), 260–269. (Cited on pp. 591, 592, 605.)

[340] A. J. Viterbi. A personal history of the Viterbi algorithm. *IEEE Signal Processing Magazine*, **23**:4 (2006), 120–142. (Cited on p. 605.)

[341] R. von Mises. *Wahrscheinlichkeitsrechnung, Statistik und Wahrheit* (Wien: Verlang von Julius Springer, 1928). (Cited on p. 19.)

[342] R. von Mises. *Probability, Statistics and Truth* (New York: MacMillan, 1954) (Translation of the 1928 publication.) (Cited on p. 9.)

[343] L. A. Wainstein and V. D. Zubakov. *Extraction of Signals from Noise* (Prentice-Hall Inc, 1962). (Cited on p. 177.)

[344] M. E. Wall, A. Rechtsteiner, and L. M. Rocha. Singular value decomposition and principal component analysis. In D. P. Berrar, W. Dubitzky, and M. Granzow, eds, *A Practical*

Approach to Microarray Data Analysis (Norwell, MA: Kluwer Academic Press, 2003) pp. 91–109. (Cited on pp. 372, 395.)

[345] J. Walrand. *An Introduction to Queueing Networks* (Englewood Cliffs, NJ: Prentice Hall, 1988). (Cited on p. 479.)

[346] L. R. Welch. The shannon lecture: Hidden Markov models and the Baum–Welch algorithms. *IEEE Informathion Theory Society Newsletter*, **53**:4 (2003), 1, 10–13. (Cited on pp. 598, 606.)

[347] P. D. Welch. The use of fast Fourier transform for the estimation of power spectra: a method based on time averaging over short, modified periodograms. *IEEE Transactions on Audio Electronics*, **AU-15** (1967), 70–73. (Cited on p. 357.)

[348] E. T. Whittaker. On the functions which are represented by the expansions of the interpolation theory. *Proceedings of the Royal Society of Edinburgh, Section A*, **35** (1915), 181–194. (Cited on p. 352.)

[349] N. Wiener. *The Fourier Integral and Certain of Its Applications* (Cambridge University Press, 1933). (Reissued in 1988.) (Cited on p. 194.)

[350] N. Wiener. *Extrapolation, Interpolation, and Smoothing of Stationary Time Series* (New York: John Wiley, 1949). (Cited on pp. 13, 645, 656, 690.)

[351] N. Wiener. *Collected Works, Vol. I (Edited by P. R. Masani)* (Cambridge, MA: MIT Press, 1976). (Cited on p. 11.)

[352] D. J. Wilkinson. Bayesian methods in bioinformatics and computational systems biology. *Briefings in Bioinformatics*, **8**:2 (2007), 109–116. (Cited on pp. 14, 643.)

[353] S. S. Wilks. *Mathematical Statistics* (New York: John Wiley & Sons, Inc., 1962). (Cited on pp. 37, 59, 60, 305.)

[354] D. Williams. *Probability with Martingales* (Cambridge University Press, 1991). (Cited on pp. 268, 293, 308.)

[355] D. Williams. *Weighing the Odds: A Course in Probability and Statistics* (Cambridge University Press, 2001). (Cited on p. 14.)

[356] I. H. Witten and E. Frank. *Data Mining: Practical Machine Learning Tools and Techniques*, 2nd edn (Morgan Kaufmann, 2005). (Cited on p. 616.)

[357] R. W. Wolff. Poisson arrivals see time averages. *Operations Research*, **30** (1982), 223–231. (Cited on pp. 413, 477.)

[358] R. W. Wolff. *Stochastic Modeling and Theory of Queues* (Englewood Cliffs, NJ: Prentice Hall, 1989). (Cited on pp. 418, 419, 730, 731.)

[359] E. Wong and B. Hajek. *Stochastic Processes in Engineering Systems* (Springer-Verlag, 1985). (Cited on p. 516.)

[360] R. A. Wooding. The multivariate distribution of complex normal variables. *Biometrika*, **43** (1956), 212–215. (Cited on pp. 177, 330.)

[361] J. M. Wozencraft and I. M. Jacobs. *Principles of Communication Engineering* (New York: John Wiley & Sons, Inc., 1965). (Cited on pp. 261, 268, 371.)

[362] C. F. J. Wu. On the convergence of the EM algorithm. *Annals of Statistics*, **11**:1 (1983), 95–103. (Cited on p. 563.)

[363] R. D. Yates and D. J. Goodman. *Probability and Stochastic Processes: A Friendly Introduction for Electrical and Computer Engineers*, 2nd edn (John Wiley & Sons, Inc., 2004). (Cited on p. 37.)

[364] I. Youn, B. L. Mark, and D. Richards. A statistical approach to geolocation of Internet hosts. In *Proceedings of 18th IEEE International Conference on Computer Communications and Networks (ICCCN'09)*, San Francisco, CA (August 2009). (Cited on pp. 150, 151.)

[365] S. Z. Yu. Hidden semi-Markov models. *Artificial Intelligence*, **174** (2010), 215–243. (Cited on p. 606.)

[366] S. Z. Yu and H. Kobayashi. A hidden semi-Markov model with missing data and multiple observation sequence for mobility tracking. *Signal Processing*, **83**:2 (2003), 235–250. (Cited on pp. 451, 606.)

[367] S. Z. Yu and H. Kobayashi. Practical implementation of an efficient forward-backward algorithm for an explicit-duration hidden Markov model. *IEEE Transactions on Signal Processing*, **54**:5 (2006), 1947–1951. (Cited on p. 606.)

[368] M. Zakai. Second-order properties of the pre-envelope and envelope processes. *IRE Transactions on Information Theory*, **IT-6** (1960), 556–557. (Cited on p. 340.)

[369] L. M. Zeger and H. Kobayashi. A simplified EM algorithm for detection of CPM signals in a fading multipath channel. *Wireless Networks*, **8** (2002), 649–658. (Cited on p. 566.)

[370] X. Zhang. Space–time diversity in multiple-antenna wireless communication systems. PhD thesis, Department of Electrical Engineering, Princeton University, Princeton, NJ (June 2004). (Cited on p. 395.)

[371] L. Zweig. Speech recognition with dynamic bayesian networks. PhD thesis, University of California (1998). (Cited on p. 624.)

Index

Printed in the United States
By Bookmasters